Handbook of Spatial Epidemiology

Chapman & Hall/CRC
Handbooks of Modern Statistical Methods

Series Editor

Garrett Fitzmaurice
Department of Biostatistics
Harvard School of Public Health
Boston, MA, U.S.A.

Aims and Scope

The objective of the series is to provide high-quality volumes covering the state-of-the-art in the theory and applications of statistical methodology. The books in the series are thoroughly edited and present comprehensive, coherent, and unified summaries of specific methodological topics from statistics. The chapters are written by the leading researchers in the field, and present a good balance of theory and application through a synthesis of the key methodological developments and examples and case studies using real data.

The scope of the series is wide, covering topics of statistical methodology that are well developed and find application in a range of scientific disciplines. The volumes are primarily of interest to researchers and graduate students from statistics and biostatistics, but also appeal to scientists from fields where the methodology is applied to real problems, including medical research, epidemiology and public health, engineering, biological science, environmental science, and the social sciences.

Published Titles

Handbook of Mixed Membership Models and Their Applications
Edited by Edoardo M. Airoldi, David M. Blei, Elena A. Erosheva, and Stephen E. Fienberg

Handbook of Markov Chain Monte Carlo
Edited by Steve Brooks, Andrew Gelman, Galin L. Jones, and Xiao-Li Meng

Handbook of Big Data
Edited by Peter Bühlmann, Petros Drineas, Michael Kane, and Mark van der Laan

Handbook of Discrete-Valued Time Series
Edited by Richard A. Davis, Scott H. Holan, Robert Lund, and Nalini Ravishanker

Handbook of Design and Analysis of Experiments
Edited by Angela Dean, Max Morris, John Stufken, and Derek Bingham

Longitudinal Data Analysis
Edited by Garrett Fitzmaurice, Marie Davidian, Geert Verbeke, and Geert Molenberghs

Handbook of Spatial Statistics
Edited by Alan E. Gelfand, Peter J. Diggle, Montserrat Fuentes, and Peter Guttorp

Handbook of Cluster Analysis
Edited by Christian Hennig, Marina Meila, Fionn Murtagh, and Roberto Rocci

Handbook of Survival Analysis
Edited by John P. Klein, Hans C. van Houwelingen, Joseph G. Ibrahim, and Thomas H. Scheike

Handbook of Spatial Epidemiology
Edited by Andrew B. Lawson, Sudipto Banerjee, Robert P. Haining, and María Dolores Ugarte

Handbook of Missing Data Methodology
Edited by Geert Molenberghs, Garrett Fitzmaurice, Michael G. Kenward, Anastasios Tsiatis, and Geert Verbeke

Chapman & Hall/CRC
Handbooks of Modern
Statistical Methods

Handbook of Spatial Epidemiology

Edited by

Andrew B. Lawson
Medical University of South Carolina
Charleston, USA

Sudipto Banerjee
UCLA Fielding School of Public Health
Los Angeles, California, USA

Robert P. Haining
University of Cambridge
United Kingdom

María Dolores Ugarte
Public University of Navarre
Pamplona, Spain

CRC Press
Taylor & Francis Group
Boca Raton London New York

CRC Press is an imprint of the
Taylor & Francis Group, an **informa** business
A CHAPMAN & HALL BOOK

CRC Press
Taylor & Francis Group
6000 Broken Sound Parkway NW, Suite 300
Boca Raton, FL 33487-2742

First issued in paperback 2020

Version Date: 20160222

ISBN 13: 978-0-367-57038-5 (pbk)
ISBN 13: 978-1-4822-5301-6 (hbk)

Library of Congress Cataloging-in-Publication Data

Names: Lawson, Andrew (Andrew B.) editor. | Banerjee, Sudipto editor. |
Haining, Robert P., editor. | Ugarte, Marâia Dolores, editor.
Title: Handbook of spatial epidemiology / editors, Andrew B. Lawson, Sudipto
Banerjee, Robert P. Haining, Maria Dolores Ugarte.
Description: Boca Raton : Taylor & Francis, 2016. | Includes bibliographical
references and index.
Identifiers: LCCN 2015041881 | ISBN 9781482253016 (alk. paper)
Subjects: LCSH: Medical geography. | Epidemiology.
Classification: LCC RA792 .H36 2016 | DDC 614.4/2--dc23
LC record available at http://lccn.loc.gov/2015041881

Visit the Taylor & Francis Web site at
http://www.taylorandfrancis.com

and the CRC Press Web site at
http://www.crcpress.com

Contents

Preface

Spatial epidemiology has seen considerable advances in recent years. Not only have methodologic advances been rapid during the late 1990s and 2000s, but the permeation and diffusion of the methods in areas of human and veterinary epidemiology have also been considerable. It is now commonplace to consider a geographic dimension to be included within a research design in studies involving nutrition, physical activity, and community intervention, as well as studies of health outcomes related to environmental and social determinants.

By definition, the focus of spatial epidemiology is the study of the geographical or spatial distribution of health outcomes. It is sometimes interchangeably known as disease mapping. Usually, it has the incidence of disease or prevalence of disease as its main focus. In addition, the parsimonious description of the distribution is often a major concern. This may involve initial visualization of the distribution and some simple summary measures, but can lead to quite complex modeling of the disease process. In the case of infectious disease, the transmission dynamics may also be of interest, and so some form of temporal change in incidence would be a focus. Inevitably, for more complex modeling, spatial statistical methodology may be needed to address the inferential questions.

The development of geographical information systems (GIS) has seen the automation of many map-related tasks and has allowed access to spatial information on a wide scale. The advent of mobile technology and GPS also allows dynamic spatial information to be recorded. GIS often allow flexible analysis of spatially distributed information, but in their current form, at least in widely available or commercial platforms, they do not include dynamic capabilities. In addition to GIS, developments in spatial statistics and fast computational methods in recent years have allowed formerly intractable problems to be modeled and inference about disease etiology to be improved. Both GIS and spatial statistics appear as tools for the analysis of spatial epidemiological data. In this volume, our focus is wide ranging, and because of the interdisciplinary nature of the focus area, our contributions come from epidemiologists, geographers, and statisticians.

Space and time often interact in health studies. The most basic form of information related to any health event is a diagnosis, and usually associated with that is a date or time. Hence, coupled with a geographic location, such as residential street address or census tract or postal or zip code, within which the sufferer lives, both a temporal and a spatial labeling of the event is available. Space–time variation in disease incidence can contain important etiological clues and provide important information that helps to better inform public health decision making.

This volume is divided into parts with relatively distinct foci. In Part I, a range of introductory chapters is found that address general issues related to epidemiology, GIS, environmental studies, clustering, and ecological analysis. In Part II, chapters deal with basic statistical methods of use in spatial epidemiology. These include basic likelihood principles and Bayesian methods, as well as testing and nonparametric approaches. Part III includes a bigger range of 12 chapters focused on special methods. These feature geostatistical models, splines, quantile regression, focused clustering, mixtures, and multivariate methods, among many others. Finally, in Part IV, some special problems and application areas are covered. These include residential history analysis, segregation, health services research, health surveys, infectious disease, and veterinary topics, as well as health surveillance and clustering.

Many of the constituent chapters in this volume have supplementary material associated with them in the form of color versions of figures, program code for models fitted, and datasets used in the work. These can be found on a dedicated handbook website hosted by the Medical University of South Carolina (MUSC) and accessible via the CRC Press website: https://www.crcpress.com/9781482253016. The contents are organized by chapter and have been deposited by the contributors.

Finally, we would like to thank the wide range of contributors who provided chapters to the volume and who made the timely submission of revisions and finalization of the volume possible. In addition, we would like to express our appreciation to Rob Calver of CRC Press/Taylor & Francis Group, who has encouraged the development and execution of the volume. Others at Taylor & Francis Group, including Alexander Edwards and Katy Buke at the production stage, were invaluable in guiding us to the finished product.

Andrew B. Lawson
Sudipto Banerjee
Robert Haining
María Dolores Ugarte

Editors

Andrew B. Lawson is a professor of biostatistics in the Division of Biostatistics, Department of Public Health Sciences, College of Medicine, Medical University of South Carolina (MUSC), Charleston, SC, and is an MUSC eminent scholar and American Statistical Association (ASA) fellow. He was previously a professor of biostatistics in the Department of Epidemiology and Biostatistics, University of South Carolina, Charleston. His PhD in spatial statistics is from the University of St. Andrews, St. Andrews, UK.

He has written more than 150 journal articles on the subjects of spatial epidemiology, spatial statistics, and related areas. In addition to a number of book chapters, he is the author of eight books in areas related to spatial epidemiology and health surveillance. As well as associate editorships on a variety of journals, he is an advisor in disease mapping and risk assessment for the World Health Organization. He is the founding editor of the new Elsevier journal *Spatial and Spatio-Temporal Epidemiology.*

Dr. Lawson has delivered many short courses in different locations over the past 10 years on Bayesian disease mapping with WinBUGS and INLA, spatial epidemiology, and disease clustering, and is the author of *Bayesian Disease Mapping*, Second Edition (CRC Press, New York, 2013).

Sudipto Banerjee is a professor and chair at the Department of Biostatistics, University of California, Los Angeles. His primary research interests focus on hierarchical modeling and Bayesian inference for spatially referenced data. He has published more than 100 peer-reviewed publications and two textbooks, one on spatial statistics and the other on linear algebra, and has overseen the development of open-source statistical software packages for fitting Bayesian hierarchical models to spatially oriented data. He is a recipient of the Mortimer Spiegelman Award from the American Public Health Association. He is also an elected fellow of the American Statistical Association, the Institute of Mathematical Statistics, and the International Statistical Institute.

Robert Haining retired as a professor of human geography from the University of Cambridge, Cambridge, UK, in September 2015. He has long-standing research interests in the quantitative analysis of geographical data. He has authored or coauthored more than 150 articles and has published two books with Cambridge University Press: *Spatial Data Analysis in the Social and Environmental Sciences* (1990) and *Spatial Data Analysis: Theory and Practice* (2003). These books have provided an overview of the challenges facing those working with spatial data, including what it means to "think spatially," problems of data collection and spatial representation, spatial sampling, exploratory data analysis, and relevant areas of statistical theory for small-area estimation and hypothesis testing. One of his primary areas of applied interest includes the geography of health. He has worked in collaboration with colleagues at the University of Sheffield, Sheffield, UK, on evaluating the impact of air pollution on health status using small-area statistics. The growing availability of small-area statistics has been made possible by the digital revolution and the development of geographical information systems, together with organizational changes that have made the collection of geocoded data by many public and private agencies a matter of routine today. But small-area statistics raise many challenges for statistical analysis and the drawing out of robust conclusions from what is often "noisy" small-area data.

In recent years, he has had the opportunity to work on the analysis of crime data in collaboration with various police forces in England and Wales and with home office funding. Most recently, Haining and colleagues have worked on the development of new methods for evaluating the effectiveness of small-area targeted police interventions. He has taught courses on spatial data analysis and spatial econometrics at the undergraduate and graduate levels at Cambridge and other universities around the world.

María Dolores Ugarte is a professor of statistics at the Public University of Navarre, Pamplona, Spain. Her research focus is mainly on spatiotemporal disease mapping and small-area estimation with applications in several fields. She has taught numerous statistics courses at the undergraduate, master, and doctoral levels. She has received a bachelor's degree in mathematics from the University of Zaragoza, Zaragoza, Spain, and a PhD in statistics from the Public University of Navarre, Pamplona, Spain. Her postdoctoral training was done at Simon Fraser University, Burnaby, Canada. In 2007, she received the INNOLEC Lectureship Award granted by the Faculty of Science, Masaryk University, Brno, Czech Republic, and in 2008, a rating of "Excellent Teacher" from the Public University of Navarre. She has written many papers in statistics and epidemiology and has also coauthored several statistical books in Spanish and English. The most recent one is *Probability and Statistics with R*, Second Edition (2015, CRC/Chapman & Hall). She has worked as an associate editor of the *Journal of the Royal Statistical Society Series A* and is currently an associate editor for *Statistical Modelling*, *TEST*, and *Computational Statistics and Data Analysis*. She is also a member of the editorial panel of the journal *Spatial and Spatio-Temporal Epidemiology*.

Contributors

Jared Aldstadt
Department of Geography
State University of New York at Buffalo
Buffalo, New York

Sudipto Banerjee
Department of Biostatistics
University of California
Los Angeles, California

Sarah E. Bauer
Department of Health Services Research, Management and Policy
University of Florida
Gainesville, Florida

Marta Blangiardo
MRC-PHE Centre for Environment and Health
Department of Epidemiology and Biostatistics
Imperial College London
London, UK

Patrick E. Brown
Analytics and Informatics
Cancer Care Ontario
and
Department of Statistical Sciences
University of Toronto
Toronto, Ontario, Canada

Catherine A. Calder
Department of Statistics
The Ohio State University
Columbus, Ohio

Michela Cameletti
Department of Management, Economics and Quantitative Methods
University of Bergamo
Bergamo, Italy

Howard H. Chang
Department of Biostatistics and Bioinformatics
Emory University
Atlanta, Georgia

Jungsoon Choi
Department of Mathematics
Hanyang University
Seoul, South Korea

Ana Corberán-Vallet
Department of Statistics and Operations Research
University of Valencia
Valencia, Spain

Adrian Dobra
Department of Statistics
Department of Biobehavioral Nursing and Health Systems
and
Center for Statistics and the Social Sciences
University of Washington
Seattle, Washington

Jaione Etxeberria
Department of Statistics and Operations Research
Institute for Advanced Materials (InaMat)
Public University of Navarra
Pamplona, Spain

Christel Faes
Interuniversity Institute for Biostatistics and Statistical Bioinformatics
Hasselt University
Hasselt, Belgium

Montserrat Fuentes
Department of Statistics
North Carolina State University
Raleigh, North Carolina

Federica Giardina
Swiss Tropical and Public Health Institute (Swiss TPH)
and
University of Basel
Basel, Switzerland

Tomás Goicoa
Department of Statistics and Operations Research
Institute for Advanced Materials (InaMat)
Public University of Navarra
Pamplona, Spain

Sue C. Grady
Department of Geography
Michigan State University
East Lansing, Michigan

Robert Haining
Department of Geography
University of Cambridge
Cambridge, UK

Martin L. Hazelton
Institute of Fundamental Sciences
Massey University
Palmerston North, New Zealand

Amy H. Herring
UNC Gillings School of Global Public Health
The University of North Carolina at Chapel Hill
Chapel Hill, North Carolina

James S. Hodges
Division of Biostatistics
School of Public Health
University of Minnesota
Minneapolis, Minnesota

Michael Höhle
Department of Mathematics
Stockholm University
Stockholm, Sweden

Md. Monir Hossain
Division of Biostatistics and Epidemiology
Cincinnati Children's Hospital and Medical Center
University of Cincinnati College of Medicine
Cincinnati, Ohio

Geoffrey Jacquez
Department of Geography
State University of New York at Buffalo
Buffalo, New York

Inkyung Jung
Department of Biostatistics and Medical Informatics
Yonsei University College of Medicine
Seoul, Korea

Athanasios Kottas
Department of Applied Mathematics and Statistics
University of California
Santa Cruz, California

Peter H. Langlois
Birth Defects Epidemiology and Surveillance Branch
Texas Department of State Health Services
Austin, Texas

Andrew B. Lawson
Division of Biostatistics and Bioinformatics
Department of Public Health Sciences
Medical University of South Carolina
Charleston, South Carolina

Duncan Lee
School of Mathematics and Statistics
University of Glasgow
Glasgow, UK

Ravi Maheswaran
School of Health and Related Research
University of Sheffield
Sheffield, UK

Marc Marí-Dell'Olmo
CIBER de Epidemiología y Salud Pública (CIBERESP)
Madrid, Spain
and
Agència de Salut Pública de Barcelona
and
Institut d'Investigació Biomèdica Sant Pau (IIB Sant Pau)
Barcelona, Spain

Miguel A. Martinez-Beneito
Fundación para el Fomento de la Investigación Sanitaria y Biomédica de la Comunidad Valenciana (FISABIO)
Valencia, Spain
and
CIBER de Epidemiología y Salud Publica (CIBERESP)
Madrid, Spain

Brian Neelon
Department of Public Health Sciences
Medical University of South Carolina
Charleston, South Carolina

Mark J. Nieuwenhuijsen
Center for Research in Environmental Epidemiology (CREAL)
Parc de Recerca Biomèdica de Barcelona – PRBB
Barcelona, Spain

Georgiana Onicescu
Department of Statistics
Western Michigan University
Kalamazoo, Michigan

Dirk U. Pfeiffer
Veterinary Epidemiology, Economics and Public Health
Department of Production and Population Health
Royal Veterinary College
London, UK

Brian J. Reich
Department of Statistics
North Carolina State University
Raleigh, North Carolina

Sara Wagner Robb
Department of Epidemiology and Biostatistics
College of Public Health
University of Georgia
Athens, Georgia

Peter A. Rogerson
Departments of Geography and Biostatistics
State University of New York at Buffalo
Buffalo, New York

Sujit K. Sahu
Mathematical Sciences
University of Southampton
Southampton, UK

Theresa R. Smith
Lancaster Medical School
Lancaster University
Lancaster, UK

Nafomon Sogoba
Université des Sciences
des Techniques et des Technologies de Bamako
Bamako, Mali

Kim B. Stevens
Veterinary Epidemiology, Economics and Public Health
Department of Production and Population Health
Royal Veterinary College
London, UK

Jeffrey M. Switchenko
Department of Biostatistics and Bioinformatics
Rollins School of Public Health
Emory University
Atlanta, Georgia

María Dolores Ugarte
Department of Statistics and Operations Research
Institute for Advanced Materials (InaMat)
Public University of Navarra
Pamplona, Spain

Yannick Vandendijck
Interuniversity Institute for Biostatistics and Statistical Bioinformatics
Hasselt University
Hasselt, Belgium

John E. Vena
Department of Public Health Sciences
Medical University of South Carolina
Charleston, South Carolina

Penelope Vounatsou
Swiss Tropical and Public Health Institute (Swiss TPH)
and
University of Basel
Basel, Switzerland

Jon C. Wakefield
Departments of Statistics and Biostatistics
University of Washington
Seattle, Washington

Lance A. Waller
Department of Biostatistics and Bioinformatics
Rollins School of Public Health
Emory University
Atlanta, Georgia

Joshua L. Warren
Department of Biostatistics
Yale School of Public Health
Yale University
New Haven, Connecticut

David C. Wheeler
Department of Biostatistics
Virginia Commonwealth University
Richmond, Virginia

Part I

Introduction

1

Integration of Different Epidemiologic Perspectives and Applications to Spatial Epidemiology

Sara Wagner Robb
Department of Epidemiology and Biostatistics
College of Public Health
University of Georgia
Athens, Georgia

Sarah E. Bauer
Department of Health Services Research, Management and Policy
University of Florida
Gainesville, Florida

John E. Vena
Department of Public Health Sciences
Medical University of South Carolina
Charleston, South Carolina

CONTENTS

1.1 Introduction

The ability of epidemiology to determine the relationships between health and various risk factors, especially environmental insults, has become exceedingly difficult. The multifactorial nature of disease and the diversity of the insults, which include biologic, physical, social, and cultural factors, combined with genetic susceptibility, suggest the need to incorporate comprehensive perspectives of multidisciplinary epidemiologic investigation. Further, there is a need to utilize tools, such as geographic information systems (GIS) and other geospatial methods, which can integrate multilevel, spatial, and temporal factors and can help limit the potential for misclassification of exposure estimates. The multifactorial nature of disease and availability of geospatial tools encourage collaboration and creativity in the field of environmental epidemiology. The recent technology advances in smart phone and GIS utilization and the diffusion of innovation now offer unprecedented opportunities for applications in public health.

1.1.1 Historical background and justification of integrated perspectives

There has been a notable amount of dialogue in recent literature about the theoretical basis for investigation surrounding current epidemiologic methods and philosophies missing the mark and perhaps even reaching their limits. Concerns have been expressed regarding the emphasis on molecular epidemiology and biology of the disease outcomes such that the macrolevel picture involving the individual in a social, cultural, and physical setting may be missed. As the disciplines involved in studies are expanded, and thus the perspectives leading to understanding increase, concern is being expressed about the existence of an integrating process enabling the construction of new descriptions of risk and disease. One of the fundamental advantages of incorporating geospatial techniques is the ability to address these population-level disease determinants, which may be ignored in standard nonspatial individual-level epidemiologic studies.

Presently, public health must confront unprecedented challenges, including dramatic global population growth, an aging population, and possibly irreversible changes in key environmental health determinants with reference to globalization and climate change. As stressed by Fielding (1999) in a recent review, advancement in epidemiologic methods has occurred, but the determinants of health at the community level have been ignored, thus leading to simplistic formulations of multiple risk factors. New tools to assess health may promise greater efficiency and effectiveness for public health. Epidemiologic investigators should marry the biopsychosocial model of disease with the environmental-social cause model to determine common final pathways, thereby understanding how the underlying environmental and genetic factors produce intermediate risks and how these translate into health, disease, and quality of life (Fielding 1999). Susser and Susser (1996a, 1996b) and Pearce (1996) have been important voices in the dialogue that criticized modern risk factor epidemiology as a discipline too focused on individual risk factors and too disconnected from examination of the broader historical and social forces that determine population disease risk (Pearce 1996; Susser and Susser 1996a, 1996b).

They also suggested that ecoepidemiology was not simply an emphasis on more ecologic studies (Schwartz et al. 1999). The Sussers stressed that epidemiology, indeed, must learn to encompass multiple levels of organization from the societal to the molecular. Rather than diminish molecular epidemiology, they propose its further development and integration into studies of higher levels of organization. If a consensus position is apparent, it would be the need to draw more of a balance between modern risk factor epidemiology and population-based approaches that are closely linked to the historical roots of the discipline (Vena et al. 1999).

The ecoepidemiology paradigm proposed by the Sussers addresses the interdependence of individuals and their connection with the biological, physical, social, and historical contexts in which they live (Schwartz et al. 1999). It encompasses the changeable contributions and effects on the individual levels of both macro and microlevels of organization. The emphasis on the time dimension implies that health and disease, in fact, involve processes. Therefore, one would aim to assess causal factors at different levels of organization, over both the life course of the individual and the history of populations. Substantial methodological and inferential barriers need to be overcome, and available research designs and analytic techniques have not been well suited to elucidating processes at multiple levels of organization.

In 2005, Christopher Paul Wild called for increased resources to be devoted to studies of the "exposome" to complement the advances that have been made in genome research (Wild 2005). Cumulative risk assessment applications have become relatively common, not only for assessments of chemicals that operate by the same mode of action, but also for community-based, population-based assessments that may include more varied stressors than just chemicals alone. The demand for more sophisticated human health risk assessments has driven the need for research into cumulative risk assessment, population-focused assessments, aggregate exposure assessment, and risk from chemical mixtures. In addition, there was a recent call from Patel and Ioannidis (2014) for a new model to discover environmental exposures associated with disease incorporating "environment-wide association studies" (EWAS) methods. Buck Louis and Sundaram (2012) noted that investigations must accommodate measurement of multiple environmental exposures at different times in the life span, particularly in development, and new analytical methods must be able to capture the complex temporal relationship between multiple exposures. Considerations of individual variation based on genetic susceptibility, life stage, timing of exposures, and interaction of nonchemical stressors are required context for both routes and holistic assessment of risk factors associated with complex environmental disease. A framework for facilitating translating information to support risk assessment for decision making in public health was outlined by Cohen Hubal et al. (2013). We propose that multidisciplinary epidemiology, which embraces and utilizes the new and innovative approaches and data resources of GIS and geospatial methods from multiple epidemiologic perspectives, can address these methodological concerns, providing an integrated process to better understand disease from a more comprehensive viewpoint (Table 1.1).

The underlying theme of this chapter is to emphasize and develop a practical model to examine applications of GIS and geospatial analytic methods in different epidemiologic studies using different perspectives. Although GIS and other spatial methods are now being used more commonly in environmental epidemiology (Briggs and Elliott 1995; Croner et al. 1996), their use tends not to encompass the entire continuum of health. The ultimate purpose of public health is to directly impact the health of a community or individual. We propose that the examination of exposure–disease relationships from multiple perspectives (traditional, acute event, community) may provide a more comprehensive knowledge base. For example, GIS analyses, considering issues such as exposure, disease risk, and population composition,

TABLE 1.1
Relationship between proposed geospatial perspectives and theoretical comparisons

Geospatial perspective model	Geospatial methods	Issues	NIH clinical translational model
Traditional	Disease mapping Disease clustering Ecological analysis Exposure assessment	Simplistic emphasis on multiple risk factors	Bench
Acute event	Rapid event assessments Environmental toxin spread Future disaster prevention	Unprecedented challenges, such as global population growth, aging, and irreversible environmental health determinants	Bedside
Community	Social structure Community-derived data Maximizing impact via intervention location	Need for methods to examine historical and social forces that determine population disease risk	Community

could be used to decide on an ideal location for placing an intervention within a community for optimizing positive health changes.

This comprehensive viewpoint may be conceptualized as comparable to a clinical translational research model (CTSA) (Figure 1.1). In basic science, this emphasizes the practical application of clinical research (in which scientists study diseases at a molecular or cellular level) and then progresses to the clinical level or the patients' "bedside" (http://www.ncats.nih.gov/research/cts/ctsa/ctsa.html). In the geospatial model, we can consider the traditional perspective as the bench; that is, in an epidemiologic sense, this is the most basic and traditional way of understanding an exposure–disease relationship, through descriptive epidemiology and analytic hypothesis testing. Likewise, in the geospatial model, the acute event perspective can be considered similar to the bedside perspective; that is, an acute event occurs and the goal becomes handling a crisis quickly and efficiently while simultaneously collecting appropriate epidemiologic data—this is similar to the occurrence of a clinical event at the bedside. Finally, the last stage in both basic science and geospatial epidemiology is community; incorporating lessons from both perspectives provides a meaningful intervention in the community. Of course, this is neither a linear nor a single-discipline process. For example, in basic research, "basic scientists provide clinicians with new tools for use with patients and for assessment of their impact, and clinical researchers make novel observations about the nature and progression of disease that often stimulate basic investigations" (http://www.ncats.nih.gov/research/cts/ctsa/ctsa.html). In a similar way, use of geospatial tools may cross-fertilize research among the three proposed perspectives and academic disciplines.

The chapter will stress the importance of fostering collaborations and understanding GIS from multiple perspectives—traditional, disaster, and community. Applications for each of these perspectives (emphasizing the use of spatial methods) are outlined. The limitations and gaps that may arise when viewing a problem from only one perspective are also highlighted. Finally, an example of an integrated approach is given. The chapter ends with an

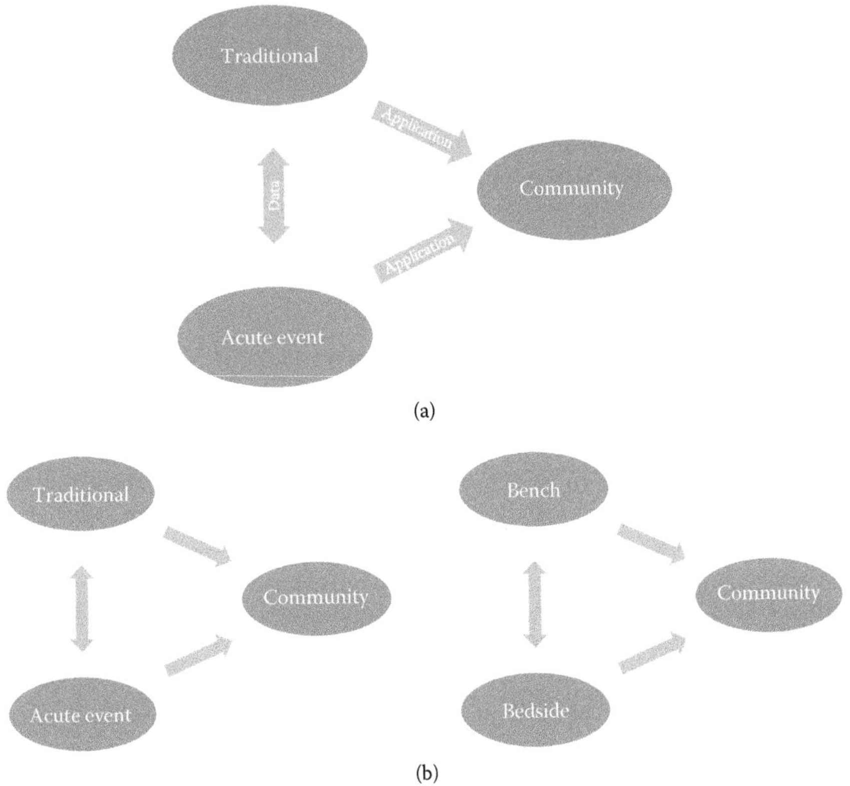

(a)

(b)

FIGURE 1.1
(a) Interconnectedness between the three proposed geospatial perspectives. At the most basic level, the traditional perspective provides quantitative data, which may be applied at the community perspective and may benefit understanding and implementation from the acute event perspective. Likewise, the acute event perspective provides unique data with different acute scenarios to strengthen understanding at the traditional perspective and which may be applied at the community perspective. (b) Relationship of geospatial perspectives to the clinical translational research model. The figure on the left represents the geospatial perspectives, which are mirrored by the figure on the right, representing the clinical translational research model.

annotated bibliography, which presents selected papers that represent each of the three perspectives.

1.2 Perspective 1: Traditional Applications

In the most traditional sense, GIS and spatial methods can be utilized as a means to understand exposure–disease relationships in descriptive epidemiology and etiologic studies. This can be compared to bench research in the clinical science translational model. In basic science, this refers to the direct efforts of scientists in the laboratory that test and yield scientific discoveries. Studies taking a traditional perspective of geospatial tools in environmental epidemiology are perhaps the most common. Studies in this category include the surveillance of infectious disease spread, electromagnetic field monitoring, and quantification of environmental risk factors and their potential impact on public health (Rytkönen 2004).

Traditional applications of GIS or spatial methods have been defined in many different ways. Lawson (2006) defines three hypotheses in spatial epidemiology: disease mapping, disease clustering, and ecological analysis. These classifications fit well within the traditional perspective as defined here. However, it should be noted that, ideally, the perspectives outlined in this chapter should not be conceptualized as compartments, but rather fluid perspectives. For the purposes of the traditional perspective, we will follow Lawson's grouping, considering

1. *Mapping.* What is the basic function of mapping in GIS? Why is mapping so fundamentally important to environmental risk assessment and environmental epidemiology in general? How has viewing GIS as a mapping tool hindered us?

2. *Clustering.* What is clustering? What can it tell us?

3. *Ecologic analysis.* Why is it important to take GIS beyond mapping? Why consider spatial modeling or geographic modeling? This section will also address the importance of using geospatial tools to improve exposure assessment in environmental epidemiology studies.

For a more detailed discussion of GIS in spatial epidemiology and public health, see Chapter 4 in this volume.

1.2.1 Disease mapping

Disease mapping can be used to provide *visual cues* about disease etiology, particularly as it relates to environmental exposures. A substantial portion of GIS applications in environmental epidemiology are cartographic (Kistenmann et al. 2002). Of course, this is a primary function of GIS, and can be a useful method for the initial identification of potential health problems (Kistenmann et al. 2002). The uses of GIS for cartographic purposes in environmental epidemiology are vast. Most fundamentally, GIS can be used to articulate "where?"—an essential question one must consider for a geographic foundation for public health (Cromley 2003). Mapping where things are allows visualization of a baseline pattern or spatial structure of disease, potential detection of disease clusters, and the initial investigation of an exposure–disease relationship (Jarup 2004). Much of disease mapping therefore focuses on using spatial methods to "clean" a map to better reveal spatial patterns (Lawson 2006). GIS may also be used to map quantities, densities, and change. For example, a continued mapping of disease over time may illuminate patterns coinciding with changes in environmental exposures (Jarup 2004). All of these uses provide a researcher with visual cues about the spatial structure of the disease, exposure, and covariates that encompass an epidemiologic hypothesis.

Although the visual power of GIS in environmental epidemiology and public health is great (Cromley 2003), the use of GIS should extend beyond cartography. Not only that, but there are true concerns about disease map misinterpretation and lack of accompanying statistical information (Lawson 2006). However, in public health, GIS continues to be conceptualized as a pure mapping tool (Jacquez 2000; Ricketts 2003; Graham et al. 2004), although more analytic applications are now being explored (Jerrett et al. 2010). After all, what purpose do visual cues serve if they are not visually honest, statistically tested, and practically useful (as the other perspectives will highlight)?

1.2.2 Disease clustering

Disease cluster detection can provide one with *relational cues* about epidemiologic problems. Clustering, in spatial epidemiology, refers to the detection of "unusual" aggregations of disease (Lawson 2006). Lawson and others define two primary types of cluster modeling: general and focused. General clustering is a nonspecific form of cluster detection where the primary purpose is to add to the descriptive understanding of a disease map. Therefore, in addition to understanding the visual nature of a disease, general clustering allows one to ask, "Is the spatial pattern of disease X statistically clustered?" Often, the output of a general clustering test is an autocorrelation parameter or some type of statistic to assess the aggregation of disease cases (Lawson 2006). Focused clustering asks, "Where in space (or space and time) are there clusters of disease X?" Focused clustering tests are important to refine locations for future study and to assess disease risk around environmental putative risk sources (Lawson 2006). Within these broad categories, general and focused, there are many different forms the cluster model can take—each with its own statistical model and estimation procedures (Lawson 2006). For a more detailed discussion of clustering, see Chapters 2 and 30 in this volume. Cluster modeling therefore is important to evaluate how disease (or exposure) data relate spatially (or spatially and temporally) to other disease (or exposure) data, but does not traditionally allow one to evaluate environmental etiologic risk factors for a particular disease (Glass 2000). Relational cues are insightful, but how functional is this type of information without statistical hypothesis testing?

1.2.3 Ecological analysis

Ecological analyses provide *etiologic cues* about the relationship between spatial disease distributions and explanatory factors. Geographic correlation studies may also be grouped within this category. Geographic correlation studies investigate geographic variations of environmental exposures, sociodemographic indicators, and lifestyle factors in relation to health outcomes at the aggregate level (Elliott and Wartenberg 2004). Ecologic studies may take advantage of specialized spatial statistical procedures to test statistical associations between exposures and disease at an aggregate level (Kistenmann et al. 2002; Lawson 2006). Exploratory ecologic or geographic correlation studies, which take advantage of aggregate-level data, allow one to develop and test spatial hypotheses of disease etiology (Elliott and Wartenberg 2004). For a more detailed discussion of regression, see Chapter 4 in this volume. Ecological analysis may therefore be conceptualized as an end product of spatial modeling—one of the mechanisms by which geospatial "bench discoveries" are made. However, despite the utility these types of analyses have in providing etiologic cues at an ecologic level, how can they be successfully translated to the individual without assessment and application at this level?

1.2.4 Geographic exposure modeling or exposure assessment

Another important application of GIS technology in epidemiology is its functionality in environmental exposure assessment. This application can be used for all types of epidemiologic studies (Vine et al. 1997). Nuckols et al. (2004) detailed several categories for which GIS could be used in exposure assessment, including defining the study population, identifying the source and potential route of exposure, estimating environmental levels of target contaminants, and estimating personal exposures. Along these lines, it is important to note that a main advantage of using GIS in exposure assessment is the ability to model exposure in a geographic sense, allowing one to save the time and resources needed to measure exposure at the individual level (Jarup 2004). Beyea (1999) terms this functionality geographic exposure modeling and emphasizes its priority in environmental epidemiology. Geographic modeling takes a study beyond using proximity as an exposure or provides predicted exposure estimates when direct measurements of exposure are not sufficient (Beyea 1999). Geographic modeling can go beyond just a GIS. Again, most tend to think of GIS only as a data storage, retrieval, and presentation system. However, a GIS is a sophisticated tool and can be used in combination with other spatial software to allow for the geographic analysis of health data (Jerrett et al. 2010).

1.2.5 Limitations of the traditional (bench) approach

Throughout this discussion, the utility of the traditional approach has been highlighted, but viewing GIS and spatial methods from only this perspective has limitations. First, as with the bench approach in basic science, if there is no application of findings, no impact will be made. Truly changing health calls for the community perspective (see Perspective 3). How can a discovered etiologic relationship, for example, be applied in the real world so that health outcomes are improved? Second, discoveries made through the traditional approach are typically limited to the data (or resources to collect data) available. Not every scenario with existing (or collected) data can possibly be tested using a traditional approach. Often, unique situations and exposures, unimagined events, and critical incidents occur. Ideally, therefore, one must possess the knowledge and skills to incorporate a disaster perspective (see Perspective 2). Finally, most traditional epidemiologic studies (especially those using ecologic analysis or those using GIS to improve exposure assessment in case-control studies) assume a chronic exposure (e.g., chronic drinking water consumption or long-term air pollution exposure). This may not always be the most important exposure, especially in disaster scenarios. Therefore, we next consider the utility of these geospatial technologies from the acute event and community perspectives.

1.3 Perspective 2: Acute Event Epidemiology Applications

Acute events, most broadly, may include infectious disease and disaster epidemiology. Acute event epidemiology is a grouping of epidemiologic studies in which the causes of and ways of controlling these types of events are examined. Understanding the causes of an event may incorporate knowledge of the mechanisms behind the event or the morbidity and mortality associated with the event. Controlling disasters and lessening the burden created by disasters incorporates preparedness, relief, and surveillance efforts. Geospatial techniques can be used to enhance all of these efforts. The perspective, as defined here, may be related to the bedside stage in the clinical translational model. In basic research, the bedside refers most commonly to the efforts of a clinician with a sick (often acutely sick) patient. Considerations in this stage include quickly assessing a patient's symptoms to decide on an appropriate course of treatment and prevention of future complications. A disaster or crisis creates an immediate need for rapid event evaluation to provide timely and appropriate care to the affected population and environment. The outbreak of Ebola in West Africa has dramatically illustrated the importance of these applications. Public health officials, such as the World Health Organization (WHO) Ebola Task Force, are using identified cases to improve surveillance and response techniques. Moreover, disaster epidemiology applications include the inhibition of future negative events, just as bedside treatments serve to deter subsequent morbidity. In this section, specific applications of geospatial methods in the acute event perspective will be highlighted: rapid event assessments, environmental toxin spread, and future event prevention.

1.3.1 Rapid event assessments

Immediately following a disaster, the effects of the acute event on the population must be rapidly assessed. GIS methods allow one to create a map of the event area for current and future planning; determine event burden and geographic extent (through spatial display of morbidity and mortality rates); and incorporate spatial data to improve assessment quality on environmental conditions, transportation, land usage, resource availability, and distance to health clinics, among other variables (Kaiser et al. 2003). Other more sophisticated geospatial analyses, such as a modified cluster sampling method, which has been used to conduct rapid needs assessment, have been used to evaluate immediate health and medical needs in an affected population quickly and efficiently (Waring et al. 2005). Furthermore, maps of population movements can be created using GIS functionality to ensure the selection of an appropriate relief location for population transfers, if necessary (Kaiser et al. 2003).

1.3.2 Environmental toxic agent and future disaster prevention

In addition to assisting with rapid disaster event assessments, geospatial tools may be used to enhance the tracking of toxic chemicals released in certain types of disasters in general and, more sophisticatedly, by incorporating environmental factors. *Toxin* is a general term and could refer to toxins in gas, air, water, radiation, oil spread, water spread due to hurricanes, or the destructive path of a tornado. For example, the Graniteville train wreck released large amounts of chlorine, and GIS plume modeling was used to model its spread to provide for a more efficient population evacuation and cleanup efforts (Svendsen et al. 2010). GIS methods were also used to plan a health registry and track participants geographically (Abara et al. 2014). GIS data layers enable the incorporation of environmental and geographical information that may affect the spread of a toxin or infectious disease, such as temperature, wind direction and speed, topography, and land cover. Incorporating these environmental and geographically specialized attributes allows for a more accurate assessment of toxin spread, which may ultimately create a more efficient disaster response. Finally, an important application of geospatial methods in disaster epidemiology is the prevention of future events through the creation of early-warning systems and disaster monitoring. GIS specifically can be used in these events to assist in site planning, enhance survey methods, establish health information systems, and integrate health data information systems (Kaiser et al. 2003).

1.3.3 Limitation of the acute event (bedside) approach

Despite the utility and importance of viewing applications of geospatial tools from a disaster perspective, there are caveats to maintaining this time-limited approach. For example, an acute event does not always provide a researcher with all the information necessary to fully understand a problem or a disease etiology. When dealing with acute events, the priority is often rapid needs assessment, and the formation of a health database or medical monitoring program often takes time and may not be complete initially. Viewing problems from the traditional perspective (see Perspective 1) allows the incorporation of data from long-standing databases to more thoroughly reach sound scientific conclusions. Additionally, it is important to examine how the information gleaned from a disaster evaluation can improve the community through the subsequent development of disease registries and through the implementation of disaster prevention interventions in spatially specific geographic locations (see Perspective 3, community).

1.4 Perspective 3: Community Approach

From the community perspective, integration of each of the first two perspectives (traditional and acute event) is paramount. From the community perspective, one can incorporate and apply what was learned from both the traditional and disaster perspectives at the community level. In this sense, a community perspective refers to epidemiologic research focused on both incorporating knowledge from the community (e.g., through community-based participatory research [CBPR]) and translating knowledge to the community (e.g., locating an optimal location for a community-based intervention). In fact, community health has been labeled the culmination of disease surveillance, risk analysis, and health access and planning (Nykiforuk and Flaman 2011). Like the other two perspectives discussed in this chapter, geospatial tools can be used to enhance community-level knowledge at a geographic perspective, and researchers are newly applying geographic methodologies to the exploration of these issues (Beyer et al. 2010). The community perspective, as defined by this chapter, is comparable to the community approach in the clinical research model. Just as a clinician may apply knowledge that was gained from the bench and bedside stages to the community itself, epidemiologists may use geospatial techniques to glean knowledge of etiologic disease risk factors in the traditional perspective and knowledge of acute disease manifestations in the disaster perspective and apply these lessons to more efficiently target the health of a community. Geospatial gradients, the joining and weighting of GIS data layers, may be used to combine many types of spatial information to make applications at the community level. Geospatial gradients can be used to consider the structure of a society, incorporate CBPR data, and prioritize locations of health interventions.

1.4.1 Social structure

An important component of the geospatial community perspective is understanding how the composition of a neighborhood or community may affect an exposure–disease relationship. Although the effects of social networks can manifest in many forms, consideration of such a phenomenon from a geographic perspective may be vital to truly understanding an epidemiologic question. Using GIS or other geospatial tools to assess geographic social structure effects may enhance understanding of an exposure–disease relationship. Relevant questions may include such issues as the following: What is the racial geographic distribution? Are health facilities located within a reachable distance to those who need them—not only as far as distance is concerned, but also as far as road networks and other public transportation routes? How are relevant disease risk factors distributed geographically? For example, are the locations where individuals can obtain healthy food items located within a reachable distance, and are they in culturally relevant locales?

Using GIS as a tool for understanding the social structure of a community and its health implications is an evolving concept. Once the physical aspects of a community have been explored, the social aspects can be examined. Recent studies have assessed elements of a community, such as feelings of belonging and social interaction with neighbors. GIS methods provide insight into the spatial scope of one's daily activities. More research is needed to elicit the personal value individuals place on their community integration (Chan et al. 2014).

1.4.2 Community-derived data

On another level, researchers have recently been attempting to engage community members in enhancing geographic information. The utilization of geospatial methods to infer social interactions and individual feelings is limited without community participation. Determining the attitudes of integration and the meaning of community is critical to understanding the dynamics of a population. Locally derived data employ individual perspectives and knowledge to reveal aspects of a community that are otherwise unobtainable. Innovative, participatory approaches to gathering locally derived data are being developed instead of relying on traditional survey methods. An example of this can be seen in participatory mapping, where individuals draw the locations they frequent and express the value these places have for them. GIS can then assess an individual's activity space and relate it to other community measures (Chan et al. 2014).

For example, there is a national movement in citizen science. Citizen science (also known as crowd science, crowd-sourced science, civic science, or networked science) is scientific research conducted, in whole or in part, by amateur or nonprofessional scientists. Formally, citizen science has been defined as "the systematic collection and analysis of data; development of technology; testing of natural phenomena; and the dissemination of these activities by researchers on a primarily avocational basis" (OpenScientist 2011). Citizen science is sometimes included in terms such as *public participation in scientific research* and *participatory action research* (Hand 2010). The Environmental Protection Agency, in a recent announcement of a funding opportunity, asked these fundamental questions:

> How can low-cost portable air pollution sensors be used by communities to understand and reduce the pollutant concentrations to which they are exposed, in outdoor and/or indoor environments? How do communities and individuals interact with low-cost portable air pollution sensors? What are effective distribution methodologies, training programs, design features, data products, etc., that maximize the value of sensors and their outputs for communities? What are effective methods for sharing and disseminating the information from low-cost portable air pollution sensors?

Substantial engagement with community groups was encouraged in this solicitation. Community-engaged research (CEnR) and community-based participatory research (CBPR) are frameworks for conducting research that supports citizen involvement in the research process, communication, and interpretation of the findings. Community engagement is the process of working collaboratively with and through groups of people that share a common geography, special interest, or similar situations to address issues affecting their overall well-being. Community engagement is one of the fundamental principles of public health that protects the right of the community to engage in making decisions about problems that affect their well-being (http://www.atsdr.cdc.gov/communityengagement/pdf/PCE_Report_508_FINAL.pdf). The most effective way to address a problem in a community is to involve the members of that community in every aspect that pertains to addressing that problem. This means involving the community from the very beginning to the end—from identifying the problem, addressing the problem, and evaluating the effectiveness of the solution to sharing results with the community. This fosters community buy-in, trust, and a sense of community empowerment, and eliminates a we–they polarity, a potential source of conflict (Abara et al. 2014). Svendsen et al. (2014) recently used these approaches to study air pollution in North Charleston, South Carolina.

1.4.3 Maximizing impact via intervention location

Perhaps one of the most important issues to consider from a community perspective is to define priority locations for resources and health-based interventions. As noted by Beyer et al. (2010), geospatial tools, specifically mapping and spatial analysis, allow researchers to evaluate the *where* in health research, by more effectively targeting interventions and resources at a geographic level. One idea is to create a geospatial gradient of where a health care clinic or other intervention (generated as a result of Perspective 1 and 2 assessments) will positively influence the community most effectively. This spatial gradient could incorporate many important pieces of geospatial information. For example, a spatial gradient may consider access to road networks and the ability to easily transport patients from residence to clinic or intervention. GIS technology allows one to consider road networks in terms of travel time, public transit, and walkability. Another important issue related to this is the placement proximity of a clinic or intervention to a community. In what type of community, considering type of social structure, race, education, and income, would the clinic or intervention be most effective? Many other geographic variables can be incorporated, including any type of environmental data and even data obtain directly from the community through something like CBPR research. For example, community members in different geographies can be asked to describe their greatest barriers or what they consider most important to them in accessing the clinic or intervention, and then the geographies or geographic attributes could be assigned accordingly.

1.4.4 Limitation of the community approach

The community perspective may appear to be the most broadly encompassing. However, there are limitations when one views the use of geospatial tools from only this perspective. Perhaps the most obvious limitation is that quantitative evidence is needed to fully understand an exposure–disease relationship before an intervention or health clinic based on some etiologic relationship can be established. This emphasizes the need for the traditional perspective. Moreover, another important source of information about an exposure–disease relationship may come through gathering of data following a disaster on the acute effects of an exposure. This makes the community perspective in GIS dependent on the foundation that the traditional and acute contexts provide. Once quantitative evidence is gathered and understood, community-derived data can be more efficiently collected and analyzed. This emphasizes the interconnectivity of different GIS perspectives and their role in multidisciplinary epidemiologic investigations.

1.5 Collaboration and Creativity

The importance of multidisciplinary efforts in epidemiology has been reinforced since the early 1990s. There was clear recognition of the need for cooperation among the disciplines, including industrial hygiene, statistics, risk assessment, meteorology, engineering, epidemiology, and biologic monitoring of both exposure and outcome. Collaborative efforts are paramount when proposing new visions for scientific research. The geospatial perspectives proposed in this chapter are an example of a creative and collaborative addition to existing epidemiologic research.

Dr. Saxon Graham, in his keynote address in 1988 as president of the Society for Epidemiologic Research, presented a paper on enhancing creativity in epidemiology (Graham 1988). It has not gotten much attention, but it offered valuable advice on how to be creative.

He suggested that creative people begin creative production early and continue later in life. He indicated that as quantity increased, so did quality. He summarized the observation with the statement, "So, the more you do, the better you are going to be at doing it." He suggested that the individual needed energy, dedication, and enthusiasm. He urged early contact with top-notch mentors. He stressed the importance of finding kindred spirits to work with. He suggested that creative people were not wary of making errors and ranged far and wide in their interests and activities. They take chances and entertain thoughts that lead to some error and are not afraid of making mistakes. The stimuli to creativity are injection of new ideas from new sources and exposure to a diversity of new concepts and fields. He summarized by emphasizing that creativity meant learning from each other. He stated that there was a need to share ideas and methods to be creative. Stimulus of the new often derives from new technology and methods.

Dr. Roberta Ness encourages "frame shifting" to maximize innovation. She advocates that creativity in science is not always linear, and that this can make us uncomfortable. Instead, we should embrace the "meandering and torture" that is inherent in creativity. Dr. Ness suggests group collaboration is powerful, but it is important to remain realistic about their benefits. Groups do not always stimulate the initiation of ideas, but often promote the elaboration of ideas. This stresses the importance of interdisciplinary collaboration for brainstorming and idea development (Ness 2012).

Creative production is generally the result of the innovative joining of two disparate elements already in the field or of elements in the primary field with elements from a new field (Graham 1988). New issues in epidemiology require new approaches. These also involve individuals with perceptions and knowledge different from traditional epidemiology. Creativity in this new setting requires utilization of these new technologies. There is risk as new procedures are included. Innovation in multidisciplinary research has progressed as the members of the team achieve understanding and sharing of ideas. New ideas, approaches, and methods must stand up to criticism and scrutiny. However, competitiveness and being supercritical will certainly reduce the amount of creative works produced. Too much criticism likely has been one of the important factors impeding the progress in spatial epidemiology. We propose that by more formal and integrative collaboration between environmental epidemiologists, geographers, and related disciplines such as geospatial scientists and statisticians and remote sensing experts, scientific innovations will be encouraged.

We also propose that multidisciplinary collaborations and scientific creativity are necessary in achieving a new frontier in spatial epidemiology, and we believe that incorporating new geospatial perspective into existing thought is an important starting point.

1.6 Integration of Perspectives

This chapter has discussed the importance of using geospatial tools from multiple epidemiologic perspectives, specifically the traditional (bench), disaster (bedside), and community (community) perspectives. To better illustrate this concept, a theoretical example is provided here, which takes an arbitrary toxin A and disease Y through the continuum of perspectives as defined by this chapter.

1.6.1 Traditional

The traditional perspective focuses on revealing etiologic cues about an exposure (toxin A) and disease (disease Y) relationship. First, GIS may be used for descriptive mapping, attempting to provide visual cues about an exposure or health problem. For example, consider toxin A as a gas, and that exposure monitors distributed throughout the study region (the state of Georgia, in this example) have collected concentration data. Concentrations of toxin A at these monitors can be incorporated in a spatial interpolation model (the procedure of estimating the values of properties at unsampled sites within an area covered by existing observations) to generate and map a predicted surface of toxin A concentrations throughout the state of Georgia. Additionally, descriptive mapping may be used to display outcome data. In this scenario, a statewide registry for disease Y may exist, and rates of disease Y may be mapped individually at geocoded point locations (pending confidentiality considerations) or by an aggregate geographic level, such as Georgia counties. As a result, visual information will help determine if areas with higher predicted concentrations of toxin A also have a higher occurrence or higher rates of disease Y.

Next, relational cues about toxin A and disease Y may be examined. First, a focused cluster analysis around a specific point source related to toxin A could be performed. The question would be, is disease Y clustered around point source X (which emits toxin A)? Results might show that a statistically elevated cluster of disease Y exists around point source X. This may give us a clue about a primary source of origin for toxin A.

A second approach is to test for general clustering, answering the question "Where, in Georgia, are disease Y and toxin A clustered together?" This type of clustering test may be used to better understand the spatial relationship of toxin A and disease Y. Another important goal of cluster analysis may be the identification of a specific location for individual-level studies of high-risk populations. Practically, resources are limited for collecting data for individual-level studies, and using geospatial tools such as cluster analysis can facilitate highlighting areas of most importance (i.e., areas with the most spatial exposure or disease overlap). A geographic area of interest is now defined, and resources for the next series of analyses can be focused.

The next step in this geospatial analysis is an evaluation of potential etiologic cues of the relationship between toxin A and disease Y through an ecological spatial model. This type of analysis allows the evaluation of a statistical relationship with the inclusion of covariates and potential adjustment for spatial autocorrelation. A potential question would be, is there more disease Y in areas with higher toxin A in the environment? Answering this type of question will lead to a firm statistical hypothesis that can be tested with an individual-level assessment.

A final step for a traditional assessment is the individual-level study toward which the analyses have been building. Geospatial tools can be applied to improve exposure assessment, determine important covariates related to the exposure–disease relationship, and assess a statistical association between exposure and disease (i.e., does toxin A effect disease Y? to what extent? and what is the strength of this relationship dependent on?). Depending on the type of exposure data available to the researcher, several options exist for how best to use GIS for improving exposure assessment. If only monitor-level data are available, spatial interpolation techniques could be used to estimate a predicted surface of toxin A concentrations. In a similar manner, environmental contamination data may have been collected near individual residential locations. If these are indirect measurements, data could be interpolated in a similar manner and used to assign exposure status to individuals based on their geocoded residential address. The GIS may also be used to assign geographic-based covariates to individuals for use in our statistical model. For example, different spatial data layers could be incorporated to determine an index of susceptibility, including environmental data affecting the concentration of toxin A, such as amount of precipitation or elevation (depending on the route of exposure). Also, using mixed modeling techniques, census variables such as population density or income may affect an individual's susceptibility and could be incorporated if one lacked individual assessment of these types of variables.

1.6.2 Acute event

The next step in this example is to examine toxin A and disease Y from the acute event perspective. Statistical knowledge about the relationship between toxin A and disease Y based on our examination of the problem using geospatial tools from the traditional perspective is now available. However, there are additional pieces of knowledge that can be gleaned from this perspective. For example, imagine a scenario where toxin A was a gas, and there was an explosion at a chemical plant containing toxin A. From the acute event perspective, geospatial tools can be used to geographically model the spread of the plume so populations most affected by its spread can be identified. With this, environmental variables that may affect the flow of toxin A, such as topology, wind direction, and speed, may be considered. This would provide us with additional information on the properties of its environmental spread. Similarly, if toxin A was a liquid, GIS and geospatial water flow models could be used to determine the toxin's spread.

Another important piece of information that can arise from examination of the problem from this perspective is how quickly toxin A may affect people. A disaster scenario creates a natural experiment that allows for the evaluation of a unique situation where toxin A spreads quickly and at high concentrations throughout an unsuspecting population. Epidemiologic evidence, if collected properly and in a timely manner, will allow the assessment of the acute effects of the toxin.

In the long term in this disaster scenario, the identification of an optimal location for a disease registry could be accomplished efficiently with geospatial tools. This is important for long-term data collection to take advantage of the natural experiment setting. This may be accomplished with geospatial tools as discussed before, taking into consideration the spatial distribution of the toxin, disease, and population.

1.6.3 Community

A final method for understanding the toxin A–disease Y problem is its examination through the community perspective. An initial question is, where is the ideal location (based on the potential reduction in disease Y) for an intervention against toxin A? This location choice will be based on several factors, including the findings from the traditional and acute event–based geospatially focused studies. In addition, community involvement and CBPR practices allow for a determination from a community perspective as to what location would be optimal for their intervention to be performed. Several working groups with predefined questions could be held, which evaluate what community members consider most important in their access to the intervention.

Spatial GIS layers of geographic variables can also be incorporated and weighted by importance to help determine an optimal intervention location. For example, GIS can be used to determine which areas (e.g., counties) have people that may be the most susceptible to toxin A based on their demographic and social characteristics (e.g., lower income and older age). Those populations and geographic areas with a higher susceptibility would be considered a better location for the intervention. In a similar manner, GIS spatial data layers could be used to incorporate information on specific areas lacking physicians or health services or health insurance coverage in general or those types of services specific to treating either toxin A or disease Y. A GIS could be used to map the density of health services by geography, or more sophisticated techniques could be used to incorporate accessibility based on ease of transportation and road network structures.

Environmental exposure to toxin A could also be refined with GIS spatial layers. This type of analysis may consider exposure based on distance to a contaminated river, for example. A contaminated river (or contaminated portion of river) could be identified, and

GIS buffers could be drawn around the river to identify the most exposed area. Buffers could be calculated from the river to consider gradations of exposure—or a buffer could be generated by more sophisticated modeling considering the environmental determinants of exposure (e.g., the people to the northwest of the river may be more exposed due to wind direction, so the buffer angle would emphasize the northwest). Soil types or air quality could also be considered. For example, one could create a map of soil types that are more likely to contain toxin A (if soil based) and could weight exposure based on the type of soils underlying a population. Then exposure could be assigned to individual residential locations or averages, or proportions of high-exposure soils could be determined by geographic area.

Another important issue to consider is the number of people that an intervention would serve. GIS could be used to calculate population density by geographic area and then target areas with the most population. Another important question is, if the intervention is using a health district or facility as a platform, who is served by that facility? GIS can be used to determine what type of population (what characteristics they have) is being served by drawing a boundary around a service area that considers its targeted population. This could be based on a radial distance or on the road network system. This would allow one to determine the extent of the population served (if more than one intervention housing center is allowable) and to better target health care services (if certain languages or races are targeted for educational materials or focus groups).

Although most often we think of geographic layers in terms of those that make a person or a particular area more susceptible to an exposure or disease, there are also layers that may increase protection against an exposure or disease (i.e., protective layers). For example, neighborhood green space or walkability may be considered if the outcome is linked with physical fitness. Other protective variables that have been discussed include a positive neighborhood composition with strong social ties. This would make one person going to a positive intervention more influential of other population members going.

1.7 Conclusion

The utilization of different epidemiologic perspectives fosters a multidisciplinary approach to understanding the relationships between health, exposures, and outcomes. Further, these perspectives can be applied through a geospatial lens to better understand population-level disease determinants. The geospatial analytic methods discussed in this chapter illustrate practical approaches to epidemiological studies. The traditional, disaster, and community perspectives ultimately cultivate both a comprehensive understanding and an efficient solution to epidemiological questions.

1.8 Annotated Bibliography Summary

The annotated bibliography provided at the end of this chapter is comprised of multiple applications of GIS in public health domains. Differing methodological approaches in spatial research were classified as traditional, acute event, or community based in nature. Traditional approaches include surveillance of infectious disease and quantification of environmental risk factors. These traditional approaches include disease mapping and clustering, exposure assessment, and ecological analysis. Acute event epidemiology focuses on the causes, assessment, and prevention of natural disasters or isolated events. The community-based approach works to incorporate knowledge from the community or translate knowledge back to the community. Multiple examples of each classification are included.

In addition, we identified a wide variety of review articles focusing on specific and general methodological approaches using GIS as a tool for spatial and temporal research. General methodological reviews highlight the usefulness of certain GIS methods in public health and also provide an overview of the expanding field of spatial epidemiology. These review articles focus on literature that use GIS as the organizing system for health data and present examples of how and where GIS-based analyses will be expanded in the future (Moore and Carpenter 1999; Rushton 2003; Jerrett et al. 2010). Specific review articles document the use of spatial epidemiological tools for emerging viral zoonoses and the control and global distribution of infectious disease (Rogers and Randolph 2003; Yang et al. 2005; Clements and Pfeiffer 2009).

Current literature corroborates the value of GIS and other geospatial tools add to the fields of epidemiology, medical geography, biostatistics and related disciplines. However, authors caution researchers and readers about the errors in the design, analysis and interpretation of geospatial models (Cromley 2003; Elliott and Wartenberg 2004; Graham et al. 2004; Nuckols et al. 2004; Rytkönen 2004). Future research incorporating GIS in public health disciplines is needed to communicate a comprehensive understanding of the data exploration and analysis strategies used by researchers in public health applications.

References

Abara, W., S. Wilson, J. Vena, L. Sanders, T. Bevington, J. M. Culley, L. Annang, L. Dalemarre, and E. Svendsen. (2014). Engaging a chemical disaster community: lessons from Graniteville. *Int J Environ Res Public Health* 11(6): 5684–5697.

Beyea, J. (1999). Geographic exposure modeling: a valuable extension of geographic information systems for use in environmental epidemiology. *Environ Health Perspect* 107(Suppl 1): 181–190.

Beyer, K. M., S. Comstock, and R. Seagren. (2010). Disease maps as context for community mapping: a methodological approach for linking confidential health information with local geographical knowledge for community health research. *J Commun Health* 35(6): 635–644.

Briggs, D. J. and P. Elliott. (1995). The use of geographical information systems in studies on environment and health. *World Health Stat Q* 48(2): 85–94.

Buck Louis, G. M., and R. Sundaram. (2012). Exposome: time for transformative research. *Stat Med* 31(22): 2569–2575.

Chan, D. V., C. A. Helfrich, N. C. Hursh, E. S. Rogers, and S. Gopal. (2014). Measuring community integration using geographic information systems (GIS) and participatory mapping for people who were once homeless. *Health Place* 27: 92–101.

Clements, A. C. and D. U. Pfeiffer. (2009). Emerging viral zoonoses: frameworks for spatial and spatiotemporal risk assessment and resource planning. *Vet J* 182(1): 21–30.

Cohen Hubal, E. A., T. de Wet, L. Du Toit, M. P. Firestone, M. Ruchirawat, J. van Engelen, and C. Vickers. (2013). Identifying important life stages for monitoring and assessing risks from exposures to environmental contaminants: results of a World Health Organization review. *Regul Toxicol Pharmacol* 2014; 69(1): 113.

Cromley, E. K. (2003). GIS and disease. *Annu Rev Public Health* 24: 7–24.

Croner, C. M., J. Sperling, and F. R. Broome. (1996). Geographic information systems (GIS): new perspectives in understanding human health and environmental relationships. *Stat Med* 15(17–18): 1961–1977.

Elliott, P. and D. Wartenberg. (2004). Spatial epidemiology: current approaches and future challenges. *Environ Health Perspect* 112(9): 998–1006.

Fielding, J. E. (1999). Public health in the twentieth century: advances and challenges. *Annu Rev Public Health* 20: xiii–xxx.

Glass, G. E. (2000). Update: spatial aspects of epidemiology: the interface with medical geography. *Epidemiol Rev* 22(1): 136–139.

Graham, A. J., P. M. Atkinson, and F. M. Danson. (2004). Spatial analysis for epidemiology. *Acta Trop* 91(3): 219–225.

Graham, S. (1988). Enhancing creativity in epidemiology. *Am J Epidemiol* 128(2): 249–253.

Hand, E. (2010). Citizen science: People power. *Nature* 466(7307): 685–687.

Jacquez, G. M. (2000). Spatial analysis in epidemiology: nascent science or a failure of GIS? *J Geograph Syst* 2: 91–97.

Jarup, L. (2004). Health and environment information systems for exposure and disease mapping, and risk assessment. *Environ Health Perspect* 112(9): 995–997.

Jerrett, M., S. Gale, and C. Kontgis. (2010). Spatial modeling in environmental and public health research. *Int J Environ Res Public Health* 7(4): 1302–1329.

Kaiser, R., P. B. Spiegel, A. K. Henderson, and M. L. Gerber. (2003). The application of geographic information systems and global positioning systems in humanitarian emergencies: lessons learned, programme implications and future research. *Disasters* 27(2): 127–140.

Kistenmann, T., F. Dangendorf, and J. Schweikart. (2002). New perspectives on the use of geographical information systems (GIS) in environmental health sciences. *Int J Hyg Environ Health* 205: 169–181.

Lawson, A. B. (2006). *Statistical Methods in Spatial Epidemiology*, 2nd ed. Hoboken, NJ: Wiley.

Moore, D. A. and T. E. Carpenter. (1999). Spatial analytical methods and geographic information systems: use in health research and epidemiology. *Epidemiol Rev* 21(2): 143–161.

Ness, R. (2012). *Innovation generation: How to Produce Creative and Useful Scientific Ideas.* Oxford: Oxford University Press.

Nuckols, J. R., M. H. Ward, and L. Jarup. (2004). Using geographic information systems for exposure assessment in environmental epidemiology studies. *Environ Health Perspect* 112(9): 1007–1015.

Nykiforuk, C. I. and L. M. Flaman. (2011). Geographic information systems (GIS) for health promotion and public health: a review. *Health Promot Pract* 12(1): 63–73.

OpenScientist. (2011). Finalizing a definition of "citizen science" and "citizen scientists." OpenScientist, September 3. http://www.openscientist.org/2011/09/finalizing-definition-of-citizen.html.

Patel, C. J. and J. P. Ioannidis. (2014). Studying the elusive environment in large scale. *JAMA* 311(21): 2173–2174.

Pearce, N. (1996). Traditional epidemiology, modern epidemiology, and public health. *Am J Public Health* 86(5): 678–683.

Ricketts, T. C. (2003). Geographic information systems and public health. *Annu Rev Public Health* 24: 1–6.

Rogers, D. J. and S. E. Randolph. (2003). Studying the global distribution of infectious diseases using GIS and RS. *Nat Rev Microbiol* 1(3): 231–237.

Rushton, G. (2003). Public health, GIS, and spatial analytic tools. *Annu Rev Public Health* 24: 43–56.

Rytkönen, M. (2004). Not all maps are equal: GIS and spatial analysis in epidemiology. *Int J Circumpolar Health* 63: 9–24.

Schwartz, S., E. Susser, and M. Susser. (1999). A future for epidemiology? *Annu Rev Public Health* 20: 15–33.

Susser, M. and E. Susser. (1996a). Choosing a future for epidemiology. I. Eras and paradigms. *Am J Public Health* 86(5): 668–673.

Susser, M. and E. Susser. (1996b). Choosing a future for epidemiology. II. From black box to Chinese boxes and eco-epidemiology. *Am J Public Health* 86(5): 674–677.

Svendsen, E. R., S. Reynolds, O. A. Ogunsakin, E. M. Williams, H. Fraser-Rahim, H. Zhang, and S. M. Wilson. (2014). Assessment of particulate matter levels in vulnerable communities in North Charleston, South Carolina prior to port expansion. *Environ Health Insights* 8: 5–14.

Svendsen, E. R., N. C. Whittle, L. Sanders, R. E. McKeown, K. Sprayberry, M. Heim, R. Caldwell, J. J. Gibson, and J. E. Vena. (2010). GRACE: public health recovery methods following an environmental disaster. *Arch Environ Occup Health* 65(2): 77–85.

Vena, J. E., C. B. Ambrosone, F. F. Kadlubar, and J. E. Vena. (1999). The authors reply. *Am J Epidemiol* 149(11): 1072–1073.

Vine, M. F., D. Degnan, and C. Hanchette. (1997). Geographic information systems: their use in environmental epidemiologic research. *Environ Health Perspect* 105: 598–605.

Waring, S., A. Zakos-Feliberti, R. Wood, M. Stone, P. Padgett, and R. Arafat. (2005). The utility of geographic information systems (GIS) in rapid epidemiological assessments following weather-related disasters: methodological issues based on the Tropical Storm Allison experience. *Int J Hyg Environ Health* 208(1–2): 109–116.

Wild, C. P. (2005). Complementing the genome with an "exposome": the outstanding challenge of environmental exposure measurement in molecular epidemiology. *Cancer Epidemiol Biomarkers Prev* 14(8): 1847–1850.

Yang, G. J., P. Vounatsou, X. N. Zhou, J. Utzinger, and M. Tanner. (2005). A review of geographic information system and remote sensing with applications to the epidemiology and control of schistosomiasis in China. *Acta Trop* 96(2–3): 117–129.

Annotated Bibliography

Perspective 1: Traditional applications

- **Bauer, S., Wagner, S., Burch, J., Bayakly, R., and Vena, J. A case-referent study: light at night and breast cancer risk in Georgia. *International Journal of Health Geographics* 2013; 12: 23.** This study examined the codistribution of light at night (LAN) and breast cancer incidence in Georgia. DMSP-OLS Nighttime Light Time Series satellite images were used to estimate LAN levels. LAN levels were extracted for each year of exposure prior to case or referent diagnosis using ArcGIS. Study results suggest positive associations between LAN and breast cancer incidence, especially among whites. The consistency of author findings with previous studies suggests that there could be fundamental biological links between exposure to artificial LAN and increased breast cancer incidence, although additional research using exposure metrics at the individual level is required to confirm or refute findings.
- **Block, J. P., Scribner, R. A., and DeSalvo, K. B. Fast food, race/ethnicity, and income—a geographic analysis. *American Journal of Preventive Medicine* 2004; 27(3): 211.** In this paper, the geographic distribution of fast food restaurants is examined relative to neighborhood sociodemographics. Using GIS software, all fast food restaurants within the city limits of New Orleans, Louisiana, were mapped. Buffers around census tracts were generated to simulate 1- and 0.5-mile "shopping areas" around and including each tract, and fast food restaurant density was calculated for each area. The link between fast food restaurants and black and low-income neighborhoods may contribute to the understanding of environmental causes of the obesity epidemic in these populations.
- **Bove, G. E., Rogerson, P. A., and Vena, J. E. Case control study of the geographic variability of exposure to disinfectant byproducts and risk for rectal cancer. *International Journal of Health Geographics* 2007; 6.** This work assessed the geographic variability in exposure to trihalomethane (THM) disinfectant by-products and rectal cancer in white males in western New York State. Using a combination of case-control methodology and spatial analysis, the spatial patterns of THMs and individual measures of tap water consumption provided estimates of the effects of ingestion of specific amounts of some by-products on rectal cancer risk. Addresses of THM sample collection sites were used, together with residential addresses from the case-control study using GIS. Case-control addresses allowed for determination of the location of each study participant in the water distribution system in relation to the water sample points. Trihalomethane levels varied spatially within the county, although risk for rectal cancer did not increase with total level of trihalomethanes.
- **Brauer, M., Hoek, G., van Vliet, P., Meliefste, K., Fischer, P., Gehring, U., et al. Estimating long-term average particulate air pollution concentrations: application of traffic indicators and geographic information systems. *Epidemiology* 2003; 14(2): 228.** The authors used a measurement and modeling procedure to estimate long-term average exposure to traffic-related particulate air pollution in communities throughout the Netherlands; in Munich, Germany; and in Stockholm

County, Sweden. At each site, fine particles and filter absorbance were measured and used to calculate annual average concentrations. Traffic-related variables were collected using GIS and used in regression models predicting annual average concentrations. A substantial fraction of the variability in annual average concentrations for all locations was explained by traffic-related variables. This approach can be used to estimate individual exposures for epidemiologic studies and offers advantages over alternative techniques relying on surrogate variables or traditional approaches that utilize ambient monitoring data alone.

- **Brooker, S. Schistosomes, snails and satellites. *Acta Tropica* 2002; 82(2): 207.** This paper gives an overview of the recent progress made in the use and application of geographical information systems (GIS) and remotely sensed (RS) satellite sensor data for the epidemiology and control of schistosomiasis in sub-Saharan Africa. Details are given for the use of GIS to collate, map, and analyze available parasitological data. The use of RS data to better understand the broad-scale environmental factors influencing schistosome distribution is defined, and examples are detailed for the prediction of schistosomiasis in unsampled areas. Finally, the current practical applications of GIS and remote sensing are reviewed in the context of national control programs.

- **Burch, J. B., Wagner Robb, S., Puett, R., Cai, B., Wilkerson, R., Karmaus, W., et al. Mercury in fish and adverse reproductive outcomes: results from South Carolina. *International Journal of Health Geographics* 2014; 13: 30.** This work is the first to examine the relationship between fish total mercury concentrations and adverse reproductive outcomes in a large population-based sample of African-American women. Geocoded residential locations for live births were linked with spatially interpolated total mercury concentrations in fish to estimate potential mercury exposure from consumption of locally caught fish. Using spatial coordinates for fish sample locations, a geostatistical model within a GIS was used to map the predicted statewide distribution of total mercury concentrations in fish. It was hypothesized that risk of low birth weight or preterm birth was greater among women living in areas with elevated total mercury in fish. The results suggest a need for more detailed investigations to characterize patterns of local fish consumption and potential dose–response relationships between mercury exposure and adverse reproductive outcomes, particularly among African-American mothers.

- **Clennon, J. A., King, C. H., Muchiri, E. M., Kariuki, H. C., Ouma, J. H., Mungai, P., et al. Spatial patterns of urinary schistosomiasis infection in a highly endemic area of coastal Kenya. *American Journal of Tropical Medicine and Hygiene* 2004; 70(4): 443.** Urinary schistosomiasis remains a major contributor to the disease burden along the southern coast of Kenya. Selective identification of transmission hot spots offers the potential for more effective disease control and reduction in transmission. In the present study, a GIS was used to integrate demographic, parasitologic, and household location data for an endemic village and neighboring households. A global spatial statistic was used to detect area-wide trends of clustering for human infection at the household level. Local spatial statistics were then applied to detect specific household clusters of infection and, as a focal spatial statistic, to evaluate clustering of infection around a putative transmission site. High infection intensities were clustered significantly around a water contact site.

- **Green, C., Yu, B. N., and Marrie, R. A. Exploring the implications of small-area variation in the incidence of multiple sclerosis. *American Journal of Epidemiology* 2013; 178(7): 1059.** This study describes the geospatial variation in the incidence of multiple sclerosis (MS) in Manitoba, Canada, and the sociodemographic characteristics associated with MS incidence. Administrative health data were used to identify all incident cases of MS and geocode cases to 230 neighborhoods in the city of Winnipeg and 268 municipalities in rural Manitoba. The application of the spatial scan statistic revealed high-rate clusters in southwestern and central Winnipeg and low-rate clusters in north-central Winnipeg and northern Manitoba. This study suggests that the causes of MS are pervasive across all population groups. Searching for local-level causes of the disease may therefore not be as productive as investigating etiological factors operating at the population level.

- **Han, D. W., Rogerson, P. A., Nie, J., Bonner, M. R., Vena, J. E., Vito, D., et al. Geographic clustering of residence in early life and subsequent risk of breast cancer (United States). *Cancer Causes and Control* 2004; 15(9): 921.** This study focused on residential geographic clustering of breast cancer cases and identified spatiotemporal clustering of cases and controls. A geographical information system was used to geocode all residential locations and the k-function difference between cases and controls was used to identify spatial clustering patterns of residence in early life. Study findings provide evidence that clustered residences at birth and at menarche were stronger than those for first birth or other time periods in adult life. This study provides evidence that early environmental exposures may be related to breast cancer risk, especially for premenopausal women.

- **Hebert, J. R., Daguise, V. G., Hurley, D. M., Wilkerson, R. C., Mosley, C. M., Adams, S. A., et al. Mapping cancer mortality-to-incidence ratios to illustrate racial and sex disparities in a high-risk population. *Cancer* 2009; 115(11): 2539.** The mortality-to-incidence rate ratio (MIR) provides a population-based indicator of survival. South Carolina Central Cancer Registry incidence data and Vital Registry death data were used to construct MIRs, and ArcGIS 9.2 was used to map cancer MIRs by sex and race for eight health regions within South Carolina. Racial differences in cancer MIRs were observed for both sexes for all cancers combined and for most individual sites. Comparing and mapping race- and sex-specific cancer MIRs provided a powerful way to observe the scope of the cancer problem. MIR mapping allows for pinpointing areas where future research has the greatest likelihood of identifying the causes of large, persistent, cancer-related disparities. Other regions with access to high-quality data may find it useful to compare MIRs and conduct MIR mapping.

- **Li, Z., Yin, W., Clements, A., Williams, G., Lai, S., Zhou, H., et al. Spatiotemporal analysis of indigenous and imported dengue fever cases in Guangdong province, China. *BMC Infectious Diseases* 2012; 12: 132.** This study aimed to explore spatiotemporal characteristics of imported and indigenous dengue fever cases in Guangdong province. The space–time scan statistic was used to determine space–time clusters of dengue fever cases at the county level, and a geographical information system was used to visualize the location of the clusters. This study demonstrated that the geographic range of imported and indigenous dengue fever cases has expanded over recent years, and cases were significantly clustered in two heavily urbanized areas of Guangdong province. This provides the foundation for further investigation of risk factors and interventions in these high-risk areas.

- **Luque Fernandez, M. A., Schomaker, M., Mason, P. R., Fesselet, J. F., Baudot, Y., Boulle, A., et al. Elevation and cholera: an epidemiological spatial analysis of the cholera epidemic in Harare, Zimbabwe, 2008–2009.**

***BMC Public Health* 2012; 12: 442.** This study analyzed the association between topographic elevation and the distribution of cholera cases in Harare during the cholera epidemic in 2008 and 2009. The study illustrated average elevation and cholera cases by suburbs using geographical information. Study findings identified a spatial pattern of the distribution of cholera cases in the Harare epidemic, characterized by a lower cholera risk in the highest elevation suburbs of Harare. This study highlights the importance of considering topographical elevation as a geographical and environmental risk factor in order to plan cholera-preventive activities linked with water and sanitation in endemic areas. Furthermore, elevation information, among other risk factors, could help to spatially orientate cholera control interventions during an epidemic.

- **Mazumdar, S., Winter, A., Liu, K. Y., and Bearman, P. Spatial clusters of autism births and diagnoses point to contextual drivers of increased prevalence. *Social Science and Medicine* 2013; 95: 87.** Identifying the spatial pattern of autism cases at birth and at diagnosis can help clarify which contextual drivers are affecting autism's rising prevalence. Authors searched for spatial clusters of autism at time of birth and at time of diagnosis by geocoding residential addresses using a GIS and applying a spatial scan approach that controls for key individual-level risk factors. Findings implicate a causal relationship between neighborhood-level diagnostic resources and spatial patterns of autism incidence, but do not rule out the possibility that environmental toxicants have also contributed to autism risk.
- **Nagel, C. L., Carlson, N. E., Bosworth, M., and Michael, Y. L. The relation between neighborhood built environment and walking activity among older adults. *American Journal of Epidemiology* 2008; 168(4): 461.** This study examined the relationship between objectively measured characteristics of the local neighborhood and walking activity among a sample of 546 community-dwelling older adults in Portland, Oregon. A GIS was used to derive measures of the built environment within a quarter-mile (0.4 km) and half-mile (0.8 km) radius around each participant's residence. Multilevel regression analysis was used to examine the association of built environment with walking behavior. These findings suggest that built environment may not play a significant role in whether older adults walk, but among those who do walk, it is associated with increased levels of activity.
- **Shyu, H. J., Lung, C. C., Ho, C. C., Sun, Y. H., Ko, P. C., Huang, J. Y., et al. Geographic patterns of hepatocellular carcinoma mortality with exposure to iron in groundwater in Taiwanese population: an ecological study. *BMC Public Health* 2013; 13: 352.** This research aimed to explore the geographical distribution of hepatocellular carcinoma (HCC) mortality rates and evaluate the association between HCC mortality, land subsidence, and iron levels in groundwater in Taiwan. Authors conducted an ecological study and calculated the HCC age-standardized mortality and incidence rates according to death certificates issued in Taiwan from 1992 to 2001 and incidence data from 1995 to 1998. Both geographical information systems and Pearson correlation coefficients were used to analyze the relationship between HCC mortality rates, land subsidence, and iron concentrations in groundwater. This study showed that HCC mortality is clustered in southwestern Taiwan and the association with the iron levels in groundwater in the Taiwanese population warrants further investigation.
- **Wagner, S. E., Bauer, S. E., Bayakly, A. R., and Vena, J. E. Prostate cancer incidence and tumor severity in Georgia: descriptive epidemiology, racial disparity, and geographic trends. *Cancer Causes and Control* 2013; 24(1): 153.** This study describes and compares the temporal and geographic trends

of prostate cancer incidence in Georgia as mapped using a geographical information system. County-level hot spots of prostate cancer incidence were analyzed with the Getis-Ord Gi* statistic in a GIS, and a census tract-level cluster analysis was performed with a discrete Poisson model and implemented in SaTScan® software. This study revealed a pattern of higher incidence and more advanced disease in northern and northwest-central Georgia, highlighting geographic patterns that need more research and investigation of possible environmental determinants.

- **Wagner, S. E., Burch, J. B., Bottai, M., Puett, R., Porter, D., Bolick-Aldrich, S., et al. Groundwater uranium and cancer incidence in South Carolina. *Cancer Causes and Control* 2011; 22(1): 41.** This ecologic study tested the hypothesis that census tracts with elevated groundwater uranium and more frequent groundwater use have increased cancer incidence. A GIS was used to combine incident cancer cases from the South Carolina Central Cancer Registry, demographic and groundwater consumption information from the U.S. Census Bureau, and groundwater uranium concentrations. Census tracts with $\geq$50% groundwater use and uranium concentrations in the upper quartile had increased risks for colorectal, breast, kidney, prostate, and total cancer compared to referent tracts.

- **Wagner, S. E., Burch, J. B., Hussey, J., Temples, T., Bolick-Aldrich, S., Mosley-Broughton, C., et al. Soil zinc content, groundwater usage, and prostate cancer incidence in South Carolina. *Cancer Causes and Control* 2009; 20(3): 345.** In this study, a GIS and Poisson regression were used to test the hypothesis that census tracts with reduced soil zinc concentrations, elevated groundwater use, or more agricultural or hazardous waste sites had elevated prostate cancer risks. Increased prostate cancer rates were associated with reduced soil zinc concentrations and elevated groundwater use, although this observation is not likely to contribute to South Carolina's racial prostate cancer disparity. Statewide mapping and statistical modeling of relationships between environmental factors, demographics, and cancer incidence can be used to screen hypotheses focusing on novel prostate cancer risk factors.

- **Wagner, S. E., Hurley, D. M., Hebert, J. R., McNamara, C., Bayakly, A. R., and Vena, J. E. Cancer mortality-to-incidence ratios in Georgia: describing racial cancer disparities and potential geographic determinants. *Cancer* 2012; 118(16): 4032.** The objective of this study was to evaluate racial cancer disparities in Georgia by calculating and comparing mortality-to-incidence ratios (MIRs) by health district and in relation to geographic factors. A GIS was used to map MIRs for each cancer site by Georgia health district. The highest MIRs were detected in west- and east-central Georgia, and the lowest MIRs were detected in and around Atlanta. Districts with better health behavior, clinical care, and social and economic factors had lower MIRs, especially among whites. More fatal cancers, particularly prostate, cervical, and oral cancer in men, were detected among blacks, especially in central Georgia, where health behavior and social and economic factors were worse. MIRs are an efficient indicator of survival and provide insight into racial cancer disparities. Additional examination of geographic determinants of cancer fatality in Georgia as indicated by MIRs is warranted.

- **Zhou, Y. B., Liang, S., Chen, G. X., Rea, C., Han, S. M., He, Z. G., et al. Spatial-temporal variations of *Schistosoma japonicum* distribution after an integrated national control strategy: a cohort in a marshland area of China. *BMC Public Health* 2013; 13: 297.** Understanding spatial variations of *Schistosoma* infections and their associated factors is crucial for the development of site-specific intervention strategies. This study reports on a GIS-based spatial analysis that was conducted to identify geographic distribution patterns of schistosomiasis infections at the household scale. The results of the spatial autocorrelation analysis revealed significant spatial clusters of human infections at the household level. The findings imply that it may be necessary to reassess risk factors of *S. japonicum* transmission over the course of control and adjust control measures in the community.

Perspective 2: Acute event

Epidemiology applications

- **Bull, M., Hall, I. M., Leach, S., and Robesyn, E. The application of geographic information systems and spatial data during Legionnaires' disease outbreak responses. *EuroSurveillance* 2012; 17(49).** A literature review was conducted to highlight the application and potential benefit of using GIS during Legionnaires' disease outbreak investigations. Relatively few published sources were identified; however, certain types of data were found to be important in facilitating the use of GIS, namely, patient data, locations of potential sources (e.g., cooling towers), demographic data relating to the local population, and meteorological data. These data were then analyzed to gain a better understanding of the spatial relationships between cases and their environment, the cases' proximity to potential outbreak sources, and the modeled dispersion of contaminated aerosols. The use of GIS in an outbreak is not a replacement for traditional outbreak investigation techniques, but it can be a valuable supplement to a response.

- **Curtis, J. W., Curtis, A., and Upperman, J. S. Using a geographic information system (GIS) to assess pediatric surge potential after an earthquake. *Disaster Medicine and Public Health Preparedness* 2012; 6(2): 163.** GIS and geospatial technology (GT) can help hospitals improve plans for postdisaster surge by assessing numbers of potential patients in a catchment area and providing estimates of special needs populations, such as pediatrics. In this study, landslide and liquefaction zones were overlaid on U.S. Census Bureau block groups. Units that intersect with the hazard zones were selected for computation of pediatric surge potential in case of an earthquake. In addition, cartographic visualization and cluster analysis were performed to identify hot spots of socially vulnerable populations. The results suggest the need for locally specified vulnerability models for pediatric populations. GIS and GT have untapped potential to contribute local specificity to planning for surge potential after a disaster. Although this case focuses on an earthquake hazard, the methodology is appropriate for an all-hazards approach. With the advent of Google Earth, GIS output can now be easily shared with medical personnel for broader application and improvement in planning.

- **Waring, S., Zakos-Feliberti, A., Wood, R., Stone, M., Padgett, P., and Arafat, R. The utility of geographic information systems (GIS) in rapid epidemiological assessments following weather-related disasters: methodological issues based on the Tropical Storm Allison experience. *International Journal of Hygiene and Environmental Health* 2005; 208(1–2): 109.** The goal of the public health response to Tropical Storm Allison was to rapidly evaluate the immediate health needs of the community. With the use of geographical information system (GIS) technology, the authors conducted a rapid needs assessment and used a modified cluster sampling facilitated by GIS methodology. Of the 420 households participating in the survey, the authors found a significant increase in illness, injuries, and immediate health needs among persons living in flooded homes compared to nonflooded homes. Study findings also were used to guide relief efforts, which underscore the usefulness of rapid needs assessment as a tool to identify actual health threats and facilitate delivery of resources to those with the greatest and most immediate need.

- **Wilson, J. L., Little, R., and Novick, L. Estimating medically fragile population in storm surge zones: a geographic information system application. *Journal of Emergency Management* (Weston, Mass.) 2013; 11(1): 9.** The objective of this study was to develop a simple, cost-effective method for determining the size and geographic distribution of medically fragile (MF) individuals at risk from tropical storm surges for use by emergency management planners. The study used GIS spatially referenced layers based on secondary data sources from both state and federal levels. The study setting included the eastern North Carolina coastal counties that would be affected by tropical storm surges. The main outcome of this study was a series of local and regional maps that portrayed the geographic distribution and estimated counts of the potentially at-risk MF population from a tropical storm surge scenario. Maps depicting the geographic distribution and potential numbers of MF individuals are important information for planning and preparedness in emergency management and potentially engaging the public.

Perspective 3: Community approach

- **Ansumana, R., Malanoski, A. P., Bockarie, A. S., Sundufu, A. J., Jimmy, D. H., Bangura, U., et al. Enabling methods for community health mapping in developing countries. *International Journal of Health Geographics* 2010; 9.** Spatial epidemiology is useful but difficult to apply in developing countries due to the low availability of digitized maps and address systems, accurate population distributions, and computational tools. A community-based mapping approach was used to demonstrate that participatory geographic information system (PGIS) techniques can provide information helpful for health and community development. The methods developed in this paper serve as a model for the involvement of communities in the generation of municipal maps and their application to community and health concerns.
- **Beyer, K. M., Comstock, S., and Seagren, R. Disease maps as context for community mapping: a methodological approach for linking confidential health information with local geographical knowledge for community health research. *Journal of Community Health* 2010; 35(6): 635.** In understanding the contextual influences on health, community health researchers have increasingly employed both geographic methodologies, including GIS, and community participatory approaches. However, the use of geographical methods and datasets to characterize community environments is lacking due in part to concerns and restrictions regarding community access to confidential health data. This study presents a method for linking confidential, geocoded health information with community-generated experiential geographical information in a GIS environment. This work opens the door for future efforts to integrate empirical epidemiological data with community-generated experiential information to inform community health research and practice.
- **Cashman, S. B., Adeky, S., Allen, A. J., Corburn, J., Israel, B. A., Montano, J., et al. The power and the promise: working with communities to analyze data, interpret findings, and get to outcomes. *American Journal of Public Health* 2008; 98(8): 1407.** Although the intent of community-based participatory research (CBPR) is to include community voice in all phases of a research

initiative, community partners appear less frequently engaged in data analysis and interpretation than in other research phases. Using four brief case studies, each with a different data collection methodology, the authors provide examples of how community members participated in data analysis, interpretation, or both, thereby strengthening community capacity and providing unique insight. The roles and skills of the community and academic partners were different from but complementary to each other. We suggest that including community partners in data analysis and interpretation, while lengthening project time, enriches insights and findings and consequently should be a focus of the next generation of CBPR initiatives.

- **Chan, D. V., Helfrich, C. A., Hursh, N. C., Rogers, E. S., and Gopal, S. Measuring community integration using geographic information systems (GIS) and participatory mapping for people who were once homeless. *Health and Place* 2014; 27: 92.** This is the first study to use GIS to document the spatial presence of individuals with disabilities and the relationship to integration outcomes. The current study documents the activity space, or area of daily experiences, of 37 individuals who were once homeless through participatory mapping and GIS. Location addresses from participant maps were identified through Google Maps© and then geocoded, or assigned the corresponding latitude and longitude coordinates. Once geocoded, locations were entered into the GIS database system for analysis using ESRI ArcGIS. Study findings suggested there was no significant relationship between activity space size and community integration measures, except a negative association with physical integration. Although the activity space measures were not associated with community integration outcomes in the expected direction, GIS provided meaningful information about an individual's spatial presence in the community and where important activities occurred as part of integration efforts. Continued research is needed using a spatial mapping approach to evaluate accessibility to resources available in the community area in conjunction with activity space measures as spatial factors that can promote or impede integration efforts.

- **Clements, A. C., Reid, H. L., Kelly, G. C., and Hay, S. I. Further shrinking the malaria map: how can geospatial science help to achieve malaria elimination? *Lancet Infectious Diseases* 2013; 13(8): 709.** Malaria is one of the biggest contributors to deaths caused by infectious disease. The spatial distribution of malaria, at all levels of endemicity, is heterogeneous. Moreover, populations living in low-endemic settings where elimination efforts might be targeted are often spatially heterogeneous. Geospatial methods, therefore, can help design, target, monitor, and assess malaria elimination programs. Rapid advances in technology and analytical methods have allowed the spatial prediction of malaria risk and the development of spatial decision support systems, which can enhance elimination programs by enabling accurate and timely resource allocation. However, no framework exists for assessment of geospatial instruments. Research is needed to identify measurable indicators of elimination progress and quantify the effect of geospatial methods in the achievement of elimination outcomes.

- **Dongus, S., Nyika, D., Kannady, K., Mtasiwa, D., Mshinda, H., Fillinger, U., et al. Participatory mapping of target areas to enable operational larval source management to suppress malaria vector mosquitoes in Dar es Salaam, Tanzania. *International Journal of Health Geographics* 2007; 6.** A simple community-based mapping procedure that requires no electronic devices in the field was developed to facilitate routine larval surveillance in Dar es Salaam, Tanzania. The mapping procedure included community-based development of sketch maps and verification of sketch maps through technical teams using laminated aerial photographs in the field, which were later digitized and analyzed using GIS. The procedure developed enabled complete coverage of targeted areas with larval control through comprehensive spatial coverage with community-derived sketch maps. The procedure is practical, affordable, and requires minimal technical skills. This approach can be readily integrated into malaria vector control programs, scaled up to towns and cities all over Tanzania, and adapted to urban settings elsewhere in Africa.

- **Jankowski, P. Towards participatory geographic information systems for community-based environmental decision making. *Journal of Environmental Management* 2009; 90(6): 1966.** This article discusses the potential of GIS to become an information technology enabling groups of people to participate in decisions shaping their communities and promoting sustainable use of natural resources. It explains the concept of participation in the context of planning and decision making. In this context, participatory GIS (PGIS) offers tools that can be used to help the public become meaningfully involved in decision-making processes affecting their communities. Following an overview of research on PGIS and its current status, the article presents two recent studies of PGIS in water resource planning: one involving the use of computer-generated maps representing simple information structures and the other involving the use of more sophisticated information tools. The synthesis of both studies provides the bases for discussing the prospects of PGIS to empower citizens in making decisions about their communities and resources.

- **Maman, S., Lane, T., Ntogwisangu, J., Modiba, P., vanRooyen, H., Timbe, A., et al. Using participatory mapping to inform a community-randomized trial of HIV counseling and testing. *Field Methods* 2009; 21(4): 368.** Participatory mapping and transect walks were used to inform the research and intervention design and to begin building community relations in preparation for Project Accept, a community-randomized trial sponsored by the National Institute of Mental Health. Results from the mapping exercises informed decisions such as defining community boundaries and identifying appropriate criteria for matching community pairs for the trial, as well as where to situate the services. The participatory methods enabled researchers at each site to develop an understanding of the communities that could not have been derived from existing data or data collected through standard data collection techniques. Furthermore, the methods lay the foundation for collaborative community research partnerships.

- **Pearce, J. Invited commentary: history of place, life course, and health inequalities—historical geographic information systems and epidemiologic research. *American Journal of Epidemiology* 2015; 181(1): 26.** In recent years, epidemiologists, geographers, and sociologists have become increasingly interested in the ways in which local geographical circumstances are related to residents' health. The emerging field of historical GIS offers possibilities to researchers interested in relationships between place and health. Integrating spatial data from various historical sources can enable the reconstruction of past urban environments. These spatial data, accrued over time and appended with detailed cohort information, can offer analytical opportunities for better understanding of how place-based factors influence health and well-being over the life course. This article discusses King and Clarke' s epidemiologic study on sociospatial inequalities in health, which is the first national-level study in the United States to emphasize that health-related neighborhood resources are socially patterned.

- **Stopka, T. J., Krawczyk, C., Gradziel, P., and Geraghty, E. M. Use of spatial epidemiology and hot spot analysis to target women eligible for prenatal women, infants, and children services. *American Journal of Public Health* 2014; 104(Suppl 1): S183.** This study used a GIS and cluster analyses to determine locations in need of enhanced Special Supplemental Nutrition Program for Women, Infants, and Children (WIC) services. Analyses focused on the density of pregnant women who were eligible for but not receiving WIC services in California's 7049 census tracts. Methods included the use of incremental spatial autocorrelation and hot spot analyses to identify clusters of WIC-eligible nonparticipants. Hot spot analyses provided a rigorous and objective approach to determine the locations of statistically significant clusters of WIC-eligible nonparticipants. Results helped inform WIC program and funding decisions, including the opening of new WIC centers, and offered a novel approach for targeting public health services.

Strengths and challenges of GIS applications in epidemiology

- **Beck, L. R., Lobitz, B. M., and Wood, B. L. Remote sensing and human health: new sensors and new opportunities. *Emerging Infectious Diseases* 2000; 6(3): 217.** Increased computing power and spatial modeling capabilities of GIS could extend the use of remote sensing beyond the research community into operational disease surveillance and control. This article illustrates how remotely sensed data have been used in health applications and assesses earth-observing satellites that could detect and map environmental variables related to the distribution of vector-borne and other diseases.

- **Bonner, M. R., Han, D., Nie, J., Rogerson, P., Vena, J. E., and Freudenheim, A. L. Positional accuracy of geocoded addresses in epidemiologic research. *Epidemiology* 2003; 14(4): 408.** Geocoding is an important step in the use of GIS in epidemiologic research, and the validity of epidemiologic studies using this methodology depends, in part, on the positional accuracy of the geocoding process. Authors conducted a study comparing the validity of positions geocoded with a commercially available program to positions determined by global positioning system (GPS) satellite receivers. Addresses ($N = 200$) were randomly selected from a recently completed case-control study in western New York State. This study indicates that the suitability of geocoding for epidemiologic research depends on the level of spatial resolution required to assess exposure. Although sources of error in positional accuracy for geocoded addresses exist, geocoding of addresses is, for the most part, very accurate.

- **Beyea, J. and Hatch, M. Geographic exposure modeling: a valuable extension of geographic information systems for use in environmental epidemiology. *Environmental Health Perspectives* 1999: 107(Suppl 1): 181.** This article presents the terminology and methodology of geographic modeling, describes applications to date in the field of epidemiology, and evaluates the potential of this relatively new tool. The author discusses the accuracy and reliability of geographic modeling, highlighting some concerns and responses to relevant issues. From a review of the literature, the article includes a list of steps on exposure reconstructions followed by analysts, and finally, the conditions appropriate for use of geographic modeling in epidemiology.

- **Clark, K. C., McLafferty, S. L., and Tempalski, B. J. On epidemiology and geographic information systems: a review and discussion of future directions. *Emerging Infectious Diseases* 1996; 2(2): 85.** The article provides insight into GIS, including functional capabilities, existing applications in epidemiology, spatial analysis, data issues, and hardware and software advances. GIS offers expanding opportunities for epidemiology because it allows the user to choose between options when geographic distributions are part of the problem. In order to capitalize on the use of GIS, rethinking and reorganizing the way data are collected should not be ignored. Education opportunities and curriculum development need to integrate GIS instruction. GIS allows the user to have a spatial perspective on disease, provides tools for analysis and decision making, and provides deep potential for public health and epidemiology.

- **Cromley, E. K. GIS and disease. *Annual Review of Public Health* 2003; 24: 7.** GIS and related technologies such as remote sensing are increasingly used to analyze the geography of disease, specifically the relationships between pathological factors (causative agents, vectors and hosts, and people) and their geographical environments. GIS applications in the United States have described the sources and geographical distributions of disease agents, identified regions in time and space where people may be exposed to environmental and biological agents, and mapped and analyzed spatial and temporal patterns in health outcomes. Although GIS show great promise in the study of disease, their full potential will not be realized until environmental and disease surveillance systems are developed that distribute data on the geography of environmental conditions, disease agents, and health outcomes over time based on user-defined queries for user-selected geographical areas.

- **Croner, C. M. Public health, GIS, and the Internet. *Annual Review of Public Health* 2003; 24: 57.** Internet access and use of georeferenced public health information for GIS application will be an important and exciting development for health agencies in the upcoming years. Technological progress toward public health geospatial data integration, analysis, and visualization of space–time events using the web portends eventual robust use of GIS by public health and other sectors of the economy. Increasing web resources from distributed spatial data portals and global geospatial libraries, and a growing suite of web integration tools, will provide new opportunities to advance disease surveillance, control, and prevention, and ensure public access and community empowerment in public health decision making.

- **Croner, C. M., Sperling, J., and Broome, F. R. Geographic information systems (GIS): new perspectives in understanding human health and environmental relationships. *Statistics in Medicine* 1996; 15(17–18): 1961.** GIS in public health is in the formative stage, and the growing need for a reliable environmental geospatial database is a fundamental concern. GIS is a much-awaited tool for public health professionals capable of improving public health strategies involving surveillance, risk assessment, analysis, and the control and prevention of human disease. The article outlines some of the recent GIS activities at the Centers for Disease Control and Prevention (CDC) and Agency for Toxic Substances and Disease Registry (ATSDR) and how GIS has advanced the mission of spatial analysis through high degrees of metric precision, and statistical predictability. The growing uses of remotely sensed imagery and satellite-facilitated global positioning systems are contributing to unprecedented surveillance of the environment and further understanding of known and suspected environmental disease associations with human and animal health. Earth science and public health monitoring GIS databases offer new analytic opportunities for disease assessment and prevention, although data privacy issues pose a serious, but resolvable restraint on the sharing of georeferenced health databases. The article contains a review of GIS properties, issues pertaining to data confidentiality and GIS in public health, and a selected glossary of spatial statistical terms related to GIS mapping and analysis.

- **Dunn, C. E. Participatory GIS—a people's GIS? *Progress in Human Geography* 2007; 31(5): 616.** Recent years have witnessed a growing number of applications of GIS that grant legitimacy to indigenous geographical knowledge, as well as to "official" spatial data. By incorporating various forms of community participation, these newer framings of geographical information systems as participatory GIS (PGIS) offer a response to the critiques of GIS, which were prevalent in the 1990s. This paper reviews PGIS in the context of the "democratization of GIS." It explores aspects of the control and ownership of geographical information, representations of local and indigenous knowledge, scale and scaling up, web-based approaches, and some potential future technical and academic directions.
- **Elliott, P. and Wartenberg, D. Spatial epidemiology: current approaches and future challenges. *Environmental Health Perspectives* 2004; 112(9): 998.** This article focuses on small-area analyses, encompassing disease mapping, geographic correlation studies, disease clusters, and clustering. Advances in GIS have created new opportunities to investigate environmental and other factors in explaining local geographic variation in disease, as well as introduce new challenges. These challenges include the large random component that may predominate disease rates across small areas. Although this can be remedied with appropriate statistical analysis, sensitivity, potential biases, confounding, and a detailed understanding of data quality are all important. Disease clusters often arise nonsystematically due to public, media, or physician concern, and one way to resolve such concerns is the replication of analyses in different areas based on routine data. With advancements in exposure mapping and surveillance of large health databases, there is promise to improve our understanding of the unique relationship between the environment and our health.
- **Graham, A. J., et al. Spatial analysis for epidemiology. *Acta Tropica* 2004; 91: 219.** This paper introduces a range of techniques used in remote sensing, GIS, and spatial analysis that are relevant to epidemiology. Remote sensing and the ecology of disease, spatial scale, and spatial statistics are covered in further detail. Possible future directions for the application of remote sensing, GIS, and spatial analysis are also suggested. The papers mentioned in this article were selected to provide insight into the opportunities offered by remote sensing, GIS, and spatial analysis for enhancing epidemiological investigations. This article advocates for the uptake of remote sensing and GIS techniques in the health sciences and states some barriers to quick uptake by health professionals.
- **Jacquez, G. M. Spatial analysis in epidemiology: nascent science or a failure of GIS? *Journal of Geographical Systems* 2000; 2: 91.** This paper summarizes contributions of GIS in epidemiology, and identifies needs required to support spatial epidemiology as science. GIS supports disease mapping, location analysis, the characterization of populations, and spatial statistics and modeling. Although laudable, these accomplishments are not sufficient to fully identify disease causes and correlates. One reason is the failure of present-day GIS to provide tools appropriate for epidemiology. Two needs are most pressing. First, we must reject the static view: meaningful inference about the causes of disease is impossible without both spatial and temporal information. Second, we need models that translate space–time data on health outcomes and putative exposures into epidemiologically meaningful measures. The first need will be met by the design and implementation of space–time information systems for epidemiology, and the second by process-based disease models.

- **Jarup, L. Health and environment information systems for exposure and disease mapping, and risk assessment. *Environmental Health Perspectives* 2004; 112: 995.** The article details the capabilities of GIS to produce maps of exposure and disease to reveal spatial patterns. As the environment is consumed with thousands of new chemicals, some with toxic properties, rapid assessment of risk associated with the use of those chemicals is essential to protect people from harmful exposure. Exposure is typically unevenly distributed geographically as well as temporally, and disease occurrence also follows similar trends. Exposure mapping using advanced GIS modeling may enhance exposure assessment in environmental epidemiologic studies. Disease maps may be useful in risk assessment used to explore changes in disease pattern, potentially linked to environmental exposures. Spatial variations in risk and trends may be studied using software tools such as the Rapid Inquiry Facility, for an initial quick evaluation of any potential health hazards associated with an environmental pollutant.

- **Kaiser, R., Spiegel, P. B., Henderson, A. K., and Gerber, M. L. The application of geographic information systems and global positioning systems in humanitarian emergencies: lessons learned, programme implications and future research. *Disasters* 2003; 27(2): 127.** Recent areas of application of GIS methods in humanitarian emergencies include hazard, vulnerability, and risk assessments; rapid assessment and survey methods; disease distribution and outbreak investigations; planning and implementation of health information systems; data and program integration; and program monitoring and evaluation. The main use of GIS in these areas is to provide maps for decision making and advocacy, which allow overlaying types of information that may not normally be linked. GIS is also used to improve data collection in the field (e.g., for rapid health assessments or mortality surveys). Development of GIS methods requires further research. Although GIS methods may save resources and reduce error, initial investment in equipment and capacity building may be substantial. Especially in humanitarian emergencies, equipment and methodologies must be practical and appropriate for field use. Add-on software to process GIS data needs to be developed and modified. As equipment becomes more user-friendly and costs decrease, GIS will become more of a routine tool for humanitarian aid organizations in humanitarian emergencies, and new and innovative uses will evolve.

- **Kandwal, R., Garg, P. K., and Garg, R. D. Health GIS and HIV/AIDS studies: perspective and retrospective. *Journal of Biomedical Informatics* 2009; 42(4): 748.** A GIS is a useful tool that aids and assists in health research, health education, and the planning, monitoring, and evaluation of health programs that are meant to control and eradicate certain life-threatening diseases and epidemics. This communication is an attempt to link and understand the health scenario in a GIS context with emphasis on HIV/AIDS. Various GIS-based functionalities for health studies and their scope in analyzing and controlling epidemiological diseases are explored. Finally, the authors conclude with the general management problems, issues, and challenges related to HIV/AIDS prevailing in India.

- **Kistenmann, T., Dangendorf, F., and Schweikart, J. New perspectives on the use of geographical information systems (GIS) in environmental health sciences. *International Journal of Hygiene and Environmental Health* 2002; 205: 169.** Within the domains of environmental health, disease ecology, and public health, GIS has become an indispensible tool for processing, analyzing, and visualizing spatial data. In the field of geographic epidemiology, GIS are used for drawing up disease maps and for ecological analysis. Advantages GIS offers for disease mapping are simplified generation and variation maps, as well as a broader variety in terms of determining areal units. When conducting ecological analysis, GIS can assist with the assessment of the distribution of health-relevant environmental factors via interpolation and modeling. Conversely, GIS-supported methods for the detection of striking spatial patterns of disease distribution require much improvement and the integration of the time dimension.

- **Nuckols, J. R., Ward, M. H., and Jarup, L. Using geographic information systems for exposure assessment in environmental epidemiology studies. *Environmental Health Perspectives* 2004; 112(9): 1007.** GIS are being used with increasing frequency in environmental epidemiology studies. Reported applications include locating the study population by geocoding addresses (assigning mapping coordinates), using proximity analysis of contaminant source as a surrogate for exposure, and integrating environmental monitoring data into the analysis of the health outcomes. This article discusses fundamentals of three scientific disciplines instrumental to using GIS in exposure assessment for epidemiologic studies: geospatial science, environmental science, and epidemiology. The authors also explore how a GIS can be used to accomplish several steps in the exposure assessment process. The authors present and discuss examples for the first three steps, and discuss potential uses of GIS and global positioning systems (GPS) in the last step. On the basis of the findings, the authors conclude that the use of GIS in exposure assessment for environmental epidemiology studies not only is feasible, but also can enhance the understanding of the association between contaminants in our environment and disease.

- **Ostfeld, R. S., Glass, G. E., and Keesing, F. Spatial epidemiology: an emerging (or re-emerging) discipline. *Trends in Ecology and Evolution* 2005; 20(6): 328.** This communication briefly describes approaches to spatial epidemiology that are spatially implicit, such as metapopulation models of disease transmission, and then focuses on research in spatial epidemiology that is spatially explicit, such as the creation of risk maps for particular geographical areas. Although the spatial dynamics of infectious diseases are the subject of intensive study, the impacts of landscape structure on epidemiological processes have so far been neglected. The few studies that demonstrate how landscape composition (types of elements) and configuration (spatial positions of those elements) influence disease risk or incidence suggest that a true integration of landscape ecology with epidemiology will be useful.

- **Ricketts, T. C. Geographic information systems and public health. *Annual Review of Public Health* 2003; 24: 1.** GIS use has become widespread and well accepted, and although GIS is not the complete solution to understanding the distribution of disease and the many problems of public health, GIS does provide more than a mapping tool, it provides an analytic system. GIS requires substantial skills on the part of specialists, immense investments in the data development and adaptation, and a structure for determining what the data are telling us when they are cast in a spatial context. Maps and geostatistical visualizations of human phenomenon are only approximations and estimates and do introduce biases that limit our ability to usefully interpret the information. Authors in the article comment on future uses of GIS, including virtual community models of how GIS-based decision support systems may be advanced into broad use and acceptance.

- **Robinson, T. P. Spatial statistics and geographical information systems in epidemiology and public health. *Advances in Parasitology* 2000; 47: 81.** This chapter surveys the principles behind spatial statistics and GIS and their application to epidemiology and public health. Like the other introductory chapters, it is aimed mainly to facilitate understanding in the chapters specific to certain diseases that follow, and to provide a short introduction to the field. A brief overview of spatial statistics and GIS is provided in the introduction. The sections that follow explore the ways in which we can map the distribution of disease, ways in which we can look for spatial patterns in the distribution of disease, and ways in which we can apply spatial statistics and GIS to the problem of identifying the causal factors of observed patterns. In the last section, the author discusses some of the ways in which these techniques have been applied to assist decision making for disease intervention, and concludes by discussing future developments in the field, and some of the issues surrounding the integration of spatial statistics and GIS.

- **Rytkönen, M. Not all maps are equal: GIS and spatial analysis in epidemiology. *International Journal of Circumpolar Health* 2004; 63: 9.** This work highlights one of GIS's most useful functions, basic mapping, although given the quality and quantity of data and the methodology used in analysis, a given map may be useful or misleading. GIS undoubtedly offers epidemiologists a new tool, but if used improperly, it can do more damage than good. The authors caution researchers to use spatial analysis techniques with care and statistical awareness. The methodology should be chosen in view of the data and research problem, and no general methodology can be recommended. The article further details the importance of spatial data and scale, mapping techniques, and insights into spatial statistics and modeling.

- **Sieber, R. Public participation geographic information systems: a literature review and framework. *Annals of the Association of American Geographers* 2006; 96(3): 491.** Public participation geographic information systems (PPGIS) pertains to the use of GIS to broaden public involvement in policy making, as well as to the value of GIS to promote the goals of nongovernmental organizations, grassroots groups, and community-based organizations. The article first traces the social history of PPGIS. It then argues that PPGIS has been socially constructed by a broad set of actors in research across disciplines and in practice across sectors. This produced and reproduced concept is then explicated through four major themes found across the breadth of the PPGIS literature: place and people, technology and data, process, and outcome and evaluation. The themes constitute a framework for evaluating current PPGIS activities and a road map for future PPGIS research and practice.

- **Tim, U. S. The application of GIS in environmental health sciences: opportunities and limitations. *Environmental Research* 1995; 71: 75.** This paper examines major issues related to the application of GIS in environmental health sciences. In detail, the paper discusses the basic principles, potential benefits, and major limitations of GIS in environmental health research. Through the discussion and included example, it is concluded that GIS can significantly add value to environmental and public health data in areas such as exploratory data analysis, hypotheses generation, confirmatory data analysis, and decision making. Widespread acceptance of these GIS tactics in the field is halted by issues such as inconsistent spatial scales of data, data quality and currency, lack of appropriate statistical function for data analysis and interpretation, and data security and confidentiality.

- **Vine, M. F., Degnan, D., and Hanchette, C. Geographic information systems: their use in environmental epidemiologic research. *Environmental Health Perspectives* 1997; 105: 598.** This paper provides an overview of some of the capabilities and limitations of GIS technology, and provides examples of several functions of a GIS, including automated address matching, distance functions, buffer analysis, spatial query, and polygon overlay. The article further discusses methods and limitations of address geocoding, often fundamental to the use of GIS in environmental epidemiologic endeavors, and suggests ways to facilitate its use in future studies.

- **Weis, B. K., Balshawl, D., Barr, J. R., Brown, D., Ellisman, M., Liov, P., et al. Personalized exposure assessment: promising approaches for human environmental health research. *Environmental Health Perspectives* 2005; 113(7): 840.** New technologies and methods for assessing human exposure to chemicals, dietary and lifestyle factors, infectious agents, and other stressors provide an opportunity to extend the range of human health investigations and advance our understanding of the relationship between environmental exposure and disease. In this article, a "toolbox" of methods for measuring external (environmental) and internal (biologic) exposure and assessing human behaviors that influence the likelihood of exposure to environmental agents was established. These methods use environmental sensors, GIS, biologic sensors, toxicogenomics, and body burden (biologic) measurements. Each method is discussed in relation to current use in human health research; specific gaps in the development, validation, and application of the methods are highlighted. Authors also present a conceptual framework for moving these technologies into use and acceptance by the scientific community. Improved methods for exposure assessment will result in better means of monitoring and targeting intervention and prevention programs.

Review articles

- **Clements, A. and Pfeiffer, D. Emerging viral zoonoses: frameworks for spatial and spatiotemporal risk assessment and resource planning. *Veterinary Journal* 2009; 182(1): 21.** Spatial epidemiological tools are increasingly being applied to emerging viral zoonoses (EVZs). This review documents applications of GIS, remote sensing (RS), and spatially explicit statistical and mathematical models to epidemiological studies of EVZ.

- **Jerret, M., Gale, S., and Gontkis, C. Spatial modeling in environmental and public health research. *International Journal of Environmental Research and Public Health* 2010; 7: 1302.** This paper aims to summarize various GIS methods and provides a review of studies that have utilized these methods. This paper highlights the usefulness of certain methods in research and also serves as an overview of the expanding field of spatial epidemiology.

- **Kalluri, S., Gilruth, P., Rogers, D., and Szczur, M. Surveillance of arthropod vector-borne infectious diseases using remote sensing techniques: a review. *PLoS Pathogens* 2007; 3(10): 1361.** Epidemiologists are adopting new remote sensing techniques to study a variety of vector-borne diseases. This paper serves as a review of the status of remote sensing studies of arthropod vector-borne diseases due to mosquitoes, ticks, blackflies, tsetse flies, and sandflies, which are responsible for the majority of vector-borne diseases in the world. Examples of simple image classification techniques that associate land use and land cover types with vector habitats, as well as complex statistical models that link satellite-derived multitemporal meteorological observations with vector biology and abundance, are discussed here. Future improvements in remote sensing applications in epidemiology are also discussed.

- **Moore, D. A. and Carpenter, T. E. Spatial analytical methods and geographic information systems: use in health research and epidemiology. *Epidemiologic Reviews* 1999; 21(2): 143.** In the last decade, computerization of spatial data, through the use of GIS, has emerged as a tool for health care research and epidemiology. No single review to date has covered both spatial analytical techniques and modern GIS. The purpose of this paper is to review both topics by highlighting some notable classic studies and some early examples of spatial analytical methods and GIS in human and animal health research. Several types of analytic techniques are described, as well as the role GIS have had in improving understanding of the spatial aspects of health care and disease research.

- **Rogers, D. J. and Randolph, S. E. Studying the global distribution of infectious diseases using GIS and RS. *Nature Reviews Microbiology* 2003; 1(3): 231.** Analytical tools that are based on geographical information systems and that can incorporate remotely sensed information about the environment offer the potential to define the limiting conditions for any disease in its native region for which there are at least some distribution data. The direction, intensity, or likelihood of its spread to new regions could then be predicted, potentially allowing disease early-warning systems to be developed.
- **Rushton, G. Public health, GIS, and spatial analytic tools. *Annual Review of Public Health* 2003; 24: 43.** The review focuses on literature that uses spatial analytic tools in contexts where GIS is the organizing system for health data or where methods discussed will likely be incorporated in GIS-based analyses in the future. The review concludes with the opinion that the literature is moving toward the development and use of systems of analysis that integrate the information geocoding and database functions of GISystems with the geoinformation processing functions of GIScience. Recent advances in the analysis of disease maps have been influenced by and benefited from the adoption of new practices for georeferencing health data and new ways of linking such data geographically to potential sources of environmental exposures, the location of the health resources, and the geodemographic characteristics of populations.
- **Rushton, G., Armstrong, M. P., Gittler, J., Greene, B. R., Pavlik, C. E., West, M. M., and Zimmerman, D. L. Geocoding in cancer research—a review. *American Journal of Preventive Medicine* 2006; 30(2): S16.** This paper reviews geocoding practice in relation to major purposes and discusses methods to improve the accuracy of geocoded cancer data. The authors conclude that selection of one particular type of geographic area as the geocode may unnecessarily constrain future work. Therefore, the longitude and latitude of each case is the superior basic geocode; all other geocodes of interest can be constructed from this basic identifier.
- **Yang, G. J., Vounatsou, P., Zhou, X. N., Utzinger, J. R., and Tanner, M. A review of geographic information system and remote sensing with applications to the epidemiology and control of schistosomiasis in China. *Acta Tropica* 2005; 96(2–3): 117.** GIS and remote sensing (RS) technologies offer new opportunities for rapid assessment of endemic areas, provision of reliable estimates of populations at risk, prediction of disease distributions in areas that lack baseline data and are difficult to access, and guidance of intervention strategies, so that scarce resources can be allocated in a cost-effective manner. Here, the article focuses on the epidemiology and control of schistosomiasis in China and review GIS and RS applications to date. The article also discusses the limitations of the previous work, and outlines potential new applications of GIS and RS techniques, namely, quantitative GIS, WebGIS, and utilization of emerging satellite information, as they hold promise to further enhance infection risk mapping and disease prediction. Finally, the authors stress current research needs to overcome some of the remaining challenges of GIS and RS applications for schistosomiasis, so that further and sustained progress can be made to control this disease in China and elsewhere.

2

Environmental Studies

Mark J. Nieuwenhuijsen
Center for Research in Environmental Epidemiology (CREAL)
Parc de Recerca Biomèdica de Barcelona – PRBB
Barcelona, Spain

CONTENTS

2.1 Introduction

Exposure is a substance or factor affecting human health, either adversely or beneficially. More precisely, in environmental epidemiology, exposure to an environmental substance is generally defined as any contact between a substance in an environmental medium (e.g., water, air, and soil) and the surface of the human body (e.g., skin and respiratory tract); after uptake into the body, it is referred to as dose. Exposure assessment is the study of distribution and determinants of substances or factors affecting human health. It consists of three components: the design of the study, data collection, and the interpretation of the data. This chapter discusses briefly some of the basic issues and introduces topics for the following chapters.

In environmental and spatial epidemiology, the focus is on chemical, biological, and physical substances in our everyday environment. In today's world, risks associated with environmental exposure are generally small, and therefore to detect a risk when there is truly a risk, the exposure assessment has to be very refined. This generally requires considerable effort and resources. Part of the exposure assessment process in environmental and spatial epidemiology is to optimize the exposure estimate with the aim of detecting a possible risk or optimizing the exposure–response relation in an epidemiological study. This can be achieved, for example, by optimizing the distribution of the variance of the exposure estimates. The main focus of this chapter is on chemical and biological substances, but the underlying principles apply to other exposures as well.

In recent years, there has been increasing interest in the field of exposure assessment, causing it to develop rapidly. We know now more than ever to what, where, and how people are exposed, and improvements have been made to methods for assessing the level of exposure, its variability, and the determinants. New methods have been developed or newly applied throughout this field, including analytical, measurement, modeling, and statistical methods. The use of geographical information systems (GIS) for the assessment of outdoor pollutants and factors has been essential; the use of remote sensing, personal sensors, and OMICS technologies is being explored to improve exposure assessment and the concept of the exposome re-energized the field (see below). All this has led to a considerable improvement in exposure assessment in epidemiological studies, and therefore improvement in the epidemiological studies themselves. Here, we will discuss the various issues, with the main focus on exposure assessment of spatial epidemiological studies.

2.2 Source–Receptor Models and Exposure Route and Pathways

The physical course a pollutant takes from the source to a subject is often referred to as exposure pathway, while the way a substance enters the body is often referred to as exposure route. Source–receptor models include the routes and pathways of exposure and are helpful in understanding how people are exposed. In this kind of model, it often becomes clear that humans create their own exposure by, for example, their activities and where they spend time, and that there is an interaction between the two. Figure 2.1 provides an example of a source–receptor model for air pollution. The source may be cars, and air pollutants such as particles, carbon monoxide, and nitrogen dioxide are emitted from exhaust pipes. Dispersion will take place into the streets and beyond, leading to environmental concentrations. Dispersion takes places into so-called microenvironments such as houses, travel routes and modes, and workplaces, where people come into contact with the pollutants, and is now referred to as exposure. People will inhale the pollutants through the lungs, leading to whole-body uptake (dose), where it may react with the lung cells, or it may be distributed to other parts of the body and react with body tissue (biologically relevant dose). The reaction with the

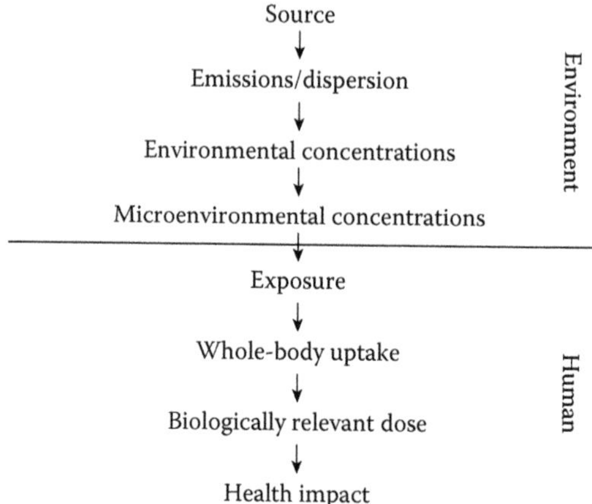

FIGURE 2.1
Source–receptor model for air pollutants.

body tissue may have an impact on health. In this case, there is only one exposure route (inhalation). There are, however, three possible exposure routes for substances:

- Inhalation through the respiratory system
- Ingestion through the gastrointestinal system
- Absorption through the skin

The exposure routes of a substance and the amount of uptake depend on, for example, the biological, chemical, and physical characteristics of the substances, location and activity of the person, and the persons themselves. Inhalation of particles through the respiratory system depends on the particle size or diameter (physical characteristic). Smaller particles are more often inhaled and penetrate deeper into the lungs. The inhalable dust fraction (particles with 50% cutoff diameter of 100 μm) is the fraction of the dust that enters the nose and mouth and is deposited anywhere in the respiratory tract. The thoracic fraction (particles with a 50% cutoff diameter of 10 μm), often referred to as PM_{10}, is the fraction that enters the thorax and is deposited within the lung airways and the gas-exchange region. The alveolar fraction (particles with a 50% cutoff diameter of 2.5 μm), often referred to as $PM_{2.5}$, is deposited in the gas-exchange region (alveoli). The smallest fraction is the ultrafine particulate, which has a diameter of less than 100 nm. Furthermore, inhalation depends on the breathing rate of the subject: those doing heavy work may inhale much more air and more deeply (20 L min^{-1} for light physical activity vs. 60 L min^{-1} for heavy physical activity) (activity of a person). And, people move through different microenvironments with different particle concentrations (location).

Skin absorption can play an important role for uptake of substances such as solvents, pesticides, and trihalomethanes. The volatile trihalomethanes are formed when water is chlorinated and the chlorine reacts with organic matter in the water. In this context, there are a number of possible exposure pathways and routes (Figure 2.2). The main pathway of ingestion is drinking tap water or tap water–based drinks (e.g., tea, coffee, and squash). Swimming, showering, bathing, and dish washing may all result in considerable uptake through inhalation and skin absorption and, for the former three, ingestion to a minor extent. Standing or flushing water in the toilet may lead to uptake via inhalation through volatilization of the trihalomethanes. The uptake of trihalomethanes may be assessed using the concentration measured in exhaled breath or serum.

In the human body, the uptake, distribution, transformation, and excretion of substances such as trihalomethanes can be modeled using physiologically based pharmacokinetic models. These models are becoming more sophisticated, although they are still rarely used in

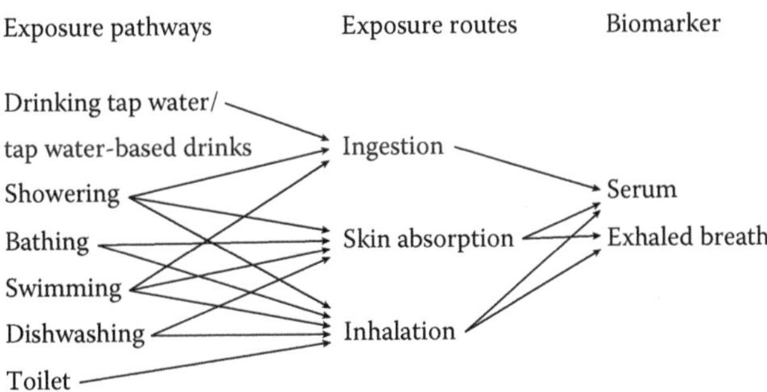

FIGURE 2.2
Examples of exposure pathways, routes, and biomarkers for trihalomethanes.

environmental epidemiology. They can be used to estimate the contribution of various exposure pathways and routes to the total uptake and model the dose of a specific target organ. For example, where trihalomethanes through ingestion may mostly be metabolized rapidly in the liver and not appear in blood, uptake through inhalation and skin increases the blood levels substantially. Furthermore, metabolic polymorphism may lead to different dose estimates under similar exposure conditions, and this can also be modeled. For a number of agents, the level of external environmental exposure may be either reduced or increased, depending on the capacity of Phase I (activation), Phase II (detoxification), and DNA repair enzymes. In this approach, genetic susceptibility markers (e.g., CYP1A1, CYP2E1, NAT1, NAT2, GSTM1, GSTT1, or DNA repair capacity) are used as if they were internal personal protective equipment. For example, low capacity of activation enzymes (e.g., CYP1A1) and high capacity of detoxification (e.g., NAT2) and DNA repair enzymes would have higher protective functions than high capacity of activation enzymes and low capacity of detoxification and DNA repair enzymes that may result in reducing cancer-causing doses of xenobiotics (Vineis 1999).

2.3 Exposure Dimensions

Besides the actual nature of the exposure, there are also three dimensions:

- Duration (e.g., in hours or days)
- Concentration (e.g., in mg m^{-3} in air or mg L^{-1} in water)
- Frequency (e.g., times per week)

In case of exposure through ingestion, the dimensions are concentration, amount (e.g., litres), and frequency. Any of these can be used as an exposure index in an epidemiological study, but they can also be combined to obtain a new exposure index, for example, by multiplying duration and concentration to obtain an index of cumulative exposure. The choice of index depends on the health effect of interest. For substances that cause acute effects such as ammonia (irritation), the short-term concentration is generally the most relevant exposure index, while for substances that cause chronic effects such as asbestos (cancer), long-term exposure indices such as cumulative exposure may be a more appropriate exposure index. However, it is rarely used in environmental epidemiology, but regularly in occupational epidemiology.

2.4 Exposure Level and Variability

The concentration of exposure varies temporally and spatially. Figure 2.3 provides the exposure levels of carbon monoxide during 1 day. Peak exposure levels, that is, exposure levels considerably higher than the overall average, are caused by smoking, cooking, and traffic.

2.5 Ecological versus Individual Exposure Estimates

To obtain exposure estimates for a population in an epidemiological study, two main approaches are available: (1) individual and (2) exposure grouping (Figure 2.4). In the first, exposure estimates are obtained at the individual level; for example, every member

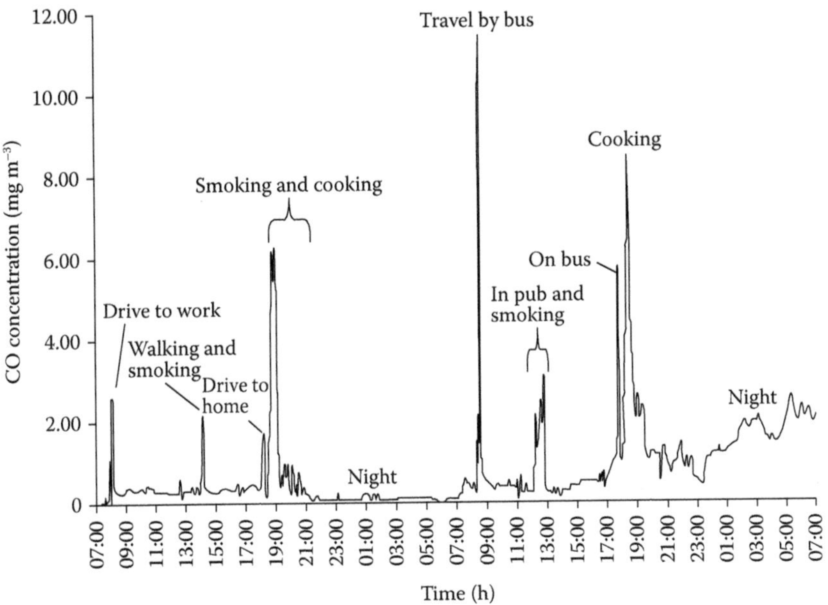

FIGURE 2.3
Variation in carbon monoxide levels over a day.

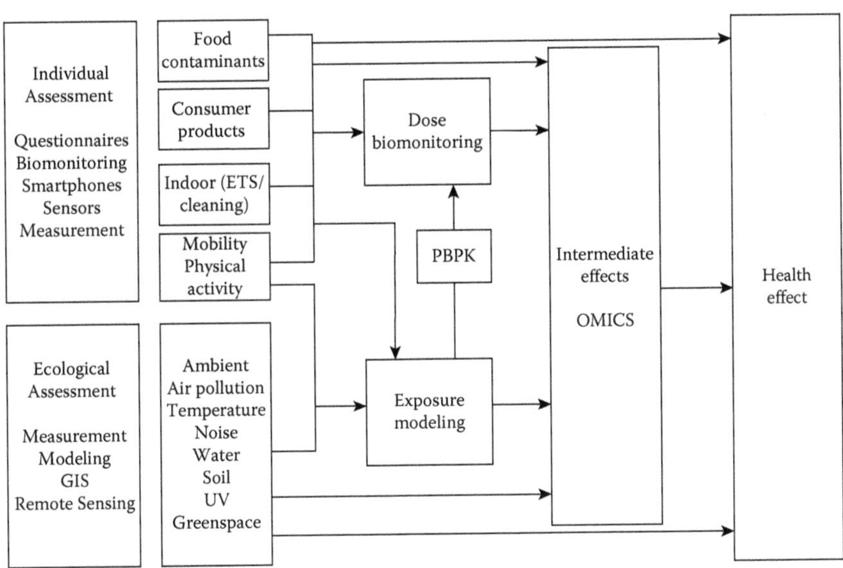

FIGURE 2.4
Individual vs. ecological assessments.

of the population is monitored either once or repeatedly. Information on individuals is collected by questionnaire (e.g., Environmental Tobacco Smoke [ETS], gas cooking, mould, and damp) (Esplugues et al. 2013; Gehring et al. 2013), biological monitoring (e.g., polychlorinated biphenyls, phthalates, phenols, and metals) (Govarts et al. 2012; Christensen et al. 2014; Llop et al. 2012; Gehring et al. 2013), and smartphones and sensors (physical activity, air pollution, and location) (de Nazelle et al. 2013; Nieuwenhuijsen et al. 2014) and estimates directly assigned to the subjects. In the second approach, the population is first split into smaller subpopulations, or more often referred to as exposure groups, based on specific determinants of exposure, and group or ecological exposure estimates are obtained for each exposure group, for example, distance from a source. This approach is often used for outdoor exposures such as air pollution, noise, temperature, and green space. GIS are used to allocate subjects, often their residences, and then assign exposure estimates based on routine monitoring stations (Wilhelm and Ritz 2005; Samoli et al. 2008), models (Beelen et al. 2014), land use characteristics such as green space (Dadvand et al. 2014a), or remote sensing data (Kloog et al. 2012; Dadvand et al. 2014b). In environmental epidemiological studies, exposure groups may be defined, for example, on the basis of distance from an exposure source (e.g., roads or factories). The underlying assumption is that subjects within each exposure group experience similar exposure characteristics, including exposure levels and variation. Exposure estimates can be assigned to the groups, for example, data from ambient air pollution monitors in the area where the subjects live, or a model to predict the exposure. Alternatively, a representative sample of members from each exposure group can be personally monitored, either once or repeatedly. If the aim is to estimate mean exposure, the average of the exposure measurements is then assigned to all the members in that particular exposure group. However, this type of approach is generally not used in environmental epidemiology. Ecological and individual estimates can be combined, for example, in the case of chlorination by-products, where routinely collected trihalomethane measurements providing ecological estimates are combined with individual estimates on actual ingestion, showering, and bathing.

Intuitively, it is expected that the individual estimates provide the best exposure estimates for an epidemiological study. This is not often true, however, because of the variability in exposure and the limited number of samples. In general, in epidemiological studies, individual estimates lead to attenuated, though more precise, health risk estimates than ecological estimates. The ecological estimates, in contrast, result in less attenuation of the risk estimates, albeit less precise. These differences can be explained by the classical and the Berkson-type error models. The between-group, between-subject, and within-subject variance can be estimated using analysis of variance models (see Chapter 5), and this information can be used to optimize the exposure–response relationship, for example, by changing the distribution of exposure groups, as has been demonstrated in occupational epidemiological studies (Kromhout and Heederik 1995; Nieuwenhuijsen 1997; van Tongeren et al. 1997). In this case, the aim is to increase the contrast in exposure between exposure groups, expressed as the ratio between the between-group variance and the sum of the between- and within-group variance, while maintaining reasonably precise exposure estimates of the groups.

2.6 Exposure Classification, Measurement, or Modeling

Exposure can be classified, measured, or modeled, and different tools are available for this, such as questionnaires, air pollution monitors, and statistical techniques, respectively. The methods are often classified as direct and indirect (Figure 2.5).

The main aim of an exposure assessment is to obtain accurate, precise, and biologically relevant exposure estimates in the most efficient and cost-effective way. The cost of the exposure assessment increases with an increase in the accuracy and precision, and therefore the assessment is often a balancing act, with cost on one side and accuracy and precision on the other (Armstrong 1996). The choice of a particular method depends on the aim of the

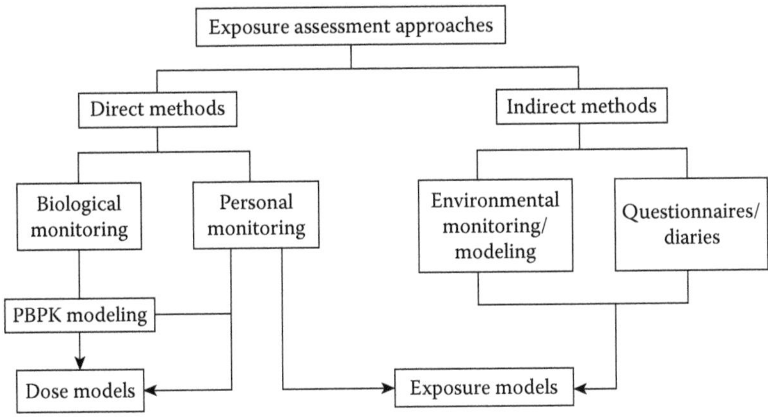

FIGURE 2.5
Different approaches to human exposure assessment.

study and, more often, on the financial resources available. Misclassification of the exposure can lead to attenuation in health risk estimates or loss of power in the epidemiological study, depending on the type of measurement error model (classical or Berkson), and should therefore be minimized.

Subjects in an epidemiological study can be classified based on a particular substance and on an ordinal scale, for example, as exposed:

- Yes or no
- No, low, medium, or high

This can be, for example, achieved by

1. *Expert assessment.* For example, a member of the research team decides, based on prior knowledge, whether the subject in the study is exposed or unexposed, for example, lives in an area with highly contaminated soil or not.
2. *Self-assessment by questionnaire.* That is, the subject in the study is asked to fill out a questionnaire in which he or she is asked about a particular substance, for example, pesticides. Questionnaires are often used to ask a subject if he or she is exposed to a particular substance and also to estimate the duration of exposure.

Questionnaires can be used to ask the subject not only to estimate the duration of their exposure, but also to obtain information related to the exposure, such as where people spent their time (time microenvironment diaries); work history, including the jobs and tasks they carried out; what they eat and drink; and where they live. These variables could be used as exposure indices in the epidemiological studies or translated into a new exposure index, for example, by multiplying the amount of tap water people drink and the contaminant level in the tap water to obtain the total ingested amount of the substance. When used on their own, they are often referred to as exposure surrogates.

Expert and self-assessment methods are generally the easiest and cheapest, but can suffer due to the lack of objectivity and knowledge and may therefore bias exposure assessment. Both experts and subjects may not know exactly what the subjects are exposed to or at what level, and therefore may misclassify the exposure, while diseased subjects may recall certain substances better than subjects without disease (recall bias) and cause differential misclassification, leading to biased health risk estimates.

A more objective way to assess exposure, particularly concentration, is through measurement. Some examples are as follows:

1. Levels of outdoor air pollution can be measured by stationary ambient air monitors (i.e., ambient or environmental air monitoring). These monitors are placed in an area, and they measure the particular substance of interest in this area. Subjects living within this area are considered to be exposed to the concentrations measured by the monitoring station. This may or may not be true, depending on, for example, where the subject in the study lives, works, or travels. The advantage of this method is that it could provide a range of exposure estimates for a large population.

2. Levels of air pollution can be measured by personal exposure monitors or sensors (i.e., personal exposure monitoring). These monitors are lightweight devices that are worn by the subject in the study. They are often used in occupational studies and are being used with increasing frequency in environmental studies too. The advantage of this method is that it is likely to estimate the subject's exposure better than, for example, ambient air monitoring. The disadvantage is that it is often labour intensive and expensive.

3. Levels of water pollutants and soil contaminants can be estimated by taking water and soil samples, respectively, and analysing these for substance of interest in the laboratory. Often these need to be combined with behavioural factors such as water intake, contaminated food intake, or hand-to-mouth contact to obtain a level of exposure.

4. Uptake levels of the substance into the body can be estimated by biomonitoring. Biomonitoring consists of taking biological samples such as urine, exhaled breath, hair, adipose tissue, or nails and measuring for, for example, lead in serum. The samples are subsequently analysed for the substance of interest itself or for a metabolite in a laboratory. Biomonitoring is expected to estimate the actual uptake (dose) of the substance of interest rather than the exposure.

The measurement of exposure is generally expansive, particularly for large populations. Modeling of exposure can be carried out preferably in conjunction with exposure measurements either to help build a model or to validate a model. It is particularly important that the model estimates be validated.

Modeling can be divided into

1. *Deterministic modeling* (i.e., physical), in which the models describe the relationship between variables mathematically on the basis of knowledge of the physical, chemical, and biological mechanisms governing these relationships (Brunekreef 1999). A deterministic model would be one where indoor air particle concentrations are explained by including in the modeling, for example, the sources, volume of rooms, air exchange rate, and settling velocity of the particles.

2. *Stochastic modeling* (i.e., statistical), in which the statistical relationships are modeled between variables. These models do not necessarily require fundamental knowledge of the underlying physical, chemical, and biological relationships between the variables. An example is the relation between land use characteristics, such as road network, traffic density, altitude, population density, and green space, and ambient air pollution levels, which is modeled using statistical regression techniques and a dataset of measurements (see Chapter 13) and is often referred to as land use regression (LUR) modeling. LUR modeling has recently

gained strong interest in air pollution epidemiology because it is relatively cheap and provides air pollution estimates for many subjects.

It is extremely important to validate the modeled estimates, and this is often not a trivial exercise. It requires substantial thought and resources, but greatly increases the validity of the study.

For example, a deterministic Gaussian-type plume model, UK-ADMS, has been applied to estimate ground-level concentrations of arsenic within 20 km of the Nováky Power Station in the Nitra Valley in Slovakia as a function of distance and direction from source and year of operation since the 1950s (Colvile et al. 2001). The power station used to burn arsenic-rich coal, resulting in high emissions of arsenic and contamination of the surrounding area. The results of modeling were used in an epidemiological study of arsenic and skin cancer around the power station (Pesch et al. 2002). The epidemiological study needed historical estimates of arsenic, but very few measurements were available. Only arsenic emission levels were available since the opening of the power station, plus details of how emissions were shared between a number of chimneys of varying height: from 100 m during the early years of operation to 300 m today. The emissions reached their peak in the 1970s, after which they declined (Figure 2.6). Air dispersion modeling was used to estimate the arsenic profiles around the power station and over time, using the arsenic emission levels and a number of assumptions.

To construct a model input that describes the local climatology, 4 years (1990–1993) of observations were used, obtained from a weather station near the plant. The relevant data for each year were in the form of monthly average temperature and wind speed at 7:00 a.m., 2:00 p.m., and 9:00 p.m.; monthly average cloud; and monthly frequency of occurrence of each wind direction by a 45° sector. Figure 2.7 shows that it is relatively rare for the wind direction to be from the west or east. This was clearly because of the valley, which runs from north to south, channeling the wind.

Apart from the wind direction, the other notable feature of the Nitra Valley climatology was the high incidence of calms, which account for 13% of the observations and as much as 32% of the calmest month, September 1990. UK-ADMS, like any Gaussian-type

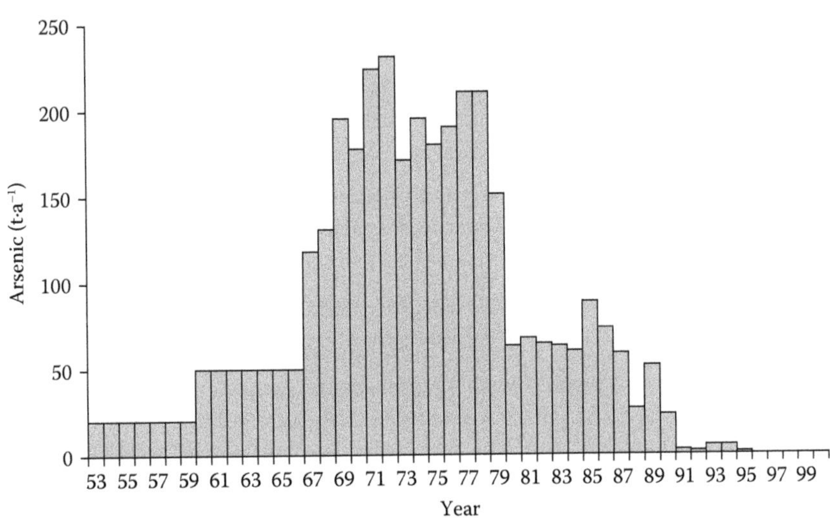

FIGURE 2.6
Arsenic emission levels of a power station in the Nitra Valley, Slovakia.

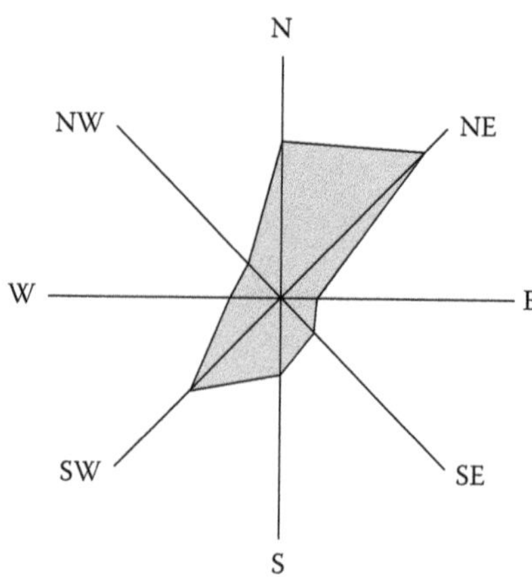

FIGURE 2.7
Wind direction climatology (1990–1993 inclusive) at Prievidza, Slovakia, showing channeling of local wind by the valley.

dispersion model, ceases to represent atmospheric physics realistically at wind speeds less than about 1 ms^{-1}. It is an unavoidable problem when using such a model that it becomes invalid at just the time when pollution concentrations can be highest. A simple box model was used to estimate the possible contribution of calms as follows. Emissions were assumed to become homogeneously mixed throughout a volume of air equal to that of the valley in the immediate vicinity of the Nováky Power Station. An average calm was assumed to last 8 h (i.e., overnight), during which time no significant removal of arsenic from the volume of air was assumed to occur. At the end of the calm, the contents of the volume of air were assumed to be removed instantly by advection. These assumptions were clearly approximate but were designed to be sufficient to carry out an order-of-magnitude estimate of the contribution of calms to the annual average atmospheric arsenic concentration.

Figure 2.8 shows the annual average ground-level concentration of arsenic modeled for the era of operation when emissions were highest, averaged over all times when wind speed was greater than 1 m^{-1}. The three main features of the concentration map are as follows:

1. The concentration bands are not circular, as has been assumed for much previous epidemiological exposure estimation work, but are stretched along the valley because of the rarity of winds blowing across.
2. The maximum of concentration is not immediately adjacent to the plant, but is displaced about 2 km to the northeast and southwest, because the pollution does not disperse vertically downward from the elevated source.
3. The concentration falls off rapidly with distance from the source, decreasing by a factor of 30 from 2 km away to 10 km away.

The modeled estimates were compared with some measurements that were available for the study period, and they showed that there was a reasonable agreement between the estimates and measurements over the years and geographically, as long as the rough calculation of the contribution of concentrations accumulating during calm weather was included (Colvile et al. 2001). Soil sampling showed that the exposure profiles for the arsenic air modeling and arsenic in soil were fairly similar, with the highest levels measured near the power station and falling to background levels approximately 10 km from the power station (Keegan et al. 2002). The epidemiological study found an association between skin cancer and cumulative

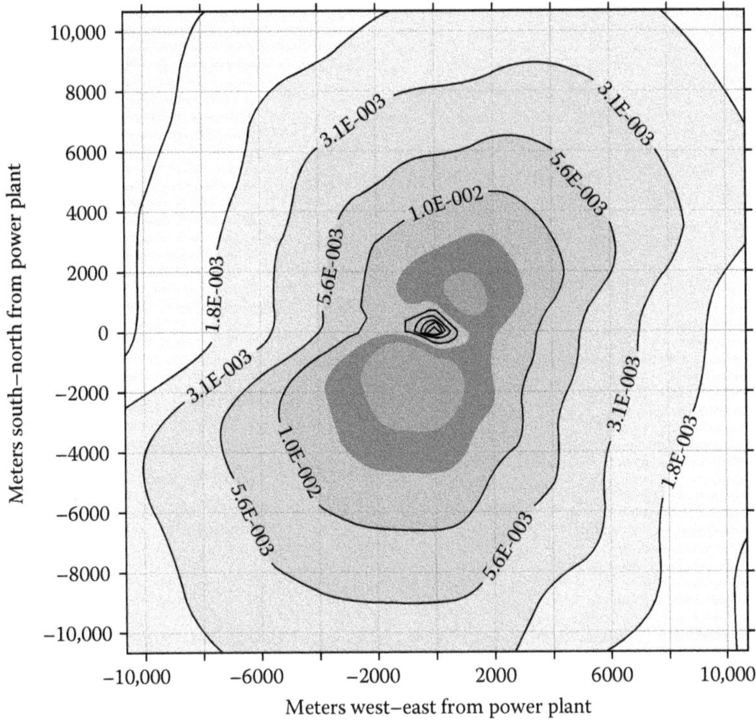

FIGURE 2.8
Profiles of annual average ground-level concentrations of arsenic (μg m^{-3}) modeled using UK-ADMS for the era of operation 1973–1975 around a power station in the Nitra Valley, Slovakia.

arsenic estimates based on the long-term air dispersion modeling estimates, taking into account changes over time, but not when simple distance measures were used, based on where subjects were living at the time of the study (Pesch et al. 2002).

An example of stochastic modeling is LUR modeling, which was recently applied in the large Europe-wide ESCAPE study (Beelen et al. 2013, 2014; Eeftens et al. 2012). In the LUR approach, measurements are taken over the study area, and information on potential determinants such as traffic network and load, green space, altitude, and so forth, are collected. A regression equation with the measurements as the dependent variable and the possible determinants as predictor variables is then developed, which can provide exposure estimates for points, that is, subjects' residences, using GIS (Figure 2.9).

More recently, the use of remote sensing has increased and is used to provide concentration levels of environmental exposures such as air pollution, temperature, and green space (Kloog et al. 2012; Dadvand et al. 2014a, 2014b).

All these different approaches are not exclusive and often are combined to obtain the best exposure index. This involves some form of modeling. At times, it may be difficult or impossible to measure the exposure to the actual substance of interest, and therefore exposure to an "exposure surrogate" is estimated. This is often the case in environmental epidemiology, where sample sizes may be large. It is important that the surrogate marker is as closely correlated as possible to the actual substance of interest. For example, the presence of a gas or electric cooker could be a surrogate for the exposure to nitrogen dioxide (yes or no exposure). Living distance from a factory could be a surrogate for the emission of pollutants from the factory. However, there is some uncertainty in these cases regarding

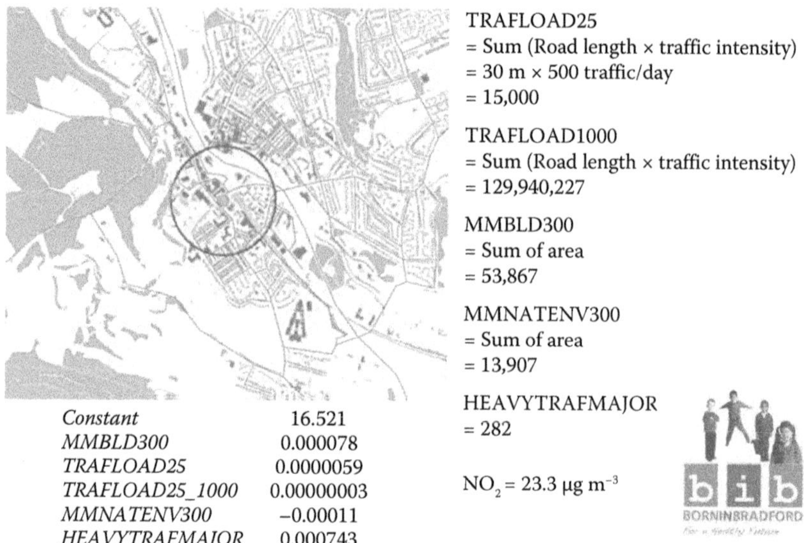

FIGURE 2.9
Example of a regression equation of a land use regression model applied to a subject's home in the Born in Bradford study. (Slide courtesy of Kees de Hoogh.)

actual exposure; for example, the gas cooker may not actually be used often, or the subject is out of the house while cooking takes place, and therefore there should be some validation.

Figure 2.10 provides an overall view of the exposure assessment approaches that were most often used in environmental epidemiological research of birth cohorts in Europe (Gehring et al. 2013). Europe has more than 35 birth cohorts that examine the relationship between a range of environmental exposures and outcomes, such as birth outcomes, growth and obesity, respiratory health, and cognitive development (Vrijheid et al. 2012).

A related issue is that people are often exposed to a number of pollutants simultaneously (a mixture), for example, to pesticides, solvents, air pollution, and disinfection by-products, although the levels may differ. Since not all the substances can be measured, a surrogate measure or exposure marker is chosen, for example, in the case of chlorination by-products in drinking water where trihalomethanes were used as a surrogate for other by-products.

Topic	Biomonitoring	Measurements	Modeling	Questionnaire
Outdoor air pollution			■	
Water contamination		■		■
Allergens and biological organisms				■
Heavy metals	■			
Pesticides				■
Persistent organic pollutants	■			
Emerging exposures	■			
Radiations				■
Smoking and ETS				■
Noise				■
Occupation				■

FIGURE 2.10
Predominant type of exposure assessment by exposure topic in European birth cohorts.

Mixtures of exposure may cause problems in the epidemiological analysis, for example, when the effects of different pollutants need to be disentangled. A possible way around this is to choose study sites where the exposure levels of the individual pollutants in the exposure mixture differ substantially and where there is a gradient of exposure.

More recently to get away from studying one exposure–one health outcome associations, a new paradigm has been developed, the exposome (see Chapter 14). The exposome encompasses the totality of human environmental (i.e., nongenetic) exposures from conception onward, complementing the genome (Wild 2005, 2012). The exposome is composed of every exposure to which an individual is subjected from conception to death. Therefore, it requires consideration of both the nature of those exposures and their changes over time. For ease of description, three broad categories of nongenetic exposures may be considered: internal, specific external, and general external. First, the exposome comprises processes internal to the body, such as metabolism, endogenous circulating hormones, body morphology, physical activity, gut microflora, inflammation, lipid peroxidation, oxidative stress, and ageing. These internal conditions will all impinge on the cellular environment and have been variously described as host or endogenous factors. Second, there is the extensive range of specific external exposures, which include radiation, infectious agents, chemical contaminants and environmental pollutants, diet, lifestyle factors (e.g., tobacco and alcohol), and occupation and medical interventions. In the past, these have been the main focus of epidemiological studies seeking to link environmental risk factors with cancer. Third, the exposome includes the wider social, economic, and psychological influences on the individual, for example, social capital, education, financial status, psychological and mental stress, urban–rural environment, and climate (Wild 2012). The dynamic nature of the exposome presents one of the most challenging features of its characterization. As a consequence, its myriad components need to be considered in relation to their temporal variation. In effect, at any given point in time, an individual will have a particular profile of exposures. Therefore, to fully characterize an individual's exposome would require either sequential measures that spanned a lifetime or a smaller number of measures that captured exposure over a series of extended periods (Wild 2012). Only because of the increase of new technologies, including GIS, sensors, remote sensing, and OMICS technologies, combined with more traditional approaches, has it become possible to start assessing the exposome, and first attempts are being made in large European projects such as HELIX (Vrijheid et al. 2014), EXPOsOMICs, and HEALS.

2.7 Validation Studies

In epidemiological studies, it is often not possible to obtain detailed exposure information on each subject in the study. For example, in a large cohort study, it is not feasible to take measurements from each subject and administer a detailed exposure questionnaire, or exposure is estimated for the subject's home, while the interest is the total exposure of the person.

In the former, it is desirable to carry out a small validation study on a representative subset of the larger population. Ideally, this will be carried out before the main study begins and can utilize information from the literature. Questions in the questionnaire could be validated with measurements, and some exposure models could be constructed. The exposure assessment in the whole population could focus on key questions that have a major influence on the exposure estimates, and thereby reduce the length of the questionnaire. Information on key determinants will also provide a better understanding of the exposure and how it may affect exposure–response relationships in epidemiological studies.

For the latter, an example is the recent VE3SPA study to evaluate the agreement between air pollution LUR estimates and personal exposure, and subjects in three different cities were asked to wear personal air pollution monitors for three periods lasting 2 weeks. The personal measurements were then compared with residential LUR estimates. The relationship varied by city and pollutant (Montagne et al. 2013, 2014). Also, new technologies, including smartphones, other GPS devices, and small sensors, may make it easier to validate spatial model estimates or at least improve existing exposure estimates. Many people nowadays have smartphones, and with the use of simple apps, they can provide information on location, mobility, physical activity, and even, to some extent, environment exposure levels, which are important features of good exposure assessment (de Nazelle et al. 2013; Donaire-Gonzalez et al. 2013; Nieuwenhuijsen et al. 2014, 2015). The smartphones, using their GPS, can be used to show objectively where people spend their time, and therefore which level of exposure they may experience, when overlaid with exposure maps (de Nazelle et al. 2013). Furthermore, the combination of assessment of personal air pollution concentrations and physical activity provides the opportunity to estimate the inhaled dose, which may be a better measure than exposure (de Nazelle et al. 2013). For example, de Nazelle et al. (2013) found, using modeled NO_2 data, that on average, time at home, which represented 51% of people's time in a day, and similarly, 54% of daily time-weighted exposures, accounted for 40% of individuals' total inhaled dose. Time at work, 33% of people's daily activity, led to 29% daily time-weighted exposures and 28% of daily inhaled NO_2. In reverse, volunteers only spent 6% of their time in transit, yet this microenvironment contributed to 11% of time-weighted exposures in a day, and 24% of daily inhaled NO_2. Nieuwenhuijsen et al. (2014) showed that using smartphone apps and a sensor, the Microaethalometer AE51 travelling routes and black carbon levels along the route could be obtained fairly easily.

Besides the validity, the reproducibility or reliability of various tools also can be evaluated in a subsample.

2.8 Retrospective Exposure Assessment

In epidemiological studies, when studying diseases with a long latency time, for example, cancer, it is not the current exposure that may be of most interest for the study, but that in the past. A reconstruction of historical exposure, often referred to as retrospective exposure assessment, is therefore. In occupational epidemiology, retrospective exposure assessment has a long tradition, but in environmental and spatial epidemiology less so (Bellander et al. 2001; Pesch et al. 2002). Retrospective exposure assessment is often difficult since there are usually many changes over time. Further work is needed in this area.

References

Armstrong B. (1996). Optimizing power in allocating resources to exposure assessment in an epidemiologic study. *American Journal of Epidemiology*, 144, 192–97.

Beelen R, Hoek G, Vienneau D, Eeftens M, Dimakopoulou K, Pedeli X, Tsai MY, et al. (2013). Development of NO_2 and NO_x land use regression models for estimating air pollution exposure in 36 study areas in Europe—the ESCAPE project. *Atmospheric Environment*, 72, 10–23.

Beelen R, Raaschou-Nielsen O, Stafoggia M, Andersen ZJ, Weinmayr G, Hoffmann B, Wolf K, et al. (2014). Effects of long-term exposure to air pollution on natural-cause mortality: An analysis of 22 European cohorts within the multicentre ESCAPE project. *Lancet*, 383, 785–95.

Bellander T, Berglind N, Gustavsson P, Jonson T, Nyberg F, Pershagen G, Järup L. (2001). Using geographic information systems to assess individual historical exposure to air pollution from traffic and house heating in Stockholm. *Environmental Health Perspectives*, 109, 633–39.

Brunekreef B. (1999). Exposure assessment. In *Environmental Epidemiology: A Text Book on Study Methods and Public Health Applications* (preliminary ed.). World Health Organization, Geneva.

Christensen K, Sobus J, Phillips M, Blessinger T, Lorber M, Tan YM. (2014). Changes in epidemiologic associations with different exposure metrics: A case study of phthalate exposure associations with body mass index and waist circumference. *Environment International*, 73C, 66–76.

Colvile RN, Stevens ES, Keegan T, Nieuwenhuijsen MJ. (2001). Atmospheric dispersion modelling for assessment of exposure to arsenic in the Nitra Valley, Slovakia. *Journal of Geophysical Research—Atmospheres*, 106, 17421–32.

Dadvand P, Ostro B, Figueras F, Foraster M, Basagaña X, Valentín A, Martinez D, et al. (2014a). Residential proximity to major roads and term low birth weight: The roles of air pollution, heat, noise, and road-adjacent trees. *Epidemiology*, 25, 518–25.

Dadvand P, Wright J, Martinez D, Basagaña X, McEachan RR, Cirach M, Gidlow CJ, de Hoogh K, Gražulevičienė R, Nieuwenhuijsen MJ. (2014b). Inequality, green spaces, and pregnant women: Roles of ethnicity and individual and neighbourhood socioeconomic status. *Environment International*, 71, 101–8.

de Nazelle A, Seto E, Donaire-Gonzalez D, Mendez M, Matamala J, Rodriguez D, Nieuwenhuijsen M, Jerrett M. (2013). Improving estimates of air pollution exposure through ubiquitous sensing technologies. *Environmental Pollution* 176, 92–99.

Donaire-Gonzalez D, de Nazelle A, Seto E, Mendez M, Rodriguez D, Nieuwenhuijsen M, Jerrett M. (2013). Comparison of physical activity measures using smartphone based CalFit and Actigraph. *Journal of Medical Internet Research*, 13, 15, e111.

Eeftens M, Beelen R, Bellander T, Cesaroni G, Cirach M, Declerq C, Dėdelė A, et al. (2012). Development of land use regression models for PM2.5, PM2.5 absorbance, PM10 and PMcoarse in 20 European study areas; results of the ESCAPE project. *Environmental Science and Technology*, 46, 11195–205.

Esplugues A, Estarlich M, Sunyer J, Fuentes-Leonarte V, Basterrechea M, Vrijheid M, Riaño I, et al. (2013). Prenatal exposure to cooking gas and respiratory health in infants is modified by tobacco smoke exposure and diet in the INMA birth cohort study. *Environmental Health*, 12, 100.

Gehring U, Casas M, Brunekreef B, Bergström A, Bonde JP, Botton J, Chévrier C, et al. (2013). Environmental exposure assessment in European birth cohorts: Results from the ENRIECO project. *Environmental Health*, 12, 8.

Govarts E, Nieuwenhuijsen M, Schoeters G, Ballester F, Bloemen K, de Boer M, Chevrier C, et al.; OBELIX; ENRIECO. (2012). Birth weight and prenatal exposure to polychlorinated biphenyls (PCBs) and dichlorodiphenyldichloroethylene (DDE): A meta-analysis within 12 European birth cohorts. *Environmental Health Perspectives*, 120, 162–70.

Keegan T, Bing Hong, Thornton I, Farago M, Jakubis P, Jakubis M, Pesch B, Ranft U, Nieuwenhuijsen M. (2002). Assessment of environmental arsenic levels in Prievidza District. *Journal of Exposure Analysis Environmental Epidemiology*, 12, 179–85.

Kloog I, Melly SJ, Ridgway WL, Coull BA, Schwartz J. (2012). Using new satellite based exposure methods to study the association between pregnancy $PM_{2.5}$ exposure, premature birth and birth weight in Massachusetts. *Environmental Health*, 11, 40.

Kromhout H, Heederik D. (1995). Occupational epidemiology in the rubber industry; implications of exposure variability. *American Journal of Industrial Medicine*, 27, 171–85.

Llop S, Guxens M, Murcia M, Lertxundi A, Ramon R, Riaño I, Rebagliato M, et al.; INMA Project. (2012). Prenatal exposure to mercury and infant neurodevelopment in a multicenter cohort in Spain: Study of potential modifiers. *American Journal of Epidemiology*, 175, 451–65.

Montagne D, Hoek G, Nieuwenhuijsen M, Lanki T, Pennanen A, Portella M, Meliefste K, et al. (2013). Agreement of land use regression models with personal exposure measurements of particulate matter and nitrogen oxides air pollution. *Environmental Science and Technology*, 47, 8523–31.

Montagne D, Hoek G, Nieuwenhuijsen M, Lanki T, Siponen T, Portella M, Meliefste K, Brunekreef B. (2014). Temporal associations of ambient PM2.5 elemental concentrations with indoor and personal concentrations. *Atmospheric Environment*, 86, 203–11.

Nieuwenhuijsen MJ. (1997). Exposure assessment in occupational epidemiology: Measuring present exposures with an example of occupational asthma. *International Archives of Occupational and Environmental Health*, 70, 295–308.

Nieuwenhuijsen MJ, Donaire-Gonzalez D, Foraster M, Martinez D, Cisneros A. (2014). Using personal sensors to assess the exposome and acute health effect. *International Journal of Environmental Research and Public Health*, 11, 7805–19.

Nieuwenhuijsen MJ, Donaire-Gonzalez D, Rivas I, Cirach M, Seto E, Jerrett M, Hoek G, Sunyer J, Querol X. (2015). Variability and agreement between modeled and personal continuously measured black carbon levels using novel smartphone and sensor technologies, 49(5), 2977–82.

Pesch B, Ranft U, Jakubis P, Nieuwenhuijsen MJ, Hergemüller A, Unfried K, Jakubis M, Miskovic P, Keegan T, EXPASCAN Study Group. (2002). Environmental arsenic exposure from a coal-burning power plant as potential risk factor for non-melanoma skin carcinoma: Results from a case-control study in the District of Prievidza, Slovakia. *American Journal of Epidemiology*, 155, 798–809.

Samoli E, Peng R, Ramsay T, Pipikou M, Touloumi G, Dominici F, Burnett R, et al. (2008). Acute effects of ambient particulate matter on mortality in Europe and North America: Results from the APHENA study. *Environmental Health Perspectives*, 116, 1480–86.

van Tongeren M, Gardiner K, Calvert I, Kromhout H, Harrington JM. (1997). Efficiency of different grouping schemes for dust exposure in the European carbon black respiratory morbidity study. *Occupational and Environmental Medicine*, 54, 714–19.

Vineis P. (ed.). (1999). *Metabolic Polymorphisms and Susceptibility.* IARC Publication 148. International Agency for Research on Cancer, Lyon, France.

Vrijheid M, Casas M, Bergström A, Carmichael A, Cordier S, Eggesbø M, Eller E, et al. (2012). European birth cohorts for environmental health research. *Environmental Health Perspectives*, 120, 29–37.

Vrijheid M, Slama R, Robinson O, Chatzi L, Coen M, van den Hazel P, Thomsen C, et al. (2014). The human early-life exposome (HELIX): Project rationale and design. *Environmental Health Perspectives*, 122, 535–44.

Wild CP. (2005). Complementing the genome with an "exposome": The outstanding challenge of environmental exposure measurement in molecular epidemiology. *Cancer Epidemiology, Biomarkers and Prevention*, 14, 1847–50.

Wild CP. (2012). The exposome: From concept to utility. *International Journal of Epidemiology*, 41, 24–32.

Wilhelm M, Ritz B. (2005). Local variations in CO and particulate air pollution and adverse birth outcomes in Los Angeles County, California, USA. *Environmental Health Perspectives*, 113, 1212–21.

3

Interpreting Clusters of Health Events

Geoffrey Jacquez
Department of Geography
State University of New York at Buffalo
Buffalo, New York

Jared Aldstadt
Department of Geography
State University of New York at Buffalo
Buffalo, New York

CONTENTS

3.1 Introduction

There has been some debate over the years regarding the utility of disease cluster analysis. While it is widely recognized that cluster analysis may be part of a public health response to an outbreak of disease, or to reports of a possible cluster by concerned citizens, the contribution and role of cluster analysis in spatial epidemiology is less clear. Applications frequently cited as successful exemplars include Snow's investigation of cholera (Snow 1855), the discovery of the link between fluoride in drinking water and dental caries (Dean 1938), and the identification of mercury poisoning as the cause of Minamata's disease (Harada 1995). Other etiological connections triggered by clustering of adverse health events include discovery of acquired immune deficiency syndrome (Friedman-Kien et al. 1981) and the association between a food allergy and tick bites (Commins et al. 2011). For chronic diseases such as cancer, the link to a putative cause is often unclear, since disease latency is often long, actual exposures are not measured, and the number of cases is often small, making a finding of a statistically significant excess difficult. Chronic disease clusters successfully linked to specific exposures include the Libby Montana cluster of lung cancer from asbestos exposures related to vermiculite mining (McDonald et al. 2004); the clustering of leukemia, lymphoma, and adverse birth outcomes in Camp Lejeune, North Carolina, due to exposures to trichloroethylene, benzene, and other carcinogens from drinking water (ATSDR 2010); and a possible cluster of brain and central nervous systems cancers in Toms River, New Jersey, possibly caused by exposures to styrene, acrylonitrile, and styrene-acrylonitrile (SAN) trimer (ATSDR 1997). Although there have been some positive cluster studies, the majority of them do not find a significant excess, leading Neutra and others to view cluster analysis as an expensive endeavor that yields little insights into the existence of clusters, let alone their underlying causes (Neutra 1990). What then are the plausible underlying causes of disease clusters? And how have cluster analyses advanced our understanding of disease processes?

This chapter provides an overview of issues central to the interpretation of disease clusters. Here, we concern ourselves not with methods (refer to Chapter 8 in this volume for statistical tests for clustering and surveillance); rather, we discuss when cluster studies have been used, their role in scientific inquiry, and relevant issues such as case ascertainment, the small numbers problem, and aspects of uncertainty. We seek to identify what can be known, and the pitfalls encountered along the way. A case study is included to demonstrate that several of the pitfalls can be avoided.

This chapter is organized around three questions or topics:

1. How are cluster studies used, and what is their function?
2. Are cluster studies science, and what is the role of cluster analysis in scientific inquiry?
3. What are the issues to consider when interpreting cluster study results?

3.2 How Are Cluster Studies Used and What Is Their Function?

Cluster studies are commonly used in three capacities: (1) as part of a public health response to a cluster allegation brought forward by concerned citizens, (2) in confirmatory studies that search for an association between a putative environmental exposure and a health outcome, and (3) in a hypothesis testing framework to systematically advance scientific knowledge. We consider each of these in turn.

3.2.1 Public health response

Kingsley et al. (2007) and others provide a useful review of cluster studies at the Centers for Disease Control and Prevention (CDC) and how they may be used within the context of a public health response to a clustering of health events identified and reported by concerned citizens (Buehler et al. 2004; CDC 1990; Kingsley et al. 2007). Attendant issues within this framework include preselection bias, known colloquially as the "Texas sharpshooter problem," which may increase the type I error (false positives). The rationale is that by constantly surveying their environment, citizens act to "preselect" apparent local disease excesses and thereby report chance clusters with higher probability. In practice, the majority of cluster studies have negative results—that is, they do not find a statistically significant excess of disease cases (Schulte et al. 1987; Warner and Aldrich 1988). This failure to result in positive findings is one of the reasons why Neutra (1990) suggested cluster studies are a wasteful enterprise for public health agencies, especially when those resources might reliably be used in vaccination, screening, and other activities that reduce disease burden. The application of clustering methods within the public health response framework may be called pre-epidemiology since a designed sample may be lacking and an epidemiologically sound study design is imposed only after a statistically significant cluster has been demonstrated.

3.2.2 Confirmatory studies (association)

Once geographically statistically significant clusters of health events have been found, it is tempting to seek possible causes by searching for associations with environmental exposures and other factors. While such an activity may suggest causal hypotheses, it does not have the power to reject hypotheses or test predictions (Jacquez 2004). From that perspective, confirmatory studies that search for association have little appeal for spatial epidemiology.

3.2.3 Hypothesis testing (advancing knowledge)

What, then, is the role of cluster studies in spatial epidemiology? We can think of spatial epidemiology as the documentation and analysis of spatial disease patterns to better understand the causes and correlates of disease. As noted above, cluster studies in the public health response framework often lack a sound sampling design, as they are by definition a post hoc response to alleged clusters. Confirmatory studies seek to identify associations between clusters and spatial patterns of environmental variables, and thus do not employ an inferential framework consistent with the scientific method put forward by Popper and others (Platt 1964; Popper 1968). The coupling of epidemiological study designs with cluster analysis techniques supports a more robust inference structure (Meliker et al. 2009a, 2009b) and may be referred to as post-epidemiology. Here, cluster analysis methods are applied to the residual risk that is not explained by the risk factors identified in a parent epidemiological study. Any clusters so identified then are above and beyond the causal factors identified in the parent epidemiological study. This approach has been successfully employed in case–control studies to identify potential clustering of testicular cancers (Sloan et al. 2013), childhood leukemia and diabetes (Schmiedel et al. 2011), and breast cancers (Jacquez et al. 2013), among others. When good-quality disease registries are available that document most, if not all, incident cases, it is possible to use cluster studies to identify statistically significant local excesses of disease, provided known explanatory (e.g., nuisance) factors have been taken into account. Spatial patterns in known explanatory factors may be accounted for using "neutral models" (Goovaerts and Jacquez 2004, 2005). This makes it possible to use disease cluster analysis to test hypotheses regarding the spatial patterns of disease, thereby increasing the utility of clustering methods in spatial epidemiology.

3.2.4 Are cluster studies science (role of cluster studies in scientific inquiry)?

The above discussion identified two components necessary in order for cluster studies to advance spatial epidemiological knowledge: (1) a sound sampling design, usually from an epidemiological study (e.g., case–control or cohort) or from a complete census of incident cases such as may be provided by a disease registry and (2) null hypotheses that support the inclusion of known factors that might explain observed clusters (e.g., neutral models). Several questions and issues must be addressed when interpreting clusters in spatial epidemiology:

1. *Is there an excess?* Most clustering methods fall under the rubric of global, local, or focused tests. Here, global clustering refers to the existence of excess risk somewhere in the study area; local clustering methods report where those clusters are found, usually with their spatial extent, population size affected, and an estimate

of the excess risk; and focused clustering assesses whether there is elevated risk in the immediate vicinity of a suspected location or focus (Besag and Newell 1991; Lawson 1989; Waller et al. 1995). Statistical clustering methods are used to identify whether there is a statistically significant excess, and where that excess may be found.

2. *Is it real?* The identification of a statistically significant excess does not mean that the disease cluster is real, in the sense of reflecting a true, underlying excess in disease risk. Spurious findings of an excess of health events are expected by chance. But other factors can play a role as well, giving rise to false cluster findings. The CDC makes a distinction between false and true clusters (CDC 1990). False clusters lack a plausible biological explanation and may be comprised of cases with symptoms or illnesses with unrelated causes. True clusters are comprised of a statistically significant excess of cases with a common, plausible biological explanation. Hence, biological plausibility, as well as statistical significance, must be considered when assessing whether a cluster is real.

3. *What does it mean?* When a real excess has been identified, the challenge of interpretation begins in earnest. This may involve the identification of the set of possible causes that may underlie the observed excess. Once these have been enumerated, these specific alternative explanations may be excluded in a systematic fashion by analysis of additional data or information. The set of explanations may include chance, case-attractor hypotheses, spatial pattern in covariates, and the action of causative exposures and disease processes. Case-attractor hypotheses describe processes that bring cases together in the absence of elevated disease risk. Examples include patients moving to be near treatment facilities, and modifications in behaviors that cause either cases or noncases to cluster. An example of the latter is the healthy worker effect, where in order to be employed, workers must be in comparatively sound health (Fornalski and Dobrzyński 2010).

4. *What's the hypothesis?* Almost all tests for clustering employ null and alternative hypotheses. The null hypothesis is a statement regarding the spatial pattern expected in the absence of a cluster process, while the alternative hypothesis is embodied in specification of the spatial weights (Griffiths 1995). Often, the meaning of the null and alternative hypotheses in terms of the underlying disease process is given little thought. One example is the use of complete spatial randomness (CSR) as the null hypothesis in many statistical tests of spatial disease clustering. CSR usually implies that the underlying disease risk, in the absence of a cluster process, will be uniform across the study area. In practice, this seldom is true, since disease risk may be associated with covariates such as age, income, and ethnicity. When working with area-based data, one solution is to use covariate-adjusted rates. Another solution is to employ randomization or Monte Carlo methods with randomization procedures that account for the covariate structure. One may also model the expected risk while incorporating known risk factors into the model, and then inspect the residuals (observed minus expected risk) for spatial pattern. But unless spatial structure in the covariates is fully accounted for, CSR may not be a reasonable null hypothesis.

   For the alternative hypothesis, the specification of spatial weights should be accomplished in a manner that reflects the underlying cluster process. For example, when diffusion is suspected, it may make sense to link adjacent locations to one another, for example, specify a spatial weight of 1 for nearby locations. But if transport on a network is suspected, one might choose to link locations using the actual road network. In practice, one thus may wish to explore a suite of alternative hypotheses, each corresponding to its own spatial weight set. This requires careful enumeration of what the set of alternative hypotheses might be, coupled with construction of the spatial weights for each alternative hypothesis. In certain instances, knowledge of the scale of the clustering may be absent or anecdotal, in which case spatial weight sets corresponding to different spatial scales may be employed, in a sensitivity analysis of the effect of spatial scale on the results of the cluster analysis.

5. *What is the sampling frame and design?* Interpretation of a cluster finding must be premised on the sampling frame and sampling design. Especially within the public health framework mentioned above, the data may be encountered, for example, cases that have been reported by concerned citizens. Here, a designed sample that is representative of an underlying study population is absent, and any inferences that may be drawn thus would apply only to the sample. When registry data are analyzed, all cases conceivably could be included in the study, and inferences would then apply to the entire population covered by the registry. When a postepidemiological framework is used, the scope of inference will apply to the study population from which the sample in the parent epidemiological study was drawn.
6. *What inferences can be drawn regarding underlying disease processes?* When the data consist of cases reported by concerned citizens, it is difficult to make inferences regarding underlying disease processes, since there usually will be several alternative explanations for an observed disease cluster. But when the data are sampled in a systematic fashion, and tests for spatial, temporal, and space–time disease clustering are used, it may be possible to construct inferences regarding an underlying disease process (see, e.g., Jacquez et al. 2013, table 1). But this is a difficult prospect, especially in the absence of additional information, since different diseases in different situations can give rise to similar space–time patterns of case occurrence.

3.3 Relevant Issues

There are several other issues that can impact the interpretation of cluster findings. These include case ascertainment, incomplete reporting, the small numbers problem, and geocoding location error.

Accurate *case ascertainment* is critical. Are cases correctly diagnosed, and to what extent are there misdiagnoses? Considerable effort has gone into describing and understanding the spatial patterns of West Nile virus (WNV) transmission in North America since its introduction in 1999 (Nash et al. 2001; Ruiz et al. 2010; Hayes et al. 2005). Most human WNV infections are asymptomatic, and many result in very common clinical features that are likely to be misdiagnosed (Davis et al. 2006). In this case, clusters of human WNV illnesses may be a result of variability in diagnoses, host characteristics, or strain virulence, as well as an indicator of infection risk. Additionally, comparative studies have shown that spatial analysis and clustering results vary considerably, depending on whether data are obtained through a notifiable disease reporting system or administrative health records (Jones et al. 2012; Yiannakoulias and Svenson 2009).

Incomplete reporting can reduce the power to detect a true cluster. And when the extent of incomplete reporting is spatially structured, it can give rise to the finding of clusters when they are, in fact, absent. One example of this is maps where the disease rate varies dramatically across administrative boundaries. While this may reflect a true difference in disease burden, it is often explained by reporting differences across administrative units. Passive disease surveillance systems are particularly susceptible to spatially heterogeneous intensity of reporting that confound cluster analysis. Economic and social barriers to obtaining proper healthcare and associated differences in treatment-seeking behavior may lie at the root of this problem. For example, incomplete reporting has been suggested as a cause of geographical clustering of autism incidence in California, and the apparent associations between parental education levels and autism incidence (Van Meter et al. 2010).

The *small numbers problem* refers to the increase in variance in disease rates (calculated as the number of incident cases, e.g., divided by the size of the at-risk population) as the size of the at-risk population in the denominator decreases. Many cluster analysis methods account for differences in population size, but not all of them do. The local Moran statistic, for example, is often applied to raw disease rates, does not account for population size, and hence can yield spurious cluster findings (false positives). Some practitioners recommend that disease rates be smoothed using an empirical Bayesian smoother prior to local Moran analysis. Such smoothing introduces spurious spatial autocorrelation into the resulting smoothed rates. Since this spurious autocorrelation is not accounted for in the null hypothesis (which in most studies is CSR), the cluster analysis results are unreliable.

Geocoding is frequently used in cluster analysis to identify the spatial coordinates of health events (Abe and Stinchcomb 2008; Goldberg 2008). That geocodes (the geographic coordinates that are the result of geocoding an address) have an associated location uncertainty is well known (Krieger et al. 2001; Bonner et al. 2003; Oliver et al. 2005; Whitsel et al. 2006). Location uncertainty impacts disease cluster analysis by decreasing statistical power (Rushton et al. 2006; Zandbergen et al. 2012) and introducing bias into exposure models that use geocoded locations (Mazumdar et al. 2008). However, at the time of this writing, geocoding location error is routinely ignored in the interpretation of cluster results (Jacquez 2012; Whitsel et al. 2006).

3.4 Case Study

Sloan et al. (2015) examine testicular cancer cases to determine whether there are spatial or spatial–temporal clusters of risk. Excess risk for testicular cancer is largely unexplained, and this study seeks to generate hypotheses by detecting locations of shared exposure among testicular cancer patients. Case–control study design, detailed residential history data, and adjustment for personal risk factors are employed to rigorously test the null hypothesis of uniform risk with nearest-neighbor-based Q-statistics (Jacquez et al. 2005).

The study is based on reliable case data and a rigorous case–control study design. The case data are taken from the Danish Cancer Registry and include 3297 cases diagnosed between 1991 and 2003. Two sets of birth-date-matched controls were obtained from the Danish Civil Registration System. The second set of controls was obtained to address the possibility that clustering results are due to choice of control group (Nordsborg et al. 2013). Residential histories were also obtained from the registration system. Inclusion of residential histories accounts for population mobility and also allows for the temporal alignment of cases by age at diagnosis or time prior to diagnosis. The temporal alignment of cases allows for an examination of important ages of exposure and accounts for the latency period between exposure and diagnosis.

Risk factors were examined first in a conditional logistic regression model that included individual-level characteristics and community-level socioeconomic status. Family history of testicular cancer was the only significant predictor of increased risk and was incorporated into the cluster detection analysis. The problem of multiple testing was addressed by performing simulation studies and checking for correspondence with the SaTScan procedure. Sloan et al. (2012) employed residential history data from the same registration system in a simulation study to provide a threshold for significant clusters in the face of multiple testing. Local space–time clusters detected with the Q-statistics were reexamined for spatial clustering using SaTScan, which is robust in the face of multiple testing.

This study did not find convincing evidence of clusters of risk for testicular cancer. This "negative" result is itself instructive and hypothesis generating. It may be that environmental exposures do not play a role in testicular cancer risk. The result may also indicate that important environmental exposures do not vary at a scale that can be detected by this method. Exposures within households or an environmental factor that is practically uniform throughout Denmark are an example. In either case, these robust results will be important when aggregating studies from different regions or in the design of future studies examining the behavioral and environmental risks for testicular cancer.

3.5 Conclusions

Disease cluster analysis consists of a spectrum of techniques, from inferential pattern analysis (e.g., tests for disease clustering) to modeling approaches, including regression, geostatistical, and Bayesian models. This chapter has concentrated primarily on inferential pattern analysis and the interpretation of clusters of health events.

While the 1980s and 1990s might be correctly typified as an era of pre-epidemiology in cluster analysis, advances in recent years, especially the use of rigorous sampling designs, have greatly strengthened the inferential structure of cluster analysis. The included case study demonstrates that available data and methodologies have moved cluster studies into the realm of post-epidemiology. Increasingly, health surveillance systems are using unstructured data streams, such as those from Twitter and Google search engines. These kinds of data tend to be heterogeneous, and the sampling method is only partially known, without a formal sampling design. An important area of future research in cluster analysis is how to impose representative sampling frames and designs on such data streams.

References

Abe, T. and D. Stinchcomb. (2008). Geocoding practices in cancer registries. In *Geocoding Health Data*, ed. G. Rushton, M. P. Armstrong, J. Gittler, B. R. Greene, C. E. Pavlik, M. M. West, and D. Zimmerman, 111–125. Boca Raton, FL: CRC Press.

ATSDR (Agency for Toxic Substances and Disease Registry). (1997). Childhood cancer incidence health consultation: A review and analysis of cancer registry data, 1979–1995 for Dover Township (Ocean County), New Jersey. Atlanta, GA: U.S. Department of Health and Human Services, ATSDR, and New Jersey Department of Health and Senior Services, Hazardous Site Health Evaluation Program, Division of Epidemiology, Environmental and Occupational Health.

ATSDR (Agency for Toxic Substances and Disease Registry). (2010). Camp Lejeune, North Carolina: Health study activities frequently asked questions (FAQs). Atlanta, GA: ATSDR, April 2.

Besag, J. and J. Newell. (1991). The detection of clusters in rare diseases. *Journal of the Royal Statistical Society Series A*, 154, 143–155.

Bonner, M. R., D. Han, J. Nie, P. Rogerson, J. E. Vena, and J. L. Freudenheim. (2003). Positional accuracy of geocoded addresses in epidemiologic research. *Epidemiology*, 14, 408–412.

Buehler, J. W., R. S. Hopkins, J. M. Overhage, D. M. Sosin, and V. Tong. (2004). Framework for evaluating public health surveillance systems for early detection of outbreaks: Recommendations from the CDC Working Group. *MMWR Recommendations and Reports*, 53, 1–11.

CDC (Centers for Disease Control and Prevention). (1990). Guidelines for investigating clusters of health events. *Mortality and Morbidity Weekly Report*, 39, 1–16.

Commins, S. P., H. R. James, L. A. Kelly, S. L. Pochan, L. J. Workman, M. S. Perzanowski, K. M. Kocan, J. V. Fahy, L. W. Nganga, and E. Ronmark. (2011). The relevance of tick bites to the production of IgE antibodies to the mammalian oligosaccharide galactose-α-1,3-galactose. *Journal of Allergy and Clinical Immunology*, 127, 1286–1293.

Davis, L. E., R. DeBiasi, D. E. Goade, K. Y. Haaland, J. A. Harrington, J. B. Harnar, S. A. Pergam, M. K. King, B. DeMasters, and K. L. Tyler. (2006). West Nile virus neuroinvasive disease. *Annals of Neurology*, 60, 286–300.

Dean, H. T. (1938). Endemic fluorosis and its relation to dental caries. *Public Health Report*, 53, 1443–1452.

Fornalski, K. W. and L. Dobrzyński. (2010). The healthy worker effect and nuclear industry workers. *Dose-Response*, 8, 125–147.

Friedman-Kien, A., L. Laubenstein, M. Marmor, K. Hymes, J. Green, A. Ragaz, J. Gottleib, F. Muggia, R. Demopoulos, and M. Weintraub. (1981). Kaposis sarcoma and *Pneumocystis pneumonia* among homosexual men—New York City and California. *MMWR Morbidity and Mortality Weekly Report*, 30, 305–308.

Goldberg, D. 2008. *A Geocoding Best Practices Guide*. Springfield, IL: North American Association of Central Cancer Registries.

Goovaerts, P. and G. M. Jacquez. (2004). Accounting for regional background and population size in the detection of spatial clusters and outliers using geostatistical filtering and spatial neutral models: The case of lung cancer in Long Island, New York. *International Journal of Health Geographics*, 3, 14.

Goovaerts, P. and G. M. Jacquez. (2005). Detection of temporal changes in the spatial distribution of cancer rates using local Moran's I and geostatistically simulated spatial neutral models. *Journal of Geographical Systems*, 7, 137–159.

Griffiths, D. 1995. Some guidelines for specifying the geographic weights matrix contained in spatial statistical models. In *Practical Handbook of Spatial Statistics*, ed. Arlinghau, S. L., D. A. Griffith, W. C. Arlinghaus, W. D. Drake, and J. D. Nystuen, 65–82. Boca Raton, FL: CRC Press.

Harada, M. (1995). Minamata disease: Methylmercury poisoning in Japan caused by environmental pollution. *CRC Critical Reviews in Toxicology*, 25, 1–24.

Hayes, E. B., N. Komar, R. S. Nasci, S. P. Montgomery, D. R. O'Leary, and G. L. Campbell. (2005). Epidemiology and transmission dynamics of West Nile virus disease. *Emerging Infectious Disease*, 11, 1167–1173.

Jacquez, G., J. Barlow, R. Rommel, A. Kaufmann, M. Rienti, G. AvRuskin, and J. Rasul. (2013). Residential mobility and breast cancer in Marin County, California, USA. *International Journal of Environmental Research and Public Health*, 11, 271–295.

Jacquez, G. M. (2004). Current practices in the spatial analysis of cancer: Flies in the ointment. *International Journal of Health Geographics*, 3, 22.

Jacquez, G. M. (2012). A research agenda: Does geocoding positional error matter in health GIS studies? *Spatial and Spatio-Temporal Epidemiology*, 3, 7–16.

Jacquez, G. M., A. Kaufmann, J. Meliker, P. Goovaerts, G. AvRuskin, and J. Nriagu. (2005). Global, local and focused geographic clustering for case-control data with residential histories. *Environmental Health*, 4, 4.

Jones, S. G., W. Conner, B. Song, D. Gordon, and A. Jayakaran. (2012). Comparing spatio-temporal clusters of arthropod-borne infections using administrative medical claims and state reported surveillance data. *Spatial and Spatio-Temporal Epidemiology*, 3, 205–213.

Kingsley, B. S., K. L. Schmeichel, and C. H. Rubin. (2007). An update on cancer cluster activities at the Centers for Disease Control and Prevention. *Environmental Health Perspectives*, 115, 165–171.

Krieger, N., P. Waterman, K. Lemieux, S. Zierler, and J. Hogan. (2001). On the wrong side of the tracts? Evaluating the accuracy of geocoding in public health research. *American Journal of Public Health*, 91, 1114–1116.

Lawson, A. B. 1989. Score tests for detection of spatial trend in morbidity data. Dundee, Scotland: Dundee Institute of Technology.

Mazumdar, S., G. Rushton, B. J. Smith, D. L. Zimmerman, and K. J. Donham. (2008). Geocoding accuracy and the recovery of relationships between environmental exposures and health. *International Journal of Health Geographics*, 7, 13.

McDonald, J., J. Harris, and B. Armstrong. (2004). Mortality in a cohort of vermiculite miners exposed to fibrous amphibole in Libby, Montana. *Occupational and Environmental Medicine*, 61, 363–366.

Meliker, J. R., P. Goovaerts, G. M. Jacquez, G. A. Avruskin, and G. Copeland. (2009a). Breast and prostate cancer survival in Michigan: Can geographic analyses assist in understanding racial disparities? *Cancer*, 115, 2212–2221.

Meliker, J. R., G. M. Jacquez, P. Goovaerts, G. Copeland, and M. Yassine. (2009b). Spatial cluster analysis of early stage breast cancer: A method for public health practice using cancer registry data. *Cancer Causes Control*, 20, 1061–1069.

Nash, D., F. Mostashari, A. Fine, J. Miller, D. O'Leary, K. Murray, A. Huang, A. Rosenberg, A. Greenberg, and M. Sherman. (2001). The outbreak of West Nile virus infection in the New York City area in 1999. *New England Journal of Medicine*, 344, 1807–1814.

Neutra, R. R. (1990). Counterpoint from a cluster buster. *American Journal of Epidemiology*, 132, 1–8.

Nordsborg, R. B., J. R. Meliker, A. K. Ersbøll, G. M. Jacquez, and O. Raaschou-Nielsen. (2013). Space-time clustering of non-Hodgkin lymphoma using residential histories in a Danish case-control study. *PLoS One*, 8, e60800.

Oliver, W. N., K. Matthews, M. Siadaty, F. Hauck, and L. Pickle. (2005). Geographic bias relating to geocoding error in epidemiologic studies. *International Journal of Health Geographics*, 4, 29.

Platt, J. (1964). Strong inference. *Science*, 146, 347–353.

Popper, K. 1968. *The Logic of Scientific Discovery*. New York: Harper and Rowe.

Ruiz, M. O., L. F. Chaves, G. L. Hamer, T. Sun, W. M. Brown, E. D. Walker, L. Haramis, T. L. Goldberg, and U. D. Kitron. (2010). Local impact of temperature and precipitation on West Nile virus infection in *Culex* species mosquitoes in northeast Illinois, USA. *Parasites & Vectors*, 3, 19.

Rushton, G., M. Armstrong, J. Gittler, B. Greene, C. Pavlik, M. West, and D. Zimmerman. (2006). Geocoding in cancer research—a review. *American Journal of Preventive Medicine*, 30, S16–S24.

Schmiedel, S., G. Jacquez, M. Blettner, and J. Schüz. (2011). Spatial clustering of leukemia and type 1 diabetes in children in Denmark. *Cancer Causes and Control*, 22, 849–857.

Schulte, P. A., R. L. Ehrenberg, and M. Singal. (1987). Investigation of occupational cancer clusters: Theory and practice. *American Journal of Public Health*, 77, 52–56.

Sloan, C. D., R. Baastrup-Norstrom, G. M. Jacquez, P. J. Landrigan, O. Raaschou-Nielsen, and J. R. Meliker. (2013). Space-time analysis of testicular cancer clusters using residential histories: A case-control study in Denmark. *PLoS One*, 10, e0120285.

Sloan, C. D., G. M. Jacquez, C. M. Gallagher, M. H. Ward, O. Raaschou-Nielsen, R. B. Nordsborg, and J. R. Meliker. (2012). Performance of cancer cluster Q-statistics for case-control residential histories. *Spatial and Spatio-Temporal Epidemiology*, 3, 297–310.

Sloan, C. D., R. B. Nordsborg, G. M. Jacquez, O. Raaschou-Nielsen, and J. R. Meliker. (2015). Space-time analysis of testicular cancer clusters using residential histories: A case-control study in Denmark. *PLoS One*, 10, e0120285.

Snow, J. (1855). *On the Mode of Communication of Cholera*. London: John Churchill, New Burlington Street.

Van Meter, K. C., L. E. Christiansen, L. D. Delwiche, R. Azari, T. E. Carpenter, and I. Hertz-Picciotto. (2010). Geographic distribution of autism in California: A retrospective birth cohort analysis. *Autism Research*, 3, 19–29.

Waller, L. A., B. W. Turnbull, G. Gustafsson, U. Hjalmars, and B. Andersson. (1995). Detection and assessment of clusters of disease: An application to nuclear power plant facilities and childhood leukaemia in Sweden. *Statistics in Medicine*, 14, 3–16.

Warner, S. C. and T. E. Aldrich. (1988). The status of cancer cluster investigations undertaken by state health departments. *American Journal of Public Health*, 78, 306–307.

Whitsel, E., P. Quilbrera, R. Smith, D. Catellier, D. Liao, A. Henley, and G. Heiss. (2006). Accuracy of commercial geocoding: Assessment and implications. *Epidemiologic Perspectives and Innovations*, 3, 8.

Yiannakoulias, N. and L. Svenson. (2009). Differences between notifiable and administrative health information in the spatial–temporal surveillance of enteric infections. *International Journal of Medical Informatics*, 78, 417–424.

Zandbergen, P. A., T. C. Hart, K. E. Lenzer, and M. E. Camponovo. (2012). Error propagation models to examine the effects of geocoding quality on spatial analysis of individual-level datasets. *Spatial and Spatio-Temporal Epidemiology*, 3, 69–82.

4

Geographic Information Systems in Spatial Epidemiology and Public Health

Robert Haining
Department of Geography
University of Cambridge
Cambridge, UK

Ravi Maheswaran
School of Health and Related Research
University of Sheffield
Sheffield, UK

CONTENTS

4.1 Introduction: What Is a Geographical Information System?

Definitions of geographical information systems (GIS) usually fall into one of three categories: toolbox-, database-, and organization-based definitions (Burrough and McDonnell, 2000). Geographical or spatial data are data about entities in the real world (physically represented as point, line, or area objects) that define their location and the attributes recorded at each location. Knowing where entities are located allows spatial relationships between them to be defined, such as distances between points, adjacency or otherwise of areas, and the proximity of one object to another. A GIS database is distinguished from most other databases by knowing where in geographical space objects (points, lines, and polygons) are located in relation to one another since most other kinds of databases capture only entities and their attributes. Database definitions of GIS emphasize this difference: "any ... computer based set of procedures used to *store and manipulate geographically referenced data*" (quoted in Burrough and McDonnell, 2000, p. 11, italics added). Geography is stored in the form of a series of discrete layers in the database (e.g., a layer for the roads, a layer for the waste sites, a layer for the forested areas, and a layer for the deprivation scores by census area), enabling different features to be switched on or off when visualizing or mapping an area.

Toolbox-based definitions, on the other hand, emphasize system functionality and the place of GIS within information technology: "*a powerful set of tools* for collecting, storing, retrieving at will, transforming and displaying spatial data from the real world for a particular set of purposes" (quoted in Burrough and McDonnell, 2000, p. 11, italics added). Burrough and McDonnell (2000, p. 15) list some of the basic operational requirements for a GIS. These include being able to show the locations of entities individually and in relation to others ("identify all the areas within a 15-minute travel distance of location x"), compute the physical size of areas, show the result of intersecting or overlaying different layers of spatial data (e.g., air pollution data, socioeconomic data, and facilities data), count the number of cases of an entity within a given distance, and determine paths of least cost or least resistance over a surface or network. The different layers in the database can be overlaid or buffered in order to respond to such queries that may be thought of as cartographic modelling. Certain forms of statistical modelling may also be available inside a GIS. For example, the ArcGIS software includes Geostatistical Analyst, which enables various forms of spatial statistical analysis to be executed within the GIS.

GIS capabilities, in terms of database management and toolbox functionality, have been exploited in many application fields (see, e.g., Burrough and McDonnell, 2000, p. 9). In spatial epidemiology, GIS capabilities have been used to capture the geographical distribution of disease (and how that distribution changes over discrete periods of time) and the relationship between the occurrence of a disease and various environmental as well as social and economic factors. In public health research, however, interest often focuses on the ways GIS can also contribute to planning public health services and interventions, improving access to healthcare services, assessing the locational impacts of health policy, and facilitating community participation (by groups or individuals) in addressing local health concerns (see de Lepper et al., 1995). When applied to these sorts of questions, a GIS meets the organization-based definition of a GIS: "*a decision support system* involving the integration of spatially referenced data in a problem solving environment" (quoted in Burrough and

McDonnell, 2000, p. 11, italics added). Cromley and McLafferty (2012, p. 14) remark: "GIS, as a means of exploring health problems and finding ways to address them, has taken its place in the conceptual and methodological foundations of public health." GIS literacy and GIS capability have become important for spatial epidemiology and public health research and practice.

In the context of this volume, however, there is another definition of a GIS: as an important enabling technology in the implementation of geographic information *science* (GISc). Here, we take GISc to refer to all those areas of scientific enquiry in which *where* events occur matters, perhaps for the purposes of description, explanation, or prediction of those events. It follows that the acquisition and processing of spatially referenced data are of fundamental importance to the progress of those areas of science, and a GIS facilitates key aspects of the handling and processing of that data. Both spatial epidemiology and public health research draw on three key aspects of GISc: spatial database management, spatial data visualization and mapping, and finally, spatial analysis. The latter includes cartographic and topological analysis (e.g., measuring distances and areas, and spatial relationships among observations, including overlay and buffering operations), mathematical modelling (e.g., network and surface analysis), and spatial statistics. Particular GIS products, of which there are a large number, are differentiated by how they manage spatial data and the functionality they possess.

A GIS is more than a computerized or automated mapping system (see Cowen, 1988) and, moreover, as emphasized by Cromley and McLafferty (2012, p. 16), should be seen as part of a "larger constellation of computer technologies for capturing and processing geographic data," which includes the global positioning system (GPS), satellite data collection systems, and digital scanners. This constellation of computer technologies used to process geographical data is therefore wider than proprietary GIS software packages (e.g., ArcGIS and MapInfo) and includes databases and statistical packages (e.g., Microsoft Access, SAS, R, Stata, and WinBUGS). Database functions within such packages may be used to link health datasets with datasets that already contain geographically located information (e.g., grid coordinates, census output areas, and environmental exposure values). Statistical modelling of data may be undertaken within statistical packages, and while this functionality may be considered to be part of what is involved when working in a GIS or GISc environment, standard GIS packages at best have only limited spatial statistical modelling capability. In several of the examples quoted in Sections 4.4 and 4.5, various computer software packages are used to process and analyse data, and these could all be regarded as "using GIS within the field of GISc."

Physically, a GIS comprises computer hardware (e.g., networked computers, large-format scanners, and printers), software (e.g., that enables data input and output, storage, and database management), and an organizational structure that includes skilled people who are able to operate the system. But the growth of the Internet means users of GIS do not need to have their own in-house, physical GIS in order to access geographical data or the results of spatial data processing, or even to undertake data analysis. Distributed GIS services have made it possible for many more users to take advantage of GIS capability. Peng and Tsou (2003) identified applications supported by distributed GIS: data sharing (both the original data and "data about data"—metadata), information sharing (online publishing of the results of data analyses and online accessing of servers running a GIS application that can process a client's request for information and return the results), data processing (providing online access to GIS analysis tools that can be applied to a client's data), and location-based services (providing information on a client's local environment or where they can best access particular health services mirroring developments in, e.g., retailing). More recently, it has become possible for users to post their own map data and annotate them online. Mashups allow data from multiple sources to be integrated, and mapping mashups using Google Maps is an example of this development (Cho, 2007). Cromley and McLafferty (2012, pp. 38–41) provide a brief overview of this evolving area of GIS. These developments have the potential to contribute to the development of public participation GIS (PPGIS)—systems that promote the participation of communities and individuals in raising and tackling issues of local concern (see, e.g., Sheppard et al., 1999, and for an example of PPGIS in health research and planning, see Beyer and Rushton, 2009). Chapter 1 of this volume contains detailed discussion of the important integrating contribution geospatial methods, and GIS in particular, can make to population-level studies of disease risk and public health.

4.2 The World as Captured in a GIS Database: Abstracting Reality

The database sits at the heart of a GIS, and for this reason, much attention has been given in the scientific literature on GIS to the relationship between geographical reality and what is stored in the spatial database. An understanding of the processes that take us from the limitless complexity of the real world to a database with a finite number of spatially referenced bits of information about that world enables the user to exploit GIS and implement GISc in a critical and hence rigorous way. In this section, we describe the two processes of conceptualization and representation that take us from geographical reality to a model of that reality that becomes an essential part of the model for the GIS database. The other two ingredients of that model are the attributes that are stored and the topological relationships. Subsequently, measurements are taken so that the database can be populated with data.

Figure 4.1 shows a two-step process in which at the first step phenomena in the world are conceptualized in terms of either a field (or continuous-surface) view or an object (or discrete-space) view. Environmental phenomena (air quality, soil type, and temperature) are usually conceptualized as fields. For example, ground-level air quality is conceptualized as a field because it is possible to go to any location on the earth's surface and measure air quality. On the other hand, a house or a hospital, a river or a road, a reservoir or a waste site—are all conceptualized as objects. Depending on the map scale, they would be conceptualized as point, line, or polygon (area) objects. For such phenomena, the real world is conceptualized as an empty space populated by these discrete, identifiable objects to which attributes are attached. These two models are how the geographical world is conceptualized, and with few exceptions, all geographical phenomena are modelled in terms of one or other of these two views. One important exception is population data that are sometimes conceptualized as fields (e.g., when constructing a population density map) or as objects (e.g., when producing a dot map). The next question is how these two conceptual models are represented for the purpose of constructing a model for the spatial dimension of the database, which of necessity must comprise a finite number of bits of spatial data.

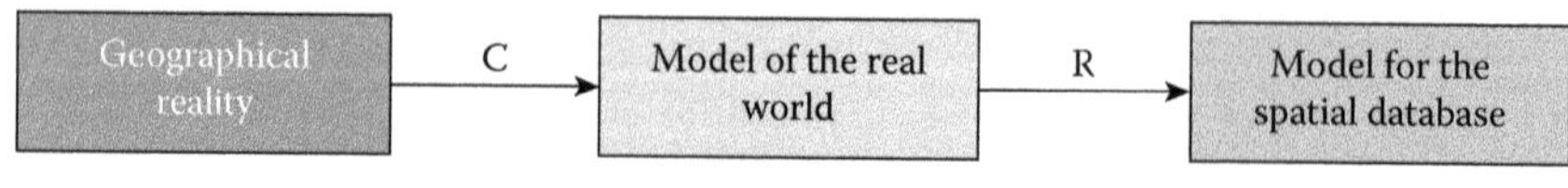

C: Conceptualization (object and field views)

R: Representation (points, lines, and polygons)

FIGURE 4.1
From geographical reality to a model for the spatial database.

In the case of the object view, each object (whether point, line, or polygon) is located on the earth's surface, and attribute values are attached to them and topological relationships defined for them. The objects, of which there are a finite number, underpin the model for the spatial database. But other issues can affect the representational choice. For example, although household data refer to point objects, for confidentiality reasons, much of this data is available only at the aggregated level, that is, for (irregular) census areas.

Field data present a different challenge. To capture field data in a (finite) spatial database, the surface needs to be sampled. This sampling may take the form of a regular or irregular sample of points (point data), a series of contour lines linking locations with the same value of the attribute (line data), or a set of polygons that partition the space. These polygons may be regular or irregular in shape. Regular polygons are constructed independently of the attribute, as in the case of a matrix of regular pixels (picture elements) of a given size from a remotely sensed image. Associated with each pixel value is a single value that provides a measure of the attribute at that location. Landsat Thematic Mapper data, for example, which provide reflectance values for each pixel, which are then classified in terms of, say, land cover features, cover a ground dimension of 30 × 30 m. This defines the data's spatial resolution and any landscape or environmental feature smaller than that will not be detected. The larger the pixel size, the larger the spatial filter and the smoother any landscape will appear. The location of each pixel is given by its position in the row and column of the matrix from which topological relations can be defined (Haining, 2003, pp. 78–79). Irregular polygons that partition an area arise when following attribute boundaries such as the edge of a forested area or the edge of a built-up area. The triangulated irregular network (TIN) is also an example of irregular polygons that partition a study area. TINs are constructed from a set of sample points on a surface. The TIN is used to capture surface variation and is often used to describe topography capturing both surface slope and aspect. Because field data are often captured by sampling, this raises the question as to what sort of sample plan to adopt (including what type of sampling design and sample point density). It also introduces sampling error into any analysis based on such data, as well as the scale and partition effects noted above in relation to aggregated object data (Haining, 2003, pp. 100–113).

Inside any GIS, map data are stored and processed in one of two main ways: the tessellation data model and the vector data model. The interested reader is referred to any standard text on GIS for a description of these two data models. At this point, we merely note that both the field and object views of geographical reality can be captured by either of these two GIS data models. That said, much, but not all, environmental data are captured using the tessellation data model, and most commonly within that set of data models, the raster model is structured as a regular array of square pixels. This is because much environmental data are obtained via remote sensing methods (aerial photographs and satellite imagery). Point- and area-based population data and network data (e.g., lines of transportation) are usually captured using the vector data model. In the case of population data, this is because reporting takes place through irregular census areas.

We now give a few examples to illustrate datasets that could be stored in a GIS database, with each geographically referenced attribute forming a layer within the GIS database.

- We have data on the set of general practice (GP) clinics for a region, represented as points with fixed locations, from which can be derived topological relations. Attached to each clinic, there are a series of attributes (patient lists with addresses and who have recently had tests for high blood pressure or been screened for a form of cancer).

- We have data on the size of the susceptible population in a region represented as counts by census areas. Attached to each census area is also the count of the number of cases of the disease during a given interval of time. There are also data recording social, economic, and demographic data for the resident population of each census area. Finally, there are data obtained from remotely sensed imagery that measure environmental (e.g., land cover) conditions across the area. Topological relationships across the set of census areas might be defined in terms of which census areas share a common boundary.

- As part of an emergency response analysis and establishing a region's vulnerability to a disaster, we have data on the location and capacity of hospitals, the regional road network (with travel times), and a map of environmental risk (arising from, say, the risk of flooding or fire), which partitions the map into areas classified by whether they have high to low levels of vulnerability. So the map objects in this case are points, lines, and areas, and topological relations between the hospitals and between the hospitals and the areas with defined levels of environmental risk are specified through the road network.

Assigning locations to objects is central to the creation of a geographical database. This creates challenges for the researcher. While the location of a hospital or a road may be fixed for the period of a study, in studies of environmental exposure, even the daily movement patterns of populations can affect exposure levels (e.g., to air pollutants), and in studies of the distribution of cases of a chronic disease, the resident histories of populations becomes important (see Chapter 34 for an extended discussion). In both cases, the analyst is forced to consider the validity of assigning populations to locations based on, for example, their current residential address. Measuring location may be problematic in other ways. Map projection issues do not need to be considered when undertaking large-scale studies (analysing small areas in high detail), but will be an issue when undertaking small-scale studies (analysing large areas, say at the scale of Europe or the continental United States, in limited detail), particularly where different datasets need to be integrated or overlaid.

In this section, we have focused on how the geographical world is conceptualized and represented for purposes of storage in a GIS database. But the database also comprises the attributes that are attached to each location. These attributes go through their own processes of conceptualization and representation. Many studies in epidemiology need to control for deprivation since deprivation often acts as a confounder in environmental exposure–response studies. There are many forms of deprivation, such as material (economic) deprivation and social deprivation. Before such terms can be used in any study, it is necessary to consider how deprivation is to be conceptualized and then how that conceptualization is best represented, that is, how it will be measured. Deprivation may be a confounder at the level of the individual, but it may also be a confounder at the area level. Some area-level attributes represent aggregates of the resident population (proportion of the population unemployed or living below the poverty line), while some attributes are only defined at the area level—for example, social cohesion or social capital, both of which need to be conceptualized, but when measured are attributes of groups of people and communities, not of individuals.

The model for the GIS database is defined by how geography is captured, how spatial relationships are defined between the finite elements representing that geography, what attributes are included in the database, and how they are to be measured. Although a GIS database cannot be used to model attributes over time, attributes will refer to points in time or intervals of time, and so time needs to be made explicit. Subsequent use of the database may produce maps comparing disease rates in a population between one time period and another.

All models are simplifications of the reality they describe and involve a trade-off between descriptive power and model complexity with diminishing returns (see Figure 4.2). We can speak of the quality of the model for the GIS database, but this cannot be assessed independently of the planned use of the database. The model is chosen on the basis of

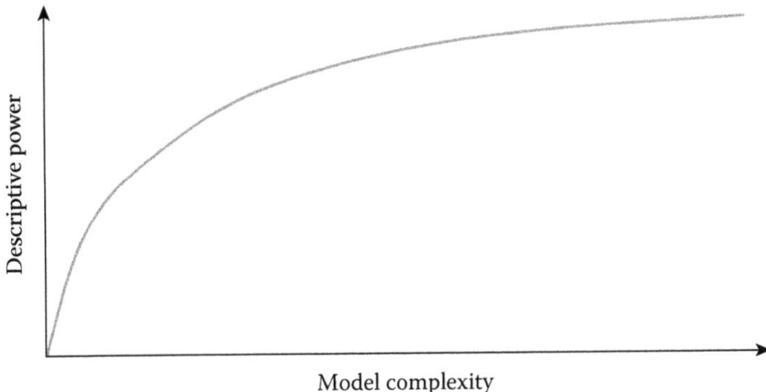

FIGURE 4.2
Schematic representation of the trade-off between model complexity and descriptive power.

whether it is fit for purpose, recognizing that the quality of the model may not be uniform across all parts of the study area (e.g., across both urban and rural areas of a study region). Model quality is assessed by such criteria as resolution (is the spatial detail adequate?), completeness (what are all the attributes necessary for undertaking a robust study of, e.g., cancer incidence by small area, and are these data available? [Swerdlow, 1992; Wakefield and Elliott, 1999]), and clarity, precision, and consistency. For extensive discussion of these terms, the reader is referred to Guptill and Morrison (1995).

We now turn from considering the model for the spatial database to issues surrounding the quality of the data that populates the spatial database.

4.3 GIS Database and Data Quality

Figure 4.3 describes the set of relationships in the construction of the GIS database and the terms often used to describe those relationships. In the case of data quality, there is a different type of trade-off from that in the case of model quality. Now the trade-off is between data quality and cost. While the costs of data acquisition may be falling on a per unit basis as more methods are routinely employed to capture data, it is still the case that if the data quality demands for a project rise, the cost of acquiring that data, particularly if it has to be primary data, increases.

A single item of geographical data comprises the triple "what, where, and when"—the attribute, the location to which the measurement refers, and the time period to which it refers. It is sometimes referred to as the geographical space–time data cube. Veregin and Hargitai (1995) identify four primary dimensions of data quality: accuracy, resolution, consistency, and completeness, thus giving rise to a 4×3 data quality matrix. Rather than try to address the full matrix, discussion will focus on the four primary dimensions, choosing examples to illustrate the point. It should be noted that as with model quality, data quality may not be uniform across the study area, particularly if the study area is large. There can be many reasons for this. In the case of mortality data, it can be due to diagnostic "fads" and other biases in specifying the cause of death (Lopez, 1992). We briefly discuss each of the four dimensions.

All measurements contain error due to the inevitable imprecision associated with the process of taking a measurement. Other types of error (inaccuracy) associated with spatial

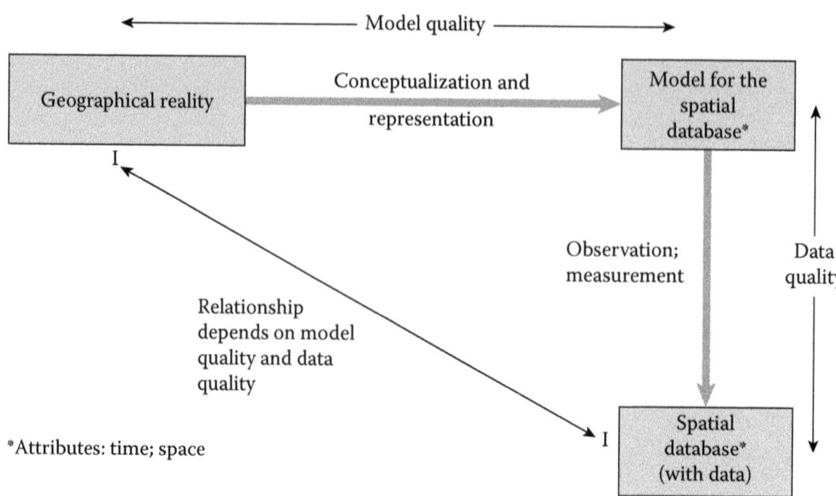

FIGURE 4.3
From geographical reality to the spatial database.

data include sampling error, operator error, and even deliberate error (as in the case of Barnardisation of UK census data). A particular concern with GIS databases is locational error when, for example, a patient is assigned to the wrong postal code or zip code. Underlying the idea of data error is the assumption that there is a true value. When drawing on socioeconomic data, it is not always clear that this is so—what is the "true" level of social capital or deprivation in an area? In large databases, it is often necessary to resort to automated methods to try to flush out possible data errors, including screening for both distributional and spatial outliers—values that are very different from their neighbours. Errors not only infect data values, but can also propagate as a result of such GIS operations as map overlaying and buffering (Haining, 2003, pp. 124–27; Burrough and McDonnell, 2000, pp. 237–39).

Resolution refers to the amount of spatial and temporal detail provided. Spatial aggregates act as filters on the real underlying variability. Large spatial units suppress variation but can make it easier to detect patterns; small spatial units retain more variation, but the resulting statistics are more affected by small errors. An important consequence of working with aggregated data is that results arising from analysing such data are conditional on the size of the census tracts (the scale effect) and their specific boundaries (the partition effect). Geographers and GI scientists refer to this as the modifiable areal unit problem. There are other challenges arising from resolution issues. Regression modelling in spatial epidemiology requires that all data are reported on the same spatial framework, but while much population data are reported by irregular census areas, environmental data are usually reported by grid squares. The challenge is to find a common spatial framework or undertake statistical modelling that recognizes the uncertainty arising when data are transferred from one reporting framework to another. Geographical data today are being collected at finer and finer spatial scales, but while improved spatial precision is to be welcomed, this raises issues of statistical precision when calculating statistics for small areas. The small numbers problem arises when working with small census units comparing and analysing rates of a rare disease. Elsewhere in this book, small-area estimation is discussed in Chapters 6 and 28, while Chapter 5 considers aggregation effects.

Consistency refers to the absence of contradictions in a database. For example, cases of a disease may be reported in a postcode where no one lives. This may be due to a geocoding error or perhaps due to the use of two databases (one health and one demographic) that do not correspond in time. Lack of completeness refers to the situation where there are missing data or there has been undercounting. The use of Internet methods that seek to engage the public in addressing public health issues may suffer from a lack of completeness. These and other issues associated with addressing GIS data quality are reviewed in, among other sources, Haining (2003, pp. 61–74), and in greater detail with reference to health data by Cromley and McLafferty (2012, pp. 43–74), who also discuss spatial databases with particular reference to U.S. data sources in their Chapter 3. Many of the issues raised in these early sections are also discussed in Maheswaran and Craglia (2004).

4.4 GIS in the Study of Disease

In this section, we illustrate the use of GIS and GISc in the study of disease. We have grouped the studies into five subsections—environmental epidemiology, communicable diseases, geographical epidemiology, exposure assessment, and disease clusters and environmental sources—but recognize that there is overlap between the subsections. We use a small number of examples but describe them in some detail in order to provide understanding of the epidemiological and public health contexts in which the scientific investigations were carried out. We have mainly used examples of work we have been involved in, supplemented by examples drawn from the work of others.

4.4.1 Environmental epidemiology

4.4.1.1 Air pollution and cardiovascular diseases in Sheffield

Existing and routinely collected data held by local and health authorities were used to investigate associations between outdoor air pollution and cardiovascular disease at the small-area level. (For further discussion of pollution fields, see Chapters 15 and 16 in this volume.) The project demonstrates the advantages and limitations of working with routinely available data. The project used air pollution estimates for three outdoor air pollutants—particulate matter, nitrogen oxides, and carbon monoxide—which had been generated by the city council using an air pollution model (Indic AirViro). The modelling process took into account a range of emission sources (represented by points, lines, and polygons) and meteorological conditions and generated a pollution surface for each pollutant at a 200 m grid square resolution. These data were imported into a GIS (ArcGIS) in order to validate model outputs and link with health outcome data. Abnormal patterns in the particulate matter pollution surface were apparent in a few quite localized areas of the city. Further investigation revealed that these were due to errors in the emissions database where old coal-fired burners that had been replaced years ago had not been updated in the city council's emissions database (Brindley et al., 2004). These areas were excluded from the analysis. We also found that pollution concentrations for nitrogen oxides were overestimated by the model when compared with measured values from monitoring stations. However, the concentrations were all overestimated at a broadly comparable level, suggesting that relative measures of pollution were probably valid (Brindley et al., 2004). Analyses therefore used relative categories by quintile and not absolute values.

The outcome data comprised deaths and hospital admissions for coronary heart disease and stroke (1994–1998), which were provided at the enumeration district level by the health authority. Enumeration district-level population estimates were based on the 1991 UK census and scaled using health authority midyear estimates for 1994–1998.

As the outcome and exposure data were available at different spatial frameworks with 1030 enumeration districts and 10,847 pollution concentration grids, GIS was used to integrate both sets of data. We calculated the average pollution level within each enumeration district using a procedure based on postcode centroids. We first assigned to each domestic postcode centroid the value of the 200 m pollution grid in which it lay. We then took the average of the values for all postcode centroids that lay within the enumeration district polygon. There are varying numbers of households within postcodes. We used the number of domestic delivery points at each postcode to weight the average value calculated for each enumeration district (Brindley et al., 2005). In order to take some account of daily local population movements, we also assigned to each postcode centroid the pollution value of the average of grid squares, which fell within a 1 km buffer of each postcode centroid based on surveys indicating that 1 km was the average walking journey length to see if this improved associations with health outcomes.

We undertook analyses using methods with increasing levels of complexity. Standard Poisson regression carried out using SAS showed some associations between air pollutants and coronary heart disease and stroke mortality and, to a lesser extent, hospital admissions (Maheswaran et al., 2005a, 2005b). We found that using the 1 km buffer made little difference to the results. Bayesian analyses carried out using WinBUGS and taking into account errors in variables and within area variation were subsequently used, and these continued to show associations between air pollutants and mortality from coronary heart disease and stroke (Maheswaran et al., 2006b; Haining et al., 2007, 2010).

4.4.1.2 Air pollution and stroke in South London

This example describes a subsequent project carried out to further investigate the association between outdoor air pollution and stroke. Point location case data and grid point resolution pollution data were used in this work. The work was carried out in an area in South London that was covered by the South London Stroke Register, a population-based stroke register set up in 1995 to capture all cases of first-ever stroke occurring among the resident population living within a defined geographical area.

The air pollution modelling for this study had been carried out at a very fine spatial resolution (20 m grid point resolution) using bespoke air pollution modelling and was available for particulate matter and nitrogen dioxide. The modelling process took into account a wide range of pollution sources and emissions, including major and minor road networks modelled with detailed information on vehicle stock, traffic flows, and speed for each road segment; pollution sources in the London Atmospheric Emissions Inventory, including large and small regulated industrial processes, boiler plants, domestic and commercial combustion sources, agriculture, rail, ships, and airports; and pollution carried into the area by prevailing winds. Model validation was carried out by comparing modelled values with measured pollution values from background monitoring stations and by visually inspecting maps showing pollution values overlaid on road networks and other pollution sources within ArcGIS. The model validated well in terms of absolute values when compared with monitored pollution values (Maheswaran et al., 2010).

We investigated two outcomes: the incidence of stroke and survival after stroke. Stroke incidence was examined using a small-area-level ecological study design, and survival after stroke was investigated using a cohort study design.

To examine incidence, a denominator population (population at risk) was needed, and we used population counts in census output areas from the UK 2001 census for this purpose. This was the smallest geographical unit at which census population counts by 5-year age band and sex were available, with approximately 300 people per output area. Stroke cases were assigned to output areas using the point-in-polygon method within ArcGIS, with a case assigned to the output area in which the postcode centroid of residence of the case was located. Observed and expected counts were calculated for each output area.

Linkage of grid point resolution pollution data to residential postcodes was also carried out within the GIS. All residential postcodes in the study area were assigned the pollution value of the grid point closest to the residential postcode centroid. Where there were equidistant points, the average value was taken. For the ecological study, an average value was then calculated for each output area, taking the average of values assigned to all postcode centroids that fell within the output area polygon, using point-in-polygon to link postcodes to output areas. A total population count by postcode was available from the 2001 census, and this was used to weight the average pollution value calculated for each output area.

The survival analysis was carried out at the individual level, using Cox regression modelling. Follow-up was for up to 11 years. Patients were assigned the pollution value attached to their residential postcode centroid. For patients who moved, we took the average of the values at the start and end of their contribution to the study. The start value was from their postcode of residence when they had the stroke. The end value was either the value from their postcode of residence at the time of death or the value from their postcode of residence at the end of the study.

We found that living in a more polluted area was associated with decreased survival after stroke (Maheswaran et al., 2010). In the ecological analysis, we found that there was no spatial structure to the output area-level incidence rates when assessed using WinBUGS, Moran's I, and visual inspection of maps, and we therefore used standard Poisson regression methods in SAS. There was no clear evidence of association between outdoor air pollutants and the incidence of stroke, although there was a suggestion of association in the 65- to 79-year age group in relation to ischemic stroke (Maheswaran et al., 2012). There was also a suggestion that the association was stronger for mild ischemic stroke (Maheswaran et al., 2014a).

4.4.2 Communicable diseases

Communicable or contagious diseases are infectious diseases involving some causative disease agent such as a virus, bacterium, or parasite that is transmitted either from person to person or via some vector or intermediate host, such as an animal. In the case of person-to-person transmission, the disease is spread by contact and the geography of that spread will depend on the geography of human interactions and may take any of several forms, including local clustering, spread following the urban hierarchy, and mixtures of the two (Cliff and Haggett, 2004). Some nonvector communicable diseases are spread through socially induced exposure to risk as in the case of HIV/AIDS (Rhodes et al., 2005); others are spread through environmental exposure, so that the geography of cases will be a function of the geography of the environmental risk—for example, the river network or water distribution system in the case of a water-borne disease (Lake et al., 2007). Modelling such communicable diseases involves the use of epidemic models that partition the population into those who are susceptible (S), those who are infected (I), and those who have recovered or been removed (R)—known as SIR models. Models may be aggregated in terms of population groups defined by their locations, which may be small areas, regions, or urban places such as the STEM model (Eclipse Foundation, 2011), or based on interactions among individuals, as in the case of agent-based models (Lee et al., 2008; Perez and Dragicevic, 2009). GIS are used to map such disease spread in space and time, to help identify disease clusters or concentrations, to map risk, and to try to predict disease spread (see, e.g., Oppong et al., 2012). Typically, GIS provides, integrates, and updates the data inputs and data layers that are used by epidemic and interaction models, which then return outputs for mapping or animation by the GIS. The GIS is loosely coupled with these models. Databases may be at many different scales, but there is growing interest in global-scale databases for communicable diseases reflecting the global nature of threats to human health arising from population mobility and other aspects of globalization. Studies that cover a large portion of the earth's surface drawing data from different countries (who may have different mapping conventions) raise map projection and other issues, and here the ability of GIS to integrate spatial data is particularly valuable.

In the case of vector-borne diseases, GIS are used in the ecological study of agent–vector–host relationships and their links to human populations (for an overview, see Cromley and McLafferty, 2012, pp. 263–302). Habitat modelling can be used to assess exposure risk, while land cover changes brought about by urban development or climate change can be used to assess whether the exposure risk is increasing or decreasing. GIS have also been used to assess the environmental characteristics of Lyme disease (Glass et al., 1992) and West Nile virus (Ruiz et al., 2007) case locations. The GIS operation of point-in-polygon can be used to look for case clusters in relation to different ecological characteristics.

The study of communicable diseases spans many disciplines involving researchers in virology, molecular biology, geography, epidemiology, and public health, so it is important to make connections. Formally integrating knowledge from different disciplines is challenging, but as Ge et al. (2012) demonstrate, GIS can help mitigate the disciplinary gaps by providing a platform for updating data inputs and layers from different disciplines, facilitating analysis, and integrating outputs for mapping (see also Chapter 26). Their study into the spatial–temporal dynamics of avian influenza H5N1 in East and Southeast Asia used GIS-based knowledge fusion. Genetic sequences were used to create phylogenetic trees to estimate and map the H5N1 virus's ability to survive and spread. Adding information about virus location together with spatial interpolation techniques produced maps of H5N1 risk. Maps of risk can also be produced by modelling social, economic, environmental, and other data (Gilbert et al., 2008) and can be obtained by analysing large concentrations of outbreaks using spatial statistics. Ge et al. (2012) used the Dempster–Shafer inference theory of evidence to integrate the three raster-layer probability maps, mapping the resulting output in a GIS.

4.4.3 Geographical epidemiology

4.4.3.1 Migration and health inequalities in Sheffield

Socioeconomic gradients in mortality at the geographical level exist across many cities and regions worldwide. These patterns may endure despite efforts by health and local government authorities to reduce inequalities by targeting appropriate interventions at deprived areas. One potential explanation for enduring inequalities at the geographical level is selective migration. This is the situation where people in poor health or those with the socioeconomic determinants of poor health, for example, unemployment, move from affluent to deprived areas, while those in good health or with the socioeconomic determinants of good health, for example, high income, move from deprived to affluent areas. Thus, although interventions may benefit individuals in deprived areas, this selective migration may perpetuate inequalities when examined at the geographical level. In this example, we describe the use of GISc in epidemiological investigation of migration and area-level mortality patterns.

We examined for evidence of selective migration and investigated the impact of selective migration on geographical inequalities in health in Sheffield, a city where there is a striking east–west gradient in area-level deprivation that is closely mirrored by gradients in life expectancy (Maheswaran et al., 2014b). The project was carried out because the local authority wanted to know if selective migration contributed to the enduring gradient in health inequalities across the city.

We used a total population cohort dataset that was provided by the local health authority in anonymized format. The dataset was created from the GP database the health authority held. This was a continually updated register of people resident in Sheffield who were registered with a GP. The health authority kept regular "snapshots" taken from this register, including the residential location of people at the snapshot time point. For this project, the health authority provided a dataset with a record for each individual who was resident in Sheffield at any point within the cohort time frame. If an individual was present in a snapshot, his or her census area (lower superoutput area from the UK 2001 census) and electoral ward of residence were provided for that snapshot time point. These census areas typically contain approximately 1500 people. Death records were also linked into this dataset, and if an individual had died, the census area and ward of residence at death were provided.

Analysis of migration can be a very complex undertaking, depending on the number of time points analysed, the number of geographical units used, and the length of time people resided in each geographical unit (see also Chapter 34 in this volume). We started out with a simplified analysis where we used two time points (residential location at the start of the study and at death or the end of the study) and divided census areas into two categories (high and low deprivation). We found clear evidence of selective migration. People moving from low- to high-deprivation areas had higher mortality than those remaining in low-deprivation areas. Conversely, people moving from high- to low-deprivation areas had lower mortality than those remaining in high-deprivation areas. The magnitude of these differentials in mortality risk diminished with increasing age. We were also provided with data on health status and socioeconomic circumstances for a sample of the population. These data had been obtained in a survey carried out before the start point of the migration analysis. Analysis of these data showed that people tended to carry their preexisting risks with them (Maheswaran et al., 2014b).

We examined the impact of migration on geographical gradients in mortality by putting people back to where they were at the start of the cohort time frame and comparing the mortality gradient across the city based on this location with the gradient based on residence at time of death. We found that selective migration made little contribution to existing socioeconomic gradients in mortality across the city (Maheswaran et al., 2014b).

The mapping, database manipulation, and analysis for this project were carried out in `R` and included use of the GIS functionalities in `R`.

4.4.3.2 Alcohol-related mortality in England

Lifestyle-related factors, which include smoking, alcohol consumption, diet, and physical activity, are key determinants of health. These lifestyle-related determinants of health are potentially modifiable and are therefore of significant public health concern. In this example, we illustrate examining the geographical epidemiology of alcohol-related mortality in relation to socioeconomic deprivation at the small-area level.

Alcohol consumption data from UK surveys suggest that alcohol consumption is marginally higher in more affluent socioeconomic groups. However, this does not appear consistent with alcohol-related mortality, which appears to be higher in lower socioeconomic groups. We investigated the association with mortality at a national scale using a small-area-level ecological correlation study (Erskine et al., 2010). We used electoral wards as the units of analysis, of which there were 8797, using an existing dataset on alcohol-related mortality that had been compiled by the Office for National Statistics for surveillance of alcohol-related mortality. The deaths had been assigned to wards using a postcode to ward lookup table. The deaths included in the dataset were those considered to be most likely directly attributable to alcohol. The predominant condition in this group was liver cirrhosis.

In addition to examining associations between alcohol-related mortality and socioeconomic deprivation at the small-area level, we also examined associations in relation to gender, age, and urban–rural location. The dataset supplied included the Carstairs index as the indicator of socioeconomic deprivation at the ward level. This is a standardized combination of four variables from the 2001 census: male unemployment, overcrowding, low social class, and lack of car ownership.

The analysis was based on 18,716 male and 10,123 female deaths over a 5-year period (1999–2003). We found a strong association between socioeconomic deprivation at the electoral ward level and alcohol-related mortality. The differential in relative risk was most pronounced in the 25- to 44-year age band. Mortality rates were higher in men than women and also higher in urban areas (Erskine et al., 2010).

The main analysis was carried out using standard Poisson regression methods in SAS due to the number of wards in the dataset and substantive analytical detail required. We also carried out Bayesian analysis on a small subset of the data to explore gender variation in the spatial pattern of alcohol-related deaths using WinBUGS (Strong et al., 2012). The adjacency matrix for this analysis was generated in ArcGIS. We initially fitted separate models for men and women and subsequently modelled male and female deaths jointly using a shared component for random effects. We investigated a range of different unstructured and spatially structured specifications for the gender-specific and shared random effects. We found significant spatial variation in ward-level alcohol-related mortality for men, but this was much less marked for women. After accounting for deprivation, there was significant unexplained elevated risk in a very small number of wards.

4.4.4 Exposure assessment

4.4.4.1 Improving estimates of air pollution exposure

Most studies examining the association between air pollution and health outcomes have used either monitored or modelled air pollution values to estimate exposure (see also Chapter 15 in this volume). Monitored values are from fixed-site air quality monitoring stations. These epidemiological studies have generally not taken daily population movements and time spent in different locations into account. Most have generally used outdoor monitored or modelled estimates, and the indoor versus outdoor concentrations have generally not been taken into account. An important element determining the dose of pollution taken in by people is the activity being undertaken, as higher energy expenditure is associated with an increased respiratory rate and depth of breathing, resulting in higher doses of pollution. These aspects are all challenging to incorporate in large-scale epidemiological studies, which are needed to examine associations with health outcomes.

The example described here is a detailed study undertaken by de Nazelle and coworkers (2013) on a small number of subjects, 36 healthy young volunteers, undertaken to accurately assess exposure. The methodology used GPS and accelerometer functions in smart phones for exposure assessment. A computer programme was developed that used accelerometer readings to estimate energy expenditure.

Estimation of air pollution exposure used modelled annual mean pollution estimates at a fine spatial scale as the start point. The estimation subsequently took into account temporal variation, both within day and by day of the week, and the microenvironment, by sampling indoor and outdoor concentrations, and measuring exposure while using different modes of transport.

A GIS platform was used to integrate the air pollution, GPS location, and activity data. Various comparisons were carried out. There was substantial variation when pollution estimates incorporating all the refinements, including energy expenditure, were compared with estimates based on home location address only, with little correlation between the two.

This example illustrates the potential for substantial exposure misclassification and also bias, with a tendency to underestimate exposure, in standard epidemiological studies. The methods described in this work are very involved, and the challenge will be to use such methods in large-scale epidemiological studies.

4.4.4.2 Assessing environmental influences on diet and exercise

Public health is concerned with the influence of environmental factors such as parks and green spaces and fast food outlets on health. Parks and green spaces provide places for physical activity, while fast food outlets may promote the consumption of foods high in saturated fat and low in fibre. Several studies have been carried out to examine the potential influence of these environmental factors on diet and physical activity, and most studies have examined exposure around the residential location. However, exposures around activity spaces away from these residential locations have been much less well studied.

The example described here is work carried out by Zenk and coworkers (2011), in which exposure in activity spaces was examined using a combination of GPS and accelerometers. GPS were programmed to record participants' position every 30 seconds over a 7-day period, and data were obtained on 120 participants. Two measures of activity space were created using the GPS information downloaded into a GIS.

The first measure was referred to as a one–standard deviation ellipse. The central location of all GPS points for a participant was calculated. An ellipse was then created around this central point, and the one–standard deviation limit meant that approximately 68% of all GPS points were included within the ellipse. The long axis of the ellipse was in the direction of maximum dispersion, while the short axis was in the direction of minimum dispersion. This measure was calculated using the spatial statistics toolbox in ArcGIS.

The second measure was referred to as the daily path area. This was created by first buffering around every GPS point for the participant using a 0.5-mile radius and then dissolving the boundaries between these buffers to create the daily path area. This measure was also created using ArcGIS.

For comparison with the standard residential location methods, a 0.5-mile street network buffer was created around the census block centroid of each participant's residential location.

The density of fast food outlets and the percentage of land that was designated as municipal park land were calculated for each of the three exposure areas. The study found very low correlations between exposures based on neighbourhood location and exposures based on activity space, especially with activity space defined as the daily path area.

The study found no association between residential neighbourhood fast food outlet density and diet. However, a positive association between fast food outlet density and an unhealthy diet was found when the daily path area was used to calculate exposure. No associations were found between physical activity and park land use in analyses using each of the three exposure space definitions.

Although the study found that fast food density in the daily path area was associated with an unhealthy diet at the individual level, there is a potential problem with interpreting this association as causal, that is, that increased exposure to fast food outlets is the cause of people eating unhealthily. This is because the daily path area is defined by the participants choosing to go along particular routes, and they may have gone along those particular routes in order to access fast food outlets.

4.4.5 Disease clusters and environmental sources

4.4.5.1 Rapid initial assessment of apparent disease clusters

Concerns about apparent clusters of disease and potentially elevated risks of disease around environmental sources of pollution, such as factories, frequently arise. These clusters, real or apparent, have the potential to cause substantial public anxiety and media interest and can result in substantial public health resources being spent in addressing these concerns if they are not handled in a timely and effective manner (Maheswaran and Staines, 1997).

Identifying disease clusters is somewhat different from examining if diseases have the general propensity for clustering. Clusters may occur in areas where there is no obvious cause, or may occur around environmental sources, typically around point sources, but also, to a lesser extent, around line and area sources. The statistical issues around clusters and clustering are covered in other chapters in this book (see Chapters 8, 9, 14, and 28). Here we describe an example, in which we were involved, of a facility set up to investigate apparent clusters of disease to support public health investigation (Aylin et al., 1999).

The Rapid Inquiry Facility was set up within the Small Area Health Statistics Unit in the UK to carry out a rapid initial assessment of apparent disease clusters. The facility is a system that combines three technical elements: a database, a GIS, and an automated statistical analytical methodology.

The database integrates datasets on health outcomes, including deaths, hospital admissions, congenital malformations, and cancer registrations. The geographical identifier for these outcome data is the postcode centroid. The database includes population denominator counts by age and sex at the small-area level. These are counts from population censuses, with population estimates for intercensal years. The database also includes information on socioeconomic deprivation at the small-area level. These are basic minimum requirements for adjusting for potential confounding variables, as disease incidence and mortality can vary substantially by age, gender, and socioeconomic status.

A GIS platform is fundamental to this system and allows integration of data from different spatial frameworks. The system offers flexibility regarding the spatial resolution at which diseases can be investigated, with the smallest unit being electoral wards and census output areas. Point source locations of environmental pollutants can be specified and different buffers created around these sources. The GIS also allows an adjacency matrix to be generated for use in smoothing risk maps.

The statistical analysis allows for automated calculation of absolute and relative risks for different buffer zones around point sources. The calculation includes confidence intervals and significance testing. The statistical methodology also includes the production of smoothed maps displaying a risk surface using Bayesian methodology utilizing the adjacency matrix created within the GIS. Areas with significantly higher or lower risks are also identified on these risk maps.

The system allows the rapid initial assessment of apparent disease clusters that have caused concern to members of the general public, media, politicians, or public health staff. It does not provide definitive answers, but can rule out clusters that do not exist statistically. If clusters or high rates in some areas are found, further investigation is needed. A key first step is to examine for artefacts and errors in the data. Incomplete data capture in some areas, inaccuracies in the data recorded, and relevance of the conditions being examined, for example, cancer clusters that comprise conditions that are aetiologically unrelated, all need to be considered.

The Rapid Inquiry Facility has been acquired for use elsewhere, for example, in Utah (Ball et al., 2008), and has undergone further enhancements (Beale et al., 2010). Chapter 2 in this volume describes environmental exposure research in detail, while cluster detection and modelling are discussed in Chapters 3, 8, 9, 14, and 28.

4.5 Examples of GIS in the Provision of Public Health Services

In this section, we give examples where GIS and GISc have been used to examine and inform the provision of public health services. We have grouped the studies into four subsections—access to services, needs assessment and health equity, variation in utilization, and planning the location of services—recognizing the overlap between sections. Health services are also discussed in Chapter 29 of this volume.

4.5.1 Access to services

4.5.1.1 Uptake of breast cancer screening in North Derbyshire

In this example, work was undertaken to inform local planning decisions (Maheswaran et al., 2006b). An increase in capacity in the provision of screening for breast cancer was needed in North Derbyshire. This was because a change in national policy meant that the age range of women invited for screening was to be increased from 50–64 years to 50–70 years. In addition, two-view mammography, which was then being undertaken only at the initial screen, was to be instituted at all screening rounds. The health authority was interested to know if there was still an issue with distance from screening site, that is, if uptake was lower among women living farther away, and if uptake was lower among women living in more socioeconomically deprived areas, in order to take these factors into consideration when reorganizing services.

Data were provided at the individual level for women invited for screening. This dataset contained the postcode of women invited for screening, whether or not they attended, and the screening location to which they were invited. A postcode to census enumeration district lookup table was used, which also contained eastings and northings for postcode centroids.

Road travel distance to a screening location was calculated from the postcode centroid to the grid location of the screening centre using 1:10,000 resolution road network data within a GIS (MapInfo). Screening was provided at a fixed site (the main district general hospital in the area) and at 12 locations throughout the health district using a mobile screening unit.

Socioeconomic deprivation was assessed using the Townsend score of the census enumeration district in which the postcode was situated. This area-level derivation indicator was a standardized combination of four 1991 census variables (unemployment, no car ownership, nonhome ownership, and overcrowding). Data were analysed on 34,868 women. The overall uptake of screening was 78%.

As this was an individual-level dataset, the analysis was carried out at the individual level using logistic regression in SAS, modelling the binary outcome of attendance or nonattendance. We found a small decrease in uptake with increasing distance from the screening location. The effect of distance on uptake, although still detectable, was likely to have been largely ameliorated through the use of the mobile unit. Deprivation, however, did have a clear effect, with lower uptake among women living in more deprived areas (Maheswaran et al., 2006b).

4.5.1.2 Walk-in centres and primary care access

This next example relates to work carried out to inform government policy on providing access to primary health care. GP surgeries were under increasing pressure due to increasing demands for their services, and walk-in centres were seen as one option for relieving pressure on these surgeries. A wave of walk-in centres had been set up in England, and the purpose of this work was to evaluate if these walk-in centres had reduced waiting times at GPs (Maheswaran et al., 2007).

Waiting times for a GP appointment were monitored by the primary care access survey, a regular monthly survey carried out nationally to assess the waiting time measured in days to the next available surgery appointment with a general practitioner. The survey was carried out on all National Health Service (NHS) GPs in England, and there was a 48-hour target set by government. We obtained these monthly survey data from the Department of Health in England.

We used two approaches to calculate exposure of GPs to walk-in centres. The first was the straight-line distance from each GP postcode centroid to the postcode centroid of the nearest walk-in centre, which was already in operation that month. This approach took into account the phased opening of walk-in centres.

The second approach used a function based on walk-in centre attendance rates by distance. For this second approach, we used attendance data, which were available for four walk-in centres, to create the function. Attendance data and population denominator counts were available by census output area. These output areas were assigned to 1 km concentric rings around the walk-in centres, and attendance rates were calculated for these distance bands. An exponential distance decay function was fitted to these rates. We used this function to calculate distance decay values for each GP by month on the basis of its distance to each walk-in centre. We then summed the values for each GP by month, which in effect took into account the effect of multiple walk-in centres in the vicinity of a GP.

We analysed data on 2509 GPs in 56 health authority areas in England and included 32 walk-in centres in the analysis. We found no evidence to suggest that walk-in centres shortened waiting times for access to primary care. As part of the project, we also examined the effect of area-level deprivation on waiting times and found clear evidence that the waiting time target was less likely to be achieved in more deprived areas.

ArcGIS was used to visualize locations of walk-in centres, GPs, and health authority boundaries. Straight-line distances were calculated using Pythagoras's theorem in a Microsoft Access database. Statistical analyses were carried out in SAS.

4.5.1.3 Renal replacement therapy in the Trent region

The purpose of this example is to illustrate the use of GISc in relation to access to health services and health outcomes (Maheswaran et al., 2003). End-stage renal failure typically results from chronic renal disease caused by a variety of medical conditions. When patients are in end-stage renal failure, they require some form of renal replacement therapy in order to survive. The three options are haemodialysis, peritoneal dialysis, and renal transplantation.

This example describes work we carried out in collaboration with the Trent Public Health Observatory in order to inform planning decisions in the Trent region. The need for more renal services was being reviewed. One of the considerations was the provision of more renal units that would improve access. Renal units are typically classified into main units and satellite units, with the latter providing mainly haemodialysis.

The observatory assembled the data and carried out descriptive analyses. Renal units within and surrounding the Trent region were identified and their locations georeferenced. Prevalence data on all patients residing in the region who were receiving any of the three forms of renal replacement therapy were obtained. Patients were assigned to census enumeration districts. Denominator populations for enumeration districts were obtained from the 1991 census and scaled to subsequent midyear estimates for health authorities within the region. The Townsend score was used as an indicator of socioeconomic deprivation at the enumeration district level. The percentage of the population of African and Asian origin at the enumeration district level was also obtained from the census. These factors were taken into account because renal disease is more common in more deprived communities and also has a higher prevalence among people of African and Asian origin. Access to renal units was assessed by calculating road travel distances from census enumeration district population centroids to the nearest renal unit.

We used Poisson regression to assess associations between travel distance, deprivation, and renal replacement therapy rates in the region. Renal replacement therapy rates were higher in more deprived areas. However, when the individual modalities of renal replacement therapy were examined, rates were higher for haemodialysis, but not for transplantation, in more deprived areas. This raises the issue of inequalities in health care, as transplantation is the preferred option for end-stage renal disease, and it would be expected that transplantation rates would also be higher in more deprived areas.

With regard to geographical access, haemodialysis rates were lower in places further from renal units. This might be expected to some extent because distance from a renal unit might be taken into consideration when decisions are made regarding whether to use haemodialysis or peritoneal dialysis. There may also be the issue of "reverse causality" for the association. The need for haemodialysis might cause people to move to live closer to renal units.

MapInfo was used to calculate road travel distances and to map and visualize locations of renal units in relation to regional geography. Assembly of the dataset was carried out using Microsoft Access, and statistical analysis was carried out in SAS.

4.5.2 Needs assessment and health equity

4.5.2.1 Health equity profiles

This example describes the use of GIS and GISc in health needs assessment and assessment of health equity. Health needs assessment may be carried out to assess the health needs of a population, people with a particular condition, or the need for a specific intervention. There is overlap with the process of health equity auditing, the first step of which is a health equity profile. The health equity profile may be assessed from spatial and social perspectives, with the latter subclassified by age, gender, class, and ethnicity.

This example is of a health equity profile that was undertaken to support planning and reconfiguration of services in a health authority in northwest England. The work used existing and routinely collected information to assess equity. The range of indicators used included mortality data, hospital admissions data, and GP-level data including data from the Quality and Outcomes Framework. The conditions for which the equity work was to be undertaken were predetermined by the health authority and included cardiovascular disease, diabetes, chronic obstructive pulmonary disease, and alcohol-related conditions.

The spatial frameworks at which the data were assembled and analysed were electoral ward level, census-based lower superoutput area level, and GP population level. The latter is not clearly defined geographically, and patients registered with a particular GP could come from a wide area. Nevertheless, the majority of patients registered with a practice live in the local area close to the practice. From the planning perspective, primary health care services are organized around practices, and this is therefore a useful level at which to investigate health needs and equity.

Data were assigned to wards and output areas by the health authority using postcode-to-area geography lookup tables. GP-level data were generated at this level. The data were used to produce a range of choropleth maps using ArcGIS. Scatterplots and other graphs were used to carry out exploratory spatial data analysis. For GP-level information, we produced a graphic that was able to show a range of practice-level indices in the same figure (Figure 4.4). This form of visualization was useful for identifying outliers and considering a range of related indicators together. This bespoke graphic was created in R and brings together a range of geographical information.

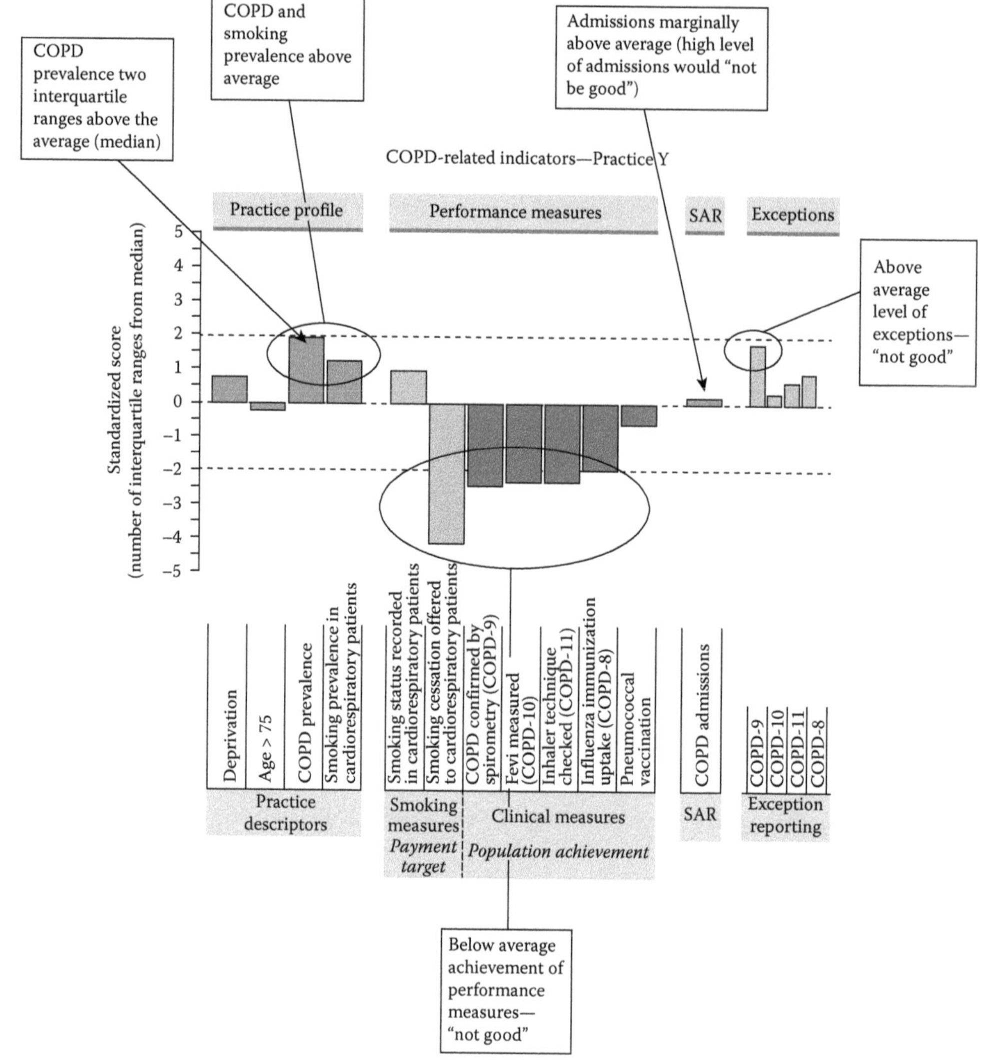

FIGURE 4.4
Simplified guide to interpretation of general practice profile charts.

The synthesis of data from different sources allowed variations in need and equity to be identified. Outlier practices were investigated further by the health authority, but it should be noted that there may be good reasons for variations in practice. Being an outlier does not automatically indicate unusual or substandard practice.

4.5.2.2 Physical activity in socioeconomically deprived areas

This example relates to a physical activity intervention that was offered as part of a multistage intervention leading up to a trial of an intervention to maintain increased physical activity (MacNab et al., 2014). Preventative services are being increasingly recognized as an important element of public health offered to communities. The prevalence of cardiovascular diseases is higher in more socioeconomically deprived communities, and attempts to reduce inequalities in health have led to services being targeted at more deprived communities.

In the initial phase of the intervention, middle-aged people living in deprived neighbourhoods in Sheffield were offered a physical activity intervention to increase their physical activity levels. The intervention comprised a motivational DVD to be sent by post, along with additional information. The DVD was sent to those who responded to an initial invitation letter asking them if they would like to receive the intervention. The health authority for Sheffield had previously characterized neighbourhoods, and people aged 40–64 in these selected most deprived neighbourhoods were to receive the intervention. This first phase was offered to all people invited, unlike the subsequent phase, in which a further intervention was given on a randomized basis to some participants.

The overall uptake was extremely low, with an overall mean of 7%. Investigation into factors associated with this low uptake included examining small-area-level factors that might be associated with low uptake. These factors were investigated using postcode areas as the units of analysis. Deprivation was assessed using the proportion of households in a postcode receiving housing benefit. This variable was not available for all postcodes, as the data were not provided for postcodes with small numbers of households or for postcodes where most or all of the households were receiving benefits. An alternative indicator, the Index of Multiple Deprivation, available at lower superoutput area level, was also used and assigned to all postcodes within the superoutput area. Other factors investigated included walking distance to the nearest gym, walking distance to the nearest swimming pool, and walking distance to the nearest municipal green space. The network distance analysis was undertaken by linking and using datasets with a fine spatial resolution within ArcGIS. A detailed description of the datasets and methodology used is provided in Goyder et al. (2014).

The spatial analysis was complicated by a number of factors. There were 2455 postcode areas analysed in the study. The postcode areas were not contiguous due to the way areas were selected. Only postcodes in selected deprived areas with one or more residents aged 40–64 years were included in the study. There were very low counts for most postcode areas. In 66 postcode areas, only one postal invitation was sent out, with no responses in 60 of the postcodes. In addition, in postcodes where more than one invitation was sent, there was a zero response from 996 postcodes.

We developed and used Bayesian hierarchical Bernoulli–binomial spatial-mixture zero-inflated binomial models to model overdispersion and separate the systematic and random variations in the noisy and mostly low response rates (MacNab et al., 2014). The models allowed for investigation of variations in patterns of mail-outs, zero responses, and response rates. We found that response rates were lower in postcodes in which a higher proportion of households received housing benefits. There was little evidence of association with the other variables examined. The postcode polygon adjacency matrix was created using ArcGIS. Spatial analysis was carried out in WinBUGS, and the statistical outputs were visualized using R.

4.5.3 Variation in utilization

4.5.3.1 Geographical variation in potentially avoidable admissions

Hospital services are coming under increasing pressure from the increasing demand for emergency hospital admissions. Not all hospital admissions are essential. While some emergency hospital admissions, for example, for a heart attack or meningococcal meningitis, are clearly urgent and essential, at the other end of the spectrum, there are admissions that could have been avoided for a number of reasons, including better social and preventative community care. In this example, we describe a project examining geographical variation in potentially avoidable admissions (O'Cathain et al., 2014).

A group of hospital admissions were used that were considered to be likely to contain a substantial number of avoidable admissions. This group of admissions included nonspecific chest pain, nonspecific abdominal pain, and chronic obstructive pulmonary disease, and was selected through a consensus process with specialists in the field. It is important to note though that not all admissions with these conditions would have been avoidable.

Using this list of defined conditions, a standardized avoidable admissions rate was calculated for health authority areas using direct standardization. The population for the whole study area (England) was used as the standard population.

The spatial framework used for the geographical areas was primary care trust boundaries. These trusts were health authorities responsible for the health of the population resident within their defined geographical boundaries. They received money from central government and are part of the NHS structure within England. The rationale for using these primary care trust boundaries was that these trusts commissioned emergency and urgent medical care (along with primary and other levels of care) for their residents, and therefore the care within a primary care trust area could be considered to comprise an emergency and urgent care system.

We analysed 152 primary care trusts in this project. There were 3.3 million admissions over a 3-year period that came under the category of defined conditions for this project, accounting for 22% of all emergency admissions. There was a 3.4-fold variation in potentially avoidable admission rates across the primary care trust areas examined.

Geographical variation was investigated by thematically mapping standardized avoidable admission rates, and there was clustering of primary care trusts with high and low rates, with noticeable clusters of high rates in northwest and northeast England. We investigated associations between a range of primary care trust-level factors and avoidable admission rates using general linear modelling with primary care trusts as the units of analysis. The list of factors examined were those that had been previously associated with geographical variation in admission rates. The large-area ecological-level regression was considered appropriate because the level of interest was systems operating at the primary care trust level. Primary care trust-level deprivation explained 72% of the observed variation across trusts. Factors related to emergency departments, ambulance services, and GP also explained some of the variation (O'Cathain et al., 2014).

A subsequent phase of the project identified trusts that were outliers, that is, with variation not explained by the factors used in the regression, and carried out in-depth case studies of a selection of these trusts using qualitative methods. Data manipulation and analysis were carried out using `R` and SPSS, and mapping was carried out in `R`.

4.5.3.2 Geographical variation in use of CT scanning services

Nixon et al. (2014) carried out a study to examine geographical variation in the use of computed tomography (CT) scanning in a region in New Zealand. CT scanning can be carried out on an emergency or a routine basis. Emergency scans may be carried out for inpatients or for patients in emergency departments before a decision to admit has been made. Routine scans may be carried out for inpatients or outpatients. CT scanning may be used purely for diagnostic purposes or for carrying out procedures under CT guidance.

The study region comprised a mix of urban and rural areas, including remote rural areas. The study area contained two large urban areas, each of which had a CT scanner at the main hospital in the area. Most of the rural areas had a rural hospital that served the local population. The spatial framework used for the analysis was the catchment area. Catchment areas were geographically defined using census-based units. The catchment areas were defined as areas from which most of the patients using the local hospital came from.

The study found that there was large variation in age-adjusted CT utilization rates. Urban areas had 63% higher CT utilization rates than remote rural areas.

There are a number of possible explanations for the variation in use. These include availability of services, variation in clinical practice, differences in population morbidity, and demography. Availability of services is most likely to be the key factor driving utilization rates. Overall, utilization rates will be the result of a mix of appropriate use (although appropriate use can be difficult to accurately define for a range of clinical situations) and potentially borderline or inappropriate use, including supply-induced demand.

Variation in clinical practice is another potentially important factor responsible for geographic variations in utilization rates. Experienced doctors may be more likely to have a higher threshold for using CT scanning, and be more likely to be able to identify patients for whom the scanning is appropriate. However, in one rural area where CT was subsequently introduced, a clear increase in utilization rates was observed, suggesting that availability of CT is the overriding factor driving utilization rates.

Referral rates were also lower for outpatient specialist CT diagnostic and procedure purposes in rural areas, even though these areas were served by specialists running clinics in the rural hospitals. A possible explanation is variation in clinical practice, with the specialists, for example, taking travel distances and inconvenience for rural patients into account if they arranged CT scanning for these patients.

This study highlights the importance of geography when investigating variations in utilization. Mapping was used to visually display hospital catchment areas by type, and the objectives of the study in relation to geography were achieved without the need to resort to more complex GIS functionality. The study was designed to be descriptive, and further investigation is needed to identify potential explanations. In addition, it was not designed to examine outcomes, and variation in utilization needs to be accompanied by subsequent study of outcomes, as lower utilization does not automatically mean poorer outcomes.

4.5.4 Planning the location of services

4.5.4.1 Optimizing access to stroke care

Ischemic stroke is a condition caused by blockage of an artery supplying blood to the brain. This acute event is most commonly caused by a blood clot. Treatment in the acute phase includes administration of a treatment (recombinant tissue-type plasminogen activator), which breaks down the clot to reestablish the blood supply. This treatment needs to be given as soon as possible after the onset of ischemic stroke, but the diagnosis needs to be confirmed first using imaging techniques.

In the United States, a hospital that has the facilities for emergency investigation and treatment of stroke patients may be certified as a primary stroke centre. The process of achieving certification is relatively onerous and is initiated by hospitals, typically larger hospitals in urban areas. This voluntary self-initiation process has led to an uneven distribution of these primary stroke centres, and substantial proportions of the population are not covered by these centres.

Leira et al. (2012) set out to quantify the percentage of the population not covered by primary stroke centres in the state of Iowa. The state comprises a mix of urban and rural areas and had 12 certified primary stroke centres. The authors used a location–allocation model to examine the percentage of the population that would have been covered in a hypothetical situation where the 12 centres were allocated de novo. They also examined how many additional centres would be needed to achieve 75% population coverage using the location–allocation model compared with a weighted random selection of additional centres. The weighted random selection was set up to mimic the current situation where larger hospitals were more likely to self-initiate the certification process. The additional sites were selected from 108 hospitals in the state that had the requirements to be designated as primary stroke centres.

There is a range of location–allocation models (see Cromley and McLafferty [2012] for a description of models) that can be used in different situations, and Leira et al. (2012) used the maximal coverage model for their investigation. They constructed a time–distance matrix from zip code tabulation area postcode centroids to potential locations, used population counts in zip code tabulation areas, and used prespecified maximum time–distance thresholds (15, 30, and 45 minutes) in their calculations. GIS tools used in their work included mapping software for visualization purposes, Microsoft's Bing Maps API for calculating travel times, and a web-based maximal coverage model calculator implemented using Java and PHP.

The authors found that the 12 existing centres only covered 37% of the Iowa population when a 30-minute maximum threshold was used for defining access. The hypothetical assignment of the 12 centres starting de novo would have covered 47.5% of the population using the maximal coverage model. A further 54 primary stroke centres would be needed to reach 75% population coverage if selected using the weighted random selection process, but only a further 31 would be needed if selected using the maximal coverage location–allocation model.

While the authors acknowledge a number of limitations to the theoretical optimization modelling approach, it nevertheless is useful in informing the debate about extending population coverage for rapid access to thrombolysis following ischemic stroke.

4.6 Concluding Remarks

Many of the applications discussed in this chapter have used regression together with other statistical modelling tools. More can be found on these methods in Chapters 5–7 and 13. In this overview, we have adopted a broad conceptualization of GIS, embedding it within GISc, thereby stressing the wider contribution it makes to how we work with geographically referenced data. The definition of what constitutes a GIS and its functionality has evolved and will continue to do so. As the domain of spatial analysis has expanded (particularly the field of spatial statistics), it is no longer reasonable (if it ever was) to expect that any single GIS product will include within it all the tools a spatial epidemiologist or public health analyst working in the field of GISc might wish to call on. Interaction between a GIS and other software systems has become increasingly important, as has the interaction between GIS and the Internet.

GIS and GISc make important contributions to all those areas of scientific investigation and policy making where elements of geography are integral—where place (from the global to the neighbourhood and community scales) and spatial relationships matter (see Chapter 33 for other examples). As will be evident from this overview and other chapters in this volume, while advanced spatial statistical methodology has an important role to play in research, even basic GIS functionalities such as data integration, mapping, and the implementation of simple spatial queries often provide important insights, placing the study of population disease and the delivery of health services in their broader social and environmental contexts.

4.7 Appendix

At the time of writing (March 2015), there is an entry in Wikipedia, "List of geographic information system software," giving both open-source and commercial GIS products.

Notable in the former category are GRASS GIS and QGIS. Notable in the latter category are ERDAS IMAGINE, ESRI (which includes ArcMap, ArcGIS, and ArcIMS), Intergraph, and MapInfo. However there are many more. See http://en.wikipedia.org/wiki/List_of_geographic_information_systems_software#cite_note-sstfoss4g-4.

GIS functionality falls into the following broad categories (Cromley and McLafferty, 2012, p. 30):

1. Measurement (e.g., distance, length, perimeter, area, centroid, buffering, volume, and shape)
2. Topology (e.g., adjacency, polygon overlay, point and line in polygon, dissolve, and merge)
3. Network and location analysis (e.g., connectivity, shortest path, routing, service areas, location–allocation modelling, and accessibility modelling)
4. Surface analysis (e.g., slope, aspect, filtering, line of sight, viewsheds, contours, and watersheds)
5. Statistical analysis (e.g., spatial sampling, spatial weights, exploratory data analysis, nearest neighbour analysis, spatial autocorrelation, spatial interpolation, geostatistics, and trend surface analysis)

GeoDa is free software for undertaking some forms of (frequentist) spatial statistical analysis, including exploratory spatial data analysis. It is user-friendly and employs a drop-down menu style. It also has some normal spatial regression modelling capability (the so-called spatial error and spatial lag regression models with likelihood-based diagnostics to help the user select). The software has some nice features, including a linked windows capability that allows the user to link database spreadsheet rows with the corresponding locations on a map and on graphs. It contains both global and local statistics, such as local and global measures of spatial autocorrelation. The software is available from https://geodacenter.asu.edu/, where tutorials can also be found. It is particularly useful for teaching (especially at the undergraduate level) and could be used in laboratory classes for a course in spatial epidemiology. The software was developed by Luc Anselin. For more advanced spatial modelling, the researcher needs to investigate, among others, the following: WinBUGS and GeoBUGS (Bayesian modelling), STATA, S-PLUS, and the R library. See other chapters in this volume. In this era of "big data," potential users need to be aware that many of these softwares encounter difficulties when used to fit models to large datasets.

References

Aylin P, Maheswaran R, Wakefield J, Cockings S, Jarup L, Arnold R, Wheeler G, Elliott P. (1999). A national facility for small area disease mapping and rapid initial assessment of apparent disease clusters around a point source: The UK Small Area Health Statistics Unit. *Journal of Public Health Medicine* 21:289–93.

Ball W, LeFevre S, Jarup L, Beale L. (2008). Comparison of different methods for spatial analysis of cancer data in Utah. *Environmental Health Perspectives* 116:1120–24.

Beale L, Hodgson S, Abellan J, LeFevre S, Jarup L. (2010). Evaluation of spatial relationships between health and the environment: The Rapid Inquiry Facility. *Environmental Health Perspectives* 118:1306–12.

Beyer K, Rushton G. (2009). Mapping cancer for community engagement. *Public Health Research, Practice and Policy* 6(1):1–8.

Brindley P, Maheswaran R, Pearson T, Wise S, Haining RP. (2004). Using modelled outdoor air pollution data for health surveillance. In Maheswaran, R, Craglia M. (eds.), *GIS in Public Health Practice*, 125–49. Boca Raton, FL: CRC Press.

Brindley P, Wise SM, Maheswaran R, Haining RP. (2005). The effect of alternative representations of population location on the areal interpolation of air pollution exposure. *Computers, Environment and Urban Systems* 29:455–59.

Burrough PA, McDonnell RA. (2000). *Principles of Geographical Information Systems.* Oxford: Oxford University Press.

Cho G. (2007). *Geographic Information Science: Mastering the Legal Issues.* Chichester, UK: Wiley.

Cliff A, Haggett P. (2004). Time, travel and infection. *British Medical Bulletin* 69(1):87–99.

Cowen DJ. (1988). GIS versus CAD versus DBMS: What are the differences? *Photogrammetric Engineering and Remote Sensing* 54:1551–54.

Cromley EK, McLafferty SL. (2012). *GIS and Public Health*, 2nd ed. New York: Guilford Press.

de Lepper MJC, Scholten HJ, Stern RM, eds. (1995). *The Added Value of Geographic Information Systems in Public and Environmental Health.* Dordrecht, Netherlands: Kluwer Academic.

de Nazelle A, Seto E, Donaire-Gonzalez D, Mendez M, Matamala J, Nieuwenhuijsen MJ, Jerrett M. (2013). Improving estimates of air pollution exposure through ubiquitous sensing technologies. *Environmental Pollution* 176:92–99.

Eclipse Foundation. (2011). The Spatiotemporal Epidemiological Modeller (STEM) Project. Ottawa: Eclipse Foundation. Retrieved August 11, 2014, from www.eclipse.org/stem.

Erskine S, Maheswaran R, Pearson T, Gleeson D. (2010). Socioeconomic deprivation, urban–rural location and alcohol-related mortality in England and Wales. *BMC Public Health* 10:99.

Ge E-J, Haining R, Li C-P, Yu Z-G, Waye MMY, Chu K-H, Leung Y. (2012). Using knowledge fusion to analyze avian influenza H5N1 in East and Southeast Asia. *PLoS One*, doi: 10.1371/journal.pone.0029617.

Gilbert M, Xiao X, Pfeiffer DU, Epprecht M, Boles S, Czarnecki C, Chaitaweesub P, et al. (2008). Mapping H5N1 highly pathogenic avian influenza risk in Southeast Asia. *Proceedings of the National Academy of Sciences of United States of America* 105(12):4769–74.

Glass G, Morgan J, Johnson D, Noy P, Israel E, Schwartz B. (1992). Infectious disease epidemiology and GIS: A case study of Lyme disease. *Geographical Information Systems* 3(3):65–69.

Goyder E, Hind D, Breckon J, Dimairo M, Minton J, Everson-Hock E, Read S, et al. (2014). A randomised controlled trial and cost-effectiveness evaluation of “booster” interventions to sustain increases in physical activity in middle-aged adults in deprived urban neighbourhoods. *Health Technology Assessment* 18:1–210.

Guptill S, Morrison J. (1995). *Elements of Spatial Data Quality*. Oxford: Elsevier Science.

Haining R. (2003). *Spatial Data Analysis: Theory and Practice*. Cambridge: Cambridge University Press.

Haining R, Law J, Maheswaran R, Pearson T, Brindley P. (2007). Bayesian modelling of environmental risk: Example using a small area ecological study of coronary heart disease mortality in relation to modelled outdoor nitrogen oxide levels. *Stochastic Environmental Research and Risk Assessment* 21:501–9.

Haining R, Li G, Maheswaran R, Blangiardo M, Law J, Best N, Richardson S. (2010). Inference from ecological models: Estimating the relative risk of stroke from air pollution exposure using small area data. *Spatial and Spatio-Temporal Epidemiology* 1:123–31.

Lake I, Harrison F, Chalmers R, Bentham G, Nichols G, Hunter P, Kovats RS, Grundy C. (2007). Case–control study of environmental and social factors influencing cryptosporidiosis. *European Journal of Epidemiology* 22(11):805–11.

Lee B, Bedford V, Roberts M, Carley K. (2008). Virtual epidemic in a virtual city: Simulating the spread of influenza in a U.S. metropolitan area. *Translational Research* 151(6):275–87.

Leira EC, Fairchild G, Segre AM, Rushton G, Froehler MT, Polgreen PM. (2012). Primary stroke centers should be located using maximal coverage models for optimal access. *Stroke* 43:2417–22.

Lopez A. (1992). Mortality data. In PJ Elliott, J Cuzick, D English, R Stern (eds.), *Geographical and Environmental Epidemiology: Methods for Small Area Studies*, 37–50. Oxford: Oxford University Press.

MacNab YC, Read S, Strong M, Pearson T, Maheswaran R, Goyder E. (2014). Bayesian hierarchical modelling of noisy spatial rates on a modestly large and discontinuous irregular lattice. *Statistical Methods in Medical Research* 23:552–71.

Maheswaran R, Craglia M. (2004). *GIS in Public Health Practice*. Boca Raton, FL: CRC Press.

Maheswaran R, Haining RP, Brindley P, Law J, Pearson T, Fryers PR, Wise S, Campbell MJ. (2005a). Outdoor air pollution and stroke in Sheffield, United Kingdom—a small-area level geographical study. *Stroke* 36:239–43.

Maheswaran R, Haining RP, Brindley P, Law J, Pearson T, Fryers PR, Wise S, Campbell MJ. (2005b). Outdoor air pollution, mortality and hospital admissions from coronary heart disease in Sheffield, United Kingdom—a small-area level ecological study. *European Heart Journal* 26:2543–49.

Maheswaran R, Haining RP, Pearson T, Law J, Brindley P, Best NG. (2006a). Outdoor NOx and stroke mortality—adjusting for small area level smoking prevalence using a Bayesian approach. *Statistical Methods in Medical Research* 15:499–516.

Maheswaran R, Payne N, Meechan D, Burden RP, Fryers PR, Wight J, Hutchinson A. (2003). Socioeconomic deprivation, travel distance and renal replacement therapy in the Trent region, United Kingdom 2000: An ecological study. *Journal of Epidemiology and Community Health* 57:523–24.

Maheswaran R, Pearson T, Beevers SD, Campbell MJ, Wolfe CD. (2014a). Outdoor air pollution, subtypes and severity of ischemic stroke—a small-area level ecological study. *International Journal of Health Geographics* 13:23.

Maheswaran R, Pearson T, Jordan H, Black D. (2006b). Socioeconomic deprivation, travel distance, location of service and uptake of breast cancer screening in North Derbyshire, UK. *Journal of Epidemiology and Community Health* 60:208–12.

Maheswaran R, Pearson T, Munro J, Jiwa M, Campbell MJ, Nicholl J. (2007). Impact of NHS walk-in centres on primary care access times: Ecological study. *British Medical Journal* 334:838–41.

Maheswaran R, Pearson T, Smeeton NC, Beevers SD, Campbell MJ, Wolfe CD. (2010). Impact of outdoor air pollution on survival after stroke: Population-based cohort study. *Stroke* 41:869–77.

Maheswaran R, Pearson T, Smeeton NC, Beevers SD, Campbell MJ, Wolfe CD. (2012). Outdoor air pollution and incidence of ischemic and hemorrhagic stroke: A small-area level ecological study. *Stroke* 43:22–27.

Maheswaran R, Pearson T, Strong M, Clifford P, Brewins L, Wight J. (2014b). Assessing the impact of selective migration and care homes on geographical inequalities in health—A total population cohort study in Sheffield. *Spatial and Spatio-Temporal Epidemiology* 10:85–97.

Maheswaran R, Staines A. (1997). Cancer clusters and their origins. *Chemistry and Industry* 7:254–56.

Nixon G, Samaranayaka A, de Graaf B, McKechnie R, Blattner K, Dovey S. (2014). Geographic disparities in the utilisation of computed tomography scanning services in southern New Zealand. *Health Policy* 118:222–28. doi: 10.1016/j.healthpol.2014.05.002.

O'Cathain A, Knowles E, Maheswaran R, Pearson T, Turner J, Hirst E, Goodacre S, Nicholl J. (2014). A system-wide approach to explaining variation in potentially avoidable emergency admissions: National ecological study. *BMJ Quality and Safety* 23:47–55.

Oppong JR, Tiwarri C, Rucktongsook W, Huddleston J, Arbona S. (2012). Mapping late testers for HIV in Texas. *Health and Place* 18:568–575.

Peng ZR, Tsou MH. (2003). *Internet GIS: Distributed Geographic Information Services for the Internet and Wireless Networks.* Hoboken, NJ: Wiley.

Perez L, Dragicevic S. (2009). An agent based approach for modelling dynamics of contagious disease spread. *International Journal of Health Geographics* 8:50. doi: 10.1186/1476-072X-8-50.

Rhodes T, Singer M, Bourgois P, Friedman S, Strathdee S. (2005). The social structural production of HIV risk among injecting drug users. *Social Science and Medicine* 61(5):1026–44.

Ruiz M, Walker E, Forster E, Haramis L, Kitron U. (2007). Association of West Nile virus illness and urban landscapes in Chicago and Detroit. *International Journal of Health Geographics* 5:10. doi: 10.1186/1476-072X-6-10.

Sheppard E, Couclelis H, Graham S, Harrington JW, Onsrud H. (1999). Geographies of the information society. *International Journal of Geographical Information Science* 13(8):797–823.

Strong M, Pearson T, MacNab YC, Maheswaran R. (2012). Mapping gender variation in the spatial pattern of alcohol-related mortality: A Bayesian analysis using data from South Yorkshire, United Kingdom. *Spatial and Spatio-Temporal Epidemiology* 3:141–49.

Swerdlow A. (1992). Cancer incidence data for adults. In PJ Elliott, J Cuzick, D English, R Stern (eds.), *Geographical and Environmental Epidemiology: Methods for Small Area Studies*, 51–62. Oxford: Oxford University Press.

Veregin H, Hargitai P. (1995). An evaluation matrix for geographical data quality. In Guptill, S, Morrison J. (eds.), *Elements of Spatial Data Quality*, 167–88. Oxford: Elsevier Science.

Wakefield J, Elliott P. (1999). Issues in the statistical analysis of small area health data. *Statistics in Medicine* 18:2377–99.

Zenk SN, Schulz AJ, Matthews SA, Odoms-Young A, Wilbur J, Wegrzyn L, Gibbs K, Braunschweig C, Stokes C. (2011). Activity space environment and dietary and physical activity behaviors: A pilot study. *Health and Place* 17:1150–61.

5

Ecological Modeling: General Issues

Jon C. Wakefield
Departments of Statistics and Biostatistics
University of Washington
Seattle, Washington

Theresa R. Smith
Lancaster Medical School
Lancaster University
Lancaster, UK

CONTENTS

5.1 Introduction

Ecological studies are based on grouped data, with the groups in a spatial context corresponding to geographical areas. In addition to epidemiology and public health [1], ecological studies have a long history in many disciplines, including political science [2], geography [3], and sociology [4]. Due to aggregation, ecological studies are susceptible to unique challenges, in particular the potential for *ecological bias*, which is the difference between estimated associations based on ecological- and individual-level data.

Ecological data may be used for a variety of purposes, including mapping (the geographical summarization of a health outcome; see Chapter 6) and cluster detection (in which anomalous areas are flagged; see Chapter 9). In this chapter, we focus on spatial regression, in which the aim is to investigate the association between an outcome and exposures, here broadly defined to include environmental variables and personal health behaviors. In mapping studies, the prediction of area-level outcome summaries is the objective, and ecological bias is not a great problem, though within-area features (lows and highs) may be obscured by the process of aggregation [5]. With respect to cluster detection, regression analysis is not the aim, although small-area anomalies may be "washed away" when data are aggregated.

5.2 Motivating Example

Throughout this chapter, we will discuss the limitations of ecological data and illustrate ecological modeling using data on breast cancer incidence diagnosed in the period 2006–2010 over the 39 counties of Washington State. The data were obtained from the Washington State Cancer Registry, with each case having an associated age and gender, along with information on race and ethnicity. Population information were obtained for each year by county, with counts by gender, age, and race (non-Hispanic white or nonwhite) from the National Center for Health Statistics. We use counts on women over 25 only, and because of data sparsity, we do not stratify by race. The data are based on intercensal estimates of the July 1 population using information on migration, births, and deaths available from the Federal State Cooperative for Population Estimates, the Internal Revenue Service, and the Social Security Administration.

We will examine the association between female breast cancer incidence and breast cancer screening rates at the county level. Information on screening is obtained from the Behavioral Risk Factor Surveillance System (BRFSS), which is an annual telephone survey that gathers information on personal risk factors as well as demographic information. We will use data from the years 2006, 2008, and 2010 (since in these years the relevant question was asked), and a woman will be counted as "screened" if she reported having received a mammogram in the previous 2 years. Within BRFSS in Washington State, a disproportionate stratified random sample scheme is implemented with stratification by county and "phone likelihood." In each county, based on previous surveys, blocks of 100 telephone numbers are classified into strata that are either "likely" or "unlikely" to yield residential numbers. Telephone numbers in the "likely" strata are sampled at a higher rate than their "unlikely" counterparts. Once a number is reached, the number of eligible adults (aged 18 or over) is determined, and one of these is randomly selected for interview. Sample (design) weights reflect the nonrandom sampling and are available for each sampled woman, along with an indicator of whether the woman has been screened for breast cancer.

We let y_{ic} represent the number of breast cancer incident cases over 2006–2010 in age (confounder) band c and county i, $i = 1, \ldots, m$, $c = 1, \ldots, C$, with $y_i = \sum_{c=1}^{C} y_{ic}$ representing the total count. The age bands are $\{25\text{–}29, 30\text{–}34, \ldots, 80\text{–}84, 85+\}$. The population in county i and confounder band c is n_{ic}, and expected numbers of cases are defined as $e_i = \sum_{c=1}^{C} n_{ic} p_c$, where p_c is the risk of breast cancer in confounder band c in 2005 (so that we are using the year before the study period to obtain the expected numbers). An age-standardized rate of screening in county i is $\hat{x}_i = \frac{1}{N_i} \sum_{c=1}^{C} n_{ic} \hat{x}_{ic}$, where x_{ic} is the rate of screening in county i and age band c. We exclude age bands below age 40 in calculating this standardized screening rate because routine mammography is not generally carried for women younger than 40 years old. We obtain an estimate $\hat{x}_{ic}$ by first calculating the weighted Horvitz–Thompson estimator of the screening prevalence $\hat{x}_{ic}$ in each county and age band. These estimates are unstable, since the data are sparse, and so we smooth the logits of these prevalences across space. Specifically, we take as response $z_{ic} = \text{logit}\ (\hat{x}_{ic})$ and assume the likelihood $z_{ic} | \lambda_{ic} \sim N(\lambda_{ic}, \hat{V}_{ic})$, where $\hat{V}_{ic}$ is the design-based variance estimate. Smoothing of λ_{ic} is then carried out using the spatial convolution model (as described in Besag et al. [6]), and with independent normal random effects for age. Further details of the method, implemented in a different context, are available elsewhere [7].

The question of interest we examine is the ecological association between screening rates and breast cancer incidence. A naive quasi-likelihood log linear ecological model is

$$E[y_i|x_i] = e_i \exp(\alpha^e + \beta^e x_i), \tag{5.1}$$

with $\text{var}(y_i|x_i) = \kappa \times E[y_i|x_i]$, where $\kappa > 0$ allows for excess-Poisson variation. The presence of excess-Poisson variability is common in spatial epidemiological applications. The parameters α^e, β^e have been superscripted with e to stress that they are ecological association parameters.

Figure 5.1 provides a map of the standardized incidence rates (SIRs) y_i/e_i, and we see large variability across counties, with the range of SIRs being (0.41, 1.09). The estimated rates in 2005 are higher than in the study period, and so the expected numbers are in general larger than the observed counts, with an average SIR of 0.88. Figure 5.2 shows a map of the estimated age standardized screening rates for women over the age of 40, with the rates ranging between 0.59 and 0.80. Figure 5.3 plots the SIRs (on the log scale) versus the standardized screening rates, with a nonparametric smoother superimposed. There is a tendency for areas with higher screening rates to have higher breast cancer incidence rates, but there is a lot of variability. We would expect higher incidence in areas with higher screening rates due to early detection of breast cancer. Fitting model (5.1) gives an estimated *ecological* (i.e., group-level) relative risk of $\exp(\widehat{\beta}) = 6.2$ with 95% confidence interval (1.48, 25.7). For this model, the overdispersion was estimated as $\widehat{\kappa} = 6.3$, which is a large amount of excess-Poisson variability and suggests (not surprisingly) that there are unobserved risk factors or errors in numerators (observed counts) and denominators (expected counts), for example, due to differential diagnosis rates across counties. For a 10% increase in screening for an area, the relative risk is 1.20, which, if interpreted literally, would suggest that for two women in areas with differences in screening rates of 10%, the woman in the area with the

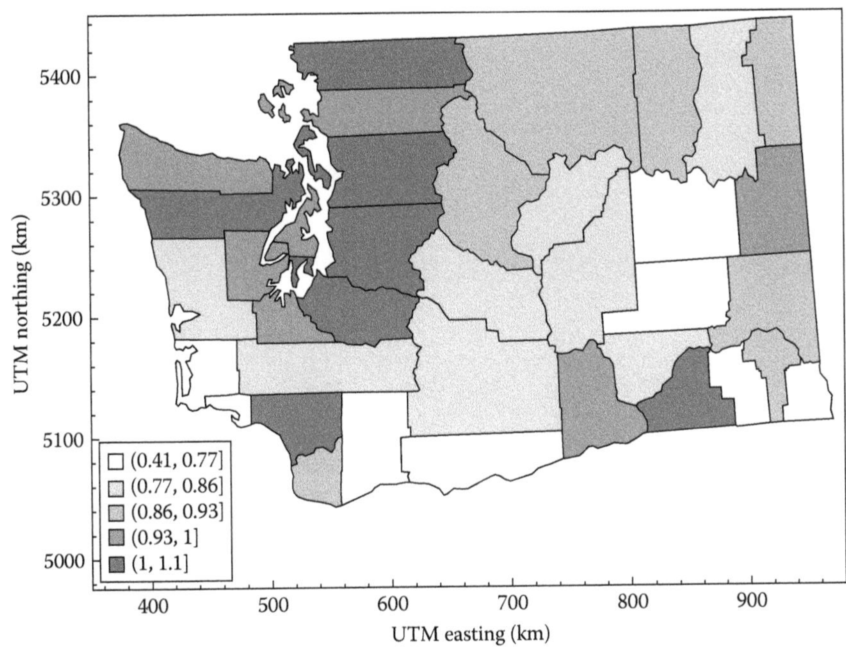

FIGURE 5.1
Standardized incidence rates (SIRs) for breast cancer across counties in Washington State, with diagnosis occurring in the period 2006–2010. UTM, Universal Transverse Mercator.

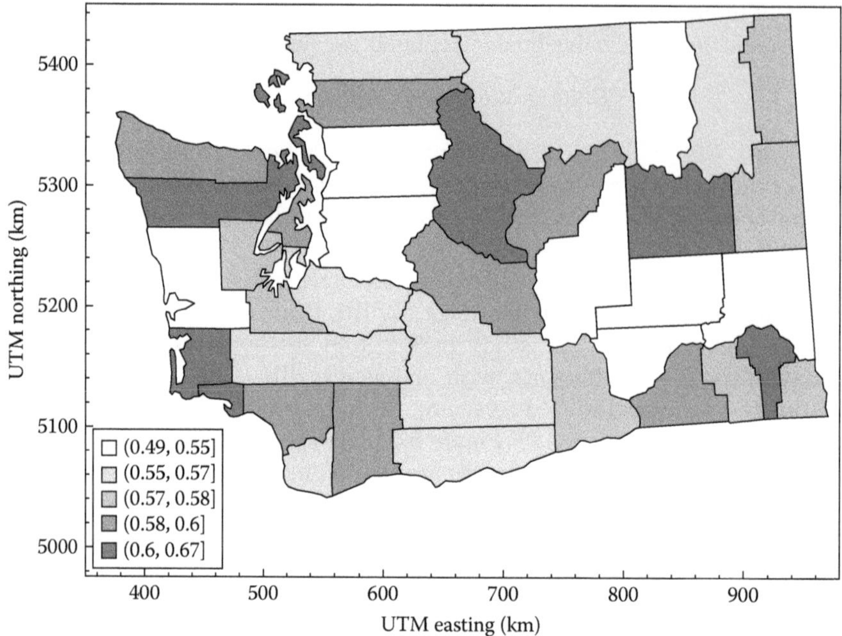

FIGURE 5.2
Breast cancer mammography rates for women over 40, across counties in Washington State.

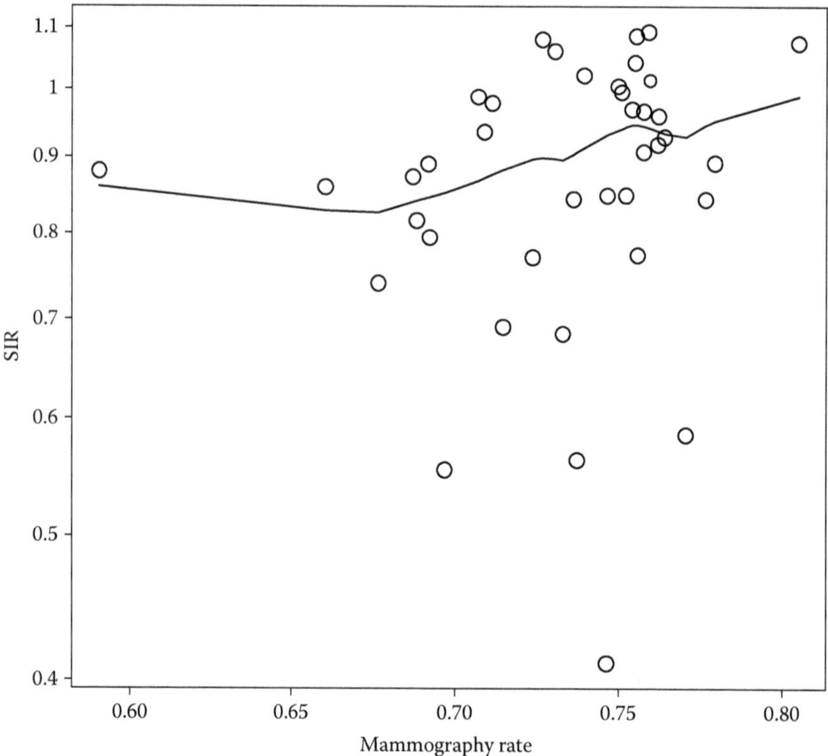

FIGURE 5.3
SIRs (on the log scale) for breast cancer vs. estimated mammography rates.

higher rate would have 20% greater risk. We now discuss methodological issues associated with ecological inference, in order to refine the above model.

5.3 Ecological Bias: Pure Specification Bias

There is a vast literature describing sources of ecological bias [1, 5, 8–19]. The fundamental problem with ecological inference is that the process of aggregation reduces information, and this information loss usually prevents identification of parameters of interest in the underlying individual-level model. When trying to understand ecological bias, it is often beneficial to specify an individual-level model, and then examine the form of the model under aggregation, in order to determine the consequences [19–21].

If there is no within-area variability in exposures and confounders, then there will be no ecological bias; hence, ecological bias occurs due to within-area variability in exposures and confounders. There are a number of distinct consequences of this variability, as we now describe.

Pure specification bias [22], which is also referred to as model specification bias [20], arises because a nonlinear risk model changes its form under aggregation. We demonstrate this bias, and discuss its implications, through a series of models that are commonly encountered in epidemiology. We emphasize that this type of bias has nothing to do with confounding. Throughout, we assume there are m areas with n_i individuals in area i. The outcome and exposure for individual j within area i are denoted Y_{ij} and x_{ij}, $j = 1, \ldots, n_i$, $i = 1, \ldots, m$.

We initially assume a single exposure x and the *linear* individual-level model

$$E[Y_{ij}|x_{ij}] = \alpha + \beta x_{ij}, \tag{5.2}$$

which might be suitable for a continuous-outcome measure such as blood pressure. If Y_{ij} is a binary disease indicator, this model corresponds to a linear risk model. In an ecological setting, the individual-level data are unavailable, and rather, we observe the aggregate data that correspond to the average outcome $\overline{y}_i = \frac{1}{n_i}\sum_{j=1}^{n_i} y_{ij}$ and average exposure $\overline{x}_i = \frac{1}{n_i}\sum_{j=1}^{n_i} x_{ij}$. On aggregation of (5.2) we obtain

$$E[\overline{Y}_i|\overline{x}_i] = \alpha + \beta \overline{x}_i \tag{5.3}$$

so that, in this very specific scenario of a linear model, aggregation has not changed the form of the model.

Unfortunately, a linear model is often inappropriate for the modeling of risk (in particular), and for rare diseases, the individual-level *log-linear* model,

$$E[Y_{ij}|x_{ij}] = \exp(\alpha + \beta x_{ij}) \tag{5.4}$$

is more appropriate. In this model, $\exp(\alpha)$ is the baseline risk associated with $x = 0$ and $\exp(\beta)$ is the relative risk corresponding to an increase in x of one unit. Aggregation of (5.4), and now taking the expectation of the total count, rather than the average, yields

$$E[Y_i|x_{ij}, j = 1, \ldots, n_i] = \sum_{j=1}^{n_i} \exp(\alpha + \beta x_{ij}). \tag{5.5}$$

Dividing the left- and right-hand sides of (5.5) by n_i illustrates that the ecological risk is the average of the risks of the constituent individuals. A naive ecological model, as considered earlier, (5.1), would assume

$$E[Y_i|\overline{x}_i] = n_i \exp(\alpha^e + \beta^e \overline{x}_i). \tag{5.6}$$

Model (5.6) is a *contextual effects* model since risk depends on the proportion of exposed individuals in the area (contextual variables are summaries of a shared environment). Interpreting $\exp(\beta^e)$ as an individual association would correspond to a belief that it is average exposure that is causative, and that individual exposure is effectively irrelevant (since it makes a small contribution to the average exposure).

The difference between (5.5) and (5.6) is clear: while the former, individual-level model, sums the risks across all *constituent* exposures, the latter, ecological-level model, is n_i times the risk function evaluated at the *average* exposure. Without further knowledge of the within-area exposure distribution (e.g., as a minimum, the within-area variances), we cannot examine the implications of ecological bias. When there is no within-area variability in exposure, so that $x_{ij} = \overline{x}_i$ for all $j = 1, \ldots, n_i$ individuals in area i, there will be no ecological bias. An intuitive implication is that pure specification bias is reduced if areas are smaller, since the heterogeneity of exposures within areas is decreased. It would be beneficial to choose geographical areas so that areas with constant exposure are obtained, but unfortunately, the level of data aggregation is usually imposed by the administrative groupings that are available, for example, from the census.

Binary exposures are the simplest to study analytically. Such exposures may correspond to an individual being below or above a pollutant threshold, or to a binary measure of a lifestyle characteristic such as smoker or nonsmoker or an indicator of screening. For a binary exposure, (5.4) can be written as

$$\exp(\alpha + \beta x_{ij}) = (1 - x_{ij}) \exp(\alpha) + x_{ij} \exp(\alpha + \beta), \tag{5.7}$$

which is linear in $\exp(\alpha)$ and $\exp(\alpha + \beta)$. This form makes it clear that the risk for an unexposed individual ($x = 0$) is $\exp(\alpha)$, while for an exposed individual ($x = 1$), the risk is $\exp(\alpha + \beta)$. Aggregation of (5.7) yields the form

$$E[Y_i|\overline{x}_i] = n_i \left[(1 - \overline{x}_i) \exp(\alpha) + \overline{x}_i \exp(\alpha + \beta)\right], \tag{5.8}$$

where $\overline{x}_i$ is the proportion exposed in area i. Hence, with a linear risk model, there is no pure specification bias so long as model (5.8) is fitted using the binary proportion, $\overline{x}_i$. If model (5.6) is fitted, there will be no correspondence between $\exp(\beta)$ and $\exp(\beta^e)$ since they are associated with completely different comparisons. The extension to general categorical exposures is straightforward, and the parameters of the disease model are identifiable so long as we have observed the aggregate proportions in each category.

We now demonstrate that for a continuous exposure, pure specification bias is dominated by the within-area relationship between the mean and the variance. In an ecological regression context, a normal within-area exposure distribution $N(x|\overline{x}_i, s_i^2)$, and the log-linear model (5.4), has been considered by a number of authors [13, 15, 21]. We assume that n_i is large so that the summation in (5.5) can be approximated by an integral, that is,

$$E[Y_i|x_{ij}, j = 1, \ldots, n_i] = \sum_{j=1}^{n_i} \exp(\alpha + \beta x_{ij}) \approx n_i \int_x \exp(\alpha + \beta x) f(x|\phi_i)\, dx, \tag{5.9}$$

where $f(x|\phi_i)$ represents the distribution, with parameters ϕ_i, of the exposure x within area i. If the exposure follows a normal distribution with mean $\overline{x}_i$ and variance s_i^2, we have, from (5.9),

$$E[Y_i|\overline{x}_i, s_i^2] = n_i \exp(\alpha + \beta\overline{x}_i + \beta^2 s_i^2/2), \tag{5.10}$$

which may be compared with the naive ecological model $n_i \exp(\alpha^e + \beta^e \overline{x}_i)$. To gain intuition as to the extent of the bias, we observe that in (5.10) the within-area variance s_i^2 is acting like a confounder, and consequently, there is no pure specification bias if the exposure is constant within each area or if the variance is independent of the mean exposure in the area. The expression (5.10) also allows us to characterize the direction of bias, if we are willing to assume a particular form for the mean–variance relationship. For example, suppose that $s_i^2 = a + b\overline{x}_i$ with $b > 0$ so that the variance increases with the mean (as is often observed with environmental exposures). In this case, the parameter we are estimating from the ecological data is

$$\beta^e = \beta + \beta^2 b/2.$$

If $\beta > 0$, then overestimation will occur using the ecological model, and if $\beta < 0$, the ecological association, β^e, may reverse sign when compared to β.

In general, there is no pure specification bias if the disease model is linear in x or if all the moments of the within-area distribution of exposure are independent of the mean. In practice, one can never check that this assumption holds, but the mean–variance will dominate, however. Unfortunately, the mean–variance relationship is impossible to assess without individual-level data on the exposure, though samples of individual exposures within areas may be available. If β is close to zero, pure specification bias is also likely to be small (since then the exponential model will be approximately linear, which, as we have seen, produces no pure specification bias), though in this case, confounding is likely to be a serious worry (Section 5.4). If the exposure is heterogeneous within areas, we need information on the variability within each area in order to control the bias. The logistic model, which is often used for nonrare outcomes in epidemiology, is unfortunately not amenable to analytical study, and so the effects of aggregation are difficult to discern [23].

5.4 Ecological Bias: Confounding

We now consider the more realistic situation in which there are confounders z that we would like to include in the model, in addition to the exposure x. As we now discuss, the key challenge is how to characterize the within-area *joint* distribution of exposures and confounders. A within-area confounder z varies within an area, in contrast to a between-area confounder that is constant within an area. We assume a single exposure x_{ij}, a single confounder z_{ij}, and the individual-level model

$$E[Y_{ij}|x_{ij}, z_{ij}] = \exp(\alpha + \beta x_{ij} + \gamma z_{ij}) \tag{5.11}$$

for individual j in area i, $j = 1, \ldots, n_i$, $i = 1, \ldots, m$. As with pure specification bias, the key to understanding sources of, and correcting for, ecological bias is to aggregate the individual-level model to give

$$E[Y_i|x_{ij}, z_{ij}, j = 1, \ldots, n_i] = \sum_{j=1}^{n_i} \exp(\alpha + \beta x_{ij} + \gamma z_{ij}). \tag{5.12}$$

To see why controlling for confounding is in general impossible with ecological data, we consider the simplest case of a binary exposure (unexposed/exposed) and a binary confounder, which for ease of explanation we assume is gender. Table 5.1 shows the distribution of the exposure and confounder within area i. The complete within-area distribution of exposure and confounder can be described by three frequencies (e.g., p_{i00}, p_{i01}, and p_{i10}), but the ecologic data usually consist of two quantities only, the proportion exposed, $\overline{x}_i$, and the proportion male, $\overline{z}_i$. From (5.12), the aggregate form is

$$E[Y_i|p_{i00},p_{i01},p_{i10},p_{i11}] = n_i\left[(1-\overline{x}_i-\overline{z}_i+p_{i11})\exp(\alpha)+(\overline{x}_i-p_{i11})\exp(\alpha+\beta)\right.$$
$$\left.+(\overline{z}_i-p_{i11})\exp(\alpha+\gamma)+p_{i11}\exp(\alpha+\beta+\gamma)\right],$$

which is simply the weighted sum of the four risks, corresponding to each combination of exposures and confounders. This expression shows that the marginal prevalences, $\overline{x}_i, \overline{z}_i$, alone, are not sufficient to characterize the joint distribution unless x and z are independent, in which case z is not a within-area confounder. This scenario has been considered in detail elsewhere [24], where it was argued that if the proportion of exposed males (p_{i11}) is missing, it should be estimated by the marginal prevalences ($\overline{x}_i \times \overline{z}_i$). It is not possible to determine the accuracy of this approximation without individual-level data, however, and we would not recommend this approach. A recurring theme in the analysis of ecological data is that bias is reduced under special cases, but estimation is crucially dependent on the appropriateness of the restrictions, which are uncheckable without individual-level data.

We now examine the situation in which we have a binary exposure and a continuous confounder. Let the confounders in the unexposed be denoted z_{ij}, $j = 1,\ldots,n_{i0}$, and the confounders in the exposed be denoted z_{ij}, $j = n_{i0}+1,\ldots,n_{i0}+n_{i1}$, with $n_{i0}+n_{i1} = n_i$. The ecological form corresponding to (5.11) is

$$E[Y_i|q_{i0},q_{i1}] = n_i\left[q_{i0}\times r_{i0}+q_{i1}\times r_{i1}\right],$$

where $q_{i0} = n_{i0}/n_i$ and $q_{i1} = n_{i1}/n_i$ are the probabilities of being unexposed and exposed, and

$$r_{i0} = \frac{\exp(\alpha)}{n_{i0}}\sum_{j=1}^{n_{i0}}\exp(\gamma z_{ij}), \quad r_{i1} = \frac{\exp(\alpha+\beta)}{n_{i1}}\sum_{j=n_{i0}+1}^{n_{i0}+n_{i1}}\exp(\gamma z_{ij})$$

are the (aggregated) risks in the unexposed and exposed groups. The important message here is that we need the confounder distribution within each exposure category, unless z is not a within-area confounder. Again, it is clear that if we fit the ecological model,

$$E[Y_i|\overline{x}_i,\overline{z}_i] = n_i\exp(\alpha^e+\beta^e\overline{x}_i+\gamma^e\overline{z}_i),$$

TABLE 5.1
Exposure and gender distribution in area i

	Female	**Male**	
Unexposed	p_{i00}	p_{i01}	$1-\overline{x}_i$
Exposed	p_{i10}	p_{i11}	$\overline{x}_i$
	$1-\overline{z}_i$	$\overline{z}_i$	1.0

Note: $\overline{x}_i$ is the proportion exposed and $\overline{z}_i$ is the proportion male; $p_{i00},p_{i01},p_{i10},p_{i11}$ are the within-area cross-classification frequencies.

where $\overline{z}_i = \frac{1}{n_i}\sum_{j=1}^{n_i} z_{ij}$, then it is not possible to equate the ecological coefficient β^e with the individual-level parameter of interest β. Clearly, the above development is also relevant to the situation in which the exposure is continuous and the confounder is binary. In this case, we need the exposure distribution in both of the confounder stratum.

We now extend our discussion to multiple exposure strata and show the link with the use of expected numbers (as defined in Section 5.2). Consider the continuous exposure x_{icj} for the $j = 1,\ldots,n_{ic}$ individuals in stratum c and area i, and suppose the individual-level model is

$$E[Y_{icj}|x_{icj},\ \text{strata } c] = \exp(\alpha + \beta x_{icj} + \gamma_c),$$

for $c = 1,\ldots,C$ stratum levels with relative risks $\exp(\gamma_c)$ (with, e.g., $\sum_{c=1}^{C}\gamma_c = 0$, imposed for identifiability) and with $j = 1,\ldots,n_{ic}$. For ease of exposition, suppose c indexes age categories (as in the motivating example; Section 5.2). Let $Y_{ic} = \sum_{j=1}^{n_c} Y_{icj}$ be the number of individuals with the disease in area i and stratum c. Then

$$E[Y_{ic}|x_{icj}, j = 1,\ldots,n_{ic}] = \exp(\alpha + \gamma_c)\sum_{j=1}^{n_{ic}}\exp(\beta x_{icj}).$$

Summing over stratum and letting $Y_i = \sum_{c=1}^{C} Y_{ic}$ be the number with the disease in area i,

$$E[Y_i|x_{icj}, j = 1,\ldots,n_{ic}, c = 1,\ldots,C] = \sum_{c=1}^{C} n_{ic}\left\{\exp(\alpha + \gamma_c)\sum_{j=1}^{n_{ic}}\exp(\beta x_{icj})\right\}. \tag{5.13}$$

If we assume a common exposure distribution across confounder stratum and let x_{ij}, $j = 1,\ldots,m_i$ be a representative exposure sample, then we could fit the model

$$\begin{aligned} E[Y_i|x_{ij}, j = 1,\ldots,m_i] &= \sum_{c=1}^{C} n_{ic}\exp(\alpha + \gamma_c)\sum_{j=1}^{m_i}\exp(\beta x_{ij}) \\ &= e_i \times \exp(\alpha)\sum_{j=1}^{m_i}\exp(\beta x_{ij}), \end{aligned} \tag{5.14}$$

where $e_i = \sum_{c=1}^{C} n_{ic}\exp(\gamma_c)$ are the expected numbers. Model (5.14) attempts to correct for pure specification bias (by summing the individual-level risks) but assumes common exposure variability across confounder stratum. Hence, we see that in this model (which has been previously used [25]), we have standardized for age (via indirect standardization), but for this approach to provide meaningful inference, we need to assume that the exposure is constant across confounder groups (so that the confounder is not a within-area confounder). This can be compared with the model that is frequently fitted:

$$E[Y_i|\overline{x}_i] = e_i \times \exp(\alpha^e + \beta^e\overline{x}_i). \tag{5.15}$$

The validity of model (5.15) goes beyond the requirement of a constant exposure distribution across stratum within each area. In addition, it also requires no within-area variability in exposure.

If the exposure is continuous then, recalling our earlier discussion, fitting model (5.15) and obtaining meaningful inference requires the exposure distribution to be approximately constant across confounder groups c and the exposure variance to be approximately independent of the mean.

This discussion is closely related to the idea of *mutual standardization* in which, if the response is standardized by age, for example, the exposure variable must also be standardized for this variable [26]. The correct model is given by (5.13) and requires the exposure distribution by age group, or at least a representative sample of exposures from each age group. The above discussion makes it clear that we need *individual-level data* to characterize the within-area distribution of confounders and exposures.

The extension to general exposure and confounder scenarios can be carried out in a similar fashion to the above derivations. If we have confounders that are constant within areas (e.g., access to health care), then they are analogous to conventional confounders since the area is the unit of analysis, and so the implications are relatively easy to understand and adjustment is straightforward.

Without an interaction between exposure and confounder, the parameters of a linear model are estimable from marginal information only, though if an interaction is present, within-area information is required [18].

5.5 Example Revisited

Breast cancer counts Y_{ic} are available by C age bands, and we have estimates of screening rates $\hat{x}_{ic}$ for county i and age group c. Let x_{icj} denote the screening status of individual j in age band c and area i with $j = 1, \ldots, n_{ic}$. The log-linear individual-level model is

$$E[Y_{icj}|x_{icj}] = \exp(\alpha + \gamma_c + x_{icj}\beta).$$

Summing over individuals leads to the form

$$E[Y_{ic}|x_{ic}] = n_{ic}\exp(\alpha + \hat{\gamma}_c)(1 - x_{ic} + x_{ic}\beta), \tag{5.16}$$

where we have substituted age estimates from the 2005 cancer incidence data, $\hat{\gamma}$. We fit a quasi-likelihood model with mean model (5.16) and obtain estimate (95% confidence interval) $\exp(\hat{\beta}) = 1.25$ (1.11, 1.39). We see that the relative risk bears no relation to the estimate obtained earlier in the log-linear ecological model (5.1). The estimate of the overdispersion $\hat{\kappa} = 2.0$, which is greatly reduced when compared with the naive ecological model.

5.6 Spatial Dependence and Hierarchical Modeling

When data are available as counts from a set of contiguous areas, we might expect residual dependence in the counts, particularly for studies in which the areas are small, due to the presence of unmeasured variables with spatial structure. The use of the word *residual* here acknowledges that variables known to influence the outcome have already been adjusted for in the mean model. If dependence is present, analysis methods that ignore the dependence are strictly not applicable, with inappropriate standard errors being the most obvious manifestation. A great deal of work has focused on models for spatial dependence [6, 27–33]; Richardson [34] provides an excellent review of this literature. A number of chapters in this volume discuss models for spatial dependence, but see in particular Chapters 7, 11, and 22.

Computation for hierarchical models has traditionally been carried out using Markov chain Monte Carlo, and there are now a variety of computing resources, including `WinBUGS`,

BayesX, and Stan; see Chapters 7 and 24. More recently [35, 36], the integrated nested Laplace approximation (INLA) approach has been developed as a quick and powerful alternative based on a combination of analytic approximation and numerical integration, and with an R implementation; see Chapter 24. We use INLA for our Bayesian analyses in this chapter.

With respect to ecological bias, an important modeling message is that greater effort should be placed on specification of the mean model because, unless the mean model is correct, adjustment for spatial dependence is a pointless exercise [5].

Spatial smoothing models have also been proposed to control for "confounding by location" [29]. A subtle but extremely important point is that such an endeavor is fraught with pitfalls since the exposure of interest usually has spatial structure, and so one must choose an appropriate spatial scale for smoothing. If the scale is chosen to be too small, the exposure effect may be attenuated; whereas if too large a scale is chosen, the signal that is due to confounding may be absorbed into the exposure association estimate. Practically, one can obtain estimates from models with and without spatial smoothing, and with a variety of spatial models, to address the sensitivity of inference concerning parameters of interest. Further discussion of confounding by location may be found elsewhere [37, 38]. Similar issues arise in time-series analysis when one must control trends by selecting an appropriate level of temporal smoothing [39]. Temporal analyses are more straightforward since time is one dimensional, the data are generally collected at regular intervals (e.g., daily), and the data are also abundant, perhaps containing many years worth.

In a much-cited book [2], a hierarchical model was proposed for the analysis of ecologic data in a political science context, as "a solution to the ecological inference problem." Identifiability in this model is imposed through the random effects prior, however, and it is not possible to check the appropriateness of this prior from the ecological data alone [19, 40].

5.7 Example Revisited

We now assume a hierarchical model with a Poisson first stage with mean model (5.16) and various assumptions for the intercept. The first model assumes

$$n_{ic}\exp(\alpha+\hat{\gamma}_c+\epsilon_i)[1-\hat{x}_{ic}+\hat{x}_{ic}\exp(\beta)], \tag{5.17}$$

with $\epsilon_i|\sigma^2_\epsilon \sim_{\text{iid}} N(0,\sigma^2_\epsilon)$. As an alternative, we also fit the convolution model [6] with mean model

$$n_{ic}\exp(\alpha+\hat{\gamma}_c+\epsilon_i+S_i)[1-\hat{x}_{ic}+\hat{x}_{ic}\exp(\beta)], \tag{5.18}$$

with, as before, $\epsilon_i|\sigma^2_\epsilon \sim_{\text{iid}} N(0,\sigma^2_\epsilon)$ and the spatial residual terms S_i following an intrinsic conditional autoregressive (ICAR) model [41], which we denote $S_i|\sigma^2_s \sim \text{ICAR}(\sigma^2_s)$.

The posterior median (95% interval) for $\exp(\beta)$ is 1.42 (1.20, 1.67) under both models (5.17) and (5.18). When compared to the quasi-likelihood model (5.16), we see that the estimate is increased. The marginal model corresponding to (5.17) has a variance that is quadratic in the mean, whereas the quasi-likelihood model has a variance that is linear in the mean, and so one should not be surprised to see differences in point and interval estimates. However, we would not expect the correspondence of the estimates under the nonspatial and spatial models to hold in general. To investigate, we examine the variance components under the two models. We find that the estimates of the variance of the nonspatial random effects are very similar in models (5.17) and (5.18) ($\hat{\sigma}^2_\epsilon = 0.0132$ vs. $\hat{\sigma}^2_\epsilon = 0.0131$).

Further, the conditional variance of the spatial random effects is quite small ($\hat{\sigma}_S^2 = 8.42 \times 10^{-5}$), confirming that including spatial random effects yields only minor changes in the overall model fit. Figure 5.4 shows the nonspatial and spatial random effects for model (5.18). The spatial random effects are small compared to the nonspatial random effects but show a high degree of spatial correlation.

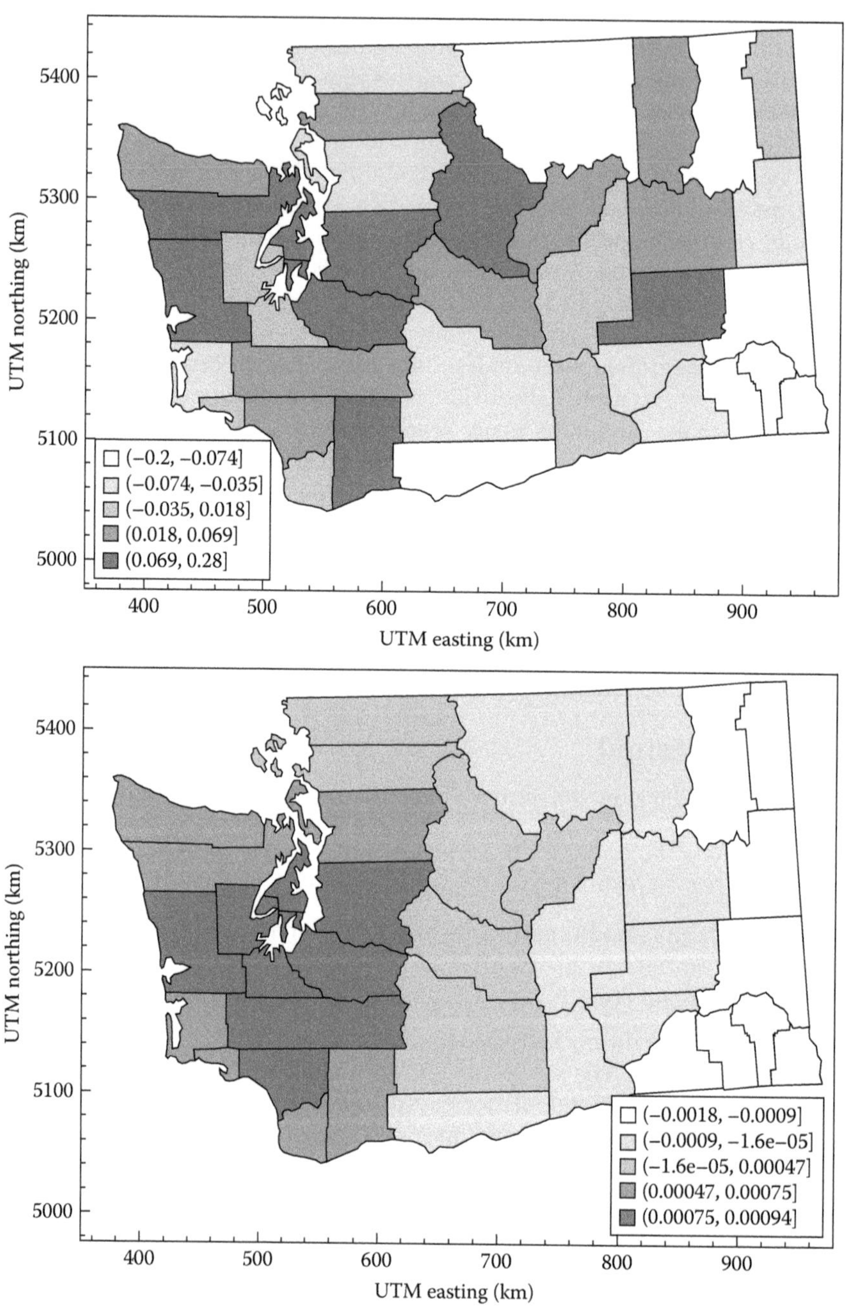

FIGURE 5.4
Estimated nonspatial (top) and spatial (bottom) random effects from model (5.18).

5.8 Related Study Designs

The only solution to the ecologic inference problem that does not require uncheckable assumptions is to add individual-level data to the ecological data. Here, we briefly review some of the proposals for such an endeavor. Another perspective is that ecological data can supplement already available individual data, in order to improve power.

Table 5.2 summarizes four distinct scenarios in terms of data availability [11, 20]. The obvious approach to adding individual-level data is to collect a random sample of individuals within areas, and this has been considered for continuous data [42, 43] and for a binary nonrare outcome [19, 44].

For a rare disease, few cases will be present in the individuals within the sample, and so only information on the distribution of exposures and confounders will be obtained via a random sampling strategy (which is therefore equivalent to using a sample survey of covariates only). This prompted the derivation of the so-called *aggregate data* method of Prentice and Sheppard [45–47]. Inference proceeds by constructing an estimating function based on the sample of $m_i \leq n_i$ individuals in each area. For example, with samples for two variables, $\{x_{ij}, z_{ij}, j = 1, \dots, m_i\}$, we have the mean function

$$E\left[\overline{Y}_i | x_{ij}, z_{ij}, j = 1, \dots, m_i\right] = \frac{1}{n_i} \frac{n_i}{m_i} \sum_{j=1}^{m_i} \exp(\alpha + \beta x_{ij} + \gamma z_{ij}).$$

There is bias involved in the resultant estimator since the complete set of exposures are unavailable ($m_i < n_i$), but Prentice and Sheppard give a finite sample correction to the estimating function based on survey sampling methods. This is an extremely powerful design since estimation is not based on any assumptions with respect to the within-area distribution of exposures and confounders, though this distribution may not be well characterized when m_i is small [48].

An alternative approach is to assume a particular distribution for the within-area variability in exposure and fit the implied model [15, 21, 49–51]. The normal model is usually assumed, in which case (for a single exposure) the mean model is (5.10). This method implicitly assumes that a sample of within-area exposures is available since the within-area moments need to be available. Approaches based on explicit likelihood models that build on the aggregate data method have also been proposed [48, 52].

A different approach to adding individual data in the context of a rare response is outcome-dependent sampling, which avoids the problems of zero cases that is encountered in random sampling. For the situation in which ecologic data are supplemented with individual case–control information gathered within the constituent areas, inferential approaches have been developed [53–55]. The case-control data remove ecological bias, while the ecological data provide increased power and constraints on the sampling distribution of the case–control data, which improves the precision of estimates.

TABLE 5.2
Study designs by level of outcome and exposure data

		Exposure	
		Individual	**Ecological**
Outcome	Individual	Individual	Semiecological
	Ecological	Aggregate	Ecological

Two-phase methods have a long history in statistics and epidemiology [56–59] and are based on an initial cross-classification by outcome and confounders and exposures; this classification provides a sampling frame within which additional covariates may be gathered via the sampling of individuals. Such a design may be used in an ecological setting, where the initial classification is based on one or more of area, confounder stratum, and possibly error-prone measures of exposure [60]. Recently, Bayesian approaches to two-phase sampling in an epidemiological setting have been proposed [61, 62].

In all of these approaches, it is clearly vital to avoid response bias in the survey samples, or selection bias in outcome-dependent sampling, and establishing a relevant sampling frame is essential.

In a semiecological study, sometimes more optimistically referred to as a "semi-individual study" [11], individual-level data are collected on outcome and confounders, with exposure information arising from another source. The Harvard six-city study [63] provides an example in which the exposure was a city-specific average exposure from pollution monitors over the follow-up of the study. These studies are still susceptible to pure specification bias, however [64].

5.9 Concluding Remarks

In this chapter, we have illustrated the pitfalls of ecological data. However, the analysis of ecological data should not be completely shunned since they can add to the totality of evidence. A useful starting point for all ecological analyses is to write down an individual-level model for the outcome–exposure association of interest, including known confounders. Ecological bias will be small when within-area variability in exposures and confounders is small, and for studies with areas of small population in particular, this may be approximately true. A serious source of bias is that due to confounding, ecological data on exposure are rarely stratified by confounder strata within areas. If a study with small population areas has been carried out with a correctly aggregated individual-level model, then parameter estimates can be cautiously interpreted at the individual level and compared with other studies at the individual level, and hence add to the totality of evidence for a hypothesis.

Less well-designed ecological studies can be suggestive of hypotheses to investigate if strong ecological associations are observed. An alternative to the pessimistic outlook expressed above is that when a strong ecological association is observed, an attempt should be made to explain how such a relationship could have arisen, if it is not due to the ecological predictor.

There are a number of issues that we have not discussed. Care should be taken in determining the effects of measurement error in an ecological study since the directions of bias may not be predictable. For example, in the absence of pure specification and confounder bias for linear and log-linear models, if there is nondifferential measurement error in a binary exposure, there will be overestimation of the effect parameter, in contrast to individual-level studies [65]. We refer interested readers to alternative sources [66, 67] for other issues, such as consideration of migration, latency periods, and the likely impacts of inaccuracies in population and health data.

Studies that investigate the acute effects of air pollution are another common situation in which ecological exposures are used. For example, daily disease counts in a city are often regressed against daily or lagged concentration measurements taken from a monitor, or the average of a collection of monitors, to estimate the acute effects of air pollution. If day-to-day exposure variability is greater than within-city variability, then we would expect ecological bias to be relatively small. We have not considered ecological bias in a space–time context; little work has been done in this area, though see Wakefield [68] for a brief development.

With respect to data availability, exposure information is generally not aggregate in nature (unless the exposure is a demographic or socioeconomic variable), and in an environmental epidemiological setting, the modeling of pollutant concentration surfaces will undoubtedly grow in popularity. However, it may be better in a health exposure modeling context to use measurements from the nearest monitor, rather than model the concentration surface, since the latter approach may be susceptible to large biases, particularly when, as is usually the case, the monitoring network is sparse [52]. A remaining challenge is to diagnose when the available data are of sufficient abundance and quality to support the use of complex models.

In Section 5.8, we described a number of proposals for the combination of ecological and individual data. Such endeavors will no doubt increase and will hopefully allow the reliable exploitation of ecological information.

Acknowledgments

The first author was supported by R01-CA095994 from the National Institutes of Health.

References

[1] H. Morgenstern. Ecologic study. In P. Armitage and T. Colton, editors, *Encyclopedia of Biostatistics*, vol. 2, 1255–76. John Wiley, New York, 1998.

[2] G. King. *A Solution to the Ecological Inference Problem.* Princeton University Press, Princeton, 1997.

[3] S. Openshaw. *The Modifiable Areal Unit Problem.* CATMOG no. 38. Geo Books, Norwich, CT, 1984.

[4] W.S. Robinson. Ecological correlations and the behavior of individuals. *American Sociological Review*, 15:351–57, 1950.

[5] J.C. Wakefield. Disease mapping and spatial regression with count data. *Biostatistics*, 8:158–183, 2007.

[6] J. Besag, J. York, and A. Mollié. Bayesian image restoration with two applications in spatial statistics. *Annals of the Institute of Statistics and Mathematics*, 43:1–59, 1991.

[7] L. Mercer, J. Wakefield, C. Chen, and T. Lumley. A comparison of spatial smoothing methods for small area estimation with sampling weights. *Spatial Statistics*, 8:69–85, 2014.

[8] S. Greenland. Divergent biases in ecologic and individual level studies. *Statistics in Medicine*, 11:1209–1223, 1992.

[9] S. Greenland and H. Morgenstern. Ecological bias, confounding and effect modification. *International Journal of Epidemiology*, 18:269–274, 1989.

[10] S. Greenland and J. Robins. Ecological studies: biases, misconceptions and counterexamples. *American Journal of Epidemiology*, 139:747–760, 1994.

[11] N. Künzli and I.B. Tager. The semi-individual study in air pollution epidemiology: a valid design as compared to ecologic studies. *Environmental Health Perspectives*, 10:1078–1083, 1997.

[12] S. Piantadosi, D.P. Byar, and S.B. Green. The ecological fallacy. *American Journal of Epidemiology*, 127:893–904, 1988.

[13] M. Plummer and D. Clayton. Estimation of population exposure. *Journal of the Royal Statistical Society, Series B*, 58:113–126, 1996.

[14] S. Richardson and C. Montfort. Ecological correlation studies. In P. Elliott, J.C. Wakefield, N.G. Best, and D. Briggs, editors, *Spatial Epidemiology: Methods and Applications*, 205–220. Oxford University Press, Oxford, 2000.

[15] S. Richardson, I. Stucker, and D. Hémon. Comparison of relative risks obtained in ecological and individual studies: some methodological considerations. *International Journal of Epidemiology*, 16:111–20, 1987.

[16] D.G. Steel and D. Holt. Analysing and adjusting aggregation effects: the ecological fallacy revisited. *International Statistical Review*, 64:39–60, 1996.

[17] J. Wakefield. Ecologic studies revisited. *Annual Review of Public Health*, 29:75–90, 2008.

[18] J.C. Wakefield. Sensitivity analyses for ecological regression. *Biometrics*, 59:9–17, 2003.

[19] J.C. Wakefield. Ecological inference for 2 $\times$ 2 tables (with discussion). *Journal of the Royal Statistical Society, Series A*, 167:385–445, 2004.

[20] L. Sheppard. Insights on bias and information in group-level studies. *Biostatistics*, 4:265–278, 2003.

[21] J.C. Wakefield and R.E. Salway. A statistical framework for ecological and aggregate studies. *Journal of the Royal Statistical Society, Series A*, 164:119–137, 2001.

[22] S. Greenland. A review of multilevel theory for ecologic analyses. *Statistics in Medicine*, 21:389–95, 2002.

[23] R.A. Salway and J.C. Wakefield. Sources of bias in ecological studies of non-rare events. *Environmental and Ecoloigcal Statistics*, 12:321–347, 2005.

[24] V. Lasserre, C. Guihenneuc-Jouyaux, and S. Richardson. Biases in ecological studies: utility of including within-area distribution of confounders. *Statistics in Medicine*, 19:45–59, 2000.

[25] K. Guthrie, L. Sheppard, and J. Wakefield. A hierarchical aggregate data model with spatially correlated disease rates. *Biometrics*, 58:898–905, 2002.

[26] P.R. Rosenbaum and D.B. Rubin. Difficulties with regression analyses of age-adjusted rates. *Biometrics*, 40:437–443, 1984.

[27] N.G. Best, K. Ickstadt, and R.L. Wolpert. Ecological modelling of health and exposure data measured at disparate spatial scales. *Journal of the American Statistical Association*, 95:1076–1088, 2000.

[28] O.F. Christensen and R. Waagepetersen. Bayesian prediction of spatial count data using generalised linear mixed models. *Biometrics*, 58:280–286, 2002.

[29] D. Clayton, L. Bernardinelli, and C. Montomoli. Spatial correlation in ecological analysis. *International Journal of Epidemiology*, 22:1193–1202, 1993.

[30] N. Cressie and N.H. Chan. Spatial modelling of regional variables. *Journal of the American Statistical Association*, 84:393–401, 1989.

[31] P.J. Diggle, J.A. Tawn, and R.A. Moyeed. Model-based geostatistics (with discussion). *Applied Statistics*, 47:299–350, 1998.

[32] J.E. Kelsall and J.C. Wakefield. Modeling spatial variation in disease risk: a geostatistical approach. *Journal of the American Statistical Association*, 97:692–701, 2002.

[33] B.G. Leroux, X. Lei, and N. Breslow. Estimation of disease rates in small areas: a new mixed model for spatial dependence. In M.E. Halloran and D.A. Berry, editors, *Statistical Models in Epidemiology, the Environment and Clinical Trials*, 179–192. Springer, New York, 1999.

[34] S. Richardson. Spatial models in epidemiological applications. In P.J. Green, N.L. Hjort, and S. Richardson, editors, *Highly Structured Stochastic Systems*, 237–259. Oxford Statistical Science Series, Oxford, 2003.

[35] Y. Fong, H. Rue, and J.C. Wakefield. Bayesian inference for generalized linear mixed models. *Biostatistics*, 11:397–412, 2010.

[36] H. Rue, S. Martino, and N. Chopin. Approximate Bayesian inference for latent Gaussian models using integrated nested Laplace approximations (with discussion). *Journal of the Royal Statistical Society: Series B*, 71:319–392, 2009.

[37] J. Hughes and M. Haran. Dimension reduction and alleviation of confounding for spatial generalized linear mixed models. *Journal of the Royal Statistical Society, Series B*, 75:131–159, 2013.

[38] B.J. Reich, J.S. Hodges, and V. Zadnik. Effects of residual smoothing on the posterior of the fixed effects in disease-mapping models. *Biometrics*, 62:1197–1206, 2006.

[39] F. Dominici, L. Sheppard, and M. Clyde. Health effects of air pollution: a statistical review. *International Statistical Review*, 71:243–276, 2003.

[40] D.A. Freedman, S.P. Klein, M. Ostland, and M.R. Roberts. A solution to the ecological inference problem (book review). *Journal of the American Statistical Association*, 93:1518–1522, 1998.

[41] H. Rue and L. Held. *Gaussian Markov Random Fields: Theory and Application.* Chapman & Hall/CRC Press, Boca Raton, FL, 2005.

[42] A. Glynn, J. Wakefield, M. Handcock, and T. Richardson. Alleviating linear ecological bias and optimal design with subsample data. *Journal of the Royal Statistical Society, Series A*, 171:179–202, 2008.

[43] T.E. Raghunathan, P.K. Diehr, and A.D. Cheadle. Combining aggregate and individual level data to estimate an individual level correlation coefficient. *Journal of Educational and Behavioral Statistics*, 28:1–19, 2003.

[44] D.G. Steele, E.J. Beh, and R.L. Chambers. The information in aggregate data. In G. King, O. Rosen, and M. Tanner, editors, *Ecological Inference: New Methodological Strategies*, 51–68. Cambridge University Press, Cambridge, 2004.

[45] R.L. Prentice and L. Sheppard. Aggregate data studies of disease risk factors. *Biometrika*, 82:113–25, 1995.

[46] L. Sheppard, R.L. Prentice, and M.A. Rossing. Design considerations for estimation of exposure effects on disease risk, using aggregate data studies. *Statistics in Medicine*, 15:1849–1858, 1996.

[47] L. Sheppard and R.L. Prentice. On the reliability and precision of within- and between-population estimates of relative rate parameters. *Biometrics*, 51:853–863, 1995.

[48] R. Salway and J. Wakefield. A hybrid model for reducing ecological bias. *Biostatistics*, 9:1–17, 2008.

[49] N. Best, S. Cockings, J. Bennett, J. Wakefield, and P. Elliott. Ecological regression analysis of environmental benzene exposure and childhood leukaemia: sensitivity to data inaccuracies, geographical scale and ecological bias. *Journal of the Royal Statistical Society, Series A*, 164:155–174, 2001.

[50] C. Jackson, N. Best, and S. Richardson. Hierarchical related regression for combining aggregate and individual data in studies of socio-economic disease risk factors. *Journal of the Royal Statistical Society, Series A*, 171:159–178, 2008.

[51] C.H. Jackson, N.G. Best, and S. Richardson. Improving ecological inference using individual-level data. *Statistics in Medicine*, 25:2136–2159, 2006.

[52] J. Wakefield and G. Shaddick. Health-exposure modelling and the ecological fallacy. *Biostatistics*, 7:438–455, 2006.

[53] S. Haneuse and J. Wakefied. Hierarchical models for combining ecological and case-control data. *Biometrics*, 63:128–136, 2007.

[54] S. Haneuse and J. Wakefield. The combination of ecological and case-control data. *Journal of the Royal Statistical Society, Series B*, 70:73–93, 2008.

[55] S. Haneuse and J. Wakefield. Geographic-based ecological correlation studies using supplemental case-control data. *Statistics in Medicine*, 27:864–887, 2008.

[56] N.E. Breslow and K.C. Cain. Logistic regression for two-stage case-control data. *Biometrika*, 75:11–20, 1988.

[57] N.E. Breslow and N. Chatterjee. Design and analysis of two-phase studies with binary outcome applied to Wilms tumour prognosis. *Applied Statistics*, 48:457–468, 1999.

[58] A.M. Walker. Anamorphic analysis: sampling and estimation for covariate effects when both exposure and disease are known. *Biometrics*, 38:1025–1032, 1982.

[59] J.E. White. A two stage design for the study of the relationship between a rare exposure and a rare disease. *American Journal of Epidemiology*, 115:119–128, 1982.

[60] J. Wakefield and S. Haneuse. Overcoming eological bias using the two-phase study design. *American Journal of Epidemiology*, 167:908–916, 2008.

[61] M.E. Ross and J.C. Wakefield. Bayesian inference for two-phase studies with categorical covariates. *Biometrics*, 69:469–477, 2013.

[62] M.E. Ross and J.C. Wakefield. Bayesian hierarchical models for smoothing in two-phase studies, with application to small area estimation. *Journal of the Royal Statistical Society, Series A*, 178:1009–23, 2015.

[63] D. Dockery, C.A. Pope III, X. Xiping, J. Spengler, J. Ware, M. Fay, B. Ferris, and F. Speizer. An association between air pollution and mortality in six U.S. cities. *New England Journal of Medicine*, 329:1753–1759, 1993.

[64] J. Wakefield and H. Lyons. Spatial aggregation and the ecological fallacy. In A. Gelfand, P. Diggle, P. Guttorp, and M. Fuentes, editors, *Handbook of Spatial Statistics*, 541–558. CRC Press, Boca Raton, FL, 2010.

[65] H. Brenner, D. Savitz, K.-H. Jockel, and S. Greenland. Effects of non-differential exposure misclassification in ecologic studies. *American Journal of Epidemiology*, 135:85–95, 1992.

[66] P. Elliott and J.C. Wakefield. Small-area studies of environment and health. In V. Barnett, A. Stein, and K.F. Turkman, editors, *Statistics for the Environment 4: Health and the Environment*, 3–27. John Wiley, New York, 1999.

[67] J. Wakefield and P. Elliott. Issues in the statistical analysis of small area health data. *Statistics in Medicine*, 18:2377–2399, 1999.

[68] J.C. Wakefield. A critique of statistical aspects of ecological studies in spatial epidemiology. *Environmental and Ecological Statistics*, 11:31–54, 2004.

Part II

Basic Methods

6

Case Event and Count Data Modeling

Andrew B. Lawson
Division of Biostatistics and Bioinformatics
Department of Public Health Sciences
Medical University of South Carolina
Charleston, South Carolina

CONTENTS

6.1 Introduction

In this chapter, the basic forms of data arising in disease mapping studies are considered and common models for the variation in disease risk are described. The chapter is broken into consideration of case event data that arises when residential addresses of cases are available, and count data that is usually found as an aggregation of case events into small areas and fixed time periods. With case event data, the data takes the form of geographic coordinates, such as latitude and longitude, of a residence. Often the (x, y) residence coordinate may not be available due to medical confidentiality restrictions, however. In that case, aggregated count data is often the only available form. For some studies, in particular those dealing with putative health hazards (see Chapter 14), a fine spatial resolution is the focus, and so address locations may be at the resolution level essential for the study to proceed. In what follows, I examine common likelihood models for both case events and count data and also consider how these can be simply extended to include extra heterogeneity in disease risk.

6.1.1 Case events: the Poisson process model

Define a study area as T and within that area m events of disease occur. Assume also a fixed time window within which the cases are observed (e.g., days, weeks, and years).

These events are usually address locations of the cases. The case could be an incident or prevalent case or could be a death certificate address. We assume at present that the cases are geocoded down to a point (with respect to the scale of the total study region). Hence, they form a point process in space. Define $\{s_i\}, i = 1, \ldots, m$, as the set of all cases within T. This is called a *realization* of the disease process, in that we assume that all cases within the study area are recorded. This is a common form of data available from government agencies. Subsamples of the spatial domain, where incomplete realizations are taken, are not considered at this point.

The basic point process model assumed for such data within disease mapping is the heterogeneous Poisson process with first-order intensity $\lambda(s)$. The basic assumptions of this model are that points (case events) are independently spatially distributed and governed by the first-order intensity. Due to the independence assumption, we can derive a likelihood for a realization of a set of events within a spatial region. For the study region defined above, the unconditional likelihood of m events is just

$$L(\{\mathbf{s}\}|\psi) = \frac{1}{m!}\prod_{i=1}^{m}\lambda(s_i|\psi)\exp\{-\Lambda_T\}, \tag{6.1}$$

where

$$\Lambda_T = \int_T \lambda(u|\psi)du.$$

The function Λ_T is the integral of the intensity over the study region, ψ is a parameter vector, and $\lambda(s_i|\psi)$ is the first-order intensity evaluated at the case event location s_i. Denote this likelihood as $PP[\{\mathbf{s}\}|\psi]$. This likelihood can be maximized with respect to the parameters in ψ and likelihood-based inference could be pursued. The only difficulty in its evaluation is the estimation of the spatial integral. However, a variety of approaches can be used for numerical integration of this function, and with suitable weighting schemes, this likelihood can be evaluated even with conventional linear modeling functions within software packages (such as glm in R) (see, e.g., Berman and Turner, 1992; Lawson, 2006, appendix C). An example of such a weighted log-likelihood approximation is

$$l(\{\mathbf{s}\}|\psi) = \sum_{i=1}^{m}\ln\lambda(s_i|\psi) - \Lambda_T, \tag{6.2}$$

where $\Lambda_T \approx \sum_{i=1}^{m} w_i\lambda(s_i|\psi)$ and w_i is an integration weight. This scheme per se is not accurate and more weights are needed. In the more general scheme of Berman and Turner, a set of additional mesh points (of size m_{aug}) are added to the data. The augmented set ($N = m + m_{\text{aug}}$) is used in the likelihood with a indicator function, I_k:

$$l(\{\mathbf{s}\}|\psi) = \sum_{k=1}^{N} w_k\left\{\frac{I_k}{w_k}\ln\lambda(s_k|\psi) - \lambda(s_k|\psi)\right\},$$

where

$$\int_T \lambda(u|\psi)du = \sum_{k=1}^{N} w_k\lambda(s_k|\psi).$$

This has the form of a weighted Poisson likelihood, with $I_k = 1$ for a case and 0 otherwise. Diggle (1990) gives an example of the use of a likelihood such as (6.1) in a spatial health data problem.

In disease mapping applications, it is usual to parameterize $\lambda(s|\psi)$ as a function of two components. The first component makes allowance for the underlying population in the study region, and the second component is usually specified with the modeled components (i.e., those components describing the excess risk within the study area).

A typical specification would be

$$\lambda(s|\psi) = \lambda_0(s|\psi_0)\lambda_1(s|\psi_1). \tag{6.3}$$

Here, it is assumed that $\lambda_0(s|\psi_0)$ is a spatially varying function of the population at risk of the disease in question. It is parameterized by ψ_0. The second function, $\lambda_1(s|\psi_1)$, is parameterized by ψ_1 and includes any linear or nonlinear predictors involving covariates or other descriptive modeling terms thought appropriate in the application. Often we assume, for positivity, that $\lambda_1(s_i|\psi_1) = \exp\{\eta_i\}$, where η_i is a parameterized linear predictor allowing a link to covariates measured at the individual level. The covariates could include spatially referenced functions as well as case-specific measures. Note that $\psi : \{\psi_0, \psi_1\}$. The function $\lambda_0(s|\psi_0)$ is a nuisance function, which must be allowed for, but which is not usually of interest from a modeling perspective.

6.1.2 Conditional logistic model

When a bivariate realization of cases and controls is available, it is possible to make conditional inference on this joint realization. Define the case events as $s_i : i = 1, \ldots, m$ and the control events as $s_i : i = m+1, \ldots, N$, where $N = m+n$ is the total number of events. Associated with each location is a binary variable (y_i) that labels the event as either a case $(y_i = 1)$ or a control $(y_i = 0)$. Assume also that the point process models governing each event type (case or control) are a heterogeneous Poisson process with intensity $\lambda(s|\psi)$ for cases and $\lambda_0(s|\psi_0)$ for controls. The superposition of the two processes is also a heterogeneous Poisson process with intensity $\lambda_0(s|\psi_0) + \lambda(s|\psi) = \lambda_0(s|\psi_0)[1 + \lambda_1(s|\psi_1)]$. Conditioning on the joint realization of these processes, then it is straightforward to derive the conditional probability of a case at any location as

$$\begin{aligned} \Pr(y_i = 1) &= \frac{\lambda_0(s_i|\psi_0).\lambda_1(s_i|\psi_1)}{\lambda_0(s_i|\psi_0)[1 + \lambda_1(s_i|\psi_1)]} \\ &= \frac{\lambda_1(s_i|\psi_1)}{1 + \lambda_1(s_i|\psi_1)} = p_i \end{aligned} \tag{6.4}$$

and

$$\Pr(y_i = 0) = \frac{1}{1 + \lambda_1(s_i|\psi_1)} = 1 - p_i. \tag{6.5}$$

The important implication of this result is that the background nuisance function $\lambda_0(s_i|\psi_0)$ drops out of the formulation, and further, this formulation leads to a standard logistic regression if a linear predictor is assumed within $\lambda_1(s_i|\psi_1)$. For example, a log-linear formulation for $\lambda_1(s_i|\psi_1)$ leads to a logit link to p_i, that is,

$$p_i = \frac{\exp(\eta_i)}{1 + \exp(\eta_i)},$$

where $\eta_i = x_i'\beta$ and x_i' is the ith row of the design matrix of covariates and β is the corresponding p-length parameter vector. Note that slightly different formulations can lead to nonstandard forms (see, e.g., Diggle and Rowlingson, 1994). In some applications, nonlinear links to certain covariates may be appropriate.

Further, if the probability model in (6.4) applies, then the likelihood of the realization of cases and controls is simply

$$L(\psi_1|\mathbf{s}) = \prod_{i \in \text{cases}} p_i \prod_{i \in \text{controls}} 1 - p_i$$
$$= \prod_{i=1}^{N} \left[\frac{\{\exp(\eta_i)\}^{y_i}}{1 + \exp(\eta_i)} \right]. \tag{6.6}$$

Hence, in this case, the analysis reduces to that of a logistic likelihood, and this has the advantage that the at-risk population nuisance function does not have to be estimated. This model is ideally suited to situations where it is natural to have a control and case realization, where conditioning on the spatial pattern is reasonable. A complete review of case event models in this context is found in Lawson (2012).

6.1.3 Binomial model for count data

In the case where we examine arbitrary small areas (such as census tracts, counties, postal zones, municipalities, and health districts), usually a count of disease is observed within each spatial unit. Define this count as y_i and assume that there are m small areas. We also consider that there is a finite population within each small area out of which the count of disease has arisen. Denote this as n_i $\forall_i$. In this situation, we can consider a binomial model for the count data conditional on the observed population in the areas. Hence, we can assume that given the probability of a case is p_i, then y_i is distributed independently as

$$y_i \sim \text{bin}(p_i, n_i),$$

and that the likelihood is given by

$$L(y_i|p_i, n_i) = \prod_{i=1}^{m} \binom{n_i}{y_i} p_i^{y_i} (1 - p_i)^{(n_i - y_i)}. \tag{6.7}$$

It is usual for a suitable link function for the probability p_i to a linear predictor to be chosen. The most common would be a logit link so that

$$p_i = \frac{\exp(\eta_i)}{1 + \exp(\eta_i)}.$$

Here, we envisage the model specification within η_i to include spatial and nonspatial components. Two applications that are well suited to this approach are the analysis of sex ratios of births and the analysis of birth outcomes (e.g., birth abnormalities) compared to total births. Sex ratios are often derived from the number of female (or male) births compared to the total birth population count in an area. A ratio is often formed, though this is not necessary in our modeling context. In this case, the p_i will often be close to 0.5 and spatially localized deviations in p_i may suggest adverse environmental risk (Williams et al., 1992). The count of abnormal births (these could include any abnormality found at birth) can be related to total births in an area. Variations in abnormal birth count could relate to environmental as well as health service variability (over time and space) (Morgan et al., 2004).

6.1.4 Poisson model for count data

Perhaps the most commonly encountered model for small-area count data is the Poisson model. This model is appropriate when there is a relatively low count of disease and the population is relatively large in each small area. Often the disease count y_i is assumed to have a mean μ_i and is independently distributed as

$$y_i \sim \text{Poisson}(\mu_i). \tag{6.8}$$

The likelihood is given by

$$L(\mathbf{y}|\boldsymbol{\mu}) = \prod_{i=1}^{m} \mu_i^{y_i} \exp(-\mu_i)/y_i!. \tag{6.9}$$

The mean function is usually considered to consist of two components: (1) a component representing the background population effect and (2) a component representing the excess risk within an area. This second component is often termed the *relative risk*. The first component is commonly estimated or computed by comparison to rates of the disease in a standard population and a local expected rate is obtained. This is often termed *standardization* (Inskip et al., 1983). Hence, we would usually assume that the data is independently distributed with expectation

$$E(y_i) = \mu_i = e_i\theta_i,$$

where e_i is the expected rate for the ith area and θ_i is the relative risk for the ith area. As we will be developing forms of hierarchical models, we will consider $\{y_i\}$ to be conditionally independent given knowledge of $\{\theta_i\}$. The expected rate is usually assumed to be fixed for the time period considered in the spatial example, although there is a literature on the estimation of small area rates that suggests this may be naive (Ghosh and Rao, 1994; Rao, 2003).

Usually the focus of interest will be the modeling of the relative risk. The most common approach to this is to assume a logarithmic link to a linear predictor model:

$$\log\theta_i = \eta_i.$$

This form of model has seen widespread use in the analysis of small-area count data in a range of applications (see, e.g., Stevenson et al., 2005; Waller and Gotway, 2004, chapter 9).

6.2 Specification of the Predictor in Case Event and Count Models

In all the above models, a predictor function (η_i) was specified to relate to the mean of the random outcome variable, via a suitable link function. Often the predictor function is assumed to be linear and a function of fixed covariates and also possibly random effects. We define this in a general form, for p covariates, here as

$$\begin{aligned}\eta_i &= x_i'\boldsymbol{\beta} + z_i'\boldsymbol{\xi},\\ &= \beta_1 x_{1i}\ldots\beta_p x_{pi} + \xi_1 z_{1i}\ldots\xi_q z_{qi},\end{aligned}$$

where x_i' is the ith row of a covariate design matrix $\mathbf{x}$ of dimension $m \times p$, $\boldsymbol{\beta}$ is a $(p \times 1)$ vector of regression parameters, $\boldsymbol{\xi}$ is a $(q \times 1)$ unit vector, and z_i' is a row vector of individual-level random effects, of which there are q. In this formulation, the unknown parameters are $\boldsymbol{\beta}$ and $\mathbf{z}$ is the $(m \times q)$ matrix of random effects for each unit. Note that in any given application, it is possible to specify subsets of these covariates or random effects. Covariates for case event data could include different types of specific level measures, such as an individual's age, gender, smoking status, or health provider, or could be environmental covariates that may have been interpolated to the address location of the individual (such as soil chemical measures or air pollution levels). For count data in small areas, it is likely that covariates will be obtained at the small-area level. For example, for census tracts, there is likely to be socioeconomic variables, such as poverty (percent population below an income level), car ownership, and median income level, available from the census. In addition, some variates could be included as supra-area variables, such as health district in which the tract lies. Environmental covariates could also be interpolated to be used at the census tract level. For example, air pollution measures could be averaged over the tract.

In some special applications, nonlinear link functions are used, and in others, mixtures of link functions are used. One special application area where this is found is the analysis of putative hazards (see Chapter 14), where specific distance- or direction-based covariates are used to assess evidence for a relation between disease risk and a fixed (putative) source of health hazard. One simple example of this is the conditional logistic modeling of disease cases around a fixed source. Let distance and direction from the source to the ith location be d_i and ϕ_i, respectively; then a mixed linear and nonlinear link model is commonly assumed, where

$$\eta_i = \{1 + \beta_1 \exp(-\beta_2 d_i)\}.\exp\{\beta_0 + \beta_3 \cos(\phi_i) + \beta_4 \sin(\phi_i)\}.$$

Here, the distance effect link is nonlinear, while the overall rate (β_0) and directional components are log-linear.

Fixed covariate models can be used to make simple descriptions of disease variation. In particular, it is possible to use the spatial coordinates of case events (or in the case of count data, centroids of small areas) as covariates. These can be used to model the long-range variation of risk: *spatial trend*. For example, let us assume that the ith unit x–y coordinates are (x_{si}, y_{si}). We could define a polynomial trend model such as

$$\eta_i = \beta_0 + \sum_{l=1}^{L} \beta_{xl} x_{si}^l + \sum_{l=1}^{L} \beta_{yl} y_{si}^l + \sum_{k=1}^{K} \sum_{l=1}^{L} \beta_{lk} x_{si}^l y_{si}^k.$$

This form of model can describe a range of smoothly varying nonlinear surface forms. However, except for very simple models, these forms are not parsimonious and also cannot capture the extra random variation that often exists in disease incidence data.

6.2.1 Bayesian linear model

In the Bayesian paradigm, all parameters are stochastic and are therefore assumed to have prior distributions. Hence, in the covariate model,

$$\eta_i = x_i' \boldsymbol{\beta},$$

the β parameters are assumed to have prior distributions. Hence, this can be formulated as

$$P(\boldsymbol{\beta}, \boldsymbol{\tau}_\beta | \text{data}) \propto L(\text{data} | \boldsymbol{\beta}, \boldsymbol{\tau}_\beta) f(\boldsymbol{\beta} | \boldsymbol{\tau}_\beta),$$

where $f(\boldsymbol{\beta}|\boldsymbol{\tau}_\beta)$ is the joint distribution of the covariate parameters conditional on the hyperparameter vector $\boldsymbol{\tau}_\beta$. Often, we regard these parameters as independent and so

$$f(\boldsymbol{\beta}|\boldsymbol{\tau}_\beta) = \prod_{j=1}^{p} f_j(\beta_j | \tau_{\beta_j}).$$

More generally, it is commonly assumed that the covariate parameters can be described by a Gaussian distribution, and if the parameters are allowed to be correlated, then we could have the multivariate Gaussian specification

$$f(\boldsymbol{\beta}|\boldsymbol{\tau}_{\beta}) = \mathbf{N}_p(\mathbf{0}, \boldsymbol{\Sigma}_{\beta}),$$

where under this prior assumption, $E(\boldsymbol{\beta}|\boldsymbol{\tau}_{\beta}) = \mathbf{0}$ and $\boldsymbol{\Sigma}_{\beta}$ is the conditional covariance of the parameters. The most common specification assumes prior independence and is

$$f(\boldsymbol{\beta}|\boldsymbol{\tau}_{\beta}) = \prod_{j=1}^{p} N(0, \tau_{\beta_j}),$$

where $N(0, \tau_{\beta_j})$ is a zero-mean single-variable Gaussian distribution with variance τ_{β_j}. At this point, an assumption about variation in the hyperparameters is usually made. At the next level of the hierarchy, hyperprior distributions are assumed for $\boldsymbol{\tau}_{\beta}$. The definition of these distributions could be important in defining the model behavior. For example, if a vague hyperprior is assumed for τ_{β_j}, this may lead to extra variation being present when limited learning is available from the data. This can affect computation of the deviance information criterion (DIC) and convergence diagnostics. While uniform hyperpriors (on a large positive range) can lead to improper posterior distributions, it has been found that a uniform distribution for the standard deviation can be useful (Gelman, 2006), that is, $\sqrt{\tau_{\beta_j}} \sim U(0, A)$, where A has a large positive value.

Alternative suggestions are usually in the form of gamma or inverse gamma distributions with large variances. For example, Kelsall and Wakefield (2002) proposed the use of gamma(0.2, 0.0001) with expectation 2000 and variance 20,000,000, whereas Banerjee et al. (2004) examine various alternative specifications, including gamma (0.001, 0.001). One common specification (Thomas et al., 2004) is gamma (0.5, 0.0005), which has expectation 1000 and variance 2,000,000.

While these prior specifications lead to relative uninformativeness, their use has been criticized by Gelman (2006) (see also Lambert et al., 2005), in favor of uniform prior distributions on the standard deviation.

6.3 Simple Case and Count Data Models with Uncorrelated Random Effects

In the previous section, some simple models were developed. These consisted of functions of fixed observed covariates. In a Bayesian model formulation, all parameters are stochastic, and so the extension to the addition of random effects is relatively straightforward. In fact, the term *mixed model* (linear mixed model [LMM], normal linear mixed model [NLMM], or generalized linear mixed model [GLMM]) is strictly inappropriate, as there are no fixed effects in a Bayesian model.

The simple regression models described above often do not capture the extent of variation present in count data. Overdispersion or spatial correlation due to unobserved confounders will usually not be captured by simple covariate models, and often it is appropriate to include some additional term or terms in a model that can capture such effects.

Initially, overdispersion or extravariation can be accommodated by either (1) inclusion of a prior distribution for the relative risk (such as a Poisson-gamma model) or (2) extension of the linear or nonlinear predictor term to include an extra random effect (lognormal model).

6.3.1 Gamma and beta models

6.3.1.1 Gamma models

The simplest extension to the likelihood model that accommodates extra variation is one in which the parameter of interest in the likelihood is given a prior distribution. One case event example would be where the intensity is specified at the ith location as $\lambda_0(s_i|\psi_0)\lambda_1(s|\psi_1) \equiv \lambda_{0i}\lambda_{1i}$, suppressing the parameter dependence, for simplicity. Note that λ_{1i} plays the role of a relative risk parameter. This parameter can be assigned a prior distribution, such as a gamma distribution to model extravariation. For most applications where count data are commonly found, a Poisson likelihood is assumed. We will focus on these models in the remainder of this section. The Poisson parameter θ_i could be assigned a gamma(a, b) prior distribution. In this case, the prior expectation and variance would be, respectively, a/b and a/b^2. This could allow for extra variation or overdispersion. Here we assume that parameters a and b are fixed and known. This formulation is attractive, as it leads to a closed form for the posterior distribution of $\{\theta_i\}$, that is,

$$[\theta_i|y_i, e_i, a, b] \sim \text{gamma}(a^*, b^*), \tag{6.10}$$

where

$$a^* = y_i + a, \; b^* = e_i + b.$$

Hence, conjugacy leads to a gamma posterior distribution with posterior mean and variance given by $\frac{y_i+a}{e_i+b}$ and $\frac{y_i+a}{(e_i+b)^2}$, respectively. Note that a variety of θ_i estimates are found depending on the values of a and b. (Lawson and Williams, 2001, pp. 78–79) demonstrate the effect of different values of a and b on relative risk maps. Samples from this posterior distribution are straightforwardly obtained (e.g., using the rgamma function on R). The prior predictive distribution of $\mathbf{y}^*$ is also relevant in this case, as it leads to a distribution often used for overdispersed count data: the negative binomial. Here, we have the joint distribution as

$$\begin{aligned}[\mathbf{y}^*|\mathbf{y}, a, b] &= \int f(\mathbf{y}^*|\boldsymbol{\theta}) f(\boldsymbol{\theta}|a, b) d\boldsymbol{\theta} \qquad (6.11)\\ &= \prod_{i=1}^{m} \left[\frac{b^a}{\Gamma(a)} \frac{\Gamma(y_i^* + a)}{(e_i + b)^{(y_i^* + a)}} \right].\end{aligned}$$

Hyperprior distributions. One extension to the above model is to consider a set of hyperprior distributions for the parameters of the gamma prior (a and b). Often these are assumed to also have prior distributions on the positive real line, such as gamma (a', b') with $a' > 0$, $b' = 1$, or $b' > 1$.

Linear parameterization. One approach to incorporating more sophisticated model components into the relative risk model is to model the parameters of the gamma prior distribution. For example, gamma linear models can be specified, where

$$[\theta_i|y_i, e_i, a, b_i] \sim \text{gamma}(a, b_i),$$

where

$$b_i = a/\mu_i$$

and

$$\mu_i = \eta_i.$$

In this formulation, the prior expectation is μ_i and the prior variance is μ_i^2/a. While this formulation could be used for modeling, often the direct linkage between the variance and mean could be seen as a disadvantage. As will be seen later, a lognormal parameterization is often favored for such models.

6.3.1.2 Beta models

When Bernoulli or binomial likelihood models are assumed (such as (6.6) or (6.7)), one may need to consider prior distributions for the probability parameter p_i. Commonly, a beta prior distribution is assumed for this:

$$[p_i|\alpha,\beta] \sim \text{Beta}(\alpha,\beta).$$

Here, the prior expectation and variance would be $\frac{\alpha}{\alpha+\beta}$ and $\frac{\alpha\beta}{(\alpha+\beta)^2(\alpha+\beta+1)}$. This distribution can flexibly specify a range of forms of distribution from peaked ($\alpha = \beta, \beta > 1$) to uniform ($\alpha = \beta = 1$) and U-shaped ($\alpha = \beta = 0.5$) to skewed or either monotonically decreasing or increasing. In the case of the binomial distribution, this prior distribution, with α, β fixed, leads to a beta posterior distribution, that is,

$$\begin{aligned}
[\mathbf{p}|\mathbf{y},\mathbf{n},\alpha,\beta] &= B(\alpha,\beta)^{-m}\prod_{i=1}^{m}\left[\binom{n_i}{y_i}p_i^{y_i}(1-p_i)^{(n_i-y_i)}.p_i^{\alpha-1}(1-p_i)^{\beta-1}\right] \\
&= B(\alpha,\beta)^{-m}\prod_{i=1}^{m}\left[\binom{n_i}{y_i}p_i^{y_i+\alpha-1}(1-p_{i'})^{(n_i-y_i+\beta-1)}\right].
\end{aligned}$$

This is the product of m independent beta distributions with parameters $y_i+\alpha$, $n_i - y_i + \beta$. Hence, the beta posterior distribution for p_i has expectation $\frac{y_i+\alpha}{n_i+\beta+\alpha}$ and variance $\frac{(y_i+\alpha)(n_i-y_i+\beta)}{(n_i+\beta+\alpha)^2(n_i+\beta+\alpha+1)}$.

Hyperprior distributions. The parameters α and β are strictly positive, and these could also have hyperprior distributions. However, unless these parameters are restricted to the unit interval, then distributions such as the gamma, exponential, or inverse gamma or inverse exponential would have to be assumed as hyperprior distributions.

Linear parameterization. An alternative specification for modeling covariate effects is to specify a linear or nonlinear predictor with a link to a parameter or parameters. For example, it is possible to consider a parameterization such as $\alpha_i = \exp(\eta_i)$ and $\beta_i = \psi\alpha_i$, where ψ is a linkage parameter with prior mean given by $\frac{1}{1+\psi}$. When $\psi = 1$, then the distribution is symmetric. The disadvantage with this formulation is that a single parameter is assigned to the linear predictor and a dependence is specified between α_i and β_i. One possible alternative is to model the prior mean as $\text{logit}(\frac{\alpha_i}{\alpha_i+\beta_i}) = \eta_i$. However, this also forces a dependence between α_i and β_i.

6.3.2 Lognormal or logistic-normal models

One simple device that is very popular in disease mapping applications is to assume a direct linkage between a linear or nonlinear predictor (η_i) and the parameter of interest (such as θ_i or p_i). This offers a convenient method of introducing a range of covariate effects and unobserved random effects within a simple formulation. The general structure of this formulation is $\eta_i = x_i'\beta + z_i'\gamma$. The simplest form involving uncorrelated heterogeneity would be

$$\eta_i = z_{1i},$$

where z_{1i} is an uncorrelated random effect.

References

Banerjee, S., B. P. Carlin, and A. E. Gelfand. (2004). *Hierarchical Modeling and Analysis for Spatial Data.* London: Chapman & Hall/CRC Press.

Berman, M. and T. R. Turner. (1992). Approximating point process likelihoods with GLIM. *Applied Statistics* 41, 31–38.

Diggle, P. J. (1990). A point process modelling approach to raised incidence of a rare phenomenon in the vicinity of a prespecified point. *Journal of the Royal Statistical Society A* 153, 349–362.

Diggle, P. and B. Rowlingson. (1994). A conditional approach to point process modelling of elevated risk. *Journal of the Royal Statistical Society A* 157, 433–440.

Gelman, A. (2006). Prior distributions for variance parameters in hierarchical models. *Bayesian Analysis* 1, 515–533.

Ghosh, M. and J. N. K. Rao. (1994). Small area estimation: an appraisal. *Statistical Science* 9, 55–93.

Inskip, H., V. Beral, P. Fraser, and P. Haskey. (1983). Methods for age-adjustment of rates. *Statistics in Medicine* 2, 483–493.

Kelsall, J. and J. Wakefield. (2002). Modelling spatial variation in disease risk: a geostatistical approach. *Journal of the American Statistical Association* 97, 692–701.

Lambert, P., A. Sutton, P. Burton, K. Abrams, and P. Jones. (2005). How vague is vague? A simulation study of the impact of the use of vague prior distributions in MCMC using Winbugs. *Statistics in Medicine* 24, 2401–2428.

Lawson, A. B. (2006). *Statistical Methods in Spatial Epidemiology* (2nd ed.). New York: Wiley.

Lawson, A. B. (2012). Bayesian point event modeling in spatial and environmental epidemiology: a review. *Statistical Methods in Medical Research* 21, 509–529.

Lawson, A. B. and F. L. R. Williams. (2001). *An Introductory Guide to Disease Mapping.* New York: Wiley.

Morgan, O., M. Vreiheid, and H. Dolk. (2004). Risk of low birth weight near EUROHAZCON hazardous waste landfill sites in England. *Archives of Environmental Health* 59, 149–151.

Rao, J. K. N. (2003). *Small Area Estimation.* New York: Wiley.

Stevenson, M., R. Morris, A. B. Lawson, J. Wilesmith, J. M. Ryan, and R. Jackson. (2005). Area level risks for BSE in British cattle before and after the July 1988 meat and bone meal feed ban. *Preventive Veterinary Medicine* 69, 129–144.

Thomas, A., N. Best, D. Lunn, R. Arnold, and D. Spiegelhalter. (2004). Geobugs user manual v 1.2. www.mrc-bsu.cam.ac.uk/bugs.

Waller, L. and C. Gotway. (2004). *Applied Spatial Statistics for Public Health Data.* New York: Wiley.

Williams, F., A. Lawson, and O. Lloyd. (1992). Low sex ratios of births in areas at risk from air pollution from incinerators, as shown by geographical analysis and 3-dimensional mapping. *International Journal of Epidemiology 21*, 311–319.

7

Bayesian Modeling and Inference

Georgiana Onicescu
Department of Statistics
Western Michigan University
Kalamazoo, Michigan

Andrew B. Lawson
Division of Biostatistics and Bioinformatics
Department of Public Health Sciences
Medical University of South Carolina
Charleston, South Carolina

CONTENTS

The development of Bayesian inference has as its kernel the data likelihood. The likelihood is the joint distribution of the data evaluated at the sample values. It can also be regarded as a function describing the dependence of a parameter or parameters on sample values. Hence, there can be two interpretations of this function. In Bayesian inference, it is this latter interpretation that is of prime importance. In fact, the *likelihood principle*, by which observations come into play through the likelihood function, and only through the likelihood function, is a fundamental part of the Bayesian paradigm (Bernardo and Smith, 1994, section 5.1.4). This implies that the information content of the data is entirely expressed by the likelihood function. Furthermore, the likelihood principle implies that any event that did not happen has no effect on an inference, since if an unrealized event does affect an inference, then there is some information not contained in the likelihood function.

7.1 Likelihood Models

The likelihood for data $\{y_i\}, i = 1, \ldots, m$, is defined as

$$L(\mathbf{y}|\boldsymbol{\theta}) = \prod_{i=1}^{m} f(y_i|\boldsymbol{\theta}), \tag{7.1}$$

where $\boldsymbol{\theta}$ is a p length vector $\boldsymbol{\theta} : \{\theta_1, \theta_2, \ldots, \theta_p\}$ and $f(.|.)$ is a probability density (or mass) function. The assumption made here is that the "sample" values of $\mathbf{y}$ given the parameters are independent, and hence it is possible to take the product of individual contributions in (7.1). Hence, the data are assumed to be conditionally independent. Note that in many spatial applications, the data would not be unconditionally independent and would in fact be correlated. This conditional independence is an important assumption fundamental to many disease mapping applications. The logarithm of the likelihood is also useful in model development and is defined as

$$l(\mathbf{y}|\boldsymbol{\theta}) = \sum_{i=1}^{m} \log f(y_i|\boldsymbol{\theta}). \tag{7.2}$$

7.1.1 Spatial correlation

Within spatial applications it is often found that correlation will exist between spatial units. This correlation is geographical and relates to the basic idea that locations close together in space often have similar values of outcome variables, while locations far apart are often different. This spatial correlation (or autocorrelation as it is sometimes called) must be allowed for in spatial analyses. This may have an impact on the structure and form of likelihood models that are assumed for spatial data. The assumption made in the construction of conventional likelihoods is that the individual contribution to the likelihood is independent, and this independence allows the likelihood to be derived as a product of probabilities. However, if this independence criterion is not met, then a different approach would be required.

7.1.1.1 Conditional independence

In some circumstances, it is possible to consider *conditional* independence of the data given parameters at a higher level of the hierarchy. For instance, in count data examples, y_i from the ith area might be thought to be independent of other outcomes given knowledge of the model parameters. In the simple case of dependence on a parameter vector $\boldsymbol{\theta}$, conditioning on the parameters can allow $[\, y_i|\boldsymbol{\theta}]$ to be assumed to be an independent contribution. This simply states that dependence only exists unconditionally (i.e., unobserved effects can induce dependence). This is often true in disease mapping examples where confounders that have spatial expression may or may not be measured in a study and their exclusion may leave residual correlation in the data. Note that this approach to correlation does not completely account for spatial effects, as there can be residual correlation effects after inclusion of confounders. These effects could be due to unobserved or unknown confounders. Alternatively, they could be due to intrinsic correlation in the process. Hence, the assumption of conditional independence may only be valid if correlation is accounted for somewhere within the model.

The idea of inclusion of spatial correlation at a hierarchical level *above* the likelihood is a fundamental assumption often made in Bayesian small-area health modeling. This means that the correlation appears in prior distributions rather than in the likelihood itself. Often parameters are given such priors, and it is assumed that conditional independence applies in the likelihood. This is valid for many situations and will be the focus of most of this book.

7.1.1.2 Joint densities with correlation

Situations exist where spatial correlation can be incorporated within a joint distribution of the data. For example, if a continuous spatial process is observed at measurement sites (such as air pollutants, soil chemical concentration, or water quality), then often a spatial Gaussian process (SGP) will be assumed (Ripley, 1981). This process assumes that any realization of the process is multivariate normal with spatially defined covariance, within its specification. Hence, if these data were observed outcome data, then the joint density would include spatial correlation.

Alternatively, it is possible to consider discrete outcome data where correlation is explicitly modeled. The autologistic and auto-Poisson models were developed for lattice data with spatial correlation included via dependence on a spatial neighborhood (Besag and Tantrum, 2003). In this approach, the normalization of the likelihood is computationally prohibitive, and likelihood approximation is often resorted to (see Section 7.1.1.3).

7.1.1.3 Pseudolikelihood approximation

Pseudolikelihood has been proposed as an option to exact likelihood analysis when correlation exists. It has a number of variants (composite, local, and pairwise: Lindsay, 1988; Tibshirani and Hastie, 1987; Kauermann and Opsomer, 2003; Nott and Rydén, 1999; Varin et al., 2005). Pseudolikelihood has been used for autologistic models in both space and time (most recently by Besag and Tantrum, 2003). In space, the likelihood is given by

$$L_p(\mathbf{y}|\boldsymbol{\theta}) = \prod_{i=1}^{m} f(y_i|y_{j\neq i}, \boldsymbol{\theta}).$$

For the autologistic model, with binary outcome y_i, a simple version could be

$$f(y_i|y_{j\neq i}) = \frac{\exp[m(\beta, \{y_j\}_{j\in\delta_i})]}{1+\exp[m(\beta, \{y_j\}_{j\in\delta_i})]},$$

where δ_i is a neighborhood set of the ith location or area, $m(.)$ is a specified function (such as mean or median), and β is a parameter controlling the spatial smoothing or degree of correlation. For nonlattice data, the neighborhood can be defined by adjacency (for count data, this could be adjacent regions, and for case event data, this could be tesselation neighbors). It is known that pseudolikelihood is least biased when relatively low spatial correlation exists (see, e.g., Diggle et al., 1994). While the autologistic model has seen some application, the auto-Poisson model is limited by its awkward negative correlation structure. An autobinomial model is also available for the situation, where y_i is a count of disease out of a finite local population n_i (see, e.g., Cressie, 1993, p. 431).

7.2 Prior Distributions

All parameters within Bayesian models are stochastic and are assigned appropriate probability distributions. Hence, a single parameter value is simply one possible realization of the possible values of the parameter, the probability of which is defined by the prior distribution. The prior distribution is a distribution assigned to the parameter before seeing the data. Note also that one interpretation of prior distributions is that they provide additional data for a problem, and so they can be used to improve estimation or identification of parameters. For a single parameter, θ, the prior distribution can be denoted $\mathbf{g}(\theta)$, while for a parameter vector, $\boldsymbol{\theta}$, the joint prior distribution is $\mathbf{g}(\boldsymbol{\theta})$.

7.2.1 Propriety

It is possible that a prior distribution can be *improper*. Impropriety is defined as the condition that integration of the prior distribution of the random variable θ over its range (Ω) is not finite:

$$\int_{\Omega} g(\theta)d\theta = \infty.$$

A prior distribution is improper if its normalizing constant is infinite. While impropriety is a limitation of any prior distribution, it is not necessarily the case that an improper prior will lead to impropriety in the posterior distribution. The posterior distribution can often be proper even with an improper prior specification.

7.2.2 Noninformative priors

Often prior distributions are assumed that do not make strong preferences over values of the variables. These are sometimes known as *vague* or *reference* or *flat* or *noninformative* prior distributions. Usually, they have a relatively flat form yielding close-to-uniform preference for different values of the variables. This tends to mean that in any posterior analysis (see Section 7.3), the prior distributions will have little impact compared to the likelihood of the data. *Jeffrey's* priors were developed in an attempt to find such reference priors for given distributions. They are based on the Fisher information matrix. For example, for the binomial data likelihood with common parameter p, the Jeffrey's prior distribution is $p \sim Beta(0.5, 0.5)$. This is a proper prior distribution. However, it is not completely noninformative, as it has asymptotes close to 0 and 1. Jeffrey's prior for the Poisson data likelihood with common mean θ is given by $g(\theta) \propto \theta^{-\frac{1}{2}}$, which is *improper.* This also is not particularly noninformative. The Jeffrey's prior is locally uniform, but can often be improper.

Choice of noninformative priors can often be made with some general understanding of the range and behavior of the variable. For example, variance parameters must have prior distributions on the positive real line. Noninformative distributions in this range are often in the gamma, inverse gamma, or uniform families. For example, $\tau \sim G(0.001, 0.001)$ will have a small mean (1) but a very large variance (1000), and hence will be relatively flat over a large range. Another specification chosen is $\tau \sim G(0.1, 0.1)$, with variance 10 for a more restricted range. On the other hand, a uniform distribution on a large range has been advocated for the standard deviation (Gelman, 2006) : $\sqrt{\tau} \sim U(0, 1000)$. For parameters on an infinite range, such as regression parameters, a distribution centered on zero with a large variance will usually suffice. The zero-mean Gaussian or Laplace distribution could be assumed. For example,

$$\beta \sim N(0, \tau_\beta)$$
$$\tau_\beta = 100{,}000$$

is typically assumed in applications. The Laplace distribution is favored in large-scale Bayesian regression to encourage removal of covariates (Balakrishnan and Madigan, 2008).

Of course, sometimes it is important to be informative with prior distributions. Identifiability is an issue relating to the ability to distinguish between parameters within a parametric model (see, e.g., Bernardo and Smith, 1994, p. 239). In particular, if a restricted range must be assumed to allow a number of variables to be *identified,* then it may be important to specify distributions that will provide such support. Ultimately, if the likelihood has little or no information about the separation of parameters, then separation or identification can only come from prior specification. In general, if proper prior distributions are assumed for parameters, then they are identified in the posterior distribution. However, how far they are identified may depend on the assumed variability. An example of identification that arises in disease mapping is where a linear predictor is defined to have two random effect components:

$$\log \theta_i = v_i + u_i,$$

and the components have different normal prior distributions with variances (say, τ_v, τ_u). These variances can have gamma prior distributions such as

$$\tau_v \sim G(0.001, 0.001)$$
$$\tau_u \sim G(0.1, 0.1).$$

The difference in the variability of the second prior distribution allows there to be some degree of identification. Note that this means that a priori we allow greater variability in the variance of v_i than that found in u_i.

7.3 Posterior Distributions

Prior distributions and the likelihood provide two sources of information about any problem. The likelihood informs about the parameter via the data, while the prior distributions inform via prior beliefs or assumptions. When there are large amounts of data, that is, the sample size is large, the likelihood will contribute more to the relative risk estimation. When the example is data poor, then the prior distributions will dominate the analysis.

The product of the likelihood and the prior distributions is called the posterior distribution. This distribution describes the behavior of the parameters after the data are observed and prior assumptions are made. The posterior distribution is defined as

$$p(\boldsymbol{\theta}|\mathbf{y}) = L(\mathbf{y}|\boldsymbol{\theta})\mathbf{g}(\boldsymbol{\theta})/C, \tag{7.3}$$

where

$$C = \int_p L(\mathbf{y}|\boldsymbol{\theta})\mathbf{g}(\boldsymbol{\theta})d\boldsymbol{\theta},$$

where $\mathbf{g}(\boldsymbol{\theta})$ is the joint distribution of the $\boldsymbol{\theta}$ vector. Alternatively, this distribution can be specified as a proportionality: $p(\boldsymbol{\theta}|\mathbf{y}) \propto L(\mathbf{y}|\boldsymbol{\theta})\mathbf{g}(\boldsymbol{\theta})$.

A simple example of this type of model in disease mapping is where the data likelihood is Poisson and there is a common relative risk parameter with a single gamma prior distribution:

$$p(\boldsymbol{\theta}|\mathbf{y}) \propto L(\mathbf{y}|\theta)g(\theta),$$

where $g(\theta)$ is a gamma distribution with parameters α, β, that is, $G(\alpha, \beta)$, and $L(\mathbf{y}|\theta) = \prod_{i=1}^{m}\{(e_i\theta)^{y_i}\exp(-e_i\theta)\}$ up to a normalizing constant. A compact notation for this model is

$$\begin{aligned} y_i|\theta &\sim Pois(e_i\theta) \\ \theta &\sim G(\alpha, \beta). \end{aligned}$$

This leads to a posterior distribution for fixed α, β of

$$[\theta|\{y_i\}, \alpha, \beta] = L(\mathbf{y}|\theta, \alpha, \beta)p(\theta)/C,$$

where

$$C = \int L(\mathbf{y}|\theta, \alpha, \beta)p(\theta)d\theta.$$

In this case, the constant C can be calculated directly, and it leads to another gamma distribution:

$$[\theta|\mathbf{y}, \alpha, \beta] = \frac{\beta^{*\alpha^*}}{\Gamma(\alpha^*)}\theta^{\alpha^*-1}\exp(-\theta\beta^*),$$

where

$$\alpha^* = \sum y_i + \alpha, \beta^* = \sum e_i + \beta.$$

7.3.1 Prior choice

Choice of prior distributions is very important, as it can be the case that the prior distributions of parameters can affect the posterior significantly. The balance between prior and posterior evidence is related to the dominance of the likelihood and is a sample size issue. For example, with large samples, the likelihood usually dominates the prior distributions. This effectively means that current data are given priority in their weight of evidence. Prior distributions that dominate the likelihood are informative, but have less influence as sample size increases. Hence, with additional data, the data speak more. Of course, when parameters are not identified within a likelihood, additional data are unlikely to change the importance of informative priors in identification. Propriety of posterior distributions is important, as only under propriety can the absolute statements about probability of posterior parameter values be made.

7.4 Predictive Distributions

The posterior distribution summarizes our understanding about the parameters given observed data, and plays a fundamental role in Bayesian modeling. However, we can also examine other related distributions that are often useful when prediction of new data (or future data) is required. Define a new observation of y as y^*. We can determine the predictive distribution of y^* in two ways. In general, the predictive distribution is defined as

$$p(y^*|\mathbf{y}) = \int L(y^*|\boldsymbol{\theta})p(\boldsymbol{\theta}|\mathbf{y})d\boldsymbol{\theta}. \tag{7.4}$$

Here the prediction is based on marginalizing over the parameters in the likelihood of the new data ($L(y^*|\boldsymbol{\theta})$) using the posterior distribution $p(\boldsymbol{\theta}|\mathbf{y})$ to define the contribution of the observed data to the prediction. This is termed the posterior predictive distribution. A variant of this definition uses the prior distribution instead of the posterior distribution:

$$p(y^*|\mathbf{y}) = \int L(y^*|\boldsymbol{\theta})p(\boldsymbol{\theta})d\boldsymbol{\theta}. \tag{7.5}$$

This emphasizes the prediction based only on the prior distribution (before seeing any data). Note that this distribution (7.5) is just the marginal distribution of y^*.

7.4.1 Poisson-Gamma example

A classic example of a predictive distribution that arises in disease mapping is the negative binomial distribution. Let y_i, $i = 1, \ldots, n$, be counts of disease in arbitrary small areas (e.g., census tracts, zip codes, and districts). Also define, for the same areas, expected rates $\{e_i\}$ and relative risks $\{\theta_i\}$. We assume that independently, $y_i|\theta_i \sim Poisson(e_i\theta_i)$. Assume that $\theta_i = \theta \;\; \forall i$ and that the prior distribution of θ, $p(\theta)$, is $\theta \sim Gamma(\alpha, \beta)$, where $E(\theta) = \alpha/\beta$ and $var(\theta) = \alpha/\beta^2$. The posterior distribution of θ is

$$[\theta|\mathbf{y}, \alpha, \beta] = \frac{\beta^{*\alpha^*}}{\Gamma(\alpha^*)}\theta^{\alpha^*-1}\exp(-\theta\beta^*),$$

where

$$\alpha^* = \sum y_i + \alpha, \;\; \beta^* = \sum e_i + \beta.$$

It follows that the (prior) predictive distribution is

$$\begin{aligned}[\mathbf{y}^*|\mathbf{y}, a, b] &= \int f(\mathbf{y}^*|\boldsymbol{\theta})f(\boldsymbol{\theta}|a, b)d\boldsymbol{\theta} \\ &= \prod_{i=1}^{m}\left[\frac{b^a}{\Gamma(a)}\frac{\Gamma(y_i^* + a)}{(e_i + b)^{(y_i^*+a)}}\right].\end{aligned} \tag{7.6}$$

7.5 Bayesian Hierarchical Modeling

In Bayesian modeling the parameters have distributions. These distributions control the form of the parameters and are specified by the investigator based, usually, on their prior belief concerning their behavior. These distributions are prior distributions, and I will denote such a distribution by $g(\theta)$. In the disease mapping context, a commonly assumed prior distribution for θ in a Poisson likelihood model is the gamma distribution, and the resulting model is the gamma-Poisson model.

7.6 Hierarchical Models

A simple example of a hierarchical model that is commonly found in disease mapping is where the data likelihood is Poisson and there is a common relative risk parameter with a single gamma prior distribution:

$$p(\boldsymbol{\theta}|\mathbf{y}) \propto L(\mathbf{y}|\theta)g(\theta),$$

where $g(\theta)$ is a gamma distribution with parameters α, β, that is, $G(\alpha,\beta)$, and $L(\mathbf{y}|\theta) = \prod_{i=1}^{m}\{(e_i\theta)^{y_i}\exp(e_i\theta)\}$ up to a normalizing constant. A compact notation for this model is

$$\begin{aligned} y_i|\theta &\sim Pois(e_i\theta) \\ \theta &\sim G(\alpha,\beta). \end{aligned}$$

In the previous section, a simple example of a likelihood and prior distribution was given. In that example, the prior distribution for the parameter also had parameters controlling its form. These parameters (α,β) can have assumed values, but more usually, an investigator will not have a strong belief in the prior parameter values. The investigator may want to estimate these parameters from the data. Alternatively and more formally, as parameters within models are regarded as stochastic (and thereby have probability distributions governing their behavior), these parameters must also have distributions. These distributions are known as hyperprior distributions, and the parameters are known as hyperparameters.

The idea that the values of parameters could arise from distributions is a fundamental feature of Bayesian methodology and leads naturally to the use of models where parameters arise within hierarchies. In the Poisson-gamma example, there is a two-level hierarchy: θ has a $G(\alpha,\beta)$ distribution at the first level of the hierarchy and α will have a hyperprior distribution (h_α), as will $\beta(h_\beta)$, at the second level of the hierarchy. This can be written as

$$\begin{aligned} y_i|\theta &\sim Pois(e_i\theta) \\ \theta|\alpha,\beta &\sim G(\alpha,\beta) \\ \alpha|\nu &\sim h_\alpha(\nu) \\ \beta|\rho &\sim h_\beta(\rho). \end{aligned}$$

For these types of models, it is also possible to use a graphical tool to display the linkages in the hierarchy. This is known as a directed acyclic graph (DAG). On such a graph, lines connect the levels of the hierarchy and parameters are nodes at the ends of the lines. Clearly, it is important to terminate a hierarchy at an appropriate place; otherwise, one could always assume an infinite hierarchy of parameters. Usually, the cutoff point is chosen to lie where further variation in parameters will not affect the lowest-level model. At this point, the parameters are assumed to be fixed. For example, in the gamma-Poisson model, if you assume α and β are fixed, then the gamma prior would be fixed and the choice of α and β would be uninformed. The data would not inform about the distribution at all. However, by allowing a higher level of variation, that is, hyperpriors for α and β, we can fix the values of ν and ρ without heavily influencing the lower-level variation.

7.7 Posterior Inference

When a simple likelihood model is employed, often maximum likelihood is used to provide a point estimate and associated variability for parameters. This is true for simple disease-mapping models. For example, in the model $y_i|\theta \sim Pois(e_i\theta)$, the maximum likelihood estimate of θ is the overall rate for the study region, that is, $\sum y_i / \sum e_i$. On the other hand, the SMR is the maximum likelihood estimate for the model $y_i|\theta_i \sim Pois(e_i\theta_i)$.

When a Bayesian hierarchical model is employed, it is no longer possible to provide a simple point estimate for any of the $\theta_i s$. This is because the parameter is no longer assumed to be fixed, but to arise from a distribution of possible values. Given the observed data, the parameter or parameters of interest will be described by the posterior distribution, and hence this distribution must be found and examined. It is possible to examine the expected value (mean) or the mode of the posterior distribution to give a point estimate for a parameter or parameters: for example, for a single parameter θ, say, $E(\theta|\mathbf{y}) = \int \theta\, p(\theta|\mathbf{y})d\theta$, $\arg\max_{\boldsymbol{\theta}} p(\boldsymbol{\theta}|\mathbf{y})$. Just as the maximum likelihood estimate is the mode of the likelihood, the *maximum a posteriori* estimate is that value of the parameter or parameters at the mode of the posterior distribution. More commonly, the expected value of the parameter or parameters is used. This is known as the posterior mean (or Bayes estimate). For simple unimodal symmetrical distributions, the modal and mean estimates coincide.

For some simple posterior distributions, it is possible to find the exact form of the posterior distribution and find explicit forms for the posterior mean or mode. However, it is commonly the case that for reasonably realistic models within disease mapping, it is not possible to obtain a closed form for the posterior distribution. Hence, it is often not possible to derive simple estimators for parameters such as the relative risk. In this situation, resort must be made to posterior sampling, that is, using simulation methods to obtain samples from the posterior distribution, which then can be summarized to yield estimates of relevant quantities. In the next section, we discuss the use of sampling algorithms for this purpose.

An exception to this situation where a closed-form posterior distribution can be obtained is the gamma-Poisson model where α, β are fixed. In that case, the relative risks have posterior distribution given by

$$\theta_i|y_i, e_i, \alpha, \beta \sim G(y_i + \alpha, e_i + \beta),$$

and the posterior expectation of θ_i is $(y_i + \alpha)/(e_i + \beta)$. The posterior variance is also available: $(y_i + \alpha)/(e_i + \beta)^2$, as is the modal value, which is

$$\arg\max_{\theta} p(\theta|\mathbf{y}) = \begin{cases} [(y_i + \alpha) - 1]/(e_i + \beta) & \text{if } (y_i + \alpha) \geq 1 \\ 0 & \text{if } (y_i + \alpha) < 1 \end{cases}.$$

Of course, if α and β are not fixed and have hyperprior distributions, then the posterior distribution is more complex. Clayton and Kaldor (1987) use an approximation procedure to obtain estimates of α and β from a marginal likelihood apparently on the assumption that α and β had uniform hyperprior distributions.

7.7.1 Bernoulli and Binomial example

Another example of a model hierarchy that arises commonly in the small-area health data is where a finite population exists within an area and within that population binary outcomes are observed. In the case event example, define the case events as $s_i : i = 1, \ldots, m$ and the control events as $s_i : i = m+1, \ldots, N$, where $N = m + n$ is the total number of events in the study area. Associated with each location is a binary variable (y_i) that labels the event either as a case $(y_i = 1)$ or a control $(y_i = 0)$. A conditional Bernoulli model is assumed for the binary outcome, where p_i is the probability of an individual being a case, given the location of the individual. Hence, we can specify that $y_i|p_i \sim Bern(p_i)$. Here, the probability either will usually have a prior distribution associated with it or will be linked to other parameters and covariates or random effects, possibly via a linear predictor. Assume that a logistic link is appropriate for the probability and that two covariates are available for the individual: x_1, age; x_2, exposure level (of a health hazard). Hence,

$$p_i = \frac{\exp(\alpha_0 + \alpha_1 x_{1i} + \alpha_2 x_{2i})}{1 + \exp(\alpha_0 + \alpha_1 x_{1i} + \alpha_2 x_{2i})}$$

is a valid logistic model for this data with three parameters (α_0, α_1, α_2). Assume that the regression parameters will have independent zero-mean Gaussian prior distributions. The hierarchical model is specified in this case as

$$\begin{aligned} y_i|p_i &\sim Bern(p_i) \\ logit(p_i) &= \mathbf{x}_i'\boldsymbol{\alpha} \\ \alpha_j|\tau_j &\sim N(0, \tau_j) \\ \tau_j &\sim G(\psi_1, \psi_2). \end{aligned}$$

In this case, $\mathbf{x}_i'$ is the ith row of the design matrix (including an intercept term), $\boldsymbol{\alpha}$ is the (3×1) parameter vector, τ_j is the variance for the jth parameter, and ψ_1 and ψ_2 are fixed scale and shape parameters.

In the binomial case, we would have a collection of small areas within which we observe events. Define the number of small areas as m and the total population as n_i. Within the population of each area, individuals have a binary label that denotes the case status of the individual. The number of cases are denoted as y_i, and it is often assumed that the cases follow an independent binomial distribution, conditional on the probability that an individual is a case, defined as p_i: $y_i|p_i \sim Bin(p_i, n_i)$.

The likelihood is given by $L(y_i|p_i, n_i) = \prod_{i=1}^{m} \binom{n_i}{y_i} p_i^{y_i}(1-p_i)^{(n_i - y_i)}$. Here, the probability either will usually have a prior distribution associated with it or will be linked to other parameters and covariate or random effects, possibly via a linear predictor such as $logit(p_i) = \mathbf{x}_i'\boldsymbol{\alpha} + \mathbf{z}_i'\boldsymbol{\gamma}$. In this general case, the $\mathbf{z}_i'$ are a vector of individual-level random effects and the $\boldsymbol{\gamma}$ is a unit vector. Assume that a logistic link is appropriate for the probability and that a random effect at the individual level is to be included: v_i. Hence,

$$p_i = \frac{\exp(\alpha_0 + v_i)}{1 + \exp(\alpha_0 + v_i)}$$

would represent a basic model with intercept to capture the overall rate and prior distribution for the intercept, and the random effect could be assumed to be $\alpha_0 \sim N(0, \tau_{\alpha_0})$, and $v_i \sim N(0, \tau_v)$. The hyperprior distribution for the variance parameters could be a distribution on the positive real line such as the gamma, inverse gamma, or uniform. The uniform distribution has been proposed for the standard deviation ($\sqrt{\tau_*}$) by Gelman (2006). Here, for illustration, we define a gamma distribution:

$$\begin{aligned} y_i|p_i &\sim Bin(p_i, n_i) \\ logit(p_i) &= \alpha_0 + v_i \\ \alpha_0 &\sim N(0, \tau_{\alpha_0}) \\ v_i &\sim N(0, \tau_v) \\ \tau_{\alpha_0} &\sim G(\psi_1, \psi_2) \\ \tau_v &\sim G(\phi_1, \phi_2). \end{aligned}$$

An alternative approach to the Bernoulli or binomial distribution at the second level of the hierarchy is to assume a distribution directly for the case probability p_i. This might be appropriate when limited information about p_i is available. This is akin to the assumption of a gamma distribution as prior distribution for the Poisson relative risk parameter. Here, one choice for the prior distribution could be a beta distribution:

$$p_i \sim Beta(\alpha_1, \alpha_2).$$

In general, the parameters α_1 and α_2 could be assigned hyperprior distributions on the positive real line, such as gamma or exponential. However, if a uniform prior distribution for p_i is favored, then $\alpha_1 = \alpha_2 = 1$ can be chosen.

7.8 Correlated Heterogeneity Models

Uncorrelated heterogeneity models with gamma or beta prior distributions for the relative risk are useful but have a number of drawbacks. First, as noted above, a gamma distribution does not easily provide for extensions into covariate adjustment or modeling, and second, there is no simple and adaptable generalization of the gamma distribution with spatially correlated parameters. Møller et al. (1998) provided an example of using correlated gamma field models, but these models have been shown to have poor performance under simulated evaluation (Best et al., 2005). The advantages of incorporating a Gaussian specification are many. First, a random effect that is log-Gaussian behaves in a similar way to a gamma variate, but the Gaussian model can include a correlation structure. Hence, for the case where it is suspected that random effects are correlated, it is simpler to specify a log-Gaussian form for *any* extra variation present. The simplest extension is to consider additive components describing different aspects of the variation thought to exist in the data.

For a spatial Gaussian process (Ripley, 1981, p. 10), any finite realization has a multivariate normal distribution with mean and covariance inherited from the process itself, that is, $\mathbf{x} \sim MVN(\mu, K)$, where μ is an m length mean vector and K is an $m \times m$ positive definite covariance matrix. Note that this is not the only possible specification of a prior structure to model correlated heterogeneity (CH) (see also Møller et al., 1998).

There are many ways of incorporating such heterogeneity in models, and some of these are reviewed here. First, it is often important to include a variety of random effects in a model. For example, both CH and uncorrelated heterogeneity (UH) might be included. One flexible method for the inclusion of such terms is to include a log-linear term with additive random effects. Besag et al. (1991) first suggested, for tract count effects, a rate parametrization of the form

$$\exp\{x_i'\beta + u_i + v_i\},$$

where $x_i'\beta$ is a trend or fixed covariate component, and u_i and v_i are correlated and uncorrelated heterogeneity, respectively. These components then have separate prior distributions. Often, the specification of the correlated component is considered to have either an intrinsic Gaussian (conditional autoregressive [CAR]) prior distribution or a fully specified multivariate normal prior distribution. The sum of two random effects within the same units can lead to identification issues. However, the spatial correlation prior (improper conditional autoregressive [ICAR]) ensures smoothness, and so partial identification is achieved.

7.8.1 CAR models

7.8.1.1 ICAR models

The intrinsic autoregression's improper difference prior distribution, developed from the lattice models of Kunsch (1987), uses the definition of spatial distribution in terms of differences and allows the use of a singular normal joint distribution. This was first proposed by Besag et al. (1991). Hence, the prior for $\{u\}$ is defined as

$$p(\mathbf{u}|r) \propto \frac{1}{r^{m/2}} \exp\left\{-\frac{1}{2r}\sum_i \sum_{j\in\delta_i}(u_i - u_j)^2\right\}, \tag{7.7}$$

where δ_i is a neighborhood of the ith tract. The neighborhood δ_i was assumed to be defined for the first neighbor only. Hence, this is an example of a Markov random field model (see, e.g., Rue and Held, 2005). More general weighting schemes could be used. For example, neighborhoods could consist of first and second neighbors (defined by common boundary) or a distance cutoff (e.g., a region is a neighbor if the centroid is within a certain distance of the region in question). The uncorrelated heterogeneity (v_i) was defined by Besag et al. (1991) to have a conventional zero-mean Gaussian prior distribution:

$$p(v) \propto \sigma^{-m/2} \exp\left\{-\frac{1}{2\sigma}\sum_{i=1}^{m} v_i^2\right\}. \tag{7.8}$$

Both r and σ were assumed by Besag et al. (1991) to have improper inverse exponential hyperpriors:

$$\text{prior}(r, \sigma) \propto e^{-\epsilon/2r} e^{-\epsilon/2\sigma}, \qquad \sigma, r > 0, \tag{7.9}$$

where ϵ was taken as 0.001. These prior distributions penalize the absorbing state at zero, but provide considerable indifference over a large range. Alternative hyperpriors for these parameters that are now commonly used are in the gamma and inverse gamma family, which can be defined to penalize at zero but yield considerable uniformity over a wide range. In addition, these types of hyperpriors can also provide peaked distributions if required.

The full posterior distribution for the original formulation where a Poisson likelihood is assumed for the tract counts is given by

$$\begin{aligned} P(u, v, r, \sigma|y_i) = {} & \prod_{i=1}^{m}\{\exp(-e_i\theta_i)(e_i\theta_i)^{y_i}/y_i!\} \\ & \times \frac{1}{r^{m/2}} \exp\left\{-\frac{1}{2r}\sum_i \sum_{j\in\delta_i}(u_i - u_j)^2\right\} \\ & \times \sigma^{-m/2} \exp\left\{-\frac{1}{2\sigma}\sum_{i=1}^{m} v_i^2\right\} \times \text{prior}(r, \sigma). \end{aligned}$$

This posterior distribution can be sampled using Markov chain Monte Carlo (MCMC) algorithms such as the Gibbs or Metropolis–Hastings samplers. A Gibbs sampler was used in the original example, as conditional distributions for the parameters were available in that formulation.

An advantage of the intrinsic Gaussian formulation is that the conditional moments are defined as simple functions of the neighboring values and number of neighbors ($n_{\delta i}$).

$$E(u_i|\ldots) = \overline{u_i}$$

and

$$var(u_i|\ldots) = r/n_{\delta i},$$

and the conditional distribution is defined as

$$[u_i|\ldots] \sim N(\overline{u}_i, r/n_{\delta i}),$$

where $\overline{u}_i = \sum_{j\in\delta_i} u_j/n_{\delta i}$, the average over the neighborhood of the ith region.

7.8.1.2 PCAR models

While the intrinsic CAR model introduced above is useful in defining a correlated heterogeneity prior distribution, this is not the only specification of a Gaussian Markov random field (GMRF) model available. In fact, the improper CAR is a special case of a more general formulation where neighborhood dependence is admitted, but which allows an additional correlation parameter (Stern and Cressie, 1999). Define the spatially referenced vector of interest as $\{u_i\}$. One specification of the proper CAR formulation yields

$$[u_i|\ldots] \sim N(\mu_i, r/n_{\delta i}) \tag{7.10}$$

$$\mu_i = t_i + \phi \sum_{j\in\delta_i} (u_j - t_j)/n_{\delta_i}, \tag{7.11}$$

where t_i is the trend ($=x_i'\beta$), r is the variance, and ϕ is a correlation parameter. It can be shown that to ensure definiteness of the covariance matrix, ϕ must lie on a predefined range that is a function of the eigenvalues of a matrix. In detail, the range is the smallest and largest eigenvalues ($\phi_{\min} = \eta_1^{-1}, \phi_{\max} = \eta_m^{-1}$) of $diag\{n_{\delta_i}^{1/2}\}.C.diag\{n_{\delta_i}^{-1/2}\}$, where $C_{ij} = c_{ij}$, that is, ($\phi_{\min} < \phi < \phi_{\max}$) and

$$c_{ij} = \begin{cases} \frac{1}{n_{\delta_i}} & \text{if } i \sim j \\ 0 & \text{otherwise} \end{cases}.$$

Of course, $\phi_{\min}$ and $\phi_{\max}$ can be precomputed before using the proper CAR as a prior distribution. It could simply be assumed that a (hyper)prior distribution for ϕ is $U(\phi_{\min}, \phi_{\max})$. As noted by Stern and Cressie (1999), this specification does lead to a simple form for the partial correlation between different sites. Note that in the simple case of no trend ($t_i = 0$), the model reduces to

$$[u_i|\ldots] \sim N(\mu_i, r/n_{\delta i}) \tag{7.12}$$

$$\mu_i = \phi\overline{u}_i. \tag{7.13}$$

The main advantages of this model formulation is that it more closely mimics fully specified Gaussian covariance models, as it has a variance and correlation parameter specified, does not require matrix inversion within sampling algorithms, and can also be used as a data likelihood.

7.9 Posterior sampling

Once a posterior distribution has been derived, from the product of likelihood and prior distributions, it is important to assess how the form of the posterior distribution is to be evaluated. If single summary measures are needed, then it is sometimes possible to obtain these directly from the posterior distribution either by direct maximization (mode: maximum a posteriori estimation) or analytically in simple cases (e.g., mean or variance) (see Section 7.3). If a variety of features of the posterior distribution are to be examined, then often it will be important to be able to access the distribution via posterior sampling. Posterior sampling is a fundamental tool for exploration of posterior distributions and can provide a wide range of information about their form. Define a posterior distribution for data $\mathbf{y}$ and parameter vector $\boldsymbol{\theta}$ as $p(\boldsymbol{\theta}|\mathbf{y})$. We wish to represent features of this distribution by taking a sample from $p(\boldsymbol{\theta}|\mathbf{y})$. The sample can be used to estimate a variety of posterior quantities of interest. Define the sample size as m_p. Analytically, tractable posterior distributions may be available to directly simulate the distribution. For example, the gamma-Poisson model with α, β known, in Section 7.7, leads to the gamma posterior distribution: $\theta_i \sim G(y_i+\alpha, e_i+\beta)$. Either this can be simulated directly (on R: rgamma) or sample estimation can be avoided by direct computation from known formulas. For example, in this instance, the moments of a gamma distribution are known: $E(\theta_i) = (y_i+\alpha)/(e_i+\beta)$, and so forth.

Define the sample values generated as θ^*_{ij}, $j = 1, \ldots, m_p$. As long as a sample of reasonable size has been taken, it is possible to approximate the various functionals of the posterior distribution from these sample values. For example, an estimate of the posterior mean would be $\widehat{E}(\theta_i) = \widehat{\theta}_i = \sum_{j=1}^{m_p} \theta^*_{ij}/m_p$, while the posterior variance could be estimated as $\widehat{var}(\theta_i) = \frac{1}{m_p-1}\sum_{j=1}^{m_p}(\theta^*_{ij} - \widehat{\theta}_i)^2$ the sample variance. In general, any real function of the jth parameter $\gamma_j = t(\theta_j)$ can also be estimated in this way. For example, the mean of γ_j is given by $\widehat{E}(\gamma_j) = \widehat{\gamma}_j = \sum_{j=1}^{m_p} t(\theta^*_{ij})/m_p$. Note that credibility intervals can also be found for parameters by estimating the respective sample quantiles. For example, if $m_p = 1000$, then the 25th and 975th largest values would yield an equal-tail 95% credible interval for γ_j. The median is also available as the 50% percentile of the sample, as are other percentiles.

The empirical distribution of the sample values can also provide an estimate of the marginal posterior density of θ_i. Denote this density as $\pi(\theta_i)$. A smoothed estimate of this marginal density can be obtained from the histogram of sample values of θ_i. Improved estimators can be obtained by using conditional distributions. A Monte Carlo estimator of $\pi(\theta_i)$ is given by

$$\widehat{\pi}(\theta_i) = \frac{1}{n}\sum_{j=1}^{n} \pi(\theta_i|\theta_{j,-i}),$$

where θ_{-i} is the θ vector excluding θ_i, and the $\theta_{j,-i}$ $j = 1, \ldots, n$ are a sample from the marginal distribution $\pi(\theta_{-i})$.

Often, m_p is chosen to be $\geq$500, more often 1000 or 10,000. If computation is not expensive, then large samples such as these are easily obtained. The larger the sample size, the closer the posterior sample estimate of the functional will be.

Generally, the complete sample output from the distribution is used to estimate functionals. This is certainly true in the case when independent sample values are available (such as when the distribution is analytically tractable and can be sampled from directly, such as in the gamma-Poisson case). In other cases, where iterative sampling must be used, it is sometimes necessary to sub-sample the output sample. In the next section, this is discussed more elaborately.

7.10 Markov Chain Monte Carlo Methods

Often in disease mapping, realistic models for maps have two or more levels, and the resulting complexity of the posterior distribution of the parameters requires the use of sampling algorithms. In addition, the flexible modeling of disease could require switching between a variety of relatively complex models. In this case, it is convenient to have an efficient and flexible posterior sampling method that could be applied across a variety of models. Efficient algorithms for this purpose were developed within the fields of physics and image processing to handle large-scale problems in estimation. In the late 1980s and early 1990s, these methods were developed further particularly for dealing with Bayesian posterior sampling for more general classes of problems (Gilks et al., 1993, 1996). Now posterior sampling is commonplace and a variety of packages (including WinBUGS, MlwiN, and R) have incorporated these methods. For general reviews of this area, the reader is referred to Cassella and George (1992) and Robert and Casella (2005). MCMC methods are a set of methods that use iterative simulation of parameter values within a Markov chain. The convergence of this chain to a stationary distribution, which is assumed to be the posterior distribution, must be assessed.

Prior distributions for the p components of $\boldsymbol{\theta}$ are defined as $g_i(\theta_i)$ for $i = 1, \ldots, p$. The posterior distribution of $\boldsymbol{\theta}$ and $\mathbf{y}$ is defined as

$$P(\boldsymbol{\theta}|\mathbf{y}) \propto L(\mathbf{y}|\boldsymbol{\theta}) \prod_i g_i(\theta_i). \tag{7.14}$$

The aim is to generate a sample from the posterior distribution $P(\boldsymbol{\theta}|\mathbf{y})$. Suppose we can construct a Markov chain with state space $\boldsymbol{\theta}_c$, where $\boldsymbol{\theta} \in \boldsymbol{\theta}_c \subset \Re^k$. The chain is constructed so that the equilibrium distribution is $P(\boldsymbol{\theta}|\mathbf{y})$, and the chain should be easy to simulate from. If the chain is run over a long period, then it should be possible to reconstruct features of $P(\boldsymbol{\theta}|\mathbf{y})$ from the realized chain values. This forms the basis of the MCMC method, and algorithms are required for the construction of such chains. A selection of recent literature on this area is found in Ripley (1987), Gelman and Rubin (1992), Smith and Roberts (1993), Besag and Green (1993), Cressie (1993), Smith and Gelfand (1992), Tanner (1996), Robert and Casella (2005), and Marin and Robert (2014).

The basic algorithms used for this construction are

1. The Metropolis and its extension, Metropolis–Hastings, algorithms

2. The Gibbs sampler algorithm

7.11 Metropolis and Metropolis–Hastings Algorithms

In all MCMC algorithms, it is important to be able to construct the correct transition probabilities for a chain that has $P(\boldsymbol{\theta}|\mathbf{y})$ as its equilibrium distribution. A Markov chain consisting of $\boldsymbol{\theta}^1, \boldsymbol{\theta}^2, \ldots, \boldsymbol{\theta}^t$ with state space Θ and equilibrium distribution $P(\boldsymbol{\theta}|\mathbf{y})$ has transitions defined as follows.

Define $q(\boldsymbol{\theta}, \boldsymbol{\theta}')$ as a transition probability function, such that if $\boldsymbol{\theta}^t = \boldsymbol{\theta}$, the vector $\boldsymbol{\theta}^t$ drawn from $q(\boldsymbol{\theta}, \boldsymbol{\theta}')$ is regarded as a proposed possible value for $\boldsymbol{\theta}^{t+1}$.

7.11.1 Metropolis updates

In this case, choose a symmetric proposal $q(\boldsymbol{\theta}, \boldsymbol{\theta}')$ and define the transition probability as

$$p(\boldsymbol{\theta}, \boldsymbol{\theta}') = \begin{cases} \alpha(\boldsymbol{\theta}, \boldsymbol{\theta}')q(\boldsymbol{\theta}, \boldsymbol{\theta}') & \text{if } \boldsymbol{\theta}' \neq \boldsymbol{\theta} \\ 1 - \sum_{\theta''} q(\boldsymbol{\theta}, \boldsymbol{\theta}'')\alpha(\boldsymbol{\theta}, \boldsymbol{\theta}'') & \text{if } \boldsymbol{\theta}' = \boldsymbol{\theta} \end{cases},$$

where $\alpha(\boldsymbol{\theta}, \boldsymbol{\theta}') = \min\left\{1, \frac{P(\boldsymbol{\theta}'|\mathbf{y})}{P(\boldsymbol{\theta}|\mathbf{y})}\right\}$.

In this algorithm, a proposal is generated from $q(\boldsymbol{\theta}, \boldsymbol{\theta}')$ and is accepted with probability $\alpha(\boldsymbol{\theta}, \boldsymbol{\theta}')$. The acceptance probability is a simple function of the ratio of posterior distributions as a function of $\boldsymbol{\theta}$ values. The proposal function $q(\boldsymbol{\theta}, \boldsymbol{\theta}')$ can be defined to have a variety of forms but must be an irreducible and aperiodic transition function. Specific choices of $q(\boldsymbol{\theta}, \boldsymbol{\theta}')$ lead to specific algorithms.

7.11.2 Metropolis–Hastings updates

In this extension to the Metropolis algorithm, the proposal function is not confined to symmetry and

$$\alpha(\boldsymbol{\theta}, \boldsymbol{\theta}') = \min\left\{1, \frac{P(\boldsymbol{\theta}'|\mathbf{y})q(\boldsymbol{\theta}', \boldsymbol{\theta})}{P(\boldsymbol{\theta}|\mathbf{y})q(\boldsymbol{\theta}, \boldsymbol{\theta}')}\right\}.$$

Some special cases of chains are found when $q(\boldsymbol{\theta}, \boldsymbol{\theta}')$ has special forms. For example, if $q(\boldsymbol{\theta}, \boldsymbol{\theta}') = q(\boldsymbol{\theta}', \boldsymbol{\theta})$, then the original Metropolis method arises, and further, with $q(\boldsymbol{\theta}, \boldsymbol{\theta}') = q(\boldsymbol{\theta}')$ (i.e., when no dependence on the previous value is assumed),

$$\alpha(\boldsymbol{\theta}, \boldsymbol{\theta}') = \min\left\{1, \frac{w(\boldsymbol{\theta}')}{w(\boldsymbol{\theta})}\right\},$$

where $w(\boldsymbol{\theta}) = P(\boldsymbol{\theta}|\mathbf{y})/q(\boldsymbol{\theta})$ and $w(.)$ are importance weights. One simple example of the method is $q(\boldsymbol{\theta}') \sim \text{Uniform}(\boldsymbol{\theta}_a, \boldsymbol{\theta}_b)$ and $g_i(\theta_i) \sim \text{Uniform}(\theta_{ia}, \theta_{ib})\ \forall i$; this leads to an acceptance criterion based on a likelihood ratio. Hence, the original Metropolis algorithm with uniform proposals and prior distributions leads to a stochastic exploration of a likelihood surface. This, in effect, leads to the use of prior distributions as proposals. However, in general, when the $g_i(\theta_i)$ are not uniform, this leads to inefficient sampling. The definition of $q(\boldsymbol{\theta}, \boldsymbol{\theta}')$ can be quite general in this algorithm, and in addition, the posterior distribution only appears within a ratio as a function of $\boldsymbol{\theta}$ and $\boldsymbol{\theta}'$. Hence, the distribution is only required to be known up to proportionality.

7.11.3 Gibbs updates

The Gibbs sampler has gained considerable popularity, particularly in applications in medicine, where hierarchical Bayesian models are commonly applied (see, e.g., Gilks et al., 1993). This popularity is mirrored in the availability of software that allows its application in a variety of problems (e.g., WinBUGS, MLwiN, JAGS, and MCMCpack). This sampler is a special case of the Metropolis–Hastings algorithm where the proposal is generated from the conditional distribution of θ_i given all other $\boldsymbol{\theta}$'s, and the resulting proposal value is accepted with probability 1.

More formally, define

$$q(\theta_j, \theta_j') = \begin{cases} p(\theta_j^*|\theta_{-j}^{t-1}) & \text{if} \quad \theta_{-j}^* = \theta_{-j}^{t-1} \\ 0 & \text{otherwise} \end{cases},$$

where $p(\theta_j^*|\theta_{-j}^{t-1})$ is the conditional distribution of θ_j given all other $\boldsymbol{\theta}$ values (θ_{-j}) at time $t-1$. Using this definition, it is straightforward to show that

$$\frac{q(\boldsymbol{\theta},\boldsymbol{\theta}')}{q(\boldsymbol{\theta}',\boldsymbol{\theta})} = \frac{P(\boldsymbol{\theta}'|\mathbf{y})}{P(\boldsymbol{\theta}|\mathbf{y})},$$

and hence $\alpha(\boldsymbol{\theta},\boldsymbol{\theta}') = 1$.

7.11.4 Special methods

Alternative methods exist for posterior sampling when the basic Gibbs or Metropolis–Hastings updates are not feasible or appropriate. For example, if the range of the parameters is restricted, then slice sampling can be used (Robert and Casella, 2005, chapter 7; Neal, 2003). When exact conditional distributions are not available but the posterior is log-concave, adaptive rejection sampling (ARS) algorithms can be used. The most general of these algorithms (ARS algorithm; Robert and Casella, 2005, pp. 57–59) has wide applicability for continuous distributions, although it may not be efficient for specific cases. Block updating can also be used to effect in some situations. When generalized linear model components are included, then block updating of the covariate parameters can be effected via multivariate updating.

7.11.5 Convergence

MCMC methods require the use of diagnostics to assess whether the iterative simulations have reached the equilibrium distribution of the Markov chain. Sampled chains need to be run for an initial burn-in period until they can be assumed to provide approximately correct samples from the posterior distribution of interest. This burn-in period can vary considerably between different problems. In addition, it is important to ensure that the chain manages to explore the parameter space properly, so that the sampler does not "stick" in local maxima of the surface of the distribution. Hence, it is crucial to ensure that a burn-in period is adequate for the problem considered. Judging convergence has been the subject of much debate and can still be regarded as art rather than science: a qualitative judgment has to be made at some stage as to whether the burn-in period is long enough.

There are a wide variety of methods now available to assess convergence of chains within MCMC. Robert and Casella (2005) and Liu (2001) provide recent reviews. The available methods are largely based on checking the distributional properties of samples from the chains. In general, define an output stream for a parameter vector $\boldsymbol{\theta}$ as $\{\boldsymbol{\theta}^1,\boldsymbol{\theta}^2,\ldots,\boldsymbol{\theta}^m,\boldsymbol{\theta}^{m+1},\ldots,\boldsymbol{\theta}^{m+m_p}\}$. Here, the mth value is the end of the burn-in period and a (converged) sample of size m_p is taken. Hence, the converged sample is $\{\boldsymbol{\theta}^{m+1},\ldots,\boldsymbol{\theta}^{m+m_p}\}$. Define a function of the output stream as $\gamma = t(\boldsymbol{\theta})$ so that $\gamma^1 = t(\boldsymbol{\theta}^1)$.

The Brooks–Gelman–Rubin (BGR) statistic (Brooks and Gelman, 1998) focuses on the between- and within-chain variability. They developed a statistic R that is a function of the ratio of between- and within-chain variances, and they also extended this diagnostic to a multiparameter situation. On R, the statistic is available in the CODA package as gelman.diag. On WinBUGS, the BGR statistic is available in the Sample Monitor Tool. On WinBUGS, the width of the central 80% interval of the pooled runs and the average width of the 80% intervals within the individual runs are color-coded (green, blue), and their ratio R is red—for plotting purposes, the pooled and within-interval widths are normalized to have an overall maximum of 1. On WinBUGS, the statistics are calculated in bins of length 50. R would generally be expected to be greater than 1 if the starting values are suitably over-dispersed. Brooks and Gelman (1998) emphasize that one should be concerned both with convergence of R to 1 and with convergence of both the pooled and within-interval widths to stability. One caveat should be mentioned concerning the use of between- and within-chain diagnostics. If the posterior distribution being approximated were to be highly multimodal, which could be the case in many mixture and spatial problems, then the variability across chains could be large even when close to the posterior distribution, and it could be that very large bins would need to be used for computation.

There is some debate about whether it is useful to run one long chain, as opposed to multiple chains with different start points. The advantage of multiple chains is that they provide evidence for the robustness of convergence across different subspaces. However, as long as a single chain samples the parameter space adequately, then these have benefits. The reader is referred to (Robert and Casella, 2005, chapter 8) for a thorough discussion of diagnostics and their use.

7.12 Posterior and Likelihood Approximations

From the point of view of computation, it is now straightforward to examine a range of posterior distributional forms. This is certainly true for most applications of disease mapping where relative risk is estimated. However, there are situations where it may be easier or more convenient to use a form of approximation to the posterior distribution or to the likelihood itself. Some approximations have been derived originally when posterior sampling was not possible and where the only way to obtain fully Bayesian estimates was to approximate (Bernardo and Smith, 1994). However, other approximations arise due to the intractability of spatial integrals (e.g., in point process models).

7.12.1 Pseudolikelihood and other forms

In Section 7.1.1.3, the idea of pseudolikelihood is introduced, and this idea is extended here. In certain spatial problems, found in imaging and elsewhere, normalizing constants arise that are highly multidimensional. A simple example is the case of a Markov point process. Define the realization of m events within a window T as $\{\mathbf{s}_1, \ldots, \mathbf{s}_m\}$. Under a Markov process assumption, the normalized probability density of a realization is

$$f_\theta(\mathbf{s}) = \frac{1}{c(\theta)} h_\theta(\mathbf{s}),$$

where

$$c(\theta) = \sum_{k=0}^{\infty} \frac{1}{k!} \int_{T^k} h_\theta(\mathbf{s}) \lambda^k (d\mathbf{s}).$$

Conditioning on the number of events (m), the normalization of $f_m(\mathbf{s}) \propto h_m(\mathbf{s})$ is over the m-dimensional window:

$$c(\theta) = \int_T \cdots \int_T h_m(\{\mathbf{s}_1, \ldots, \mathbf{s}_m\}) d\mathbf{s}_1, \ldots, d\mathbf{s}_m.$$

For a conditional Strauss process, $f_m(\mathbf{s}) \propto \gamma^{n_R(\mathbf{s})}$ and $n_R(\mathbf{s})$ is the number of R-close pairs of points to $\mathbf{s}$.

It is also true that a range of lattice models developed for image processing applications also have awkward normalization constants (auto-Poisson and autologistic models and Gaussian Markov random field models: Besag and Tantrum, 2003; Rue and Held, 2005).

This has led to the use of approximate likelihood models in many cases. For example, for Markov point processes, it is possible to specify a conditional intensity (Papangelou) that is independent of the normalization. This conditional intensity $\lambda^*(\xi, \mathbf{s}|\theta) = h(\boldsymbol{\xi} \cup \mathbf{s})/h(\boldsymbol{\xi})$ can be used within a pseudolikelihood function. In the case of the above Strauss process, this is just $\lambda^*(\xi, \mathbf{s}|\theta) = \lambda^*(\mathbf{s}|\theta) = \gamma^{n_R(\mathbf{s})}$ and the pseudolikelihood is

$$L_p(\{\mathbf{s}_1, \ldots, \mathbf{s}_m\}|\theta) = \prod_{i=1}^{m} \lambda^*(\mathbf{s}_i|\theta) \exp\left(-\int_T \lambda^*(\mathbf{u}|\theta) d\mathbf{u}\right).$$

As this likelihood has the form of an inhomogeneous Poisson process likelihood, it is relatively straightforward to evaluate. The only issue is the integral of the intensity over the window T. This can be handled via special numerical integration schemes (Berman and Turner, 1992, Lawson, 1992a, 1992b). Bayesian extensions are generally straightforward. Note that once a likelihood contribution can be specified, this can be incorporated within a posterior sampling algorithm such as Metropolis–Hastings. This can be implemented on WinBUGS via a zeroes trick if the Berman–Turner weighting is used. For example, the model with the ith likelihood component, $l_i = \log \lambda^*(\mathbf{s}_i|\theta) - w_i \lambda^*(\mathbf{s}_i|\theta)$, can be fitted using this method, where the weight w_i is based on the Dirichlet tile area of the ith point or a function of the Delauney triangulation around the point (see Berman and Turner, 1992; Baddeley and Turner, 2000; Lawson, 2006, appendix C.5.3).

In application to lattice models, Besag and Tantrum (2003) give the example of a Markov random field of m dimension, where the pseudolikelihood $L_p = \prod_{i=1}^{m} p(y_i^0|y_{-i}^0; \theta)$ is the product of the full conditional distributions. In the (auto)logistic binary case,

$$p(y_i^0|y_{-i}^0; \theta) = \frac{\exp(f(\alpha_0, \{y_{-i}^0\}_{\in \partial i}, \theta)}{1 + \exp(f(\alpha_0, \{y_{-i}^0\}_{\in \partial i}, \theta)},$$

where ∂_i denotes the adjacency set of the ith site.

Other variants of these likelihoods have been proposed. Local likelihood (Tibshirani and Hastie, 1987) is a variant where a contribution to likelihood is defined within a local domain of the parameter space. In spatial problems, this could be a spatial area. This has been used in a Bayesian disease mapping setting by Hossain and Lawson (2005). Pairwise likelihood (Nott and Rydén, 1999; Heagerty and Lele, 1998) has been proposed for image restoration and for general spatial mixed models (Varin et al., 2005). All these variants of full likelihoods will lead to models that are approximately valid for real applications. It should be born in mind, however, that they ignore aspects of the spatial correlation, and if these are not absorbed in some part of the model hierarchy, then this may affect the appropriateness of the model.

7.12.2 Asymptotic approximations

It is possible to approximate a posterior distribution with a simpler distribution that is found asymptotically. The use of approximations lies in their often common form and also the ease with which parameters may be estimated under the approximation. Often, the asymptotic approximating distribution will be a normal distribution. Here, two possible approaches are examined: the asymptotic quadratic form approximation and the integral approximation via Laplace's method.

7.12.2.1 Asymptotic quadratic form

Large sample convergence in the form of the likelihood or posterior distribution is considered here. In many cases, the limiting form of a likelihood or posterior distribution in large samples can be used as an approximation. The Taylor series expansion of the function $f(.)$ around vector $\mathbf{a}$ is

$$f(\mathbf{a})+U(\mathbf{a})^T(\mathbf{x}-\mathbf{a})+\frac{1}{2}(\mathbf{x}-\mathbf{a})^T H(\mathbf{a})(\mathbf{x}-\mathbf{a})+R,$$

where $U(\mathbf{a})$ is the score vector evaluated at $\mathbf{a}$, R is a remainder, and $H(\mathbf{a})$ is the Hessian matrix of second derivatives of $f(.)$ evaluated at $\mathbf{a}$. For an arbitrary log-likelihood with p length vector of parameters $\boldsymbol{\theta}$, an expansion around a point is required. Usually, the mode of the distribution is chosen. Define the modal vector as $\boldsymbol{\theta}^{mo}$ and $l(\mathbf{y}|\boldsymbol{\theta}) \equiv l(\boldsymbol{\theta})$ for brevity. The expansion is defined as

$$l(\boldsymbol{\theta}) = l(\mathbf{y}|\boldsymbol{\theta}^{mo})+U(\boldsymbol{\theta}^{mo})(\boldsymbol{\theta}-\boldsymbol{\theta}^{mo})-\frac{1}{2}(\boldsymbol{\theta}-\boldsymbol{\theta}^{mo})^T H(\boldsymbol{\theta}^{mo})(\boldsymbol{\theta}-\boldsymbol{\theta}^{mo}).$$

Here, $U(\boldsymbol{\theta}^{mo}) = \mathbf{0}$, as we have expanded around the maxima, and so this reduces to

$$l(\boldsymbol{\theta}) = l(\mathbf{y}|\boldsymbol{\theta}^{mo})-\frac{1}{2}(\boldsymbol{\theta}-\boldsymbol{\theta}^{mo})^T H(\boldsymbol{\theta}^{mo})(\boldsymbol{\theta}-\boldsymbol{\theta}^{mo}). \tag{7.15}$$

Note that $H(\boldsymbol{\theta}^{mo})$ describes the local curvature of the likelihood at the maxima and is defined by

$$H(\boldsymbol{\theta}^{mo}) = \left(-\frac{\partial^2 l(\boldsymbol{\theta})}{\partial\theta_i\partial\theta_j}\right)\Bigg|_{\boldsymbol{\theta}=\boldsymbol{\theta}^{mo}}.$$

This approximation, given $\boldsymbol{\theta}^{mo}$, consists of a constant and a quadratic form around the maxima. In a likelihood analysis, the $\boldsymbol{\theta}^{mo}$ might be replaced by maximum likelihood estimates $\widehat{\boldsymbol{\theta}}^{mo}$.

For a posterior distribution, it is possible to also approximate the prior distribution with a Taylor expansion, in which case a full posterior approximation would be obtained. In the case of the joint prior distribution, defined by $p(\boldsymbol{\theta}|\boldsymbol{\Gamma})$, Γ is a parameter vector or matrix. Assuming that Γ is fixed, the approximation around the modal vector $\boldsymbol{\theta}^p$, again assuming the score vector is zero at the maxima, is given by

$$\log p(\boldsymbol{\theta}|\boldsymbol{\Gamma}) = \log p(\boldsymbol{\theta}^p|\boldsymbol{\Gamma})-\frac{1}{2}(\boldsymbol{\theta}-\boldsymbol{\theta}^p)^T H_p(\boldsymbol{\theta}^p)(\boldsymbol{\theta}-\boldsymbol{\theta}^p)+R_0,$$

where R_0 is the remainder term and

$$H_p(\boldsymbol{\theta}^p) = \left(-\frac{\partial^2 \log p(\boldsymbol{\theta}|\boldsymbol{\Gamma})}{\partial\theta_i\partial\theta_j}\right)\Bigg|_{\boldsymbol{\theta}=\boldsymbol{\theta}^p}.$$

Again, given $\boldsymbol{\theta}^p$, this is simply a quadratic form around the maxima. There are then two posterior approximations that might be considered:

1. Likelihood approximation only:

$$p(\boldsymbol{\theta}|\mathbf{y}) \propto p(\boldsymbol{\theta}|\boldsymbol{\Gamma})\exp\left\{-\frac{1}{2}(\boldsymbol{\theta}-\boldsymbol{\theta}^{mo})^T H(\boldsymbol{\theta}^{mo})(\boldsymbol{\theta}-\boldsymbol{\theta}^{mo})\right\}.$$

2. Full posterior approximation:

$$p(\boldsymbol{\theta}|\mathbf{y}) \propto \exp\left\{-\frac{1}{2}(\boldsymbol{\theta}-\boldsymbol{\theta}^p)^T H_p(\boldsymbol{\theta}^p)(\boldsymbol{\theta}-\boldsymbol{\theta}^p) - \frac{1}{2}(\boldsymbol{\theta}-\boldsymbol{\theta}^{mo})^T H(\boldsymbol{\theta}^{mo})(\boldsymbol{\theta}-\boldsymbol{\theta}^{mo})\right\}$$
$$\propto \exp\left\{-\frac{1}{2}(\boldsymbol{\theta}-\mathbf{m}^n)^T H_n(\boldsymbol{\theta}-\mathbf{m}^n)\right\},$$

where $H_n = H_p(\boldsymbol{\theta}^p) + H(\boldsymbol{\theta}^{mo})$ and $\mathbf{m}^n = H_n^{-1}(H_p(\boldsymbol{\theta}^p)\boldsymbol{\theta}^p + H(\boldsymbol{\theta}^{mo})\boldsymbol{\theta}^{mo})$.

Note that $H(\boldsymbol{\theta}^{mo})$ is the observed information matrix. As the sample size increases, this quadratic form approximation improves in its accuracy and two important results follow:

1. The posterior distribution tends toward a normal distribution, that is,

$$\text{as } m \to \infty \text{ then } p(\boldsymbol{\theta}|\mathbf{y}) \to N_p(\boldsymbol{\theta}|\mathbf{m}^n, H_n).$$

2. The information matrix tends toward the Fisher (expected) information matrix in the sense that $H(\boldsymbol{\theta}^{mo}) \to mI(\boldsymbol{\theta}^{mo})$, where the ijth element is

$$I(\boldsymbol{\theta})_{ij} = \int p(y|\boldsymbol{\theta})\left(-\frac{\partial^2 l(\boldsymbol{\theta})}{\partial\theta_i\partial\theta_j}\right)dy.$$

   This means that it is possible to consider further asymptotic distributional forms. For instance, if the variability in the prior distribution is negligible compared to the likelihood, then

$$p(\boldsymbol{\theta}|\mathbf{y}) \to N_p(\boldsymbol{\theta}|\boldsymbol{\theta}^{mo}, H(\boldsymbol{\theta}^{mo}))$$

   or

$$p(\boldsymbol{\theta}|\mathbf{y}) \to N_p(\boldsymbol{\theta}|\boldsymbol{\theta}^{mo}, mI(\boldsymbol{\theta}^{mo})).$$

   Often the maximum likelihood (ML) estimates would be substituted for $\boldsymbol{\theta}^{mo}$. If $\boldsymbol{\theta}^{mo}$ are given or estimated via ML, the posterior distribution will be multivariate normal in large samples.

Hence, a normal approximation to the posterior distribution is justified at least asymptotically (as $m \to \infty$). This approximation should be reasonably good for continuous likelihood models and may be reasonable for discrete models when the rate parameter (Poisson) is large or the binomial probability is not close to 0 or 1. Of course, this is likely not to hold when there is sparseness in the count data, as can arise when rare diseases are studied. Further discussion of different asymptotic results can be found in Bernardo and Smith (1994).

An example of such a likelihood approximation would be where a binomial likelihood has been assumed and $y_i|p_i \sim Bin(p_i, n_i)$, with $p_i \sim Beta(2,2)$. In this case, assume $p(\boldsymbol{\theta}|\mathbf{y}) \sim N_p(\boldsymbol{\theta}|\mathbf{m}^n, H_n)$ and $\mathbf{m}^n = \frac{\widehat{p}_i(1-\widehat{p}_i)}{n_i}\left[\frac{n_i}{\widehat{p}_i(1-\widehat{p}_i)}\widehat{p}_i\right] = \widehat{p}_i$, $H_n = 0 + \frac{n_i}{\widehat{p}_i(1-\widehat{p}_i)}$, where $\widehat{p}_i = \frac{y_i}{n_i}$, and so the distribution is $N_m(p_i|\widehat{p}_i, diag\{\frac{n_i}{\widehat{p}_i(1-\widehat{p}_i)}\})$. Hence, the approximate distribution is centered around the saturated maximum likelihood estimator. In this case, the prior distribution has little effect on the mean or the variance of the resulting Gaussian distribution. If, on the other hand, an asymmetric prior distribution favoring low rates of disease were assumed, such as $p_i \sim Beta(1.5, 5)$, then the approximation is given by $\mathbf{m}^n = (H_p(\boldsymbol{\theta}^p) + \frac{n_i}{\widehat{p}_i(1-\widehat{p}_i)})^{-1}\ [H_p(\boldsymbol{\theta}^p)\boldsymbol{\theta}^p + \frac{n_i\widehat{p}_i}{\widehat{p}_i(1-\widehat{p}_i)}]$ and $H_n = H_p(\boldsymbol{\theta}^p) + \frac{n_i}{\widehat{p}_i(1-\widehat{p}_i)}$, where $H_p(\boldsymbol{\theta}^p) = 81.383$ and $\boldsymbol{\theta}^p = 0.11$. Here, the mean and variance are influenced considerably.

Note that it is also possible to approximate posterior distributions with mixtures of normal distributions, and this could lead to closer approximation to complex (multimodal) distributions. Hierarchies with more than two levels have not been discussed here. However, in principle, if a normal approximation can be made to each prior in turn (perhaps via mixtures of normals), then a quadratic form would result with a more complex form.

7.12.2.2 Laplace integral approximation

In some situations, ratios of integrals must be evaluated and it is possible to employ the integral approximation method suggested by Laplace (Nott and Rydén, 1999). For example, the posterior expectation of a real-valued function $g(\boldsymbol{\theta})$ is given by

$$E(g(\boldsymbol{\theta})|\mathbf{y}) = \int g(\boldsymbol{\theta})p(\boldsymbol{\theta}|\mathbf{y})d\boldsymbol{\theta}.$$

This can be considered a ratio of integrals, given the normalization of the posterior distribution. The approximation is given by

$$\widehat{E}(g(\boldsymbol{\theta})|\mathbf{y}) \approx \left(\frac{\sigma^*}{\sigma}\right) \exp\{-m[h^*(\boldsymbol{\theta}^*) - h(\boldsymbol{\theta})]\},$$

where $-mh(\boldsymbol{\theta}) = \log p(\boldsymbol{\theta}) + l(\mathbf{y}|\boldsymbol{\theta})$ and $-mh^*(\boldsymbol{\theta}) = \log g(\boldsymbol{\theta}) + \log p(\boldsymbol{\theta}) + l(\mathbf{y}|\boldsymbol{\theta})$ and $-h(\widehat{\boldsymbol{\theta}}) = \max_{\boldsymbol{\theta}}\{-h(\boldsymbol{\theta})\}$, $-h^*(\boldsymbol{\theta}^*) = \max_{\boldsymbol{\theta}}\{-h^*(\boldsymbol{\theta})\}$, $\widehat{\sigma} = |m\nabla^2 h(\widehat{\boldsymbol{\theta}})|^{-1/2}$ and $\widehat{\sigma} = |m\nabla^2 h^*(\boldsymbol{\theta}^*)|^{-1/2}$, where

$$[\nabla^2 h(\widehat{\boldsymbol{\theta}})]_{ij} = \left.\frac{\partial^2 h(\boldsymbol{\theta})}{\partial\theta_i \partial\theta_j}\right|_{\boldsymbol{\theta}=\widehat{\boldsymbol{\theta}}}.$$

7.12.2.3 INLA and R-INLA

A recent development in the use of approximations to Bayesian models has been proposed in a sequence of papers by Rue and coworkers (Rue et al., 2009; Lindgren et al., 2011). The basic idea is that a wide range of models that have a latent Gaussian structure can be approximated via integrated nested Laplace approximation (INLA). These approximations can be seen as successive approximations of functions within integrals. The integrals are then approximated by fixed integration schemes. If we consider a Poisson model for observed counts, y_i, $i = 1, \ldots, m$, then with the set of hyperparameters given by $\boldsymbol{\phi}$ and a log link to an additive set of effects (random effects),

$$\begin{aligned} y_i|\boldsymbol{\lambda}_i &\sim Pois(e_i\theta_i) \\ \theta_i &= \exp\{\alpha + v_i + u_i\} \\ \text{where } \boldsymbol{\lambda}_i &= \{\alpha, v_i, u_i\}^T. \end{aligned}$$

Note that the parameters in $\boldsymbol{\lambda}$ all have Gaussian distributions and prior distribution $P(\boldsymbol{\lambda}|\boldsymbol{\phi})$. In this case, it is possible to approximate the posterior marginal distribution $P(\boldsymbol{\lambda}_i|y)$ by

$$P(\boldsymbol{\lambda}_i|\, y) = \int_{\boldsymbol{\phi}} P(\boldsymbol{\lambda}_i|y, \boldsymbol{\phi})\boldsymbol{P}(\boldsymbol{\phi} \mid y)d\boldsymbol{\phi}$$

for each component $\boldsymbol{\lambda}_i$ of the latent fields. The terms $P(\boldsymbol{\lambda}_i|y, \boldsymbol{\phi})$ and $\boldsymbol{P}(\boldsymbol{\phi}|y)$ can each be approximated by Laplace approximation. The simplest of these is the Gaussian approximation where matching of the mode and curvature to a normal distribution is used. Finally, the integral approximation leads to

$$\widetilde{P}(\boldsymbol{\lambda}_i|\, \boldsymbol{y}) = \sum_k \triangle_k \widetilde{P}(\boldsymbol{\lambda}_i|\boldsymbol{y}, \boldsymbol{\phi}_k)\widetilde{P}(\boldsymbol{\phi}_k|\boldsymbol{y}).$$

The sum is over values of $\boldsymbol{\phi}$ with area weights $\triangle_k$. This approximation approach is now available in R (package R-inla: www.r-inla.org) and can be used for a wide variety of applications. Application of these approximations has been made by Schrodle et al. (2011) to veterinary spatial surveillance data. In that work, they demonstrate the closeness of the final estimates to that achieved using posterior sampling within MCMC. However, they do not show any simulated comparisons where a ground truth is compared. Further demonstration of the capabilities in spatiotemporal modeling is given by Schrodle and Held (2011) and Ugarte et al. (2014).

In a recent extension, the parallel with finite element solutions to differential equations was exploited by Lindgren et al. (2011), whereby the spatial field is a solution to a stochastic partial differential equation (SPDE) with form $\lambda(s_i) = \sum_k \phi_k(s)w_k$, where the $\phi_k(s)$ are basis functions and w_k are weights. This is formally close in form to the kernel process convolution models of Higdon (2002). A comparison is made by Simpson et al. (2012).

7.13 Case Study: Models with Correlated Heterogeneity

We performed an analysis of the fetal death pregnancy outcome in 46 counties of South Carolina during the time period 2008–2013, using a convolution model with both uncorrelated heterogeneity (UH) and correlated heterogeneity (CH) random effects (see, e.g., Lawson, 2013, chapter 5). The spatial dependency was modeled using a conditional autoregressive (CAR) prior distribution.

Data were downloaded from the South Carolina Community Assessment Network (SCAN). The form of the data includes county-level fetal death and gestational delivery (fetal death and live birth) counts. The expected fetal deaths were calculated by multiplying the gestational deliveries for each county with the state rate. Specifically,

$$E_i = p_i \frac{\sum y_i}{\sum p_i}, \quad i = 1, \dots, 46,$$

where E_i denotes the expected count for county i, p_i is the number of gestational deliveries in county i, and y_i is the number of fetal deaths in county i.

County	Observed cases O_i	Expected cases E_i	SMR $\frac{O_i}{E_i}$	Adjacent counties
1	10	14.07	0.71	33, 30, 24, 23, 4
2	107	101.08	1.06	41, 38, 32, 19, 6
...	...	...	...	...
46	111	155.76	0.71	44, 29, 12, 11

Standardized mortality rate (SMR) is computed as the ratio between the observed number of deaths and the number of deaths that would be expected. We smooth the raw SMRs by fitting a Poisson model allowing for spatial correlation using a CAR model. We implement the model in WinBUGS, INLA, and the R CARBayes package.

7.13.1 WinBUGS

The convolution model can be implemented in WinBUGS or OpenBUGS as follows:

```
model {
# Likelihood
for (i in 1 : N) {
fetaldeath[i] ~ dpois(mu[i])
log(mu[i]) <- log(E[i]) + alpha0+v[i]+u[i]
vusum[i] <- v[i]+u[i]
}
# CAR prior distribution for random effects:
u[1:N] ~ car.normal(adj[], weights[], num[], tau.u)
for(k in 1:sumNumNeigh) {
weights[k] <- 1
}
# Other priors:
alpha0 ~ dnorm(0,tau.alpha0)
tau.alpha0<-1/(sigma.alpha0*sigma.alpha0)
sigma.alpha0~ dunif(0,100)
tau.v<-1/(sigma.v*sigma.v)
sigma.v~ dunif(0,100)
tau.u<-1/(sigma.u*sigma.u)
sigma.u ~ dunif(0,100)
}
```

Our model was run for 40,000 iterations, the first 30,000 being discarded as burn-in. The samples were monitored for convergence using the Brooks–Gelman–Rubin (BGR) diagnosis, which examines the between- and within-sample variances. The analysis of posterior samples was performed using the coda R package. Estimated means of posterior sums of unstructured and structured random effects are displayed in Figure 7.1 and show larger values in the South-East of the state.

7.13.2 INLA

In order to fit spatial models in INLA, we need to have a graph with neighborhood adjacency information. This can be created using the R spdep library.

```
library(spdep)
adjSC=poly2nb(SC)
nb2INLA("SCgraph.txt",adjSC)
```

There are two ways of specifying a convolution model in INLA. One option is to add a spatially correlated CAR term to the uncorrelated random effects as follows:

```
formula=y ~ f(county, model = "iid") +
f(county2,model="besag",graph="SC.txt")+offset(log(data$pop))
```

where county is the variable in the dataset denoting the geographical indexing variable, in our case the county number. Notice that it is required to use an identical duplicate of this variable in the second `f` function. This is needed because there is a separate summary output for each part of the model.

An alternative to the above formulation is to specify a convolution model having only one `f` function. In this case, the results are concatenated into a single output accessed by `summary.random$county`.

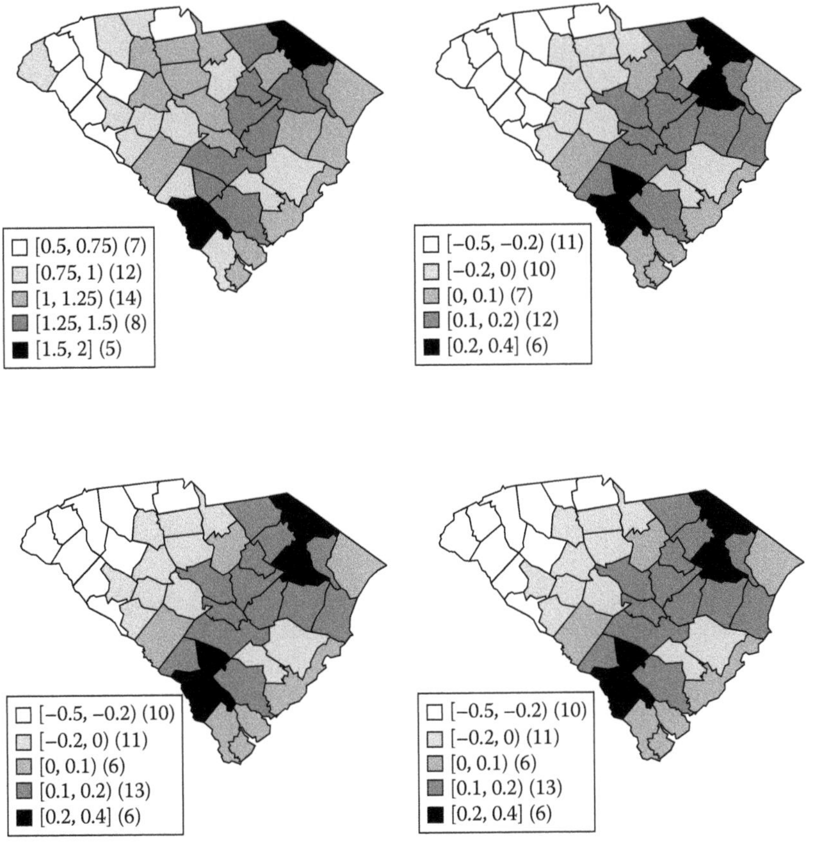

FIGURE 7.1
Display of standardized mortality ratio (SMR) and estimated mean of posterior sum of unstructured and structured random effects; row-wise from top left: SMR, posterior mean estimates fitted in WinBUGS, INLA, and the CARBayes R package for the same dataset; South Carolina county-level fetal deaths, 2008–2013.

The estimated posterior marginal of the sum of the unstructured and spatially correlated random effects is listed first (UH + CH), followed by summary estimates of the unstructured random effects (UH).

```
formula=y ~ f(county,model="bym",graph="SCgraph.txt")
+offset(log(data$pop))
result=inla(formula,family="poisson",data=data)
INLAmeanuvsum=as.numeric(unlist(result$summary.random
$county[1:46,][2]))
```

Estimated means of posterior sum of unstructured and structured random effects are displayed in Figure 7.1.

7.13.3 R CARBayes package

CARBayes is an add-on package to the statistical software R and can fit both independent and spatially correlated random effects, where the data likelihood can be binomial, Gaussian, or Poisson. Using the CARBayes package, inference is based on MCMC simulations using a combination of Gibbs sampling and Metropolis steps. The convolution model can be fit using the `bymCAR.re` function, for which the spatial neighborhood information is provided in a n_c x n_c neighborhood matrix W, where n_c is in number of counties. The binary adjacency matrix can be constructed using the spdep R package. The linear predictor is specified as an R formula object.

```
library(spdep)
library(CARBayes)
adjSC=poly2nb(SC)
weight <- listw2mat(nb2listw(adjSC, style = "B"))
Wmat<- as(weight, "CsparseMatrix")
y=data$fetaldeath
pop=data$pop
formulaR= y~1+offset(log(data$pop)) #intercept only
modelR=S.CARbym(formulaR,family="poisson",W=as.matrix(Wmat),burnin=30000,
        n.sample=40000,thin=1,verbose=TRUE)
REsum=modelR$samples$re
REsummean=colMeans(REsum)
```

We have displayed in Figure 7.1 the SMR and estimated mean of posterior sum of unstructured and structured random effects fitted in WinBUGS, INLA, and the CARBayes R package. We notice that the estimated random effect maps are very similar despite the fact that the default values for priors and hyperparameters may differ across packages.

References

Baddeley, A. and R. Turner. (2000). Practical maximum pseudolikelihood for spatial point patterns. *Australia and New Zealand Journal of Statistics* 42, 283–322.

Balakrishnan, S. and D. Madigan. (2008). Algorithms for sparse linear classifiers in the massive data setting. *Journal of Machine Learning Research* 9, 313–337.

Berman, M. and T. R. Turner. (1992). Approximating point process likelihoods with GLIM. *Applied Statistics* 41, 31–38.

Bernardo, J. M. and A. F. M. Smith. (1994). *Bayesian Theory.* New York: Wiley.

Besag, J. and P. J. Green. (1993). Spatial statistics and Bayesian computation. *Journal of the Royal Statistical Society B* 55, 25–37.

Besag, J. and J. Tantrum. (2003). Likelihood analysis of binary data in space and time. In Green, P., N. Hjort, and S. Richardson (eds.), *Highly Structured Stochastic Systems*, pp. 289–295. Oxford: Oxford University Press.

Besag, J., J. York, and A. Mollié. (1991). Bayesian image restoration with two applications in spatial statistics. *Annals of the Institute of Statistical Mathematics* 43, 1–59.

Best, N., S. Richardson, and A. Thomson. (2005). A comparison of Bayesian spatial models for disease mapping. *Statistical Methods in Medical Research* 14, 35–59.

Brooks, S. and A. E. Gelman. (1998). General methods for monitoring convergence of iterative simulations. *Journal of Computational and Graphical Statistics* 7, 434–455.

Cassella, G. and E. I. George. (1992). Explaining the Gibbs sampler. *American Statistician* 46, 167–174.

Clayton, D. G. and J. Kaldor. (1987). Empirical Bayes estimates of age-standardised relative risks for use in disease mapping. *Biometrics* 43, 671–691.

Cressie, N. A. C. (1993). *Statistics for Spatial Data* (revised ed.). New York: Wiley.

Diggle, P., T. Fiksel, Y. Ogata, D. Stoyan, and M. Tanemura. (1994). On parameter estimation for pairwise interaction processes. *International Statistical Review* 62, 99–117.

Gelman, A. (2006). Prior distributions for variance parameters in hierarchical models. *Bayesian Analysis* 1, 515–533.

Gelman, A. E. and D. Rubin. (1992). Inference from iterative simulation using multiple sequences [with discussion]. *Statitsical Science* 7, 457–511.

Gilks, W. R., D. G. Clayton, D. J. Spiegelhalter, N. G. Best, A. J. McNeil, L. D. Sharples, and A. J. Kirby. (1993). Modelling complexity: Applications of Gibbs sampling in medicine. *Journal of the Royal Statistical Society B* 55, 39–52.

Gilks, W. R., S. Richardson, and D. J. Spiegelhalter (eds.). (1996). *Markov Chain Monte Carlo in Practice.* London: Chapman & Hall.

Heagerty, P. and S. Lele. (1998). A composite likelihood approach to binary spatial data. *Journal of the American Statistical Association* 93, 1099–1111.

Higdon, D. (2002). Space and space-time modeling using process convolutions. In Anderson, E. A. C. (ed.), *Quantitative Methods for Current Environmental Issues.* London: Springer.

Hossain, M. and A. B. Lawson. (2005). Local likelihood disease clustering: Development and evaluation. *Environmental and Ecological Statistics* 12, 259–273.

Kauermann, G. and J. D. Opsomer. (2003). Local likelihood estimation in generalized additive models. *Scandinavian Journal of Statistics* 30, 317–337.

Kunsch, H. (1987). Intrinsic autoregressions and related models on the two-dimensional lattice. *Biometrika* 74, 517–524.

Lawson, A. B. (1992a). GLIM and normalising constant models in spatial and directional data analysis. *Computational Statistics and Data Analysis* 13, 331–348.

Lawson, A. B. (1992b). On fitting non-stationary Markov point processes on glim. In Dodge, Y. and J. Whittacker (eds.), *Computational Statistics I.* Berlin: Physica Verlag.

Lawson, A. B. (2006). *Statistical Methods in Spatial Epidemiology* (2nd ed.). New York: Wiley.

Lawson, A. B. (2013). *Bayesian Disease Mapping: Hierarchical Modeling in Spatial Epidemiology* (2nd ed.). New York: CRC Press.

Lindgren, F., H. Rue, and J. Lindstrom. (2011). An explicit link between Gaussian fields and Gaussian Markov random fields: The stochastic partial differential equation approach. *Journal of the Royal Statistical Society B* 74, 423–498.

Lindsay, B. G. (1988). Composite likelihood methods. *Contemporary Mathematics* 80, 221–238.

Liu, J. S. (2001). *Monte Carlo Strategies in Scientific Computing.* New York: Springer.

Marin, J. M. and C. Robert. (2014). *Bayesian Essentials with R* (2nd ed.). New York: Springer.

Møller, J., A. Syversveen, and R. P. Waagepetersen. (1998). Log Gaussian Cox processes. *Scandinavian Journal of Statistics* 25, 451–482.

Neal, R. M. (2003). Slice sampling. *Annals of Statistics* 31, 1–34.

Nott, D. J. and T. Rydén. (1999). Pairwise likelihood methods for inference in image models. *Biometrika* 86, 661–676.

Ripley, B. D. (1981). *Spatial Statistics.* New York: Wiley.

Ripley, B. D. (1987). *Stochastic Simulation.* New York: Wiley.

Robert, C. and G. Casella. (2005). *Monte Carlo Statistical Methods* (2nd ed.). New York: Springer.

Rue, H. and L. Held. (2005). *Gaussian Markov Random Fields: Theory and Applications.* New York: Chapman & Hall/CRC.

Rue, H., S. Martino, and N. Chopin. (2009). Approximate Bayesian inference for latent Gaussian models by using integrated nested Laplace approximations. *Journal of the Royal Statistical Society B* 71, 319–392.

Schrodle, B. and L. Held. (2011). Spatio-temporal disease mapping using INLA. *Environmetrics* 22, 725–734.

Schrodle, B., L. Held, A. Rieber, and J. Danuser. (2011). Using integrated nested Laplace approximations for the evaluation of veterinary surveillance data from Switzerland: A case-study. *Applied Statistics* 60, 261–279.

Simpson, D., F. Lindgren, and H. Rue. (2012). In order to make spatial statistics computationally feasible, we need to forget about the covariance function. *Environetrics* 23, 65–74.

Smith, A. F. M. and A. E. Gelfand. (1992). Bayesian statistics without tears: A sampling-resampling perspective. *American Statistician* 46, 84–88.

Smith, A. F. M. and G. Roberts. (1993). Bayesian computation via the Gibbs sampler and related Markov chain Monte Carlo methods. *Journal of the Royal Statistical Society B* 55, 3–23.

Stern, H. S. and N. A. C. Cressie. (1999). Inference for extremes in disease mapping. In Lawson, A. B., A. Biggeri, D. Boehning, E. Lesaffre, J. F. Viel, and R. Bertollini (eds.), *Disease Mapping and Risk Assessment for Public Health.* New York: Wiley, chapter 5.

Tanner, M. A. (1996). *Tools for Statistical Inference* (3rd ed.). New York: Springer Verlag.

Tibshirani, R. and T. Hastie. (1987). Local likelihood estimation. *Journal of the American Statistical Association* 82, 559–568.

Ugarte, M. D., A. Adin, T. Goicoa, and A. F. Militino. (2014). On fitting spatio-temporal disease mapping models using approximate Bayesian inference. *Statistical Methods in Medical Research* 23(6), 507–530.

Varin, C., G. Høst, and Ø. Skare. (2005). Pairwise likelihood inference in spatial generalized linear mixed models. *Computational Statistics and Data Analysis* 49, 1173–1191.

8

Statistical Tests for Clustering and Surveillance

Peter A. Rogerson
Departments of Geography and Biostatistics
State University of New York at Buffalo
Buffalo, New York

Geoffrey Jacquez
Department of Geography
State University of New York at Buffalo
Buffalo, New York

CONTENTS

8.1 Introduction

A common question that arises in spatial epidemiology is whether a mapped disease pattern exhibits any significant deviation from a pattern of uniform disease risk throughout the study area. When temporal data such as onset of disease is available in addition to data on geographic location, additional questions may be asked. Has the spatial pattern changed over time? How might we detect an emergent cluster of cases as quickly as possible?

In this chapter, we review several classic statistical approaches to these questions. The approaches are largely based in the context of statistical methods for testing hypotheses of complete spatial randomness, and they depend on both the nature of the data and the nature of the question. Locations of disease cases are most often given in terms of point locations (e.g., x–y or latitude–longitude coordinates) or the number of cases (often together with the number of people at risk of becoming a case) for a set of mutually exclusive and collectively exhaustive subregions that comprise the study region. A third type of data comes under the heading of "case-control" data—here, the locations of cases are compared with the locations of controls (e.g., the locations of healthy individuals without the disease), and the spatial patterns of the two events are examined to determine whether there is any spatial clustering of cases, relative to the clustering of controls. A fourth type of data is a disease rate calculated as the number of health events in a given area divided by the population at risk within that area. A fifth type of data is the space–time paths traced by individuals (e.g., cases and controls) as they move across their life course. These have been referred to as "geospatial lifelines" and "mobility histories" (Miller 2005; Sinha and Mark 2005).

There are several types of clustering that may be considered: temporal clustering, spatial clustering, space–time clustering, and interaction. Temporal clustering refers to an excess of cases in specific time periods, and may be the signature of an outbreak or seasonal disease pattern. Spatial clustering is defined by an excess of cases at specific locations (e.g., local or focused clustering, below) or by an unusual pattern over the entire study area (e.g., global clustering, below). Space–time clustering arises when cases are clustered in both space and time. This is expected, for example, when there is an outbreak of an infectious disease, or when chronic diseases, such as cancer, are associated with a localized exposure, such as a release of a carcinogen from a specific point source that varies through time. Interaction occurs when nearby cases occur at about the same time. This is expected for infectious diseases that require contact between susceptible and infectious individuals for infection transmission. It also may be found for acute physiological responses to localized irritants, such as emergency room visits for asthma arising from exposure to airborne dusts from grain loading activities. This chapter deals primarily with spatial clustering and disease surveillance methods.

The question arises as to what can be learned or inferred from statistical tests for disease clustering. Is there a mapping, for example, of disease processes to specific space–time case patterns? What kinds of null and alternative hypotheses are available, and do they correspond to disease patterns expected for different transmission dynamics? Questions such as these are the topic of Chapter 3 in this handbook.

Besag and Newell (1991) provide a convenient classification scheme for the nature of statistical questions related to clustering. *Global* or *general* tests use a single statistic to decide whether to reject the null hypothesis of no raised risk in the study region. If the statistic is sufficiently large to reject the null hypothesis, the user is not provided with additional information about the size, shape, or location of the region or regions of raised risk. *Local* tests search for clusters anywhere in the study region and will, for each candidate cluster, report their locally elevated risk, spatial extent, and statistical significance. *Focused* tests consider the question of clustering from the perspective of a prespecified location of interest. We may wish to know, for example, whether disease risk is higher near a hazardous waste site than it is in locations far from the site. Focused tests are distinguished from local tests in that one must specify the location (called the focus) where disease risk is hypothesized to be elevated. Finally, Besag and Newell consider what they term "tests for the detection of clustering." These tests not only result in a decision regarding the null hypothesis of no clustering, but also provide the analyst with specific information on the location and size of clusters. This broad group of *scan tests* may be thought of as a set of many focused or local tests, where the entire map is scanned to see whether clustering exists around *any* possible location on the map. Because scan tests consist of a large number of correlated local tests, the statistical significance of the results must be adjusted for the multiple testing that occurs. Waller and Jacquez (1995) provide a useful framework for thinking about the components of a statistical test for disease clustering. They defined five components: the null hypothesis, the null spatial model, the alternative hypothesis, the cluster statistic, and the distribution of the cluster statistic under the null spatial model.

The *null hypothesis* is a statement regarding the spatial pattern of cases expected in the absence of a disease process. Most frequently employed is the null hypothesis that disease risk is uniform across the study region.

The *null spatial model* may be of two types, computational and theoretical. Randomization algorithms may be used by computers to generate realizations of the distribution of cases under the null hypothesis, for example, by "sprinkling" the cases across the study region in a manner consistent with the underlying sizes of the at-risk population. This is often referred to as a distribution-free approach. Theoretical distributions may be used when certain assumptions about the data are met (e.g., a sufficient number of observations). Here, the distribution of the test statistic under the null hypothesis is assumed to follow a theoretical distribution, such as the Gaussian or inverse exponential.

The *alternative hypothesis* describes the spatial pattern expected under the cluster process. This may be the omnibus "not the null hypothesis" or a more specific alternative, such as "clustering about a specific location" (e.g., a focused test). Notice when the omnibus null is used, the entire universe of possible patterns is covered under the null and alternative hypotheses; this may not be true when a more specific alternative is employed, in which case the statistical test will have poor power to detect departures from the null not covered by that specific alternative hypothesis.

The *cluster statistic* is a statistic designed to be sensitive to a specific geographic pattern of cases. For nearest-neighbor statistics, this might be based on the distance from a given case to its nearest neighbor that is also a case (and not a control). Hence, the cluster test one employs should be based on a cluster statistic that will be large when case clustering exists.

The *distribution of the cluster statistic under the null spatial model* is obtained either through randomization (also referred to as Monte Carlo techniques) or based on distribution theory. The statistical significance of the cluster test is found through comparison of the observed value of the test statistic to its distribution under the null spatial model. When randomization is used, the probability of the observed value of the cluster statistic is found by counting the number of simulation runs for which the value of the cluster statistic under randomization is larger than or equal to the observed value of the cluster statistic. This is then divided by the number of randomization runs using:

$$P(s^*|H_0) = \frac{NGE+1}{Nruns+1}.$$

Here, $P(s^*|H_0)$ is the probability of the observed value of the test statistic under the null hypothesis, NGE is the number of randomization runs for which the value of the test statistic under simulation is greater than or equal to $s*$, and Nruns is the total number of simulation runs conducted. When using distribution theory, one compares the observed value of the test statistic to its theoretical distribution, and its probability is given by that portion of the distribution that is greater than or equal to that observed value. In this chapter, we primarily describe statistical tests for clustering that rely on distribution theory.

The remainder of this chapter is organized as follows. Section 8.2 provides an overview of global tests. Section 8.3 summarizes local or focused tests, and scan tests are discussed in

Section 8.4. Section 8.5 is devoted to a review of statistical surveillance, Section 8.6 provides an example, and Section 8.7 provides a summary.

8.2 Global Tests

8.2.1 Historical antecedents: Nearest-neighbor and quadrat tests

Global statistical tests for testing the null hypothesis of spatial randomness have their antecedents in two tests developed in the field of species ecology and biogeography.

8.2.1.1 Quadrat tests

Quadrat tests assess spatial patterns by first overlaying a grid of cells on top of a study area that contains a set of points representing observations (cases), and then counting the number of cases within each cell. Student (1907) (better known for Student's t-distribution, and lesser known as a quality control employee of the Guinness Brewing Company whose actual name was William Gossett) first developed the methods while studying counts of cells in blood using a hemocytometer. Gleason (1920) and Blackman (1935) were early adopters of the approach in the field of species ecology, and Fisher (1925) suggested the statistical test that is widely used today:

$$\chi^2 = \frac{\sum_{i=1}^{m}(x_i - \bar{x})^2}{\bar{x}}.$$

Here, m is the number of quadrats, x_i is the frequency observed in quadrat i, and $\bar{x}$ is the mean number of observations per quadrat. Under the null hypothesis that the observations occur randomly across quadrats, this statistic has a chi-square distribution with $m-1$ degrees of freedom. One question that arises in the use of this test is the optimal size of the quadrat. Quadrats that are too small will result in many cells with no observations, while quadrats that are too large are not ideal since the spatial pattern of observations within the quadrat is ignored. One guideline is to choose a quadrat size that is in line with the best educated guess that can be made about the size of any potential cluster. The classic version of the quadrat test outlined above assumes that the expected number of cases in each cell does not vary from cell to cell, but this is easily generalized to allow for the more realistic situation where say the observed number of disease cases in each of the m subregions (which also do not necessarily have to be quadrats, cells, or regions of the same size or shape) is compared with the expected number for that cell (e_i)—a simple use of the common chi-square goodness-of-fit test:

$$\chi^2 = \frac{\sum_{i=1}^{m}(x_i - e_i)^2}{e_i}.$$

Although it is commonly assumed that the expected frequency in each cell should be at least five for the test to be valid, much work has shown that this is overly conservative (see, e.g., Koehler and Larntz 1980); more realistic assumptions are that the values of e_i exceed about 1, and that the value of $(\sum e_i)^2/m$ is greater than or equal to 10.

8.2.1.2 Nearest-neighbor tests

The nearest-neighbor statistic (Clark and Evans 1954) tests the null hypothesis that a point pattern is random. This is achieved by comparing the average distance from a point to its nearest neighboring point ($\bar{d}$) with the distance expected between nearest neighbors in a random pattern. Clark and Evans give this latter quantity as $0.5/\sqrt{\rho}$, where ρ is the average density of points in the study area (i.e., the number of points n, divided by the size of the area, A). Since Clark and Evans also provide the variance of the distance between points and their nearest neighbors, a z-test can be carried out as follows:

$$z = \frac{\bar{d} - 1/(2\sqrt{\rho})}{0.261/\sqrt{n\rho}}.$$

This statistic is affected by the presence of boundary effects. The expressions for the expected value and variance of nearest-neighbor distance assume an infinite plane, and in practice, the observed distance between a point and its nearest neighbor will be, on average, larger than $0.5/\sqrt{\rho}$. One way to address boundary effects is to use Monte Carlo testing to simulate the null hypothesis—in this case, deriving the average distance to nearest neighbors in random patterns so that the unusualness of the observed value may be assessed. Another is to retain the z-statistic described above, but to add a guard area around the study region—points falling in the guard area can then be potential nearest neighbors of the points within the original study region.

The statistic is also affected by the shape of the study area—randomly distributed points in long, rectangular regions have nearest neighbors that are on average closer than a distance of $0.5/\sqrt{\rho}$.

It should also be noted that since it focuses on nearest neighbors, the statistic is designed to detect deviations from spatial randomness at small spatial scales. Failure to reject the null hypothesis does not mean that deviations from randomness may occur at other scales. To address this, extensions of the statistics to second, third, and so forth, closest neighbors have been developed.

Perhaps the biggest limitation is that a homogeneous background—a uniform density of the at-risk population—is assumed. In epidemiological applications, one is often not interested in whether the spatial pattern of points is random. The spatial pattern of disease cases is most often *not* spatially random—the more relevant question is whether it is random relative to the location of the typically nonuniform distribution of the population. In numerical ecology, the term *neutral models* was posed to incorporate more complex underlying distributions into the null hypothesis. In spatial epidemiology, the term *neutral models* has been used to describe realistic distributions of cases expected in the absence of the disease process. Hence, a neutral model in spatial epidemiology conceivably could account for variation in the size of the underlying at-risk population, relevant covariates such as age and socioeconomic status, and other factors that might impact the distribution of cases one might plausibly expect in the absence of a spatial disease process (refer to Liebisch et al. [2002] for simulation algorithms that account for spatial pattern and Goovaerts and Jacquez [2004] for a discussion of neutral models in spatial epidemiology). The Cuzick–Edwards test (see below) addresses the problem of nonhomogeneous geographic distribution of the at-risk population.

8.2.2 Moran's *I*

One of the most commonly used measures of spatial patterns (spatial autocorrelation) is Moran's I, which focuses on the values observed in pairs of regions:

$$I = \frac{n \sum_{i=1}^{n} \sum_{j=1}^{n} w_{ij}(x_i - \bar{x})(x_j - \bar{x})}{\left(\sum_{i=1}^{n} \sum_{j=1}^{n} w_{ij}\right) \sum_{i=1}^{n} (x_i - \bar{x})^2},$$

where there are n regions, and where w_{ij} is a weight describing the relationship between regions i and j. Most typically, this definition is one of binary adjacency: w_{ij} is equal to 1 if regions i and j are adjacent, and is equal to zero otherwise. The numerator of this equation provides its conceptual core—if pairs of regions are simultaneously above (or below) average, this contributes to a positive value of I; if one region is above average and the other below, this contributes negatively to the numerator. The denominator serves to scale the measure so that its value lies typically between -1 and $+1$, although it is possible for I to fall outside this range. A pattern such that nearby (i.e., connected by positive weights) locations have a similar value results in positive I values and is called positive spatial autocorrelation. When nearby locations are dissimilar in value, Moran's I is negative, and this is called negative spatial autocorrelation. In a random spatial pattern, I is near zero (the expectation of the statistic is actually slightly negative and equal to $-1/(n-1)$). The null hypothesis of no spatial association may be tested using the observed value of I, its expectation, and expressions for the variance of I (see, e.g., Griffith 1987). Application of Moran's I is usually made by assuming that $w_{ii} = 0$; thus, it is usually employed as a measure of spatial association, to determine, for example, whether high (low) values are surrounded by high (low) values.

It should be noted that Moran's I, as written above, is not written explicitly for epidemiology, in the sense that it does not include either a numerator or a denominator. Oden (1995) has modified the statistic to address this issue. Specifically, his *Ipop* statistic explicitly accounts for variation in population sizes across regions, and also provides assessment of what occurs within regions. These modifications make it more suitable for testing geographic clustering in the context of spatial epidemiology.

8.2.3 Tango's statistic and spatial chi-square statistics

Tango (1995) introduced the global statistic

$$C_G = \sum_{i=1}^{n}\sum_{j=1}^{n} w_{ij}(r_i - p_i)(r_j - p_j),$$

where r_i and p_i represent, respectively, the observed and expected proportions of cases falling in region i. Again, pairs of regions that are simultaneously above (or below) their expectations will contribute to a larger value of the test statistic, C_G. In addition, there is an internal contribution made by the squared deviation between observed and expected proportions in each region (weighted by w_{ii}).

Tango gives the expectation and variance of this statistic under the null hypothesis that cases are distributed at random, according to their regional expectation. This allows for a hypothesis test based on the assumption that the null distribution is approximately normal. Tango also gives a more accurate chi-square approximation to the null distribution.

Rogerson (1997) gives as a variant of this a spatial chi-square statistic:

$$R = \sum_{i=1}^{n}\sum_{j=1}^{n} w_{ij}\frac{(r_i - p_i)}{\sqrt{p_i}}\frac{(r_j - p_j)}{\sqrt{p_j}}.$$

Note that this is simply Tango's statistic where Tango's weights are divided by $\sqrt{p_j p_j}$, and thus Tango's expressions and approximations may be used to test the null hypothesis of no spatial clustering. If $w_{ij} = 0$, when $i \neq j$, R reduces to the aspatial chi-square statistic used for the quadrat test. In the general case, it is the contributions made by both individual regions (as in the case of the quadrat method) and pairs of regions (as in Moran's I) that ultimately determine the significance of these global statistics.

8.2.4 Cuzick–Edwards test

We have seen that a drawback of the nearest-neighbor test is that it does not account for spatial variation in the background population—disease cases may be more clustered than random, but this could simply be due to the fact that the population is more clustered than random. The Cuzick–Edwards test compares the point pattern of cases with the point pattern of "controls"—these controls, for example, could be healthy individuals with relevant characteristics (such as age, sex, and race) that are similar to the cases. The null hypothesis is now one that states that cases are no closer to other cases than they are to controls.

The Cuzick–Edwards statistic is easy to state—it is simply a count of the nearest neighbors of a case that are also cases. More generally, though, the statistic is defined not necessarily for nearest neighbors, but for neighbors that allow for larger spatial scales. Let $w_{ij} = 1$ if point j is a k-nearest neighbor of point i. Let δ_i equal 1 if location i is a case and zero if it is a control. Then the statistic is

$$T_k = \sum_{i=1}^{n}\sum_{j=1}^{n} w_{ij}\delta_i\delta_j,$$

and this is equal to the total number of cases that are k-nearest neighbors of cases. Its expectation, under the null hypothesis that the case and control locations are labeled as cases and controls randomly, is

$$E[T_k] = \frac{kn_o(n_0 - 1)}{n - 1},$$

where n_0 is the number of cases, and n is the total number of cases and controls. Under the null hypothesis, the variance of T_k is approximately equal to the expectation (but Cuzick and Edwards give a more precise expression for the variance), and this allows hypothesis testing using a z-statistic, assuming that the null distribution is approximately normal. This assumption may be relaxed by using randomization approaches, and Cuzick and Edwards' test also can be modified to account for uncertainty in the spatial locations of cases and controls (Jacquez 1994).

A conceptual problem with Cuzick and Edwards' test is that it assumes a static geography; that is, the cases and controls are represented by single, spatial locations. In many instances, a more realistic representation would be a space–time thread, such that each person can travel over the study duration—for example, changing residences over their life course. A number of statistical tests and modeling approaches have been developed specifically for residential mobility histories of cases and controls. These include Q-statistics (Jacquez et al. 2006), generalized additive models (Vieira et al. 2005), risk surface modeling (Han et al. 2005), and exposure opportunity models (Sabel et al. 2000).

8.3 Focused Tests

8.3.1 Aspatial: Local quadrat test

Focal tests are designed to test the null hypothesis of no raised incidence around a prespecified location. For example, there may be interest in whether disease incidence is raised around a toxic site. In perhaps the simplest case, an aspatial focused test could be carried out by examining the observations in a single region or quadrat. For a sufficiently large expected number of cases, the normal approximation

$$z = \frac{x_i - e_i}{\sqrt{e_i}}$$

may be used; an alternative approach for small expectations would be to make use of the Poisson distribution. Note here that the sum of the squared z-values across regions is equal to the global chi-square statistic. The fact that local statistics (or a function of them) sum to a corresponding global statistic (up to a constant) is a desirable characteristic of local statistics (Anselin 1995).

8.3.2 Stone's test

An early focused test was suggested by Stone (1988). Here subregions are ordered in terms of distance away from the prespecified location. For each subregion, the ratio of the cumulative number of cases observed to the cumulative number expected is formed, and the statistic is simply the maximum of these ratios. Its statistical significance can be assessed with some effort analytically, but is more often evaluated by comparing it with the results of a simulation of the null hypothesis.

8.3.3 Score statistics

The score statistic is widely used to evaluate the possibility of raised incidence around a source location, with some of the earliest descriptions attributed to Lance Waller and Andrew Lawson (Lawson 1989; Waller et al. 1992). It is defined as

$$U_i = N \sum_{j=1}^{n} w_{ij}(r_j - p_j),$$

where N is the total number of cases and the other terms are defined as in Section 8.2.3. When the null hypothesis holds, this statistic has a distribution that is approximately normal, with mean zero and variance

$$V[U_i] = N \left\{ \sum_{j=1}^{n} w_{ij}^2 p_j - \left(\sum_{j=1}^{n} w_{ij} p_j \right)^2 \right\}.$$

This test has the desirable characteristic that it is uniformly most powerful against the alternative where the expected number of cases is $Np_j(1 + w_{ij}\varepsilon)$ instead of the Np_j that characterizes the null hypothesis. Lawson (1993) devotes attention to alternative specifications of the weights that may be useful in these applications.

8.4 Scan Tests

8.4.1 Introduction

Scan tests are motivated by the desire to find spatial clusters on a map when the analyst has no a priori idea about the location, size, and shape of potential clusters. Unlike global statistics, scan tests result in an estimate of cluster location. Operationally, the tests may be thought of as a set of tests consisting of multiple local or focused tests. Because scan tests look for clusters in many locations, some adjustment for the multiple testing should be made. In the most straightforward case, this is achieved quite simply via a Bonferroni adjustment (explained in more detail in Section 8.4.2). More commonly, the adjustment is complicated by the fact that the local tests are often correlated. Adjustment also has been accomplished by using the number of significant local clusters as a single test statistic to assess the overall probability of clustering (see, e.g., Besag and Newell 1991).

8.4.2 Aspatial: Multiple local quadrat tests

In the search for subregions with higher than expected rates, it is tempting to examine the z-scores described in Section 8.3.1 for each region. If this is done for many such regions, some of the statistics are likely to be significant by chance alone. Thus, if the probability of a type I error is set at 0.05, we would expect one region of every 20 examined to yield a significant statistic. The Bonferroni adjustment is employed to ensure that the overall type I error probability remains at the desired level. It is implemented by dividing the desired level, α, by the number of tests (say, m) to be carried out, and then this result is used in conjunction with each of the z-tests. The probability of finding at least one of the m tests to be significant, when the null hypothesis is true, is approximately equal to α. However, it is widely acknowledged that the Bonferroni adjustment is overly conservative.

8.4.3 Some early scan tests

Openshaw et al. (1987) developed their geographical analysis machine to harness the increasing power of computing. This scan test is implemented by creating circles of many sizes around each grid point in a fine-resolution set of such grid points overlaid on the study area. For each circle, a Poisson test is used (via simulation of the null hypothesis) to determine how unusual the count of events is inside of the circle. This is repeated for a large number of circles; circles associated with p-values of less than 0.002 are left on the map. Clearly, the choice of 0.002, meant to adjust for the multiple testing, is arbitrary.

Besag and Newell (1991) improved upon this by noting that there was no need for simulation of events within each circle, and that the Poisson distribution could be used directly. For each focal location, they used as their statistic the number of surrounding subregions it would take to collect k cases, where k is chosen prior to the analysis. Rejection of the null hypothesis occurs when this number is smaller than would be expected when using the Poisson distribution. They also address the issue of multiple testing by comparing the number of significant focal regions with that expected when using a type I error probability of α.

Turnbull et al. (1990) adopt a similar approach, but in their case, they collect around each of the m focal locations or subregions a fixed, prechosen population size (say, P). This standardizes the denominator for each local test, and their statistic is simply the maximum number of cases that are found among the m neighborhoods that have been so constructed. They assess statistical significance by comparing this observed statistic with the distribution of statistics found via simulation of the null hypothesis.

Several other approaches to scan tests have been suggested, including the methods set forth by Fotheringham and Zhan (1996) and Rushton and Lolonis (1996). In the next two subsections, we review two other tests.

8.4.4 Spatial scan test

By far the most commonly employed scan test is the spatial scan statistic of Kulldorff and Nagarwalla (1994). There are now many variants of this test, including those that allow for the possibility of elliptical scanning windows and clusters (Kulldorff et al. 2006).

The spatial scan statistic is based upon examining the likelihood ratio within a window placed over the study area; the statistic itself is the maximum ratio found after windows of different sizes have scanned the study area. The ratio of the likelihood of the alternative hypothesis (that incidence is raised within the window) to the likelihood of the null hypothesis (that the probability that an individual has the disease in question is constant throughout the population) is

$$\frac{p_z^{c_z}(1-p_z)^{n_z-c_z}q_z^{C-c_z}(1-q_z)^{(N-N_z)-(C-c_z)}}{p_0^C(1-p_0)^{N-C}},$$

where there are C cases, N is the size of the population, and c_z and n_z are the number of cases and size of the population within the window, respectively. Furthermore, $p_0 = C/N$, $p_z = c_z/n_z$, and $q_z = (C-c_z)/(N-n_z)$.

This particular specification is based upon the assumption that under the null hypothesis, cases are distributed according to the binomial distribution; other distributions, such as the Poisson and multinomial, are also commonly employed.

8.4.5 Geometrical probability

Rogerson (2001) has suggested that the statistical work carried out in the literature on brain activation studies to detect regions of increased blood flow in the three-dimensional brain be modified for use in two-dimensional studies of geographic clustering. This work has its immediate antecedents in the work of Worsley (1996), who used results from the field of geometrical probability to find p-values associated with clustering for one, two, and three dimensions.

Regional data (y_j) are first smoothed using a Gaussian kernel:

$$z = \sum_j^n w_{ij} y_j$$

where the Gaussian weights are

$$w_{ij} = \frac{e^{-d_{ij}^2/2\sigma^2}}{\sqrt{\pi}\sigma}.$$

The distance between regions i and j is given by d_{ij}^2, and the bandwidth of the kernel is denoted by σ. To test the null hypothesis using a type I error probability of α, the maximum value of z_i across the n regions is compared with the critical value

$$z_{crit} = \sqrt{\sqrt{\pi}\ln\left(\frac{4\alpha(1+0.81\sigma^2)}{n}\right)}.$$

8.5 Statistical Surveillance in Spatial Epidemiology

A key goal in public health surveillance is to detect new trends as quickly as possible. Statistical approaches to quick detection have been suggested and used since the early work of Hill et al. (1968) and Weatherall and Haskey (1976); see Barbujani (1987) for a review.

These contributions were made primarily in the context of temporal surveillance, with the objective of spotting, for example, increases in congenital malformations quickly. More recent attention has been given to surveillance in a spatial context. In this case, spatial statistics of the type discussed in previous sections are themselves monitored over time, and if increases are detected, this is suggestive of temporal changes in spatial patterns. One might consider simply employing the methods discussed above time period after time period, but clearly some form of adjustment for the multiple testing is necessary or else type I errors would increase substantially, as some statistics would be "significant" by chance alone as more and more statistics were observed.

8.5.1 Shewart charts

A straightforward way to implement surveillance is to simply plot the z-scores of observations as they occur. Limits or thresholds are set, and if an observation occurs that is outside this limit, an "alarm" is declared. Suppose, for example, that each month we observe a new map of disease cases. For each map, we find Moran's I and its corresponding z-score. If we set our threshold at ± 3, then we would expect, under the null hypothesis, to witness an observation outside of these limits once every 370 observations (this corresponds to the fact that a standard normal table reveals that $0.0027 = 1/370$ of all observations have absolute value greater than 3). The threshold determines the false alarm rate. If we had used wider limits (e.g., ± 4), we would have fewer false alarms, but it would also take longer to detect a true change when it occurred.

These plots of z-scores over time, termed Shewart charts, are ultimately limiting because of their dependence on, and sensitivity to, single, outlying observations. Other methods allow for quicker detection of change when it occurs, and we turn to these in the next subsection.

8.5.2 Cumulative sum surveillance

Cumulative sum (or cusum) methods are designed to minimize the time to detection when change occurs. They were developed primarily within the context of industrial process control, where the goal is often to find defects in industrial machinery or processes quickly (see, e.g., Wetherill and Brown 1991; Montgomery 1996; Hawkins and Olwell 1998).

For the most commonly employed cases, where a standard normal random variable (z_t) is being monitored, and quick detection of an increase in the mean is desired, the one-sided cumulative sum after observation t is

$$S_t = \max(0, S_{t-1} + z_t - k),$$

where $S_0 = 0$ and k is a parameter chosen to be equal to one-half the size of the deviation one would like to detect quickly. The parameter k is commonly set equal to 1/2; this minimizes the time that the cumulative sum approach takes to find a change in the process from a mean of zero to a mean of 1.

Like Shewhart charts, alarms are declared when the cumulative sum exceeds a threshold. The threshold, h, is given by (Rogerson and Yamada 2009)

$$h \approx \frac{2k^2 \mathrm{ARL}_0 + 2}{2k^2 \mathrm{ARL}_0 + 1} \ln\left(\frac{2k^2 \mathrm{ARL}_0}{2k} + 1\right) - 1.166,$$

where ARL_0 is the desired number of observations between false alarms (analogous to the type I error probability in hypothesis testing). When $k = 1/2$, this reduces to

$$h \approx \frac{\mathrm{ARL}_0 + 4}{\mathrm{ARL}_0 + 2} \ln\left(\frac{\mathrm{ARL}_0}{2} + 1\right) - 1.166.$$

8.6 Example: Cluster Morphology Analysis of Pancreatic Cancer in Michigan

The question often arises regarding which test is best. Not surprisingly, this is a complex question, as each method has specific null hypotheses, assumptions regarding the underlying probability distribution, and specific spatial patterns they are sensitive to. Additionally, there are peculiarities specific to the geography and disease pattern being analyzed. Depending on how they are implemented, different techniques may or may not ignore geographic discontinuities such as rivers and lakes. One must consider, then, the statistical performance of different methods given the specifics of the at-risk population. This suggests undertaking a performance comparison premised on the disease outcome, geography being studied, distribution of the at-risk population, and population sizes for the disease under consideration. This is the motivation behind cluster morphology analysis (CMA), which may be thought of as "a meta-analysis of the results of clustering approaches found to have the best statistical performance for a given disease, geography, and at-risk population" (Jacquez 2009, p. 22). Here, we summarize the approach using the example of pancreatic cancer in southeastern Michigan. For details, refer to Jacquez (2009).

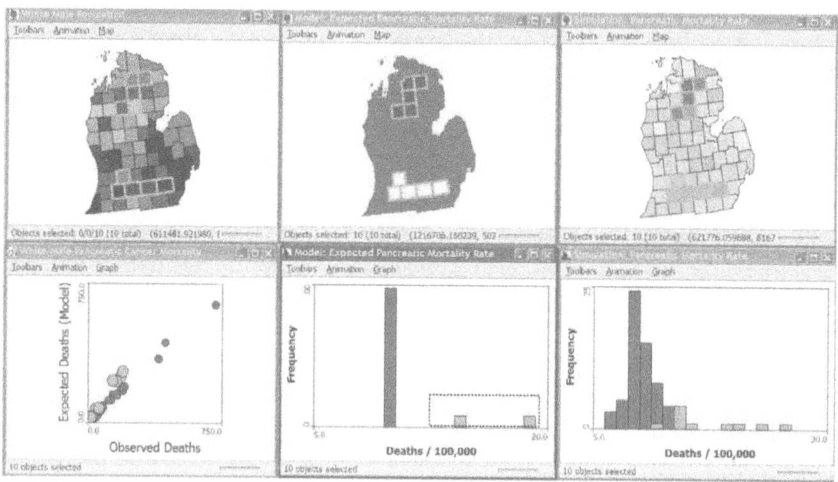

FIGURE 8.1
Cluster model construction for the example of pancreatic cancer mortality in Michigan counties. The at-risk population is defined as all white males from 1970 to 1994 (upper left). Two clusters were imposed, the S-shaped one in the north and the backward L shape in the south. The cluster model (center top) shows the observed background rate, with relative risks of 2.0 in the north cluster and 1.5 in the south cluster; also see center lower pane for a histogram of the risk model. The simulation model introduces stochasticity as a Poisson sampling process; see lower right pane. This provided sufficient information to simulate disease rates with the map from one simulation shown in the upper right pane. The relationships between modeled and observed deaths by county are in the lower left pane, with the counties in the simulated clusters highlighted. This validates the relationship between observed and expected deaths under the model. (From Jacquez, G. M., *Spatial and Spatio-Temporal Epidemiology*, 1, 19–29, 2009.)

There are six steps when conducting a CMA:

1. *Construct a realistic cluster model*: Use the disease geography under scrutiny to construct a cluster model incorporating the observed and expected numbers of cases, the size of the at-risk population, and the observed background risk for that cancer. Model clusters comprised of cases or contiguous areas (depending on whether the analysis is for case locations or rates in areas) with relative risks the researcher wishes to be able to detect (refer to Chapter 28 in this handbook for details on disease and cluster modeling). This example uses pancreatic cancer in Michigan counties (Figure 8.1).

2. *Power comparison*: Conduct a power analysis of several appropriate clustering techniques, using the cluster model from Step 1. This example evaluated 10 methods and a range of parameter values, for a total of 23 comparisons (Table 8.1).

TABLE 8.1
Results of CMA steps 2 and 3

Test	Parameter	Power	False negative	False positive	Specificity	Accuracy
B		1.000	0.000	0.069	0.931	0.714
FlexScan	$k = 3$	1.000	0.000	0.086	0.914	0.667
FlexScan	$k = 5$	1.000	0.000	0.086	0.914	0.667
Circular scan		1.000	0.000	0.103	0.897	0.625
FlexScan	$k = 7$	1.000	0.000	0.138	0.862	0.556
Kernel	$r = 10$	1.000	0.000	0.138	0.862	0.556
FlexScan	$k = 9$	1.000	0.000	0.224	0.776	0.435
Turnbull	$R = 2{,}000{,}000$	1.000	0.000	0.276	0.724	0.385
ULS scan RR		1.000	0.000	0.276	0.724	0.385
ULS scan 1-p(RR)		1.000	0.000	0.310	0.690	0.357
Kernel	$r = 15$	0.900	0.100	0.017	0.983	0.900
FlexScan	$k = 2$	0.900	0.100	0.017	0.983	0.900
Besag and Newell	$k = 210$ (north cluster size)	0.700	0.300	0.052	0.948	0.700
Kernel	$r = 25{,}000$	0.500	0.500	0.000	1.000	1.000
Turnbull	$R = 400{,}000$	0.500	0.500	0.034	0.966	0.714
Turnbull	$R = 800{,}000$	0.500	0.500	0.103	0.897	0.455
Local Moran		0.400	0.600	0.000	1.000	1.000
Kernel	$r = 50$	0.400	0.600	0.000	1.000	1.000
Kernel	$r = 20{,}000$	0.400	0.600	0.000	1.000	1.000
Turnbull	$R = 250{,}000$	0.400	0.600	0.017	0.983	0.800
G*		0.400	0.600	0.069	0.931	0.500
G		0.400	0.600	0.069	0.931	0.500
Besag and Newell	$k = 1034$ (south cluster size)	0.200	0.800	0.138	0.862	0.200

Source: Jacquez, G. M., *Spatial and Spatio-Temporal Epidemiology*, 1, 19–29, 2009.
Note: Repeated runs of the simulation model where the modeled clusters are known a priori allowed calculation of statistical power and proportion of false negatives and false positives. Results were then ranked by power from highest to lowest, resulting in the $m = 5$ best cluster methods.

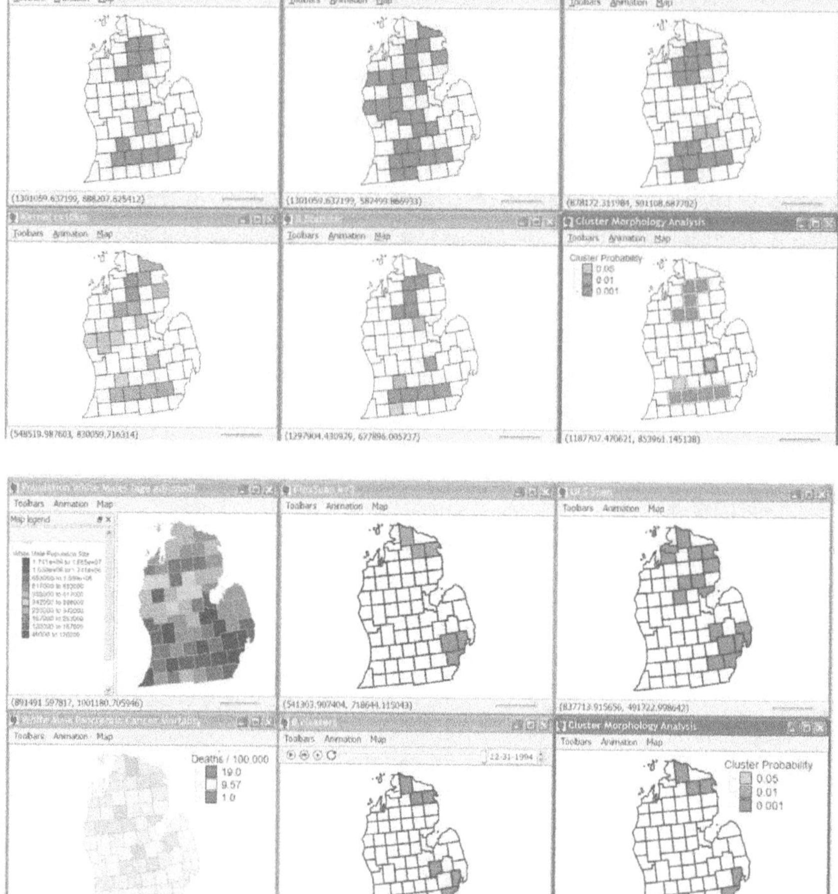

FIGURE 8.2
Cluster morphology analysis of pancreatic cancer mortality in white males, 1970–1994. Simulated clusters (top) and observed data (bottom). CMA correctly found the 10 counties in simulated clusters in the north and in the south, with one false positive. Applied to the real data, CMA found two clusters of high pancreatic cancer mortality in white males, one in the north and one in the southeast (bottom panes). The southeast cluster in Macomb and Wayne Counties was later validated using Surveillance, Epidemiology, and End Results (SEER) data. Screen capture from SpaceStat. (From Jacquez, G. M., *Spatial and Spatio-Temporal Epidemiology*, 1, 19–29, 2009.)

3. *Rank the methods*: Now rank the methods in terms of performance, first by power to detect true clusters and second by proportion of false positives. Use the top m methods (this example uses the top 5; $m = 5$) in the remaining steps. In this example, methods called the B statistic (Jacquez et al. 2008), upper level set (ULS) scan statistic (Patil et al. 2006), FlexScan (Takahashi et al. 2004), Kernel (Rushton and Lolonis 1996), and circular scan (Kulldorff 1997) had power = 1, with false positives from 0.069 to 0.276 (Table 8.1).

4. *Define cluster sets*: Define the cluster set $\mathbf{C}_{\mathrm{q}}$ to be the clusters found by method q. So should method q yield a cluster comprised of five counties, the cluster set would be those five counties.

5. *Combine the results of the best methods*: Combining the cluster results from several different methods may be accomplished in several ways; we think of this as a meta-analysis where each of the best methods casts a vote. In this example, we combine results from the five best methods by using geographic set intersection defined as $\mathbf{C}_{\mathrm{CMA}} = \mathbf{C}_1 \cap \mathbf{C}_2 \cap \ldots \cap \mathbf{C}_m$ (Figure 8.2).

The rationale is as follows. We recognize that the results of any clustering methods consist of two types: true cluster members and false positives. When several alternative clustering methods are employed, we might expect the false positives found by each method are not necessarily the same for the other methods—they are founded on different null and alternative hypotheses, and this should be reflected in the false positives for each technique. Combining information across the different "best" methods seeks to reduce the overall false positive rate while maintaining statistical power to find true clusters. This example found CMA had power = 1 and proportion of false positives = 0.017, which is lower than the false positive rate for any of the individual methods themselves.

8.7 Summary

Spatial epidemiologists are often fascinated by the complexity and, in some instances, the elegance of disease maps. When conducting cluster analysis and health surveillance, it can be easy to forget that every observation is a person, and that every case and control has his or her own story. This reinforces the need for analysts to keep individual-level location-referenced data confidential, and to treat each record as if it were our own health history. This chapter has presented an overview of spatial clustering and surveillance methods and is meant by no means to be exhaustive. We have focused primarily on well-known techniques with established distribution theory that allows for easy implementation and description. By focusing on a methodological overview, we have necessarily given little coverage to issues of cluster interpretation (what does a positive finding mean?) and to practical problems such as how to account for location uncertainty, incomplete data, incomplete reporting, sources of bias in the underlying data, and the propagation and visualization of uncertainty in the cluster statistics themselves. Should these topics be of interest, we suggest looking at Chapter 3 in this handbook, which deals with issues of cluster interpretation. We have provided no coverage of related clustering techniques that frequently are employed when exploring a cluster; these include temporal, space–time, and tests for interaction.

The history of clustering and surveillance methods is now relatively long, and dozens of methods have been proposed. Data availability, volume, and timeliness are increasing dramatically, and it is likely that several of the existing methods to date will need to be reexamined, modified, and even scrapped. For example, the question of "spatial-only" clustering is not as interesting as it used to be, because almost all disease data are now collected through time. Hence, questions of cluster persistence, and location and cluster size changes through time are growing in importance. Similarly, the data models themselves are becoming increasingly complex. It seems a spatial point pattern is not as relevant as it once was, since it ignores how that pattern changes through time. And would not a thread representation, which accounts for each individual's residential and daily movement history, be more appropriate when considering where and when exposures and health outcomes occurred? These issues are invigorating disease clustering and surveillance as a research area.

References

Anselin, L. 1995. Local indicators of spatial association—LISA. *Geographical Analysis* 27: 93–115.

Barbujani, G. 1987. A review of statistical methods for continuous monitoring of malformation frequencies. *European Journal of Epidemiology* 3: 67–77.

Besag, J. and Newell, J. 1991. The detection of clusters in rare diseases. *Journal of the Royal Statistical Society: Series A* 154: 143–155.

Blackman, G. E. 1935. A study by statistical methods of the distribution of species in grassland associations. *Annals of Botany* 49: 749–777.

Clark, P. J. and Evans, F. C. 1954. Distance to nearest neighbor as a measure of spatial relationships in populations. *Ecology* 35: 445–453.

Cuzick, J. and Edwards, R. 1990. Spatial clustering for inhomogeneous populations [with discussion]. *Journal of the Royal Statistical Society: Series B* 52: 73–104.

Fisher, R. A. 1925. *Statistical Methods for Research Workers.* Edinburgh, Oliver and Boyd, pp. 239.

FlexScan: Software for the flexible spatial scan statistic. National Institute of Public Health, Saitama, Japan.

Fotheringham, A. S. and Zhan, F. B. 1996. A comparison of three exploratory methods for cluster detection in spatial point patterns. *Geographical Analysis* 28: 200–218.

Gleason, H. A. 1920. Some applications of the quadrat method. *Bulletin of the Torrey Botanical Club* 47: 21–33.

Goovaerts, P. and Jacquez, G. M. 2004. Accounting for regional background and population size in the detection of spatial clusters and outliers using geostatistical filtering and spatial neutral models: The case of lung cancer in Long Island, New York. *International Journal of Health Geographics* 3: 14.

Griffith, D. 1987. Spatial autocorrelation: a primer. Washington, DC: Association of American Geographers.

Han, D., Rogerson, P. A., Bonner, M. R., Nie, J., Vena, J. E., Muti, P., Trevisan, M., and Freudenheim, J. L. 2005. Assessing spatio-temporal variability of risk surfaces using residential history data in a case control study of breast cancer. *International Journal of Health Geographics* 4: 9.

Hawkins, D. M. and Olwell, D. H. 1998. *Cumulative Sum Charts and Charting for Quality Improvement.* New York: Springer-Verlag.

Hill, G. B., Spicer, C. C., and Weatherall, J. A. C. 1968. The computer surveillance of congenital malformations. *British Medical Journal* 24: 215–218.

Jacquez, G. M. 1994. Cuzick and Edwards' test when exact locations are unknown. *American Journal of Epidemiology* 140: 58–64.

Jacquez, G. M. 2009. Cluster morphology analysis. *Spatial and Spatio-Temporal Epidemiology* 1: 19–29.

Jacquez, G. M., Kaufmann, A., and Goovaerts, P. 2008. Boundaries, links and clusters: A new paradigm in spatial analysis? *Environmental and Ecological Statistics* 15: 403–419.

Jacquez, G. M., Meliker, J. R., Avruskin, G. A., Goovaerts, P., Kaufmann, A., Wilson, M. L., and Nriagu, J. 2006. Case-control geographic clustering for residential histories accounting for risk factors and covariates. *International Journal of Health Geographics* 5: 32.

Koehler, K. and Larntz, K. 1980. An empirical investigation of goodness-of-fit statistics for sparse multinomials. *Journal of the American Statistical Association* 75: 336–344.

Kulldorff, M. 1997. A spatial scan statistic. *Communications in Statistics* 26: 1481–1496.

Kulldorff, M., Huang, L., Pickle, L., and Duczmal, L. 2006. An elliptical spatial scan statistic. *Statistics in Medicine* 25: 3929–3943.

Kulldorff, M. and Nagarwalla, N. 1994. Spatial disease clusters: Detection and inference. *Statistics in Medicine* 14: 799–810.

Lawson, A. B. 1989. Score tests for detection of spatial trend in morbidity data. Dundee, Scotland: Dundee Institute of Technology.

Lawson, A. B. 1993. On the analysis of mortality events associated with a prespecified fixed point. *Journal of the Royal Statistical Society: Series A* 156: 363–377.

Liebisch, N., Jacquez, G. M., Goovaerts, P., and Kaufmann, A. 2002. New methods to generate neutral images for spatial pattern recognition. In *Geographic Information Science*, ed. M. J. Egenhofer and D. M. Mark, vol. 2478 of Lecture Notes in Computer Science, 181–195. Berlin: Springer-Verlag.

Miller, H. J. 2005. What about people in geographic information science? In *Re-Presenting Geographical Information Systems*, ed. P. Fisher and D. Unwin, 215–242. New York: John Wiley.

Montgomery, D. 1996. *Introduction to Statistical Quality Control*. New York: John Wiley.

Oden, N. 1995. Adjusting Moran's *I* for population density. *Statistics in Medicine* 14: 17–26.

Openshaw, S., Charlton, M., Wymer, C., and Craft, A. 1987. A mark 1 geographical analysis machine for the automated analysis of point data sets. *International Journal of Geographical Information Systems* 1: 335–358.

Patil, G. P., Modarres, R., Myers, W. L., and Patankar, A. P. 2006. Spatially constrained clustering and upper level set scan hotspot detection in surveillance geoinformatics. *Environmental and Ecological Statistics* 13: 365–377.

Rogerson, P. 1997. Surveillance methods for monitoring the development of spatial patterns. *Statistics in Medicine* 16: 2081–2093.

Rogerson, P. 2001. A statistical method for the detection of geographic clustering. *Geographical Analysis* 33: 215–227.

Rogerson, P. and Yamada, I. 2009. *Statistical Detection and Surveillance of Geographic Clusters*. Boca Raton, FL: Taylor & Francis.

Rushton, G. and Lolonis, P. 1996. Exploratory spatial analysis of birth defect rates in an urban population. *Statistics in Medicine* 15: 717–726.

Sabel, C. E., Gatrell, A. C., Loytonen, M., Maasilta, P., and Jokelainen, M. 2000. Modelling exposure opportunities: Estimating relative risk for motor neurone disease in Finland. *Social Science and Medicine* 50: 1121–1137.

Sinha, G. and D. Mark. 2005. Measuring similarity between geospatial lifelines in studies of environmental health. *Journal of Geographical Systems* 7: 115–136.

Stone, R. 1988. Investigation of excess environmental risks around putative sources: Statistical problems and a proposed test. *Statistics in Medicine* 7: 649–660.

"Student." 1907. On the error of counting with a haemocytometer. *Biometrika* 5: 351–360.

Takahashi, K., Yokoyama, T. and Tango, T., 2004. FleXScan: Software for the flexible spatial scan statistic. National Institute of Public Health, Japan.

Tango, T. 1995. A class of tests for detecting "general" and "focused" clustering of rare diseases. *Statistics in Medicine* 7: 649–660.

Turnbull, B. W., Iwano, E. J., Burnett, W. S., Howe, H. L., and Clark, L. C. 1990. Monitoring for clusters of disease: Application to leukemia incidence in upstate New York. *American Journal of Epidemiology* 132: S136–S143.

Vieira, V., Webster, T., Weinberg, J., Aschengrau, A., and Ozonoff, D. 2005. Spatial analysis of lung, colorectal, and breast cancer on Cape Cod: An application of generalized additive models to case-control data. *Environmental Health* 4: 11.

Waller, L. A. and Jacquez, G. M. 1995. Disease models implicit in statistical tests of disease clustering. *Epidemiology* 6: 584–590.

Waller, L. A., Turnbull, B. W., Clark, L. C., and Nasca, P. 1992. Chronic disease surveillance and testing of clustering of disease and exposure: Application to leukemia incidence and TCE-contaminated dumpsites in upstate New York. *Environmetrics* 3: 281–300.

Weatherall, J. A. C. and Haskey, J. C. 1976. Surveillance of malformations. *British Medical Journal* 32: 39–44.

Wetherill, G. W. and Brown, D. W. 1991. *Statistical Process Control: Theory and Practice.* New York: Chapman & Hall.

Worsley, K. J. 1996. The geometry of random images. *Chance* 9(1): 27–40.

9

Scan Tests

Inkyung Jung
Department of Biostatistics and Medical Informatics
Yonsei University College of Medicine
Seoul, Korea

CONTENTS

9.1 Introduction

Scan statistics have been extensively studied since Naus first stated the problem of clustering points on a line and in two dimensions in 1965 (Naus 1965a, 1965b). Various topics on scan statistics are thoroughly covered in the books by Glaz and Balakrishnan (1999) and Glaz et al. (2001, 2009). In spatial epidemiology, we are often concerned with the location, size, and intensity of clusters when studying local area clustering (Pfeiffer et al. 2008). The spatial scan statistic proposed by Kulldorff (1997) is one of the most popular methods for identifying local clusters. This method imposes a large number of scanning windows with variable sizes on a study region and compares the areas inside versus those outside the scanning windows. The procedure of finding the most likely cluster is formulated using a hypothesis testing framework with a test statistic based on the likelihood ratio test.

Spatial scan statistics have been developed for various probability models, including Bernoulli and Poisson (Kulldorff 1997), ordinal (Jung et al. 2007), multinomial (Jung et al. 2010), exponential (Huang et al. 2007), and normal (Kulldorff et al. 2009; Huang et al. 2009). The choice of model depends on the application. The method may be applied in purely spatial (two-dimensional) or spatiotemporal (three-dimensional) settings. A scanning window is circular shaped in two dimensions and cylindrical shaped in three dimensions, with the base representing space and the height representing time. Other shapes, such as elliptical (Kulldorff et al. 2006) and irregular (Patil and Taillie 2004; Duczmal and Assunção 2004; Tango and Takahashi 2005) shapes, can also be considered. The standard method of inference conducted on the detected clusters is Monte Carlo (MC) hypothesis testing (Dwass 1957). Recently, a p-value approximation using an extreme value distribution has been proposed to reduce computational burden and obtain precise p-values (Abrams et al. 2010; Read et al. 2013, Jung and Park 2015).

Applications of spatial scan statistics have been successfully conducted in various areas, including not only spatial epidemiology, but also medical imaging, astronomy, and forestry. In cancer epidemiology, geographical variation may exist in the incidence, prevalence, mortality, or survival rates of cancer. Spatial scan statistics can be used to identify spatial (or spatiotemporal) cancer clusters and evaluate the statistical significance of the clusters.

Section 9.2 introduces the various probability models of spatial scan statistics. Section 9.3 reviews the inference procedures, and Section 9.4 presents some extensions of the method regarding the shapes of the scanning windows and covariate adjustments. Section 9.5 provides examples of applications of spatial scan statistics in cancer epidemiology, infectious disease epidemiology, and syndromic surveillance, and details a case study using Korean colorectal cancer mortality data. Section 9.6 presents the summary and discussion.

9.2 Models and Test Statistics

The spatial scan statistic imposes a large number of scanning windows with variable sizes on a study region. At each data point location, overlapping circular scanning windows are constructed. Theoretically, the radius of the scanning windows varies continuously from zero to some upper limit. In reality, the number of scanning windows is finite because the number of data points is finite. The exact locations of the data points may be available, and the data may be available in an aggregated form, and hence the data point locations are centroids of geographical areas, such as census tracts or counties. When dealing with data in an aggregated form, a scanning circle includes all areas with their centroids within the circle.

The spatial scan statistic is based on the likelihood ratio test statistic and is formulated as follows:

$$\lambda = \frac{\max_{z \in Z, H_a} L(\theta|z)}{\max_{z \in Z, H_0} L(\theta|z)} = \max_{z \in Z} LR(z),$$

where Z represents the collection of scanning windows constructed on the study region, H_a is an alternative hypothesis, H_0 is a null hypothesis, and $L(\theta|z)$ is the likelihood function with parameter θ given window z. The null hypothesis says that there is no spatial clustering on the study region, and the alternative hypothesis is that there is a certain area with high (or low) rates of outcome variables. The null and alternative hypotheses and the likelihood function may be expressed in different ways depending on the probability model under consideration. Likelihood $L(\theta|z)$ does not depend on z under the null hypothesis, and test statistic λ is the maximum over $z \in Z$ of the likelihood ratio test statistics $LR(z)$ comparing inside and outside scanning window z. The most likely cluster is defined as the area associated with the value of λ.

9.2.1 Poisson model

The Poisson model is used to compare cases against the underlying population at risk, for example, using disease incidence or mortality data. The null hypothesis is written as $H_0 : p = q$ for all z, and the alternative hypothesis is $H_a : p > q$ for some z, where p and q are the intensities of the outcome variable (e.g., incidence rates) inside and outside scanning window z, respectively. The alternative hypothesis can be $H_a : p < q$ if we want to search for clusters with low rates. The likelihood ratio test statistic $LR(z)$ given z is expressed as

$$LR(z) = \left(\frac{c_z}{e_z}\right)^{c_z} \left(\frac{C - c_z}{C - e_z}\right)^{C-c_z} I\left(\frac{c_z}{e_z} > \frac{C - c_z}{C - e_z}\right),$$

where c_z is the observed number of cases within window z, e_z is the expected number of cases within z, C is the total number of cases in the whole study region, and $I()$ denotes an indicator function. More details on the derivation of likelihood ratio test statistic $LR(z)$ is given in the paper by Kulldorff (1997). The expected number of cases is calculated proportional to the population size when there are no covariates. The covariate-adjusted expected number of cases can be calculated using indirect standardization when we have categorical covariates to be controlled for.

The Poisson model has been extensively used in various areas. In geographical disease surveillance, the most common outcome is disease incidence or mortality. Jennings et al. (2005) used the Poisson-based scan statistic to determine whether statistically significant geographic clusters of high-prevalence gonorrhea cases can be located after controlling for race and ethnicity in Baltimore, Maryland. Other examples include studies of childhood mortality in rural Burkina Faso in West Africa (Sankoh et al. 2001) and of regional variation in the incidence of symptomatic pesticide exposure in Oregon (Sudakin et al. 2002).

9.2.2 Bernoulli model

When we have dichotomous outcome variables, such as cases and noncases of certain diseases, the Bernoulli model is used. The null hypothesis is written as $H_0 : p = q$ for all z, and the alternative hypothesis is $H_a : p > q$ for some z, where p and q are the outcome probabilities (e.g., the probability of being a case) inside and outside scanning window z, respectively. Given window z, the test statistic is

$$LR(z) = \frac{\left(\frac{c_z}{n_z}\right)^{c_z} \left(\frac{n_z - c_z}{n_z}\right)^{n_z-c_z} \left(\frac{C - c_z}{N - n_z}\right)^{C-c_z} \left(\frac{(N - n_z) - (C - c_z)}{N - n_z}\right)^{(N-n_z)-(C-c_z)}}{\left(\frac{C}{N}\right)^{C} \left(\frac{N - C}{N}\right)^{N-C}}$$

$$\cdot I\left(\frac{c_z}{n_z} > \frac{C - c_z}{N - n_z}\right),$$

where c_z and n_z are the numbers of cases and observations (cases and noncases) within z, respectively, and C and N are the total numbers of cases and observations in the whole study region, respectively. Kulldorff (1997) applied the Bernoulli model to sudden infant death syndrome (SIDS) data in North Carolina and found two significant clusters of higher incidence. The Poisson model was also applied to the same data, and the same clusters were found. The Poisson model is a good approximation to the Bernoulli model when the total number of cases C is very small compared to the total number of observations N, as in the North Carolina SIDS data.

9.2.3 Ordinal model

Categorical data with order information, such as cancer stage or grade, can be analyzed using the ordinal model. Jung et al. (2007) first proposed the ordinal model based on the likelihood ratio ordering alternative hypothesis. That is, the null and alternative hypotheses are written as $H_0 : p_1 = q_1, \ldots, p_k = q_k$ for all z versus $H_a : p_1/q_1 \leq \cdots \leq p_k/q_k$ for some z with at least one strict inequality. Here, p_k and q_k $(k = 1, \ldots, K)$ are probabilities that an observation belongs to category k inside and outside scanning window z, respectively. The alternative hypothesis ensures that detected clusters represent an area with high rates of more severe stage than the surrounding area. Using the opposing inequalities in the alternative hypothesis, we can search for clusters with low rates of more severe stage.

Given scanning window z, the likelihood ratio test statistic is expressed as

$$LR(z) = \frac{\prod_k(\prod_{i \in z} \hat{p}_k^{c_{ik}} \prod_{i \notin z} \hat{q}_k^{c_{ik}})}{\prod_k \left(\frac{C_k}{C}\right)^{C_k}}, \tag{9.1}$$

where $\hat{p}_k$ and $\hat{q}_k$ are the maximum likelihood estimates (MLEs) of p_k and q_k under the alternative hypothesis, c_{ik} is the number of observations in location i and category k, C_k is the total number of observations in category k, and C is the total number of observations in the whole study region. The MLEs $\hat{p}_k$ and $\hat{q}_k$ can be obtained using an isotonic regression or the pool-adjacent-violators algorithm. More details are provided in the paper by Jung et al. (2007).

Jung et al. (2007) showed that the proposed ordinal model has good power and detects correct clusters compared to the Bernoulli model with dichotomized categories. They successfully applied this model to Maryland prostate cancer data of stage and grade. However, the alternative hypothesis can be restrictive in that it does not include all situations in which the probabilities of more severe disease categories are higher. Jung and Lee (2012) proposed an ordinal model based on the stochastic ordering alternative hypothesis $H_a : \sum_{k=1}^{j} p_k \leq \sum_{k=1}^{j} q_k$, for all $j = 1, \ldots, K$ for some z, which includes the likelihood ratio ordering as a special case. Jung and Lee (2012) showed that the stochastic ordering-based method can detect spatial clusters with high rates of more severe disease status in more general situations that may be missed by the likelihood ratio ordering-based method.

9.2.4 Multinomial model

The multinomial model is used for spatial cluster detection for categorical data without intrinsic order information. The null hypothesis is the same as the ordinal model, and the alternative hypothesis is that p_k and q_k $(k = 1, \ldots, K)$ are not all the same for some z. Unlike the other models, the multinomial model does not search for clusters with "high" or "low" rates, but rather for clusters with different category distributions. The test statistic given window z can be written in the same way as in (9.1), but MLEs $\hat{p}_k$ and $\hat{q}_k$ are much simpler. They are simply the proportions of the number of cases in category k to the number of total cases inside and outside window z, respectively:

$$\hat{p}_k = \sum_{i\in z} c_{ik} \Big/ \sum_k \sum_{i\in z} c_{ik}$$

and

$$\hat{q}_k = \sum_{i\notin z} c_{ik} \Big/ \sum_k \sum_{i\notin z} c_{ik},$$

with the same notation for c_{ik} as in the previous section.

The multinomial model was applied to meningitis data in Nottingham and Derbyshire Counties in the United Kingdom to search for spatial clusters where the meningitis-type distribution statistically significantly differs from the remaining regions in the two counties (Jung et al. 2010).

9.2.5 Exponential model

Determining whether there are geographical clusters of people with shorter (or longer) than expected survival time may be of interest. Survival-type data can be analyzed using the exponential model proposed by Huang et al. (2007). The null and alternative hypotheses are written as $H_0 : \theta_{in} = \theta_{out}$ for all z and $H_a : \theta_{in} < \theta_{out}$ for some z, respectively, where θ_{in} and θ_{out} are the means of survival time inside and outside scanning window z, respectively. Survival time for each individual is assumed to follow an exponential distribution, which can deal with censored observations. The likelihood ratio statistic for a given window z is given by

$$LR(z) = \frac{\left(\frac{r_{in}}{\sum_{i\in z} t_i}\right)^{r_{in}} \left(\frac{r_{out}}{\sum_{i\notin z} t_i}\right)^{r_{out}}}{\left(\frac{R}{\sum_i t_i}\right)^R} I\left(\frac{r_{in}}{\sum_{i\in z} t_i} < \frac{r_{out}}{\sum_{i\notin z} t_i}\right),$$

where r_{in} and r_{out} are the number of noncensored individuals inside and outside window z, respectively, t_i is the observed survival time (i.e., the minimum of survival and censoring time) of individual i ($i = 1, \ldots, N$), and $R = r_{in} + r_{out}$ is the total number of noncensored individuals.

Huang et al. (2007) applied the method to prostate cancer survival data in Connecticut, adjusting for covariates. The authors showed that the method performs well for different survival distribution functions, such as gamma and lognormal distributions.

For survival-type data, a spatial scan statistic based on Cox's proportional hazard model has been proposed by Cook et al. (2007). The hypotheses are expressed using a regression parameter of Cox's model, and the test statistic is formulated using a score test statistic. The authors applied the method to the Home Allergens and Asthma prospective cohort study to find clusters of disease in the first 4 years of life for time to asthma, time to allergic rhinitis or hay fever, and time to eczema.

9.2.6 Normal model

For continuous outcome data such as birth weight, the normal model can be used. Kulldorff et al. (2009) proposed a scan statistic where the likelihood is calculated using the normal probability model. The null and alternative hypotheses are respectively written as $H_0 : \mu = \eta$ for all z and $H_a : \mu > \eta$ for some z, where μ and η are the means of outcome variables inside and outside scanning window z, respectively. The normal distribution has two parameters of mean and variance. Kulldorff et al. (2009) assumed a common variance inside and outside the scanning window under the alternative hypothesis. Given window z, the log-likelihood ratio test statistic $LLR(z)$, equivalent to $LR(z)$, is given by

$$LLR(z) = \log LR(z) = N \ln(\hat{\sigma}) + \sum_i \frac{(x_i - \hat{\mu})^2}{2\hat{\sigma}^2} - \frac{N}{2} - N \ln\left(\sqrt{\hat{\sigma}_z^2}\right),$$

where N is the total number of observations, x_i are the continuous observations $(i = 1, \ldots, N)$, $\hat{\mu} = \Sigma_i x_i/N$ and $\hat{\sigma}^2 = \sum_i (x_i - \hat{\mu})^2/N$ are the MLEs of the mean and variance under the null hypothesis, respectively, and $\hat{\sigma}_z^2$ is the MLE of the common variance under the alternative hypothesis, which is given by

$$\hat{\sigma}_z^2 = \frac{1}{N}\left\{\sum_{i \in z}(x_i - \hat{\mu}_z)^2 + \sum_{i \notin z}(x_i - \hat{\eta}_z)^2\right\}.$$

Here, $\hat{\mu}_z = \sum_{i \in z} x_i/n_z$ and $\hat{\eta}_z = \sum_{i \notin z} x_i/(N - n_z)$ are the MLEs of the mean parameters under the alternative hypothesis, where n_z is the number of observations inside window z. The $LLR(z)$ depends on z only through the last term, and hence, the most likely cluster is the area that minimizes the variance under the alternative hypothesis.

Kulldorff et al. (2009) applied the method for New York City birth weight data, finding two statistically significant clusters of low birth weight that corresponded to areas with high infant mortality. The authors mentioned that the normal model can be used for a wide variety of continuous data that may not be normally distributed, but do not recommend it for exponential or other types of survival data.

The normal model described above is used when individual continuous outcomes at each location are available. To evaluate spatial heterogeneity of continuous measures in population data such as incidence, survival, and mortality rates at the county level, a weighted normal scan statistic has been proposed by Huang et al. (2009). The method can investigate clusters of geographic units with unusually high or low continuous regional measures, where the weights reflect the uncertainty of the regional measures or the sample sizes in the geographic units. Huang et al. (2009) applied the method to census tract–level lung cancer survival rate data in Los Angeles County and county-level breast cancer mortality rate data in the United States. The standard normal model can be treated as a special case of the weighted normal model with homogeneous weights.

9.3 Inference Procedure

9.3.1 Monte Carlo hypothesis testing

To obtain a p-value of the most likely cluster, we need to know the null distribution of the test statistic, which is not the case for the spatial scan statistics. Instead, MC hypothesis testing (Dwass 1957) can be used. The MC method is the standard method for calculating p-values of detected clusters using spatial scan statistics. Under the null hypothesis, a large number of random data sets are generated and test statistics are computed for each random data set. The value of the test statistic from the original data set is compared with those from the randomly generated data sets. The p-value based on MC hypothesis testing is defined as $p = r/(R+1)$, where r is the rank of the test statistic from the original data set among all data sets, and R is the number of replicates (i.e., the number of random data sets). Often, $R = 999$ or 9999 is used.

9.3.2 Gumbel-based p-value

Although the MC hypothesis testing method is known to maintain the correct significance level regardless of the number of replications, a drawback is that we must increase the number of replications to obtain more precise p-values. Further, as Abrams et al. (2010) mentioned, statistical power is higher with more replicates. For data sets with a small number of locations, a large number of replications may not be a problem. In some cases, however, the MC hypothesis testing method can be computationally very intensive with a large number of replications.

Abrams et al. (2010) showed that p-value approximation based on the method of fitting a Gumbel distribution to random data sets generated under the null can be advantageous to MC hypothesis testing. The Gumbel-based p-value is defined as $p_G = 1 - F_G(\lambda)$, where $F_G(x) = \exp\{-\exp(-(x-\mu)/\sigma)\}$ is the cumulative density function of the Gumbel distribution with location parameter μ and scale parameter σ. These parameters can be estimated using the test statistics calculated from the randomly generated data sets under the null distribution using the method of moments or the maximum likelihood estimation method. Although this method also requires replications under the null, Abrams et al. (2010) and Read et al. (2013) showed that the method can produce relatively accurate p-values with a smaller number of replicates for spatial scan statistics with Poisson and Bernoulli data. Jung and Park (2015) evaluated the method for other probability models and showed that the Gumbel approximation also works well for multinomial and ordinal models.

9.3.3 Secondary clusters

In addition to the most likely cluster, there may exist secondary clusters with high-likelihood-ratio test statistic values. It is useful to report those secondary clusters as well when they have no geographical overlap with another reported cluster with a higher likelihood ratio. The statistical significance of a secondary cluster can be evaluated irrespectively of other clusters by comparing its likelihood ratio value with the maximum likelihood ratios from the generated random data sets, which is somewhat conservative (Kulldorff 1997). Alternatively, a sequential method for evaluating the statistical significance of secondary clusters can be used by removing the effect caused by the previously detected stronger clusters (Zhang et al. 2010). Through a simulation study, Zhang et al. (2010) showed that the sequential version maintains the type I error probability close to the nominal level and detects secondary clusters with higher power than the standard version.

9.3.4 Software

All spatial scan statistic models introduced in Section 9.2 are implemented into the SaTScan^{TM} software (Kulldorff and Information Management Services, 2015), which is freely available at www.satscan.org. Analyses can be conducted for purely spatial, purely temporal, and spatiotemporal settings in retrospective and prospective ways. MC hypothesis testing is the default inference procedure, and Gumbel-based p-values are provided for some models.

9.4 Extensions of Spatial Scan Statistics

9.4.1 Scanning window shapes

While circular-shaped scanning windows are the most commonly used, other shapes for scanning windows, such as squares, triangles, or ellipses, can be considered. When a true cluster is noncircular, the circular scan statistics may have difficulty correctly detecting the cluster. It is also shown that the circular spatial scan statistics tend to detect a cluster larger than the true cluster by swallowing neighboring areas with nonelevated risk (Tango 2010).

Patil and Taillie (2004) proposed the upper level set (ULS) scan statistic using irregularly shaped scanning windows consisting of geographically connected areas. Because the collection of all connected areas can be very large, it may be infeasible to explore all windows. Thus, Patil and Taillie (2004) defined the ULS to reduce the parameter space of the scanning windows. Duczmal and Assunção (2004) also considered irregularly shaped windows and proposed a simulated annealing (SA) algorithm to examine only the most promising windows in the configuration space by abandoning the directions that seem uninteresting, instead of calculating the likelihood ratio test statistics for all scanning windows.

The flexibly shaped (FS) spatial scan statistic proposed by Tango and Takahashi (2005) also considers irregularly shaped windows, but it makes a thorough search of all possible windows. The set of windows for a given area i is composed of the nearest k areas (of "length" k) of a given area for k from 1 to the preset maximum length K, usually set as 15 or 20 (by the authors). Takahashi et al. (2009) provided FlexScan software for this method.

Kulldorff et al. (2006) proposed an elliptic version of the spatial scan statistic, which uses an elliptic scanning window of variable location, shape (eccentricity), angle, and size, with and without an eccentricity penalty. The authors concluded that in terms of power, the elliptic scan statistic performs well for circular clusters, and that the circular scan statistic performs well for elliptic clusters. The elliptic scan statistic may detect clusters more accurately than the circular version if the true cluster is an elongated one.

9.4.2 Covariate adjustment

Searching for spatial clusters for a certain disease may require adjusting for covariates if the covariates are related to the disease and not geographically randomly distributed. Otherwise, clusters that are not interesting may be found and their interpretation may be erroneous. In cancer data, for example, age is often related to the disease risk, and hence we may find cancer clusters that are only areas consisting of older ages if proper adjustment for a confounding covariate is not performed.

For the Poisson model, covariate adjustment relies on calculating the covariate-adjusted expected number of cases using an indirect standardization method for categorical covariates (Kulldorff 1997) and using estimates from regression models (Klassen et al. 2005). For the exponential model, the survival and censoring time can be adjusted based on the risk estimates for each covariate by first fitting an exponential regression model without spatial information (Huang et al. 2007). For the normal model, covariate adjustment can be done by replacing the observed values with their residuals from linear regression modeling (Kulldorff et al. 2009).

For the Bernoulli, ordinal, and multinomial models, covariate adjustment is rather restricted. It can be accomplished using multiple data sets stratified by categorical covariates (Kulldorff et al. 2007). However, the method is still limited in that only categorical covariates are allowed, and that a large number of covariate categories may render the procedure infeasible. While the method cannot be applied to the ordinal model based on likelihood ratio ordering, it may be applied to the ordinal model based on stochastic ordering (Jung and Lee 2012).

Jung (2009) proposed using a generalized linear model approach to construct spatial scan statistics, which is readily in a form for covariate adjustment and formulated in a single framework for different probability models. The approach can address both the categorical and continuous covariates easily and even interaction terms among covariates. Jung (2009) applied the method for the Bernoulli model to Texas female breast cancer data to find spatial clusters with high rates of late-stage cancer cases adjusting for covariates of race or ethnicity and age group. The method has also been applied to the multinomial model (Jung et al. 2010).

9.4.3 Spatial scan statistics based on other types of statistics

Neill et al. (2006) developed the Bayesian spatial scan statistic, which computes the posterior probability that an event has occurred in each geographical area. The method does not require randomization testing, and the interpretation of results is based on the posterior probability that each area has been affected. This method was extended to a multivariate version (Neill et al. 2007; Neill and Cooper 2010).

A spatial scan statistic based on the Wald statistic in log-linear models was proposed by Zhang and Lin (2009). The scanning windows as well as the covariates are incorporated in the model, and the Wald statistic is constructed using the regression parameter estimate and standard error for scanning windows. The test statistic is defined as the maximum of the Wald statistic over all scanning windows. The authors showed that the model-based approach is useful for adjustments for covariates and overdispersion, and that the proposed method is comparable with the original spatial scan statistic in terms of power and location specificity through simulation.

9.5 Applications

9.5.1 Cancer epidemiology

In cancer epidemiology, we may be interested in geographical variation in the incidence, prevalence, or mortality of certain cancers. The spatial scan statistic has been applied to a number of studies to identify local cancer clusters of high rates of incidence, prevalence, or mortality. Pollack et al. (2006) used the Bernoulli-based scan statistic to identify geographic variation in colorectal cancer by stage at diagnosis in California. They found two clusters where the observed number of late-stage cancer occurrences differed from the number expected from the distribution over the rest of the state. Lin et al. (2015) identified clusters of village-level women's lung cancer mortality rates in Xuan Wei, China, using the Poisson-based scan statistic. Liu-Mares et al. (2013) identified pancreatic cancer clusters in Florida using the Poisson-based scan statistic and found that cases living within 1 mile of known arsenic-contaminated wells were significantly more likely to be diagnosed within a cluster relative to cases living more than 3 miles away from known sites.

In addition to cancer incidence, prevalence, or mortality, geographic variation in cancer screening and treatment can provide important information in cancer epidemiology. Sheehan et al. (2000) identified geographical areas with a high proportion of late-stage diagnosis cases in Massachusetts, which may reflect gaps in screening efforts. Gregorio et al. (2001) evaluated geographical differences across Connecticut in the proportions of cases with early-stage breast cancer treated by partial mastectomy.

9.5.2 Infectious disease epidemiology

The examples for cancer epidemiology in the previous section used a retrospective analysis, in which cluster detection analysis is performed once for historical data. In infectious disease epidemiology, prospective analyses may also be performed to quickly detect an emerging infectious disease outbreak. The space–time scan statistic can be used in a prospective manner with repeated periodic analyses to detect currently active geographical clusters of disease (Kulldorff 2001). Jones et al. (2006) applied the prospective space–time Poisson-based scan statistic to Chicago's 2002 shigellosis surveillance data to evaluate its utility in objectively describing clusters and assisting in the prioritization of the investigation. Fifty-two separate space–time analyses, one for each week of 2002, were run, and 12 live clusters were detected. Examples of retrospective analyses include Osei and Duker (2008), who identified four significant spatial clusters of cholera prevalence in Ghana for the year 2005; Hixson et al. (2011), who studied spatial clustering of HIV prevalence in Atlanta, Georgia; and Kammerer et al. (2013), who used the Poisson-based spatial scan statistic to detect outbreaks of tuberculosis during 2008–2009 in the United States.

9.5.3 Syndromic surveillance

The fundamental objective of syndromic surveillance is to identify illness clusters early, before diagnoses are confirmed and reported to public health agencies, and to mobilize a rapid response, thereby reducing morbidity and mortality (Henning 2004). Syndromic surveillance systems monitor routine healthcare contact, such as emergency room visits, outpatient visits, telephone calls, over-the-counter drug sales, and absenteeism data. The space–time permutation scan statistic (Kulldorff et al. 2005) is very useful when only the number of cases is available without suitable population-at-risk data. It makes minimal assumptions about the time, geographical location, or size of the outbreak, and it adjusts for natural purely spatial and purely temporal variation. Besculides et al. (2005) evaluated school absenteeism data for early outbreak detection in New York City using the prospective space–time scan statistic. van den Wijngaard et al. (2010) used the space–time permutation scan statistic to evaluate whether syndromic surveillance detects local outbreaks of lower-respiratory infection without swamping true signals by false alarms in the Netherlands using hospitalization data collected from the Dutch National Medical Register for 1999–2006.

9.5.4 A case study: Korean colorectal cancer mortality data

We obtained Korean colorectal cancer mortality data from Statistics Korea for the years 2010–2012. The total number of deaths reported for the 3-year period in South Korea was 13,167 males and 10,014 females. The data were aggregated into 250 Si–Gun–Gu (city–county–ward) administrative districts, grouped in 5-year age intervals. For the population data of the 250 administrative districts, we used the 2010 Population and Housing Census data from Statistics Korea. To search for clusters of high or low rates of colorectal cancer mortality, the Poisson-based spatial scan statistic was used with age group adjusted using indirect standardization. Analyses were conducted separately for each gender using SaTScan.

Figure 9.1 shows a choropleth map of age-standardized mortality rates (ASRs) per 100,000 people for colorectal cancer in South Korea (2010–2012), overlaid with the most likely and secondary clusters identified for male (a) and female (b). The circles without lines represent clusters with high mortality rates, and the circles with hatched lines represent those with low mortality rates. One low-rate cluster in the south region and one high-rate cluster, including some areas of Seoul (the capital city of South Korea) and Incheon (one of

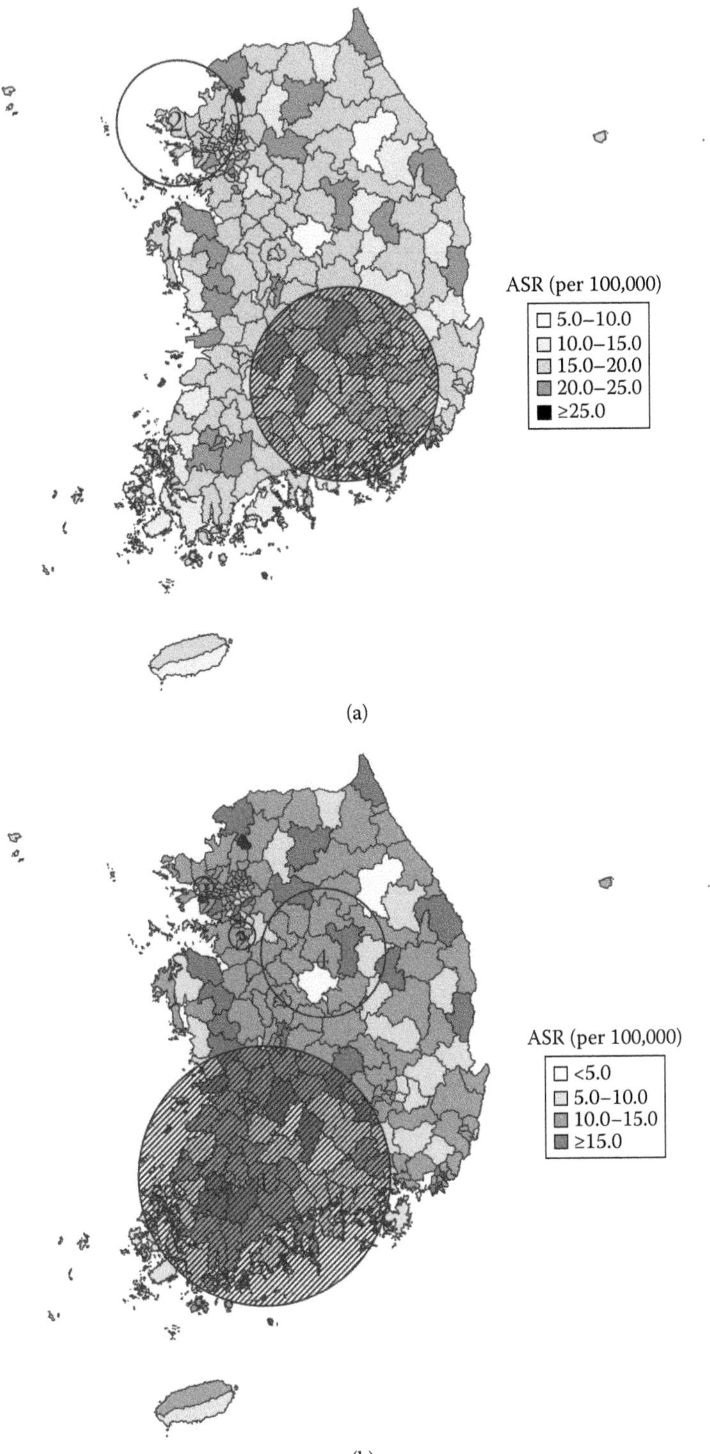

FIGURE 9.1
Clusters of high (circles without lines) or low (circles with hatched lines) rates of colorectal cancer mortality detected using the Poisson-based spatial scan statistic in South Korea (2010–2012): (a) male; (b) female.

TABLE 9.1
Most likely and secondary clusters of high or low rates of colorectal cancer mortality in South Korea (2010–2012) for males

Cluster	Number of areas	Number of observed cases	Number of expected cases	Relative risk	p-Value
1 (low)	48	1826	2053.70	0.87	0.00013
2 (high)	43	3626	3407.52	1.09	0.028

TABLE 9.2
Most likely and secondary clusters of high or low rates of colorectal cancer mortality in South Korea (2010–2012) for females

Cluster	Number of areas	Number of observed cases	Number of expected cases	Relative risk	p-Value
1 (low)	61	1763	2003.86	0.85	<0.00001
2 (high)	2	90	47.55	1.90	0.00022
3 (high)	4	292	213.18	1.38	0.001
4 (high)	16	583	487.86	1.21	0.039

six metropolitan cities), were detected for males. For females, one low-rate cluster was detected in the southwest region, and three high-rate clusters were found, one small cluster each in Gyeonggi province and Incheon and one cluster in the central region. Tables 9.1 (male) and 9.2 (female) list the clusters ordered by their statistical significance.

9.6 Summary and Discussion

In this chapter, we reviewed the methodology of spatial scan statistics covering several probability models, inference procedures based on MC hypothesis testing and Gumbel approximation for p-values, and shapes of scanning windows. Examples of applications in spatial epidemiology were also listed.

The spatial scan statistic is an important tool for spatial cluster detection analysis. We should mention that there are other useful tools for spatial cluster detection, such as Besag and Newell's method (1991) and the local Moran test (Anselin 1995), termed local indicators of spatial association (LISA). Such methods of cluster detection tests, including spatial scan statistics, are distinguished from focused tests such as Stone's test (1988), Lawson and Waller's score test (Lawson 1993; Waller and Lawson 1995), and Tango's score test (1995). Cluster detection tests are used when we want to explore spatial clusters without a priori knowledge of the location and size of clusters, whereas focused tests are used when we want to examine local clustering around a prespecified fixed-point source. There is another type of test for global clustering, which is used to assess if clustering exists in a study region without picking out the location of clusters, such as Moran's I (1950), Cuzick and Edwards' k-nearest neighbor test (1990), and Tango's MEET (2000). The three different types of tests are used for different purposes and are complementary to each other.

Research on spatial scan statistics has considerably grown in recent years and is expected to continue to widen. While spatial epidemiology is the most suitable area for their application, spatial scan statistics will be increasingly used in diverse areas.

References

Abrams, A., Kleinman, K., and Kulldorff, M. (2010). Gumbel based p-value approximations for spatial scan statistics. *International Journal of Health Geographics*, 9:61.

Anselin, L. (1995). Local indicators of spatial association—LISA. *Geographical Analysis*, 27:93–115.

Besag, J. and Newell, J. (1991). The detection of clusters in rare diseases. *Journal of the Royal Statistical Society: Series A*, 154:143–55.

Besculides, M., Heffernan, R., Mostashari, F., and Weiss, D. (2005). Evaluation of school absenteeism data for early outbreak detection, New York City. *BMC Public Health*, 5:105.

Cook, A. J., Gold, D. R., and Li, Y. (2007). Spatial cluster detection for censored outcome data. *Biometrics*, 63(2):540–49.

Cuzick, J. and Edwards, R. (1990). Spatial clustering for inhomogeneous populations. *Journal of the Royal Statistical Society: Series B*, 52:73–104.

Duczmal, L. and Assunção, R. (2004). Simulated annealing strategy for the detection of arbitrarily shaped spatial clusters. *Computational Statistics and Data Analysis*, 45:269–86.

Dwass, M. (1957). Modified randomization tests for nonparametric hypotheses. *Annals of Mathematical Statistics*, 28:181–87.

Glaz, J. and Balakrishnan, N. (1999). *Scan Statistics and Applications*. Birkhäuser, Boston.

Glaz, J., Naus, J., and Wallenstein, S. (2001). *Scan Statistics*. Springer, New York.

Glaz, J., Pozdnyakov, V., and Wallenstein, S. (2009). *Scan Statistics: Methods and Applications*. Birkhäuser, Boston.

Gregorio, D. I., Kulldorff, M., Barry, L., Samociuk, H., and Zarfos, K. (2001). Geographical differences in primary therapy for early-stage breast cancer. *Annals of Surgical Oncology*, 8(10):844–49.

Henning, K. J. (2004). What is syndromic surveillance? *Morbidity and Mortality Weekly Report*, 53:7–11.

Hixson, B. A., Omer, S. B., del Rio, C., and Frew, P. M. (2011). Spatial clustering of HIV prevalence in Atlanta, Georgia and population characteristics associated with case concentrations. *Journal of Urban Health*, 88(1):129–41.

Huang, L., Kulldorff, M., and Gregorio, D. (2007). A spatial scan statistic for survival data. *Biometrics*, 63:109–18.

Huang, L., Tiwari, R. C., Zuo, Z., Kulldorff, M., and Feuer, E. J. (2009). Weighted normal spatial scan statistic for heterogeneous population data. *Journal of the American Statistical Association*, 104:886–98.

Jennings, J. M., Curriero, F. C., Celentano, D., and Ellen J. M. (2005). Geographic identification of high gonorrhea transmission areas in Baltimore, Maryland. *American Journal of Epidemiology*, 161(1):73–80.

Jones, R. C., Liberatore, M., Fernandez, J. R., and Gerber, S. I. (2006). Use of a prospective space-time scan statistic to prioritize shigellosis case investigations in an urban jurisdiction. *Public Health Reports*, 121:133–39.

Jung, I. (2009). A generalized linear models approach to spatial scan statistics for covariate adjustment. *Statistics in Medicine*, 28(7):1131–43.

Jung, I., Kulldorff, M., and Klassen, A. C. (2007). A spatial scan statistic for ordinal data. *Statistics in Medicine*, 26(7):1594–607.

Jung, I., Kulldorff, M., and Richard, O. J. (2010). A spatial scan statistic for multinomial data. *Statistics in Medicine*, 29(18):1910–18.

Jung, I. and Lee, H. (2012). Spatial cluster detection for ordinal outcome data. *Statistics in Medicine*, 31(29):4040–48.

Jung, I. and Park, G. (2015). *p*-Value approximations for spatial scan statistics using extreme value distributions. *Statistics in Medicine*, 34(3):504–14.

Kammerer, J. S., Shang, N., Althomsons, S. P., Haddad, M. B., Grant, J., and Navin, T. R. (2013). Using statistical methods and genotyping to detect tuberculosis outbreaks. *International Journal of Health Geographics*, 12:5.

Klassen, A. C., Kulldorff, M., and Curriero, F. (2005). Geographical clustering of prostate cancer grade and stage at diagnosis, before and after adjustment for risk factors. *International Journal of Health Geographics*, 4:1.

Kulldorff, M. (1997). A spatial scan statistic. *Communication in Statistics—Theory and Methods*, 26(6):1481–96.

Kulldorff, M. (2001). Prospective time periodic geographical disease surveillance using a scan statistic. *Journal of the Royal Statistical Society: Series A*, 164:61–72.

Kulldorff, M., Heffernan, R., Hartman, J., Assunção, R. M., and Mostashari, F. (2005). A space-time permutation scan statistic for disease outbreak detection. *PLoS Medicine*, 2:216–24.

Kulldorff, M., Huang, L., and Konty, K. (2009). A scan statistic for continuous data based on the normal probability model. *International Journal of Health Geographics*, 8:58.

Kulldorff, M., Huang, L., Pickle, L., and Duczmal, L. (2006). An elliptic spatial scan statistic. *Statistics in Medicine*, 25:3929–43.

Kulldorff, M. and Information Management Services, Inc. (2015). SaTScan™ v9.4: Software for the spatial and space-time scan statistics. http://www.satscan.org/.

Kulldorff, M., Mostashari, F., Duczmal, L., Yih, K., Kleinman, K., and Platt, R. (2007). Multivariate scan statistics for disease surveillance. *Statistics in Medicine*, 26:1824–33.

Lawson, A. B. (1993). On the analysis of mortality events associated with a prespecified fixed point. *Journal of the Royal Statistical Society: Series A*, 156:363–77.

Lin, H., Li, J., Ho, S. C., Huss, A., Vermeulen, R., and Tian, L. (2015). Lung cancer mortality among women in Xuan Wei, China: A comparison of spatial clustering detection methods. *Asia-Pacific Journal of Public Health*, 27(2):NP392–401.

Liu-Mares, W., MacKinnon, J. A., Sherman, R., Fleming, L. E., Rocha-Lima, C., Hu, J. J., and Lee, D. J. (2013). Pancreatic cancer clusters and arsenic-contaminated drinking water wells in Florida. *BMC Cancer*, 13:111.

Moran, P. A. P. (1950). Notes on continuous stochastic phenomena. *Biometrika*, 37:17–23.

Naus, J. (1965a). Clustering of random points in two dimensions. *Biometrika*, 52:263–67.

Naus, J. (1965b). The distribution of the size of the maximum cluster of points on a line. *Journal of the American Statistical Association*, 60:532–38.

Neill, D. B. and Cooper, G. F. (2010). A multivariate Bayesian scan statistic for early event detection and characterization. *Machine Learning*, 79:261–82.

Neill, D. B., Moore, A. W., and Cooper, G. F. (2006). A Bayesian spatial scan statistic. *Advances in Neural Information Processing Systems*, 18:1003–10.

Neill, D. B, Moore, A. W., and Cooper, G. F. (2007). A multivariate Bayesian scan statistic. *Advances in Disease Surveillance*, 2:60.

Osei, F. B. and Duker, A. A. (2008). Spatial dependency of *V. cholera* prevalence on open space refuse dumps in Kumasi, Ghana: A spatial statistical modeling. *International Journal of Health Geographics*, 7:62.

Patil, G. P. and Taillie, C. (2004). Upper level set scan statistic for detecting arbitrarily shaped hotspots. *Environmental and Ecological Statistics*, 11:183–97.

Pfeiffer, D. U., Robinson, T. P., Stevenson, M., Stevens, K. B., Rogers, D. J., and Clements, A. C. A. (2008). *Spatial Analysis in Epidemiology*. Oxford University Press, Oxford.

Pollack, L. A., Gotway, C. A., Bates, J. H., Parikh-Patel, A., Richards, T. B., Seeff, L. C., Hodges, H., and Kassim, S. (2006). Use of the spatial scan statistic to identify geographic variations in late stage colorectal cancer in California (United States). *Cancer Causes and Control*, 17:449–57.

Read, S., Bath, P. A., Willett, P., and Maheswaran, R. (2013). A study on the use of Gumbel approximation with the Bernoulli spatial scan statistic. *Statistics in Medicine*, 32(19):3300–13.

Sankoh, O. A., Ye, Y., Sauerborn, R., Muller, O., and Becher, H. (2001). Clustering of childhood mortality in rural Burkina Faso. *International Journal of Epidemiology*, 30:485–92.

Sheehan, T. J., Gershman, S. T., MacDougal, L., Danley, R., Mrosszczyk, M., Sorensen, A. M., and Kulldorff, M. (2000). Geographical surveillance of breast cancer screening by tracts, towns and zip codes. *Journal of Public Health Management and Practice*, 6:48–57.

Stone, R. A. (1988). Investigations of excess environmental risks around putative sources: Statistical problems and a proposed test. *Statistics in Medicine*, 7:649–60.

Sudakin, D. L., Horowitz, Z., and Giffin, S. (2002). Regional variation in the incidence of symptomatic pesticide exposures: Applications of geographic information systems. *Clinical Toxicology*, 40:767–73.

Takahashi, K., Yokoyama, T., and Tango, T. (2009). FlexScan: Software for the flexible spatial scan statistic, v3.0. http://www.niph.go.jp/soshiki/gijutsu/index_e/html.

Tango, T. (1995). A class of tests for detecting 'general' and 'focused' clustering of rare disease. *Statistics in Medicine*, 14:2323–34.

Tango, T. (2000). A test for spatial disease clustering adjusted for multiple testing. *Statistics in Medicine*, 19:191–204.

Tango, T. (2010). *Statistical Methods for Disease Clustering*. Springer, Berlin.

Tango, T. and Takahashi, K. (2005). A flexibly shaped spatial scan statistic for detecting clusters. *International Journal of Health Geographics*, 4:11.

van den Wijngaard, C. C., van Asten, L., van Pelt, W., Doornbos, G., Nagelkerke, N. J., Donker, G. A., van der Hoek, W., and Koopmans, M. P. (2010). Syndromic surveillance for local outbreaks of lower-respiratory infections: Would it work? *PLoS One*, 5(4):e10406.

Waller, L. A. and Lawson, A. B. (1995). The power of focused tests to detect disease clustering. *Statistics in Medicine*, 14:2291–308.

Zhang, T. and Lin, G. (2009). Spatial scan statistics in loglinear models. *Computational Statistics and Data Analysis*, 53:2851–58.

Zhang, Z., Assunção, R., and Kulldorff, M. (2010). Spatial scan statistics adjusted for multiple clusters. *Journal of Probability and Statistics*, 642379.

10

Kernel Smoothing Methods

Martin L. Hazelton
Institute of Fundamental Sciences
Massey University
Palmerston North, New Zealand

CONTENTS

10.1 Introduction

This chapter is concerned with kernel methods for examining the spatial distribution and variation in disease risk for individual event data. Specifically, we assume that the geographical coordinates of disease cases (and perhaps controls) are observed. This may occur when we have very precise information on these locations (e.g., the home addresses of individuals), but the methods are also generally applicable to situations in which cases are identified within spatial units that are tiny in comparison to the study region as a whole. In such circumstances, it is typically convenient to model the data as points located at the centroids of these small areas.

The observed locations can be modelled as the realization of some random point process over the study region. Many of the interesting properties of a point process are naturally described by smooth functions. For example, an inhomogeneous Poisson process is characterized by its intensity function. Estimation of such a function from point process data is a smoothing problem.

There are a variety of statistical methodologies available for smoothing. Spline-based methods are very popular, particularly for regression problems. See, for example, Ruppert et al. (2003) and also Chapter 12 of this volume. However, kernel smoothing methods are most widely used for estimation of density and intensity functions (and functionals thereof) when the data are of low dimension. In particular, kernel smoothing is preeminent for bivariate density estimation, and so is well suited to many of the types of smoothing problem that are encountered in spatial epidemiology.

Kernel smoothing techniques can play a number of roles. Perhaps the simplest of these is as a means of data visualization. For example, a raw plot of disease data over the region can be difficult to interpret, particularly when there are many overlapping points. A kernel estimate of the intensity function can provide a clearer picture of the overall spatial distribution of the disease. An estimate of the ratio of case-to-control densities can display the spatial variation in relative risk (Bithell, 1990, 1991).

A second use of kernel smoothing is to account for the background spatial distribution of the population at risk in semiparametric models. Diggle et al. (2005), for example, combined a kernel estimate of the marginal spatial distribution of cases with a parametric description of temporal variation in order to develop a model for disease surveillance. More generally, kernel smoothing can aim to describe the large-scale spatial variations in disease, allowing local anomalies to be more readily identified as disease clusters. This idea underpins the use of clustered point process models, such as the log-Gaussian Cox and Neyman–Scott processes, in the presence of an inhomogeneous background population. See Davies and Hazelton (2013) for an example.

A third use of kernel smoothing methods is as the basis for tests and the specification of confidence regions and related quantities. For example, kernel methods have been used to provide tolerance contours, which indicate areas of the study region for which the relative risk of disease is significantly elevated (Kelsall and Diggle, 1995a). Kernel methods can also be employed as an alternative to scan statistics (cf. Chapter 9) for testing the null hypothesis of constant risk over the study region. See, for example, the tests proposed by Kelsall and Diggle (1995a) and Hazelton and Davies (2009).

In Section 10.2 of this chapter, we examine the fundamental problem of estimating density and intensity functions. We also describe methods for selecting the bandwidth, which controls the degree of smoothing employed. In Section 10.3, we focus on kernel methods for modelling spatial relative risk. This is a challenging problem because it involves estimation of a ratio of density functions, but a number of recent developments have led to substantial improvements in methodology. We look at the use of kernel-based methods for inference in Section 10.4.

10.2 Kernel Intensity Estimation and Related Problems

10.2.1 The basics of kernel estimation of intensity and density functions

Let $\boldsymbol{s}_1, \ldots, \boldsymbol{s}_m$ denote the geographical location of disease cases (or controls) over a study region $\mathcal{R}$. A simple statistical model is to assume that these data are generated by an inhomogeneous Poisson process with intensity function $\lambda(\boldsymbol{s})$. For almost any conceivable real-life application, it will not be possible to specify a credible parametric form for λ. Estimation must therefore be by nonparametric means. Kernel smoothing is such a technique, and can be regarded as a generalization of the two-dimensional histogram (see, e.g., Wand and Jones, 1995).

The kernel estimate of the spatial intensity function is defined by

$$\hat{\lambda}(\boldsymbol{s}) = \sum_{i=1}^{m} K_h(\boldsymbol{s} - \boldsymbol{s}_i), \tag{10.1}$$

where $K_h(\boldsymbol{s}) = h^{-2}K(\boldsymbol{s}/h)$. Here K is the unscaled kernel function, usually taken to be a spherically symmetric probability density function. The standard bivariate normal density is a very popular choice for kernel function, often referred to as the Gaussian kernel. The spherically symmetric bivariate biweight kernel $K(\boldsymbol{s}) = \frac{3}{\pi}(1 - \boldsymbol{s}^\top \boldsymbol{s})^2 \mathbf{1}_{\{||\boldsymbol{s}||<1\}}$ is an alternative with compact support. The kernel K_h is a scaled version of K. The scaling parameter h controls the overall degree of smoothness of the final intensity estimate and is termed the bandwidth. Note that $\int_{\mathcal{R}} \hat{\lambda}(\boldsymbol{s})\, d\boldsymbol{s} = m$, so that kernel estimate has mean intensity $m/|\mathcal{R}|$ over the region, where $|\mathcal{R}|$ denotes the area of $\mathcal{R}$.

We illustrate kernel estimation of the intensity function on a dataset comprising the geographical locations of $m = 62$ cases of childhood leukaemia and lymphoma in North Humberside, UK, for the period 1974–1986. See Cuzick and Edwards (1990) for details. The estimate displayed in Figure 10.1 was constructed using the Gaussian kernel with a bandwidth of $h = 1.5$ km. The estimated intensity function reflects the underlying population density, reaching a peak in the south of the region around the city of Hull.

A closely related problem is kernel estimation of the spatial density function, f, of the data. This function is simply a version of the intensity function that is scaled so as to integrate to one across the region. It can be estimated by the kernel density estimator

$$\hat{f}(\boldsymbol{s}) = \frac{1}{m} \sum_{i=1}^{m} K_h(\boldsymbol{s} - \boldsymbol{s}_i). \tag{10.2}$$

The functions $\hat{f}$ and $\hat{\lambda}$ provide exactly the same information about the spatial variation of the data in a relative sense. For example, a plot of $\hat{f}$ for the North Humberside childhood cancer data would look identical to the right-hand panel of Figure 10.1, except that the calibration of the colour scale would be changed (by a constant multiple).

10.2.2 Bandwidth selection

It is widely agreed that the choice of kernel function is of relatively minor importance in kernel density (or intensity) estimation, but that the choice of bandwidth is critical. If the bandwidth h is too small, then the density (or intensity) estimate will look extremely rough. It will then be very difficult to distinguish interesting structure in the estimate from noise. If the bandwidth h is too large, then the estimate will be oversmoothed, obscuring potentially important fine detail.

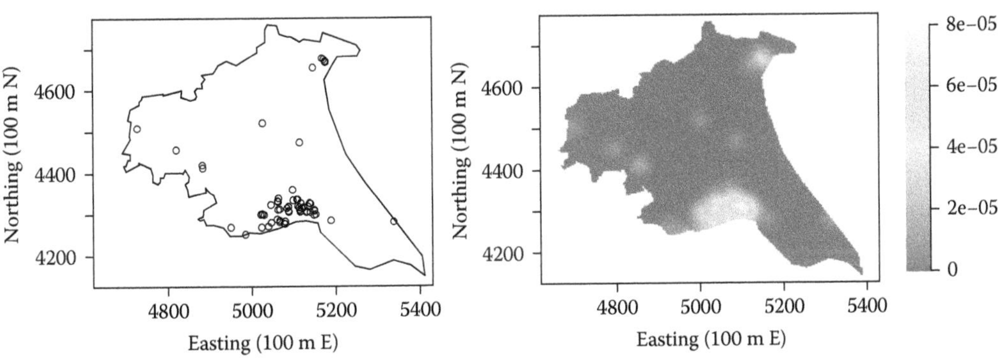

FIGURE 10.1
Plots of cases of childhood leukaemia and lymphoma in North Humberside for the period 1974–1986. The left-hand panel shows the locations of the cases; the right-hand panel displays a heat map of a kernel estimate of the intensity function (in cases/km^2 over the time window).

Considerable attention has been directed to methods for selecting h. The simplest is to choose a value by eye, in order to obtain an aesthetically pleasing estimate of the density or intensity function. A less subjective methodology is to attempt to find the value of h that minimizes some measure of the error of the function estimate. A common goal is to attempt to minimize the mean integrated squared error (MISE), defined by

$$\mathsf{MISE}(\hat{\lambda}) = \mathsf{E}\left[\int_{\mathcal{R}} \left\{\hat{\lambda}(\boldsymbol{s}) - \lambda(\boldsymbol{s})\right\}^2 d\boldsymbol{s}\right] \tag{10.3}$$

for the intensity function estimate (having conditioned on the observed number of data, m). For density estimation, $\mathsf{MISE}(\hat{f})$ is defined in a corresponding manner. The problem of selecting a bandwidth to minimize MISE is the same regardless of whether we are focused on f or λ.

Direct minimization of $\mathsf{MISE}(\hat{\lambda})$ is not possible, because this error criterion depends on the unknown target function λ. There are a number of ways forward. Plug-in approaches work by taking an asymptotic expansion of the MISE as a function of h, under the assumption that $h \to 0$ and $m \to \infty$. This expansion is a function of λ (or f) and its derivatives. Substituting pilot estimates of these functionals produces a data-driven estimate of $\mathsf{MISE}(\hat{\lambda})$, which can be minimized to give the plug-in bandwidth. See, for example, Wand and Jones (1994).

A second approach to data-driven bandwidth selection is to use leave-one-out cross-validation. In its most basic form, this methodology works by selecting h so as to minimize the cross-validation criterion

$$\mathsf{CV}(\hat{\lambda}) = \int_{\mathcal{R}} \left\{\hat{\lambda}(\boldsymbol{s})\right\}^2 d\boldsymbol{s} - 2\sum_{i=1}^{m}\sum_{\substack{j=1\\ j\neq i}}^{m} K_h(\boldsymbol{s}_i - \boldsymbol{s}_j), \tag{10.4}$$

which can be shown to be an unbiased estimate of $\mathsf{MISE}(\hat{\lambda})$ modulo, a constant independent of h. A refined version of this methodology is described by Diggle (1985) and Berman and Diggle (1989). A third approach is to use the maximal smoothing principle of Terrell (1990). This produces the so-called oversmoothing bandwidth, which provides the maximum amount of smoothing that could be required to minimize the MISE for any target density (or intensity) function.

We illustrate application of cross-validation and oversmoothing bandwidth selectors on the North Humberside childhood cancer data (see Figure 10.2). Cross-validation methods have a reputation for having somewhat variable performance, and in particular for producing bandwidths that are too small (see, e.g., Wand and Jones, 1995). This is arguably reflected in the left-hand panel of Figure 10.2, where the variation in intensity across the city of Hull (in the south of the region) may simply be due to noise. The oversmoothing bandwidth, as expected, smooths out all this noise, but also obscures most of the detail in the intensity function. See the right-hand panel of Figure 10.2.

The kernel estimators that we have considered above all employ a single scalar bandwidth h to control the degree of smoothing. With radially symmetric kernel functions, this leads to isotropic smoothing. Such a smoothing regimen seems very reasonable with geographical data, and has the advantage of reducing the problem of bandwidth selection to a search for a single value. Nonetheless, it is worth noting that Equations 10.1 and 10.2 can be generalized to incorporate a bandwidth matrix. This allows the kernel functions to take general elliptical shapes at arbitrary orientations. Duong and Hazelton (2003, 2005) describe data-driven methods for selecting bandwidth matrices for bivariate density estimation.

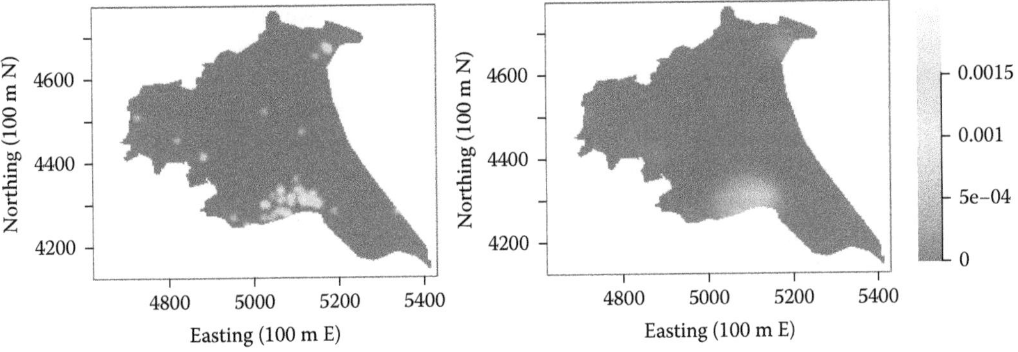

FIGURE 10.2
Estimates of the intensity function for cases of childhood leukaemia and lymphoma in North Humberside for the period 1974–1986. Both were computed using a Gaussian kernel. The estimate in the left-hand panel employs a bandwidth of $h = 5.79$ (chosen by cross-validation), and the one in the right-hand panel a bandwidth of $h = 21.92$ (chosen by the oversmoothing principle). The heat maps for the intensity estimates use a common colour scale.

10.2.3 Dealing with edge effects

Much of the research on kernel density estimation has focused on situations where the support of the data is unbounded. However, an important characteristic of event data in spatial epidemiology is that they are necessarily restricted to some finite study region. Ignoring this leads to severe bias in the estimates of intensity or density functions near the boundary of the region. The cause of the problem is that for a data point close to the edge of the region, some of the mass of the kernel will spill over the boundary. What is more, there are no observed data outside the region to compensate in a reciprocal fashion. If the kernel estimate is truncated at the boundary, this loss of mass leads to severe underestimation of the target function near the edge of $\mathcal{R}$. For estimation on the boundary itself, the bias is so severe that it does not disappear even asymptotically as $m \to \infty$.

The most straightforward way of addressing this problem is to apply a correction factor to estimates of the density or intensity function near the boundary. Specifically, for estimation at a point $\boldsymbol{s} \in \mathcal{R}$, we may define a correction factor (dependent on h) by

$$q_h(\boldsymbol{s}) = \int_{\mathcal{R}} K_h(\boldsymbol{u} - \boldsymbol{s})\, d\boldsymbol{u}. \tag{10.5}$$

Then the scaled estimators $\hat{f}(\boldsymbol{s})/q_h(\boldsymbol{s})$ and $\hat{\lambda}(\boldsymbol{s})/q_h(\boldsymbol{s})$ are asymptotically unbiased for density and intensity functions, respectively. See Hazelton and Marshall (2009). These authors also discuss more sophisticated methods of dealing with edge effects, which in theory reduce estimation error near the borders of $\mathcal{R}$, but simulation studies have shown that the practical benefits of these more complex methods are marginal.

10.2.4 Spatially adaptive kernel smoothing methods

The kernel smoothing methods that we have considered to date have employed a constant degree of smoothing across the region $\mathcal{R}$. However, it is well known that the optimal amount of smoothing for estimation at a particular point $\boldsymbol{s} \in \mathcal{R}$ depends on the local density f. Intuitively, in areas where the density (or intensity) is relatively large, there will be more data, and hence more scope to describe fine detail in f. As a consequence, one would prefer to use quite a small value of h for estimation at such a point. However, in areas where the data are sparse, a much larger value of the bandwidth is desirable in order to smooth the contributions from individual isolated data points.

One of the challenges in kernel smoothing in spatial epidemiology is that highly heterogeneous spatial distributions are the norm rather than the exception. We are typically faced with situations in which the optimal smoothing requirements in population centres (towns and cities, etc.) are very different from those in sparsely populated rural areas. Fixed-bandwidth kernel estimates like (10.1) and (10.2) are doomed to seek a compromise value of h, not quite small enough to reveal all the desirable detail in areas of high density, yet not quite large enough to smooth out stochastic bumps and troughs in regions of low density.

Spatially adaptive kernel smoothers address this problem by allowing the bandwidth h to vary across the region. This is a very attractive idea in principle for applications in spatial epidemiology, although a variety of issues concerning the practical implementation of this technique remain open research questions. The most successful adaptive kernel density estimator to date is due to Abramson (1982). He proposed that the bandwidth change with the data point, so that the density function estimate (for example) becomes

$$\breve{f}(\boldsymbol{s}) = \sum_{i=1}^{m} K_{h_i}(\boldsymbol{s} - \boldsymbol{s}_i), \tag{10.6}$$

where $h_i = h(\boldsymbol{s}_i) = h_0/[f(\boldsymbol{s}_i)^{1/2}\gamma]$, in which h_0 is the global bandwidth (controlling the overall degree of smoothing), and $\gamma = \exp\{-n^{-1}\sum_{i=1}^{m}\log[f(\boldsymbol{s}_i)]/2\}$ is a scaling constant introduced so that h_0 is directly comparable to the value of h for fixed-bandwidth estimators. Note that the form of the adaptive bandwidth function $h(\boldsymbol{s}_i)$ is intuitive, setting the amount of smoothing to be inversely proportional to the (square root of) the underlying density.

It can be shown that Abramson's (1982) adaptive estimator has excellent theoretical properties. In particular, its bias is of an asymptotically lower order than that of the fixed-bandwidth kernel estimator. Moreover, simulation studies have demonstrated that this theory is generally reflected by strong practical performance.

There are a number of issues that must be address for practical implementation of (10.6). First, the theoretical bandwidth function depends on the unknown density f. We therefore replace this by a pilot estimate. Typically, a fixed-bandwidth estimate is used for this purpose, and is sufficient to ensure that the resultant adaptive estimator retains its desirable theoretical properties (see Hall and Marron, 1988). The second matter that needs attention is the choice of global bandwidth h_0. In principle, cross-validation can be employed for this purpose, but the search for better methods remains a research challenge. Finally, adaptive kernel estimators required edge correction in order to avoid boundary bias. The general ideas mirror those for fixed-bandwidth methods, although the details are more complicated. See Marshall and Hazelton (2010).

10.3 Kernel Methods for Describing Spatially Varying Risk

Kernel estimates of intensity or density functions for case data alone do not describe the spatial variation of disease risk directly, since they take no account of the geographical distribution of the underlying population. This motivated Bithell (1990, 1991) to propose the spatial relative risk function, which is defined by

$$r(\boldsymbol{s}) = \frac{f(\boldsymbol{s})}{g(\boldsymbol{s})}, \tag{10.7}$$

where f and g are, respectively, the probability density functions for the spatial distribution of cases and controls for the disease of interest. (It would be more precise to refer to r as a relative odds function, as noted by Wakefield and Elliott [1999], but for a rare disease the distinction can be ignored.) The relative risk function is defined so that its average value (with respect to the control density) is 1: that is, $\int_{\mathcal{R}} r(\boldsymbol{s})g(\boldsymbol{s})\,d\boldsymbol{s} = 1$. It follows that r does not provide any information regarding the absolute risk of disease, only the relativities between risk a different spatial locations. Some authors (e.g., Lawson and Williams, 1993) have referred to estimation and plotting of r as extraction mapping, since it involves removal of the density g from $f(\boldsymbol{s}) = r(\boldsymbol{s})g(\boldsymbol{s})$.

The spatial relative risk function is designed so that it can be estimated from case-control data. With that in mind, we update the previous notation so that $\boldsymbol{s}_1, \ldots, \boldsymbol{s}_{m_1}$ denote the locations of m_1 disease cases, and $\boldsymbol{s}_{m_1+1}, \ldots, \boldsymbol{s}_m$ denote that locations of $m_2 = m - m_1$ randomly selected controls. A natural estimator of the relative risk function is $\hat{r}(\boldsymbol{s}) = \hat{f}(\boldsymbol{s})/\hat{g}(\boldsymbol{s})$, where $\hat{f}$ and $\hat{g}$ are kernel density estimates of the case and control densities, respectively. See Bithell (1990). For display purposes, it is common to plot an estimate $\hat{\rho}(\boldsymbol{s})$ of the log-relative risk function, $\rho(\boldsymbol{s}) = \log\{r(\boldsymbol{s})\}$.

Kernel estimation of $\rho(\boldsymbol{s})$ is a particularly challenging smoothing problem. In essence, this is because of the inevitable instability in estimating f and g in areas of low density, where the data will be sparse and estimation errors greatest in a relative sense. For estimation of the densities themselves, this will be of little importance, because even a large relative error of a very small density will be barely noticeable in a plot. However, when estimating a log-relative risk function, large relative errors in $\hat{f}$ and $\hat{g}$ translate to highly visible absolute errors in $\hat{\rho}(\boldsymbol{s})$.

The previous discussion implies that the issue of bandwidth selection is particularly critical for estimation of $\rho(\boldsymbol{s})$. An immediate question is whether one should use different or common bandwidths for computing $\hat{f}$ and $\hat{g}$. There are certain theoretical advantages in using separate bandwidths (see Kelsall and Diggle, 1995a), but these seem very difficult to realize in practice. Indeed, the difficulties inherent in choosing a pair of bandwidths (rather than a singleton) are such that a common bandwidth is typically preferable (Hazelton and Davies, 2009). Even selecting a single (common) bandwidth is challenging. Kelsall and Diggle (1995b) and Hazelton (2008) suggested leave-one-out cross-validation methodologies, but neither has proven particularly reliable in practice (Davies and Hazelton, 2010).

It can be argued that the lack of success of data-driven bandwidth selection for $\hat{\rho}(\boldsymbol{s})$ merely reflects the fact that the problem is, in a sense, too difficult. Specifically, a fixed-bandwidth kernel estimate is completely ill-equipped to cope with the contrasting smoothing demands across $\mathcal{R}$ as the population density varies, particularly because the errors in low-density areas can have such a large impact when working with log densities. This observation motivates the use of adaptive kernel estimation of f and g. Davies and Hazelton (2010) proposed such a methodology, based on Abramson's (1982) adaptive estimator (as described above). This technique is implemented in the software `R` (R Core Team, 2013) using the package sparr (Davies et al., 2011).

We illustrate this methodology on data for 761 cases and 3020 controls for primary biliary cirrhosis, collected between 1987 and 1994 in northeast England. See Prince et al. (2001) for details. The raw case and control data are plotted, respectively, in the left-hand and central panels of Figure 10.3, while an estimate of the spatial relative risk function appears in the right-hand panel. This last plot suggests tangible spatial variation in the relative risk of primary biliary cirrhosis, with elevated levels of the disease in the cities of Newcastle and Gateshead (just under halfway up on the eastern side of the region) and lower risk in rural western areas.

An alternative method of describing the spatial variation in risk is to plot the probability of disease as a function of location (possibly conditional on the observed case and control

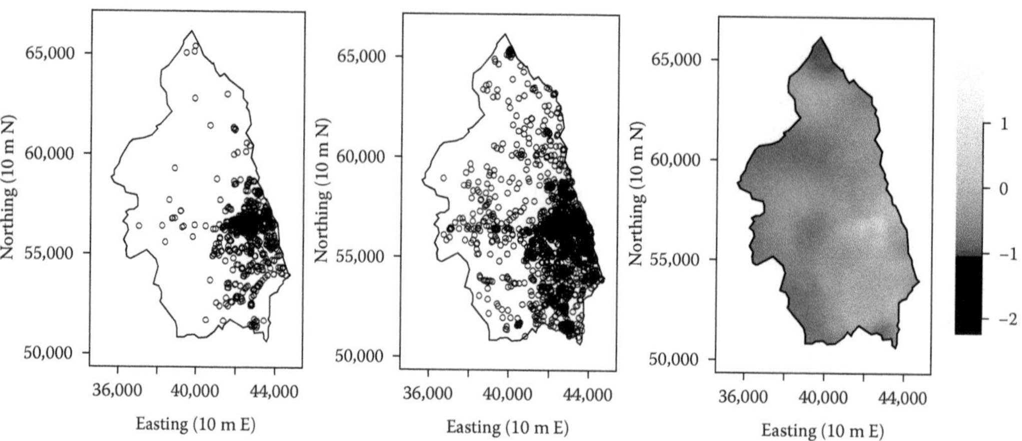

FIGURE 10.3
Analysis of the spatial variation in risk for primary biliary cirrhosis based on case-control data collected between 1987 and 1994 in northeast England. The data are plotted in the left-hand (cases) and central (controls) panels, while the right-hand panel displays the spatial relative risk function, plotted on the log scale.

sample sizes). To develop this idea further, let y_i denote the status of the observation at $\boldsymbol{s}_i$, so that $y_i = 1$ for a case and $y_i = 0$ for a control. The conditional probability of disease $p(\boldsymbol{s}) = \mathbb{P}(y = 1|\, \boldsymbol{s})$ is then given by

$$p(\boldsymbol{s}) = \frac{\lambda_f(\boldsymbol{s})}{\lambda_f(\boldsymbol{s}) + \lambda_g(\boldsymbol{s})}, \tag{10.8}$$

where λ_f and λ_g are, respectively, the spatial intensity functions for disease cases and controls.

A direct estimator, which we denote by $\hat{p}_0$, can be obtained by replacing the intensity functions in (10.8) by kernel estimates thereof. An alternative is to use kernel regression techniques (see, e.g., Wand and Jones, 1995). Specifically, if we employ kernel binary regression on the logit scale, then the estimate $\hat{p}(\boldsymbol{s})$ is defined by using the model

$$\mathsf{logit}(p(\boldsymbol{z})) = \log\left\{\frac{\hat{p}(\boldsymbol{z})}{1 - \hat{p}(\boldsymbol{z})}\right\} = \hat{Q}(\boldsymbol{z} - \boldsymbol{s}) \tag{10.9}$$

for points $\boldsymbol{z}$ in the neighbourhood of $\boldsymbol{s}$, where $\hat{Q}$ is an rth-order polynomial fitted so as to maximize the kernel weighted likelihood

$$L(Q, \boldsymbol{s}) = \sum_{i=1}^{m} \left(y_i Q(\boldsymbol{s}_i - \boldsymbol{s}) - \log\left\{1 + \exp[Q(\boldsymbol{s}_i - \boldsymbol{s})]\right\}\right) K_h(\boldsymbol{s}_i - \boldsymbol{s}). \tag{10.10}$$

The estimate is then defined by $\hat{p}(\boldsymbol{s}) = \hat{Q}(\boldsymbol{0})$. See Tibshirani and Hastie (1987).

Setting $r = 0$ implies use of a constant polynomial and produces a so-called local constant regression estimator. It can be shown that this is identical to $\hat{p}_0$. A popular alternative is to set $r = 1$ when we obtain the local linear binary regression estimator, $\hat{p}_1(\boldsymbol{s})$. This estimator was studied in detail by Clark and Lawson (2004). Based on results from a simulation study, these authors found that the local linear estimator improved markedly on the local constant version for estimation of p.

The conditional probability $p(\boldsymbol{s})$ and spatial relative risk function $r(\boldsymbol{s})$ are closely linked. Specifically, it can be shown that

$$\mathsf{logit}(p(\boldsymbol{s})) = \rho(\boldsymbol{s}) + c,$$

where c is a constant that does not depend on location. It follows that for understanding the (relative) spatial variation of disease, it is purely a matter of taste as to whether one plots $p(\boldsymbol{s})$ on the raw probability scale or $\rho(\boldsymbol{s})$ (equivalent to plotting this probability on the logit scale, modulo an additive constant). However, the choice of scale has a significant impact on the preferred choice of kernel smoothing methodology. As Fernando et al. (2014) showed, the use of local linear regression typically produces a worse estimate of ρ than is obtained using the direct density ratio method described at the start of this section. We conclude that for estimation and display of the conditional probability of disease $p(\boldsymbol{s})$, one should prefer local linear regression, while the density ratio method is preferable when one wishes to plot an estimate of the log-relative risk $\rho(\boldsymbol{s})$.

10.4 Kernel-Based Inference

Kernel smoothed estimates of the spatial relative risk function are frequently used as a method of data visualization. For example, it is difficult to come to any conclusions about the spatial variation of risk based on the plots of the raw data in the left-hand and central panels of Figure 10.3, whereas the geographical relative risk function is easy to interpret. Nevertheless, use of kernel estimates of the relative risk function or conditional probability are not limited to such exploratory data analysis activities. They can also be employed for more formal testing purposes.

Tests for disease risk can be of two types: local or global. A local test addresses the question of whether the relative risk of disease is elevated at a given spatial location $\boldsymbol{s}$. A global test is concerned with whether there is significant variation in the relative risk across the study region as a whole.

We first consider local testing, where the null and alternative hypotheses are, respectively, $H_0\colon r(\boldsymbol{s}) = 1$ and $H_1\colon r(\boldsymbol{s}) > 1$. Note that this is equivalent to testing $H_0\colon \rho(\boldsymbol{s}) = 0$ against $H_1\colon \rho(\boldsymbol{s}) > 0$. One approach is to conduct a permutation test, working on the premise that if H_0 holds, then the locations of cases and controls can be exchanged at random. We proceed as follows. First, obtain an estimate of the relative risk $\hat{\rho}(\boldsymbol{s})$ from the observed case-control data. Next, we remove the case and control labels from the original data, and then reassign them entirely at random. We construct a new estimate $\hat{\rho}^*(\boldsymbol{s})$ from this relabelled dataset. We repeat this relabelling process N times, producing N estimates of ρ built from datasets that are entirely consistent with the null hypothesis. A p-value is then given by the proportion of the estimates $\hat{\rho}^*(\boldsymbol{s})$ from the relabelled data that are larger than the observed value. See Kelsall and Diggle (1995b) and Lawson et al. (2007), for example.

A more computationally efficient alternative is to construct a z-test statistic, $\hat{\rho}/\hat{\sigma}_{\hat{\rho}}$, where $\hat{\sigma}_{\hat{\rho}}$ is a standard error for $\hat{\rho}$ derived using asymptotic theory for kernel density estimation. This approach was explored by Hazelton and Davies (2009), who showed that the results were generally very comparable to those from the permutation test, but at a fraction of the computational cost. Indeed, this methodology can give more reliable p-values in areas of low density.

An interesting use of local tests of risk is for the production of tolerance contours (Kelsall and Diggle, 1995a, 1995b). These work by computing p-values for the local test of $H_0\colon \rho(\boldsymbol{s}) = 0$ against $H_1\colon \rho(\boldsymbol{s}) > 0$ for all locations $\boldsymbol{s} \in \mathcal{R}$. A tolerance contour at the α significance level is then obtain by plotting the α contour for this field of p-values. Any point $\boldsymbol{s}$ lying interior to such a contour will correspond to a location for which we reject H_0 in favour of the alternative hypothesis of elevated risk. In practice, this idea is implemented by conducting the tests over a grid of locations within the study region. Tolerance contours are often used to embellish plots of the relative risk function, to help distinguish areas of truly elevated risk from noisy artefacts in the estimate $\hat{\rho}$.

We illustrate these ideas on the primary biliary cirrhosis introduced earlier. See Figure 10.4. In the left-hand panel, we display a filled 5% tolerance contour, while in the right-hand panel, the tolerance contour (again at the $\alpha = 0.05$ significance level) is added as an embellishment to the log-relative risk function. It is noticeable that only for the Newcastle area on the east coast do we have clear evidence of elevated risk of disease. The other (smaller) peaks in the relative risk function are indistinguishable from noise at a 5% significance level.

Kernel estimates of the relative risk function can also be used to test the global hypothesis $H_0\colon r = 1$ uniformly across $\mathcal{R}$ against the alternative $H_0\colon r$ is not constant over $\mathcal{R}$. Natural test statistics for this purpose are the global maximum log-relative risk,

$$\hat{\rho}_{\max} = \sup_{\boldsymbol{s}\in\mathcal{R}} \hat{\rho}(\boldsymbol{s}), \tag{10.11}$$

and the integral,

$$\hat{t} = \int_{\mathcal{R}} \hat{\rho}(\boldsymbol{s})^2\, d\boldsymbol{s}, \tag{10.12}$$

as proposed by Kelsall and Diggle (1995a). Other (related) test statistics are discussed by Anderson and Titterington (1997). Note that $\hat{t}$ takes the value zero for a perfectly uniform

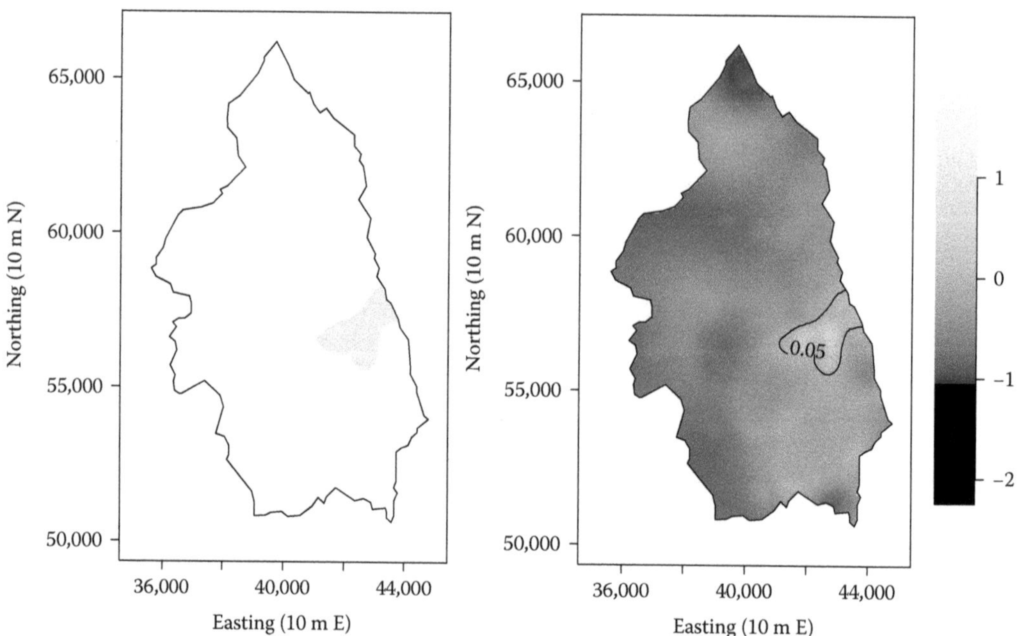

FIGURE 10.4
Tolerance contours for the risk primary biliary cirrhosis in northeast England. The left-hand panel displays a filled 5% tolerance contour; the right-hand panel superimposes this tolerance contour on the estimated log-relative risk function.

relative risk function, whereas sufficiently large values of this statistic will indicate tangible evidence for the failure of the null hypothesis. Hazelton and Davies (2009) examined the utility of these methods, with a particular emphasis on their comparative performance against the scan statistics of Kulldorff (1997, 2006). They concluded that the kernel test statistic $\hat{t}$ is typically preferable to $\hat{\rho}_{\max}$. What is more, $\hat{t}$ appears to have higher power than standard scan statistics for detecting departures from the null hypothesis involving gentle spatial variation in ρ. However, the scan statistic is preferable for detecting tight local clusters.

10.5 Conclusions

Kernel smoothing methods can be used to describe the spatial variation in disease, either by direct estimation of the underlying intensity function for cases or through the spatial relative risk function. An important use of such estimates is for data visualization. Plots of kernel smoothed spatial relative risk functions are becoming increasingly common in the spatial epidemiology literature, for instance. Kernel estimates can also be used for more formal inferential purposes, such as testing for significant local elevation, or global variation, of disease risk.

Whatever the intents and purposes of kernel estimation, it is critical to select an appropriate degree of smoothing. This can be particularly problematic when using fixed-bandwidth methods in epidemiological applications, because the spatial distributions of populations of interest (whether human or animal) are typically hugely heterogeneous. Spatially adaptive kernel smoothers show promise, although more work is needed to refine current techniques.

Our focus throughout this chapter has been on kernel methods applied to just the spatial locations for case and case-control data. However, in practice, we will often have information on covariates for each case and control. Fernando and Hazelton (2014) describe a methodology for generalizing the relative risk function so as to show how the geographical pattern of risk alters with changes to other variables. This provides an enhanced tool for data visualizing. However, if one's primary aim is to test covariate significance adjusted for general spatial trends, then a natural approach is to use kernel regression within the framework of generalized additive models, as discussed by Clark and Lawson (2004).

References

Abramson, I. (1982). On bandwidth variation in kernel estimation—a square root law. *Annals of Statistics* 10, 1217–1223.

Anderson, N. and Titterington, D. (1997). Some methods of investigating spatial clustering, with epidemiological applications. *Journal of the Royal Statistical Society: Series A* 169, 87–105.

Berman, M. and Diggle, P. (1989). Estimating weighted integrals of the second-order intensity of a spatial point process. *Journal of the Royal Statistical Society: Series B (Methodological)* 51, 81–92.

Bithell, J. F. (1990). An application of density estimation to geographical epidemiology. *Statistics in Medicine* 9, 691–701.

Bithell, J. F. (1991). Estimation of relative risk functions. *Statistics in Medicine* 10, 1745–1751.

Clark, A. B. and Lawson, A. B. (2004). An evaluation of non-parametric relative risk estimators for disease maps. *Computational Statistics and Data Analysis* 47(1), 63–78.

Cuzick, J. and Edwards, R. (1990). Spatial clustering for inhomogeneous populations. *Journal of the Royal Statistical Society: Series B* 52, 73–104.

Davies, T. M. and Hazelton, M. L. (2010). Inference based on kernel estimates of the relative risk function in geographical epidemiology. *Statistics in Medicine* 29, 2423–2437.

Davies, T. and Hazelton, M. (2013). Assessing minimum contrast parameter estimation for spatial and spatiotemporal log-Gaussian Cox processes. *Statistica Neerlandica* 67(4), 355–389.

Davies, T., Hazelton, M., and Marshall, J. (2011). Sparr: analyzing spatial relative risk using fixed and adaptive kernel density estimation in R. *Journal of Statistical Software* 39(i01).

Diggle, P. (1985). A kernel method for smoothing point process data. *Applied Statistics* 34(2), 138–147.

Diggle, P., Rowlingson, B., and Su, T. (2005). Point process methodology for on-line spatio-temporal disease surveillance. *Environmetrics* 16, 423–434.

Duong, T. and Hazelton, M. L. (2003). Plug-in bandwidth matrices for bivariate kernel density estimation. *Journal of Nonparametric Statistics* 15(1), 17–30.

Duong, T. and Hazelton, M. L. (2005). Cross-validation bandwidth matrices for multivariate kernel density estimation. *Scandinavian Journal of Statistics, Theory and Applications* 32(3), 485–506.

Fernando, S. and Hazelton, M. (2014). Generalizing the spatial relative risk function. *Spatial and Spatio-Temporal Epidemiology* 8, 1–10.

Fernando, W., Ganesalingam, S., and Hazelton, M. (2014). A comparison of estimators of the geographical relative risk function. *Journal of Statistical Computation and Simulation* 84(7), 1471–1485.

Hall, P. and Marron, J. S. (1988). Variable window kernel estimates of probability densities. *Probability Theory and Related Fields* 80, 37–49.

Hazelton, M. L. (2008). Kernel estimation of risk surfaces without the need for edge correction. *Statistics in Medicine* 27, 2269–2272.

Hazelton, M. L. and Davies, T. M. (2009). Inference based on kernel estimates of the relative risk function in geographical epidemiology. *Biometrical Journal* 51, 98–109.

Hazelton, M. L. and Marshall, J. C. (2009). Linear boundary kernels for bivariate density estimation. *Statistics and Probability Letters* 79, 999–1003.

Kelsall, J. E. and Diggle, P. J. (1995a). Kernel estimation of relative risk. *Bernoulli* 1(1–2), 3–16.

Kelsall, J. E. and Diggle, P. J. (1995b). Non-parametric estimation of spatial variation in relative risk. *Statistics in Medicine* 14, 2335–2343.

Kulldorff, M. (1997). A spatial scan statistic. *Communications in Statistics, Theory and Methods* 26(6), 1481–1496.

Kulldorff, M. (2006). Tests of spatial randomness adjusted for an inhomogeneity: a general framework. *Journal of the American Statistical Association* 101(475), 1289–1305.

Lawson, A. and Williams, F. (1993). Applications of extraction mapping in environmental epidemiology. *Statistics in Medicine* 12(13), 1249–1258.

Lawson, A., Williams, F., and Liu, Y. (2007). Some simple tests for spatial effects around putative sources of health risk. *Biometrical Journal* 49(4), 493–504.

Marshall, J. and Hazelton, M. (2010). Boundary kernels for adaptive density estimators on regions with irregular boundaries. *Journal of Multivariate Analysis* 101(4), 949–963.

Prince, M. I., Chetwynd, A., Diggle, P. J., Jarner, M., Metcalf, J. V., and James, O. F. W. (2001). The geographical distribution of primary biliary cirrhosis in a well-defined cohort. *Hepatology* 34, 1083–1088.

R Core Team. (2013). *R: A Language and Environment for Statistical Computing.* R Foundation for Statistical Computing, Vienna, Austria. http://www.R-project.org/.

Ruppert, D., Wand, M. P., and Carroll, R. J. (2003). *Semiparametric Regression.* Cambridge University Press, Cambridge.

Terrell, G. R. (1990). The maximal smoothing principle in density estimation. *Journal of the American Statistical Association* 85, 470–477.

Tibshirani, R., and Hastie, T. (1987). Local likelihood estimation. *Journal of the American Statistical Association* 82, 559–567.

Wakefield, J. and Elliott, P. (1999). Issues in the statistical analysis of small area health data. *Statistics in Medicine* 18, 2377–2399.

Wand, M. P. and Jones, M. C. (1994). Multivariate plug-in bandwidth selection. *Computational Statistics* 9(2), 97–116.

Wand, M. P. and Jones, M. C. (1995). *Kernel Smoothing.* Chapman & Hall, London.

Part III

Special Methods

11

Geostatistics in Small-Area Health Applications

Patrick E. Brown
Analytics and Informatics
Cancer Care Ontario
and
Department of Statistical Sciences
University of Toronto
Toronto, Ontario, Canada

CONTENTS

11.1 Introduction

The discipline of spatial statistics has traditionally been divided into three themes: geostatistics, discrete spatial models, and spatial point processes. Much of spatial epidemiology is concerned with discrete spatial models of the sort described in Chapters 6 and 7, with any residual spatial variation in risk or exposure (meaning variations beyond those caused by individual-level risk factors) being estimated at the level of health regions or census areas. Geostatistical models operate continuously in space, where the residual spatial component is different at each and every location in the study area, but varying with some degree of smoothness. Two individuals located in close proximity to one another are assumed to have values of this residual risk that are quite similar, and similarity decreases as the distance of separation gets larger. Observational data on human health outcomes tend to be made available on spatially discrete scales, usually case counts in administrative regions, and geostatistics in epidemiology has generally been confined to modelling environmental exposures. Study data, however, can often contain precise spatial information, such as full street addresses or GPS coordinates of the subjects' homes, and geostatistical models are worth consideration when location data of this sort are available.

This chapter will present two examples of geostatistical modelling, the first modelling a continuously valued exposure and the second a binary-valued response. The software used for carrying out the analyses is contained in the `geostatsp` package for R and is described in detail with worked examples in Brown (2015). Code for reproducing the analysis is included in the online supplement to this volume. The `R` packages `geoR` and `geoRglm` are also options for fitting Gaussian and non-Gaussian models, respectively, and information on these packages is contained in Section 34.2.1 of Chapter 34, as well as Brown (2015).

11.2 Gaussian Random Fields

The Gaussian random field (GRF) forms the basis of geostatistical models, and Figure 11.1 shows simulations of three different GRFs with their correlation functions. Figure 11.1d shows a reasonably smoothly varying spatial surface, whose values change gradually from one location to the next. In comparison, Figure 11.1b is a very rough surface, and Figure 11.1c is somewhere between the other two. Each of these images is characterized by its spatial correlation function, with the three correlation functions superimposed in Figure 11.1a. The curves show the correlation between a spatial surface at two locations separated by a distance specified on the x axis. The dotted line is the correlation function of the smooth plot in Figure 11.1d and has a correlation decaying very slowly with distance. The dashed and solid curves have correlations that fall more quickly with distance, with the steeper solid curve giving the roughest surface found in Figure 11.1b.

The GRFs in Figure 11.1 are *stationary*, meaning the correlation between the values of a GRF evaluated at two locations s_1 and s_2 depends only on the separation between them, $s_1 - s_2$. A stationary GRF is characterized by its variance σ^2, a correlation function $\rho(\cdot)$, and a mean parameter that for the time being will be fixed at zero. A spatial correlation function should be able to respond to changes in units of measurements (i.e., from kilometres to inches) without changing the statistical properties of the underlying process, and this

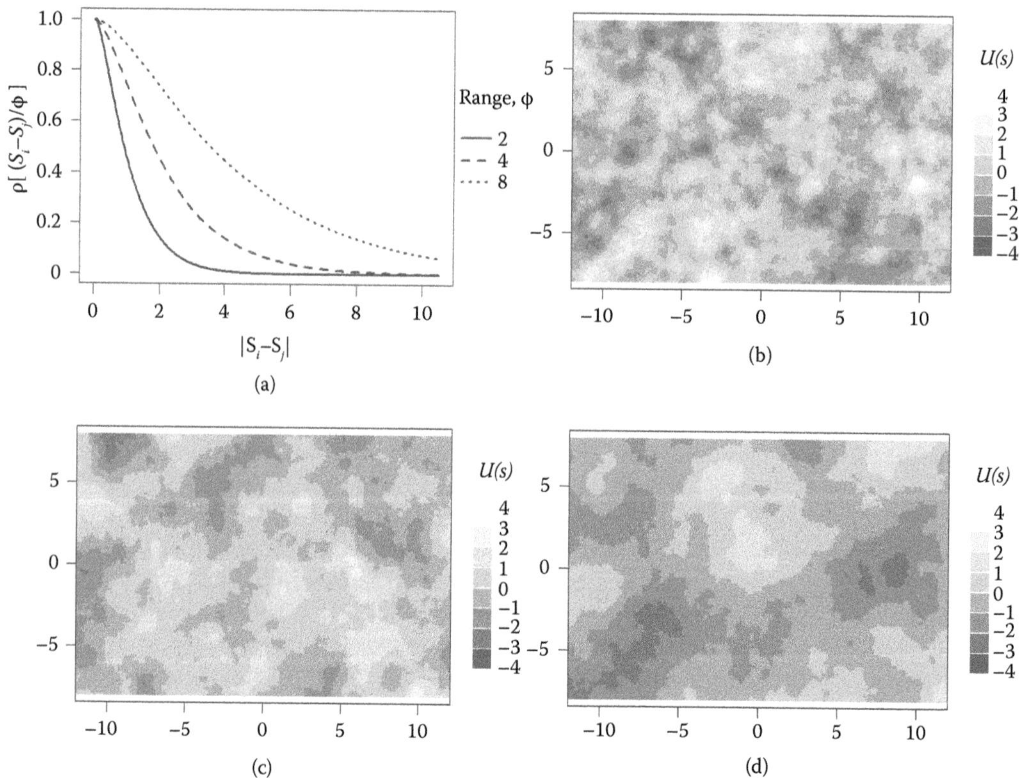

FIGURE 11.1
(a) Matérn correlation functions with shape parameter $\kappa = 1$ and different range parameters ϕ. (b) Simulated Gaussian random field with $\phi = 2$. (c) Simulation with $\phi = 4$. (d) Simulation with $\phi = 8$.

is best accomplished by scaling coordinates with a range parameter. Denoting this range parameter as ϕ, a stationary GRF can be specified as

$$\text{cov}[U(s_i), U(s_j)] = \sigma^2 \rho[(s_i - s_j)/\phi].$$

In Figure 11.1, the GRFs are *isotropic*, with their correlation depending only on the distance $|s_i - s_j|$ and not the direction from one point to the other. The three correlation functions in Figure 11.1a have distance $|s_i - s_j|$ on the horizontal axis and correlations $\rho[(s_i - s_j)/\phi]$ on the vertical axis, with ϕ taking values of 2, 4, and 8.

The Matérn (1960) function has become the most popular family of geostatistical correlation functions, although prior to the publication of Stein (1999), spherical functions and powered exponentials were more widely used. The (isotropic) Matérn correlation has the inconvenient formula

$$\rho(u; \kappa) = \frac{1}{\Gamma(\kappa)2^{\kappa-1}} \left(\sqrt{8\kappa}|u|\right)^{\kappa} K_{\kappa}\left(\sqrt{8\kappa}|u|\right),$$

where u is a spatial separation, $\Gamma(\cdot)$ is a gamma function and K_κ is a modified Bessel function of the second kind of order κ. The κ parameter controls the differentiability of the correlation function and GRF, with small κ producing surfaces with sharp spikes and dips and large κ given more gently rounded surfaces. Figure 11.1 contains Matérn functions with $\kappa = 1$, a compromise value that induces a certain degree of jaggedness in the contour lines even for the strongly correlated $\phi = 8$ surface.

Figure 11.2 shows two examples of anisotropic GRFs, where the correlation between $U(s)$ at two locations depends on the orientation as well as the length of the separation between the points. The horizontal and vertical axes (Figure 11.2a and b) represent distances in the x and y dimensions between two points s_i and s_j, and the shading shows the correlations $\rho[(s_i - s_j)/\phi]$. Figure 11.2a shows a correlation with a strong south–west to north–east directional effect, with the simulation with this correlation in Figure 11.2b having a pronounced striping in this direction. The correlation and simulation in Figure 11.2c and d, respectively, have a weaker directional effect and a less pronounced but still perceptible striping effect. These anisotropic correlation functions, called *geometric anisotropy*, contain two additional parameters controlling the angle of the preferred direction and the ratio of the rate of decay of the correlation in the x direction to the rate in the y direction. In Figure 11.2a and b, the elliptical contours of the correlations are 20° anticlockwise from the x axis, and the width of each ellipse in the direction 20° from the y axis is 0.25 of its width 20° from the x axis.

11.3 Linear Geostatistical Model

The linear geostatistical model (LGM) has foundations in Krige (1951) and was popularized in the statistical community by Cressie (1993). Stein (1999) is largely responsible for the Matérn correlation and maximum likelihood estimation having supplanted method-of-moments-based estimation with spherical or powered exponential correlations, and Diggle and Ribeiro (2006) provide a comprehensive and consolidated description of the LGM and its use. The basic premise of the LGM is to introduce a GRF $U(s)$ to the ordinary least-squares regression model $Y_i = X_i^T\beta + \epsilon_i$, with $U(s)$ accounting for spatial correlation in the residuals ϵ_i.

Writing Y_i as the value of the ith observation, s_i as the location at which this observation was taken, $X(s) = [X_1(s) \ldots X_P(s)]^T$ as a vector of explanatory variables measured at

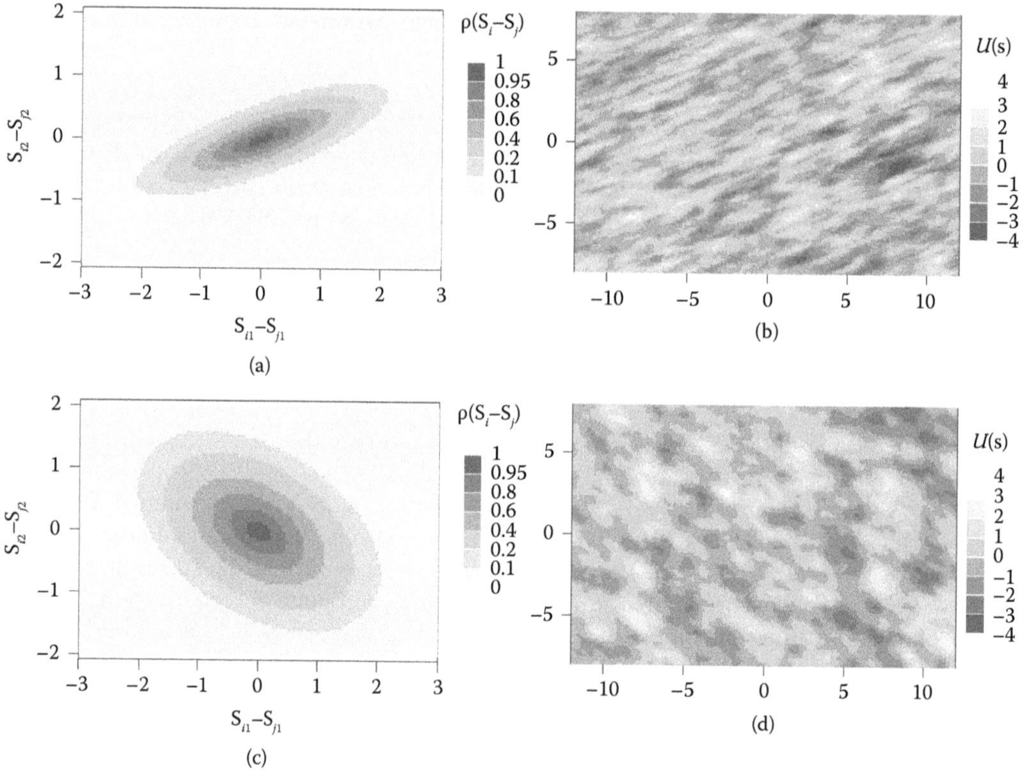

FIGURE 11.2
Correlations and simulations of Gaussian random fields with geometric anisotropy. (a) Correlation, ratio 0.25, angle −20°. (b) Simulation, ratio 0.25, angle −20°. (c) Correlation, ratio 0.6, angle 30°. (d) Simulation, ratio 0.6, angle 30°.

location s, the LGM is written as

$$\begin{aligned} Y_i|U(s_i) &\sim \mathrm{N}[\lambda(s_i), \tau^2] \\ \lambda(s) &= X(s)^T\beta + U(s) \\ \mathrm{cov}[U(s_i), U(s_j)] &= \sigma^2\rho[(s_i - s_j)/\phi, \theta]. \end{aligned} \tag{11.1}$$

Here, $U(s)$ is a GRF with variance σ^2 and a correlation function ρ, which depends on a vector of parameters θ in addition to its scale parameter ϕ. The τ^2 term is the variance of any observation errors, sometimes referred to as a "nugget effect" or microscale spatial variation. Nonspatial covariates, assigned to observations i rather than locations s, can be included by writing these covariates as W_i, their coefficients as γ, and changing the top line of (11.1) to $\mathrm{N}[\lambda(s_i) + W_i\gamma, \tau^2]$.

The normal distribution of the Y_i in the LGM is only suitable for modelling continuously defined outcomes that can be transformed to having a fairly bell-shaped density. For that reason, applications of the LGM in health sciences tend to relate to modelling exposures rather than health outcomes. Figure 11.3 shows one example of the application of an LGM to a health exposures, showing an area of California's Central Valley, with the dots in Figure 11.3a giving the locations of groundwater samples and the colours referring to concentrations of arsenic measured in these samples. The data can be obtained from the National Water Quality Monitoring Council at the Water Quality Portal (www.waterqualitydata.us) using

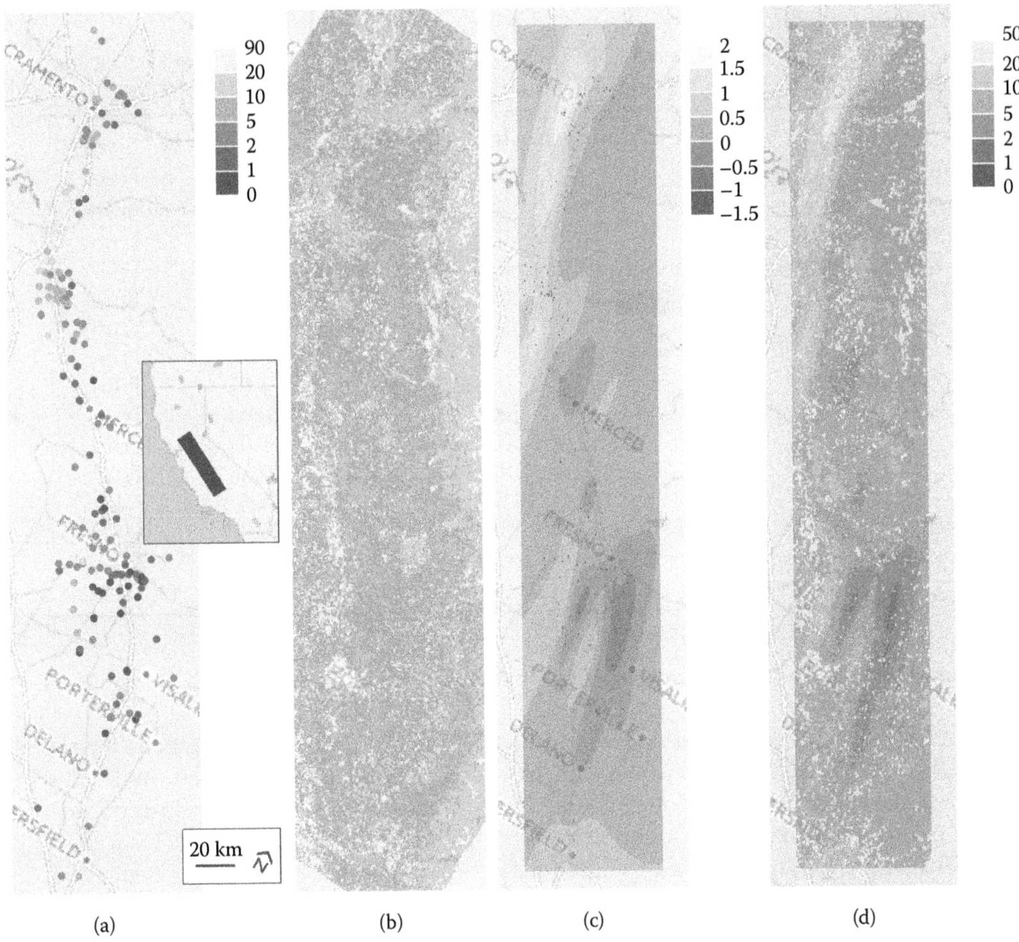

FIGURE 11.3
(a) Sample locations and observed data. (b) Land categories, a predictor variable. (c) Predicted residual spatial variation $\text{E}[U(s)|Y]$. (d) Predicted arsenic $\text{E}[\lambda(s)|Y]$ on the natural scale. Background © CartoDB.

code in the online supplement. Figure 11.3b shows a map of covariates $X(s)$, which are land usage categories with shading referring to land types given in Table 11.1. Being a categorical variable, the vector $X(s)$ at each s is an indicator variable of length 6, and the "herbaceous vegetation" category is the baseline.

As with ordinary least-squares regression, inference on model parameters in the LGM can be accomplished by maximizing the likelihood function. Writing $\mathbf{Y} = [Y_1 \ldots Y_N]^T$ and the matrix $\mathbf{X}$ having entry X_{ip} as element p of $X(s_i)$, the probability density function for an LGM (logged and times −2) is

$$-2\log\left[pr(\mathbf{Y};\beta,\sigma^2,\tau^2,\psi,\theta)\right] = \log|\Sigma| + (\mathbf{Y}-\mathbf{X}\beta)^T\Sigma^{-1}(\mathbf{Y}-\mathbf{X}\beta) + C. \tag{11.2}$$

The variance matrix $\Sigma = \text{var}(\mathbf{Y})$ has entries $\Sigma_{ij} = \sigma^2\rho(s_i - s_j/\phi,\theta)$ on the off-diagonals ($i \neq j$) and $\sigma^2+\tau^2$ on the diagonals. Maximum likelihood estimates (MLEs) of the model parameters, $\hat{\beta}$, $\hat{\sigma}^2$, $\hat{\tau}^2$, $\hat{\psi}$, and $\hat{\theta}$, maximize the probability in (11.2). There is no closed-form solution to maximizing the likelihood of an LGM, and a numerical optimizer must be used to find the MLEs. Standard errors and confidence intervals for the model parameters can

TABLE 11.1
Land type legend and model parameter estimates from fitting an LGM to the California arsenic data

	Colour	Estimate	ci0.025	ci0.975	Estimated
(Intercept)		1.03	0.46	1.60	True
Herbaceous vegetation		0.00	NA	NA	False
Mosaic vegetation		−0.11	−0.51	0.29	True
Mosaic forest or shrubland		−0.13	−0.55	0.30	True
Needle-leaved evergreen forest		0.00	−0.55	0.54	True
Mosaic cropland		−0.22	−0.70	0.26	True
Artificial surfaces		0.21	−0.38	0.81	True
Rain-fed croplands		0.39	−0.26	1.05	True
τ		0.57	0.47	0.68	True
σ		0.72	0.51	1.02	True
$\phi/1000$		19.53	9.40	40.58	True
Shape		1.00	NA	NA	False
anisoRatio		3.91	1.40	10.87	True
anisoAngleDegrees		16.68	1.50	31.86	True
Box–Cox		−0.05	−0.14	0.04	True

be obtained from the second derivative of the likelihood function (the Fisher information matrix).

Table 11.1 shows the estimated model parameters and 95% confidence intervals for the LGM applied to the California arsenic data. An anisotropic model was used, and the θ parameter in the correlation function consists of two geometric anisotropy parameters, with the shape parameter κ fixed at 1. An additional parameter included in the model is a Box–Cox transformation parameter, used to account for the skewed distribution of arsenic with most values in the 0 to 2 range and a small number of measurements above 20. The final column in Table 11.1 shows that the shape parameter and the coefficient on herbaceous vegetation were treated as fixed and not estimated from the data, in the case of the latter variable because of it being the baseline category. The range ϕ and the "anisoRatio" and "anisoAngleDegrees" parameters should be interpreted together. The 17° angle means the contours of the covariance and stripes of $U(s)$ will be oriented a small amount clockwise from the north–south axis, with the ratio of roughly 4 meaning correlation is much stronger in the north–south direction than the east–west direction. The range is 20 km at 17° from the east–west direction and 76 km at 17° from the north–south direction. The confidence intervals for these three parameters are quite large, indicating there is considerable uncertainty associated with these estimates. The standard deviations σ and τ are, in contrast, estimated fairly precisely. The Box–Cox transformation parameter has a confidence interval containing zero, equivalent to a log transform.

Figure 11.3c and d show spatial predictions of the $U(s)$ and $\lambda(s)$ surfaces produced from the parameter MLEs. The $U(s)$ surface can be interpreted as *residual spatial variation*, which in this instance is variability in arsenic not accounted for by the land category covariates. These predictions are conditional expectations $\text{E}(U(s)|\mathbf{Y};\hat{\phi},\hat{\sigma}^2,\hat{\tau}^2,\hat{\beta},\hat{\theta})$ evaluated using the multivariate normal distribution. Writing $\mathbf{U}$ as a vector of $U(g_\ell)$ with the $g_\ell, \ell = 1 \ldots L$ covering the centres of the pixels in Figure 11.3c, we have

$$\text{E}(\mathbf{U}|\mathbf{Y}) = \hat{\Sigma}(\hat{\Sigma}+\hat{\tau}^2 I)^{-1}[\mathbf{Y}-\mathbf{X}\hat{\beta}]$$

and

$$\text{var}(\mathbf{U}|\mathbf{Y}) = \hat{\Sigma} - \hat{\Sigma}(\hat{\Sigma}+\hat{\tau}^2 I)^{-1}\hat{\Sigma}.$$

Computation of the conditional expectation is often referred to as *Kriging* after Krige (1951). The predicted arsenic intensity $\mathrm{E}(\lambda(s)|\mathbf{Y})$ is made by adding $\mathbf{X}\hat{\beta}$ to $\mathrm{E}(U(s)|\mathbf{Y})$, and converting the results from the Gaussian or Box–Cox scale to the natural scale. This back transformation of the $\lambda(s)$ is nonlinear, and numerical integration is required to compute these expectations.

Were the purpose of this modelling to be able to predict arsenic exposures for individuals for whom groundwater samples are not available, the map of $\mathrm{E}(\lambda(s)|\mathbf{Y})$ in Figure 11.3d would be of interest. If the research question were to quantify the effect of land categories on groundwater arsenic, the coefficients and confidence intervals for the land categories in Table 11.1 would be the quantities to consult. Were this exercise part of a study attempting to find predictors or determinants of groundwater arsenic other than the land categories included in the model, Figure 11.3c would be the map to consult. Large values in Figure 11.3c show where arsenic is elevated above and beyond the levels predicted by the spatial distribution of land types. Figure 11.3d is rougher than Figure 11.3c, as the former includes effects of land categories that Figure 11.3b shows are not distributed evenly or smoothly. There are areas of rain-fed cropland to the east of Merced that show reasonably high predicted arsenic values (above 5) due to the coefficient on rain-fed cropland being greater than zero, and despite the GRF term $U(s)$ being close to the average.

11.4 Generalized Linear Geostatistical Model

Adapting the LGM to handle non-Gaussian observations Y_i is conceptually straightforward, but the methodology for performing inference is considerably more complex. Recent advances in Bayesian inference by Rue et al. (2009) and Gaussian Markov random fields by Lindgren et al. (2011) have, when combined with the excellent INLA software from www.r-inla.org, made the process of using non-Gaussian spatial models much easier for the end user. Information on Bayesian inference, integrated nested Laplace approximation (INLA), and Markov random field approximation of the Matérn correlation can be found in Chapter 34.

The generalized linear geostatistical model (GLGM) was first laid out by Diggle et al. (1998) and is covered extensively in Diggle and Ribeiro (2006). The model is specified in (11.3), with the top row from the LGM in (11.1) changed from a normal distribution to an arbitrary probability distribution π as follows:

$$\begin{aligned} Y_i|U(s_i) &\sim \pi\{f[\lambda(s_i)], \nu\} \\ g[\lambda(s)] &= \beta X(s) + U(s) \\ \mathrm{cov}[U(s_i), U(s_j)] &= \sigma^2 \rho[(s_i - s_j)/\phi, \theta]. \end{aligned} \tag{11.3}$$

Here, $f(\cdot)$ is a link function, and ν is a parameter or vector of parameters on which π depends. The LGM results from setting $f(x) = x$, $\nu = \tau^2$, and π being a Gaussian distribution. The scenarios most frequently encountered in the health sciences are a Bernoulli distribution for π and a logit function for $f(\cdot)$ for modelling occurrence or nonoccurrence of health outcomes. This was the case in Diggle et al. (1998) when Y_i was presence or absence of malaria in a child tested, and s_i was the child's place of residence. Jiang et al. (2014) have Y_i as the age at which individual i is diagnosed with lung cancer, a time-to-event outcome that for most individuals is censored at their age at the end of the study period. A Weibull distribution was used for π, the Weibull shape parameter was ν, and $f(\cdot)$ was an exponential function.

Figure 11.4 shows data related to political preference and coffee drinking habits, a topic of considerable controversy (see, e.g., Kelly, 2014). The 2010 mayoral election in Toronto, Canada, was won by the populist politician Rob Ford, who "fashioned his image as the defender of the little guy, the suburban strivers, against the downtown elites, with their degrees and their symphonies," according to pundit Andrew Coyne (2013). Anne Golden concisely summarized the situation: "A Ford voter was more likely [to] order a medium double-double, ... and the city people order grande, non-fat lattes" (Canadian Press, 2014), using the Canadian term for a filter coffee with twice the standard amounts of cream

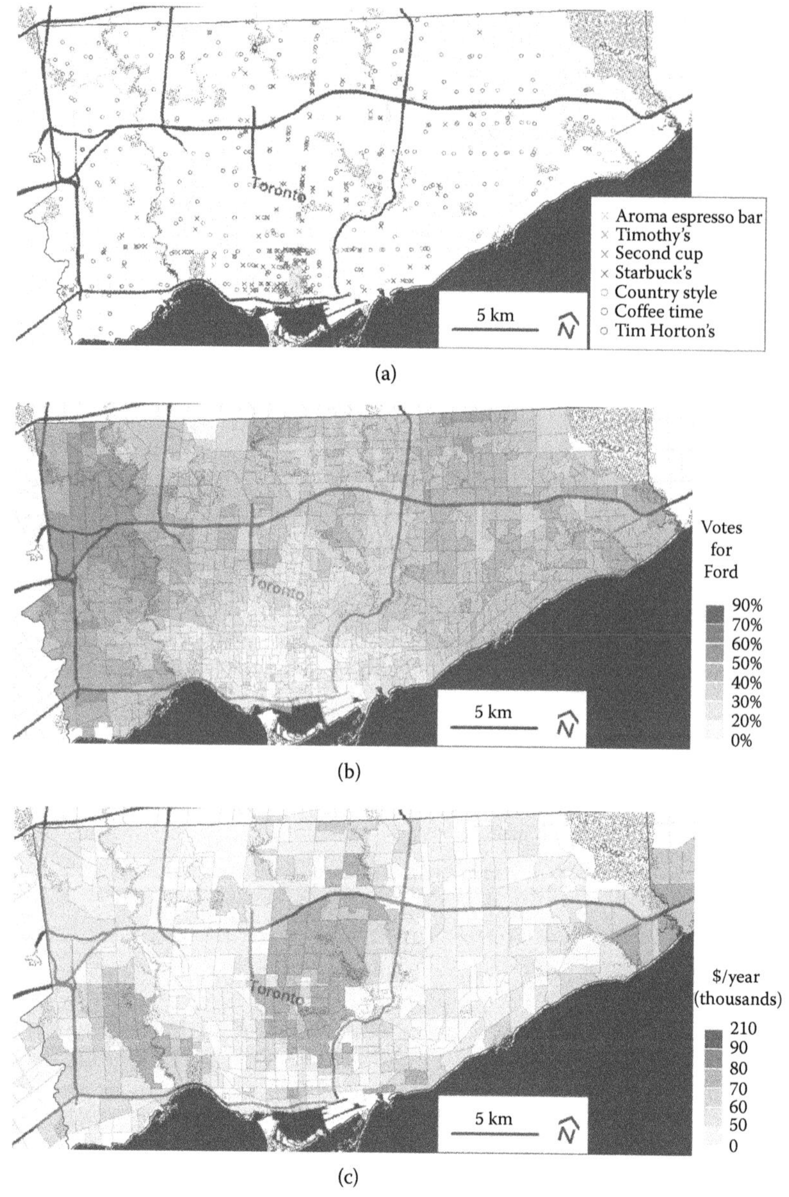

FIGURE 11.4
(a) Coffee shop names and locations. (b) Proportion of votes for Rob Ford, 2010 election. (c) Median household income, 2006 census. Background © Stamen Design.

and sugar. Figure 11.4a contains the locations of major chains of coffee shops in Toronto inspected by Toronto Public Health, with the cafés marked as X's selling lattes and cappuccinos for prices starting at C$ 4 and establishments shown as O's charging in the region of C$ 1.50 for cups of filter coffee. Figure 11.4b reports the proportion of votes that were cast for Rob Ford in each electoral subdivision, showing a clear geographical pattern mimicking the relative preponderance of the different types of coffee shops and suggesting support for Golden's hypothesis.

The model used by Diggle et al. (1998) for malaria prevalence is directly applicable to the coffee shop data, with a latte-serving coffee shop considered "diseased" and the shops serving inexpensive filter coffee being "normal." Diggle et al. (1998) were aiming to predict malaria prevalence from a satellite-derived measure of vegetation intensity, and the equivalent predictor for the coffee problem will be the proportion of voters supporting Rob Ford within 500 m of each coffee shop. The use of bed nets by the children in the malaria study was used as a confounder variable by Diggle et al. (1998), and the confounder here is median household income from the 2006 census shown in Figure 11.4c. The hypothesis to be assessed with the GLGM is that coffee shops catering to the pro-establishment anti-Ford voter are more likely to serve costly espresso-based beverages than coffee shops with a clientele of pro-Ford antielitist voters, allowing for the possibility that income is a confounder. The grey rectangular area in the lower-central area of the map is the lightly populated downtown core, which is excluded from the analysis.

In the notation of (11.3), the coffee data have $Y_i = 1$ for shop i serving lattes and $Y_i = 0$ otherwise; $X(s)$ as a length 3 vector with an intercept, the log of the median household income (centred and averaged over 500 m), and the proportion of votes against Rob Ford (scaled and centred); and $U(s)$ representing any additional determinants of coffee preference, with the ethnicity of individuals nearby being one possible contender. The link function is a logit transformation $f(x) = \exp(x)/[1+\exp(x)]$, and π is a Bernoulli distribution. The research question concerns the coefficient β on the voting covariate. Here, $U(s)$ is not of direct interest, which was also the case in the malaria application, but rather a necessary inclusion in the model in order to make valid inference on β without assuming the data are spatially independent. An isotropic model is used with the shape parameter fixed at 2. The study area is covered with 500×500 m cells on a 41×80 grid, and a Matérn correlation is approximated with a Gaussian Markov random field. The `glgm` function in the `geostatsp` package is used to carry out the analysis, which in turn calls the `inla` function in the `INLA` package.

Bayesian inference (see Chapter 7) requires specifying prior distributions for model parameters, and these prior distributions are particularly important for the spatial correlation parameters σ^2 and ϕ. The regression coefficients β can, in practice, be assigned uninformative prior distributions such as the $N(0, 1000)$ priors, which are the default in `inla`. There is unfortunately no equivalent uninformative prior for variance and range parameters. Setting a uniform $(0, 1000)$ prior for σ^2, for instance, is not equivalent to a uniform prior on the standard deviation σ or the precision $1/\sigma^2$. A broad prior on σ^2 or σ with a large upper bound (100, say) translates to a belief that the variance is almost certainly above 2, a large value when $U(s)$ operates on the log or log-odds scale. It is not uncommon for spatial data to contain relatively little information about σ^2 and ϕ, so some degree of prior sensitivity should be expected.

Prior distributions should in theory be specified using genuine subject-area knowledge without having consulted the dataset in question. A practical compromise used in the `geostatsp` package is to specify the priors via a 95% prior interval for σ and ϕ. For the coffee data, the interval used for σ is (0.2, 3), with the upper bound being large in the sense that $U(s)$ would on occassion reach values of 6, which converted from the log-odds scale would give odds ratios approaching 400. The `geostatsp` package finds prior distributions having the 95% intervals specified while being compatible with the INLA software, which is a gamma prior on $1/\sigma^2$ in this case, and Figure 11.5a shows this prior distribution as a dashed line. One might wish to use a smaller lower bound than 0 for σ, but doing so causes the mode of the prior to become more pronounced and shift leftwards. Toronto is roughly 40 km across, and the 95% prior interval chosen here for ϕ of (1, 30) km would at the upper end give $U(s)$ with very little variation. The prior distribution given to INLA, a gamma distribution on $1/\phi$, is shown in Figure 11.5b.

The posterior distributions of σ and ϕ from the GLGM fit to the coffee data appear as solid lines on Figure 11.5, and there has been a modest shift in the posteriors from the priors. Of note are the relatively light right tails of the posteriors in comparison with the heavy tails of the priors. The table of parameter estimates and posterior 95% credible intervals in Table 11.2 is easier to interpret than the posterior distribution plots. The coefficient on Ford votes is shown as the odds ratio for a 25-point decrease in the proportion voting for Rob Ford, for example, comparing a location with 60% of votes for Rob Ford to a region where Ford obtained 35% of votes cast. The income covariate was modelled on the log scale, and the odds ratio shown for a 50% increase in median household income would compare a location with an income of \$60,000 to one where the income is \$90,000. Both variables have a 95% posterior credible, which excludes 1.0 and can be regarded as significant. Nonsupport of Rob Ford is evidently strongly associated with coffee shops serving lattes and cappuccinos, even after one accounts for the fact that incomes are often higher in the areas where Ford had low levels of support. Even at the lower end of the credible interval, every 25-point reduction in the percentage of votes for Rob Ford doubles the odds of an espresso machine being present.

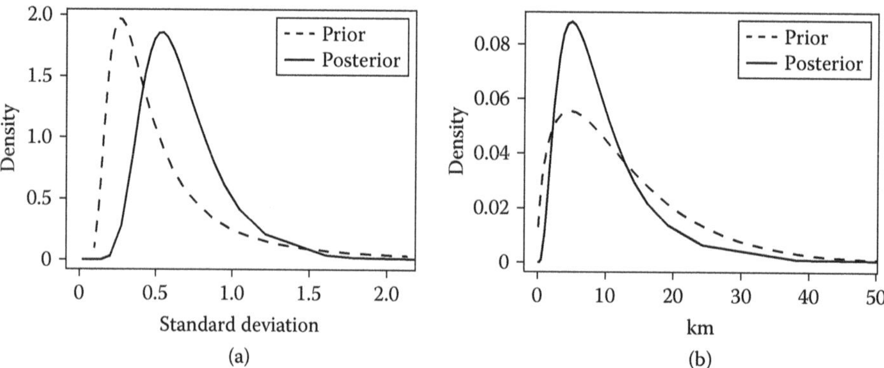

FIGURE 11.5
Prior and posterior distributions of spatial covariance parameters. (a) Standard deviation σ. (b) Range $\phi/1000$.

TABLE 11.2
Posterior means and 95% credible intervals for odds ratios of regression coefficients $\exp(\beta)$ and spatial covariance parameters

	Mean	**0.025 quant**	**0.975 quant**
(Intercept)	0.39	0.22	0.57
Ford, 25-point decrease	3.27	1.98	5.70
Income, 50% increase	2.40	1.61	3.61
$\phi/1000$	9.97	1.90	29.84
σ	0.66	0.31	1.40

Figure 11.6 contains maps related to the prevalence of latte-serving cafés $\rho(s)$ and the residual spatial variation $U(s)$. Figure 11.6a shows the predicted prevalence of lattes, or $\mathrm{E}[\rho(s)|\mathbf{Y}]$, with more than 70% of coffee shops predicted to be infected with espresso in the central area, which is both wealthy and averse to Ford. Figure 11.6b shows the predicted residual spatial variation (on the odds scale) $\mathrm{E}\{\exp[U(s)]|\mathbf{Y}\}$, with values above 1.0 at a location s, indicating there are more latte-sipping coffee shop patrons at s than what is typical given the income and Ford support at s. Four of the largest shopping malls in Toronto (Sherway Gardens, Yorkdale, Fairview Mall, and Scarborough Town Centre) and the transit interchange at Yonge and Sheppard are situated in areas with prevalences that are low to moderate yet above what would be expected given the substantial support for Rob Ford at these locations. The residents northwest of the intersection of Yonge St. and Eglinton Ave. have a risk of espresso consumption that is elevated both in absolute terms ($\mathrm{E}[\rho(s)|\mathbf{Y}] \approx 0.8$) and relative to other regions with similar levels of Ford opposition ($\mathrm{E}\{\exp[U(s)]|\mathbf{Y}\} \approx 2$).

There is always uncertainty inherent in making spatial predictions, and uncertainty is a particular concern when modelling spatially referenced binary health outcomes. The 584 coffee shops, of which 229 test positive for espresso, contain less information than a similar number of continuous observations. The coffee data have produced 95% confidence intervals for model parameters larger than the 95% confidence intervals obtained from the 199 arsenic samples in Section 11.3. A useful way of conveying uncertainty is with plots of exceedance probabilities, which are the predicted probabilities that either ρ or U is above or below a threshold. Figure 11.6c shows the probability that $\rho(s) > 0.5$, or the confidence with which it can be concluded that more than half of coffee shops serve lattes, conditional on the observed data. Not all of the areas in Figure 11.6a where prevalence is predicted to be above 0.6 can be said to exceed the 0.5 level with any degree of certainty, although in much of the central region there is more than 95% confidence that the threshold is exceeded. Figure 11.6d shows probabilities that the residual variation $U(s)$ is inducing a fairly modest 10% increase in latte prevalence in excess of what is typical given the covariates $X(s)$.

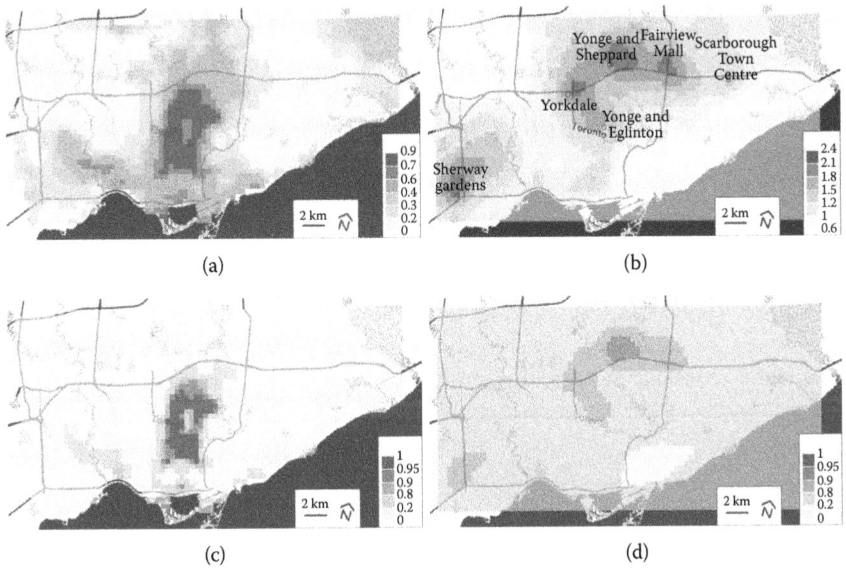

FIGURE 11.6
Latte prevalence and residual spatial variation. (a) $\mathrm{E}(\rho(s)|\mathbf{Y})$. (b) $\mathrm{E}(\exp[U(s)]|\mathbf{Y})$. (c) $pr(\rho(s) > 0.5|\mathbf{Y})$. (d) $pr(U(s) > \log(1.1)|\mathbf{Y})$. (Background © Stamen Design.)

There is considerably more uncertainty in the predicted $U(s)$ than the prevalence $\rho(s)$, with the overwhelming majority of locations having inconclusive probabilities in the 0.2–0.8 range. Only a small area east of Yonge and Sheppard reaches the 90% level of confidence that the excess of lattes is greater than 10%.

11.5 Conclusions

Although geostatistical models are used less frequently than area-level spatial models for applications in the health sciences, their use is likely to increase as advances in mobile technologies make georeferencing to point locations more common. Recent statistical innovations and increases in computer processing power have made it possible to fit geostatistical models to non-Gaussian data on most laptop computers. Area-level spatial models are often the only option for data that are provided in a spatially aggregated format, although Taylor et al. (2015) and Nguyen et al. (2012) are two counterexamples to this assertion. When point locations are available, however, geostatistical models should be considered to be the default option and fitting area-level models to aggregated points can be avoided.

Many of the problems and pitfalls inherent in drawing conclusions from area-level spatial models are amplified with geostatistical models, and the several obvious notes of caution in the coffee example are good examples of these. First, the posterior distributions of the standard deviation and range parameters are not entirely dissimilar to their respective priors. Identifiability of these parameters can be a problem with spatial models, and some sort of prior sensitivity would need to be performed were these results to be taken seriously. Second, results from geostatistical models can be sensitive to model assumptions that may or may not be appropriate. Area-level models that treat each spatial unit as homogeneous (cancer risk in a health region) might be less appropriate than a geostatistical model when the residual spatial variation is due to an environmental factor such as air quality. The discontinuities at borders in area-level models would, however, be more suitable than geostatistical models if the spatial variation is due to some social factor that may well change abruptly once a health region or school district boundary is crossed. The stationarity assumption in geostatistical models, while essential for permitting models to be identified from the modestly sized arsenic and coffee datasets, discourages abrupt changes in the spatial process, as might be expected to occur when crossing the river in the arsenic example or a major road in the coffee example.

Finally, the distinction between causality and correlation is particularly important when using spatial covariates to describe individual-level characteristics such as income or political preference. One should not rule out the possibility that relationship between political preference and coffee shop type is not causal, but rather a symptom of each being influenced by a common spatial factor, such as urban density or proximity to public transit. As spatial data become more widespread and easily attainable, a greater importance will need to be attached to selecting model response variables and covariates based on sound scientific reasoning rather than data availability.

Acknowledgements

Background maps are licensed as follows:

Figure 11.3: Map tiles by CartoDB under CC BY 3.0. Data by OpenStreetMap available under the Open Database License.

Figures 11.4 and 11.6: Map tiles by Stamen Design under CC BY 3.0. Data by OpenStreetMap available under the Open Database License.

The author is supported by a Discovery Grant from the Natural Sciences and Engineering Research Council of Canada (www.nserc-crsng.gc.ca).

References

Brown, P. E. (2015). Geostatistics the easy way. *Journal of Statistical Software*, 63.

Canadian Press (2014). Academics gather in U.S. capitol to test their intellectual mettle with a tough question: 'How is Rob Ford mayor of Toronto?' *National Post*, May 6.

Coyne, A. (2013). Rob Ford mess a monster born of divisive and condescending populism. *National Post*, November 15.

Cressie, N. (1993). *Statistics for Spatial Data.* New York: John Wiley & Sons.

Diggle, P. J., Moyeed, R. A., and Tawn, J. A. (1998). Model-based geostatistics. *Applied Statistics*, 47, 299–350.

Diggle, P. J. and Ribeiro, P. J. (2006). *Model-Based Geostatistics.* New York: Springer-Verlag.

Jiang, H., Brown, P. E., Rue, H., and Shimakura, S. (2014). Geostatistical survival models for environmental risk assessment with large retrospective cohorts. *Journal of the Royal Statistical Society A*, 177(3), 679–695.

Kelly, J. (2014). Why are lattes associated with liberals? *BBC News Magazine.*

Krige, D. G. (1951). A statistical approach to some basic mine valuation problems on the Witwatersrand. *Journal of Chemical, Metallurgical, and Mining Society of South Africa*, 52(6), 119–139.

Lindgren, F., Rue, H., and Lindström, J. (2011). An explicit link between Gaussian fields and Gaussian Markov random fields: The stochastic partial differential equation approach. *Journal of the Royal Statistical Society B*, 73(4), 423–498.

Matérn, B. (1960). Spatial variation. *Meddelanden fran statens Skogsforskningsinstitut*, 49(5).

Nguyen, P., Brown, P. E., and Stafford, J. (2012). Mapping cancer risk in southwestern Ontario with changing census boundaries. *Biometrics*, 68.

Rue, H., Martino, S., and Chopin, N. (2009). Approximate Bayesian inference for latent Gaussian models by using integrated nested Laplace approximations. *Journal of the Royal Statistical Society B*, 71(2), 319–392.

Stein, M. L. (1999). *Interpolation of Spatial Data: Some Theory for Kriging.* New York: Springer-Verlag.

Taylor, B. M., Davies, T. M., Rowlingson, B. S., and Diggle, P. J. (2015). Bayesian inference and data augmentation schemes for spatial, spatiotemporal and multivariate log-gaussian cox processes in r. *Journal of Statistical Software*, 63(7).

12

Splines in Disease Mapping

Tomás Goicoa
Department of Statistics and Operations Research
Institute for Advanced Materials (InaMat)
Public University of Navarra
Pamplona, Spain

Jaione Etxeberria
Department of Statistics and Operations Research
Institute for Advanced Materials (InaMat)
Public University of Navarra
Pamplona, Spain

María Dolores Ugarte
Department of Statistics and Operations Research
Institute for Advanced Materials (InaMat)
Public University of Navarra
Pamplona, Spain

CONTENTS

12.1 Introduction

Smoothing techniques based on splines have been recently incorporated into the disease mapping toolkit as an alternative to the widely used conditional autoregressive (CAR) models. In this chapter, we deal with areal data. Readers interested in alternative smoothing techniques for examining the spatial distribution and variation in disease risk for individual event data are referred to Chapter 10 of this handbook (Hazelton, 2015).

Although splines have become popular in a spatiotemporal setting, their use has been mainly limited to model temporal rather than spatial effects (MacNab and Dean, 2001; MacNab and Gustafson, 2007; MacNab, 2007; Silva et al., 2008). Recently, Lee and Durbán (2009) proposed a two-dimensional P-spline model to smooth risks in space. This last model has been extended to a spatiotemporal setting by Ugarte et al. (2010b), giving rise to an anisotropic and nonseparable three-dimensional model allowing a different amount of smoothing in each dimension (longitude, latitude, and time). An ANOVA-type P-spline model has also been used by Ugarte et al. (2012b) to analyze the spatiotemporal evolution of prostate cancer in Spain. These authors develop an estimator for the mean squared error of the log-risk predictor, taking into account the variability associated with the estimation of the smoothing parameters. In what follows, we focus on the P-spline approach and look into similarities and differences with classical CAR smoothing and temporal B-spline smoothing without penalties.

The rest of the chapter is organized as follows. In Section 12.2, a brief insight into general P-splines is given. Spatial P-splines in disease mapping are described in Section 12.3. Section 12.4 is devoted to spatiotemporal P-splines and forecasting. A case study is presented in Section 12.5. The chapter closes with some final remarks.

12.2 P-Spline Basics

Penalized splines or P-splines with penalties on the coefficients were popularized by Eilers and Marx (1996), and their use has increased during the last few years. They have been shown to be very useful due to their flexibility and some interesting properties. First, they do not show boundary effects, what makes them attractive for forecasting purposes. Second, they are low-rank smoothers, as the size of the basis is smaller than the dimension of the data at hand, reducing computational complexity, and third, the choice of knots is not very important because of the penalty. Another interesting feature is that they can be expressed as linear (or generalized linear) mixed models, and well-posed theory can be applied. Different penalties have been considered in the literature (see O'Sullivan, 1986, 1988), but here we focus on penalties on the coefficients. Next, we briefly review the theory of P-splines with B-spline bases. B-spline bases are constructed joining polynomial pieces of degree p at some values of x, called knots. Once the knots are chosen, B-splines are computed using a recursive algorithm (de Boor, 1978). Usually, equally spaced quantiles of x are considered to locate the knots, and its number is selected according to the rule min{number of unique values of $x/4, 40$} (Ruppert, 2002), although some authors restrict the number of knots in spatial and spatiotemporal settings to reduce computational burden.

Let us consider paired data (y_i, x_i) and the model

$$y_i = f(x_i) + \epsilon_i \approx a_0 B_0(x_i) + a_1 B_1(x_i) + \ldots + a_K B_K(x_i) + \epsilon_i, \quad i = 1, \ldots n.$$

In matrix form, the model is expressed as

$$\mathbf{Y} = \mathbf{f} + \boldsymbol{\epsilon} \approx \mathbf{Ba} + \boldsymbol{\epsilon},$$

where $\mathbf{Y}$ is the vector of sampled observations, $\mathbf{f}$ is an unknown smooth function that can be well approximated using P-splines with B-spline bases, $\mathbf{B} = (\mathbf{B}_0, \ldots, \mathbf{B}_K)$ is the matrix of the B-spline basis obtained from the covariate $\mathbf{X}$, $\mathbf{a}' = (a_0, \ldots, a_K)$ is the vector of the basis coefficients, $\boldsymbol{\epsilon}' = (\epsilon_1, \ldots, \epsilon_n)' \sim N(\mathbf{0}, \sigma^2 \mathbf{I}_n)$, and $\mathbf{I}_n$ is the identity matrix. The P-spline approach minimizes the penalized sum of squares

$$\mathbf{S}(\mathbf{a}; \mathbf{Y}, \lambda) = \sum_{i=1}^{n} (y_i - f(x_i))^2 + \lambda \mathbf{a}'\mathbf{D}'\mathbf{D}\mathbf{a} = (\mathbf{Y} - \mathbf{Ba})'(\mathbf{Y} - \mathbf{Ba}) + \lambda \mathbf{a}'\mathbf{P}\mathbf{a}, \tag{12.1}$$

where $\mathbf{P} = \mathbf{D}'\mathbf{D}$ is a penalty matrix imposing smoothness on the adjacent coefficients, and $\mathbf{D}$ is a difference matrix of order d (usually $d = 2$). The parameter λ determines the influence of the penalty matrix. A common way to optimize the smoothing parameter is using cross-validation techniques (see Hastie and Tibshirani, 1990, p. 43); however, the variability due to estimation of the smoothing parameter λ is not taken into account.

One of the more interesting properties of P-splines is that they can be reformulated as mixed-effect models (see Eilers, 1999; Currie and Durbán, 2002). To express a P-spline model with B-spline bases as a mixed model, a transformation matrix $\mathbf{T}$ is needed. The transformation matrix is not unique, and it is common to use the one based on the singular value decomposition of the penalty matrix, that is, $\mathbf{T} = [\mathbf{U}_1 : \mathbf{U}_2\widetilde{\boldsymbol{\Sigma}}^{-1/2}]$, where $\mathbf{U}_1$ and $\mathbf{U}_2$ are matrices of singular vectors corresponding to the null and nonzero eigenvalues, respectively, and $\widetilde{\boldsymbol{\Sigma}} = \text{diag}(\tau_1, \ldots, \tau_c)$ is a diagonal matrix whose elements are the nonzero eigenvalues of $\mathbf{D}'\mathbf{D}$. Then, defining $\mathbf{BT} = [\mathbf{X} : \mathbf{Z}] = [\mathbf{BU}_1, \mathbf{BU}_2\widetilde{\boldsymbol{\Sigma}}^{-1/2}]$, one can reparameterize $\mathbf{Ba} = \mathbf{X}\boldsymbol{\beta} + \mathbf{Z}\boldsymbol{\alpha}$, where $\mathbf{a} = \mathbf{T}\begin{pmatrix}\boldsymbol{\beta}\\ \boldsymbol{\alpha}\end{pmatrix}$, and $\boldsymbol{\beta}$ and $\boldsymbol{\alpha}$ are sets of fixed and random effects, respectively. Using this transformation, Equation 12.1 can be written as

$$\mathbf{S}(\boldsymbol{\beta}, \boldsymbol{\alpha}; \mathbf{Y}, \lambda) = (\mathbf{Y} - \mathbf{X}\boldsymbol{\beta} - \mathbf{Z}\boldsymbol{\alpha})'(\mathbf{Y} - \mathbf{X}\boldsymbol{\beta} - \mathbf{Z}\boldsymbol{\alpha}) + \lambda\boldsymbol{\alpha}'\boldsymbol{\alpha}. \tag{12.2}$$

Dividing Equation 12.2 by σ_e^2, setting $\lambda = \sigma_e^2/\sigma_u^2$, and differentiating with respect to $\boldsymbol{\beta}$ and $\boldsymbol{\alpha}$, the mixed-model equations corresponding to the mixed model

$$\mathbf{Y} = \mathbf{X}\boldsymbol{\beta} + \mathbf{Z}\boldsymbol{\alpha} + \boldsymbol{\epsilon}, \quad \boldsymbol{\alpha} \sim N(\mathbf{0}, \sigma_u^2\mathbf{I}), \quad \boldsymbol{\epsilon} \sim N(\mathbf{0}, \sigma_e^2\mathbf{I})$$

are attained. The smoothing parameter now becomes a variance component, and it can be estimated using restricted maximum likelihood (REML). Then, the uncertainty arising from the estimation of the variance components can be taken into account.

12.3 P-Splines in Spatial Disease Mapping

P-splines have become popular in space–time disease mapping to describe temporal trends. However, their use in disease mapping to smooth the risk surface and unveil spatial patterns of a disease has not been so common. As far as we know, the first piece of research using P-splines to describe the geographical pattern of a disease is the work by Lee and Durbán (2009). They suggest modeling the spatial variability using a two-dimensional P-spline considering the centroid of the areas as spatial locations. Here, we briefly review their work. Let us assume that the observed number of mortality or incidence cases in each area Y_i follows a Poisson distribution with mean $\mu_i = e_i\theta_i$, that is, $Y_i \sim \text{Poisson}(\mu_i = e_i\theta_i)$ for $i = 1, \ldots, n$, where θ_i is the unknown relative risk. In this context, the interest lies in estimating the relative risk θ_i for each area i. Then, the log risk is modeled as

$$\log\theta_i = f(x_{1i}, x_{2i}), \quad i = 1, \ldots, n,$$

where f is an unknown smooth function of the covariates that is assumed to be approximated sufficiently well by P-splines, and x_{1i} and x_{2i} are the coordinates of the centroid of the ith small area (longitude and latitude, respectively). In matrix form,

$$\log\boldsymbol{\theta} = \mathbf{Ba},$$

where $\mathbf{B}$ is a bivariate B-spline basis and $\mathbf{a} = (a_1, \ldots, a_k)'$ is the vector of coefficients. For scatter spatial data, the B-spline basis is defined as the row-wise Kronecker product (see Eilers et al., 2006) of the marginal bases. If $\mathbf{B}_1$ and $\mathbf{B}_2$ are $n \times k_1$ and $n \times k_2$ marginal bases for longitude and latitude, respectively, $\mathbf{B}$ is of dimension $n \times k_1 k_2$ and is given by

$$\mathbf{B} = \mathbf{B}_2 \square \mathbf{B}_1 = (\mathbf{B}_2 \otimes 1'_{k1}) \odot (1'_{k2} \otimes \mathbf{B}_1),$$

where $\odot$ represents "element-wise" matrix product and $\mathbf{1}_{k_1}$ and $\mathbf{1}_{k_2}$ are vectors of ones of lengths k_1 and k_2, and $k = k_1 k_2$. To achieve smoothness, the following penalty, based on marginal penalties for longitude and latitude, is placed on the coefficients (see Currie et al., 2004):

$$\mathbf{P} = \lambda_1 \mathbf{P}_2 \otimes \mathbf{I}_{k_1} + \lambda_2 \mathbf{I}_{k_2} \otimes \mathbf{P}_1,$$

where λ_i, $i = 1, 2$ are smoothing parameters, $\mathbf{P}_i = \mathbf{D}_i' \mathbf{D}_i$, and the matrices $\mathbf{D}_i$ are difference matrices, making "neighboring" coefficients similar. A detailed insight into the penalty is deserved to see how smoothing is working with P-splines. This is described in what follows. Let us arrange the coefficients in a lattice where rows represent latitude and columns represent longitude (see Figure 12.1), and let us consider difference matrices $\mathbf{D}_i$ of order 1. Then, coefficient a_{ij} would be influenced by $a_{i(j-1)}$ and $a_{i(j+1)}$ (longitude), and by $a_{(i-1)j}$ and $a_{(i+1)j}$ (latitude), $1 < i < k_1$ and $1 < j < k_2$, with the corresponding adjustments for coefficients on the border of the grid. However, a_{ij} is not equally affected by rows and columns because the smoothing parameters for longitude and latitude are different. This can be seen as an anisotropic model with a CAR distribution on the basis coefficients (see Rue and Held, 2005, pp. 104–107). If the smoothing parameters happened to be equal, this would be equivalent to an intrinsic conditional autoregressive (ICAR) distribution on the basis coefficients; that is, a_{ij} would be equally influenced by its first-order neighbors $a_{(i-1)j}$, $a_{(i+1)j}$, $a_{i(j-1)}$, and $a_{i(j+1)}$. From a Bayesian point of view, this would be equivalent to considering a Gaussian prior for the coefficients with a structure matrix given by the penalty (see Lang and Brezger, 2004), which in this case would be a neighborhood matrix for the coefficients. In practice, differences of order 2 on the basis coefficients have been considered, indicating that coefficient a_{ij} is influenced by $a_{i(j-1)}$, $a_{i(j-2)}$, $a_{i(j+1)}$, and $a_{i(j+2)}$ (longitude) and similarly for latitude. The reader should note that classical CAR models for disease mapping use the CAR distribution for spatial random effects, not for fixed coefficients,

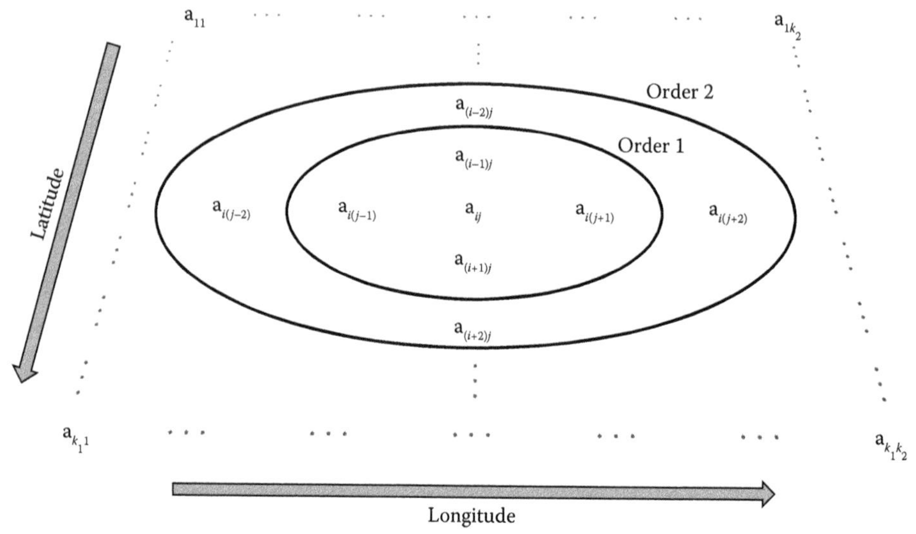

FIGURE 12.1
Coefficient grid for the two-dimensional P-spline model.

and they assume that two areas are neighbors if they share a common border. Here, the coefficients are neighbors if they correspond to adjacent B-splines. The two-dimensional P-spline model can be expressed as a mixed model

$$\log \boldsymbol{\theta} = \mathbf{Ba} = \mathbf{X}\boldsymbol{\beta} + \mathbf{Z}\boldsymbol{\alpha}, \quad \boldsymbol{\alpha} \sim N(\mathbf{0}, \mathbf{G}(\boldsymbol{\lambda})),$$

where expressions for $\mathbf{X}$, $\mathbf{Z}$, and $\mathbf{G}(\boldsymbol{\lambda})$ are based on the singular value decomposition of the marginal penalties and are detailed in Lee and Durbán (2009), and $\boldsymbol{\lambda} = (\lambda_1, \lambda_2)$. Then, the well-known penalized quasi-likelihood (PQL) technique (Breslow and Clayton, 1993) can be used for model estimation and inference. The smoothing parameters become variance components, and they are estimated using the restricted maximum likelihood equations. This in turn allows us to include this source of variability in the mean squared error (MSE) estimation of the log risk predictor, something that has been previously ignored. The MSE estimator for the log risk predictor for CAR models has been developed by Ugarte et al. (2008), borrowing ideas from the small-area estimation literature. Similarly, Goicoa et al. (2012) propose the following MSE estimator for spatial P-spline models:

$$\widehat{MSE}[\widehat{\log \theta_i}] = g_{1i}(\hat{\boldsymbol{\lambda}}) + g_{2i}(\hat{\boldsymbol{\lambda}}) + 2g_{3i}(\hat{\boldsymbol{\lambda}}), \tag{12.3}$$

where the terms g_1, g_2, and g_3 capture variability related to the estimation of the random effects, fixed effects, and variance components, respectively.

Additional models based on P-splines have been considered by Lee and Durbán (2009) to analyze the well-known example on lip cancer incidence cases in Scotland, namely, a P-spline model with an individual random effect for each area (PRIDE) (see Perperoglou and Eilers, 2010) and a P-spline model including a CAR random effect (Smooth-CAR) with different specifications: a convolution prior (Besag et al., 1991), a Leroux et al. (2000) prior, and a similar parametrization given by Dean et al. (2001). However, the authors recommend caution with the Smooth-CAR model, as identifiability issues may arise. The performance of the two-dimensional P-spline and PRIDE models has been assessed by Goicoa et al. (2012) in terms of smoothing, sensitivity (ability to detect high-risk regions), and specificity (ability to discard false positives created by noise) throughout a simulation study. The authors use confidence intervals based on the MSE estimator given by Equation 12.3 to classify regions (see Ugarte et al., 2009a) and conclude that, in general, the P-spline model detects more true positives, but at the cost of increasing the number of false positives. They also give some guidelines if the aim is to detect high-risk areas. However, their scenarios were simulated using the CAR, P-spline, and PRIDE models, and other model-free settings should be explored. In general, if the risk surface changes smoothly, the bivariate P-spline model seems to be appropriate.

12.4 P-Splines in Spatiotemporal Disease Mapping

The two-dimensional P-spline approach has been extended to include the time dimension giving rise to three-dimensional P-splines. In this section, three-dimensional P-splines for describing the spatiotemporal evolution of mortality risks are reviewed (Ugarte et al., 2010b). A first attempt to model standardized mortality rates in space and time using splines was made by van der Linde et al. (1995). However, only two time periods were finally considered in their application.

In spatiotemporal disease mapping, data are available for every region and time period, and the number of deaths Y_{it} is assumed to follow a Poisson distribution with mean

$\mu_{it} = e_{it}\theta_{it}$, that is, $Y_{it} \sim Poisson(\mu_{it} = e_{it}\theta_{it})$, $i = 1, \ldots, n$, and $t = 1, \ldots, T$. Then, the log risk is modeled as

$$\log \theta_{it} = f(x_{1i}, x_{2i}, t), \tag{12.4}$$

where x_{1i} and x_{2i} are the coordinates of the centroid of the ith small area (longitude and latitude, respectively), t is the time, and f is an unknown three-dimensional smooth function, assumed to be well approximated by P-splines. In matrix form,

$$\log \boldsymbol{\theta} = \mathbf{Ba},$$

where $\mathbf{B}$ is a three-dimensional B-spline basis and $\mathbf{a}$ is the vector of coefficients. If $\mathbf{B}_1$, $\mathbf{B}_2$, and $\mathbf{B}_3$ are $n \times k_1$, $n \times k_2$, and $n \times k_3$ marginal B-spline bases for longitude, latitude, and time, respectively, then the B-spline basis $\mathbf{B}$ is given by

$$\mathbf{B} = \mathbf{B}_3 \otimes (\mathbf{B}_2 \square \mathbf{B}_1), \tag{12.5}$$

and the penalty matrix on the coefficients is expressed in terms of the marginal penalties $\mathbf{P}_i = \mathbf{D}_i'\mathbf{D}_i$, $i = 1, 2, 3$ as

$$\mathbf{P} = \lambda_1 I_{K_3} \otimes I_{K_2} \otimes \mathbf{P}_1 + \lambda_2 I_{K_3} \otimes \mathbf{P}_2 \otimes I_{K_1} + \lambda_3 \mathbf{P}_3 \otimes I_{K_2} \otimes I_{K_1},$$

where λ_i are smoothing parameters and $\mathbf{D}_i$ are difference matrices. Similarly, as in Figure 12.1, a coefficient a_{ijt} would be influenced by its row (longitude) and column (latitude) neighbors. As we have a penalty in time, it would also be affected by the time neighbors. However, the effect of row, column, and time neighbors is not the same, as the smoothing parameters are different. The spatiotemporal P-spline model can be reformulated as a linear mixed model as

$$\mathbf{Ba} = \mathbf{X}\boldsymbol{\beta} + \mathbf{Z}\boldsymbol{\alpha}, \quad \boldsymbol{\alpha} \sim N(\mathbf{0}, \mathbf{G}(\boldsymbol{\lambda})),$$

where expressions for $\mathbf{X}$, $\mathbf{Z}$, and $\mathbf{G}(\boldsymbol{\lambda})$ can be found in Ugarte et al. (2010b). This model was used by Ugarte et al. (2010a) to analyze breast cancer mortality in Spain. Note that the covariance matrix $\mathbf{G}(\boldsymbol{\lambda})$ depends on the smoothing parameters $\boldsymbol{\lambda} = (\lambda_1, \lambda_2, \lambda_3)$. A possible limitation of this model may be that the same smoothing parameters λ_1, λ_2, and λ_3 used for longitude, latitude, and time are also used for the space–time interaction. To make the model more flexible, an ANOVA-type P-spline model (Lee and Durbán, 2011) can be used in this context (Ugarte et al., 2012b). In this model, the log risk is expressed as

$$\log \theta_{it} = \delta + f_s(x_{1i}, x_{2i}) + f_t(x_t) + f_{(st)}(x_{1i}, x_{2i}, x_t), \tag{12.6}$$

where δ is an intercept, $f_s(x_{1i}, x_{2i})$ captures the spatial effects, $f_t(x_t)$ is a trend common to all areas, and $f_{(st)}(x_{1i}, x_{2i}, x_t)$ can be seen as the specific temporal trend for each area. Here, the B-spline basis is $\mathbf{B} = [\mathbf{1}_{nT} : \mathbf{1}_T \otimes (\mathbf{B}_2 \square \mathbf{B}_1) : \mathbf{B}_3 \otimes \mathbf{1}_n : \mathbf{B}_3 \otimes (\mathbf{B}_2 \square \mathbf{B}_1)]$, where $\mathbf{1}_{nT}$, $\mathbf{1}_T$, and $\mathbf{1}_n$ are column vectors of ones of dimensions nT, T, and n, respectively. Finally, the penalty takes the form

$$\begin{aligned} \text{Blockdiag}(0, \lambda_1 I_{k_2} \otimes \mathbf{P}_1 + \lambda_2 \mathbf{P}_2 \otimes I_{k_1}, \lambda_3 \mathbf{P}_3, \tau_1 I_{K_3} \otimes I_{K_2} \otimes \mathbf{P}_1 + \tau_2 I_{K_3} \otimes \mathbf{P}_2 \otimes I_{K_1} \\ + \tau_3 \mathbf{P}_3 \otimes I_{K_2} \otimes I_{K_1}). \end{aligned}$$

Note that here three additional parameters, τ_1, τ_2, and τ_3, are used for the interaction term.

Both models (12.4) and (12.6) consider space–time interactions, and the coefficients are jointly penalized in space and time; that is, they could be classified as type IV interaction models (see Knorr-Held, 2000)—temporal trends are different from region to region, but are more likely to be similar for adjacent regions. However, there have been other approaches in the literature. All of them consider one-dimensional splines for the temporal trends and spatial random effects with a CAR distribution for space. Basically, the models take the form

$$\log \theta_{it} = \delta + f(x_t) + \phi_i + f_i(x_t),\ i = 1, \ldots, n,\ t = 1, \ldots, T,$$

where t is centered, $\delta + f(x_t)$ represents a global trend common to all areas, and $\phi_i + f_i(x_t)$ is a regional trend, where ϕ_i is a spatial effect constant along time. The main differences are the type of splines used and the distribution on the spline coefficients. If spatial dependence is placed on the coefficients, then neighboring regions will have similar temporal trends. For example, the simplest approach, consisting of an additive model with a spatial random effect ϕ_i with a CAR distribution, and a temporal trend $\delta + f(x_t)$ common to all regions modeled with B-splines without penalties, has been considered by MacNab and Dean (2001). Additionally, they proposed an interaction model given by different B-splines for each area. They use the methodology to model infant mortality data in Canada. A similar approach was developed by Silva et al. (2008) using a binomial distribution to model revascularization odds. Torabi and Rosychuk (2011) and Torabi (2013) also consider spatial random effects and B-splines for time to analyze childhood cancer incidence in Alberta and asthma visits to hospital in Manitoba. MacNab and Gustafson (2007) extend this approach from a fully Bayesian perspective considering random coefficients for the B-spline bases. They propose different spatial priors as well as spatially unstructured priors for these coefficients and evaluate different numbers of knots. In their study, they analyze iatrogenic injuries among males aged 65 or older in British Columbia and conclude that a spatial prior for the coefficients leads to the best model. Zhang et al. (2006) use smoothing splines (Kimeldorf and Wahba, 1970) to model temporal trends. To compare the performance of different smoothing methods, MacNab (2007) studies temporal smoothing with unpenalized B-splines, and penalized alternatives, namely, smoothing splines and P-splines. She analyzes different CAR priors and spatially unstructured priors for the coefficients and suggests that smoothing splines and P-splines may be less flexible, as they could be limited by the penalty (and the boundary conditions in the case of smoothing splines). Nevertheless, in her example, the P-spline approach was close to a smoothing with four-knot spatially unstructured B-splines. Very recently, Etxeberria et al. (2014) also considered spatially structured temporal P-splines to smooth risks.

12.4.1 Forecasting with spatiotemporal P-splines

Cancer mortality data are currently available with a delay of roughly 3 years due to administrative procedures to create the registries. Therefore, it seems sensible to use statistical techniques to provide risk and count predictions in different areas. Risk and count predictions have been provided using spatiotemporal models with CAR distributions for space and random walks of orders 1 and 2 for time (see Knorr-Held and Rainer, 2001; Schmid and Held, 2004), but forecasting with P-splines is not common in disease mapping, even though they have been used for forecasting purposes in other fields (see Currie et al., 2004; Ugarte et al., 2009b). Here, we briefly revise some work on forecasting mortality risks and counts with P-splines. More precisely Ugarte et al. (2012a) extend the P-spline model (12.4) to forecast prostate cancer mortality risks and counts in 50 Spanish provinces. The key point is to extend the B-spline basis for time, denoted by $\mathbf{B}_t^*$, to include the future years. This leads to an extended three-dimensional basis $\mathbf{B}^*$ and to an extended penalty $\mathbf{P}^*$. The main issue here is to look for an extended transformation matrix $\mathbf{T}^*$ preserving the original transformation matrix $\mathbf{T}$, so that the reformulation of the P-spline as a mixed model includes the future relative risks $\boldsymbol{\theta}_{it}^*$. That is,

$$\log(\boldsymbol{\theta}_{it}^*) = \mathbf{x}_{it}^* \boldsymbol{\beta} + \mathbf{z}_{it}^* \boldsymbol{\alpha}^* \quad t = T+1, T+2, \ldots, T+p,$$

where $\mathbf{x}^*_{it}$, and $\mathbf{z}^*_{it}$ are the design matrices within the mixed-model reformulation corresponding to the future years, and $\boldsymbol{\alpha}^*$ is the vector of random effects related to the extended years. To assess the predictive performance of the model, a validation study was conducted concluding that the spatiotemporal P-spline model outperforms simple temporal models fitted independently to each area in terms of goodness of fit and coverage rates of prediction intervals. Lately, Etxeberria et al. (2014) have compared the P-spline model with other alternatives based on CAR models and combinations of CAR and P-splines. They show that the P-spline model is a good alternative to provide short-term projections of counts and mortality risks. An ANOVA-type P-spline model has also been recently considered for forecasting cancer mortality by Etxeberria et al. (2015).

12.5 Illustration

In this section, the performance of the spatiotemporal interaction P-spline model given by (12.4) is considered for illustration purposes using female breast cancer mortality data in 50 Spanish provinces during the period 1990–2010. This is an interaction model appropriate to analyze mortality or incidence trends by provinces, as it allows a different trend for each region.

Data on Spanish population and breast cancer deaths were obtained from records of the Spanish Statistical Institute. A total of 121,905 deaths were observed in the whole period, varying from 5 to 847 per province and year. The number of expected counts per province and per year ranges from 16.44 to 931.70. The expected counts are calculated on the basis of age-specific mortality rates for Spain during the whole period and the age-specific population at risk for each year.

The three-dimensional B-spline basis $\mathbf{B}$ given by Equation 12.5 is constructed once the number of (equidistant) internal knots, the degree of the B-splines, and the order of the penalty are defined. Here, cubic B-splines, a second-order penalty, 10 knots for both spatial components (longitude and latitude), and 6 for the time component are chosen. The main objectives of the analysis are to detect changes in spatial risks patterns along the years and to study temporal trends in the different Spanish provinces.

Figure 12.2 displays the temporal evolution of the geographical pattern of female breast cancer mortality in the study period. Changes in the spatial distribution of risks are clear. At the beginning of the period and until 1993, there is a cluster of regions with high risks in the northeast of Spain. This cluster encompasses the provinces in Catalonia. Balearic and Canary Islands also exhibit high risks. In general, provinces located in the central north and the Mediterranean area exhibit the highest mortality risks, together with the southwestern provinces around the nineties. From then on, the geographical pattern changes, and from 2003 onward, the distribution of the relative risk is fairly homogeneous among the provinces.

Figure 12.3, top row, displays the standardized mortality ratios (SMRs) (nonsmooth lines), the smooth mortality risks, and the corresponding 95% pointwise confidence bands (grey bands) for three provinces located in different geographical spots of Spain in the period 1990–2010. The horizontal line at one is placed for interpretation purposes. Values of the true risk and the lower (upper) bound of the confidence interval greater (less) than 1 mean a significant high (low) risk when compared to Spain in the whole period.

The P-spline model (12.4) is an interaction model, but when reformulated as a mixed model, it can be easily split into the spatial, temporal, and interaction components. This can be done by grouping the columns of the design matrices corresponding to space, time, and

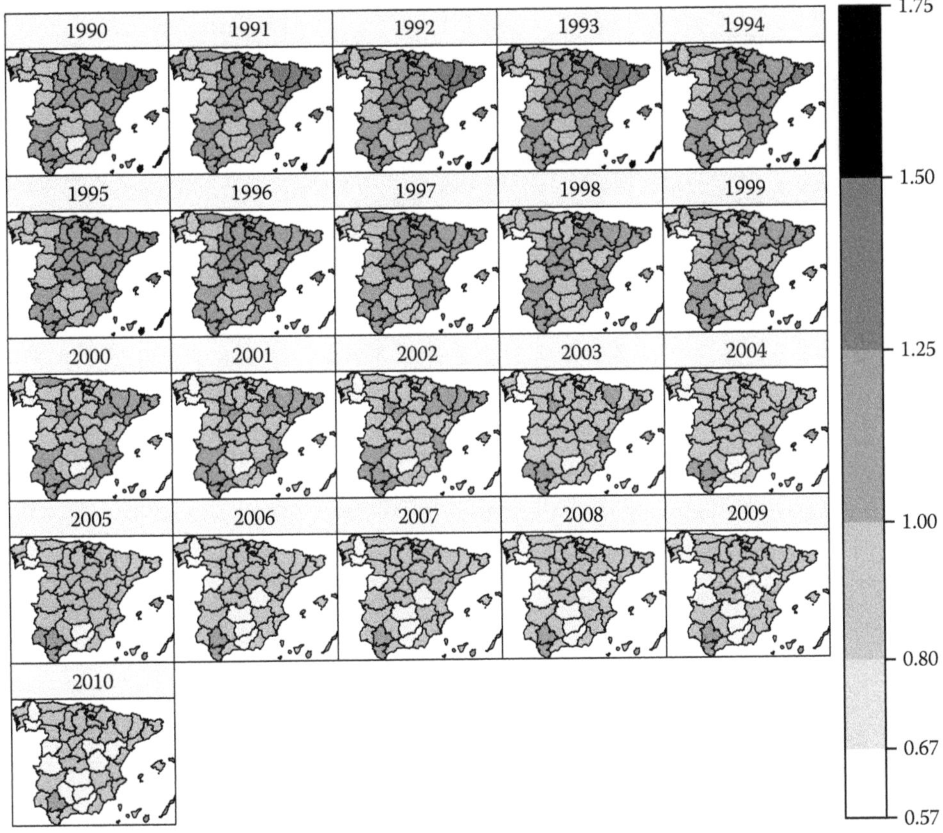

FIGURE 12.2
Spatiotemporal distribution of breast cancer mortality risks in Spain (1990–2010). The Canary Islands have been moved from their exact location; they are displayed at the bottom right of the maps.

interaction. Hence, it is very interesting to look at the different parts of this P-spline model. Figure 12.3, bottom row, displays the estimated temporal trend common to all areas, the spatial effect constant along the whole period, the area-specific temporal trend, and the total risk estimates for the same three provinces. The common temporal trend is decreasing during the period, and hence it contributes to reduce the total risk. This could be attributed to the implementation of screening programs in Spain or to general improvements in treatment. The constant spatial effect can be interpreted as the mortality risk associated with a certain region. The area-specific temporal trend makes the risk increase or decrease, depending on the area, and it could be attributed to specific factors related to a particular region.

To illustrate the predictive performance of the P-spline model (12.4), Table 12.1 displays the observed number of deaths in 2010 for the three selected provinces, and predictions for the years 2011, 2012, and 2013. Ninety-five percent prediction intervals are also provided (see Etxeberria et al., 2014, for the construction of these intervals). Note that although risks tend to decrease (see Figure 12.3), counts do not necessarily have to mimic that behavior, as they depend on the population, which may increase or decrease from one year to another.

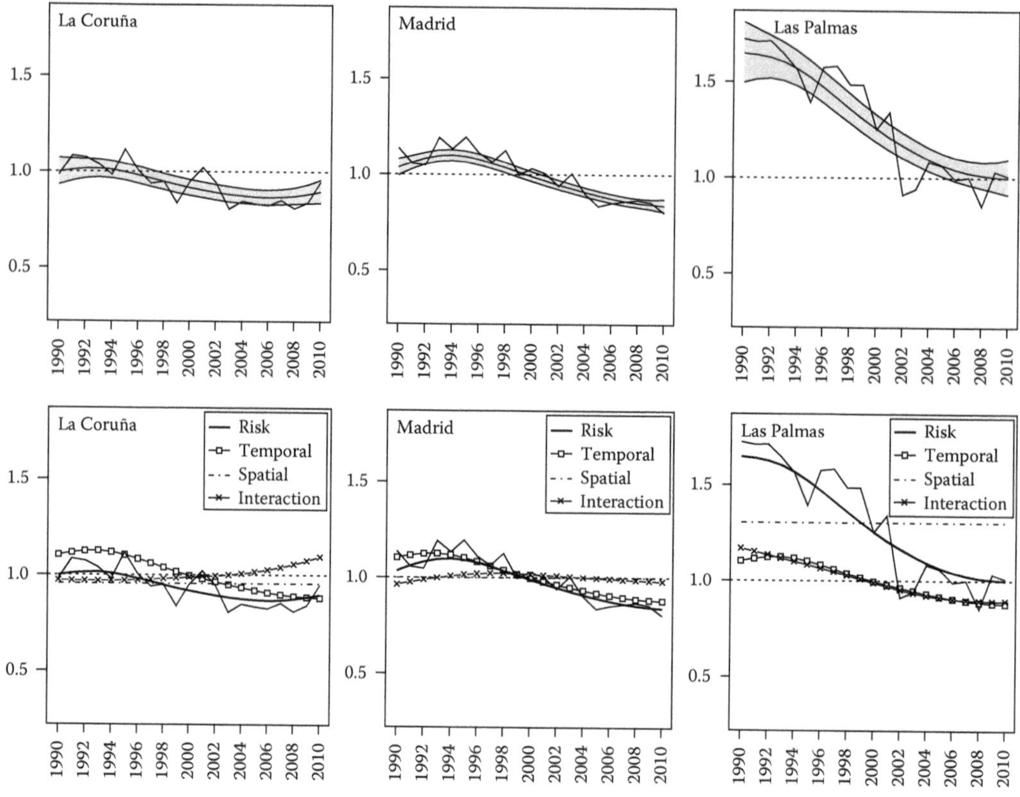

FIGURE 12.3
Smooth temporal trends obtained with the P-spline model (12.4) in the period 1990–2010. Top row: Smoothed breast cancer mortality risks and confidence bands (in grey). Nonsmooth lines represent standardized mortality ratios (SMRs). Bottom row: Temporal evolution of the different terms of the model. The smooth solid line corresponds to the total risk estimates, the boxed line represents the temporal trend common to all areas, the dash–dot line represents the spatial effect, and the X line captures the area-specific temporal trend.

TABLE 12.1
Number of deaths in 2010, and predictions for 2011, 2012, and 2013 obtained with the P-spline model (12.4)

Province	Observed deaths 2010	Predictions 2011	Predictions 2012	Predictions 2013
La Coruña	198	189.52	182.02	152.42
		[157.32, 221.72]	[141.75, 222.28]	[79.81, 225.03]
Madrid	748	742.20	752.49	755.46
		[672.87, 811.54]	[657.36, 847.61]	[558.76, 952.16]
Las Palmas	128	119.26	120.68	119.63
		[94.07, 144.45]	[92.58, 148.77]	[77.26, 161.99]

Note: Number in brackets corresponds to 95% prediction intervals.

12.6 Final Remarks

Smoothing techniques based on splines have been recently incorporated into the disease mapping toolkit to describe temporal trends as well as spatiotemporal patterns. One of the simplest models consists of a spatial random effect for space with a CAR distribution and a global temporal trend common to all areas modeled with B-splines without penalties. This model has been extended to include space–time interactions in the form of specific temporal trends for each area. These temporal trends may be both spatially structured and unstructured; that is, neighboring regions have similar temporal risk trends or trends can randomly vary, respectively. An alternative approach is to use two-dimensional P-splines to describe the underlying risk surface. This approach uses B-spline bases for longitude and latitude and places a difference penalty on the coefficients of adjacent B-splines, so that the surface varies smoothly. The amount of smoothing in each direction is controlled by two different smoothing parameters. This can be seen as an anisotropic model with a CAR distribution on the basis coefficients. Were these two smoothing parameters equal and the differences of first order, this would be equivalent to an ICAR distribution on the coefficients with a structure matrix given by the penalty. The spatial P-spline approach has been extended to the spatiotemporal setting by using three-dimensional P-splines so that smoothing is carried out jointly in space and time using a penalty with different smoothing parameters for longitude, latitude, and time. The described approaches are sensible and produce valid results. The first one mainly focuses on describing temporal trends in each region with temporal B-splines without penalties. Then, trends may or may not be spatially structured. Basically, these models consider a one-dimensional B-spline common to all areas or different B-splines for each area. They do not include penalties. The CAR distribution on the coefficients makes neighboring trends similar, but this is not a common penalty, as the CAR prior is placed on the coefficients of different B-splines. The P-spline approach jointly smooths space and time using three-dimensional B-spline bases and penalties. That is, a single three-dimensional P-spline is used, and it directly produces smooth maps and trends using penalties for space and time, unlike the previous approach, where different unpenalized B-splines for each area are considered. The P-spline approach can be seen as a kind of type IV interaction model, whereas the unpenalized approach leads to type I or type III interaction models (see Knorr-Held, 2000) because there are no penalties in time.

Another approach that deserves further research is possible. Instead of considering temporal trends and studying whether they are spatially structured, it would be possible to consider smooth surfaces and study their temporal evolution, taking into account if they are temporally structured or not. In any case, all of these models would be alternative and valid approaches to model spatiotemporal interactions using splines. If the main focus is on temporal trends, then temporal B-splines spatially structured or unstructured would be appropriate. If the goal is to monitor the evolution of spatial patterns during some years, two-dimensional B-splines temporally structured or unstructured would seem to be the election. The P-spline approach reviewed here can be seen as a combination of both approaches, as it smooths risks in space and time, and it would then be appropriate in both cases.

Acknowledgments

This work has been supported by the Spanish Ministry of Science and Innovation (project MTM2011-22664 cofunded with FEDER grants and project MTM2014-51992-R) and by the

Health Department of the Navarre government (Project 113, Res. 2186/2014). We would like to thank the National Epidemiology Center (area of Environmental Epidemiology and Cancer) for providing the data, originally created by the Spanish Statistical Office.

References

Besag, J., York, J., and Mollié, A. 1991. Bayesian image restoration, with two applications in spatial statistics. *Ann. Inst. Stat. Math.*, 43(1), 1–59. With discussion and a reply by Besag.

Breslow, N. E. and Clayton, D. G. 1993. Approximate inference in generalized linear mixed models. *J. Am. Stat. Assoc.*, 88(421), 9–25.

Currie, I. D. and Durbán, M. 2002. Flexible smoothing with *P*-splines: a unified approach. *Stat. Model.*, 2(4), 333–349.

Currie, I. D., Durbán, M., and Eilers, P. H. C. 2004. Smoothing and forecasting mortality rates. *Stat. Model.*, 4(4), 279–298.

Dean, C. B., Ugarte, M. D., and Militino, A. F. 2001. Detecting interaction between random region and fixed age effects in disease mapping. *Biometrics*, 57(1), 197–202.

de Boor, C. 1978. *A Practical Guide to Splines.* Applied Mathematical Sciences, vol. 27. Springer-Verlag, New York.

Eilers, P. H. C. 1999. Comment on "The analysis of designed experiments and longitudinal data by using smoothing splines," by Verbyla AP, Cullis BR, Kenward MG, and Welham JS. *Journal of the Royal Statistical Society: Series C (Applied Statistics)*, 48(3), 269–311.

Eilers, P. H. C., Currie, I. D., and Durbán, M. 2006. Fast and compact smoothing on large multidimensional grids. *Comput. Stat. Data Anal.*, 50(1), 61–76.

Eilers, P. H. C. and Marx, B. D. 1996. Flexible smoothing with *B*-splines and penalties. *Stat. Sci.*, 11(2), 89–121. With comments and a rejoinder by the authors.

Etxeberria, J., Goicoa, T., Ugarte, M. D., and Militino, A. F. 2014. Evaluating space-time models for short-term cancer mortality risk predictions in small areas. *Biometrical Journal*, 56(3), 383–402.

Etxeberria, J., Ugarte, M. D., Goicoa, T., and Militino, A. F. 2015. On predicting cancer mortality using ANOVA-type P-spline models. *REVSTAT*, 13, 21–40.

Goicoa, T., Ugarte, M. D., Etxeberria, J., and Militino, A. F. 2012. Comparing CAR and P-spline models in spatial disease mapping. *Environ. Ecol. Stat.*, 19(4), 573–599.

Hastie, T. J. and Tibshirani, R. J. 1990. *Generalized Additive Models.* Monographs on Statistics and Applied Probability, vol. 43. Chapman & Hall, London.

Hazelton, M. L. 2015. Kernel smoothing methods. In *Handbook of Spatial Epidemiology*, ed. Lawson, A. B., Banerjee, S., Haining, R., and Ugarte, M. D., pp. 195–208. Chapman & Hall/CRC Press, Boca Raton, FL.

Kimeldorf, G. S. and Wahba, G. 1970. A correspondence between Bayesian estimation on stochastic processes and smoothing by splines. *Ann. Math. Stat.*, 41, 495–502.

Knorr-Held, L. 2000. Bayesian modelling of inseparable space-time variation in disease risk. *Stat. Med.*, 19(17–18), 2555–2567.

Knorr-Held, L. and Rainer, E. 2001. Projections of lung cancer mortality in West Germany: a case study in Bayesian prediction. *Biostatistics*, 2(1), 109–129.

Lang, S. and Brezger, A. 2004. Bayesian P-splines. *J. Comput. Graph. Stat.*, 13(1), 183–212.

Lee, D.-J. and Durbán, M. 2009. Smooth-CAR mixed models for spatial count data. *Comput. Stat. Data Anal.*, 53(8), 2968–2979.

Lee, D.-J. and Durbán, M. 2011. *P*-spline ANOVA-type interaction models for spatio-temporal smoothing. *Stat. Model.*, 11(1), 49–69.

Leroux, B. G., Lei, X., and Breslow, N. 2000. Estimation of disease rates in small areas: a new mixed model for spatial dependence. In *Statistical Models in Epidemiology, the Environment, and Clinical Trials (Minneapolis, MN, 1997)*, IMA Volumes in Mathematics and Its Applications, vol. 116, pp. 179–191. Springer, New York.

MacNab, Y. C. 2007. Spline smoothing in Bayesian disease mapping. *Environmetrics*, 18(7), 727–744.

MacNab, Y. C. and Dean, C. B. 2001. Autoregressive spatial smoothing and temporal spline smoothing for mapping rates. *Biometrics*, 57(3), 949–956.

MacNab, Y. C. and Gustafson, P. 2007. Regression B-spline smoothing in Bayesian disease mapping: with an application to patient safety surveillance. *Stat. Med.*, 26(24), 4455–4474.

O'Sullivan, F. 1986. A statistical perspective on ill-posed inverse problems [with discussion]. *Stat. Sci.*, 1(4), 502–527. With comments and a rejoinder by the author.

O'Sullivan, F. 1988. Fast computation of fully automated log-density and log-hazard estimators. *SIAM J. Sci. Stat. Comput.*, 9(2), 363–379.

Perperoglou, A. and Eilers, P. H. C. 2010. Penalized regression with individual deviance effects. *Comput. Stat.*, 25(2), 341–361.

Rue, H. and Held, L. 2005. *Gaussian Markov Random Fields.* Monographs on Statistics and Applied Probability, vol. 104. Chapman & Hall/CRC, Boca Raton, FL. Theory and applications.

Ruppert, D. 2002. Selecting the number of knots for penalized splines. *J. Comput. Graph. Stat.*, 11(4), 735–757.

Schmid, V. and Held, L. 2004. Bayesian extrapolation of space-time trends in cancer registry data. *Biometrics*, 60(4), 1034–1042.

Silva, G. L., Dean, C. B., Niyonsenga, T., and Vanasse, A. 2008. Hierarchical Bayesian spatiotemporal analysis of revascularization odds using smoothing splines. *Stat. Med.*, 27(13), 2381–2401.

Torabi, M. 2013. Spatio-temporal modeling for disease mapping using CAR and B-spline smoothing. *Environmetrics*, 24(3), 180–188.

Torabi, M. and Rosychuk, R. J. 2011. Spatio-temporal modelling using B-spline for disease mapping: analysis of childhood cancer trends. *J. Appl. Stat.*, 38(9), 1769–1781.

Ugarte, M. D., Goicoa, T., Etxeberria, J., and Militino, A. F. 2012a. Projections of cancer mortality risks using spatio-temporal P-spline models. *Stat. Methods Med. Res.*, 21(5), 545–560.

Ugarte, M. D., Goicoa, T., Etxeberria, J., and Militino, A. F. 2012b. A P-spline ANOVA type model in space-time disease mapping. *Stochastic Environ. Res. Risk Assess.*, 26(6), 835–845.

Ugarte, M. D., Goicoa, T., Etxeberria, J., Militino, A. F., and Pollán, M. 2010a. Age-specific spatio-temporal patterns of female breast cancer mortality in Spain (1975–2005). *Ann. Epidemiol.*, 20(12), 906–916.

Ugarte, M. D., Goicoa, T., and Militino, A. F. 2009a. Empirical Bayes and fully Bayes procedures to detect high-risk areas in disease mapping. *Comput. Stat. Data Anal.*, 53(8), 2938–2949.

Ugarte, M. D., Goicoa, T., and Militino, A. F. 2010b. Spatio-temporal modeling of mortality risks using penalized splines. *Environmetrics*, 21(3-4), 270–289.

Ugarte, M. D., Goicoa, T., Militino, A. F., and Durbán, M. 2009b. Spline smoothing in small area trend estimation and forecasting. *Comput. Stat. Data Anal.*, 53(10), 3616–3629.

Ugarte, M. D., Militino, A. F., and Goicoa, T. 2008. Prediction error estimators in empirical Bayes disease mapping. *Environmetrics*, 19(3), 287–300.

van der Linde, A., Witzko, K. H., and Jöckel, K. H. 1995. Spatial-temporal analysis of mortality using splines. *Biometrics*, 51.

Zhang, S., Sun, D., He, C. Z., and Schootman, M. 2006. A Bayesian semi-parametric model for colorectal cancer incidences. *Stat. Med.*, 25(2), 285–309.

13

Quantile Regression for Epidemiological Applications

Brian J. Reich
Department of Statistics
North Carolina State University
Raleigh, North Carolina

CONTENTS

The predominant analytic approach in epidemiological studies is linear mean regression. In linear regression, it is assumed that the covariates affect only the mean of the response distribution and no other aspect, such as variance or skewness. While this approach is computationally efficient and easy to interpret, it is insufficient in many cases. For example, linear regression can miss subtle relationships such as covariate effects on the variance or the likelihood of extreme events. Also, linear regression is known to be sensitive to outlying observations.

In this chapter, we discuss an alternative approach using linear quantile regression (QR). In QR, the quantiles, not the mean, of the response distribution are assumed to be linear in the covariates. Each quantile is allowed to have different linear regression covariates, and therefore QR provides a more comprehensive analysis of the relationship between the covariates and the response. For example, QR may reveal that a covariate has a positive effect on the center of the distribution (the median is the 0.5 quantile), a negative effect on the spread (the interquartile range is 0.75–0.25 quantile), and no effect on the extremes (0.99 quantile). In addition to this modeling flexibility, QR is more robust to outliers than mean regression. For example, QR includes median regression as a special case, and just as the sample median is more robust than the sample mean, median (and other quantile levels) regression is more robust than mean regression.

An excellent example of QR for environmental epidemiological data is Lee and Neocleous (2010), who regress hospital admissions onto ambient air pollution and find a stronger effect for lower quantile levels. This leads to the subtle conclusion that ambient air pollution has a stronger effect in healthier regions. As an illustration, in this chapter we consider the birthweight data plotted in Figure 13.1a. These data consist of $n = 7093$ preterm (26–36 weeks) births in 2012 from the National Center for Health Statistics (2012).

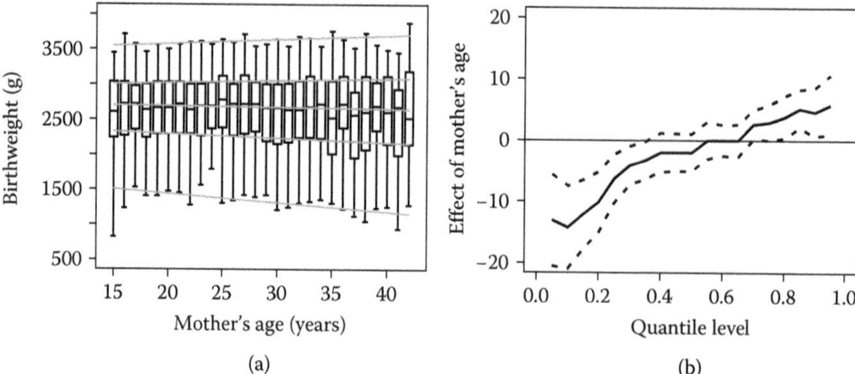

FIGURE 13.1
Plot of birthweight by maternal age (MA). (a) Birthweights for each MA are shown as a boxplot with horizontal lines at the $\tau = 0.05, 0.25, 0.50, 0.75$, and 0.95 quantiles. The dashed lines are the QR fitted lines (from the `quantreg` package) $\hat{\beta}_1(\tau) + \hat{\beta}_2(\tau)$ MA for the same quantile levels. (b) Estimated slopes $\hat{\beta}_2(\tau)$ and 95% interval (dashed lines) by quantile level.

The response variable is the baby's birthweight (BW), and the predictor is the mother's age (MA). The typical linear regression analysis would focus only on estimating the mean BW for a baby of a certain MA. However, this does not capture the entire relationship between these variables. For example, in these data it does not appear that MA has any effect on the center of the BW distribution. The only apparent trend is that the likelihood of very low BW decreases with MA.

Figure 13.1 summarizes a QR analysis of these data. A different slope is plotted for each quantile level in Figure 13.1a, and the slopes are shown with 95% intervals for several quantile levels in Figure 13.1b. We see that the effect of MA on BW varies considerably by quantile level. The slope is negative for low quantiles and near zero at and above the median. These comparisons of slopes communicate the complex relationship between MA and BW. Trends in the quantiles (or equivalently percentiles) such as these are easily communicated in many fields, such as those that use growth charts routinely to summarize an individual's place in the population or to compare two populations.

Quantile regression is certainly not new. It was first discussed by Koenker and Bassett (1978) and is discussed in depth in Koenker (2005). However, due to the ever-increasing size and complexity of data and ever-improving computational tools, QR is gaining in popularity in many fields. In this chapter, we present the essential concepts and methods of QR. We then discuss recent extensions that are especially relevant for epidemiological applications, including methods to deal with clustered and spatially referenced data. Throughout the chapter, we discuss implementation using the free open-source software `R` (R Core Team, 2013). Code is available at http://www4.stat.ncsu.edu/~reich/code/.

13.1 Definitions and Notation

Denote the continuous response variable as Y and the covariate vector as $\mathbf{X} = (X_1, \ldots, X_p)^T$, with $X_1 = 1$ for the intercept. For quantile level $\tau \in (0, 1)$, denote the conditional quantile of $Y|\mathbf{X}$ as $q(\tau|\mathbf{X})$, so that

$$\text{Prob}[Y < q(\tau|\mathbf{X})] = \tau. \tag{13.1}$$

Linear quantile regression assumes that $q(\tau|\mathbf{X})$ is linear in the covariates,

$$q(\tau|\mathbf{X}) = \sum_{j=1}^{p} X_j\beta_j(\tau). \tag{13.2}$$

The vector $\boldsymbol{\beta}(\tau) = [\beta_1(\tau), \ldots, \beta_p(\tau)]^T$ is the regression coefficient for quantile level τ. These coefficients are interpreted in the usual way, with $\beta_j(\tau)$ representing the increase in the τth quantile corresponding to a unit increase in X_j while holding all other covariates fixed.

A key modeling step is to allow the regression coefficients to vary by quantile level. This provides the flexibility to permit covariates to have different effects on different aspects of the conditional distribution. For example, $\beta_j(0.5)$ is the covariate's effect on the center of the distribution (with $\tau = 0.5$, $q(0.5|\mathbf{X})$ is the conditional median of Y given $\mathbf{X}$). The covariates are also allowed to affect the spread of the conditional distribution. For example, the interquartile range is

$$q(0.75|\mathbf{X}) - q(0.25|\mathbf{X}) = \sum_{j=1}^{p} X_j[\beta_j(0.75) - \beta_j(0.25)]. \tag{13.3}$$

Thus, $\beta_j(0.75) - \beta_j(0.25)$ controls the effect of covariate j on the distribution's spread. The covariates can also affect extremes via $\beta_j(\tau)$ for τ near zero or 1.

While quantile regression can be applied more generally, it contains location-scale models as special cases. For example, if $\beta_1(\tau) = \beta_1 + \sigma\Phi^{-1}(\tau)$, where Φ is the standard normal distribution function, and $\beta_j(\tau) \equiv \beta_j$ for all $j > 1$ and all $\tau \in (0,1)$, then the quantile regression model reduces to the usual linear regression model:

$$Y|\mathbf{X} \sim \mathrm{N}\left(\sum_{j=1}^{p} X_j\beta_j, \sigma^2\right). \tag{13.4}$$

In general, if a covariate's effect is constant across quantile level, $\beta_j(\tau) \equiv \beta_j$, then the covariate's effect is to simply shift the distribution and not alter the distribution's shape.

If a covariate effect is not constant over quantile level, then the relationship between the covariate and the response is more complex than a simple location shift. Perhaps the simplest example is a location-scale relationship. The heteroskedastic normal model,

$$Y|\mathbf{X} \sim \mathrm{N}\left[\sum_{j=1}^{p} X_j\beta_j, \left(\sum_{j=1}^{p} X_j\sigma_j\right)^2\right], \tag{13.5}$$

allows the covariates to affect both the conditional mean and standard deviation. For this model, the quantile functions are $\beta_j(\tau) = \beta_j + \sigma_j\Phi^{-1}(\tau)$. Clearly, in this example restrictions on the σ_j are required to obtain a valid model; this will be revisited in Section 13.3.

13.2 Model-Free Estimation

The most common estimation approach was proposed by Koenker and Bassett (1978). The estimate for quantile level τ is

$$\hat{\boldsymbol{\beta}}(\tau) = \underset{\boldsymbol{\beta}\in\mathcal{R}^p}{\operatorname{argmin}} \sum_{i=1}^{n} \rho_\tau(Y_i - \mathbf{X}_i^T\boldsymbol{\beta}), \tag{13.6}$$

where $\rho_\tau(e)$ is the check function

$$\rho_\tau(e) = \begin{cases} \tau|e| & e > 0 \\ (1-\tau)|e| & e \leq 0. \end{cases} \tag{13.7}$$

The intuition behind using this check-loss function is that it balances the penalty above and below the quantile level of interest. For example, with $\tau = 0.9$, the loss is nine times larger for positive errors ($Y_i - \mathbf{X}_i^T\boldsymbol{\beta} > 0$) than negative errors. By downweighting errors below zero, the loss places more emphasis on the upper tail of the conditional distribution, which is appropriate for estimating the 0.9 quantile.

The estimator has been shown to have excellent frequenstist properties, including consistency and asymptotic normality under mild regulatory conditions. The asymptotic distribution is

$$\sqrt{n}\left[\hat{\boldsymbol{\beta}}(\tau) - \boldsymbol{\beta}(\tau)\right] \to \mathrm{N}\left[0, \tau(1-\tau)D^{-1}\Omega D^{-1}\right], \tag{13.8}$$

where $D = E[f(\mathbf{X}^T\boldsymbol{\beta})\mathbf{X}^T\mathbf{X}]$, $\Omega = E(\mathbf{X}^T\mathbf{X})$, and $f(e)$ is the residual density. Standard errors can be computed in several ways. The simplest approach is a nonparametric bootstrap. To avoid tedious resampling, kernel density estimates of f have been used to approximate the asymptotic covariance. This estimator and inferential tools are appealing because they make no assumptions about the parametric form of the residuals (i.e., they do not require normality) and estimates are robust to outliers (Koenker, 2005).

Computing the solution (13.6) is a linear programming problem and is therefore computationally efficient even for large datasets. Because the check-loss function in nondifferentiable at zero, the solution is not guaranteed to be unique. This can cause numerical issues for small datasets, especially with categorical covariates. However, this is typically not an issue for moderately large datasets. The solution is provided by the `rq` function in the `R` package `quantreg`, as shown in Figure 13.1.

13.3 Model-Based Estimation

13.3.1 Single-quantile methods

An advantage for the check-loss estimator (13.6) is that no model on the residuals is assumed. However, in some settings a carefully chosen parametric model may be preferred. For example, a Bayesian analysis requires a proper likelihood be specified. Also, accounting for complex features such as random effects, missing data, and censoring is often simplified if a parametric model is assumed.

The asymmetric Laplace (ASL) likelihood (Koenker and Machado, 1999; Yu and Moyeed, 2001) is a natural choice. The ASL density function for $Y|\mathbf{X}$ is

$$f_\tau(y|\mathbf{X}) = \frac{\tau(1-\tau)}{\lambda}\exp[-\rho_\tau(y - \mathbf{X}^T\boldsymbol{\beta})/\lambda], \tag{13.9}$$

where $\lambda > 0$ the scale parameter. The ASL's mode and τth quantile are $\mathbf{X}^T\boldsymbol{\beta}$, and clearly the maximum likelihood estimator equals the check-loss estimate (13.6). Therefore, the MLE from this model will share many of the attractive properties of the check-loss estimate, including consistency and robustness. However, standard errors computed assuming the ASL likelihood (e.g., those based on the observed information) will not be the same as the standard errors discussed in Section 13.2. Typically, the standard errors are reduced when assuming a parametric likelihood, which improves inference if the model fits the data well.

Often the method is applied separately for a grid of quantile levels to produce results analogous to Figure 13.1b. These results must be interpreted with caution; the ASL distribution is right-skewed for small τ and left-skewed for large τ, and therefore not likely to be appropriate for all τ simultaneously. This leads to questionable inferential properties (undercoverage of interval estimates) for at least some quantile levels.

More general likelihood functions have been proposed to overcome the limitations of the ASL. Kottas and Gelfand (2001), Hanson and Johnson (2002), Kottas and Krnjajic (2009), and Reich et al. (2010) use nonparametric Bayesian methods to estimate the residual distribution. Yang and He (2010) use an empirical likelihood to avoid specifying a parametric residual distribution.

13.3.2 Simultaneous quantile regression

As discussed in Section 13.3.1, assigning the same parametric family to the residuals can lead to questionable statistical inference when many quantile levels are considered. In addition, all methods that analyze several quantile levels separately, including the model-free methods discussed in Section 13.2, are susceptible to crossing quantile levels. That is, it may be that for some $\mathbf{X}$, the estimated quantiles are not increasing in τ. Of course, for prediction it is awkward to estimate that, say, the median is larger than the 0.75 quantile. Many post hoc adjustments or methods to ensure noncrossing for a finite number of quantile levels have been proposed (He, 1997; Neocleous and Portnoy, 2008; Wu and Liu, 2009; Bondell et al., 2010).

An alternative to analyzing the quantile levels separately is a simultaneous analysis of all quantile levels (Dunson and Taylor, 2005; Todkar and Kadane, 2011; Reich et al., 2011; Reich, 2012; Reich and Smith, 2013). In this approach, we specify a valid statistical model for the data that preserves the linear quantile relationship for all quantile levels. This ensures that prediction made from this model will not have crossing quantiles because they are generated from a valid statistical model. A simultaneous model sacrifices some of the robustness properties of model-free QR, but has the added benefit of pooling information across quantile levels. Assuming $\beta_j(\tau)$ varies smoothly in τ, a joint model can provide substantially smaller uncertainty than separate fits, especially for τ near 0 or 1 where data are sparse.

To ensure a valid model, the quantile function must be an increasing function of τ for all $\mathbf{X}$. Equivalently, the derivative of the quantile function with respect to τ must be positive. Reich et al. (2011) and Reich and Smith (2013) satisfy these restrictions by expressing the quantile functions using a finite basis expansion

$$\beta_j(\tau) = \theta_{0l} + \sum_{l=1}^{L} B_l(\tau)\theta_{jl}, \tag{13.10}$$

where $B_l(\tau)$ are known functions of τ and $\boldsymbol{\theta}_l = (\theta_{1l}, \ldots, \theta_{pl})^T$ are unknown parameters that determine the shape of the quantile function. Reich et al. (2011) use Bernstein basis functions for $B_l(\tau)$. Reich and Smith (2013) use the piecewise distribution function of a known parametric distribution $q_0(\tau)$.

The model of Reich and Smith (2013) is motivated by the heteroskedastic model in (13.5). For a location-scale model with location $\mathbf{X}^T\boldsymbol{\theta}_0$, scale $\mathbf{X}^T\boldsymbol{\theta}_1$, and base distribution $q_0(\tau)$ (e.g., $\Phi^{-1}(\tau)$), the quantile functions are $q(\tau|\mathbf{X}) = \mathbf{X}^T\boldsymbol{\theta}_0 + [\mathbf{X}^T\boldsymbol{\theta}_1]q_0(\tau)$ and $q'(\tau|\mathbf{X}) = [\mathbf{X}^T\boldsymbol{\theta}_1]q_0'(\tau)$. To provide added flexibility and allow covariates to have a different effect on different aspects of the distribution, Reich and Smith (2013) assume different scale parameters for different values of τ. Let $0 = \kappa_1 < ... < \kappa_{L+1} = 1$ be a fixed grid of quantile levels, and then $q'(\tau) = [\mathbf{X}^T\boldsymbol{\theta}_l]q_0'(\tau)$ for $\tau \in (\kappa_l, \kappa_{l+1})$. Basis functions that lead to this derivative are

$$B_1(\tau) = \begin{cases} q_0(\tau), & \tau \le \kappa_1 \\ q_0(\kappa_1), & \tau > \kappa_1 \end{cases} \tag{13.11}$$

and

$$B_l(\tau) = \begin{cases} 0, & \tau \le \kappa_{l-1} \\ q_0(\tau) - q_0(\kappa_{l-1}), & \kappa_{l-1} < \tau \le \kappa_l \\ q_0(\kappa_l) - q_0(\kappa_{l-1}), & \tau > \kappa_l \end{cases}$$

for $l > 1$. Since $q_0'(\tau) > 0$, the derivative $q'(\tau)$ is positive if and only if $\mathbf{X}^T\boldsymbol{\theta}_l > 0$ for all $\mathbf{X}$ and $l > 0$. This provides a manageable expression to ensure a valid quantile process. For example, Reich and Smith (2013) assume the covariates are transformed so that $X_1 = 1$, $X_j \in [-1, 1]$ for all $j > 1$, and then use a prior for the $\boldsymbol{\theta}_l$ that has $\theta_{1l} - \sum_{j=2}^{p} |\theta_{jl}| > 0$ with probability 1.

This approach has several advantages. It has a closed-form expression for the likelihood,

$$f(y|\mathbf{X}) = \sum_{l=1}^{L} \frac{I\left[q(\kappa_{l-1}|\mathbf{X}_i) < y < q(\kappa_l|\mathbf{X}_i)\right]}{\mathbf{X}_i^T\boldsymbol{\theta}_l} f_0\left[\frac{y - \mathbf{X}_i^T\boldsymbol{\theta}_0 - I(l > 1)\mathbf{X}_i^T\boldsymbol{\theta}_l q_0(\kappa_l)}{\mathbf{X}_i^T\boldsymbol{\theta}_l}\right], \tag{13.12}$$

where f_0 is the density corresponding to q_0. Therefore, implementing either a likelihood or Bayesian approach is computationally convenient. From a modeling perspective, this framework balances parametric and semiparametric approaches. If $\theta_{j2} = \ldots = \theta_{jL}$, then the model reduces to the parametric heteroskedastic model in (13.5); further, if $\theta_{j2} = \ldots = \theta_{jL} = 0$, then $\beta_j(\tau) = \theta_{j1}$ for all τ and the effects of the jth covariate are simply additive. Therefore, if the parametric model fits well, this method will be efficient if the prior or the penalty encourages shrinkage to common θ_{jl}. On the other hand, Reich and Smith (2013) show that by increasing the number of basis functions, the model can approximate any valid linear quantile function, and therefore provides a very rich modeling framework.

The method is implemented in the `R` package `Bsquare` (Smith and Reich, 2014). Figure 13.2a plots the posteriors of the quantile coefficients for the birthweight data, taking q_0 to be the standard normal quantile function and $L = 4$ basis functions. The estimated

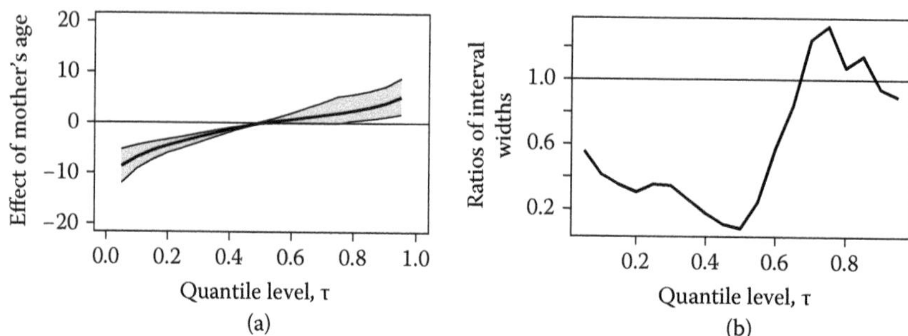

FIGURE 13.2
Simultaneous QR of birthweight and maternal age. (a) Posterior mean and 95% interval for the estimated slopes $\beta_2(\tau)$ from the simultaneous analysis using `BSquare`. (b) Ratio of widths of 95% intervals from the simultaneous (`BSquare`) and separate (`quantreg`) analyses.

slopes are similar to the separate fits in Figure 13.1; both analyses show a negative relationship for lower quantiles, no relationship in the center of the distribution, and a slightly positive effect in the upper tail. The two fits differ in that the estimated slopes vary more smoothly over τ for the simultaneous fit, and the 95% intervals are generally narrower for the simultaneous fit (Figure 13.2b), especially for low quantile levels.

13.4 Clustered Data

Epidemiologic data are often collected in clusters. For example, there may be repeated measurements within a subject, clinic, or county. Due to shared and unmeasured variables, observations within a cluster may be dependent. It is therefore important to account for clustering in the analysis to obtain valid standard errors and interval estimates for regression coefficients $\boldsymbol{\beta}(\tau)$. In some cases, the objective of the analysis is actually to identify clusters with unusual features, such as extremely large quantiles.

A common method to account for clustering is to include a random effect for the cluster. Denote Y_{ij} as the ith observations for cluster j, $\mathbf{X}_{ij}$ as the corresponding covariates, and u_j as the random effect for cluster j. Koenker (2004) proposes to incorporate the random effects into (13.6) via the penalized loss function

$$[\hat{\boldsymbol{\beta}}(\tau), \hat{\mathbf{u}}(\tau)] = \underset{\boldsymbol{\beta}\in\mathcal{R}^p, \mathbf{u}\in\mathcal{R}^m}{\text{argmin}} \sum_{j=1}^{m}\sum_{i=1}^{n_j} \rho_\tau(Y_{ij} - \mathbf{X}_{ij}^T\boldsymbol{\beta} - u_j) + \phi \sum_{j=1}^{m} |u_j|. \tag{13.13}$$

The tuning parameter $\phi > 0$ in this penalized regression controls shrinkage of the random effects and prevents overfitting that could be caused by adding many parameters $u_1, \ldots, u_m$ to the model. If $\phi = 0$, then the cluster random effects are essentially fixed effects capturing the differences by cluster, and if $\phi = \infty$, then the random effects are shrunk to zero and the estimate is equivalent to (13.6).

Geraci and Bottai (2007) propose a parametric version of this random effects approach. They assume ASL likelihood $Y_{ij} = \mathbf{X}_{ij}^T\boldsymbol{\beta}(\tau) + u_j + \epsilon_{ij}$, where the errors ϵ_{ij} are independent with ALS distribution with the τth quantile equal to zero and scale parameter λ. The random effects distribution is also taken to be parametric, for example, $u_i \sim \mathrm{N}(0, \sigma^2)$. The advantage of a parametric model for clustered QR is to avoid selecting the tuning parameter ϕ and that it permits estimation of and inference on the random effects u_j. Estimation is carried out using a Monte Carlo EM algorithm, which is available in the `lqmm` package in `R` (Geraci, 2014).

A drawback of the random effects approach to accounting for within-cluster dependence is that adding random effects changes the interpretation of the regression coefficients, $\boldsymbol{\beta}(\tau)$. In a random effects model, $\boldsymbol{\beta}(\tau)$ are interpreted in terms of the quantile of one subject conditioned on the random effect. However, this is often not the desired interpretation. Instead, we might want to interpret $\boldsymbol{\beta}(\tau)$ in terms of the population's quantile, marginally over the subject random effects, and these two interpretations can be substantially different. Reich et al. (2010) propose a random effects approach that gives this population interpretation of the regression coefficients while accounting for clustering. Another limitation of the approaches mentioned thus far is that the observations are considered to be independent conditioned on the random effects, leading to an exchangeable dependence structure. This is not sufficient when the observations within a cluster are ordered in time, such as daily blood levels collected over a study period. Smith et al. (2015) propose a QR model that explicitly accounts for autocorrelation in longitudinal data.

13.5 Spatial Data

Large epidemiological datasets are often spatially referenced. For example, in the birthweight example illustrated in Figure 13.1, it may be of interest to conduct a QR for each county to account for unmeasured confounders that vary by county, compare trends across counties, or identify anomalous counties. Define $\beta_{jk}(\tau)$ as the effect of covariate j on quantile level τ in spatial region $k \in \{1, \ldots, m\}$. Here, we assume areal data with m spatial regions of interest (e.g., $m = 100$ counties of North Carolina), although most of the methods below extend naturally to geostatistical data with an uncountable number of potential spatial locations (e.g., the latitude and longitude coordinates of the mother's residence).

The simplest spatial analysis is to conduct a separate quantile regression in each region. Denote $\hat{\beta}_{jk}(\tau)$ as the check-loss estimate (13.6) using data only from region k, and denote its variance as $s_{jk}^2(\tau)$ (the jth diagonal element of the covariance in (13.8)). This may be sufficient for large datasets, but when there are many predictors or data are limited, a spatial model that borrows strength across nearby regions is more powerful. Reich et al. (2011) propose a two-stage procedure that first computes estimates separately by spatial region in the first stage, and then smooths these estimates using a spatial model in the second stage. This two-stage approach resembles a meta-analysis, where the first-stage estimates are combined in the second stage using a spatial random effects model. To define the second-stage spatial model for a single covariate and single quantile level, we drop notational dependence on j and τ and let $\beta_{jk}(\tau) = \beta_k$, $\hat{\beta}_{jk}(\tau) = \hat{\beta}_k$, and $s_{jk}(\tau) = s_k$. The second-stage spatial model then treats the first-stage estimates $\hat{\beta}_k$ as data and models the estimates as

$$\hat{\beta}_k \overset{indep}{\sim} \text{N}(\beta_k, s_k^2), \tag{13.14}$$

where β_k has a spatial model.

For areal data, the most common spatial model is the intrinsic conditionally autoregressive model (Gelfand et al., 2010). In this model, the conditional distribution of β_k given β_l for all regions $l \neq k$ is normally distributed with mean $\rho\bar{\beta}_k$ and variance τ^2/m_k, where $\bar{\beta}_k$ is the mean of β over the m_k regions that neighbor region k and $\rho \in (0, 1)$ and τ^2 are covariance parameters. The joint prior for $\boldsymbol{\beta} = (\beta_1, \ldots, \beta_n)^T$ is the multivariate normal with mean vector zero and covariance matrix $\tau^2 Q^{-1}$, where the diagonal elements of Q are m_k and the off-diagonals are $-\rho$ if regions k and l are neighbors and zero otherwise. Analysis of these models is straightforward using `OpenBUGS` (Thomas et al., 2006). Reich et al. (2011) extend this two-stage approach to handle multiple covariates and several quantile levels simultaneously, and to estimate the quantile functions in a single stage using a Bayesian hierarchical model. They show that spatial smoothing can provide a substantial improvement over separate regressions by spatial location and quantile level.

More sophisticated approaches have been proposed for a single quantile level. For example, Yang and He (2015) use an empirical likelihood approach to spatial QR. Taking a parametric approach, Lum and Gelfand (2012) propose an ASL process model. Extending the ASL representation of Kuzobowski and Podgorski (2000) to the spatial setting, they model

$$Y_k = \mathbf{X}_k^T \boldsymbol{\beta} + \sqrt{\frac{2\varepsilon_k}{\lambda\tau(1-\tau)}} Z_k + \frac{1-2\tau}{\tau(1-\tau)}\varepsilon_k, \tag{13.15}$$

where Z_k is a Gaussian random effect with spatial covariance (e.g., CAR) and ε_k are independent (though other options were discussed) exponentially distributed errors. Under this hierarchical specification, the marginal distribution of Y_k is ASL with the τth quantile equal to $\mathbf{X}_k^T\boldsymbol{\beta}$. On the other hand, the conditional distribution given the ε is Gaussian. Assuming $\mathbf{Z} = (Z_1, ..., Z_m)^T$ follows the CAR model, then the ASL process likelihood can be written

$$\mathbf{Y}|\mathbf{X}, \boldsymbol{\varepsilon} \sim \mathrm{N}\left[\mathbf{X}^T\boldsymbol{\beta} + c\boldsymbol{\varepsilon}, \tau^2\mathbf{D}(\boldsymbol{\varepsilon})Q^{-1}\mathbf{D}(\boldsymbol{\varepsilon})\right], \tag{13.16}$$

where $c = (1-2\tau)/[\tau(1-\tau)]$, $\boldsymbol{\varepsilon} = (\varepsilon_1, \ldots, \varepsilon_n)^T$, and $\mathbf{D}(\boldsymbol{\varepsilon})$ is diagonal with diagonal elements $\sqrt{2\varepsilon_k/[\lambda\tau(1-\tau)]}$. This facilitates Bayesian computing in standard packages such as `OpenBUGS` (Thomas et al., 2006).

Simultaneous QR models have also been proposed for spatial data. Reich (2012) extends the basis expansion in (13.10) by allowing the basis coefficients θ_{jl} to vary spatially following Gaussian process priors. This allows the quantile function and covariate effects to vary smoothly across space and quantile level, while ensuring a valid quantile function for spatial prediction. In addition to allowing for spatial variation in the marginal distribution via the quantile function, he accounts for residual spatial dependence using a Gaussian copula (Nelsen, 1999).

13.6 Discussion

This chapter introduces the emerging field of quantile regression and discusses applications to epidemiological data. QR is a valuable tool for revealing complex relationships between predictors and a response and performing statistical inference on associations other than the mean. The literature on QR has been fleshed out in recent years to cover most data types seen in practice. In this chapter, we have presented methods for clustered and spatial data, but methods are also available to accommodate missing data (e.g., Wei et al., 2012), censoring (e.g., Portnoy, 2003), discrete data (e.g., Lee and Neocleous, 2010; Machado and Silva, 2005), and many others. In addition, open-source code is now available for many of these methods. Therefore, QR is poised to play a prominent role in epidemiological studies.

References

Bondell, H. D., Reich, B. J., and Wang, H. (2010). Non-crossing quantile regression curve estimation. *Biometrika*, 97, 825–838.

Dunson, D. B. and Taylor, J. A. (2005). Approximate Bayesian inference for quantiles. *Journal of Nonparametric Statistics*, 17, 385–400.

Gelfand, A. E., Diggle, P. J., Fuentes, M., and Guttorp, P. (2010). *Handbook of Spatial Statistics*. New York: Chapman & Hall/CRC.

Geraci, M. (2014). Linear quantile mixed models: The lqmm package for Laplace quantile regression. *Journal of Statistical Software*, 54, 1–29.

Geraci, M. and Bottai, M. (2007). Quantile regression for longitudinal data using the asymmetric Laplace distribution. *Biostatistics*, 8, 140–154.

Hanson, T. and Johnson, W. (2002). Modeling regression error with a mixture of Polya trees. *Journal of the American Statistical Association*, 97, 1020–1033.

He, X. (1997). Quantile curves without crossing. *American Statistician*, 51, 186–192.

Koenker, R. (2004). Quantile regression for longitudinal data. *Journal of Multivariate Analysis*, 91, 74–89.

Koenker, R. W. (2005). *Quantile Regression.* Boston: Cambridge University Press.

Koenker, R. W. and Bassett, G. W. (1978). Regression quantiles. *Econometrica*, 46, 33–50.

Koenker, R. and Machado, J. (1999). Goodness of fit and related inference processes for quantile regression. *Journal of the American Statistical Association*, 94, 1296–1309.

Kottas, A. and Gelfand, A. (2001). Bayesian semiparametric median regression modeling. *Journal of the American Statistical Association*, 96, 1458–1468.

Kottas, A. and Krnjajic, M. (2009). Bayesian nonparametric modeling in quantile regression. *Scandinavian Journal of Statistics*, 36, 297–319.

Kuzobowski, T. J. and Podgorski, K. (2000). Multivariate and asymmetric generalization of Laplace distribution. *Computational Statistics*, 15, 531–540.

Lee, D. and Neocleous, T. (2010). Bayesian quantile regression for count data with application to environmental epidemiology. *Journal of the Royal Statistical Society: Series C*, 59, 905–920.

Lum, K. and Gelfand, A. E. (2012). Spatial quantile multiple regression using the asymmetric Laplace process. *Bayesian Analysis*, 7, 235–258.

Machado, J. A. F. and Silva, J. S. (2005). Quantiles for counts. *Journal of the American Statistical Association*, 100, 1226–1237.

National Center for Health Statistics. (2012). Data file documentations, natality. Hyattsville, MD: National Center for Health Statistics. http://www.nber.org/data/vital-statistics-natality-data.html.

Nelsen, R. (1999). *An Introduction to Copulas.* New York: Springer-Verlag.

Neocleous, Y. and Portnoy, S. (2008). On monotonicity of regression quantile functions. *Statistics and Probability Letters*, 78, 1226–1229.

Portnoy, S. (2003). Censored quantile regression. *Journal of American Statistical Association*, 98, 1001–1012.

R Core Team. (2013). *R: A Language and Environment for Statistical Computing.* Vienna, Austria: R Foundation for Statistical Computing. http://www.R-project.org/.

Reich, B. J. (2012). Spatiotemporal quantile regression for detecting distributional changes in environmental processes. *Journal of the Royal Statistical Society: Series C*, 64, 535–553.

Reich, B. J., Bondell, H. D., and Wang, H. (2010). Flexible Bayesian quantile regression for independent and clustered data. *Biostatistics*, 11, 337–352.

Reich, B. J., Fuentes, M., and Dunson, D. B. (2011). Bayesian spatial quantile regression. *Journal of the American Statistical Association*, 106, 6–20.

Reich, B. J. and Smith, L. B. (2013). Bayesian quantile regression for censored data. *Biometrics*, 69, 651–660.

Smith, L. B., Fuentes, M., Gordan-Larsen, P., and Reich, B. J. (2015). Quantile regression for mixed models with an application to examine blood pressure trends in China. *Annals of Applied Statistics*, 9, 1226–1246.

Smith, L. B. and Reich, B. J. (2014). BSquare: Bayesian simultaneous quantile regression. R package version 1.1. http://CRAN.R-project.org/package=BSquare.

Thomas, A., Hara, B. O., Ligges, U., and Sturtz, S. (2006). Making BUGS open. *R News*, 6, 12–17.

Todkar, S. T. and Kadane, J. B. (2011). Simultaneous linear quantile regression: A semi-parametric Bayesian approach. *Bayesian Analysis*, 6, 1–12.

Wei, Y., Ma, Y., and Carroll, R. J. (2012). Multiple imputation in quantile regression. *Biometrika*, 101, 1–16.

Wu, Y. and Liu, Y. (2009). Stepwise multiple quantile regression estimation using non-crossing constraints. *Statistics and Its Interface*, 2, 299–310.

Yang, Y. and He, X. (2010). Bayesian empirical likelihood for quantile regression. *Annals of Statistics*, 40, 1102–1130.

Yang, Y. and He, X. (2015). Quantile regression for spatially correlated data: An empirical likelihood approach. *Statistica Sinica*, 25, 261–274.

Yu, K. and Moyeed, R. A. (2001). Bayesian quantile regression. *Statistics and Probability Letters*, 54, 437–447.

14

Focused Clustering: Statistical Analysis of Spatial Patterns of Disease around Putative Sources of Increased Risk

Lance A. Waller
Department of Biostatistics and Bioinformatics
Rollins School of Public Health
Emory University
Atlanta, Georgia

David C. Wheeler
Department of Biostatistics
Virginia Commonwealth University
Richmond, Virginia

Jeffrey M. Switchenko
Department of Biostatistics and Bioinformatics
Rollins School of Public Health
Emory University
Atlanta, Georgia

CONTENTS

14.1 Introduction

Dr. John Snow's studies of cholera mortality in the vicinity of the Broad Street water pump of the Soho neighborhood of London in 1854 represent an early example of assessing how the *pattern* of disease incidence with respect to suspected sources of risk can reveal insight into the *processes* driving underlying morbidity or mortality. The rapid proliferation of georeferenced data and advances in geographic information systems offer great opportunities for the quantification and analysis of spatial patterns in epidemiologic data, especially with respect to assessing the potential impact of putative sources of excess risk. Examples include (but certainly are not limited to) assessments of leukemia risk around hazardous waste sites (Lagakos et al. 1986; Waller et al. 1992, 1994) or nuclear sites (Stone 1988; Bithell et al. 1994; Waller et al. 1995; National Research Council 2012), or respiratory outcomes near industrial sites (Lawson and Williams 1994; Diggle et al. 1999) or highways (English et al. 1999).

Assessing statistical significance of an observed pattern compared to patterns expected under a (null) hypothesis of no excess risk in areas near the suspected source (or sources) of risk is termed a test to detect *focused clusters* (Besag and Newell 1991), and numerous approaches appear in the statistical and epidemiologic literature (for recent comprehensive reviews, see Waller and Gotway 2004; Lawson 2006; Tango 2010). We use the terms *foci* and *putative sources* interchangeably throughout the sections below to denote the points, lines, or areas hypothesized as sources of exposures generating increased risk. In addition to hypothesis tests assessing the significance of any clusters, model-based assessments provide estimates of the functional nature of associations (if any) between local observed counts of disease and exposures to the foci or location-based exposure surrogates (when measured exposures are not available), while allowing for more comprehensive adjustment for known risk factors and potential confounders or effect modifiers in the population at risk.

In this chapter, we review the typical types of disease incidence and putative risk exposure data used in focused cluster studies and the statistical framework for focused cluster analysis. We provide illustrations of methods and references to additional statistical details and applications in the literature, building on the general introduction by Rogerson and Jacquez (2016, Chapter 8 of this volume).

14.2 Basic Elements: Data Types and Inferential Framework

Most focused cluster studies involve either point-referenced locations of cases and controls (noncases) or counts of cases and the number of at-risk individuals in small enumeration districts (e.g., counties or census tracts in the United States and postcodes in the United Kingdom). Building on the available data, analytic methods typically fall into two classes: methods comparing statistical descriptions of spatial patterns of case events to those of control events, with special attention to areas near the foci of putative risk, and methods comparing observed small-area counts to numbers expected under a model of no excess risk associated with the foci of putative excess risk.

14.2.1 Framework for case-control point data

For point data, statistical inference proceeds from an assumed stochastic spatial point process defining the local probability of observing a case event or a control event.

Typically, we assume locations follow a heterogeneous spatial Poisson process where events are independent of one another, events are distributed in space according to a smooth spatial probability *density* function, and the probability of two events occurring at exactly the same location is zero. Details regarding the mathematics of spatial Poisson processes appear in Møller and Waagepetersen (2004), Waller and Gotway (2004, chapter 5), Lawson (2006), and Diggle (2014).

The spatial density function, $f(\boldsymbol{s})$, is typically a continuous function for all locations $\boldsymbol{s}$ in the study area. The density is proportional to the spatial *intensity* function, $\lambda(\boldsymbol{s})$, a smooth function whose integral over a small area defines the number of cases expected in that small area. Many point-based tests to detect focused clusters compare estimates (e.g., via kernels or splines) of the underlying intensity functions of cases $\lambda_1(\boldsymbol{s})$ to those of controls $\lambda_0(\boldsymbol{s})$. Local variations in risk due to known risk factors unrelated to exposure to the foci (e.g., population size, age, race, and income) yield hills and valleys in these intensity functions (corresponding to areas with higher and lower expected numbers of cases or controls, respectively), and we seek to detect local deviations in $\lambda_1(\boldsymbol{s})$ away from $\lambda_0(\boldsymbol{s})$, representing location deviations in expected case events from the expected numbers of control events due to local variations in known risk factors. To detect focused clusters, we direct (via weighting or even truncation) methods to examine such deviations in locations with exposure to (or proximity to) the foci (our putative sources of increased risk) and pay less attention to deviations in low exposure locations.

14.2.2 Framework for count data

For count data, we again typically base inference on an underlying (but unobserved) heterogeneous spatial Poisson point process aggregated to provide counts of events for each small area. We then draw on the property that the number of events occurring in a small area A follows a Poisson distribution with expected value defined by the integral over A of $\lambda_1(\boldsymbol{s})$ for cases and $\lambda_0(\boldsymbol{s})$ for controls. We next compare the Poisson small-area counts to the numbers we would expect in the absence of a focus of increased risk, via either a hypothesis test or a Poisson regression model, to estimate associations with small-area exposure measures (or surrogates). The Poisson distribution provides an adequate approximation of an underlying binomial model of disease occurrence among residents of the area, provided the overall risk of disease is small. Analyses of nonrare outcomes often use a binomial model directly, but both Poisson and binomial models are subject to ecological biases when analyzing aggregate count data with a goal of inference regarding individual risk (Salway and Wakefield 2005; Wakefield and Smith 2016, Chapter 5 of this volume).

Both point data and areal count data require some assessment of exposure to the foci. Local exposure data may consist of (in decreasing order of accuracy): monitored values at fixed locations (e.g., residences) or for individuals (e.g., via personal monitors); exposure fields interpolated between monitored values (e.g., via kriging); modeled exposure fields based on release data and mathematical diffusion models, for example, the U.S. Environmental Protection Agency's Community Multiscale Air Quality (CMAQ) model (Byun and Schere 2006); reported releases from the focus (e.g., Toxic Release Inventory data from the U.S. Environmental Protection Agency, www.epa.gov/tri/); or simply the location of the focus itself (where proximity to the focus defines an exposure surrogate). As noted in National Research Council (1991, 2008), these different levels of exposure accuracy have implications on the strength of conclusions that can be drawn regarding the nature of the exposure–risk relationship.

With these basic building blocks of intensity functions for point case-control data, Poisson expectations for areal counts, and some measure or surrogate of exposure, we next define hypothesis test-based assessment of focused clusters in Section 14.3 and model-based approaches in Section 14.4.

14.3 Hypothesis Tests to Detect Focused Clusters around a Putative Source of Increased Risk

We begin by considering a hypothesis test framework for detecting clusters of disease around a prespecified location (or locations) hypothesized to be a source of excess risk. We note that tests to detect focused clusters differ from tests to detect general disease clusters in that we wish to "focus" the test to assess whether the excess risk is located in the areas near the putative sources of risk; we generally are less concerned with identifying and assessing clusters elsewhere.

To begin, we define the null hypothesis to represent a lack of excess risk near the foci, that is, a model of spatially homogeneous risk regardless of exposure to (or proximity to) the foci. Waller and Gotway (2004) note that this null hypothesis typically is operationalized in one of two ways: the first is a hypothesis of *constant risk*, wherein individuals are subject to the same risk of disease regardless of location. This may be generalized to account for known risk factors that are unassociated with the foci. The second is a hypothesis, often used for point data, of *random labeling*, wherein the spatial intensity function of cases matches that of controls, up to a multiplicative constant (reflecting the ratio of cases to controls). The term *random labeling* refers to fixing the observed case and control locations and simply randomly assigning "case" and "control" labels following the observed numbers of cases and controls. This random labeling generates patterns of cases among the full set of locations and provides a convenient recipe for Monte Carlo–based hypothesis testing; that is, the random labeling hypothesis defines how to generate replicate data sets under the null hypothesis, conditioning on the observed locations of events and the observed numbers of cases and controls.

The two null hypotheses are almost identical, one difference occurring when the value of the constant risk is estimated from data external to the analysis. In such a case, hypothesis tests could reject a hypothesis of constant risk simply because the entire study area experiences an overall increase or decrease in incidence; that is, the entire study area is a "cluster" of increased risk. In contrast, random labeling is always defined based on the observed case-control ratio and will not be sensitive to such differences. We note that tests of general clustering use the same two implementations of the null hypothesis. The defining feature of a test to detect *focused* clusters is the alternative hypothesis that defines where we look for deviations, that is, which deviations from the null are of primary interest. With these concepts in place, we next review several categories of focused tests to detect disease clusters.

14.3.1 Scan statistics

Scan statistics provide a popular approach to assessing significance of either general or focused clusters, introduced by Kulldorff (1997) and widely applied via the SaTScan™ software package (Kulldorff and Information Management Services 2009). In the focused cluster setting, we consider windows of varying size (and possibly shape) centered on areas suspected of having high exposure to the foci. These windows may be simple circular buffers around point foci, or could be adjusted to be ellipses (or even more general shapes) for wind direction or stream flow, although relatively few applications use windows outside of basic circular or elliptical buffers. Each window defines a potential cluster of cases.

The basic structure of the scan statistic approach compares the disease risk observed inside the potential cluster to that observed outside and defines a score measuring the unusualness of the potential cluster to what would be expected under the null hypothesis. SaTScan, the most popular implementation of spatial scan statistics, uses a likelihood ratio for this comparison of each potential cluster, and the "most likely cluster" is the one yielding the most extreme ratio. Inference follows from comparing the most extreme ratio observed across all potential clusters to the distribution of the most extreme values obtained from Monte Carlo simulations of either the case-control assignment (for point data) or case counts under the null hypothesis (random labeling for point data, and constant risk for small-area counts). The most likely cluster in each simulation may be at a different location.

A distinguishing feature between a *focused* scan statistic and a *general* scan statistic is the set of potential clusters under evaluation. General scan statistics center potential clusters on all locations (all case and control locations for point data, all small-area centroids for small-area count data), while focused scan statistics limit attention to potential clusters in areas with exposure to the foci or to potential clusters consistent with locally increased risk.

We note that the scan statistic can be applied to either point-based case-control data or small-area count data. For count data, one observes n_i cases in region i, where n_i is assumed to follow a Poisson distribution with mean $E[\mu_i]$. Let N denote the total number of observed cases. For any potential cluster, the likelihood ratio comparing the risk of disease inside any potential cluster Z to the risk outside of Z is

$$\left(\frac{n(Z)}{E[n(Z)]}\right)^{n(Z)}\left(\frac{n-n(Z)}{n-E[n(Z)]}\right)^{n-n(Z)}. \tag{14.1}$$

To restrict attention to potential clusters where the observed risk of disease is greater inside the cluster, one can multiply this by an indicator of whether the observed rate inside the potential cluster is greater than that outside the cluster. For case-control point data, one replaces the Poisson likelihood by a Bernoulli likelihood and defines the likelihood ratio in a similar manner. Again, for a simple test of focused clustering, one limits attention to the set of potential clusters centered on the foci of interest.

Variants of the scan statistic expand the set to include more general sets of potential clusters, including the work of Patil and Taillie (2004), Tango and Takahashi (2005), and Assunçao et al. (2006). As the set of potential clusters becomes more general, computation time associated with the Monte Carlo test often increases and efficient implementation becomes a priority.

14.3.2 Isotonic spatial scan statistic

We note that the simple extension of general scan statistics to focused testing above limits attention to potential focused clusters clearly defined by a boundary (the scan statistic radius) within which we anticipate elevated risk compared to the area outside of the potential cluster. This binary high- or low-risk delineation represents an abrupt local change in risk which we can generalize by considering a broader class of potential focused clusters defined by rings of monotonically decreasing risk as one moves farther away from the focus. The idea of *isotonic* (monotonically decreasing) risk as one moves farther from a putative source of higher risk motivated Stone (1988) to develop a test statistic based on *isotonic regression*, where estimated regression slopes are constrained to reflect the isotonic assumption.

Stone's (1988) approach directly motivated a number of tests for focused clustering, including the *isotonic spatial scan statistic* (Kulldorff 1999), which differs from the standard spatial scan statistic by defining a specific set of isotonic alternative hypotheses. In the standard spatial scan statistic described above, the alternative model of clustering posits a higher rate inside the cluster and a lower rate outside the cluster. In the isotonic spatial scan, under the alternative hypothesis, the cluster is modeled via isotonic regression with successively decreasing risk with increasing distance from the cluster center (Kulldorff 1999). Instead of one circular window, the isotonic spatial scan statistic defines the window via a central circle and a set of concentric rings centered on the same point with different radii. The alternative (focused clustering) model suggests that the rate is highest within the innermost circle and lower in each subsequent larger circle. The alternative model of the isotonic spatial scan mirrors a focused cluster around a point source where exposure decreases gradually as a function of distance (but still in a series of discrete steps). While motivated by a desire to detect focused clusters, in most applications of the approach, one considers the isotonic alternatives in a very general way and considers all isotonic potential clusters without prespecifying locations for the foci. We note that such an application defines a "general" test by considering the set of isotonic potential clusters, but the structure of each potential cluster (increasing risk toward a central peak) can point to likely locations of unknown foci of higher risk.

More specifically, with the isotonic spatial scan, the number and sizes of the circles are not defined in advance of the analysis, with the exception of the size of the largest ring, whose maximum size is set in advance. The number and sizes of the rings are determined from the method to maximize the likelihood ratio statistic, where many possible ring centroids are considered as the search window moves A over space. More formally, consider a set $\mathbf{C}$ of centroids that cover a study region A. Kulldorff (1999) initially proposed using an areal set of census regions and, for a particular circle centroid $C \in \mathbf{C}$, ordering the census areas by increasing distance from the centroid. Let i index the ranks of the distances from the circle centroid to the census areas, and let n_i, $i = 1, \ldots, J$ denote the number of cases in census area i. As with the spatial scan statistic, the likelihood can be calculated using either a Poisson or Bernoulli model, depending on how cases are represented in the data.

For the Bernoulli model, let p_i denote the probability that an individual in area i is a case and m_i is the total number of cases and controls in the area, respectively. The likelihood is defined as

$$L(C, p_1, \ldots, p_J) = \prod_{i=1}^{J} p_i^{n_i}(1-p_i)^{m_i - n_i}. \tag{14.2}$$

The maximum likelihood is defined as

$$L(C) = \max_{p_1 \geq p_2 \geq \ldots \geq p_J} L(C, p_1, \ldots, p_J), \tag{14.3}$$

and the likelihood function under the null hypothesis is

$$L_0 = \max_{p_1 = p_2 = \ldots = p_J} L(C, p_1, \ldots, p_J). \tag{14.4}$$

The test statistic is the maximum likelihood ratio over all possible circle centroids $C \in \mathbf{C}$,

$$S_C = \max_{C \in \mathbf{C}} \frac{L(C)}{L_0}, \tag{14.5}$$

where L_0 does not depend on C and has the same value as with the spatial scan statistic.

For the Poisson model, we define μ_i as the expected number of cases in census area i under the null hypothesis. The test statistic becomes

$$S_C = \max_{\lambda_1 \geq \lambda_2 \geq \ldots \geq \lambda_J} \frac{\prod_{i=1}^{J} (\lambda_i \mu_i)^{n_i}}{\left(\sum_{j=1}^{J} \lambda_j \mu_j\right)^{N}}, \tag{14.6}$$

where λ_i is the relative risk in area i and, again, N is the total number of cases.

The test statistic in Equation 14.6 identifies the most likely disease cluster. The p-value for the most likely cluster is calculated through Monte Carlo hypothesis testing. The testing process conditions on the observed number of cases and a large number K (e.g., 9999) of random replications of the data are generated under the null hypothesis of constant risk. The maximum likelihood ratio is calculated for each replicated data set with the same procedure applied to the observed data, yielding a distribution of the test statistic under the null hypothesis. The maximum likelihood ratio values from the $K + 1$ data sets, including the observed data, are ranked and the rank for the observed data is denoted R. The p-value is calculated as the fraction of statistics in the simulated data that exceed the observed statistic, that is, $p = R/(K + 1)$.

Typically, a large number of circle centroids will be considered. If there is only one circle centroid C evaluated, then the isotonic spatial scan is the focused cluster test proposed by Stone (1988), which is the maximum likelihood ratio test T_l for an isotonic regression problem with a simple ordering. Other extensions include that of Morton-Jones et al. (1999), who propose an extension to Stone's test for Poisson data that allows for covariate adjustment via a log-linear Poisson regression model. Stone's original test is implemented in the R package DCluster (Gómez-Rubio et al. 2005).

14.3.3 Example: Isotonic spatial scan statistic and non-Hodgkin lymphoma

As a demonstration of using the isotonic local spatial scan statistic, we apply it to a case-control study of non-Hodgkin lymphoma (NHL). The National Cancer Institute (NCI) Surveillance, Epidemiology and End Results (SEER) NHL study is a case-control study of 1321 cases aged 20–74 years that were diagnosed between July 1, 1998 and June 30, 2000, in four SEER cancer registries, including Detroit, Iowa, Seattle, and Los Angeles County. The study has been described previously (Wheeler et al. 2011). Briefly, population controls (1057) were selected from residents of the SEER areas using random digit dialing (<65 years of age) or Medicare eligibility files (65 and over) and were frequency matched to cases by age (within 5-year groups), sex, race, and SEER area. Among eligible subjects contacted for an interview, 76% of cases and 52% of controls participated in the study. The goal of the NCI-SEER NHL study was to investigate potential genetic and environmental risk factors for NHL.

Computer-assisted personal interviews were conducted during a visit to each subject's home to obtain lifetime residential and occupational histories, medical history, and information on demographics and risk factors. Written informed consent was obtained during the home visit, and human subject review boards approved the study at the NCI and at all participating institutions. Historic addresses were collected in a residential history section of an interviewer-administered questionnaire. Participants were mailed a residential calendar in advance of the interview and were requested to provide the complete address of every home in which they lived from birth to the current year, listing the years when they moved in and out of each address (De Roos et al. 2010). Interviewers reviewed the residential calendar with respondents and probed to obtain missing address information. Subjects reported residential addresses for the time period 1923–2001. Residential addresses were matched to geographic address databases to yield geographic coordinates.

We applied the isotonic spatial scan to the Los Angeles County study area and the Detroit study area of the NCI-SEER NHL study using freely available SaTScan software. These two study areas previously had the most significant unexplained spatial risk of NHL after adjusting for covariates in a generalized additive model (GAM) (Wheeler et al. 2011, 2012). Using residential histories, the most significant unexplained risk occurred at 20 years before diagnosis among long-term residents of each study area. For the spatial scan analysis, we conducted a purely spatial retrospective analysis at 20 years before diagnosis, set the maximum spatial cluster size as 50% of the population at risk, used a Bernoulli probability model, and scanned for areas with high NHL rates.

The isotonic spatial scan identified a significant area of high risk in both Los Angeles County (Figure 14.1) and Detroit (Figure 14.2). The steps in the cluster are shaded in decreasing grayscale, and the residential locations have been jittered in the maps to mask the true locations. The p-value was 0.037 for the isotonic spatial scan statistic in Los Angeles County and 0.038 in Detroit. As the figures show, there were three steps in the risk function for the cluster in Los Angeles County and two in Detroit. In Detroit, the relative risks by step in order were 1.78 and 1.18, and the cluster relative risk was 1.35. In Los Angeles,

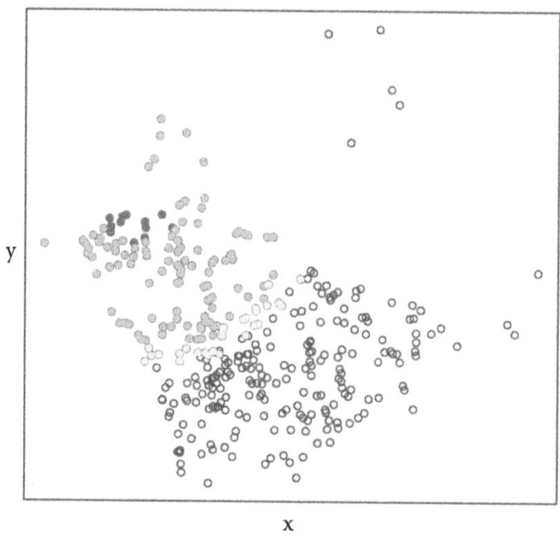

FIGURE 14.1
Isotonic spatial scan result for NCI-SEER NHL study in Los Angeles County.

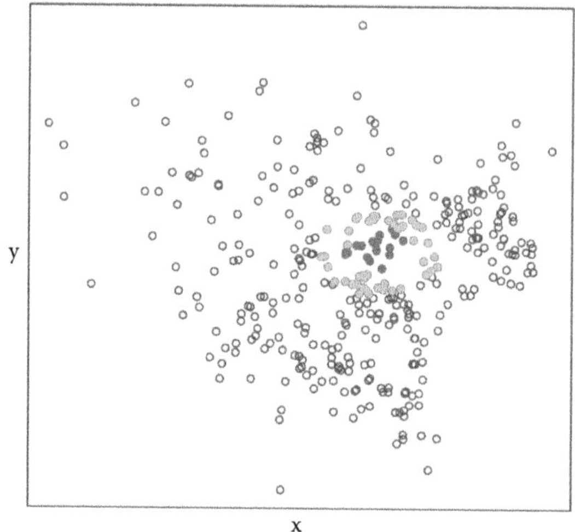

FIGURE 14.2
Isotonic spatial scan result for NCI-SEER NHL study in Detroit.

the relative risks by step in order were 2.08, 1.30, and 1.00, and the cluster relative risk was 1.31. Inside the Los Angeles County cluster, 91 of the 145 observations were cases (62.8%), and the ratio of observed/expected cases was $91/78.49 = 1.16$. Inside the Detroit cluster, 48 of the 63 observations were cases (76.2%), and the ratio of observed/expected cases was $48/37.66 = 1.27$. For both Los Angeles and Detroit, the cluster areas overlap with the significantly elevated risk areas previously found using a GAM (Wheeler et al. 2011).

In summary, scan statistics identify the most likely cluster from a family of potential clusters for a particular data set. If these potential clusters are all focused, the approach identifies the most likely focused cluster. Note that while the isotonic scan statistic identifies the most likely isotonic cluster and a circular area of highest risk, it does not identify the precise location of a focus of increased risk. Indeed, while it identifies where the local risk is high, the isotonic scan statistic approach does not provide inference on an a priori reason for the local increase, and analysts must take care not to fall victim to the so-called Texas sharpshooter approach of shooting the barn and painting a bull's-eye around the bullet hole by identifying a local region with significantly high risk, and then identifying some feature within this region and claiming to have assessed the significance of clustering around that point alone. The assumptions and questions addressed by any approach define the appropriate context for interpretation, an issue of critical importance in focused cluster studies.

14.3.4 Score tests to detect focused clusters

We next consider methods that move beyond binary or isotonic stepwise definitions of exposure where individuals within a given radius are considered to share the same exposure.

For small-area count data, suppose we have an exposure measure or surrogate, say g_i, for each small area $i = 1, \ldots, I$. Lawson (1993) and Waller et al. (1992, 1994) independently proposed a score test based on a focused alternative hypothesis where risk increases with exposure. One such alternative hypothesis is

$$\mathrm{H_A} : \mathrm{E}[n_i] = pop_i \lambda (1 + g_i \varepsilon),$$

where n_i denotes the observed number of cases in small area i, pop_i is the population size in area i, λ is the shared constant risk of disease assumed in the null hypothesis, and ε denotes a small, positive constant. The null hypothesis is defined when $\varepsilon = 0$, that is,

$$\mathrm{H_0} : \mathrm{E}[n_i] = pop_i \lambda.$$

Assuming a Poisson distribution for Y_i, the locally most powerful test of $\mathrm{H_A}$ is defined by the statistic

$$\Sigma g_i (n_i - pop_i \lambda).$$

This statistic is intuitive in that it is a weighted sum of the observed cases minus the expected number of cases (under the null hypothesis), weighted by exposure. That is, the statistic gives more weight to observed deviations in highly exposed areas, precisely what we would want a test to detect focused clusters to do. The test statistic is asymptotically normally distributed (as the number of areas increases) with mean zero and variance $\Sigma g_i^2 pop_i \lambda$ (for known λ). For small numbers of areas where the asymptotic distribution may not be accurate, a Monte Carlo implementation provides a practical and accurate method of assessing significance. Extensions to adjust for covariates are possible by replacing the null expectation $pop_i \lambda$ by the expectation adjusted for the covariates in area i.

14.3.5 Distance–decay functions as exposure surrogates

Specific applications of the score tests often include the use of distance–decay functions as exposure surrogates. Distance–decay functions assume that exposure is a function of distance from a source. Increasing distance is assumed to lead to decreasing exposure, typically taking a linear, inverse square, curvilinear, inverse curvilinear, or exponential form. Decay functions commonly appear as exposure surrogates in the Lawson–Waller score test above.

Distance–decay functions have seen extensive application in racial environmental disparity studies. Using the Environmental Protection Agency's Toxics Release Inventory (TRI) database, Pollock and Vittas (1995) estimated local exposure to a specified toxic chemical using a log-transformed distance to TRI as an exposure proxy, and Downey (2006) incorporated total air releases as a proxy for facility size in an effort to generate relative cumulative effects of multiple hazard sources. Challenges to broader implementation include selecting the correct functional form of the distance–decay function as the actual rate at which the chemical's hazard decline is typically not well known.

Expanded use of distance–decay functions has included pollution plume modeling, which incorporates not only distance to sources, but also other factors, such as wind speed, wind direction, other meteorological influences, and smokestack height. Morello-Frosch et al. (2001) estimate health risk scores using air pollution and toxicity data, taken from the EPA's Cumulative Exposure Project, and Ash and Fetter (2004) built similar estimates using the Risk-Screening Environmental Indicators project.

14.3.6 Extensions to the score test for detecting focused clusters

The basic score test has been extended to other families of focused alternatives as well. Bithell (1995) provides a very general definition of score tests, and Tango (2002) considers two families of focused alternatives for a given focus location. First, suppose the relative risk of disease for individuals living in region i compared to the background rate is denoted θ_i. If $\theta_{(i)}$ denotes the relative risk in the ith nearest region to the focus, the family of monotonically decreasing risk with distance is defined by

$$\mathrm{H_A} : \theta_{(1)} \leq \theta_{(2)} \leq \ldots \leq \theta_{(I)}.$$

Tango (2002, 2010) also defined score tests for the family of alternatives with a peak risk occurring some distance from the focus, say at the kth farthest region from the focus,

$$\mathrm{H_A} : \theta_{(1)} \leq \theta_{(2)} \leq \ldots \leq \theta_{(k)} \geq \ldots \geq \theta_{(I)}.$$

The test statistics are provided in detail in Tango (2002) and Tango (2010, section 8.4.4). As one would expect from the nature of score tests, in simulations under each of the families, the score test defined for that particular family tends to provide the best performance in terms of statistical power.

14.3.7 Example: Score tests to detect focused clusters of leukemia around hazardous waste sites

To illustrate the application and performance of score tests of focused clustering, we consider reports of leukemia (all types) occurring in the period 1978–1982 within 281 census tracts in an eight-county region of upstate New York (Waller and Gotway 2004). The data originally appear at the census block group resolution for seven of the eight counties (census tract resolution for the eighth) in Waller et al. (1992, 1994). Our observed data include observed case counts $(n_1, n_2, \ldots, n_{281})$ and population counts from the 1980 U.S. census $(pop_1, pop_2, \ldots, pop_{281})$. We consider a constant risk null hypothesis where the overall risk of disease is assumed known and equal to the overall rate defined by the total number of cases divided by the total number of individuals at risk, that is, $592/1{,}057{,}673 \approx 0.00056$. Our foci of putative risk include 11 inactive hazardous waste sites registered with the New York Department of Environmental Conservation as containing trichloroethylene (TCE). As an exposure surrogate, we define the inverse distance of each census tract centroid to the location of each waste site. For a full analysis of the data, see Waller et al. (1992, 1994), Waller and Gotway (2004), and Bivand et al. (2013), among others.

Here, we focus on the components of the score test. Figure 14.3 illustrates the centroid locations of the 281 census tracts and the location of the hazardous waste sites. The density of tract centroids roughly mirrors population density with the city of Syracuse in the north-central part of the map and the city of Binghamton in the south-central part of the map. For illustrative purposes, we consider Site 1, located in Binghamton, and Site 10, located in the smaller town of Ithaca. Figures 14.4 and 14.5 illustrate the components of the score test to detect focused clusters. The top plot in each illustrates the difference between observed disease counts and the value expected under a constant risk null hypothesis, ordered by the distance to each site. (We connect the points to aid in visualization.) We note that identical values appear in the top plots in Figures 14.4 and 14.5; the only difference is the ordering by distance to Site 1 in Figure 14.4 and to Site 10 in Figure 14.5. The middle

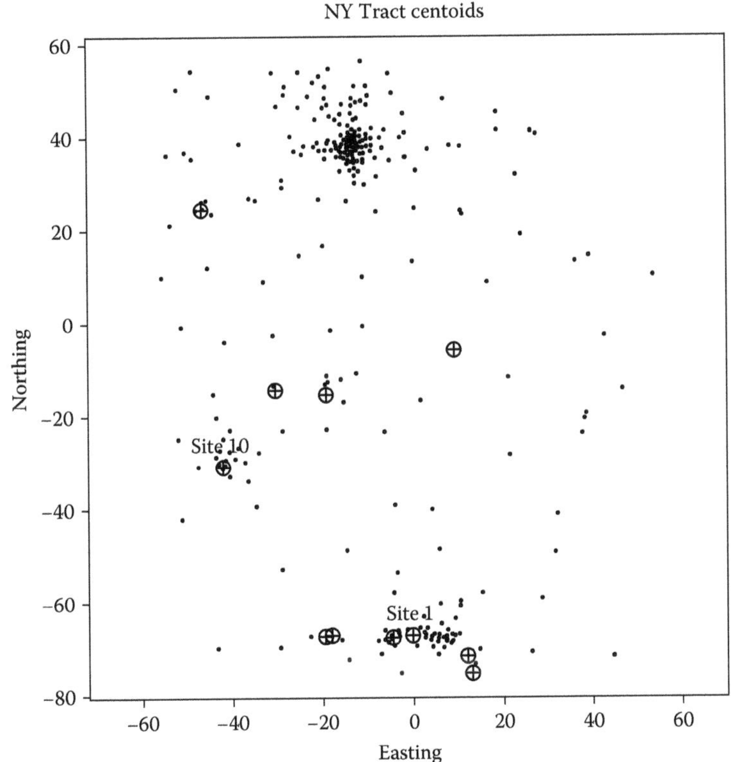

FIGURE 14.3
Location of 281 census tract centroids in central New York (small dots) and the location of 11 inactive hazardous waste sites containing trichloroethylene (large dots). We explore two sites as potential foci of increased risk, one in the city of Binghamton (Site 1) and the other in the smaller town of Ithaca (Site 10).

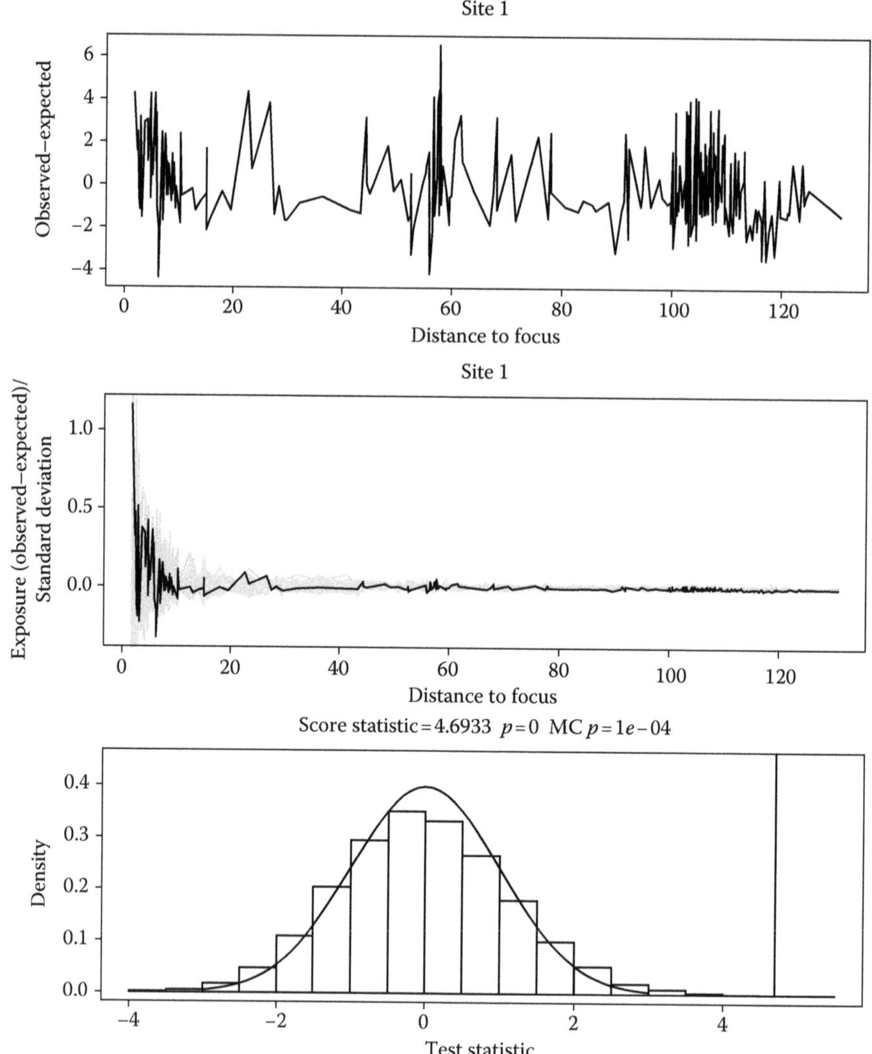

FIGURE 14.4
Components of the score test for Site 1. The top figure shows the difference between observed and expected counts ordered by distance to Site 1. The middle figure weights these values by inverse distance (as an exposure surrogate) and standardizes by the asymptotic variance of the statistic under the null hypothesis. The grey lines represent 50 simulations under the constant risk hypothesis. The bottom figure shows the histogram of test statistic values under the constant risk hypothesis and the asymptotic normal distribution, along with p-values associated with each.

plot in Figures 14.4 and 14.5 illustrates the same values multiplied by the inverse distance exposure surrogate and rescaled by dividing by the asymptotic variance of the statistic. The score test statistic is found by summing the values displayed in the middle plots. Here, we see that by defining the focus and weighting by our exposure surrogate, we increase the impact of deviations near the focus and decrease the impact of observations farther away.

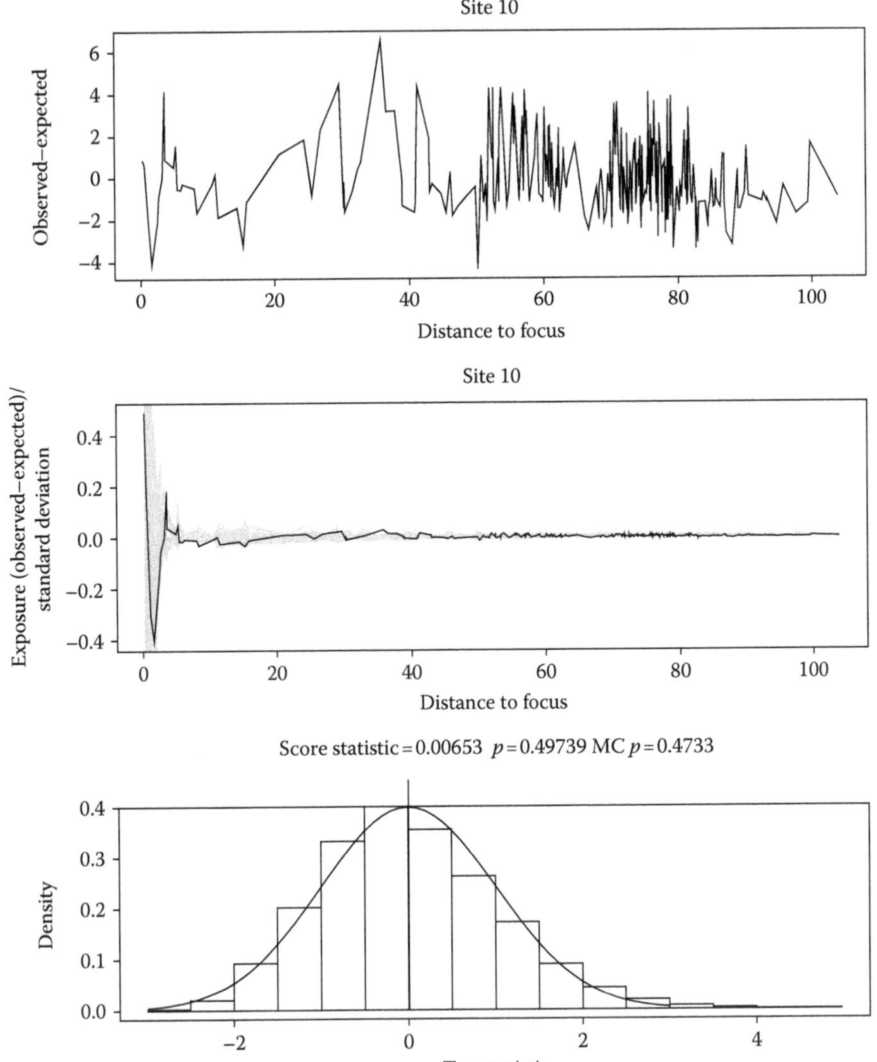

FIGURE 14.5
Components of the score test for Site 10. The top figure shows the difference between observed and expected counts ordered by distance to Site 10. The middle and bottom figures follow those in Figure 14.4, but for Site 10.

To illustrate the variability under the null hypothesis, grey lines in the middle plot indicate 50 simulations under the null hypothesis. For Site 1, excess observations occur near the site and receive high weight, resulting in a significant statistic. For Site 2, we observe one large positive deviation and one large negative deviation very close to the site, resulting in an observed statistic value near zero and no significant evidence of clustering around Site 10. The histograms in the bottom plots of Figures 14.4 and 14.5 illustrate that the asymptotic normal distribution is fairly accurate for this sample size, but that some skewness toward lower values remains apparent in the simulations.

14.4 Statistical Modeling to Detect Focused Clusters around a Putative Source of Increased Risk

The hypothesis testing approaches above build inference relating to detecting (or failing to detect) significant clusters of disease around the foci. We next describe two model-based extensions, one for point data and the other for count data.

14.4.1 Raised incidence model

The raised incidence model explores whether there appears to be raised incidence of disease around a point source or sources of pollution. For case-control point data, Diggle (1990) proposed the use of an inhomogeneous Poisson point process for cases where the intensity accounts for the distance to the pollution point source(s). The case intensity is specified as

$$\lambda_1(x) = \rho\lambda_0(x)f(x - x_0;\theta), \tag{14.7}$$

where λ_0 is the spatial variation of the underlying population, ρ measures the overall number of events per unit area, and $f(x - x_0;\theta)$ is a function of distance from the location of the source x_0 to point x and parameters θ. The spatial variation of the underlying population is independent of the effect of the point source. Diggle (1990) uses a distance–decay function,

$$f(x - x_0;\alpha,\beta) = 1 + \alpha\exp(-\beta||x - x_0||^2), \tag{14.8}$$

with a square of distance. The parameters α, β, ρ can be estimated by maximizing the likelihood of the Poisson point process, assuming that kernel smoothing is used to estimate the intensity λ_0 with a specified kernel bandwidth. The selection of the bandwidth can be difficult and substantially impact the results (Bivand et al. 2013).

Another approach does not require kernel bandwidth estimation. Using the previous scenario and conditioning on the location of cases and controls, Diggle and Rowlingson (1994) model the probability of being a case at location x as

$$p(x) = \frac{\lambda_1(x)}{\lambda_1(x) + \lambda_0(x)} = \frac{\rho f(x - x_0;\alpha,\beta)}{1 + \rho f(x - x_0;\alpha,\beta)}, \tag{14.9}$$

which factors out the intensity for the underlying population and reduces the problem to a nonlinear binary regression, an adjustment similar to proportional hazard modeling in survival analysis (Cox 1972). The model parameters can be estimated by maximizing the log-likelihood

$$L(\rho,\theta) = \sum_{i=1}^{n_1}\log(p(x_i)) + \sum_{j=1}^{n_0}\log(1 - p(x_i)), \tag{14.10}$$

where n_1 is the number of cases and n_0 is the number of controls. In addition to parametric modeling of exposure to the point source, the approach allows a natural way to include covariate effects in the definition of f via

$$f = \prod_{i=1}^{N_s}(1 + \alpha_i\exp(-\beta_i d_i^2))\prod_{j=1}^{N_c}\exp(\theta_j z_j), \tag{14.11}$$

where N_s is the number of sources of suspected raised incidence, d_i is the distance to the ith source, N_c is the number of covariates, z_j is the value of the jth covariate, and θ_j is its effect. When there are multiple putative sources, it is possible to reduce the number of parameters to estimate by using the same parameter pair (α_i, β_i) for different sources, assuming in advance that the sources would have the same effect on incidence (Diggle and Rowlingson 1994).

One can evaluate the significance of the distance to point sources or covariates by comparing model log-likelihoods. For example, one can compare the maximized log-likelihood for the null model and the model that includes the distance function for a point source. The test statistic is $D = 2$(log-likelihood for alternative model – log-likelihood for null model), which approximately follows a chi-square distribution with degrees of freedom equal to the difference in the number of model parameters. The raised incidence model can be fitted using function tribble from the R package splancs (Rowlingson and Diggle 2013).

This modeling framework provides a broad base allowing extensions to accommodate more detailed epidemiologic design, for example, matched case-control data to reduce the impact of potential confounding factors such as age, race, and sex. Diggle et al. (2000) provide the initial extension of the modeling framework to assess the impact of point-source exposures on disease risk in matched case-control data. Wakefield and Morris (2001) expand the approach to a hierarchical Bayesian setting, allowing for residual spatial correlation via random effects (as in Lawson 2016, Chapter 6 of this volume), and Li et al. (2012) provide further extensions allowing for multiple, ordered classes of disease severity (rather than a simple binary case-control label).

14.4.2 Example: Raised incidence model assessing childhood asthma in North Derbyshire, England

To illustrate the raised incidence modeling approach, we briefly review the application to one of the original motivating data sets, with additional discussion and detail provided by Diggle and Rowlingson (1994), Bivand et al. (2013), and Diggle (2014). Diggle and Rowlingson (1994) introduce a data set for a case-control study of asthmatic symptoms in a set of 10 elementary schools in north Derbyshire, England. In 5 of the 10 schools, the head teacher had expressed concern regarding the level of asthma among students. The data consist of questionnaire responses from parents, including the location of residence, the child's school (classified as among the five with prior expressed concerns regarding asthma), and binary (yes or no) responses as to whether the child has suffered from asthma, whether the child suffered from hay fever, and whether the child lived with individuals who smoked. The data also include the primary road network and three potential foci: a coking facility ("coking works"), a chemical plant, and a waste treatment facility. Figure 14.6 illustrates the residential locations of cases and controls and the location of the three foci. The data include 215 case locations (filled circles) and 1076 control locations (open circles).

Using the maximum likelihood approach outlined in the preceding section, Diggle and Rowlingson (1994) fitted several models with different subsets of the covariates and compared fit via the model deviance, that is, twice the maximized log-likelihood. The analysis provides a mechanism to assess the impact of residential proximity to the focus and simultaneous adjustment for additional, nonspatial covariates. The analysis revealed a significant effect of hay fever, followed by a moderately significant effect of proximity to the coking works (details provided in Diggle 2014, p. 186), a two-factor effect that would be difficult to assess in simple hypothesis tests to detect focused clusters.

The modeling perspective in this example moves from the spatial descriptive analysis of the scan statistic through the distance–decay assessment of the score test toward a fuller inclusion of potentially confounding effects (e.g., hay fever and household smoking) in a familiar regression-type setting. More recent and complex spatial point process models, such as log-Gaussian Cox processes (Møller and Waagepetersen 2004; Diggle 2014) further extend these ideas and offer additional directions for further expansion.

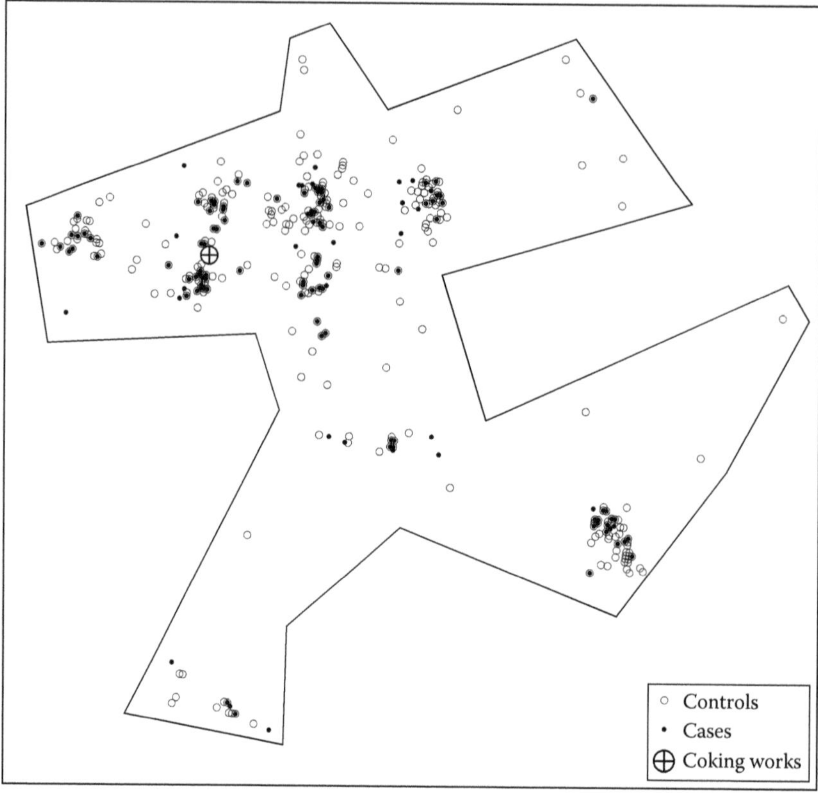

FIGURE 14.6
Location of residences for cases of pediatric asthma (filled circles) and 1076 control schoolchildren (open circles) in north Derbyshire, England. The circled cross represents the location of coke works. (Original data from Diggle, P.J. and Rowlingson, B.S., *Journal of the Royal Statistical Society: Series A*, 157, 433–440, 1994.)

14.5 Discussion

In the sections above, we transitioned from scan statistics to evaluate the most unusual collection of cases around prespecified putative sources or foci of excess risk, to identification of the most unusual collection of cases in concentric, isotonic decreases in risk (without prespecified foci), to definitions of families of score tests for particular focused cluster alternatives, to models of spatially varying intensity functions for point and regional count data. The approaches provide statistical tools ranging from descriptive to model based with increasingly specific inference and increasing ability to adjust for other potential risk factors. All build on elements of probability models and statistical inference, but are tuned to address the specific questions at hand. That said, the examples above also illustrate that each tool accomplishes a specific job and answers specific questions. A scan statistic identifies and evaluates the most likely cluster from a family (often a large family) of potential clusters. A score test assesses local deviations from the null hypothesis consistent with a specific family of alternative hypotheses. While score tests have attractive theoretical properties, these properties often rely on some assumptions, such as independent observations that may not be realistic in small-area studies of health data. Simulation-based power studies

(Waller and Lawson 1995; Tango 2010) provide some quantitative comparisons of performance for the alternatives driving the simulation. However, for most applications where local population sizes vary across the study area, power and other measures of performance are themselves spatial quantities; that is, the power to detect a cluster depends not only on the type and strength of the cluster, but also on the *location* of a cluster relative to the local population at risk (Waller et al. 2006; Waller 2014). Such dependencies work against single overall comparative rankings of methods by power, and stress the importance of careful, data set–specific power assessments to provide appropriate context for comparing results of multiple methods applied to the same data set.

Future work will continue to expand the toolbox by refining existing methods and extending the bridge from hypothesis testing to more comprehensive spatial modeling of disease patterns, providing a link to developing methods in spatial point process modeling (Møller and Waagepetersen 2004; Diggle 2014) and disease mapping for count data (Lawson 2013). The assessment of focused clusters in space and time provides additional analytic challenges and provides an additional frontier for further development. The analytic tools for spatiotemporal point and count data continue to develop (Cressie and Wikle 2011; Diggle 2014) and provide the theoretic base for extended work in this area. As an example, the extension of models to accommodate temporal variation in spatial exposures (e.g., through definition of *episodes* of high exposure) represents a particularly challenging and relevant issue in model-based detection of unusual spatiotemporal *focused* clusters (Al-Hadhrami and Lawson 2011; Blangiardo et al. 2011).

In summary, the detection and assessment of focused clusters within disease surveillance data provide multiple challenges in analysis and interpretation. Spatial statistical tools provide valuable insight and pave the way for deeper understanding of observed patterns to give clues to the underlying processes driving disease.

Acknowledgments

We thank the following collaborators for providing access to the data from the National Cancer Institute Surveillance, Epidemiology, and End Results Interdisciplinary Case-Control Study of NHL: Mary Ward, Lindsay Morton, Patricia Hartge, Anneclaire De Roos, James Cerhan, Wendy Cozen, and Richard Severson. Research reported in this publication was supported by the National Cancer Institute of the National Institutes of Health under award number R03CA173823 (Wheeler).

References

Al-Hadhrami, A. and Lawson, A.B. 2011. Bayesian hierarchical modeling of latent period switching in small-area putative health hazard studies. *Statistical Methods in Medical Research* 20: 5–28.

Ash, M. and Fetter, T. 2004. Who lives on the wrong side of the environmental tracks? Evidence from the EPA's Risk-Screening Environmental Indicators model. *Social Science Quarterly* 85: 441–462.

Assunçao, R., Costa, M., Tavares, A., and Ferreira, S. 2006. Fast detection of arbitrarily shaped disease clusters. *Statistics in Medicine* 25: 723–742.

Besag, J. and Newell, J. 1991. The detection of clusters in rare diseases. *Journal of the Royal Statistical Society: Series A* 154: 143–155.

Bithell, J.F. 1995. The choice of test for detecting raised disease risk near a point-source. *Statistics in Medicine* 14: 2309–2322.

Bithell, J.F., Dutton, S.J., Draper, G.J., and Neary, N.M. 1994. Distribution of childhood leukaemias and non-Hodgkin's lymphomas near nuclear installations in England and Wales. *British Medical Journal* 309: 501–505.

Bivand, R.S., Pebesma, E., and Gómez-Rubio, V. 2013. *Applied Spatial Data Analysis with R*. Springer, Berlin.

Blangiardo, M., Richardson, S., Gulliver, J., and Hansell, A. 2011. A Bayesian analysis of the impact of air pollution episodes on cardio-respiratory hospital admissions in the Greater London area. *Statistical Methods in Medical Research* 20: 69–80.

Byun, D. and Schere, K.L. 2006. Review of the governing equations, computational algorithms, and other components of the Models-3 Community Multiscale Air Quality (CMAQ) modeling system. *Applied Mechanics Reviews* 59: 51–77.

Cox, D.R. 1972. Regression models and life-tables (with discussion). *Journal of the Royal Statistical Society: Series B* 34: 187–220.

Cressie, N. and Wikle, C.K. 2011. *Statistics for Spatio-Temporal Data*. Wiley, Hoboken, NJ.

De Roos, A.J., Davis, S., Colt, J.S., Blair, A., Airola, M., Severson, R.K., Cozen, W., Cerhan, J.R., Hartge, P., Nuckols, J.R., and Ward, M.H. 2010. Residential proximity to industrial facilities and risk of non-Hodgkin lymphoma. *Environmental Research* 110: 70–78.

Diggle, P.J. 1990. A point process modelling approach to raised incidence of a rare phenomenon in the vicinity of a prespecified point. *Journal of the Royal Statistical Society: Series A* 153: 349–362.

Diggle, P.J. 2014. *Statistical Analysis of Spatial and Spatio-Temporal Point Patterns*. Chapman & Hall/CRC, Boca Raton FL.

Diggle, P., Morris, S., and Morton-Jones, T. 1999. Case-control isotonic regression for investigation of elevation in risk around a point source. *Statistics in Medicine* 18: 1605–1613.

Diggle, P.J., Morris, S.E., and Wakefield, J. 2000. Point-source modeling using matched case-control data. *Biostatistics* 1: 89–105.

Diggle, P.J. and Rowlingson, B.S. 1994. A conditional approach to point process modelling of elevated risk. *Journal of the Royal Statistical Society: Series A* 157: 433–440.

Downey, L. 2006. Environmental racial inequality in Detroit. *Social Forces* 85: 771–796.

English, P., Neutra, R., Scalf, R., Sullivan, M., Waller, L., and Zhu, L. 1999. Examining associations between pediatric asthma and traffic flow using a geographic information system. *Environmental Health Perspectives* 107: 761–767.

Gómez-Rubio, V., Ferrándiz-Ferragud, J., and López-Quílez, A. 2005. Detecting clusters of disease with R. *Journal of Geographical Systems* 7(2): 189–206.

Kulldorff, M. 1997. A spatial scan statistic. *Communications in Statistics: Theory and Methods* 26: 1481–1496.

Kulldorff, M. 1999. An isotonic spatial scan statistic for geographical disease surveillance. *Journal of the National Institute of Public Health* 48: 94–101.

Kulldorff, M. and Information Management Services, Inc. 2009. SaTScan^{TM} v8.0: software for the spatial and space-time scan statistics. http://www.satscan.org/.

Lagakos, S.W., Wessen, B.J., and Zelen, M. 1986. An analysis of contaminated well water and health effects in Woburn, Massachusetts [with discussion]. *Journal of the American Statistical Association* 81: 583–614.

Lawson, A.B. 1993. On the analysis of mortality events associated with a prespecified fixed point. *Journal of the Royal Statistical Society: Series A* 156: 363–377.

Lawson, A.B. 2006. *Statistical Methods for Spatial Epidemiology*, 2nd ed. Wiley, New York.

Lawson, A.B. 2013. *Bayesian Disease Mapping: Hierarchical Modeling in Spatial Epidemiology*, 2nd ed. Chapman & Hall/CRC, Boca Raton, FL.

Lawson, A.B. 2015. Case event and count modeling in disease mapping. In *Handbook of Spatial Epidemiology*, ed. A.B. Lawson, S. Banerjee, R. Haining, and M.D. Ugarte. Chapman & Hall/CRC, Boca Raton FL, pp. 121–131.

Lawson, A.B. and Williams, F.L.R. 1994. Armadale: a case-study in environmental epidemiology. *Journal of the Royal Statistical Society: Series A* 157: 285–298.

Li, S., Mukherjee, B., and Batterman, S. 2012. Point source modeling of matched case-control with multiple disease subtypes. *Statistics in Medicine* 31: 3617–3637.

Møller, J. and Waagepetersen, R. 2004. *Statistical Inference and Simulation for Spatial Point Processes*. Chapman & Hall/CRC, Boca Raton, FL.

Morello-Frosch, R., Pastor, M., and Sadd, J. 2001. Environmental justice and Southern California's 'Riskscap': the distribution of air toxic exposures and health risks among diverse communities. *Urban Affairs Review* 36: 551–578.

Morton-Jones, T., Diggle, P., and Elliott, P. 1999. Investigation of excess environmental risk around putative sources: Stone's test with covariate adjustment. *Statistics in Medicine* 18: 189–197.

National Research Council. 1991. *Environmental Epidemiology*, vol. 1: *Public Health and Hazardous Wastes*. National Academy Press, Washington, DC.

National Research Council. 2008. *The Utility of Proximity-Based Herbicide Exposure Assessment in Epidemiologic Studies of Vietnam Veterans*. National Academies Press, Washington, DC.

National Research Council. 2012. *Analysis of Cancer Risks in Populations Near Nuclear Facilities: Phase I*. National Academies Press, Washington, DC.

Patil, G.P. and Taillie, C. 2004. Upper level set scan statistic for detecting arbitrarily shaped hotspots. *Environmental and Ecological Statistics* 11: 183–197.

Pollock, P. and Vittas, M. 1995. Who bears the burden of environmental pollution? Race, ethnicity, and environmental equity in Florida. *Social Science Quarterly* 76: 294–310.

Rogerson, P. and Jacquez, G. 2015. Statistical tests for clustering and surveillance. In *Handbook of Spatial Epidemiology*, ed. A.B. Lawson, S. Banerjee, R. Haining, and M.D. Ugarte. Chapman & Hall/CRC, Boca Raton, FL, pp. 161–178.

Rowlingson, B. and Diggle, P. 2013. splancs: spatial and space-time point pattern analysis. R package version 2.01-33. http://CRAN.R-project.org/package=splancs.

Salway, R. and Wakefield, J. 2005. Sources of bias in ecological studies of non-rare events. *Environmental and Ecological Statistics* 12: 321–347.

Stone, R.A. 1988. Investigations of excess environmental risks around putative sources: statistical problems and a proposed test. *Statistics in Medicine* 7(6): 649–60.

Tango, T. 2002. Score tests for detecting excess risks around putative sources. *Statistics in Medicine* 21: 497–514.

Tango, T. 2010. *Statistical Methods for Disease Clustering.* Springer, New York.

Tango, T. and Takahashi, K. 2005. A flexibly shaped spatial scan statistic for detecting clusters. *International Journal of Health Geographics* 4: 11.

Wakefield, J.C. and Morris, S.E. 2001. The Bayesian modeling of disease risk in relation to a point source. *Journal of the American Statistical Association* 96: 77–91.

Wakefield, J.C. and Smith, T.R. 2015. Ecological modeling: general issues. In *Handbook of Spatial Epidemiology*, ed. A.B. Lawson, S. Banerjee, R. Haining, and M.D. Ugarte. Chapman & Hall/CRC, Boca Raton FL, pp. 99–117.

Waller, L.A. 2014. Putting spatial statistics (back) on the map. *Spatial Statistics* 9: 4–19.

Waller, L.A. and Gotway, C.A. 2004. *Applied Spatial Statistics for Public Health Data.* Wiley, Hoboken, NJ.

Waller, L.A., Hill, E.G., and Rudd, R.A. 2006. The geography of power: statistical performance of tests of clusters and clustering in heterogeneous populations. *Statistics in Medicine* 25: 853–865.

Waller, L.A. and Lawson, A.B. 1995. The power of focused tests to detect disease clustering. *Statistics in Medicine* 14: 2291–2308.

Waller, L.A., Turnbull, B.W., Clark, L.C., and Nasca, P. 1992. Chronic disease surveillance and testing of clustering of disease and exposure: application to leukemia incidence and TCE-contaminated dumpsites in upstate New York. *Environmetrics* 3: 281–300.

Waller, L.A., Turnbull, B.W., Clark, L.C., and Nasca, P. 1994. Spatial pattern analyses to detect rare disease clusters. In *Case Studies in Biometry*, ed. N. Lange, L. Ryan, L. Billard, D. Brillinger, L. Conquest, and J. Greenhouse. Wiley, New York, pp. 3–23.

Waller, L.A., Turnbull, B.W., Gustavsson, G., Hjalmars, U., and Andersson, B. 1995. Detection and assessment of clusters of disease: an application to nuclear power plant facilities and childhood leukemia in Sweden. *Statistics in Medicine* 14: 3–16.

Wheeler, D.C., De Roos, A.J., Cerhan, J.R., Morton, L.M., Severson, R., Cozen, W., and Ward, M.H. 2011. Spatial-temporal cluster analysis of non-Hodgkin lymphoma in the NCI-SEER NHL study. *Environmental Health* 10: 63.

Wheeler D.C., Ward, M.H., and Waller, L.A. 2012. Spatial-temporal analysis of cancer risk in epidemiologic studies with residential histories. *Annals of the Association of American Geographers* 102(5): 1049–1057.

15

Estimating the Health Impact of Air Pollution Fields

Duncan Lee
School of Mathematics and Statistics
University of Glasgow
Glasgow, UK

Sujit K. Sahu
Mathematical Sciences
University of Southampton
Southampton, UK

CONTENTS

15.1 Introduction

The health impacts of many facets of the natural and built environment have been well studied in recent years, including air pollution (Krall et al., 2013), green space (Richardson and Mitchell, 2010), and water quality (Wymer and Dufour, 2002). This chapter focuses on quantifying the health impact of air pollution, although the environmental, epidemiological, and statistical challenges discussed are applicable in the wider environmental context. For a more in-depth discussion of environmental studies, see Chapter 2. Quantifying the impact of air pollution is an inherently spatial as well as temporal problem, because air pollution concentrations vary at fine spatiotemporal scales. Furthermore, individuals move through this spatiotemporal pollution field, which makes quantifying both their exposure to air pollution and its resulting health impact a difficult modelling challenge. Nevertheless, this has been an active research topic since the 1990s, with one of the first studies quantifying the effect of short-term increases in concentrations in London (Schwartz and Marcus, 1990). Since then, a truly voluminous literature has developed, which has collectively quantified the health effects resulting from exposure to air pollution in both the short and the long term. This literature has included both single-site studies and large multicity studies, the latter being advantageous because of the comparability of the results across multiple locations due to unified data and analysis protocols. Collectively, these studies have helped to drive and shape legislation limiting pollution concentrations around the world, with examples being the 1990 Clean Air Act in the United States; the 2007 Air Quality Strategy for England, Scotland, Wales, and Northern Ireland; and the 2008 European Parliament directive on ambient air quality and cleaner air for Europe.

Three main study designs have been used to estimate the health impact of air pollution: time series studies, cohort studies, and areal unit studies. Time series studies are used to estimate the health impact of short-term exposure to pollution, that is, exposure to a few days of elevated concentrations, often termed an air pollution episode. The disease data used in such studies are population-level summaries of disease prevalence rather than individual disease cases, meaning that it is an ecological association study and cannot be used to determine individual-level cause and effect. However, due to the routine availability of population-level disease summaries, time series studies are inexpensive, quick to implement, and the most commonly used study design. Prominent examples include the large multicity studies entitled Air Pollution and Health: A European Approach (APHEA-2) (Samoli et al., 2007) in Europe and the National Morbidity, Mortality and Air Pollution Study (NMMAPS) (Dominici et al., 2002) in the United States. In contrast, cohort studies quantify the health effects resulting from long-term exposure to pollution, that is, prolonged exposure over months or years. They utilize individual-level data, and as a result, individual-level cause and effect can be established. However, they are costly and time-consuming to implement, due to the large amount of data collection required and the length of time needed for the follow-up period to assess disease incidence in the cohort. Examples of cohort studies include the Six Cities Study (Dockery et al., 1993), the Multi-Ethnic Study of Atherosclerosis (MESA) (Kaufman et al., 2012), and the European Study of Cohorts for Air Pollution Effects (ESCAPE) (Cesaroni et al., 2014).

As a result of the high cost of cohort studies, areal unit study designs have also been used to quantify the long-term health impact of air pollution. These studies are the spatial analogue of time series studies, and estimate the effects of air pollution based on spatial contrasts in disease risk and pollution concentrations across a set of contiguous areal units. Like time series studies, they utilize population-level rather than individual-level disease data, and cannot be used to quantify individual-level cause and effect. However, the areal unit data required to implement such studies have become widely available in recent times, with an example being the Health and Social Care Information Centre (https://indicators.ic.nhs.uk/) in the UK. Therefore, areal unit studies are quick and inexpensive to implement, and they contribute to and independently corroborate the evidence from cohort studies. These studies have been implemented using both spatial (Jerrett et al., 2005; Lee et al., 2009) and spatiotemporal (Greven et al., 2011; Lawson et al., 2012) designs, and the latter have also been used to estimate the short-term impact of pollution (see, e.g., Zhu et al., 2003; Fuentes et al., 2006; Choi et al., 2009).

This chapter provides a critique of the statistical and epidemiological challenges faced by researchers conducting areal unit studies, reviews the literature in this area to date, and provides an illustrative example of such a study. Although we focus on the spatial modelling challenges inherent in areal unit studies, there are similar challenges to be encountered when conducting time series and cohort studies. For simplicity, we discuss these challenges in the context of a spatial rather than a spatiotemporal study, but we note that similar challenges exist in the latter design. An in-depth discussion of spatiotemporal modelling is given in Chapter 19. The layout of this chapter is as follows. The next two sections describe the study design and data used in areal unit studies, as well as the statistical models commonly used in the literature to analyse these data. This review is followed by two sections highlighting the main statistical challenges facing researchers in this area, including modelling residual spatial autocorrelation in the disease data after accounting for known covariates, and accounting for the spatial misalignment between the pollution and disease data. These discussions are followed by an example that illustrates the issues discussed so far, and the chapter ends with a section providing the main conclusions and a discussion of future work needed in this area.

15.2 Areal Unit Studies

The study region $\mathcal{A}$ is a large geographical region such as a city, state, or country, and is partitioned into n areal units $\mathcal{A} = \{\mathcal{A}_1, \ldots, \mathcal{A}_n\}$ such as local authorities or census tracts. The areal units are typically defined by administrative boundaries, and the populations living in each one will be of different sizes and have different demographic structures. The disease data are denoted by $\mathbf{y} = (y_1, \ldots, y_n)$ and are counts of the total numbers of disease cases observed for each areal unit during an extended period of time, such as a year. Differences in the population sizes and demographic structures between areal units are accounted for by computing the expected numbers of disease cases, which are denoted here by $\mathbf{e} = (e_1, \ldots, e_n)$. For this calculation, the population in each areal unit is split into R strata based on their age, sex, and possibly ethnicity, and N_{ik} denotes the size of the population in strata k in areal unit i. Let r_k denote the risk of disease for the kth strata from a reference population, and then $e_i = \sum_{k=1}^{R} N_{ik} r_k$, which is known as indirect standardization. The standardized morbidity/mortality ratio (SMR) for area k is given by $\hat{\theta}_i = y_i/e_i$, which is the maximum likelihood estimator for the risk of disease from the simple Poisson model, $y_i \sim \text{Poisson}(e_i \theta_i)$. A SMR value of 1 represents an average risk relative to e_i, while a SMR value of 1.2 means an area has a 20% increased risk of disease.

A vector of representative air pollution concentrations for the n areal units is denoted by $\mathbf{x} = (x_1, \ldots, x_n)$ and is typically measured in micrograms per cubic metre (μgm^{-3}). For simplicity of exposition, we work with a single pollutant here, which can also be taken as a continuous index of air quality, but the methodology can be easily extended to include multiple pollutants. The health impacts of numerous different pollutants have been investigated in areal unit studies, including carbon monoxide (Maheswaran et al., 2005), nitrogen dioxide (Haining et al., 2010), ozone (Young et al., 2009), particulate matter (Jerrett et al., 2005), and sulphur dioxide (Elliott et al., 2007). Particulate matter is a mixture of small solid and liquid particles in the air, and particles with diameters of less than 10 microns (PM_{10}) and 2.5 microns ($PM_{2.5}$) have been associated with ill health in existing areal unit studies (see Lee, 2012, and Janes et al., 2007, respectively). However, estimating $\mathbf{x}$ is a challenging task, and two different data types can be used. The first are measured concentrations from a pollution monitoring network, with examples being the State and Local Air Monitoring Stations (SLAMS) network run by the U.S. Environmental Protection Agency (USEPA), and the Automatic Urban and Rural Network (AURN) maintained by the Department for Environment, Food and Rural Affairs (DEFRA) of the UK government. The second type of data are modelled concentrations from an air pollution dispersion model, with examples being the Community Multi-Scale Air Quality Model (CMAQ) and the Air Quality in the Unified Model (AQUM) (see Savage et al., 2013) developed by the UK Met Office. Monitoring networks measure air pollution concentrations with little error, but they typically do not have good spatial coverage of the study region, and a number of the n areal units may not contain any pollution monitors. In contrast, computer dispersion models estimate pollution concentrations on a regular grid and give complete spatial coverage of the study region without missing observations. However, modelled concentrations are known to contain errors and biases and are less accurate than the measured values.

Finally, there are a number of confounding factors that must be adjusted for when estimating the health impact of air pollution using an areal unit study design, the most prominent of which is the differential rates of smoking across the n areal units. However, reliable smoking data are often hard to obtain, so almost all studies use measures of socioeconomic deprivation as a proxy for smoking prevalence, due to the likely high correlation between smoking and deprivation (Kleinschmidt et al., 1995). However, socioeconomic deprivation is multidimensional, and variables that have been used to account for it include individual measures of income (Lawson et al., 2012), unemployment (Lee and Mitchell, 2014), and house price (Jerrett et al., 2005), as well as composite deprivation indices such as the Carstairs (Elliott et al., 2007) and Townsend (Haining et al., 2010) indices. Let $\mathbf{U} = (\boldsymbol{u}_1^{\mathrm{T}}, \ldots, \boldsymbol{u}_n^{\mathrm{T}})$ denote the matrix of p confounders, where the values relating to areal unit $\mathcal{A}_i$ are denoted by $\boldsymbol{u}_i^{\mathrm{T}} = (u_{i1}, \ldots, u_{ip})$.

15.3 Modelling

Poisson log-linear models are typically used to estimate the health impact of air pollution, and both classical (e.g., Jerrett et al., 2005; Greven et al., 2011) and Bayesian (e.g., Fuentes et al., 2006; Lee et al., 2009) approaches have been used for inference. For the latter, Markov chain Monte Carlo (MCMC) (for details, see Robert and Casella, 2010) simulation and integrated nested Laplace approximation (INLA) (for details, see Rue et al., 2009) have both been used; for further details, see Chapter 7. The use of INLA is rare, however, in this air pollution and health context, with one of the few example studies being Lee and Mitchell (2014). A Bayesian approach is the most popular inferential framework in these studies, because the models used are typically hierarchical in nature and include spatial autocorrelation and different levels of variation. The first stage of a general Bayesian hierarchical model for these data is given by

$$\begin{aligned} y_i &\sim \text{Poisson}(e_i\theta_i) \quad \text{for } i = 1, \ldots, n, \\ \ln(\theta_i) &= \beta_0 + x_i\beta_x + \boldsymbol{u}_i^{\mathrm{T}}\boldsymbol{\beta}_u + \phi_i, \\ \boldsymbol{\beta} = (\beta_0, \beta_x, \boldsymbol{\beta}_u) &\sim \mathrm{N}(\boldsymbol{\mu}_\beta, \Sigma_\beta), \end{aligned} \tag{15.1}$$

and similar models are used in disease mapping applications, as discussed in Chapters 5 through 7. Here the expected value of the disease count y_i is the product $e_i\theta_i$, where θ_i is the risk of disease in areal unit i. Here, a value of θ_i greater (less) than 1 indicates that areal unit $\mathcal{A}_i$ has a higher (lower) than average disease risk relative to e_i, and $\theta_i = 1.15$ corresponds to a 15% increased risk of disease. The log risks for all n areal units are modelled as a linear combination of an overall intercept term β_0, air pollution concentrations $(x_i\beta_x)$, confounding factors $(\boldsymbol{u}_i^{\mathrm{T}}\boldsymbol{\beta}_u)$, and a vector of random effects $\boldsymbol{\phi} = (\phi_1, \ldots, \phi_n)$. The latter account for any residual spatial autocorrelation remaining in the disease data after the covariate effects have been removed, as well as any overdispersion resulting from the restrictive Poisson assumption that $\mathrm{Var}[y_i] = \mathbb{E}[y_i]$. One possible cause of this residual autocorrelation and overdispersion is unmeasured confounding, which occurs when an important spatially autocorrelated covariate is either unmeasured or unknown. The spatial structure in this covariate induces spatial autocorrelation and overdispersion into the disease data, which cannot be accounted for in a regression model. Other possible causes of residual spatial autocorrelation are neighbourhood effects, where, in general, a subject's behaviour is influenced by that of neighbouring subjects, and grouping effects, where subjects choose to be close to similar subjects.

The regression parameter β_x quantifies the relationship between air pollution and disease risk on the log scale, and is transformed to a relative risk for the purposes of interpretation. The relative risk for a ν (say) unit increase in pollution concentrations measures the proportional increase in risk from increasing pollution by ν and is calculated as

$$\text{RR}(\beta_x, \nu) = \frac{e_i \exp(\beta_0 + (x_i + \nu)\beta_x + \boldsymbol{u}_i^{\text{T}} \boldsymbol{\beta}_u + \phi_i)}{e_i \exp(\beta_0 + x_i \beta_x + \boldsymbol{u}_i^{\text{T}} \boldsymbol{\beta}_u + \phi_i)} = \exp(\nu \beta_x). \tag{15.2}$$

Hence, a relative risk of 1.05 corresponds to a 5% increase in disease risk when pollution concentrations increase by ν μgm^{-3}. The posterior distribution and hence 95% credible intervals for the relative risk can be computed by applying the transformation given by (15.2) to the posterior distribution for β_x, and there is substantial evidence of a relationship between air pollution and disease risk if the 95% credible interval does not include 1.

A number of approaches have been proposed for modelling the residual spatial autocorrelation and overdispersion in the disease data unaccounted for by the covariates, including geographically weighted regression (Young et al., 2009), geostatistical models (Elliott et al., 2007), and simultaneous autoregressive models (Jerrett et al., 2005). However, the most common approach is to model the vector of random effects $\boldsymbol{\phi}$ with a conditional autoregressive (CAR) prior (see Maheswaran et al., 2005; Fuentes et al., 2006; Lee et al., 2009, 2014; Lawson et al., 2012), which is a special case of a Gaussian Markov random field (GMRF). This prior can be written as $\boldsymbol{\phi} \sim \text{N}(\mathbf{0}, \tau^2 \mathbf{Q}(\mathbf{W})^{-1})$, where $\mathbf{Q}(\mathbf{W})_{n \times n}$ is a potentially singular, precision matrix, τ^2 is a variance parameter, and $\mathbf{0}$ is an $n \times 1$ mean vector of zeros. Spatial autocorrelation is induced into this joint distribution via a $n \times n$ neighbourhood matrix $\mathbf{W}$, which determines the spatial adjacency structure of the n areal units. Typically, the elements of $\mathbf{W}$ are taken to be binary, and element $w_{ij} = 1$ if areal units $(\mathcal{A}_i, \mathcal{A}_j)$ are spatial neighbours and share a common border (denoted $i \sim j$), while $w_{ij} = 0$ otherwise (denoted $i \nsim j$).

The intrinsic model (ICAR) (Besag et al. 1991) is the simplest CAR prior and has a singular precision matrix (its row sums equal zero) given by $\mathbf{Q}(\mathbf{W}) = \text{diag}(\mathbf{W1}) - \mathbf{W}$, where $\text{diag}(\mathbf{W1})$ is an $n \times n$ diagonal matrix containing the row sums of $\mathbf{W}$. The spatial autocorrelation structure implied by this prior is more easily observed from its set of univariate full conditional distributions, that is, from $f(\phi_i | \boldsymbol{\phi}_{-i})$ for $i = 1, \ldots, n$, where $\boldsymbol{\phi}_{-i} = (\phi_1, \ldots, \phi_{i-1}, \phi_{i+1}, \ldots, \phi_n)$. The Markov nature of this model means that the conditioning is in fact only on the random effects in geographically adjacent areal units, which induces spatial autocorrelation into $\boldsymbol{\phi}$. Using standard multivariate Gaussian theory, the full conditional distribution $f(\phi_i | \boldsymbol{\phi}_{-i})$ is given by

$$\phi_i | \boldsymbol{\phi}_{-i}, \tau^2, \mathbf{W} \sim \text{N}\left(\frac{\sum_{j=1}^{n} w_{ij} \phi_j}{\sum_{j=1}^{n} w_{ij}}, \frac{\tau^2}{\sum_{j=1}^{n} w_{ij}} \right). \tag{15.3}$$

The Bayesian hierarchical model defined by (15.1) and (15.3) is completed by assuming $\tau^2 \sim \text{inverse gamma}(a, b)$, where the hyperparameters (a, b) are typically chosen to make the prior proper but weakly informative, such as $(a = 2, b = 1)$ (Gelman, 2006), to avoid controversy regarding the use of a noninformative prior, which may lead to an improper posterior distribution. The ICAR prior is a natural model for strong spatial autocorrelation, because its conditional expectation is the mean of the random effects in neighbouring areal units, while its conditional variance is inversely proportional to the number of neighbouring units. The rationale for the latter is that the more neighbouring units an area has, the more information there is about the value of its random effect—hence its variance is smaller. However, the ICAR model can only capture strong spatial autocorrelation, because it does not have a spatial autocorrelation parameter. Note that if $\boldsymbol{\phi}$ is multiplied by a positive constant, then the spatial autocorrelation structure will remain unchanged, but τ^2 will increase.

Therefore, a number of different approaches have been proposed for allowing for varying levels of spatial autocorrelation in $\boldsymbol{\phi}$, the most popular of which is the *convolution* or *BYM* model proposed by Besag et al. (1991). This model augments the linear predictor in (15.1) with a second set of random effects, say $\boldsymbol{\gamma} = (\gamma_1, \dots, \gamma_n)$, which are modelled independently as $\gamma_i \sim \text{N}(0, \sigma^2)$ for $i = 1 \dots, n$. However, identifiability problems may arise in this model, as it has n data points and $2n$ random effects. Therefore, Leroux et al. (1999) proposed a CAR prior with a spatial autocorrelation parameter ρ, which has full conditional distributions given by

$$\phi_i|\boldsymbol{\phi}_{-i}, \tau^2, \rho, \mathbf{W} \sim \text{N}\left(\frac{\rho\sum_{j=1}^n w_{ij}\phi_j}{\rho\sum_{j=1}^n w_{ij} + 1 - \rho}, \frac{\tau^2}{\rho\sum_{j=1}^n w_{ij} + 1 - \rho}\right). \tag{15.4}$$

Here, $\rho = 1$ corresponds to the intrinsic CAR model for strong spatial autocorrelation, while $\rho = 0$ corresponds to independent random effects with a constant variance. A uniform prior on the unit interval is typically specified for ρ, and the joint distribution for $\boldsymbol{\phi}$ is multivariate Gaussian with a mean of zero, a variance of τ^2, and a precision matrix given by $\mathbf{Q}(\mathbf{W}) = \rho[\text{diag}(\mathbf{W1}) - \mathbf{W}] + (1 - \rho)\mathbf{I}_n$, where $\mathbf{I}_n$ is the $n \times n$ identity matrix. The models discussed in this section can be easily implemented in the freely available R package *CARBayes* (Lee, 2013), and further details are available from http://cran.r-project.org. See Chapter 24 for an in-depth discussion of R packages for spatial data analysis.

15.4 Controlling for Residual Spatial Autocorrelation

CAR priors such as (15.3) or (15.4) force the random effects $\boldsymbol{\phi}$ to be globally spatially smooth, which is illustrated by their assumed partial autocorrelation between ϕ_i and ϕ_j conditional on the remaining random effects $\boldsymbol{\phi}_{-ij}$. For model (15.4), this partial autocorrelation is given by

$$\text{Corr}[\phi_i, \phi_j|\boldsymbol{\phi}_{-ij}] = \frac{\rho w_{ij}}{\sqrt{(\rho\sum_{k=1}^n w_{ik} + 1 - \rho)(\rho\sum_{l=1}^n w_{jl} + 1 - \rho)}}, \tag{15.5}$$

where $w_{ij} = 1$ for all pairs of adjacent areal units. Therefore, if ρ is close to 1 all pairs of adjacent random effects are spatially autocorrelated, leading to a globally smooth surface for $\boldsymbol{\phi}$. Conversely, if ρ is close to zero, then all pairs of adjacent random effects are close to being conditionally independent given all other random effects, leading to no spatial autocorrelation anywhere in the random effects surface. In either case, the random effects exhibit a single global level of spatial smoothness throughout the study region, which is likely to be inappropriate. This is because the residual spatial autocorrelation in the disease data is unlikely to be globally spatially smooth, which is evidenced empirically for our England study in the left panel of Figure 15.1. The figure shows that some pairs of adjacent areal units exhibit similar disease risks, while between other adjacent pairs, there are large step changes. As shown by (15.5), such localized spatial autocorrelation cannot be accurately represented

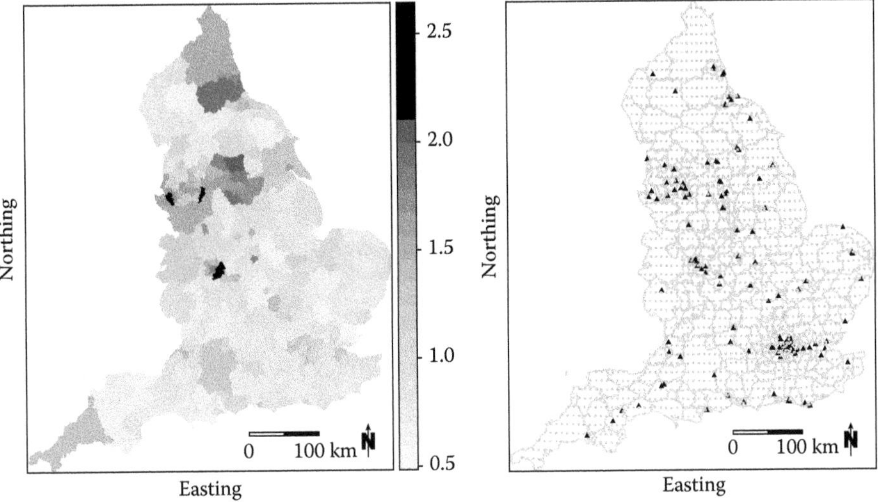

FIGURE 15.1
The left map displays the SMR for hospital admissions due to respiratory disease in 2010, while the right map shows the locations of the pollution monitors (black triangles) and the corners of the 12-kilometre-square grid cells (grey dots).

by CAR priors with a single spatial smoothing parameter ρ. Furthermore, research by Clayton et al. (1993) and Reich et al. (2006) has shown the potential for collinearity between globally smooth random effects and any covariate in the model that is also spatially smooth, such as air pollution. This potential collinearity can lead to variance inflation and instability in the estimation of the air pollution effect, and the simple solution of omitting the random effects from (15.1) is not appropriate, as ignoring residual spatial autocorrelation can lead to similar problems. Therefore, a number of approaches to solving these problems have been proposed in the literature to date, and a review is given here.

15.4.1 Two-stage modelling of autocorrelation

Lawson et al. (2012) utilize a spatiotemporal rather than a spatial study design, and propose estimating the covariate effects and the residual spatiotemporal autocorrelation in the disease data separately using a two-stage approach. They suggest first computing the residuals, on the linear predictor scale, from fitting a simple Poisson log-linear model similar to (15.1) to the disease and covariate data without the random effects. These residuals are then modelled in the second stage by a spatiotemporal mixture model, which allows complex spatiotemporal autocorrelation structures to be estimated. Finally, the fitted values from this second-stage residual model are included in a modified disease model as a fixed offset term, and the health effects of air pollution are reestimated, allowing for this residual spatiotemporal autocorrelation structure.

15.4.2 Orthogonal autocorrelation

Hughes and Haran (2013) propose replacing the random effects $\boldsymbol{\phi}$ with a set of spatially smooth basis functions that are orthogonal to the covariates. These basis functions are based on the residual projection matrix from a Gaussian linear model given by

$$\mathbf{P} = \mathbf{I}_n - \tilde{\mathbf{U}}(\tilde{\mathbf{U}}^{\mathrm{T}}\tilde{\mathbf{U}})^{-1}\tilde{\mathbf{U}}^{\mathrm{T}},$$

where $\tilde{\mathbf{U}} = (\mathbf{x}, \mathbf{U})$ is the complete covariate matrix. The basis functions used by Hughes and Haran (2013) are the eigenvectors of the matrix product **PWP**, and they show that these eigenvectors correspond to all possible mutually distinct patterns of spatial autocorrelation orthogonal to the covariates $\tilde{\mathbf{U}}$. Hughes and Haran (2013) suggest choosing the first $q << n$ eigenvectors corresponding to positive and decreasing eigenvalues, which are collectively denoted by $\mathbf{M}$, where $\mathbf{m}_i$ denotes the ith row of $\mathbf{M}$. The resulting model is similar to (15.1) except that its linear predictor has the form

$$\begin{aligned} \ln(\theta_i) &= \beta_0 + x_i\beta_x + \boldsymbol{u}_i^{\mathrm{T}}\boldsymbol{\beta}_u + \mathbf{m}_i^{\mathrm{T}}\boldsymbol{\delta}, \\ \boldsymbol{\delta} &\sim \mathrm{N}(\mathbf{0}, \tau^2\mathbf{Q}(\mathbf{W})_s^{-1}). \end{aligned} \quad (15.6)$$

Here, the precision matrix for $\boldsymbol{\delta}$ is given by $\mathbf{Q}(\mathbf{W})_s = \mathbf{M}^{\mathrm{T}}\mathbf{Q}(\mathbf{W})\mathbf{M}$, and $\mathbf{Q}(\mathbf{W})$ is defined as for the intrinsic CAR prior. This model can be implemented by the *ngspatial* package in the statistical software R, and further details can be found in Hughes and Haran (2013).

15.4.3 Localized autocorrelation

Brezger et al. (2007), Lu et al. (2007), Ma et al. (2010), Lee and Mitchell (2014), and Lee et al. (2014) have extended the class of CAR priors to allow for localized spatial autocorrelation, by modelling the elements of the neighbourhood matrix $\mathbf{W}$ corresponding to geographically adjacent areal units as binary random quantities, rather than being fixed equal to 1. Denoting this set by $\mathcal{W} = \{w_{jk}|j \sim k\}$, Equation (15.5) shows that if $w_{ij} \in \mathcal{W}$ equals 1 then (ϕ_i, ϕ_j) are spatially autocorrelated and are smoothed over in the modelling process, while if $w_{ij} \in \mathcal{W}$ is estimated as zero, then they are conditionally independent and no such spatial smoothing is enforced.

Lee et al. (2014) propose a localized conditional autoregressive (LCAR) model, which specifies a joint prior distribution $f(\tilde{\boldsymbol{\phi}}, \mathcal{W}) = f(\tilde{\boldsymbol{\phi}}|\mathcal{W})f(\mathcal{W})$ for an extended vector of random effects $\tilde{\boldsymbol{\phi}} = (\boldsymbol{\phi}, \phi_*)$ and the adjacency elements $\mathcal{W}$. Here, $f(\tilde{\boldsymbol{\phi}}|\mathcal{W})$ is a CAR prior conditional on a fixed neighbourhood structure $\mathcal{W}$, but because the elements of $\mathcal{W}$ can be estimated as zero, then (15.3) would be an inappropriate choice, as $\sum_{j=1}^{n} w_{ij}$ could equal zero, leading to an infinite mean and variance. Therefore, an additional random effect ϕ_* is included in the model, which prevents the infinite mean and variance problem from occurring. Let $\tilde{\mathbf{W}}$ be an $(n+1) \times (n+1)$ neighbourhood matrix for $\tilde{\boldsymbol{\phi}}$, where there is a one-to-one relationship between a particular set of values in $\mathcal{W}$ and its matrix representation $\tilde{\mathbf{W}}$. The remaining unspecified adjacency relations in $\tilde{\mathbf{W}}$ are between (ϕ_i, ϕ_*), and are denoted by w_{i*} for $i = 1, \ldots, n$. Here, $w_{i*} = 0$ if all of the adjacency elements in $\mathcal{W}$ relating to areal unit $\mathcal{A}_i$ equal one, otherwise $w_{i*} = 1$. Based on $\tilde{\mathbf{W}}$, Lee et al. (2014) propose a zero-mean multivariate Gaussian prior for $\tilde{\boldsymbol{\phi}}$ with precision matrix $\mathbf{Q}(\tilde{\mathbf{W}}, \epsilon) = \mathrm{diag}(\tilde{\mathbf{W}}\mathbf{1}) - \tilde{\mathbf{W}} + 0.001\mathbf{I}$, where $0.001\mathbf{I}$ is added to make the precision matrix diagonally dominant and hence invertible. The full conditional distribution $f(\phi_i|\tilde{\boldsymbol{\phi}}_{-i})$ corresponding to this joint distribution is given by

$$\phi_i|\tilde{\boldsymbol{\phi}}_{-i}, \tau^2, \tilde{\mathbf{W}} \sim \mathrm{N}\left(\frac{\sum_{j=1}^{n} w_{ij}\phi_j + w_{i*}\phi_*}{\sum_{j=1}^{n} w_{ij} + w_{i*} + 0.001}, \frac{\tau^2}{\sum_{j=1}^{n} w_{ij} + w_{i*} + 0.001}\right). \quad (15.7)$$

Thus, if $\sum_{j=1}^{n} w_{ij} = 0$, then $w_{i*} = 1$ and the prior mean and variance simplify to $\phi_*/(1+\epsilon)$ and $\tau^2/(1+\epsilon)$, respectively, which corresponds to ϕ_i being independent of its neighbouring random effects. The next part of the model is the prior distribution $f(\mathcal{W})$, which is specified via a discrete uniform prior on its neighbourhood matrix representation $\tilde{\mathbf{W}}$:

$$\tilde{\mathbf{W}} \sim \text{Discrete Uniform}(\tilde{\mathbf{W}}^{(0)}, \tilde{\mathbf{W}}^{(1)}, \ldots, \tilde{\mathbf{W}}^{(N_{\mathcal{W}})}). \quad (15.8)$$

This discrete uniform prior is specified to simplify the sample space for $\mathcal{W}$ to being one-dimensional and having size $N_{\mathcal{W}}+1$, where $N_{\mathcal{W}} = |\mathcal{W}| = \mathbf{1}^T\mathbf{W}\mathbf{1}/2$ is the number of binary adjacency elements in $\mathbf{W}$. This simplification is undertaken because the number of elements to estimate $N_{\mathcal{W}}$ is much larger than the number of data points n, and existing research (Li et al., 2011) has shown that these elements are only weakly identifiable from the data. Therefore, the set of allowable adjacency structures $(\tilde{\mathbf{W}}^{(0)}, \tilde{\mathbf{W}}^{(1)}, \ldots, \tilde{\mathbf{W}}^{(N_{\mathcal{W}})})$ in (15.8) have a natural ordering, with $\tilde{\mathbf{W}}^{(j)}$ having j elements in $\mathcal{W}$ equal to 1 and $N_{\mathcal{W}} - j$ elements equal to zero. Thus, $\tilde{\mathbf{W}}^{(N_{\mathcal{W}})}$ corresponds to the intrinsic CAR model for strong spatial smoothing, while $\tilde{\mathbf{W}}^{(0)}$ has all elements of $\mathcal{W}$ equal to zero and corresponds to independence. The set of candidate values $(\tilde{\mathbf{W}}^{(1)}, \ldots, \tilde{\mathbf{W}}^{(N_{\mathcal{W}}-1)})$ are elicited from disease data prior to the study period using a Gaussian approximation, and further details are given by Lee et al. (2014). The model proposed by Lee et al. (2014) is given by (15.1), (15.7), and (15.8), and software to implement this model is provided in the supplementary material accompanying Lee et al. (2014).

15.5 Estimating Fine-Scale Pollution Concentrations and Their Association with Disease Risk

The disease and pollution data are spatially misaligned, because the geographical scales at which these data are measured are different. The disease data are single summaries for each areal unit $\mathcal{A}_i$, which are irregular in size and shape and are typically administrative geographies such as local authorities or census tracts. In contrast, the measured pollution data relate to an irregular set of point locations throughout the study region, while the modelled pollution concentrations are available on a regular grid covering the entire study region. This spatial misalignment has been termed the *change of support problem* by Gelfand et al. (2001), and two main solutions have been proposed for addressing it. The first is to aggregate the pollution concentrations to the same spatial scale as the disease data, while the second is to allow for the inherent spatial variation in the pollution concentrations within an areal unit. Both these approaches are discussed below.

15.5.1 Estimating a spatially representative pollution concentration

The most common approach to accounting for the spatial misalignment between the disease and pollution data is to create a spatially representative pollution concentration for each areal unit $\mathcal{A}_i$, which Gelfand et al. (2001) argue is given by

$$x_i = \frac{1}{|\mathcal{A}_i|}\int_{\mathbf{s}\in\mathcal{A}_i} x(\mathbf{s})d\mathbf{s}. \tag{15.9}$$

Here, $x(\mathbf{s})$ is the true unobserved concentration at location $\mathbf{s}$, and Equation 15.9 is the average concentration across areal unit $\mathcal{A}_i$. Alternatively, a population weighted average could be used, which is given by

$$x_i = \int_{\mathbf{s}\in\mathcal{A}_i} p(\mathbf{s})x(\mathbf{s})d\mathbf{s}, \tag{15.10}$$

where the population density at point $\mathbf{s}$, $p(\mathbf{s})$, is scaled so that $\int_{\mathcal{A}_i} p(\mathbf{s})d\mathbf{s} = 1$. This latter measure adjusts for varying population density within an areal unit, so that the pollution measure is representative of the average concentration to which the population might be exposed. However, the integrals in (15.9) and (15.10) are unknown, and their estimation can be based on either the measured or modelled pollution data. The simplest approach to estimating (15.9) or (15.10) is to average the modelled pollution concentrations within each areal unit, as they have complete spatial coverage of the study region. This approach was adopted by Lee et al. (2009) and Haining et al. (2010), who respectively estimate (15.9) and (15.10), but does not use the measured pollution data, which are more accurate than the modelled concentrations. We note that simple averaging of the measured pollution data is unlikely to be possible, as these data are spatially sparse and may not be available in many of the n areal units at which the disease data are recorded. Alternatively, Zhu et al. (2003), Fuentes et al. (2006), Choi et al. (2009), and Lawson et al. (2012) estimate these unknown integrals using Monte Carlo integration, based on a regular grid of prediction points $(\mathbf{s}^*_{i1}, \ldots, \mathbf{s}^*_{iN_i})$ within areal unit $\mathcal{A}_i$. Using this approach (15.9) can be estimated by

$$x_i = \frac{1}{N_i}\sum_{j=1}^{N_i} x(\mathbf{s}^*_{ij}), \tag{15.11}$$

where $x(\mathbf{s}^*_{ij})$ is a prediction of the pollution concentration at location $\mathbf{s}^*_{ij}$. A number of models have been proposed in the literature for modelling and hence predicting spatially referenced air pollution data $\mathbf{z} = (z(\boldsymbol{s}_1), \ldots, z(\boldsymbol{s}_J))$, where $(\boldsymbol{s}_1, \ldots, \boldsymbol{s}_J)$ are the J locations at which measured pollution data are available. These models have the general form

$$\begin{aligned} z(\mathbf{s}_j) &= x(\mathbf{s}_j) + \epsilon(\mathbf{s}_j), \quad \epsilon(\mathbf{s}_j) \sim \mathrm{N}(0, \sigma^2_\epsilon), \\ x(\mathbf{s}_j) &= \boldsymbol{v}(\mathbf{s}_j)^{\mathrm{T}}\boldsymbol{\alpha} + \eta(\mathbf{s}_j), \end{aligned} \tag{15.12}$$

where the measured concentration $z(\mathbf{s}_j)$ is modelled as being equal to the true concentration $x(\mathbf{s}_j)$ plus measurement error $\epsilon(\mathbf{s}_j)$. The true concentration is represented by a spatial process $\eta(\mathbf{s}_j)$ and covariates, $\boldsymbol{v}(\mathbf{s}_j)$, the latter of which include the modelled concentration in the grid square containing location $\mathbf{s}_j$, measures of meteorology, spatial trend terms, and any other relevant covariates. The spatial process at all the J spatial locations is denoted by $\boldsymbol{\eta} = (\eta(\mathbf{s}_1), \ldots, \eta(\mathbf{s}_J))$ and can be modelled as a zero-mean Gaussian process

$$\boldsymbol{\eta} \sim \mathrm{N}(\mathbf{0}, \sigma^2_\eta S_\eta(\rho)). \tag{15.13}$$

Here, $\mathbf{0}$ is a vector of zeros, and $S_\eta(\rho)$ is an exponential spatial autocorrelation matrix that has elements $S_\eta(\rho)_{jk} = \exp(-\rho||\mathbf{s}_j - \mathbf{s}_k||)$, where $||.||$ denotes Euclidean distance. Bayesian Kriging is used to sample from the posterior predictive distribution $f(x(\mathbf{s}^*_{ij})|\mathbf{z})$ for each prediction location $(x(\mathbf{s}^*_{i1}), \ldots, x(\mathbf{s}^*_{iN_i}))$, which thus allows (15.11) to be estimated. This process can be repeated for G MCMC samples, yielding a posterior predictive distribution for (15.11). Further discussion of geostatistical models can be found in Chapter 11.

Two-stage approaches are typically adopted to estimate the health impact of air pollution (see, e.g., Lawson et al., 2012), where in the first stage the pollution model given by (15.12) and (15.13) is fitted and used to estimate (15.11), before a disease model such as (15.1) and (15.4) is fitted separately in the second stage. Alternatively, both models could be combined into a joint Bayesian hierarchical model, but this would allow the disease data to inform on the pollution concentrations, whereas it is the relationship in the opposite direction that is of primary interest. In a two-stage approach, one is cutting the feedback between the pollution and disease models, which is advantageous not only for the reason above, but also because it is computationally simpler to implement.

The simplest two-stage approach thus estimates (15.11) by its posterior predictive mean or median, and assumes this value is fixed when fitting the disease model in stage 2. However, this approach ignores the fact that the pollution concentrations $(x(\mathbf{s}^*_{i1}), \ldots, x(\mathbf{s}^*_{iN_i}))$ and hence (15.11) are unknown, and their uncertainty could be propagated into the disease model. This is achieved by modelling $\mathbf{x} = (x_1, \ldots, x_n)$, with x_i defined as in (15.11), as unknown parameters in the disease model, rather than being fixed quantities. The latter could be achieved by the extended disease model

$$\begin{aligned} y_i &\sim \text{Poisson}(e_i\theta_i) \quad \text{for } i = 1, \ldots, n, \\ \ln(\theta_i) &= \beta_0 + x_i\beta_x + \boldsymbol{u}_i^{\mathrm{T}}\boldsymbol{\beta}_u + \phi_i, \\ \boldsymbol{\beta} = (\beta_0, \beta_x, \boldsymbol{\beta}_u) &\sim \mathrm{N}(\boldsymbol{\mu}_\beta, \Sigma_\beta), \\ x_i &\sim f(x_i|\mathbf{z}), \end{aligned} \tag{15.14}$$

where the random effects can be modelled using any of the specifications discussed in this chapter. The prior distribution for x_i is simply the posterior predictive distribution from the first-stage pollution model, which allows the uncertainty in (15.11) to be propagated into the disease model. However, if the posterior uncertainty in $f(x_i|\mathbf{z})$ is large, say of the same order of magnitude as the spatial variation in $(x_1, \ldots, x_n)$ across the n areal units, then the above model may not give good inference for β_x.

15.5.2 Allowing for spatial variation in pollution within each areal unit

The disease models (15.1) and (15.14) both assume there is a single spatially representative pollution concentration x_i for the ith areal unit, which can be estimated using a number of different approaches. However, pollution concentrations vary within each areal unit, and this within-area variability in concentrations has so far been ignored. We note that the variation discussed in the previous section relates to posterior uncertainty in a spatially representative measure of pollution such as (15.11), whereas the variation discussed here is spatial variation in the pollution concentrations within an areal unit. For simplicity of exposition in the following discussion, we assume that the pollution concentrations $(x(\mathbf{s}_{i1}^*), \ldots, x(\mathbf{s}_{iN_i}^*))$ at N_i locations within the ith areal unit are known, but note that their uncertainty could be accounted for using a similar approach to that described above. The statistical challenge is then how to relate the N_i pollution concentrations $(x(\mathbf{s}_{i1}^*), \ldots, x(\mathbf{s}_{iN_i}^*))$ to one disease count y_i that accounts for the within–areal unit variation in the concentrations.

Wakefield and Salway (2001) show that in the presence of within-area variation in pollution concentrations, the naive disease model (15.1) has a different algebraic form than what one would obtain by aggregating an individual-level risk model to the population scale. The bias in the health effect estimate resulting from this difference is known as *ecological bias*, and a full discussion is given by Wakefield and Salway (2001). This bias occurs because a nonlinear risk model changes its form under aggregation from the individual to the population level. Consider the ideal situation of having individual-level data on disease presence or absence and pollution exposure for all individuals in the study region. Let y_{ik}, for $k = 1, \ldots, n_i$, be the binary observation denoting whether individual k in areal unit $\mathcal{A}_i$ has the disease under study, and let x_{ik} denote that individual's pollution exposure. Then a simple individual-level risk model is given by

$$\begin{aligned} y_{ik} &\sim \text{Bernoulli}(p_{ik}) \quad \text{for } k = 1, \ldots n_i, \ i = 1, \ldots, n, \\ \ln(p_{ik}) &= \beta_0 + \boldsymbol{u}_i^{\mathrm{T}}\boldsymbol{\beta}_u + \phi_i + x_{ik}\beta_I, \end{aligned} \tag{15.15}$$

where a log rather than logit link function is used because the probability of disease in any single individual is small. Here, $(\boldsymbol{u}_i^{\mathrm{T}}\boldsymbol{\beta}_u, \phi_i)$, respectively, model the effects of known and unknown areal unit–level covariates on disease prevalence. However, for areal unit studies, the individual y_{ik}'s are unknown, and only the total number of disease cases from the population living in each areal unit $y_i = \sum_{k=1}^{n_i} y_{ik}$ is known. Computing the expectation of the aggregated disease counts y_i under the individual-level model gives

$$\begin{aligned}\mathbb{E}[y_i] &= \mathbb{E}\left[\sum_{k=1}^{n_i} y_{ik}\right] \\ &= \exp(\beta_0 + \boldsymbol{u}_i^{\mathrm{T}}\boldsymbol{\beta}_u + \phi_i)\frac{1}{n_i}\sum_{k=1}^{n_i}\exp(x_{ik}\beta_I),\end{aligned} \tag{15.16}$$

where a factor $\ln(n_i)$ has been absorbed into the intercept term β_0 to counterbalance the division by n_i. Comparing (15.1) and (15.16) shows that ecological bias occurs because $\frac{1}{n_i}\sum_{k=1}^{n_i}\exp(\beta_I x_{ik}) \neq \exp(x_i\beta_x)$, where in this context $x_i = \frac{1}{n_i}\sum_{k=1}^{n_i} x_{ik}$. The implication is that the commonly estimated β_x is not equal to the desired β_I in general; thus, the pollution–health effect estimated from (15.1) may be subject to bias. However, the individual pollution exposures $(x_{i1}, \ldots, x_{in_i})$ are unknown, and instead, only a sample of point predictions $(x(\mathbf{s}_{i1}^*), \ldots, x(\mathbf{s}_{iN_i}^*))$ from a pollution model such as (15.12) is available. Two main approaches have been proposed for approximating (15.16), the first of which uses the available point predictions $(x(\mathbf{s}_{i1}^*), \ldots, x(\mathbf{s}_{iN_i}^*))$ for the ith areal unit directly to approximate (15.16). This approach was proposed by Wakefield and Shaddick (2006), who propose changing the risk model for θ_i to

$$\theta_i = \exp(\beta_0 + \boldsymbol{u}_i^{\mathrm{T}}\boldsymbol{\beta}_u + \phi_i)\frac{1}{N_i}\sum_{k=1}^{N_i}\exp(x(\mathbf{s}_{ik}^*)\beta_I), \tag{15.17}$$

where the available point predictions $(x(\mathbf{s}_{i1}^*), \ldots, x(\mathbf{s}_{iN_i}^*))$ have replaced the unknown exposures $(x_{i1}, \ldots, x_{in_i})$. The second approach was first proposed by Richardson et al. (1987), who suggested representing the within-area distribution of exposures $(x_{i1}, \ldots, x_{in_i})$ by a parametric distribution. Let X_i denote the random variable characterizing the within-area exposure distribution for areal unit $\mathcal{A}_i$. Then equating the sample average in (15.16) with expectation over X_i shows that the desired quantity is $\mathbb{E}[\exp(X_i\beta_I)]$, the moment-generating function of X_i. Assuming that the within-area exposure distribution $X_i \sim \mathrm{N}(\mu_i, \sigma_i^2)$ leads to the risk model

$$\theta_i = \exp(\beta_0 + \boldsymbol{u}_i^{\mathrm{T}}\boldsymbol{\beta}_u + \phi_i + \mu_i\beta_I + \sigma_i^2\beta_I^2/2), \tag{15.18}$$

where (μ_i, σ_i^2) can be estimated by their sample equivalents from the N_i samples $(x(\mathbf{s}_{i1}^*), \ldots, x(\mathbf{s}_{iN_i}^*))$. However, if the within-area exposure distribution is skewed, then a normal approximation may be inappropriate, and a lognormal distribution could be used instead. The moment-generating function of a lognormal distribution does not exist, so Salway and Wakefield (2008) propose approximating it with a three-term Taylor expansion leading to the risk model

$$\theta_i = \exp(\beta_0 + \boldsymbol{u}_i^{\mathrm{T}}\boldsymbol{\beta}_u + \phi_i + \mu_i\beta_I + \sigma_i^2\beta_I^2/2 + \eta_i^3\beta_I^3/6), \tag{15.19}$$

where η_i^3 is the third central moment and is given by $\eta_i^3 = \sigma_i^2/(\mu_i[(\sigma_i^2/\mu_i^2) + 3])$. However, as the pollution–health effect β_I is typically small, (β_I^2, β_I^3) will both be close to zero, so the bias that will result in using (15.1) in place of (15.18) or (15.19) should be negligible.

15.6 Illustrative Example

We illustrate the models discussed in this chapter by investigating the impact of long-term exposure to $PM_{2.5}$ on respiratory hospitalization risk in England in 2010. Here, England is partitioned into $n = 323$ local and unitary authorities (LUAs), which are typically either individual cities or larger rural areas. The disease data are counts of the numbers of hospital admissions with a primary diagnosis of respiratory disease in 2010, and the spatial pattern in the standardized morbidity ratio ($SMR_i = y_i/e_i$) is displayed in the left panel of Figure 15.1. The figure shows that the LUAs with the highest SMRs are cities in the north and central parts of England, such as Liverpool, Birmingham, and Manchester. In contrast, the lowest risks occur in the larger rural LUAs, such as Rutland, West Somerset, and Richmondshire. The SMR map shows evidence of localized spatial smoothness, with some pairs of neighbouring LUAs exhibiting similar risks, while other neighbouring pairs have vastly different values.

The pollution metric considered in this study is the annual average $PM_{2.5}$ concentrations in 2009, where the measured data come from the AURN network of sites, while the modelled concentrations come from the AQUM model. Their locations are displayed in the right panel of Figure 15.1, where the black triangles represent the measured data locations, while the grey dots are the corners of the 12-kilometre-square grid cells for which the modelled data are available. The figure shows that the measured data are clustered mainly in the cities, while the rural areas, such as the southwest of England, are very sparsely covered. In contrast, the modelled concentrations are calculated on a regular grid of size 12 kilometres, and thus provide complete spatial coverage of the study region. Finally, the confounding effects of socioeconomic deprivation on disease risk are accounted for by including the English index of multiple deprivation as a covariate in the disease model.

A two-stage approach was taken to modelling these data, where in the first-stage model (15.12) was fitted to annual mean $PM_{2.5}$ concentrations measured at the $J = 166$ monitoring sites shown in Figure 15.1. This model was used to predict $PM_{2.5}$ concentrations for each of the 12-kilometre grid cells displayed in the right panel of Figure 15.1. Five thousand predictions were made at each grid cell, and following the majority of the existing literature, these predictions were aggregated to the LUA scale using (15.11). The posterior predictive means and standard deviations for each LUA are displayed in Figure 15.2. The figure shows

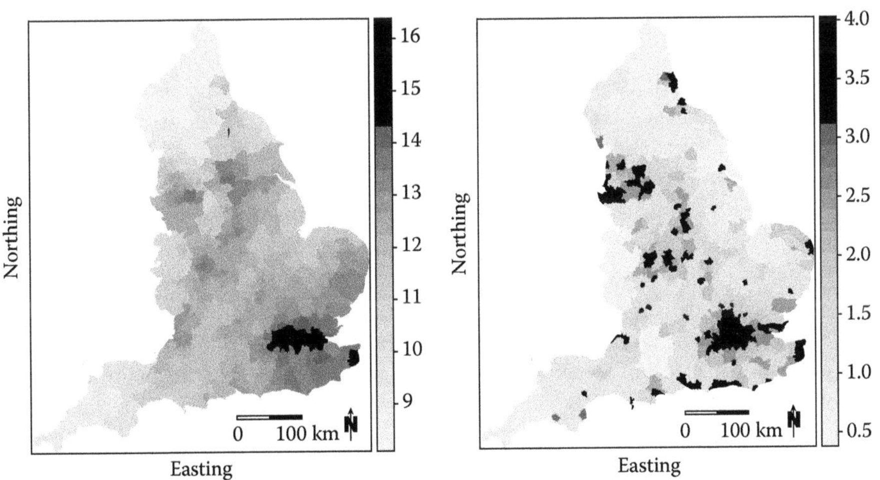

FIGURE 15.2
The maps display the posterior mean (left panel) and standard deviation (right panel) of the annual average $PM_{2.5}$ concentrations in 2009 for each LUA in England.

that the highest concentrations are observed in the city of London in the southeast, as well as the large cities of Birmingham and Manchester in the centre of the country, while the rural areas have the lowest pollution concentrations, particularly in the far southwest and north of England, where population density is relatively low. The posterior predictive standard deviations in the right panel of Figure 15.2 show there is substantial posterior uncertainty in the average concentrations within an LUA relative to the mean levels, with an average standard deviation of 2.33.

We fit two different disease models to our data, where in both cases, the posterior mean $PM_{2.5}$ concentrations estimated in stage 1 are assumed to be fixed. Model A is given by (15.1) in conjunction with the CAR prior (15.4), and is routinely used in studies of this type. Model B extends Model A by using a more flexible spatial autocorrelation model for the random effects, namely, the LCAR prior given by (15.7) and (15.8). The estimated relative risks and 95% credible intervals corresponding to a 1 μgm^{-3} increase in $PM_{2.5}$ concentrations are as follows Model A, 1.032 (1.001, 1.079), and Model B, 1.040 (1.011, 1.067). Both models suggest that areal units with higher concentrations of $PM_{2.5}$ exhibit higher risks of respiratory disease, with increases ranging between 3.2% and 4%. Replacing a globally smooth set of random effects (Model A) with a locally smooth set (Model B) has inflated the risk by 0.8%, while the width of the 95% credible interval from Model A is wider than the corresponding interval from Model B. Both these differences may be due to the collinearity between the globally smooth random effects in Model A and the globally smooth $PM_{2.5}$ covariate.

15.7 Discussion

This chapter has critiqued the statistical challenges involved in estimating the long-term health impact of air pollution using an ecological areal unit study design, and has illustrated the two-stage modelling approach commonly used in the existing literature. Although such two-stage approaches are commonplace (see Chang et al., 2011; Lawson et al., 2012), Szpiro and Paciorek (2013) highlight that they have the potential to induce their own biases into the estimated health effects. Therefore, future research in this area will need to consider a single integrated model that simultaneously estimates the spatiotemporal pattern in air pollution concentrations and its resulting effects on disease risk. An important issue in an integrated Bayesian model is that of feedback; namely, should information in the disease counts be allowed to affect the estimated pollution concentrations; when it is the relationship in the opposite direction that is of primary interest? In time series studies, this feedback has been prevented (see Lee and Shaddick, 2010), but an interesting area of work would be to examine the impact of this in an areal unit study. The other future research direction is to extend the methodology discussed here into the spatiotemporal domain, as a number of studies are now utilizing spatiotemporal data (Janes et al., 2007; Greven et al., 2011; Lawson et al., 2012). In addition to the challenges outlined here for spatial studies, spatiotemporal studies are likely to throw up a number of additional modelling challenges, for which methodological development is required. The most obvious of these is developing a model for spatiotemporal autocorrelation, which is flexible enough to allow for nonstationarity, nonseparability, and varying levels of autocorrelation in both space and time. The use of a spatiotemporal study also naturally leads to questions about lag times of air pollution effects, a subject that has only been investigated in a time series context to date.

Acknowledgements

This work was funded by the Engineering and Physical Sciences Research Council (EPSRC) grant numbers EP/J017442/1 and EP/J017485/1, and the data were provided by the UK Met Office; the Department for Environment, Food and Rural Affairs (DEFRA); and the National Health Service Health and Social Care Information Centre.

References

Besag, J., J. York, and A. Mollie. (1991). Bayesian image restoration with two applications in spatial statistics. *Annals of the Institute of Statistical Mathematics* 43, 1–59.

Brezger, A., L. Fahrmeir, and A. Hennerfeind. (2007). Adapative Gaussain Markov random fields with applications in human brain mapping. *Journal of the Royal Statistical Society: Series C* 56, 327–345.

Cesaroni, G., F. Forastiere, M. Stafoggia, Z. Andersen, C. Badaloni, R. Beelen, B. Caracciolo, et al. (2014). Long term exposure to ambient air pollution and incidence of acute coronary events: prospective cohort study and meta-analysis in 11 European cohorts from the ESCAPE Project. *British Medical Journal* 348, f7412.

Chang, H., R. Peng, and F. Dominici. (2011). Estimating the acute health effects of coarse particulate matter accounting for exposure measurement error. *Biostatistics* 12, 637–652.

Choi, J., M. Fuentes, and B. Reich. (2009). Spatial-temporal association between fine particulate matter and daily mortality. *Computational Statistics and Data Analysis* 53, 2989–3000.

Clayton, D., L. Bernardinelli, and C. Montomoli. (1993). Spatial correlation in ecological analysis. *International Journal of Epidemiology* 22, 1193–1202.

Dockery, D., C. Pope, X. Xu, J. Spengler, J. Ware, M. Fay, B. Ferris, and F. Speizer. (1993). An association between air pollution and mortality in six U.S. cities. *New England Journal of Medicine* 329, 1753–1759.

Dominici, F., M. Daniels, S. Zeger, and J. Samet. (2002). Air pollution and mortality: estimating regional and national dose-response relationships. *Journal of the American Statistical Association* 97, 100–111.

Elliott, P., G. Shaddick, J. Wakefield, C. Hoogh, and D. Briggs. (2007). Long-term associations of outdoor air pollution with mortality in Great Britain. *Thorax* 62, 1088–1094.

Fuentes, M., H. Song, S. Ghosh, D. Holland, and J. Davis. (2006). Spatial association between speciated fine particles and mortality. *Biometrics* 62, 855–863.

Gelfand, A., L. Zhu, and B. Carlin. (2001). On the change of support problem for spatio-temporal data. *Biostatistics* 2, 31–45.

Gelman, A. (2006). Prior distributions for variance parameters in hierarchical models. *Bayesian Analysis* 1, 515–534.

Greven, S., F. Dominici, and S. Zeger. (2011). An approach to the estimation of chronic air pollution effects using spatio-temporal information. *Journal of the American Statistical Association* 106, 396–406.

Haining, R., G. Li, R. Maheswaran, M. Blangiardo, J. Law, N. Best, and S. Richardson. (2010). Inference from ecological models: estimating the relative risk of stroke from air pollution exposure using small area data. *Spatial and Spatio-Temporal Epidemiology* 1, 123–131.

Hughes, J. and M. Haran. (2013). Dimension reduction and alleviation of confounding for spatial generalized linear mixed models. *Journal of the Royal Statistical Society: Series B* 75, 139–160.

Janes, H., F. Dominici, and S. Zeger. (2007). Trends in air pollution and mortality: an approach to the assessement of unmeasured confounding. *Epidemiology* 18, 416–423.

Jerrett, M., M. Buzzelli, R. Burnett, and P. DeLuca. (2005). Particulate air pollution, social confounders, and mortality in small areas of an industrial city. *Social Science and Medicine* 60, 2845–2863.

Kaufman, J., S. Adar, R. Allen, G. Barr, M. Budoff, G. Burke, A. Casillas, et al. (2012). Prospective study of particulate air pollution exposures, subclinical atherosclerosis, and clinical cardiovascular disease. *American Journal of Epidemiology* 176, 825–837.

Kleinschmidt, I., M. Hills, and P. Elliott. (1995). Smoking behaviour can be predicted by neighbourhood deprivation measures. *Journal of Epidemiology and Community Health* 49, S72–S77. DOI: 10.1136/jech.49.Suppl2.S72.

Krall, J., G. Anderson, F. Dominici, M. Bell, and R. Peng. (2013). Short-term exposure to particulate matter constituents and mortality in a national study of U.S. urban communities. *Environmental Health Perspectives* 121, 1148–1153.

Lawson, A., J. Choi, B. Cai, M. Hossain, R. Kirby, and J. Liu. (2012). Bayesian 2-stage space-time mixture modeling with spatial misalignment of the exposure in small area health data. *Journal of Agricultural, Biological and Environmental Statistics* 17, 417–441.

Lee, D. (2012). Using spline models to estimate the varying health risks from air pollution across Scotland. *Statistics in Medicine* 31, 3366–3378.

Lee, D. (2013). Carbayes: an R package for Bayesian spatial modelling with conditional autoregressive priors. *Journal of Statistical Software* 55, 13.

Lee, D., C. Ferguson, and R. Mitchell. (2009). Air pollution and health in Scotland: a multicity study. *Biostatistics* 10, 409–423.

Lee, D. and R. Mitchell. (2014). Controlling for localised spatio-temporal autocorrelation in long-term air pollution and health studies. *Statistical Methods in Medical Research* 23, 488–506.

Lee, D., A. Rushworth, and S. Sahu. (2014). A Bayesian localized conditional autoregressive model for estimating the health effects of air pollution. *Biometrics* 70, 419–429.

Lee, D. and G. Shaddick. (2010). Spatial modeling of air pollution in studies of its short term health effects. *Biometrics* 66, 1238–1246.

Leroux, B., X. Lei, and N. Breslow. (1999). Estimation of disease rates in small areas: a new mixed model for spatial dependence. In *Statistical Models in Epidemiology, the Environment and Clinical Trials*, Halloran, M., and Berry, D. (eds.), pp. 135–178. Springer-Verlag, New York.

Li, P., S. Banerjee, and A. McBean. (2011). Mining boundary effects in areally referenced spatial data using the Bayesian information criterion. *Geoinformatica* 15, 435–454.

Lu, H., C. Reilly, S. Banerjee, and B. Carlin. (2007). Bayesian areal wombling via adjacency modelling. *Environmental and Ecological Statistics* 14, 433–452.

Ma, H., B. Carlin, and S. Banerjee. (2010). Hierarchical and joint site-edge methods for Medicare hospice service region boundary analysis. *Biometrics* 66, 355–364.

Maheswaran, R., R. Haining, P. Brindley, J. Law, T. Pearson, P. Fryers, S. Wise, and M. Campbell. (2005). Outdoor air pollution and stroke in Sheffield, United Kingdom. *Stroke* 36, 239–243.

Reich, B., J. Hodges, and V. Zadnik. (2006). Effects of residual smoothing on the posterior of the fixed effects in disease-mapping models. *Biometrics* 62, 1197–1206.

Richardson, E., and R. Mitchell. (2010). Gender differences in relationships between urban green space and health in the United Kingdom. *Social Science and Medicine* 71, 568–575.

Richardson, S., I. Stucker, and D. Hemon. (1987). Comparison of relative risks obtained in ecological and individual studies: some methodological considerations. *International Journal of Epidemiology* 16, 111–120.

Robert, C. and G. Casella. (2010). *Introducing Monte Carlo Methods with R* (1st ed.). Springer, New York.

Rue, H., S. Martino, and N. Chopin. (2009). Approximate Bayesian inference for latent Gaussian models using integrated nested Laplace approximations [with discussion]. *Journal of the Royal Statistical Society: Series B* 71, 319–392.

Salway, R. and J. Wakefield. (2008). A hybrid model for reducing ecological bias. *Biostatistics* 9, 1–17.

Samoli, E., G. Touloumi, J. Schwartz, R. Anderson, C. Schindler, B. Forsberg, M. Vigotti, J. Vonk, M. Kosnik, J. Skorkovsky, and K. Katsouyanni. (2007). Short-term effects of carbon monoxide on mortality: an analysis within the APHEA project. *Environmental Health Perspectives* 115, 1578–1583.

Savage, N. H., P. Agnew, L. S. Davis, C. Ordonez, R. Thorpe, C. E. Johnson, F. M. O'Connor, and M. Dalvi. (2013). Air quality modelling using the met office unified model (AQUM OS24-26): model description and initial evaluation. *Geoscientific Model Development* 6, 353–372.

Schwartz, J. and A. Marcus. (1990). Mortality and air pollution in London: a time series analysis. *American Journal of Epidemiology* 131, 185–194.

Szpiro, A. and C. Paciorek. (2013). Measurement error in two-stage analyses, with application to air pollution epidemiology. *Environmetrics* 24, 501–517.

Wakefield, J. and R. Salway. (2001). A statistical framework for ecological and aggregate studies. *Journal of the Royal Statistical Society: Series A* 164, 119–137.

Wakefield, J. and G. Shaddick. (2006). Health-exposure modeling and the ecological fallacy. *Biostatistics* 7, 438–455.

Wymer, L. and A. Dufour. (2002). A model for estimating the incidence of swimming-related gastrointestinal illness as a function of water quality indicators. *Environmetrics* 13, 669–678.

Young, L., C. Gotway, J. Yang, G. Kearney, and C. DuClos. (2009). Linking health and environmental data in geographical analysis: it's so much more than centroids. *Spatial and Spatio-Temporal Epidemiology* 1, 73–84.

Zhu, L., B. Carlin, and A. Gelfand. (2003). Hierarchical regression with misaligned spatial data: relating ambient ozone and pediatric asthma ER visits in Atlanta. *Environmetrics* 14, 537–557.

16

Data Assimilation for Environmental Pollution Fields

Howard H. Chang
Department of Biostatistics and Bioinformatics
Emory University
Atlanta, Georgia

CONTENTS

16.1 Introduction

Understanding the health effects of environmental pollutants, especially those due to anthropogenic sources, has become increasingly important in public health research. Toxic contaminants in air, water, and land can affect large populations, and exposures are often involuntary. Recently, the field of environmental epidemiology has benefited greatly from our ability to leverage health databases developed for administrative or billing purposes. Examples include vital certificates, medical records, educational testing data, and registries for cancer, birth defects, and neurological diseases. These databases allow researchers to conduct large-scale population-based studies, often covering large geographical regions. Advances in geographic information systems further enhance our ability to obtain detailed spatial and temporal information, enabling environmental exposures to be ascertained. Results from these population studies have played an important role in setting regulatory standards and protecting public health.

Exposure assessment is a crucial component in any environmental health study. Pollutant fields typically vary both spatially and temporally. One major limitation of retrospective population studies is that exposure levels are not collected with the health data, and they must be estimated from additional data sources. Past studies routinely utilize measurements from existing networks of monitors set up by government agencies. However, reliance on these networks leads to several analytic challenges. First, because of the high maintenance cost, monitoring networks are typically spatially sparse and lack continuous temporal data. This is a particular concern for pollution fields with high spatial heterogeneity that cannot be captured by the monitors. Second, because the networks are designed for regulatory purposes, monitors are often preferentially located in urban areas where the at-risk population is large and the pollution level is high (Shaddick and Zidek, 2014). Hence, when linking health data to monitoring measurements, the complex spatiotemporal missing data pattern will not only restrict the study population, but also can result in exposure measurement error that impacts the accuracy of health studies.

In order to improve the availability and resolution of environmental pollution data, one approach is to supplement monitoring measurements with additional data sources that reflect pollution levels. Two specific data sources that have received growing attention are satellite images and simulations from computer models. In this chapter, these data sources are referred to as proxy data. Satellite images have been used to measure environmental processes such as temperature, wild fire (Wan et al., 2010), and air pollution (Hoff and Christopher, 2009). Remotely sensed data are publicly available, and they often provide global coverage. Computer models aim to simulate a pollutant's creation and dispersion using information on pollutant sources and state-of-the-art knowledge on the chemical and physical processes. Despite their high computational demand, computer models can provide three-dimensional deterministic outputs that have complete spatiotemporal coverage for the study domain. Computer models have been utilized extensively for weather forecast (Lynch, 2008) and climate research (Shine and Henderson-Sellers, 1983; Giorgi and Mearns, 1991).

While the use of satellite images and computer simulations overcomes spatiotemporal missing data, these proxy data cannot directly replace monitoring measurements in health analyses. For example, the associations between pollutant levels and remotely sensed parameters often vary across meteorological conditions, land cover, and pollution composition. Satellite data are also subject to retrieval errors. For numerical model simulations, errors can arise from incorrect input data on sources, incorrect representation of the underlying complex processes with partial differential equations, and discretization of the continuous environmental field in space and time.

In environmental epidemiology, statistical methods for data assimilation have two simultaneous objectives: (1) combining monitoring measurements and proxy data to improve exposure assessment, while (2) accounting for the errors in proxy data. Data assimilation has also been referred to as data fusion and data blending. Recent approaches can be categorized into two groups: *melding* and *calibration*. In melding, both measurements and proxy data are viewed as error-prone realizations of a latent true pollution field, whereas in calibration, the proxy serves as a predictor for the measurements. Both melding and calibration encounter several common modeling challenges. First, satellite images and numerical model outputs represent areal spatial data over contiguous grid cells. When linked to point-referenced monitoring locations, a spatial change of support is encountered (Gotway and Young, 2002). Second, the discrepancy between monitoring measurements and proxy data can exhibit complex spatial and temporal structures.

The modeling approaches presented in this chapter are largely drawn from air quality research because of the rich literature on exposure modeling and health effect estimation. See Chapter 16 for a summary of approaches for estimating air pollution health effects. However, similar approaches have also been developed for applications in climate science (Berrocal et al., 2012a; Zhou et al., 2012), oceanography (Foley and Fuentes, 2008), and hydrology (Teutschbein and Seibert, 2012). In air quality research, recent work has focused predominantly on two pollutants: fine particulate pollution $PM_{2.5}$ (particulate matter less than 2.5 μm in aerodynamic diameter) and ground-level ozone. Both $PM_{2.5}$ and ozone have been linked to health outcomes such as premature mortality, asthma exacerbation, cardio-respiratory morbidity, and adverse birth outcomes. As a remotely sensed proxy, satellite-derived aerosol optical depths (AODs) have been used to monitor $PM_{2.5}$ concentrations. AOD measures light extinction due to airborne particles in the atmospheric column, and previous studies have found positive associations between $PM_{2.5}$ level and AOD at different spatial and temporal scales (Liu et al., 2009; Kloog et al., 2013). Similarly, several global and regional numerical models have been developed to simulate $PM_{2.5}$ and ozone concentrations. Examples include the Community Multi-Scale Air Quality Model (CMAQ) (Byun and Schere, 2006) and GEOS-Chem (Bey et al., 2001).

This chapter aims to provide a review of data assimilation methods for environmental pollution fields. A hierarchical Bayesian framework is often employed to flexibly describe the various data components. We will first introduce the general framework of Bayesian melding, followed by statistical calibration. To simplify notation, the spatial version of the approach will be presented with noted spatiotemporal extensions when they are available. Several recent advances, such as multipollutant models, multiscale fusion, and quantile calibration, will be examined. Finally, we will discuss methods for using fusion products in health studies with an emphasis on uncertainty quantification.

16.2 Bayesian Melding

Let $Y(\mathbf{s})$ denote the pollutant concentration measurement from an air quality monitor at point location $\mathbf{s}$ with coordinates (s_1, s_2). The gridded proxy data are denoted by $X(B)$, where B indexes the contiguous grid cells. The spatial resolution of the proxy data varies based on satellite retrieval algorithms and whether the numerical simulation is performed on a global or a regional scale. In air quality applications, the spatial resolution of satellite-derived AOD ranges from 10 km to 1 km; the spatial resolution ranges from 100 km to 4 km for numerical model simulations.

In Bayesian melding, we assume observations $Y(\mathbf{s})$ and $X(B)$ arise from a common latent process $Z(\mathbf{s})$, representing the true pollutant field. The hierarchical model is formulated as follows:

$$Y(\mathbf{s}) = Z(\mathbf{s}) + e(\mathbf{s}) \tag{16.1}$$

$$Z(\mathbf{s}) = \mu(\mathbf{s}) + \epsilon(\mathbf{s}) \tag{16.2}$$

$$X(B) = \frac{1}{|B|}\int_B \tilde{X}(\mathbf{s})\, ds \tag{16.3}$$

$$\tilde{X}(\mathbf{s}) = a(\mathbf{s}) + b(\mathbf{s})Z(\mathbf{s}) + \delta(\mathbf{s}). \tag{16.4}$$

Equation 16.1 treats the observed monitoring data $Y(\mathbf{s})$ as an error-prone realization of the latent process with measurement error $e(\mathbf{s})$. The latent process $Z(\mathbf{s})$ has a spatial trend $\mu(s)$ and a residual component $\epsilon(\mathbf{s})$. Equation 16.3 introduces a conceptual point-referenced proxy data $\tilde{X}(\mathbf{s})$, and its spatial average over grid cell B results in the observed grid-level proxy value. Finally, the point-level proxy $\tilde{X}(\mathbf{s})$ is linked to the latent true pollutant field $Z(\mathbf{s})$ via a linear regression model. Coefficients $a(\mathbf{s})$ and $b(\mathbf{s})$ are often interpreted as the additive and multiplicative calibration parameters for the proxy, and component $\delta(\mathbf{s})$ represents random proxy error.

The above framework was first described by Fuentes and Raftery (2005) for assessing spatial bias in CMAQ. Subsequent applications in predicting pollution fields have found that Bayesian melding consistently outperforms kriging (Berrocal et al., 2010a; Liu et al., 2011), but not for long-term $PM_{2.5}$ levels using AOD (Paciorek and Liu, 2009). Several features are worth noting. The main advantage of melding is the use of a latent continuous field that allows the spatially misaligned measurements and proxy data to jointly provide information on $Z(\mathbf{s})$, which is the quantity of interest. Conceptually, the latent variable approach offers straightforward extensions to multiple proxies (Crooks and Isakov, 2013) and multiple pollutants (Choi et al., 2009; Sahu et al., 2010). However, the model is highly parameterized and identifiability is a frequent concern. Specifically, we first need to decompose the residual variation in $Y(\mathbf{s})$ into two error components $e(\mathbf{s})$ and $\epsilon(\mathbf{s})$. The structures of the spatial calibration parameters $a(\mathbf{s})$ and $b(\mathbf{s})$ also need to be selected with care, as they determine how much variation in the proxy can be attributed to the true pollutant field versus output bias. Specifically, $a(\mathbf{s})$ and $b(\mathbf{s})$ are usually parametrized as fixed effects, instead of spatial random fields, to avoid identifiability problems with estimating $Z(\mathbf{s})$.

Using the CMAQ proxy to estimate weekly SO_2 concentration in the eastern United States, Fuentes and Raftery (2005) have the following parameterizations for the three independent variance components: $e(\mathbf{s}) \overset{iid}{\sim} N(0, \sigma_e^2)$, $\epsilon(\mathbf{s})$ is a Gaussian process with non-stationary covariance function, and $\delta(\mathbf{s}) \overset{iid}{\sim} N(0, \sigma_\delta^2)$. For the structural components, $\mu(\mathbf{s})$ and $a(\mathbf{s})$ are polynomial functions of $\mathbf{s}$, and $b(\mathbf{s})$ is assumed to be an unknown constant. For modeling ozone concentration and numerical model outputs, a similar model is employed by Berrocal et al. (2010a) and Liu et al. (2011), with the exception that the covariance function of $\epsilon(\mathbf{s})$ is assumed to be exponential. One important observation from Berrocal et al. (2010a) is that the predictive surface obtained from melding closely follows the spatial gradients of the proxy, with values closer to CMAQ, especially in unmonitored areas. In melding, the proxy can dominate for two reasons. First, there is considerably more proxy data than monitoring measurements. Second, by specifying $a(\mathbf{s})$ as a smooth spatial trend, fine-scale spatial variation in the proxy is assumed to reflect the true pollutant field. The issue of disentangling spatial scales between $a(\mathbf{s})$ and $Z(\mathbf{s})$ is further investigated by Paciorek (2012). Through a simulation study and an application of predicting $PM_{2.5}$ levels using CMAQ or AOD, Paciorek finds that modeling $a(\mathbf{s})$ flexibly significantly reduces the usefulness of the proxy.

Bayesian melding often involves considerable computational effort because of the change-of-support integral in Equation 16.3 and a large number of spatial points need to be evaluated for $Z(\mathbf{s})$. Typically, the integral is approximated using Monte Carlo integration. For example, Berrocal et al. (2010a) use a systematic sample of four points for each CMAQ grid cell (12 km resolution). Liu et al. (2011) consider randomly selecting a fixed number of points to avoid an ill-conditioned spatial covariance matrix. Several approaches to decrease computational burden have been proposed. First, Sahu et al. (2010) introduce an areal true pollutant process $\tilde{Z}(B)$ that has the same grid as the proxy. This latent discrete spatial variation can be efficiently estimated using a conditionally autoregressive (CAR) model (Besag, 1974). A measurement error model, Equation 16.5, is then used to resolve the mismatch between point-referenced and areal true pollutant processes:

$$\begin{aligned} Y(\mathbf{s}) &= Z(\mathbf{s}) + e(\mathbf{s}) \\ Z(\mathbf{s}) &= \tilde{Z}(B) + \nu_v(\mathbf{s}) \\ X(B) &= \gamma_0 + \gamma_1 \tilde{Z}(B) + \psi(B), \end{aligned} \tag{16.5}$$

where $e(\mathbf{s})$, $\nu_v(\mathbf{s})$, and $\psi(B)$ are normal independent errors. An additional simplification is taken by McMillan et al. (2010) to model $PM_{2.5}$ levels and CMAQ outputs by eliminating the latent variable $Z(\mathbf{s})$. Under this model, monitoring measurements are linked to the areal latent process directly as in Equation 16.6:

$$\begin{aligned} Y(\mathbf{s}) &= \tilde{Z}(B) + e(\mathbf{s}) \\ X(B) &= \gamma_0 + \gamma_1 \tilde{Z}(B) + \psi(B). \end{aligned} \tag{16.6}$$

Here, $e(\mathbf{s})$ incorporates both measurement error in monitoring data and error due to spatial misalignment. One important consequence of Equation 16.6 is that the model can only provide gridded pollution predictions because $\tilde{Z}(B)$ is modeled as a discrete spatial process. This is often referred to as *upscaling*, as the original point-referenced monitoring data have been coarsened to areal level.

Because of the gain in computational speed, both Sahu et al. (2010) and McMillan et al. (2010) are able to perform Bayesian melding in a spatiotemporal setting with an additional dynamic time series model for the latent process. Their models provide daily pollution predictions that are useful for health studies that require short-term exposure assessment. Choi et al. (2009) also extend the melding framework of Fuentes and Raftery (2005) to a temporal setting by modeling the latent pollution field on day t as $Z(\mathbf{s},t) = \mu(\mathbf{s},t) + \epsilon(\mathbf{s},t)$. The trend $\mu(\mathbf{s},t)$ includes time-varying meteorological variables, and $\epsilon(\mathbf{s},t)$ is modeled as an autoregressive Gaussian process.

16.3 Statistical Calibration

Motivated by the increasing amount of proxy data being generated and the limitations of Bayesian melding, Berrocal et al. (2010a) propose a statistical calibration approach for data assimilation. Under a calibration framework, the proxy is viewed as a predictor for measurements. Let $X(B_{\mathbf{s}})$ denote the proxy grid cell linked to a monitor at point location $\mathbf{s}$. The regression model is given by

$$Y(\mathbf{s}) = \alpha_0(\mathbf{s}) + \alpha_1(\mathbf{s})X(B_{\mathbf{s}}) + e(\mathbf{s}), \quad e(\mathbf{s}) \sim N(0, \sigma_e^2). \tag{16.7}$$

Here, $\alpha_0(\mathbf{s})$ and $\alpha_1(\mathbf{s})$ can be interpreted as the spatially varying additive and multiplicative calibration parameters of the error-prone proxy data. $\alpha_0(\mathbf{s})$ and $\alpha_1(\mathbf{s})$ are modeled as a bivariate continuous spatial process via a linear model of coregionalization (LMC) (Gelfand et al., 2004). Briefly, two independent mean-zero, unit-variance Gaussian processes $W_1(\mathbf{s})$ and $W_2(\mathbf{s})$ are introduced. Correlation between $\alpha_0(\mathbf{s})$ and $\alpha_1(\mathbf{s})$ is induced by letting $\alpha_0(\mathbf{s}) = c_1 W_1(\mathbf{s})$ and $\alpha_1(\mathbf{s}) = c_2 W_1(\mathbf{s}) + c_3 W_1(\mathbf{s})$, where c_1, c_2, c_3 are constants that determine the marginal variance for each random effect and their correlation.

The idea of using proxy data as a predictor arises from the extensive literature on land use regression modeling (Ryan and LeMasters, 2007; Hoek et al., 2008). Land use regression models utilize a large number of predictors, such as population density, distance to roadway, and traffic density, that can be derived from geographical information systems. One challenge with land use regression models is that a dense monitoring network is required to capture fine-scale spatial variation associated with the land use variables. Often, additional measurement campaigns are conducted to enrich the observation dataset (Clougherty et al., 2013). Consequently, predictions from land use regression models have been predominantly used to support studies examining the health effects of long-term pollution exposures. For statistical calibration using computer model simulations, since the proxy directly reflects pollution level, additional predictors are often not considered.

There are several advantages that calibration offers compared to Bayesian melding. First, note that the only proxy data used for model fitting are those linked to a monitor. Proxy data not linked to a monitor are only used for predictions. This reduces the computational effort considerably, as the number of monitoring locations is usually much smaller than the number of proxy grid cells. Second, by treating $\alpha_0(\mathbf{s})$ and $\alpha_0(\mathbf{s})$ as continuous processes, they can be interpolated smoothly in space. This allows point-level predictions even though the proxy predictor represents an areal average. This feature is often referred to as *downscaling*, in contrast to the upscaler by McMillan et al. (2010) in Equation 16.6. Statistical downscaling has been used to combine CMAQ data (Berrocal et al., 2010a), as well as AOD for predicting $PM_{2.5}$ fields (Chang et al., 2013). When compared to Bayesian melding, Berrocal et al. (2010a) found that downscaling CMAQ for daily ozone level results in smaller spatial prediction error in cross-validation experiments. Similar findings are observed by Paciorek (2012) in a simulation study. More recently, outputs from dispersion models that provide air quality simulations at the point level have also been considered (Lindstrom et al., 2013; Priani et al., 2014).

One major limitation for using the proxy as a predictor is that it cannot contain missing values. For example, satellite-retrieved AOD can be missing due to cloud cover and highly reflective surfaces such as snow. Also, the calibration framework assumes that the observed measurements $Y(\mathbf{s})$ are the gold standard, even though instrumental error is likely to be present.

Extension to spatiotemporal data is straightforward by allowing the calibration parameters in Equation 16.7 to be time varying: $\alpha_0(\mathbf{s},t)$ and $\alpha_1(\mathbf{s},t)$. For computational efficiency, the space–time processes can be decomposed into additive components, that is, $\alpha_j(\mathbf{s},t) = \alpha_j(\mathbf{s}) + \alpha_j(t)$, for $j = 1, 2$. The temporal component is then assumed to evolve dynamically in time via an autoregressive model. In combining $PM_{2.5}$ and satellite-derived AOD, Chang et al. (2013) demonstrate the importance of considering temporal dependence in the calibration parameters because missing AOD data can result in days with no linked AOD–measurement pair. With sufficient monitoring locations, one can assume that $\alpha_j(\mathbf{s},t)$ evolves dynamically over time. However, for combining ozone and CMAQ outputs in the eastern United States, Berrocal et al. (2010a) find that the model allowing the spatial calibration parameters to be independent across days has the best prediction performance. Finally, in Zidek et al. (2012), the authors assume spatially varying calibration parameters, but model the residuals $e(\mathbf{s},t)$ to be autoregressive temporal processes with spatially varying autoregressive parameters. The motivation is to model the temporal variation in the measurements via the residuals, instead of that inherent in the proxy data.

16.4 Model Extensions

16.4.1 Multivariate assimilation

Humans are exposed to multiple pollutants simultaneously, and different pollutants can share the same sources. Consequently, there is growing interest in developing data fusion methods for multiple pollutants to support health research. A multipollutant approach may also improve prediction performance, as we can exploit the dependence between pollutants. This is particularly advantageous when the pollutant monitors are not colocated or have different measurement schedules.

Choi et al. (2009) present an interesting application of Bayesian melding for five constituents of $PM_{2.5}$. Here, the proxy data are measurements from another network that only provides the sum of the five constituents. The objective is to model individual pollutant concentration at location $\mathbf{s}$ as a function of a latent pollutant sum field. Let $Y_k(\mathbf{s})$ denote the measured kth pollutant's concentration, and let $S(\mathbf{s}) = \sum_{k=1}^{5} Y_k(\mathbf{s})$ denote the unobserved latent sum. Similarly, let $X(\mathbf{s})$ be the sum measured from the proxy network. The hierarchical model is given by

$$\begin{aligned} Y_k(\mathbf{s}) &= \theta_k(\mathbf{s})\, S(\mathbf{s}) + e_k(\mathbf{s},t) \\ S(\mathbf{s}) &= \mu(\mathbf{s}) + \epsilon(\mathbf{s}) \\ X(\mathbf{s}) &= a(\mathbf{s}) + S(\mathbf{s}) + \delta(\mathbf{s}). \end{aligned} \qquad (16.8)$$

Equation 16.8 expresses the observed pollutant as a proportion of the latent sum. The pollutant-specific proportion $\theta_k(\mathbf{s})$ is allowed to vary spatially and is specified as

$$\theta_k(\mathbf{s}) = \frac{\exp(\alpha_k(\mathbf{s}))}{\sum_{j=1}^{5}\exp(\alpha_j(\mathbf{s}))},$$

where for $j = 1, \ldots, 4$, $\alpha_j(\mathbf{s})$ is a Gaussian process, independent across j. $\alpha_5(\mathbf{s})$ is set to 0 for all $\mathbf{s}$ for identifiability purposes. To account for the correlated measurement error, $e_k(\mathbf{s})$ is modeled jointly using LMC. Note that here a change-of-support calculation is not needed because the proxy is available at the point level. Spatiotemporal data are accommodated by replacing $\theta_k(\mathbf{s})$ with a dynamic Gaussian process.

Multipollutant approaches have also been proposed under the downscaling framework. First, Berrocal et al. (2010b) extend their original model to a bipollutant setting for ozone and $PM_{2.5}$ using CMAQ outputs as the proxy. Following previous notation in Equation 16.7, the bipollutant model is given by

$$\begin{aligned} Y_1(\mathbf{s}) &= \alpha_{10}(\mathbf{s}) + \alpha_{11}(\mathbf{s})X_1(B_\mathbf{s}) + \alpha_{12}(\mathbf{s})X_2(B_\mathbf{s}) + e_1(\mathbf{s}), \quad e(\mathbf{s}) \sim N(0, \sigma^2_{e_1}) \\ Y_2(\mathbf{s}) &= \alpha_{20}(\mathbf{s}) + \alpha_{21}(\mathbf{s})X_2(B_\mathbf{s}) + \alpha_{22}(\mathbf{s})X_1(B_\mathbf{s}) + e_2(\mathbf{s}), \quad e(\mathbf{s}) \sim N(0, \sigma^2_{e_2}). \end{aligned}$$

We note that the proxy variables $X_1(B_\mathbf{s})$ and $X_2(B_\mathbf{s})$ are used to model each outcome, maximizing potential information in the proxy data. Again, the six calibration parameters are modeled jointly using LMC where various between-pollutant or between-proxy dependence structures can be investigated. To handle sites where only one of the pollutants is observed, a data augmentation step for the LMC latent variables is included in the Bayesian estimation algorithm. The above model involves numerous parameters, and extension to more than two pollutants has yet to be examined.

Finally, we describe a calibration approach by Crooks and Ozkaynak (2014) that includes a sum constraint for modeling $PM_{2.5}$ constituents. The motivating application entails simultaneously combining monitoring data and CMAQ outputs for five $PM_{2.5}$ constituents and the total $PM_{2.5}$ mass. The sum constraint is accomplished by modeling the kth pollutant concentration using a gamma distribution:

$$Y_k(\mathbf{s}) \sim \text{Gamma}\ \left(\tau^{-1} \times [\alpha_{k0}(\mathbf{s}) + \alpha_1(\mathbf{s})X_k(B_\mathbf{s})],\ \tau^{-1}\right).$$

Note that the gamma rate parameter τ and the multiplicative calibration parameter $\alpha_1(\mathbf{s})$ do not vary across pollutants. Assuming the pollutants are independent, the observed sum also follows a gamma distribution. This allows a mass conservation requirement using the observed total $PM_{2.5}$ mass. Despite these distributional assumptions, in cross-validation experiments, the authors find that both mass conservation and the multipollutant approach improve prediction accuracy.

16.4.2 Multiple-scale assimilation

In calibration, at each monitoring location, only the single linked proxy grid cell is used to provide information on the observed measurements. Because of the spatial dependence in the pollutant field, it is reasonable to also consider whether neighboring grid cells are useful for prediction. Berrocal et al. (2012b) propose two approaches to borrow proxy information across multiple grid cells. First, they replace the single grid cell predictor $X(B_\mathbf{s})$ in Equation 16.7 by a *smoothed* version, $\tilde{X}(B_\mathbf{s})$:

$$Y(\mathbf{s}) = \alpha_0(\mathbf{s}) + \alpha_1\tilde{X}(B_\mathbf{s}) + e(\mathbf{s}), \quad e(\mathbf{s}) \sim N(0, \sigma^2_e). \tag{16.9}$$

The new predictor $\tilde{X}(B_\mathbf{s})$ is taken as the spatial grid-level random effects with a CAR structure: $X_B = \tilde{X}(B) + \psi(\mathbf{s})$, where $\psi(\mathbf{s})$ is an independent normal residual error. Note that the above multiplicative calibration parameter α_1 is constant in space to avoid identifiability problems. The use of smoothed proxy data offers two advantages. First, this approach avoids the spatial misalignment due to a monitor's location within a proxy grid cell. Specifically, if a monitor is near the boundary of a grid cell, then it is natural to also consider proxy values at the closer grid cell. Second, it provides a flexible framework to utilize all of the proxy data in predicting $Y(\mathbf{s})$. This mimics the Bayesian melding approach where all the proxy data are used to estimate the latent pollutant field. However, the downscaler enables the measurement data to decide how much local smoothing is required to achieve optimal prediction.

Berrocal et al. (2012b) also consider an alternative smoothing approach by deriving a point-referenced proxy that represents a weighted-average across all proxy grid cells:

$$\tilde{X}(\mathbf{s}) = \sum_{g=1}^{G} w_g(\mathbf{s}) X(B_k).$$

Let $\mathbf{r}_g$ be the centroid of grid cell g; the weights are defined as

$$w_g(\mathbf{s}) = \frac{K(\mathbf{s}-\mathbf{r}_g)\,\exp(Q(\mathbf{r}_g))}{\sum_{l=1}^{G} K(\mathbf{s}-\mathbf{r}_l)\,\exp(Q(\mathbf{r}_l))},$$

where $K(\cdot)$ is a Gaussian kernel with bandwidth covering three grid cells in each direction, and $Q(\cdot)$ is a mean-zero Gaussian process approximated using a predictive process (Banerjee et al., 2008). By including $Q(\cdot)$, the weight $w_g(\mathbf{s})$ is allowed to be asymmetrical among the grid cells around location $\mathbf{s}$. In their application to daily ozone concentration, the authors find that the use of a smoothed CMAQ proxy provides better prediction power, especially at locations farther from the rest of the monitors.

Reich et al. (2014) propose a *spectral* downscaler that extends Berrocal et al. (2012b) further by using multiple smoothed proxies at different scales. The conceptual framework begins by considering the spectral representation of the continuous processes associated with the measurement and the proxy:

$$Y(\mathbf{s}) = \int \exp(-i\boldsymbol{\omega}^t\mathbf{s}) H_1(\boldsymbol{\omega})\, d\boldsymbol{\omega}$$

and

$$X(\mathbf{s}) = \int \exp(-i\boldsymbol{\omega}^t\mathbf{s}) H_2(\boldsymbol{\omega})\, d\boldsymbol{\omega},$$

where $H_1(\boldsymbol{\omega})$ and $H_2(\boldsymbol{\omega})$ are mean-zero Gaussian processes. The correlation between $H_1(\boldsymbol{\omega})$ and $H_2(\boldsymbol{\omega})$ is assumed to vary across frequency $\boldsymbol{\omega}$. Assuming $X(\mathbf{s})$ is observed everywhere, the conditional distribution of $Y(\mathbf{s})$ is given by

$$E[Y(\mathbf{s}) \,|\, X(\mathbf{r}) \text{ for all } \mathbf{r}] = \int \exp(-i\boldsymbol{\omega}^t\mathbf{s})\, \alpha(\boldsymbol{\omega}) H_2(\boldsymbol{\omega})\, d\boldsymbol{\omega}.$$

The above equation describes a scenario where the usefulness of the proxy to predict $Y(s)$ differs across spatial scales as captured by parameter $a(\boldsymbol{\omega})$. To estimate $a(\boldsymbol{\omega})$, Reich et al. (2014) parameterized it using basis expansion where $a(\boldsymbol{\omega}) = \sum_{l=1}^{L} A_l(\boldsymbol{\omega})\theta_l$. The standard downscaler in Equation 16.7 now takes the form

$$Y(\mathbf{s}) = \alpha_0(\mathbf{s}) + \sum_{l=1}^{L} \theta_l \tilde{X}_l(\mathbf{s}) + e(\mathbf{s}), \quad e(\mathbf{s}) \sim N(0, \sigma_e^2), \tag{16.10}$$

where

$$\tilde{X}_l(\mathbf{s}) = \int A_l(\boldsymbol{\omega}) \exp(-i\boldsymbol{\omega}^t \mathbf{s}) H_2(\boldsymbol{\omega})\, d\boldsymbol{\omega}.$$

Since the proxy data are observed completely over a grid, $\tilde{X}_l(\mathbf{s})$ can be constructed efficiently using fast Fourier transform.

The spectral downscaler is similar to using a smoothed proxy as predictor because the decomposed proxy signal at each frequency is driven by more than one individual grid cell linked to the monitor. By considering the entire range of frequencies, the spectral downscaler also provides unique insights into the utility of the proxy at different scales. When applied to CMAQ ozone simulations, the spectral downscaler showed that CMAQ outputs at 12 km resolution have low correlation for features with a period less than 24 km. This may highlight the deterministic model's limited resolution at this scale due to the coarse meteorological inputs. However, the associations between CMAQ outputs and measurements increase with increasing period, likely because ozone concentration tends to exhibit strong regional trends.

16.4.3 Rank and quantile-based calibration

All the models we have presented focus on modeling the mean of the pollutant fields. However, environmental exposures can exhibit extreme tails, and these extreme values are often more detrimental to health (e.g., extreme heat). In the United States, the air quality standards for ozone pollution use the annual fourth-highest daily 8-hour maximum concentration as the metric to determine nonattainment status. Berrocal et al. (2014) consider a downscaler that models the annual largest kth-order statistic based on the generalized extreme value distribution, which contains three parameters: location, scale, and shape. The same calibration approach in Equation 16.7 is then applied to the location parameter.

Zhou et al. (2011) propose an alternative approach that aims to characterize the entire distribution of the pollutant field at each location. This is accomplished by estimating a one-to-one mapping between the quantile functions of the measurements and the proxy data using monotonic splines. This approach also falls under the framework of nonparameteric density estimation. Unlike Berrocal et al. (2014), to address the tails of the distribution, Zhou et al. (2011) assume the 10% tail of the pollutant distribution at each extreme follows a generalized Pareto distribution and the central 80% of the distribution is determined flexibly by the splines.

16.5 Using Fusion Products in Health Studies

As fusion products become more readily available, researchers face the challenge of how to utilize them for epidemiological research. Specifically, a statistical model provides both predictions and the associated uncertainties; however, the majority of recent health studies using fusion products have treated the fitted values as the true exposures. While this naive "plug-in" approach is straightforward, it may underestimate the uncertainty associated with the estimated health effects, as well as lead to inaccurate estimates due to measurement error.

The fitted values from a statistical model contain two sources of error. First, a Berkson type error arises because the population average is used instead of the individual exposures. Second, a classical measurement error will be present from sampling variability (i.e., estimation uncertainty of the model parameters). Szpiro and Paciorek (2014) provide a detailed description of the impacts and magnitude of the Berkson and classical error components due to spatial kriging. The presence of both Berkson and classical measurement errors means that the exposure model with the best prediction performance does not necessarily give the best health effect estimate (Szpiro et al., 2011b).

Several approaches have been proposed to account for the uncertainty in exposure estimates from statistical models. Different approaches are developed to accommodate different modeling demands in the exposure model or in the health model. In cases where the computational demand for both the exposure model and the health model are low, a joint model of the exposure and health outcome is possible, for example, using a fully Bayesian hierarchical model (Gryparis et al., 2009). Models for data assimilation are obviously complex, especially if both spatial and temporal components are involved. Moreover, predicted pollutant concentrations often need to be spatially aggregated to match the aggregated health outcomes or temporally aggregated to define specific exposure windows. This additional misalignment between health and exposure data adds to the challenge for joint modeling. Finally, in epidemiological research, sensitivity analyses are often conducted to examine the robustness of the estimated health associations. Sensitivity analysis may include alternative specifications of the health model, stratified analysis, or inclusion of additional potential confounders. Typically, a large number of sensitivity analyses need to be conducted efficiently.

A *two-stage* approach where the exposure model and health model are fitted separately has been more commonly adopted in population-based environmental epidemiology. Szpiro et al. (2011a) describe several bootstrap methods to account for measurement errors. Briefly, naive parameter estimates are first obtained from fitting the exposure and the health models. At each bootstrap iteration, observed pollutant measurements and the corresponding health outcomes are simulated. The health effects obtained from the simulated data then serve as the bootstrap samples for bias correction and for deriving standard error estimation. Hence, bootstrap simulation considers outcome and exposure sampling variation and parameter estimation variation. The above algorithm involves fitting the exposure and health model, although separately, at each bootstrap iteration. A more efficient bootstrap approach is proposed where the unknown parameters are simulated from their asymptotic sampling distribution at each iteration.

In a Bayesian two-stage analysis, the posterior predictive distribution of the exposures can serve as a prior distribution of the exposures when fitting the health model. Spatial-temporal models estimated under a Bayesian framework using Markov chain Monte Carlo (MCMC) methods provide posterior predictive samples of the exposures such that posterior distributions of different spatial and temporal averages can be obtained. However, the joint posterior distribution of all the exposures is likely to exhibit complex dependence, and for complex health models, approximate Bayesian computation may be needed to reduce the computation burden. For example, Chang et al. (2011) use exposure posterior samples as the proposal distribution of a Metropolis–Hasting update where the acceptance probability is evaluated using the profile likelihood. This shares similarity with the bootstrap approach where (1) the health model data likelihood is evaluated at every iteration, and (2) the exposures are drawn from the posterior predictive distribution instead of simulations from the exposure model.

One important difference between the Bayesian and the bootstrap approach is that the Bayesian framework allows *feedback* where the health data provide information on the exposures. While feedback can lead to smaller posterior interval, it relies on the assumption that the health model and the exposure–health relationship are correctly specified. In previous applications, the impacts of health data feedback have been minimal when the health associations are small. However, the appropriateness of feedback when there exits strong exposure–health association warrants further scrutiny. It is possible to avoid feedback by fitting the health model repeatedly with posterior samples of the exposures, mimicking a multiple imputation scheme (Si and Reiter, 2011; Zhou and Reiter, 2010).

16.6 Concluding Remarks

Improved exposure assessment methods will continue to be valuable for epidemiological research and health impact studies. The prospect of combining different sources of data to assess environmental pollution is well recognized. For example, in the most recent annual global burden of disease published by the World Health Organization, global ambient $PM_{2.5}$ estimates were derived from monitoring measurements, satellite-derived AOD, and numerical model simulations (Brauer et al., 2012). Practical methods and tools for data assimilation will play an important role in environmental epidemiology as environmental engineers continue to develop state-of-the-art proxy data products and as health researchers continue to leverage novel health data sources. Techniques for modeling massive spatio-temporal datasets will also be indispensable because environmental data are being collected with better spatial resolution and computer model simulations are being generated with greater ease. Finally, as novel pollutant products are being utilized in health research, the ability to quantify and account for various sources of uncertainties in the exposure estimates may help health studies to be more accurate and reproducible.

References

Banerjee S, Gelfand AE, Finley AO, Sang H. (2008). Gaussian predictive process models for large spatial data sets. *Journal of the Royal Statistical Society: Series B* 70, 825–848.

Berrocal VJ, Craigmile PF, Guttorp P. (2012a). Regional climate model assessment using statistical upscaling and downscaling techniques. *Environmetrics* 23, 482–492.

Berrocal VJ, Gelfand AE, Holland DM. (2010a). A spatio-temporal downscaler for output from numerical models. *Journal of Agricultural, Biological, and Environmental Statistics* 15, 176–197.

Berrocal VJ, Gelfand AE, Holland DM. (2010b). A bivariate space-time downscaler under space and time misalignment. *Annals of Applied Statistics* 4, 1942–1975.

Berrocal VJ, Gelfand AE, Holland DM. (2012b). Space-time data fusion under error in computer model outputs: an application to modeling air quality. *Biometrics* 68, 837–848.

Berrocal VJ, Gelfand AE, Holland DM. (2014). Assessing exceedance of ozone standards: a space-time downscaler for fourth-highest ozone concentrations. *Environmetrics* 25, 279–291.

Besag J. (1974). Spatial interaction and the statistical analysis of lattice systems. *Journal of the Royal Statistical Society: Series B* 136, 192–236.

Bey I, Jacob DJ, Yantosca RM, Logan JA, Field BD, Fiore AM, Li QB, Liu HGY, Mickley LJ, Schultz MG. (2001). Global modeling of tropospheric chemistry with assimilated meteorology: model description and evaluation. *Journal of Geophysical Research—Atmospheres* 106, 23073–23095.

Brauer M, Amann M, Burnett RT, Cohen A, Dentener F, Ezzati M, Henderson SB, et al. (2012). Exposure assessment for estimation of the global burden of disease attributable to outdoor air pollution. *Environmental Sciences and Technology* 46, 652–660.

Byun DJ, Schere KL. (2006). Review of the governing equations, computational algorithms, and other components of the Model-3 Community Multiscale Air Quality (CMAQ) modeling system. *Applied Mechanics Review* 59, 51–77.

Chang HH, Peng PD, Dominici F. (2011). Estimating the acute health effects of coarse particulate matter accounting for exposure measurement error. *Biostatistics* 12, 637–653.

Chang HH, Xu X, Liu Y. (2013). Calibrating MODIS aerosol optical depth for predicting daily $PM_{2.5}$ concentrations via statistical downscaling. *Journal of Exposure Science and Environmental Epidemiology* 24, 398–404.

Choi J, Reich BJ, Fuentes M, Davis JM. (2009). Multivariate spatial-temporal modeling and prediction of speciated fine particles. *Journal of Statistical Theory and Practice* 3, 407–418.

Clougherty JE, Kheirbek I, Eisl HM, Ross Z, Pezeshki G, Gorczynski JE, Johnson S, Markowitz S, Kass D, Matte T. (2013). Intra-urban spatial variability in wintertime street-level concentrations of multiple combustion-related air pollutants: the New York City Community Air Survey (NYCCAS). *Journal of Exposure Science and Environmental Epidemiology* 23, 232–240.

Crooks J, Isakov V. (2013). A wavelet-based approach to blending observations with deterministic computer models to resolve the intraurban air pollution field. *Journal of the Air and Waste Management Association* 63, 1369–1385.

Crooks JL, Ozkaynak H. (2014). Simultaneous statistical bias correction of multiple $PM_{2.5}$ species from a regional photochemical grid model. *Atmospheric Environment* 95, 126–141.

Foley K, Fuentes M. (2008). A statistical framework to combine multivariate spatial data and physical models for hurricane surface wind prediction. *Journal of Agricultural, Biological, and Environmental Statistics* 13, 37–59.

Fuentes M, Raftery AE. (2005). Model evaluation and spatial interpolation by Bayesian combination of observations with outputs from numerical models. *Biometrics* 61, 36–45.

Gelfand AE, Schmidt AM, Banerjee S, Sirmans CF. (2004). Nonstationary multivariate process modeling through spatially varying coregionalization. *Test* 13, 263–312.

Giorgi F, Mearns LO. (1991). Approaches to the simulation of regional climate change: a review. *Review of Geophysics* 29, 191–216.

Gotway CA, Young LJ. (2002). Combining incompatible spatial data. *Journal of the American Statistical Association* 97, 632–648.

Gryparis A, Paciorek C, Zeka A, Schwartz J, Coull BA. (2009). Measurement error caused by spatial misalignment in environmental epidemiology. *Biostatistics* 10, 258–274.

Hoek G, Beelen R, de Hoogh K, Vienneau D, Gulliver J, Fischer P, Briggs D. (2008). A review of land-use regression models to assess spatial variation of outdoor air pollution. *Atmospheric Environment* 42, 7561–7578.

Hoff RM, Christopher SA. (2009). Remote sensing of particulate pollution from space: have we reached the promised land? *Journal of the Air and Waste Management Association* 59, 645–675.

Kloog I, Nordio F, Coull BA, Schwartz J. (2013). Incorporating local land use regression and satellite aerosol optical depth in a hybrid model of spatiotemporal $PM_{2.5}$ exposures in the Mid-Atlantic states. *Environmental Sciences and Technology* 46, 11913–11921.

Lindstrom J, Szpiro AA, Sampson PD, Oron AP, Richards M, Larson TV, Sheppard L. (2013). A flexible spatio-temporal model for air pollution with spatial and spatio-temporal covariates. *Environmental and Ecological Statistics* 21, 411–433.

Liu Y, Paciorek CJ, Koutrakis P. (2009). Estimating regional spatial and temporal variability of $PM_{2.5}$ concentrations using satellite data, meteorology, and land use information. *Environmental Health Perspectives* 117, 886–892.

Liu Z, Le ND, Zidek JV. (2011). An empirical assessment of Bayesian melding for mapping ozone pollution. *Environmetrics* 22, 340–353.

Lynch P. (2008). The origins of computer weather prediction and climate modeling. *Journal of Computational Physics* 227, 3431–3444.

McMillan NJ, Holland DM, Morara M, Feng J. (2010). Combining numerical model output and particulate data using Bayesian space-time modeling. *Environmetrics* 21, 48–65.

Paciorek CJ. (2012). Combining spatial information source while accounting for systematic errors in proxies. *Journal of the Royal Statistical Society: Series C* 61, 429–451.

Paciorek CJ, Liu Y. (2009). Limitations of remotely sensed aerosol as a spatial proxy for fine particulate matter. *Environmental Health Perspectives* 117, 904–909.

Priani M, Gulliver J, Fuller GW, Blangiardo M. (2014). Bayesian spatiotemporal modelling for the assessment of short-term exposure to particle pollution in urban areas. *Journal of Exposure Science and Environmental Epidemiology* 24, 319–317.

Reich BJ, Chang HH, Foley KM. (2014). A spectral method for spatial downscaling. *Biometrics* 70, 932–942.

Ryan PH, LeMasters GK. (2007). A review of land-use regression models for characterizing intraurban air pollution exposure. *Inhalation Toxicology* 19, 127–133.

Sahu SK, Gelfand AE, Holland DM. (2010). Fusing point and areal level space-time data with application to wet deposition. *Journal of the Royal Statistical Society: Series C* 59, 77–103.

Shaddick G, Zidek JV. (2014). A case study in preferential sampling: long term monitoring of air pollution in the UK. *Spatial Statistics* 9, 51–65.

Shine KP, Henderson-Sellers A. (1983). Modelling climate and the nature of climate models: a review. *International Journal of Climatology* 3, 81–94.

Si Y, Reiter JP. (2011). A comparison of posterior simulation and inference by combining rules for multiple imputation. *Journal of Statistical Theory and Practice* 5, 335–347.

Szpiro AA, Paciorek CJ. (2014). Measurement error in two-stage analyses, with application to air pollution epidemiology. *Environmetrics* 24, 501–517.

Szpiro AA, Paciorek CJ, Sheppard L. (2011b). Does more accurate exposure prediction improve health effect estimates? *Epidemiology* 22, 680–685.

Szpiro AA, Sheppard L, Lumley T. (2011a). Efficient measurement error correction with spatially misaligned data. *Biostatistics* 12, 610–623.

Teutschbein C, Seibert J. (2012). Bias correction of regional climate model simulations for hydrological climate-change impact studies: review and evaluation of different methods. *Journal of Hydrology* 456–457, 12–29.

Wan V, Braun WJ, Dean C, Henderson S. (2010). A comparison of classification algorithms for the identification of smoke plumes from satellite images. *Statistical Methods in Medical Research* 20, 131–156.

Zhou J, Chang HH, Fuentes M. (2012). Estimating the health impact of climate change with calibrated climate model output. *Journal of Agricultural, Biological, and Environmental Statistics* 17, 377–394.

Zhou J, Fuentes M, Davis J. (2011). Calibration of numerical model output using nonparametric spatial density functions. *Journal of Agricultural, Biological, and Environmental Statistics* 16, 531–553.

Zhou X, Reiter JP. (2010). A note on Bayesian inference after multiple imputation. *American Statistician* 64, 159–163.

Zidek JV, Le ND, Liu Z. (2012). Combining data and simulated data for space-time fields: application to ozone. *Environmental and Ecological Statistics* 19, 37–56.

17

Spatial Survival Models

Sudipto Banerjee
Department of Biostatistics
University of California
Los Angeles, California

CONTENTS

17.1 Introduction

Epidemiological and biomedical studies often require modeling and analysis for time-to-event data, where a subject is followed up to an event (e.g., death or onset of a disease) or is "censored," whichever comes first. Survival models (see, e.g., Cox and Oakes [1]) are widely used in biostatistics and epidemiology for analyzing time-to-event data, including, perhaps, several censored observations. If the event does not occur for a subject during the period of the study, then the subject's time to event is censored at the study end point. This situation is described as "right censoring" and is, perhaps, the most common. Analogously, certain study designs can produce "left-censored" or "interval-censored" data. As opposed to modeling disease incidence and mortality, survival modeling focuses on how many are expected to survive after a certain period of time, the rate of failure, and the factors driving shortened or prolonged survival; all of these may be influenced by several factors such as gender, race, age, type of cancer, treatment obtained, and access to health care facilities.

If T is a nonnegative random variable representing a subject's waiting time until an event (e.g., disease onset, relapse, or death) occurs, then we define the subject's *survival function* as $S(t) = P(T \geq t)$ and the *hazard function* as $h(t) = f(t)/S(t)$, where $f(t)$ is the probability density function of T. Survival data are often grouped into strata (or clusters), such as clinical sites and geographic regions. For this chapter, we consider the strata as geographical regions. Let (i, j) index the jth subject in region i and let $\{(t_{ij}, \delta_{ij}) : i = 1, 2, \ldots, I;\ j = 1, 2, \ldots, n_i\}$ be observations from n subjects in a study, where t_{ij} indicates the time at which either subject (i, j) experienced the event or it was censored. Associated with each t_{ij} is an "event" indicator, δ_{ij}, where $\delta_{ij} = 1$ if the subject experienced the event before the termination of the study, and $\delta_{ij} = 0$ if the subject was censored.

For right-censored data, we have the likelihood

$$\prod_{j=1}^{n_i} f(t_{ij})^{\delta_{ij}} S(t_{ij})^{1-\delta_{ij}} = \prod_{j=1}^{n_i} h(t_{ij})^{\delta_{ij}} S(t_{ij}) \,. \tag{17.1}$$

If $\delta_{ij} = 1$, then subject j contributes $f(t_{ij}) = h(t_{ij})S(t_{ij})$ to the likelihood, while if $\delta_{ij} = 0$, then it contributes $S(t_{ij})$ to the likelihood. Analogous expressions for left-censored or interval-censored data can be found in, for example, [1].

To account for heterogeneity in the population, often we observe explanatory variables, say as a $p \times 1$ vector $\mathbf{x}_{ij}$, associated with subject (i, j). Further modeling incorporates such information in (17.1). For example, an accelerated failure time (AFT) distribution is obtained by specifying the survival function for subject (i, j) as $S(t) = S_0(t/\gamma_{ij})$, where $S_0(t)$ is any parametric survival function and $\gamma_{ij} = \exp\{\mathbf{x}_{ij}^T\boldsymbol{\beta}\}$. The corresponding hazard function for subject (i, j) is $h(t) = h_0(t/\gamma_{ij})/\gamma_{ij}$, where $h_0(t)$ is the hazard derived from $S_0(t)$.

An alternative to the AFT is the proportional hazards model, which stipulates the hazard function as

$$h(t_{ij}; \mathbf{x}_{ij}) = h_0(t_{ij}) \exp(\mathbf{x}_{ij}^T\boldsymbol{\beta}), \tag{17.2}$$

where $h_0(t)$ represents the *baseline hazard* function affected only multiplicatively by the exponential term involving the covariates. There are parametric or nonparametric choices for modeling the baseline hazard. Yet another option is a proportional odds model ([2]), which requires the survival function for subject (i, j) to satisfy

$$\frac{S(t \,|\, \mathbf{x}_{ij})}{1 - S(t \,|\, \mathbf{x}_{ij})} = \frac{S_0(t)}{1 - S_0(t)} \exp(\mathbf{x}_{ij}^T\boldsymbol{\beta}), \tag{17.3}$$

where the baseline survival function $S_0(t)$ is modeled using either parametric or nonparametric approaches.

The data analytic settings where the above specifications are appropriate, or not, have been comprehensively explored and documented in the survival analysis literature. For example, the proportional odds model is clearly distinct from the proportional hazards model because the former posits that the hazard ratio approaches unity over time; that is, the covariate effects on the hazards disappear over time. While the regression estimates may be similar between the two models, the interpretation of the regression component significantly differs. The term $\exp\{\mathbf{x}^T\boldsymbol{\beta}\}$ in the proportional odds model is interpreted as the change in the odds of survival (or failure, depending on the parametrization) given the observed covariates or risk factors.

17.2 Survival Models with Spatial Frailties

Survival data grouped into strata are often analyzed using hierarchical models with stratum-specific *frailties* (e.g., Vaupel et al. [3]). Frailties are random effects assumed to capture variation specific to different strata. Conceptually, this is straightforward. We simply replace $\mathbf{x}_{ij}^T\boldsymbol{\beta}$ with $\mathbf{x}_{ij}^T\boldsymbol{\beta} + \eta_i$ in the survival regression models discussed earlier. For instance, frailty models extend the hazard function in (17.2) to

$$h(t_{ij}; \mathbf{x}_{ij}) = h_0(t_{ij})\, \omega_i \exp(\boldsymbol{\beta}^T\mathbf{x}_{ij}) = h_0(t_{ij}) \exp(\boldsymbol{\beta}^T\mathbf{x}_{ij} + \eta_i), \tag{17.4}$$

where η_i is the region-specific frailty term capturing extraneous variation in the ith region.

When the strata are spatially arranged, such as clinical sites or geographical regions, we might suspect that random effects corresponding to proximate regions will be similar in magnitude. Models for spatially arranged survival data customarily introduce *spatial frailties* (e.g., Banerjee et al. [4]). Precisely how these spatial frailties will be modeled depends on the type of spatial referencing. Spatial referencing, broadly speaking, can be classified into one of two different groups: (1) *point referenced*, where each spatial unit's exact geographic coordinates (e.g., latitude and longitude from a GPS device) are recorded and (2) *areally referenced*, where each spatial unit corresponds to an areal unit or region (e.g., state, county, zip codes, and census tracts). Spatial models depend on the type of referencing. For point-referenced data, an underlying *spatial process* is assumed and the desired inference is to measure the extent of spatial correlation (usually described in terms of a *spatial range* measured in terms of Euclidean or geodesic distances) and interpolate the outcomes at arbitrary locations. For areally referenced data, spatial autocorrelations are measured using only the positions of the strata relative to each other (e.g., which counties are adjacent to others).

For point-referenced survival data, we assume that the frailties $\eta_i \equiv \eta(\mathbf{s}_i)$ arise as finite-dimensional realizations of an underlying spatial process $\{\eta(\mathbf{s}) : \mathbf{s} \in \mathcal{D}\}$, where $\mathbf{s}_i$ references the ith strata (e.g., centroid of a region), and then the $I \times 1$ vector $\boldsymbol{\eta} = (\eta(\mathbf{s}_1), \eta(\mathbf{s}_2), \ldots \eta(\mathbf{s}_I))^T$ follows $N(\mathbf{0}, \mathbf{C}(\boldsymbol{\theta}))$, where $\mathbf{C}(\boldsymbol{\theta})$ is an $I \times I$ spatial covariance matrix whose (i, j)th element gives the covariance between $\eta(\mathbf{s}_i)$ and $\eta(\mathbf{s}_j)$. For further details on covariance function, see Chapter 11. For areal models, rather than a continuously indexed spatial process $\eta(\mathbf{s})$ over the domain, we assume that the frailty distribution is specified only over the discretely indexed regions constituting the study domain. Such models, customarily, incorporate information about the adjacency of regions rather than any type of continuous distance metric. Popular examples include the conditionally autoregressive (CAR) model and its variants (see Chapter 23).

Bayesian spatial survival models, where the spatial associations are introduced as a prior on the random effects in the hierarchy, are simple to specify and can be estimated using standard Markov chain Monte Carlo (MCMC) methods. Apart from the spatial distribution for the frailties, one needs to model the hazard function with the understanding that "expected" survival times (or hazard rates) will be more similar in neighboring regions, due to underlying factors (access to care, willingness of the population to seek care, etc.) that vary spatially. This is in contrast to the observed survival times from subjects in proximate regions being similar, which is not necessarily implied by spatially associated frailties.

There exists an extensive literature on spatial survival models. Li and Ryan [5] were among the first to establish identifiability and legitimacy of likelihood-based inference from semiparametric spatial survival models. They modeled the hazard function nonparametrically and the spatially correlated frailties using different spatial covariance functions. The proposed methods were applied to the East Boston Asthma Study to detect prognostic factors leading to childhood asthma. Henderson et al. [6] incorporated spatially dependent frailties using multivariate gamma distributions to investigate spatial variation in the survival of acute myeloid leukemia patients in northern England. Banerjee et al. [4] proposed a Bayesian hierarchical framework to introduce spatially correlated frailties and compared performances between frailties modeled using Markov random field and geostatistical covariance functions. Data from a large infant mortality study in the state of Minnesota was analyzed. Subsequent papers explored Bayesian semiparametric modeling [7], spatiotemporal modeling [8,9], semiparametric proportional odds models with spatial frailties [10], joint survival and longitudinal modeling with frailties [11], and parametric accelerated failure time models [12]. Finally, we recognize that all spatial survival models do not necessarily employ frailties (see, e.g., Lawson et al. [13]).

17.3 Spatial Cure Rate Models

With significant progress in medical and health sciences, scientists and health professionals increasingly encounter datasets where patients are expected to be *cured*. Models accounting for cure are important for understanding prognosis in potentially terminal diseases. Traditional parametric survival models such as Weibull or gamma (see, e.g., Cox and Oakes [1]) do not account for cure, instead assuming that individuals who do not experience the event are *censored*. The subtle distinction between censoring and cure is worth noting: a subject who does not fail within the time window of the experiment is considered censored, while a subject is cured if he will *never* relapse. Clearly, the latter is a more abstract concept in that we are never able to observe a cure, yet there is interest in estimating the probability of such an outcome, especially in various disease-relapse settings.

Cure models, like survival models, also enjoy a rich literature too vast to be comprehensively reviewed here. See Ibrahim et al. [14] for a methodological introduction, while Othus et al. [15] offer a more recent review and practical introduction. Cooner et al. [16] build on a flexible framework proposed in Cooner et al. [17] (see also Hurtado Rúa and Dey [18]) to introduce spatial frailties in cure models for geographically referenced data. Banerjee and Carlin [19] propose a spatial extension of earlier work by Chen et al. [20], assuming that some latent biological process is generating the observed data. Suppose that subject (i, j) has N_{ij} potential latent (unobserved) risk factors, the presence of any of which (i.e., $N_{ij} \geq 1$) will ultimately manifest the event. Chen et al. [20] consider the case of multiple latent factors, assuming that the N_{ij} are distributed as independent Poisson random variables with mean θ_{ij}; that is, $p(N_{ij}|\theta_{ij})$ is $Poi(\theta_{ij})$. For example, in cancer settings, these factors may correspond to metastasis-competent tumor cells within the individual. Subjects who do not experience the event during the observation period are considered censored. Thus, if U_{ijk}, $k = 1, 2, \ldots, N_{ij}$ is the time to an event arising from the kth latent factor for subject (i, j), the observed time to event for an uncensored individual is generated by $T_{ij} = \min\{U_{ijk}, k = 1, 2, \ldots, N_{ij}\}$.

Given N_{ij}, the U_{ijk}'s are independent with survival function $S(t|\Psi_{ij})$ and corresponding density function $f(t|\Psi_{ij})$. The parameter Ψ_{ij} is a collection of all the parameters (including possible regression parameters) that may be involved in a parametric specification for the survival function S. In this section, we will work with a two-parameter Weibull distribution specification for the density function $f(t|\Psi_{ij})$, where we allow the Weibull scale parameter ρ to vary across the regions, and η, which may serve as a link to covariates in a regression setup, to vary across individuals. Therefore, $f(t|\rho_i, \eta_{ij}) = \rho_i t^{\rho_i - 1} \exp(\eta_{ij} - t^{\rho_i} \exp(\eta_{ij}))$.

Banerjee and Carlin [19] analyze smoking cessation data using interval-censored spatial cure rate models. The outcome of interest is the time for a subject to relapse into smoking. In studies such as these, one is only able to determine patient status at the office visits themselves, meaning we observe only a time *interval* (t_{ijL}, t_{ijU}) within which the event (smoking relapse) is known to have occurred. For patients who did not resume smoking prior to the end of the study, we have $t_{ijU} = \infty$, returning us to the case of right censoring at time point t_{ijL}. Thus, we now set $\nu_{ij} = 1$ if subject ij is interval censored (i.e., experienced the event), and $\nu_{ij} = 0$ if the subject is right censored.

Following [21], the general interval-censored cure rate likelihood is given by

$$\prod_{i=1}^{I}\prod_{j=1}^{n_i} [S(t_{ijL}|\rho_i, \eta_{ij})]^{N_{ij}-\nu_{ij}} \{N_{ij}[S(t_{ijL}|\rho_i, \eta_{ij}) - S(t_{ijU}|\rho_i, \eta_{ij})]\}^{\nu_{ij}}$$

$$= \prod_{i=1}^{I}\prod_{j=1}^{n_i} [S(t_{ijL}|\rho_i, \eta_{ij})]^{N_{ij}} \left\{N_{ij}\left(1 - \frac{S(t_{ijU}|\rho_i, \eta_{ij})}{S(t_{ijL}|\rho_i, \eta_{ij})}\right)\right\}^{\nu_{ij}}.$$

If $N_{ij} \overset{iid}{\sim} \text{Ber}(\theta_{ij})$, then the marginal likelihood obtained by summing over the N_{ij}'s is $L\left(\{(t_{ijL}, t_{ijU})\} \mid \{\rho_i\}, \{\theta_{ij}\}, \{\eta_{ij}\}, \{\nu_{ij}\}\right)$ and can be written as

$$\prod_{i=1}^{I}\prod_{j=1}^{n_i} S^*\left(t_{ijL}|\theta_{ij}, \rho_i, \eta_{ij}\right)\left\{1 - \frac{S^*\left(t_{ijU}|\theta_{ij}, \rho_i, \eta_{ij}\right)}{S^*\left(t_{ijL}|\theta_{ij}, \rho_i, \eta_{ij}\right)}\right\}^{\nu_{ij}}. \tag{17.5}$$

As with the covariates, we will introduce the frailties ϕ_i through the Weibull link as intercept terms in the log relative risk; that is, we set $\eta_{ij} = \mathbf{x}_{ij}^T\boldsymbol{\beta} + \phi_i$. Here, we allow the ϕ_i to be spatially correlated across the regions; similarly, we would like to permit the Weibull baseline hazard parameters, ρ_i, to be spatially correlated. A natural approach in both cases is to use a univariate CAR prior. While one may certainly employ separate, independent CAR priors on $\boldsymbol{\phi}$ and $\boldsymbol{\zeta} \equiv \{\log \rho_i\}$, Banerjee and Carlin [19] allow these two spatial priors to themselves be correlated. In other words, we may want a bivariate spatial model for the $\delta_i = \{\phi_i, \zeta_i\} = \{\phi_i, \log \rho_i\}$. One way to achieve this is using a multivariate CAR (MCAR) distribution. The MCAR prior for $\boldsymbol{\delta} = \{\boldsymbol{\phi}, \boldsymbol{\zeta}\}$ is Gaussian with mean $\mathbf{0}$ and precision matrix $\Lambda^{-1} \otimes (\text{Diag}\,(m_i) - \rho W)$, where Λ is a 2×2 symmetric and positive definite matrix, $\rho \in (0, 1)$, and m_i and W remain as above. In the current context, we may also wish to allow different smoothness parameters (say, ρ_1 and ρ_2) for $\boldsymbol{\phi}$ and $\boldsymbol{\zeta}$, respectively. Henceforth, in this section we will denote the proper MCAR with a common smoothness parameter by $MCAR(\rho, \Lambda)$, and the multiple smoothness parameter generalized MCAR by $MCAR(\rho_1, \rho_2, \Lambda)$. Combined with independent (univariate) CAR models for $\boldsymbol{\phi}$ and $\boldsymbol{\zeta}$, these offer a broad range of potential spatial models.

17.4 Spatial Survival Models with Multiple Cancers

Simultaneous (joint) modeling of time-to-event data from different diseases, such as cancers, from the same patient could provide useful insights as to how these diseases behave together. Survival models for *multiple cancers* are sought for practical reasons, such as to assess survival from multiple primary cancers simultaneously and to adjust the survival rates from a specific primary cancer in the presence of other primary cancers [22,23]. Diva et al. [24,25] construct Bayesian hierarchical survival models for capturing spatial correlations within the proportional hazards and proportional odds frameworks. They exploit the Surveillance, Epidemiology, and End Results (SEER) database, which is available from the National Cancer Institute, along with a statistical software SEER-STAT and list cases of multiple cancers by cancer types diagnosed. Together with Medicare data, one can extract an individual patient's complete medical history, say, starting from his visits to the health clinics (from a time when he might have been perfectly healthy), and information about his interim (uneventful) visits between the diagnoses of cancers.

Diva et al. [24] consider a relatively simple version of the SEER data containing survival information on patients suffering from multiple cancers. Along with survival information, patient-specific and patient–cancer combination–specific covariates, including gender, race, marital status, number of primary cancers diagnosed, and their county of residence, are also available. Patient–cancer-specific information relates to age at diagnosis, the stage of the cancer (in situ, local, distant, or regional), the type (radiation or surgery) of treatment that each patient underwent for each type of cancer, and so on. As an example, consider a patient, say with ID 00171, who first contracted cancer of the small intestine (the first primary cancer) in 1986, developed colon cancer in 1991 and pancreatic cancer in 1995, and eventually died in 1998. The data for this particular patient are shown in Table 17.1. We extract information about this patient from the SEER database, where each entry

TABLE 17.1
Sample database entry of a single patient with several cancers

County	Patient ID	Gender	Cancer type	Age at diagnosis	Status (end point)	Time (months)
7	00171	Female	Small intestine	57	Dead	143
7	00171	Female	Colon	63	Dead	78
7	00171	Female	Pancreas	67	Dead	25

corresponding to the patient would list the type of cancer, the age at diagnosis, the stage of the cancer, the status at end point (dead or alive), and the survival time (or censorship time) in months. Note that the status is labeled as "dead" for all entries, as it corresponds to the status at the end point.

Along with the above information, we also know the patient's county of residence. Assume there are I counties, and in the ith county we observe n_i patients. Thus, the total number of patients in our study is $N = \sum_{i=1}^{I} n_i$, and we uniquely identify every individual through the ordered pair (i, j), as the jth individual $(j = 1, 2, \ldots, n_i)$ from the ith county. Suppose each of these patients has been diagnosed with at least one of K possible cancer types. Let t_{ijk} denote this time, which we will refer to as the *survival* time for the (i, j)th individual from his diagnosis with the kth type of cancer. Since not all patients develop all K types of cancer, we list the possible cancers as $\{1, 2, \ldots, K\}$, and form the subset $C_{(i,j)} \subseteq \{1, 2, \ldots, K\}$ as the indices of the cancers developed by the (i, j)th patient. Thus, if patient 00171 in Table 17.1 is the 10th individual from the seventh county, and if colorectal, stomach, and pancreas are cancer types 1, 3, and 5, in $\{1, 2, \ldots, K\}$, then $C_{(7,10)} = \{1, 3, 5\}$. Indeed, when referring to t_{ijk}, $k \in C_{(i,j)}$. Clearly, these survival times will be correlated for similar county–patient–cancer combinations. We capture these correlations by introducing appropriate frailties.

Let $u_{(i,j)}$ denote the frailty for the (i, j)th patient (these may be looked upon as patient effects nested within counties), v_k be the frailty for the kth cancer type, and ϕ_{ik} be the frailty for the kth cancer type nested within the ith county. Consider a proportional hazards model incorporating main effects for patients and cancers and nested effects for cancers within counties, which we write as

$$h\left(t_{ijk}\right) = h_0\left(t_{ijk}\right) \exp\left(\mathbf{x}_{ijk}^T \boldsymbol{\beta} + u_{(i,j)} + v_k + \phi_{ik}\right) \tag{17.6}$$

for $i = 1, 2, \ldots, I$, $j = 1, 2, \ldots, n_i$, and $k = 1, 2, \ldots, K$. Here, $\mathbf{x}_{ijk}$ denotes the patient–cancer-specific covariates (includes covariates such as age at diagnosis of primary cancer, stage of the individual cancers, and indicator of whether the cancer is the first primary cancer), and $\boldsymbol{\beta}$ is the corresponding vector of regression coefficients. Also, $h_0(t)$ is the baseline hazard function, which can be cancer specific (in which case we write $h_{0k}(t)$) or county specific (in which case we write $h_{0i}(t)$). In principle, one can also include subject–cancer interaction terms of the form $w_{(ij)k}$ in (17.6), which may reveal cancer-specific variation in patient frailties. However, in practice, reliably estimating these higher-order interaction terms becomes difficult, especially with data in which many subjects yield less than three observations. Since we already had space–cancer interaction terms in Equation 17.6 (the ϕ_{ik}'s), adding the $w_{(ij)k}$'s vastly exacerbates the identifiability problems.

Using the above notation and letting $\delta_{(i,j)}$ be a death indicator (status at end point) for patient (i, j), we obtain the likelihood

$$\prod_{i=1}^{I} \prod_{j=1}^{n_i} \prod_{k \in C_{(i,j)}} \left[S\left(t_{ijk} | \mathbf{x}_{ijk}, \boldsymbol{\beta}, \boldsymbol{\eta}\right)\right] \left[h_0\left(t_{ijk}; \boldsymbol{\eta}\right) \exp\left(\mathbf{x}_{ijk}^T \boldsymbol{\beta} + u_{(i,j)} + v_k + \phi_{ik}\right)\right]^{\delta_{ij}}, \tag{17.7}$$

where $S(t_{ijk}|\mathbf{x}_{ijk}, \boldsymbol{\beta}, \boldsymbol{\eta})$ is the survival function evaluated at t_{ijk} conditional on the parameters and observed covariates. Turning to the priors and hyperpriors, we would like to assume the patient frailties to be independent and normally distributed (zero centered) with county-specific variances, that is, $u_{(i,j)} \overset{indep}{\sim} N(0, \sigma_i^2)$. Collecting the K cancers into a vector $\mathbf{v}$, we assume $\mathbf{v} = (v_1, \ldots, v_K)^T \sim N(\mathbf{0}, \Lambda)$, where Λ is a $K \times K$ (unknown) covariance matrix. Turning to the spatial effects, one can assign any version of the multivariate CAR distribution, as discussed in Chapter 23 for $\{\phi_{ik}\}$. Further details on these approaches, along with details on the MCMC algorithms for fitting them, can be found in Diva et al. [24,25].

17.5 Illustration

We present some of the data analysis from Banerjee and Carlin [19] in the study of smoking cessation, which is of particular relevance to studies of lung health and primary cancer control. Murray et al. (1998) [26] describe the study and the entire database in greater detail. For our illustration here, we restrict attention to 223 subjects from 54 zip codes in southeastern Minnesota. These subjects were all smokers at study entry and were randomized into either a smoking intervention (SI) group or a usual care (UC) group that received no antismoking intervention. Based on a consecutive 5-year monitoring period, each of these subjects was known to have quit smoking at least once during these 5 years. The event of interest is whether they relapse into smoking (resume smoking) or not. The raw data revealed that 29.7% resumed smoking, producing an empirical cure fraction of 0.703. Additional information available for each subject includes sex, years as a smoker, and the average number of cigarettes smoked per day prior to the quit attempt.

As is not unusual in spatial datasets, the 54 zip codes that contributed the data were not contiguous and make it apparently difficult to fit neighborhood-based models. Banerjee and Carlin [19] considered 81 contiguous zip codes, shown in Figure 17.1, which included the 54 dark-shaded regions that had patients in the dataset, and the 27 regions not contributing patients were treated as if data were missing. The spatial effects are treated as latent random variables, and fairly standard Markov chain Monte Carlo algorithms deliver inference on these random effects for *all* 81 contiguous zip codes, including the 27 unshaded regions in

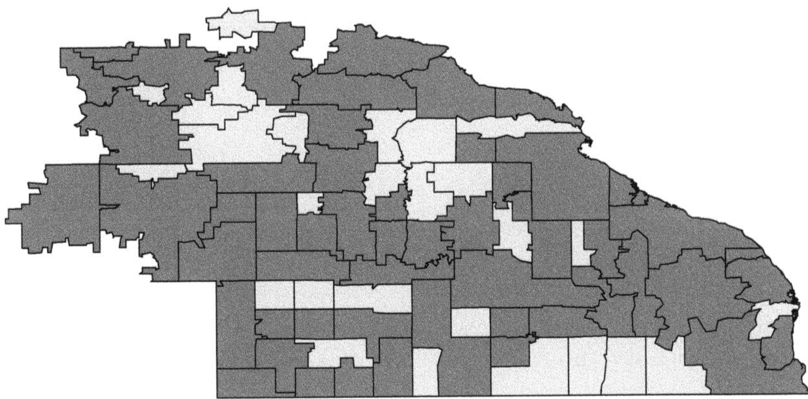

FIGURE 17.1
Map showing missingness pattern for the smoking cessation data: lightly shaded regions are those having no responses.

which no study patients actually resided. Markov chain Monte Carlo algorithms were run with five initially overdispersed sampling chains, each for 20,000 iterations. Convergence was diagnosed using correlation plots, sample trace plots, and Gelman–Rubin statistics. In every case, a burn-in period of 15,000 iterations appeared satisfactory. Retaining the remaining 5000 samples from each chain yielded a final sample of 25,000 for posterior summarization.

Table 17.2 presents deviance information criterion (DIC) scores for several interval-censored cure rate models with different random effect specifications. Models 1 and 2 have only random frailty terms ϕ_i with independent and identically distributed (i.i.d.) and CAR priors, respectively; Models 3 and 4 add random Weibull shape parameters $\zeta_i = \log \rho_i$, again with i.i.d. and CAR priors, respectively; while Models 5 and 6 consider the full MCAR structure for the (ϕ_i, ζ_i) pairs, assuming common and distinct spatial smoothing parameters, respectively. The DIC scores seem to suggest little benefit from the more complex models, and the low p_D scores indicate a high degree of shrinkage in the random effects. Here, we present results for Model 6 in order to maintain complete generality.

Table 17.3 presents estimated posterior quantiles for the fixed effects $\boldsymbol{\beta}$, cure fraction θ, and hyperparameters from Model 6. Smoking intervention does, expectedly, produce a decrease in the log relative risk of relapse, and women seem to be more likely to relapse than men. This result is often attributed to the (real or perceived) risk of weight gain following smoking cessation. The number of cigarettes smoked per day seems to be less significant,

TABLE 17.2
DIC and p_D values for various competing interval-censored models

Model	Log relative risk	pD	DIC
1	$\mathbf{x}_{ij}^T\boldsymbol{\beta} + \phi_i;\ \phi_i \overset{iid}{\sim} N(0, \tau_\phi),\ \rho_i = \rho\ \forall\ i$	10.3	438
2	$\mathbf{x}_{ij}^T\boldsymbol{\beta} + \phi_i;\ \{\phi_i\} \sim CAR(\lambda_\phi),\ \rho_i = \rho\ \forall\ i$	9.4	435
3	$\mathbf{x}_{ij}^T\boldsymbol{\beta} + \phi_i;\ \phi_i \overset{iid}{\sim} N(0, \tau_\phi),\ \zeta_i \overset{iid}{\sim} N(0, \tau_\zeta)$	13.1	440
4	$\mathbf{x}_{ij}^T\boldsymbol{\beta} + \phi_i;\ \{\phi_i\} \sim CAR(\lambda_\phi),\ \{\zeta_i\} \sim CAR(\lambda_\zeta)$	10.4	439
5	$\mathbf{x}_{ij}^T\boldsymbol{\beta} + \phi_i;\ (\{\phi_i\}, \{\zeta_i\}) \sim MCAR(\rho, \Lambda)$	7.9	434
6	$\mathbf{x}_{ij}^T\boldsymbol{\beta} + \phi_i;\ (\{\phi_i\}, \{\zeta_i\}) \sim MCAR(\rho_\phi, \rho_\zeta, \Lambda)$	8.2	434

TABLE 17.3
Posterior quantiles, full model, and interval-censored case

Parameter	Median	(2.5%, 97.5%)
Intercept	−2.720	(−4.803, −0.648)
Sex (male = 0)	0.291	(−0.173, 0.754)
Duration as smoker	−0.025	(−0.059, 0.009)
SI/UC (usual care = 0)	−0.355	(−0.856, 0.146)
Cigarettes smoked per day	0.010	(−0.010, 0.030)
θ (cure fraction)	0.694	(0.602, 0.782)
ρ_ϕ	0.912	(0.869, 0.988)
ρ_ζ	0.927	(0.906, 0.982)
Λ_{11} (spatial variance component, ϕ_i)	0.005	(0.001, 0.029)
Λ_{22} (spatial variance component, ζ_i)	0.007	(0.002, 0.043)
$\Lambda_{12}/\sqrt{\Lambda_{11}\Lambda_{22}}$	0.323	(−0.746, 0.905)

Note: The (2.5%, 97.5%) interval is the 95% Bayesian credible interval providing the 2.5th and 97.5th quantiles for the posterior distribution of the parameters.

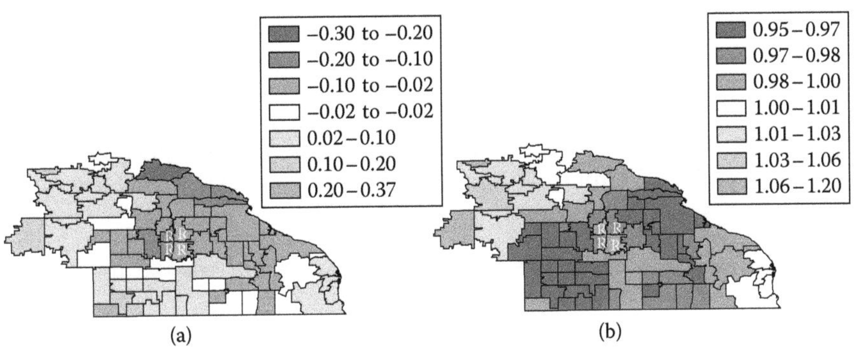

FIGURE 17.2
Maps of posterior means for the ϕ_i (a) and the ρ_i (b) in the full spatial MCAR model, assuming the data are interval censored.

but what is perhaps somewhat counterintuitive is that shorter-term smokers relapse sooner. This may be due to the fact that people are better able to quit smoking as they age (and are thus confronted more clearly with their own mortality).

The estimated cure fraction in Table 17.3 is roughly 0.70, indicating that roughly 70% of smokers who attempt to quit are "cured" and never relapse. The spatial smoothness parameters ρ_ϕ and ρ_ζ are both close to 1, suggesting little is lost by simply setting them both equal to 1 (as in the intrinsic CAR model). Finally, only a moderate correlation is seen between the two random effects, again consistent with the rather weak case for including them in the model at all.

Figure 17.2 maps the posterior medians of the spatial random effects (ϕ_i) and shape (ρ_i) parameters in Model 6. While the magnitudes of the differences appear relatively small across zip codes, the maps reveal some interesting spatial patterns. The south-central region seems to be of some concern, with its high values for both ϕ_i (high overall relapse rate) and ρ_i (increasing baseline hazard over time). By contrast, the four zip codes comprising the city of Rochester, Minnesota (home of the Mayo Clinic, and marked with an R in each map), suggest slightly better than average cessation behavior. Finally, while a nonspatial model cannot impute anything other than the null values ($\phi_i = 0$ and $\rho_i = 1$) for any zip code contributing no data (all of the unshaded regions in Figure 17.1), the spatial models are able to impute nonnull values here, in accordance with the observed values in neighboring regions.

References

[1] Cox DR and Oakes D. *Analysis of Survival Data.* Chapman & Hall, London, 1984.

[2] Bennett S. Analysis of survival data by the proportional odds model. *Statistics in Medicine*, 2:273–277, 1983.

[3] Vaupel JW, Manton KG, and Stallard E. The impact of heterogeneity in individual frailty on the dynamics of mortality. *Demography*, 16:439–454, 1979.

[4] Banerjee S, Wall M, and Carlin BP. Frailty modelling for spatially correlated survival data with application to infant mortality in Minnesota. *Biostatistics*, 4:123–142, 2003.

[5] Li Y and Ryan L. Modeling spatial survival data using semiparametric frailty models. *Biometrics*, 58:287–297, 2002.

[6] Henderson R, Shikamura S, and Gorst D. Modeling spatial variation in leukemia survival data. *Journal of the American Statistical Association*, 97:965–972, 2002.

[7] Banerjee S and Carlin BP. Spatial semi-parametric proportional hazards models for analyzing infant mortality rates in Minnesota counties. In Carriquiry A, Gelman A, Higdon D, Pauler DK, Verdinelli I, Gatsonis C, and Kass RE, eds, *Case Studies in Bayesian Statistics*, vol. VI. Springer, New York, 2002.

[8] Bastos L and Gamerman D. Dynamical survival models with spatial frailty. *Lifetime Data Analysis*, 12:441–460, 2006.

[9] Banerjee S and Carlin BP. Semiparametric spatiotemporal frailty modeling. *Environmetrics*, 14:523–535, 2003.

[10] Banerjee S and Dey DK. Semiparametric proportional odds model for spatially correlated survival data. *Lifetime Data Analysis*, 11:175–191, 2005.

[11] Zhou H, Lawson AB, Hebert J, Slate E, and Hill E. Joint spatial survival modelling for the date of diagnosis and the vital outcome for prostate cancer. *Statistics in Medicine*, 27:3612–3628, 2008.

[12] Zhang J and Lawson AB. Bayesian parametric accelerated failure time spatial model and its application to prostate cancer. *Journal of Applied Statistics*, 38:591–603, 2011.

[13] Lawson AB, Choi J, and Zhang J. Prior choice in discrete latent modeling of spatially referenced cancer survival. *Statistical Methods in Medical Research*, 23:183–200, 2014.

[14] Ibrahim JG, Chen MH, and Sinha D. *Bayesian Survival Analysis*. Springer-Verlag, New York, 2001.

[15] Othus M, Barlogie B, LeBlanc ML, and Crowley JJ. Cure models as a useful statistical tool for analyzing survival. *Clinical Cancer Research*, 18:3731–3736, 2012.

[16] Cooner F, Banerjee S, and McBean AM. Modelling geographically referenced survival data with a cure fraction. *Statistical Methods in Medical Research*, 15:307–324, 2006.

[17] Cooner F, Banerjee S, Carlin BP, and Sinha D. Flexible cure rate modeling under latent activation schemes. *Journal of the American Statistical Association*, 102:560–572, 2007.

[18] Hurtado Rúa SM and Dey D. A transformation class for spatio-temporal survival data with a cure fraction. *Statistical Methods in Medical Research*, April 18, 2012.

[19] Banerjee S and Carlin BP. Parametric spatial cure rate models for interval-censored time-to-relapse data. *Biometrics*, 60:268–275, 2004.

[20] Chen M-H, Ibrahim JG, and Sinha D. A new Bayesian model for survival data with a surviving fraction. *Journal of the American Statistical Association*, 94:909–919, 1999.

[21] Finkelstein DM. A proportional hazards model for interval-censored failure time data. *Biometrics*, 42:845–854, 1986.

[22] van Houwelingen HC and Putter H. *Dynamic Prediction in Clinical Survival Analysis*. CRC Press, Boca Raton, FL, 2011.

[23] Sankila R and Hakulinen T. Survival of patients with colorectal carcinoma: Effect of prior breast cancer. *Journal of the National Cancer Institute*, 90:63–65, 1998.

[24] Diva UA, Dey DK, and Banerjee S. Parametric models for spatially correlated survival data for patients with multiple cancers. *Statistics in Medicine*, 27:2127–2144, 2008.

[25] Diva UA, Banerjee S, and Dey DK. Modeling spatially correlated survival data for individuals with multiple cancers. *Statistical Modeling*, 7:1–23, 2007.

[26] Murray R, Anthonisen NR, Connett JE, Wise RA, Lindgren PG, Greene PG, and Nides MA (for the Lung Health Study Research Group). Effects of multiple attempts to quit smoking and relapses to smoking on pulmonary function. *Journal of Clinical Epidemiology*, 51:1317–1326, 1998.

18

Spatial Longitudinal Analysis

Andrew B. Lawson
Division of Biostatistics and Bioinformatics
Department of Public Health Sciences
Medical University of South Carolina
Charleston, South Carolina

CONTENTS

18.1 General Issues

In many biostatistical applications, there is a need to consider temporal variation. The most common examples are often found in clinical or behavioral intervention trials where a state can be reached by a patient. The time at which the state is reached could be of primary interest. The end point could be a vital outcome (death) or could disease remission, cure, or cessation of a behavior. In all cases, the time of the event is the important random variable. This is the typical scenario where *survival analysis* is employed (see Chapter 17). On the other hand, in some clinical or intervention studies, the variation of response over time is monitored. For example, cholesterol concentrations in blood might be monitored under different treatments, and the effect of these treatments over time is to be examined. In an intervention trial for diet change, food intake might be repeatedly measured via self-report questionnaire. In these cases, the time of measurement is usually fixed and the measurement itself is the random variable.

In clinical trial applications, there is not often a need, or interest, in examining the residential address of the patient or the even the neighborhood or county of residence. As clinical trials are designed experiments, and usually randomized, there should be less need to consider location of residence or area of residence as a factor affecting outcome. Two factors should be considered, however, that could impact a decision to ignore spatial effects. First, at the design stage, there could be a geographical bias in the areas displaying risk for the disease. This bias could manifest itself by differential recruitment. Any strong geographical differences might need to be represented in the study design if the study is to be representative of the population concerned. Even after design, in the analysis phase, there could be strong reasons to examine spatial effects. First, it could be important to know if there were any spatial effects, such as confounding between space and intervention groups, in relation to outcome. Second, all studies, whether designed or observational, have confounders that could be unknown or known but not included in the study. Hence, if the researcher wants to (1) explore factors affecting the study and (2) make sure that confounding is allowed for in the analysis, then the use of spatial information can help with both these tasks. Clayton et al. (1993) have stressed the usefulness of including spatial correlation terms in models to make allowance for confounder and ecological biases, and so there is reasonably strong arguments for always including contextual terms that have spatial structure, at least at the population level.

18.2 Spatial Longitudinal Analysis

There are many situations when variation in an individual response is time dependent, and the focus is the monitoring over time of the associated outcome. The areas of application for these methods are manifold, and in particular, they are commonly applied in clinical trials and community-based behavioral intervention trials.

In many clinical settings, it is possible to conduct trials of new treatments. These trials are essentially designed experiments where patient outcomes are compared between treatment groups. Often, these are monitored over time to establish whether a treatment has been effective. In the simplest case, when only two time points are employed, this leads to simple comparisons. However, if more than two time points are monitored, then considerable complication can arise. In what follows, I will emphasize a general framework for longitudinal analysis that incorporates spatial referencing. Recent general references to this area are Diggle et al. (2002), Verbeke and Molenberghs (2000), Molenberghs and Verbeke (2005), Fitzmaurice et al. (2008), and Fahrmeir and Kneib (2011). Note that there is a close connection between spatiotemporal disease modeling (see Chapter 19) and longitudinal modeling. In the former, the unit is usually a small area, which is monitored over time, whereas in the latter, the unit is usually an individual. In what follows, a Bayesian approach is assumed in the model specification and implementation. Non-Bayesian approaches are of course possible (Fitzmaurice et al., 2008), but the availability of accessible Bayesian software that allows for the incorporation of spatial effects within longitudinal analyses makes the Bayesian choice very convenient.

Define an outcome for the ith individual at a given time j as y_{ij}. Usually, in designed studies the time period is fixed and measurements are made at these fixed times. Hence, the time label denotes a fixed time period. Further define there to be $i = 1, \ldots, m$ individuals or individual observation units and $j = 1, \ldots, J$ time periods. The periods could be units of time, such as minutes, hours, days, weeks, and years, depending on the study design. Usually, recruitment to the study will be randomized in a clinical or intervention trial setting. However, in observational studies, where such control is not possible, it is likely that there will be considerable extra variation and potential for imbalance. Hence, there are strong reasons for inclusion of confounder or random effects when observational designs are used.

A simple general approach is to consider a linear model formulation whereby the outcome of interest is assumed to depend on underlying parameters, and these parameters can be time dependent. For example, serum cholesterol (*ldl*) levels could be measured in patients in a trial for a new cholesterol drug at two time points (baseline and 6 weeks). There are two groups: old drug, fixed dose, and new drug, fixed dose. These groups are labeled $l = 1, 2$, respectively. The outcome of interest is y_{ilj} at given times $j = 1, 2$, and groups $l = 1, 2$. A statistical model for this situation will depend on the nature of the outcome. If the outcome is continuous, then a Gaussian or gamma error model might be assumed. In the cholesterol trial, a continuous Gaussian variate could be assumed, as a first model, and so

$$y_{ilj} \sim N(\mu_{ilj}, \tau_y),$$

where μ_{ilj} is the mean for the ith person at time j in group l, and τ_y is a variance.

The mean parameter would then be specified as a function of available covariates and random effects. For example, it might be important to allow for age of the individual. Define this as x_{1i}. One simple regression model might be

$$\mu_{ilj} = \alpha_l + \beta_1 x_{1i} + \beta_2 t_j,$$

where t_j is the time of the jth period. Here, each group has a separate intercept. The regression parameters would be commonly assumed to have zero-mean Gaussian prior distributions with small precisions. In this model, a group effect is specified, with a covariate effect and constant linear effect over time. There is assumed to be no correlation in the error. More complex models could of course be envisaged. One possibility that is commonly introduced is a susceptibility or frailty term to allow a random response variation between individuals or units. Addition of random effects can include such frailty terms, and can also allow a random temporal dependence term (which can substitute for the regression on t_j):

$$\mu_{ilj} = \alpha_l + \beta_1 x_{1i} + v_i + \eta_j.$$

This temporal dependence term can be assumed to have a lagged dependence via, for example, a random walk prior distribution:

$$\eta_j \sim N(\eta_{j-1}, \tau_\eta), \ j > 1.$$

For the frailty term, a zero-mean Gaussian prior distribution is often assumed:

$$v_i \sim N(0, \tau_v).$$

18.2.1 Spatial effects

If inference about spatial effects is important, then a variety of possibilities exist. If individual-level information is available, such as location of residence, then there are two approaches to the inclusion of space. First, it may be possible to estimate an individual-level effect directly for residential location data. Denote a spatially structured effect as w_i. Also assume that $w_i \equiv w(s_i)$, where s_i is the ith residential address. A model could be assumed of the form

$$\mu_{ilj} = \alpha_l + \beta_1 x_{1i} + w_i + \eta_j,$$

where w_i can be assumed to have a multivariate Gaussian prior distribution and so

$$\mathbf{w} \sim \mathbf{N}(\mathbf{0}, \mathbf{\Gamma}),$$

where $\mathbf{\Gamma}$ is a covariance matrix. A number of options exist for the specification of $\mathbf{\Gamma}$. It is possible to assume a singular normal conditional autoregressive (CAR) specification whereby a Markov random field distribution (improper CAR) is defined. This could be based on the definition of adjacencies obtained from natural neighborhoods (Sibson, 1980). Alternatively, the elements of $\mathbf{\Gamma}$ could be explicitly defined as functions of distance between locations as $\mathbf{\Gamma}_{ij} = \tau_w \rho(d_{ij})$ and $d_{ij} = ||s_i - s_j||$. Suitable forms of the covariance function $\rho(.)$ are discussed in detail elsewhere in this volume, but two commonly adopted forms are the power exponential or Matérn (Diggle and Ribeiro, 2007). This approach provides for a spatially continuous spatial effect. Both of these approaches assume a reasonably dense spatial array of locations, based on which a random field is a reasonable approximation. Sparse locations such as may be found in random sampling of a population may not be dense enough to justify this approach. Second, instead of considering a spatiually continuous effect, it may be useful to consider a contextual effect, which is *inherited* from a stratification of the study region.

18.2.1.1 Contexual effects

This second approach is a specific example of the use of ancillary information from groupings that individuals belong to in the modeling of individual outcomes. For example, a nonspatial grouping could be *family* grouping, that is, where common family members are thought to inherit a family effect. In a spatial example, one such grouping could be the census tract in which someone lives. Censuses usually produce aggregate information about small areas and their demographic and socioeconomic characteristics. Hence, the socioeconomic features of a tract could affect the individuals living there, either directly via availability of resources locally or indirectly via the grouping of like people and their behaviors. Often, contextal effects are aggregated above the level of the individual in spatial studies, and they can be nested. For example, in the United States, a person lives in a census block within a census tract within a county within a state. Each of these levels could have a contextual effect on the individual.

To denote contextual effects, we assume the notation $w_{i(i \in l)}$, which means that the ith person is assigned the contextal effect of the lth grouping that he or she lives in. The simplest form for such a spatial contextual effect would be a measured covariate that is available at the aggregate level and is assigned to the ith person, the value of the covariate in the lth area. An example would be assigning the lth county average income as a predictor to the ith person who lives in the lth county. In that case, $w_{i(i \in l)} = \alpha x_l$, where α is a regression parameter and x_l is the average income for the lth county. In this way, the contextual effect can be considered a form of *neighborhood* effect (as it encompasses a larger area) (see, e.g., Kawachi and Berkman, 2003), although the neighborhood is not predefined. In addition, of course, once the individual is geocoded to a small area, it is then possible to add adjacent small areas to the neighborhood of the individual. For example, if the individual lies in the lth tract, then the data in adjacent tracts can be used also, that is, $w_{i(i \in l)} = \alpha\{f(\sum_{k \in \delta_l} x_k + x_l)\}$, where δ_l denotes the set of neighbors of the lth tract. In this way, uncertainty in the definition of neighborhoods can be partially allowed for.

A more sophisticated form of contextual effect that can be considered is where a random effect at an aggregate level is inherited by the individual. For example, we could consider a set of counties in South Carolina (46) and observe within the state individuals with health outcomes $\{y_i\}$. For the ith individual, we could conceive that he or she inherits a random effect from the lth county (that he or she lives in). Hence, $w_{i(i \in l)}$ could be the value of a random effect for the lth county so that, say, $w_{i(i \in l)} = u_l$. Now the $\{u_i\}$ could be spatially correlated or uncorrelated as required. In fact, more generally, we could assume a convolution model (Chapter 7) to describe the individuals's inherited effect:

$$w_{i(i \in l)} = v_l + u_l,$$

where v_l is an uncorrelated county effect and u_l is a correlated effect. For example, if we have m counties we could assume, for $j = 1, \ldots, m$, that

$$v_j \sim N(0, \tau_v)$$
$$u_j \sim ICAR(\tau_u),$$

where $ICAR(\tau_u)$ represents a conditional autoregressive prior distribution with variance τ_u. This type of model is directly available on free Markov chain Monte Carlo (MCMC) packages such as WinBUGS and OpenBUGS and also on the free R package INLA. Note that this type of hierarchical context modeling can be effected by using *nested indexing*. On WinBUGS, the relevant code to assign an area effect to an individual would be

$$w[label[i]],$$

where label is a variable that holds the area label (l) associated with the ith individual. Outside the i loop, the ICAR model would be

$$w[1:m] \sim car.normal(\ldots).$$

This approach can easily be extended to multiple contextual effects and can be nested so that hierarchies of spatial units can be included.

18.2.2 General model

A linear mixed model for a continuous outcome can be specified where the vector $\mathbf{y} : \{\mathbf{y}_1, \ldots, \mathbf{y}_m\}$, with $\mathbf{y}_i = (y_{i1}, \ldots, y_{iJ})$, is a realization of a random vector from a multivariate normal distribution of the form

$$\mathbf{y}_i = \mathbf{x}_i'\boldsymbol{\beta} + \mathbf{z}_i'\boldsymbol{\gamma} + \mathbf{e}_i, \tag{18.1}$$

where $\mathbf{x}_i'\boldsymbol{\beta}$ is a linear predictor (which can include a group indicator), $\mathbf{z}_i$ is a vector of g random effects for the ith individual, $\boldsymbol{\gamma}$ is a $g \times J$ unit matrix, and the model for $\mathbf{e}_i$ is $\mathbf{e} \sim \mathbf{N}(\mathbf{0}, \tau\boldsymbol{\Sigma})$, where $N = mJ$, $\mathbf{x}'$ is an $N \times p$ matrix of covariates, $\boldsymbol{\beta}$ is a parameter vector, τ is a variance, and $\boldsymbol{\Sigma}$ is a block-diagonal matrix with $J \times J$ blocks each representing the variance matrix of the vector measurements on a single subject. Various assumptions about the structure of the covariance can be made depending on the focus of the study. I do not pursue this here, but refer those interested to Diggle et al. (2002).

An extension of the approach can be easily made to generalized linear mixed models (GLMMs), whereby (18.1) is modified to allow different distributional assumptions at the data level and a link function is introduced to connect the mean to the linear predictor:

$$\mathbf{y}_i \sim f(\boldsymbol{\mu}_i),$$

where

$$E(\mathbf{y}_i) = \boldsymbol{\mu}_i$$

and

$$g(\boldsymbol{\mu}_i) = \boldsymbol{\eta}_i = \mathbf{x}_i'\boldsymbol{\beta} + \mathbf{z}_i'\boldsymbol{\gamma}.$$

Suitable models for $f(\boldsymbol{\mu}_i)$ could be the Poisson, binomial, gamma, and others in the standard exponential family.

18.2.3 Missing data

Missing data are a very important aspect of longitudinal studies. Apart from general forms of missing data found in all studies, missingness can often arise in longitudinal studies due to individuals failing to remain in the study. This *dropout*, as it is known, is common in clinical or intervention trials, and often by the end of a study, 20%–30% of the participants may have left. There are different forms of missingness mechanisms. Verbeke and Molenberghs (2000) define three basic forms,whereas Lesaffre and Lawson (2012, chapter 15) discuss outcome missingness types and predictor missingness. See also Daniels and Hogan (2008) for a more detailed examination of issues. While it may be important in given applications to consider which of these apply, here I will focus simply on the implications of georeferencing on missingness. In particular, the effect of georeferencing on dropout will be discussed. Usually a geographical delimitation of a study area is made in the design of a trial. During the course of a trial, it is possible for participants to move their residence. This is more likely to occur when longer-term studies, such as behavioral intervention trials, are considered. These trials can last up to 2–3 years. Ultimately, the mobility of participants could lead to a move outside the study area. If there is a geographical limit on the study in terms of recruitment, then those who are removed from the study area would be regarded as dropout. With intention-to-treat approaches to trials (Lachin, 2000), it may be important to use all data for those entered into the trial, and so it may be important to consider this geographical dropout. Denote the outcome for the ith participant at time j as y_{ij}. Consider residential history defined as $s_i(t)$, where t is continuous time and s denotes a spatial location (residence address). A study lasts for $j = 1, \ldots, J$ time points. Outcome measurements are made at the J time points. Often, the residential history will be discretized into the study time periods so that s_{ij} may be censored between time points. Besides location shifts, it would be useful to define a dropout indicator R whereby r_{ij} denotes presence (1) or absence (0) at the j measurement. While r_{ij} could be 0 for a range of reasons, one such could be related to s_{ij}. Hence, it might be important to consider (y_{ij}, r_{ij}, s_{ij}) jointly in modeling a given outcome. This area appears to have been little explored. In Chapter 15 of Lesaffre and Lawson (2012), a full discussion of missingness can be found.

18.2.4 Malaria intervention example

In this example, an approach to longitudinal spatial analyis is descibed within a malaria intervention in the city of Dar es Salaam, Tanzania. A full report on this intervention and related analysis has been given by Lawson et al. (2014), and other analyses by Maheu-Giroux and Castro (2013). The intervention was part of the Urban Malaria Control Program (UMCP) coordinated by the Dar es Salaam City Council, and made use of microbial larviciding to target larval stages of *Anopheles* mosquitoes, the malaria vector. The UMCP was chosen because the program collected both longitudinal and cross-sectional data, with more than 64,000 observations gathered for more than 4 years. Previous evaluations have shown that the intervention significantly reduced the prevalence of malaria infection in intervention areas. The UMCP covered 15 wards in Dar es Salaam, 5 in each of the three municipalities that comprise the city. Administratively, the smallest unit is 10 cells (TCU), which has an average of 20 households, although some can be much larger. The TCU was used as the UMCP sampling unit and is the spatial unit of analysis for this study. Figure 18.1 displays the spatial distribution of these units in the city.

UMCP data were assembled through household interviews, conducted between May 2004 and December 2008. Briefly, each survey round randomly sampled 10 TCUs in each ward (based on a roster of TCU numbers) and interviewed all members of households in the sampled TCUs who agreed to participate, aiming at a sample size of 400–450 individuals.

FIGURE 18.1
Map of 10-cell units in Dar es Salaam, Tanzania.

Upon consent, individuals were tested for malaria through microscopy. All data collected were georeferenced.

Starting on the second survey round, all households previously interviewed were followed up, and a new cross-sectional survey was conducted in different TCUs. A total of six survey rounds were done, gathering information on 64,537 people. Only 99 individuals were followed up in all six survey rounds, and 38,661 individuals were interviewed only once. Refusal during cross-sectional surveys was almost null. There was some refusal in the first follow-up, addressed with sensitization efforts.

The UMCP commenced weekly larvicide application in March 2006 in 3 out of the 15 wards, scaled-up to 9 wards in May 2007, and to all 15 wards in April 2008. Therefore, intervention periods did not exactly match survey round periods. To accommodate that, we created a variable that considered the date of the interview to indicate if the individual was in a location under the larviciding intervention.

The UMCP collected varied information on household and individual characteristics, which facilitates a comprehensive assessment of the potential impact of the larval control intervention, controlling for socioeconomic, environmental, and behavioral aspects.

18.2.4.1 Model development

The outcome of interest is whether an individual tested positive for malaria. We considered all observations longitudinally, and we denoted these for an individual as y_{ij}, where i is the individual index and j is the survey round. Here, the ith individual can appear at different times and have between one and six observations. Cross-sectional data were also analyzed in the original analyses, but we do not present those results here.

Bayesian modeling formulations considered the different ways of using the UMCP data described above. Our choice of models was determined by their relevance to the analysis of malaria and their goodness of fit. We measured goodness of fit via the deviance information criterion (DIC) and marginal likelihood.

18.2.4.2 Dropout and missingness

We adopted a strategy of utilizing complete case for individuals in the study. Therefore, any individual observation with missingness in predictors or in the outcome variable was excluded from the analysis. Using this criterion, 632 observations (0.01% of the total observations) were excluded, resulting in a dataset with 63,905 observations. Complete case can lead to bias in estimates when substantial proportions of cases are missing. In our case, a very small number of observations were excluded, and we did not observe any strong correlation between missing observations and patterns in the outcome.

In the case of loss to follow-up, we do not have information on the mechanisms that could affect failure to attend over time. We have assumed an MCAR (missing completely at random) situation and so have not anticipated there to be any systematic correlation between outcome or intervention and participation.

18.2.4.3 Longitudinal models

The outcome for the general model is a binary variable representing positive test for malaria (0/1), and our data model is of the form

$$\begin{aligned} y_{ij} &\sim Bern(p_{ij}) \\ f(p_{ij}) &= x_{ij}^T\beta + z_{ij}^T\gamma_j, \end{aligned}$$

where we assumed for simplicity that $f(.)$ is a logit link. The (possibly time-dependent) predictors are defined in vector x_{ij}^T with regression parameters in vector β. In this formulation, we considered a range of predictors within the fixed design matrix. For the random component, we assumed that $z_{ij}^T\gamma_j$ is represented by a sum of terms, with z_{ij}^T representing random effects with individual or possible time dependence, and is a binary indicator matrix of dimension. First, we considered temporal effects and then spatial contextual effects.

Individuals were sampled over time, but the times of sampling differed. Hence, denote t_{ij} as the sampled times for the ith individual at the jth survey round. Our temporal model consists of fixed predictors x_{ij}^T, so that a basic form is given by $f(p_{ij}) = x_{ij}^T\beta + g(t_{ij})$, where $g(t_{ij})$ is a link function and can be formatted in a variety of ways. The time on the study in months and the start date on the study for an individual are known.

Initially, we considered a temporal dependency case:

$$\begin{aligned} f(p_{ij}) &= x_{ij}^T\beta + \lambda_{ij}(t_{ij}) \\ \lambda_{ij}(t_{ij}) &\sim N(f_t(\phi d_{ij})\lambda_{i,j-1}(t_{i,j-1}), \tau_t), \end{aligned}$$

where the second term represents the effects of time of screening and is a function of the actual time t_{ij} for the ith individual and jth round. The actual temporal model assumed is an Orstein–Uhlenbeck (O-U) process (Guttorp, 1995) in which the time dependence is scaled by the time difference from the last observation. This process is derived from the stochastic differential equation where $\lambda_{ij}(t_{ij})$ is a function of a zero-mean Weiner process, with $\phi > 0$. The O-U process can be defined via a differential equation where $d\lambda = \mu(\lambda)dt + \sigma(\lambda)dB$. Here, B is the Weiner independent increment process, and so λ is Markovian. This can be expressed as a difference formulation whereby the mean of $\lambda_{ij}(t_{ij})$ is a scaled function of the preceding value with a time-dependent variance component. This is the continuous time analog of the AR1 model. In particular, in our model the conditional expectation and precision are

$$\begin{aligned} E(\lambda_{ij}(t_{ij})) &= \exp(-\phi d_{ij}).\lambda_{i,j-1}(t_{i,j-1}) \\ \tau_\lambda &= \tau_0(1 - \exp(-2\phi d_{ij}))^{-1} \\ d_{ij} &= t_{ij} - t_{i,j-1}. \end{aligned}$$

It is important to employ this construction as individuals, when linked in time, have different observation times and gaps between observations. Note that the ϕ parameter acts as a scaling effect on the dependence on time. An extension to this could be to consider grouped or individual scaling parameters ϕ_i, so that dependencies could be allowed to vary across groups or individuals, but I do not pursue this here.

A spatial extension to this model using longitudinal data (models L1–L4 in Figure 18.2) considered the inclusion of spatial contextual effects. We considered that the UMCP area in Dar es Salaam consists of a number of administrative units: municipalities (3), wards (15), and TCUs (913 sampled, out of 3243), as well as 6796 unique households. We sought to make allowance for the neighborhood effects that could arise from the nested administrative regions, and thus added spatial effects (fixed and random) to the basic temporal model.

For example,

$$\text{logit}(p_{ij}) = x_{ij}^T\beta + \lambda_{ij}(t_{ij}) + \nu_{i\in k} + \cdots + u_{i\in l},$$

where $\nu_{i\in k}$ denotes an effect for the kth contextual neighborhood (ward) as for $u_{i\in l}$ with the lth neighborhood (TCU). A number of these effects could be included, for example, uncorrelated heterogeneity, in which case the noninformative prior specification could be assumed (L1 and L2). Alternatively, a spatially structured effect could be assumed, allowing for spatial correlation at the contextual level (L3 and L4). In the latter case, we made the conventional assumption that the effect has a conditional autoregressive prior distribution, which is specified conditional on effects in neighboring areas as $u_l|\{u_p\}_{p\neq l} \sim N(\rho\bar{u}_{\delta_i}, \tau_u/n_{\delta_i})$.

This assumption provides for a Markov dependence on immediate neighboring areas (those that are adjacent) defined for the ith neighborhood as δ_i, with number of neighbors as n_{δ_i}. In our models, we also assumed that $\rho = 1$, which is the improper CAR (ICAR) model specification. Our overall O-U spatial model is then specified as

$$\begin{aligned}
\text{logit}(p_{ij}) &= x_{ij}^T\beta + \lambda_{ij}(t_{ij}) + \nu_{i\in k} + \cdots + u_{i\in l} \\
\lambda_{ij}(t_{ij}) &\sim N(f_t(\phi d_{ij})\lambda_{i,j-1}(t_{i,j-1}), \tau_t) \\
\tau_\lambda &= \tau_0(1 - \exp(-2\phi d_{ij}))^{-1} \\
d_{ij} &= t_{ij} - t_{i,j-1},
\end{aligned}$$

where the collection $\nu_{i\in k} + \cdots + u_{i\in l}$ consists of a range of random spatial effects (indexed by TCU, ward, municipality, and household).

Prior distributions for parameters were derived from considerations of noninformativeness and parsimony. We assumed zero-mean Gaussian prior distributions for regression parameters $\{\beta_j\}$, so that $\beta_j \sim N(0, \tau_{\beta_j})$, where the precisions are assumed to be fixed.

Parameters		Longitudinal models			
		L1	L2	L3	L4
O-U month random effect		x	x	x	x
Round fixed effect			x		x
Net treatment					
	All	x	x	x	x
Five or more symptoms					
	All	x	x	x	x
Household uncorrelated random effect		x	x	x	x
TCU uncorrelated random effect		x	x	x	x
TCU correlated random effect		x	x	x	x
DIC		34,779	34,160	34,757	34,353

FIGURE 18.2
Tabular comparison of four longitudinal models for the malaria intervention data with different key ingredients. The DIC of each model is provided.

We also assumed a lognormal prior distribution for ϕ, $\log(\phi) \sim N(0, 0.2)$. For the marginal O-U process precision τ_0^{-1}, we assumed $\tau_0^{-1} \sim Ga(1, 5e-05)$, while for the precisions on the uncorrelated or spatially structured random effects, we assumed $\tau_*^{-1} \sim Ga(1, 5e-05)$.

Overall, the best model based on DIC was model L2, which included the spatially unstructured effect only at the TCU level. The wide range of covariates were fitted under model L2, and of these, the significantly reduced odds ratios were found for variables related to mosquito screening, treatment of bed nets, and sleeping outside among others, whereas those with five or more symptoms tended to display elevated odds ratios. Figure 18.3 displays the posterior mean estimates and 95% credible intervals of hyperparameters for the model L2. The uncorrelated 10-cell unit effect for model L2 is displayed in Figure 18.4.

Hyperparameters	Model L2
(Household)	5.54 (4.13, 7.60)
(TCU uncorrelated)	2.87 (2.37, 3.44)
τ_0 (O-U) model	0.27 (0.09, 0.58)
ϕ (O-U) model	0.03 (0.01, 0.07)

FIGURE 18.3
Posterior hyperparameter mean estimates and 95% credible intervals for model L2.

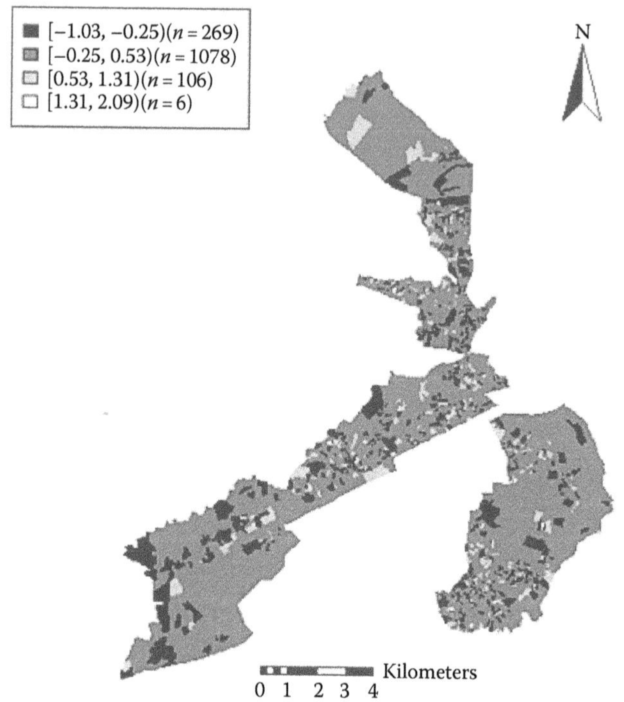

FIGURE 18.4
Ten-cell unit–level uncorrelated random effect for model L2. (Reproduced with permission of Sage Publications.)

18.3 Extensions to Repeated Events

There are many other application areas where spatial context can be fruitfully introduced or allowed for. One such area that is closely related to longitudinal analysis is the analysis of repeated events. This is sometimes called repeated event analysis or event history analysis. For a recent review, see, for example, Cook and Lawless (2007). This is an extension of longitudinal and survival analysis, where instead of making one measurement at different times or observing a single time to end point, the time period is fixed and within that period a sequence of events is observed on an individual. In the simplest case, the sequence could consist of just a single type of event, and its repeated occurrence is observed. Hence, for a single person, this is a point process in time, assuming the event does not have finite duration. An example of a simple sequence would be a sequence of doctors' visits where the time of visit is recorded. If we are only concerned about the times of the visits (and not their nature or duration), then this can be considered a point process at the unit (patient) level.

At this point, it is worth noting that there is an immediate connection between the linear or generalized linear modeling approach of longitudinal analysis and the time-based analysis for survival data. With repeated events, time is important, but when fixed time periods are used and observations are collected within these time periods, the resulting counts of events can be considered within the framework of the conventional hierarchical model. For example, for simple doctor visits, the count of visits in time periods might be treated at the first level of hierarchy as a Poisson random variable, whereas if the times are recorded directly, then a (heterogeneous) Poisson process model might be appropriate. This, of course, mirrors the duality of the point process and count model when binning of events takes place in a spatial context.

18.3.1 Simple repeated events

Assume first that a patient resides at an address and makes repeated doctor visits. Denote the address as s_i and the sequence of visits as $\{t_{i,1}, \ldots, t_{i,n_i}\}$, where t denotes visit time within a study period (t_0, t_T) and n_i is the number of visits. If some basic assumptions are made concerning conditional independence (given knowledge of all confounding and event history, independence of the events might be a reasonable starting model. If a modulated (heterogeneous) Poisson process (PP) were assumed for the event times, then a conditional PP likelihood could be assumed whereby the first-order intensity for the ith individual could be defined as

$$\lambda_i(t) = \lambda_{i0}(t)\exp(\mathbf{x}_i'\boldsymbol{\beta} + g_i(t)).$$

Note that here the covariates are included via the fixed vector $\mathbf{x}_i'$, and the $g_i(t)$ function could be a smooth function of time. The baseline function $\lambda_{i0}(t)$ is also unit specific. The associated unconditional likelihood is given by

$$\prod_{i=1}^{m}\prod_{j=1}^{n_i}\lambda_i(t_{ij})\exp\{\Lambda_i\}, \tag{18.2}$$

where

$$\Lambda_i = \int_{t_0}^{t_T}\lambda_i(u)du. \tag{18.3}$$

For fixed covariates, $\Lambda_i = \exp(\mathbf{x}_i'\boldsymbol{\beta})\int_{t_0}^{t_T}\lambda_{i0}(u)\exp(g_i(u))du$. Various methods can be used to include time-varying effects. Discretizing to allow piecewise linear terms in the baseline and $g_i(t)$ are possible, while semiparametric models could also be assumed. Of course, gamma or Dirichlet processes could also be used (Ibrahaim et al., 2000). Cook and Lawless (2007, chapter 3) discuss various possibilities for Poisson process models.

18.3.2 More complex repeated events

Two generalizations are immediate from the simple event case. First, multiple types of events could occur. Second, a feature associated with the event could be important and could vary with time or type of event. An example of the first situation could easily arise with the progression of a disease. Visits to doctors could be interspersed with hospital visits or nurse visits. In fact, complex patterns of repeated events are more usual when making observational studies on disease progression than in clinical settings. The observed data could then be of the form $\{t^s_{i,1}, \ldots, t^s_{i,n_i}\}$, where s denotes the event type. For a fixed number of event types (L), s could be defined as $s = 1, \ldots, L$. The second situation arises when the event has attached a mark. For instance, a severity score or biomarker might be measured at a given visit. Alternatively, the visit itself could have a duration. In the first case, the mark or measurement could be jointly modeled with the event time, and the correlation between the mark and time could be directly modeled. In this case, the observed data for the ith individual would be $\{t_{i,1}, x_{i,1}, \ldots, t_{i,n_i}, x_{i,n_i}\}$, where x is the mark value. When duration is associated with event time, there can be a complication, as this type of mark directly affects the subsequent event times (as duration is time based). In this case, the observed data would be $\{t_{i,1}, d_{i,1}, \ldots, t_{i,n_i}, d_{i,n_i}\}$, where d is the event duration. This might be appropriate where hospital visits involve stays of different lengths.

Ultimately, we might have a mixture of these situations where multitype events also have marks or durations, and so for the mark case, we would observe $\{t^s_{i,1}, x_{i,1} \ldots, t^s_{i,n_i}, x_{i,n_i}\}$.

Modeling approaches for these different situations depend on the observed data and also the study purpose. For example, if known times are observed, it is often convenient to conditionally model the marks given the times so that we have the joint model

$$[x, t] = [x|t][t]. \tag{18.4}$$

In this case, we consider the times to be governed by a point process model. In addition, the model for the marks could be a simple Gaussian distribution: $[x|t]\tilde{}N(\mu(t), \tau)$. Here, dependence on time could be made explicit in the mean parameterization ($\mu(t)$). This mean function could be specified to include covariates (individual or contextual) as well as time dependence, for example, $\mu(t) = x'\beta + \gamma(t)$. Note also that the point process model could also have covariate dependence, and so it is debatable whether there should be multiple entries of the same covariate in each model. For example, patient age could affect both the mark (e.g., blood pressure) and doctor visit times.

On the other hand, if we do not observe directly the visit time but simply the number of visits within a fixed time period, then usually the mark would not be any longer associated directly with the individual (as the information has been averaged). The mark either would be based on the time period or is an average mark for all events within the period. As the exact times have been lost, the resulting data would consist of counts of visits and an average mark or the value of (say) a contextual variable pertaining to the time period. In this case, the observed data would be y_{ij}, where there are $j = 1, \ldots, J$ time periods. We might be interested in the joint distribution $[x, y] = [x|y][y]$. In effect, the model for the mark is conditioned on the count and a separate count model is specified. If the focus were on the counts per se, then the alternative formulation of $[x, y] = [y|x][x]$ could be considered, and often the visit frequency is modeled conditionally, that is, via $[y|x]$ treating the mark as a covariate.

These formulations do not include any explicit spatial dependence. In Section 18.3.2.1 some proposals for how this dependence could be incorporated are proposed. As in Section 18.2, the spatial effects can be regarded as contextual.

18.3.2.1 Known times

Single events. As a first pass, we could assume that the times follow a heterogeneous Poisson process (hPP) so that

$$f(t) = \lambda(t) \exp\left(-\int_{t_0}^{t_T} \lambda(u)du\right)$$

and is conditional on n_i; then the likelihood element for the ith individual would be

$$\prod_{i=1}^{m}\prod_{j=1}^{n_i} \lambda_i(t_{ij}) \exp\{-\Lambda_i\}, \tag{18.5}$$

where

$$\Lambda_i = \int_{t_0}^{t_T} \lambda_i(u)du,$$

and $\lambda_i(t) = \lambda_{i0}(t)\exp(\mathbf{x}_i'\boldsymbol{\beta} + g_i(t))$. Note that it is possible to include within a Bayesian hierarchy the spatial effect, especially if it is not time dependent. For example, a contextual effect at the individual level could be included as $\Lambda_i = \exp(\mathbf{x}_i'\boldsymbol{\beta} + w_i)\int_{t_0}^{t_T} \lambda_{i0}(u)\exp(g_i(u))du$, where $w_i = w(\underset{i\in j}{c_j})$, as defined in Section 18.2.4.3. More generally, the spatial location of the individual (s_i) could be incorporated in the PP model and a space–time formulation could be considered directly

$$\lambda_i(s,t) = \lambda_{i0}(s,t)\exp(\mathbf{x}_i'\boldsymbol{\beta} + g_i(t) + h_i(s)),$$

and the resulting likelihood would be

$$\prod_{i=1}^{m}\prod_{j=1}^{n_i} \lambda_i(s_i, t_{ij}) \exp\{-\Lambda_i\},$$

where

$$\Lambda_i = \int_{t_0}^{t_T}\int_W \lambda_i(v,u)dvdu,$$

and W is the study region area. In addition, variants of the specification could allow for nonparametric specification of the $g_i(t)$, $h_i(s)$ functions, with a simple alternative being piecewise discretization in time and space. For space, it may be possible to define a common surface, say $h(s)$, and to form piecewise constant components from a tiling of the distribution of individuals. However, if the spatial component is zero centered, then further approximations may be available. Of course, integral approximations such as proposed by Berman and Turner (1992) could also be considered.

Multiple event types. Alternatively, simpler intensity-based methods can be pursued. Denote the intensity of the ith individual and jth type as $\lambda_{ij}(t|H_i(t))$, where $H_i(t)$ is the event history. A different period of observation is allowed for each individual: $[0, \tau_i]$. Denote the history of the ith subject as $H_i(t) = \{N_i(s),\ 0 < s < t\}$, where $N_{ij}(t)$ is the number of event of jth type occurring on the ith individual in the interval $[0, t]$ and $N_i(t) = \{N_{i1}(t), \ldots, N_{iJ}(t)\}'$. Hence, the history of the process concerns the preceding count accumulation. The resulting likelihood is given for the times t_{ijk}, $k = 1, \ldots, N_{ij}(t)$ so that

$$\prod_{j=1}^{J}\left\{\prod_{k=1}^{n_{ij}} \lambda_{ij}(t_{ijk}|H_i(t_{ijk}))\exp\left(-\int_0^{\tau_i}\lambda_{ij}(u|H_i(u))du\right)\right\}. \tag{18.6}$$

This, of course, assumes that the distribution of events is functionally independent. In many situations, this would of course not be reasonable. For example, a doctor's visit might precipitate a hospital visit. When multiple events can arise with dependence over time, it is often useful to consider transition models for the occurrence of events of different types. These types of models specify the probability that an event of a given type will occur in an interval of time given the preceding event. Hence, they are naturally specified conditioned on the preceding event, its type and time, or the type of preceding events and their times. Alternatively, a competing risk approach could be envisaged. I do not consider these further here.

Incorporation of random effects can be effected via redefinition of (18.6): $\lambda_{ij}(t_{ijk}|H_i(t_{ijk}), R_{ij}) = \lambda_{ij}(t_{ijk}|H_i(t_{ijk}), \exp\{W_{ij}\})$, where W_{ij} would represent the individual random effect for the jth event type. This could be decomposed into a number of individual specific or contextual random effects. For example, it would be possible to consider a set of L counties labeled c_l, $l = 1, \ldots, L$. Then we could consider as a first example the hierarchical random effect model: $W_{ij} = (u_{ij} + v_i + w_i)$, where $w_i = w(\underset{i \in l}{c_l})$ and $v_i = v(\underset{i \in l}{c_l})$, and these are spatial contextual effects, and $u_{ij} = u_{1i} + u_{2j}$, where u_{1i} is a general individual frailty and u_{2j} is a type-specific effect. The prior distributions for the effects could be overdispersed zero-mean Gaussian for u_{1i}, u_{2j}, v_i, while for w_i, a CAR formulation would be possible. A variety of other formulations would be possible, of course. As before, it would be possible to consider the intensity as a function of both time and space, and hence to include a specific spatial component in its definition. However, it is probably simpler and more convenient to consider conditioning on the spatial component defined at a higher level of the hierarchy via a random effect.

18.3.3 Fixed time periods

When fixed time periods are observed, it is usual to collect events within the time periods into counts of events. Of course, when this is done, information about the time sequence of events is lost within the time period. Hence, even if the residential location of individuals is known and known at a fine spatial resolution level, the resolution level in time is aggregated. In general, we denote the count within the jth time period as y_{ijl}, where i denotes the individual, $j = 1, \ldots, J$ denotes the time periods, and l denotes the event type ($l = 1, \ldots, L$).

18.3.3.1 Single events

In the case of a single outcome, we observe $\{y_{ij}\}$ for a sequence of J times and $i = 1, \ldots, m$. The simplest modeling approach is to assume a generalized linear model for the counts with some form of time dependence. For example, it could be assumed that $y_{ij} \sim Pois(\mu_{ij})$ and a log-linear model could be assumed for the mean:

$$\log \mu_{ij} = \mathbf{x}_i'\boldsymbol{\beta} + v_i + \gamma_j, \tag{18.7}$$

where $\mathbf{x}_i'$ is a vector of individual fixed covariates, $\boldsymbol{\beta}$ is the corresponding parameter vector, v_i is an individual frailty, and γ_j is a common temporal effect. The temporal effect could have a variety of specifications:

1. An uncorrelated prior distribution (e.g., $\gamma_j \sim N(0, \tau_\gamma)$)
2. A correlated prior distribution (e.g., a random walk: $\gamma_1 \sim N(0, \tau_\gamma)$, $\gamma_j \sim N(\gamma_{j-1}, \tau_\gamma)$ $j > 1$)
3. A trend regression on time $\gamma_j = \beta t_j$ (or a higher-order polynomial), where t_j is the time of the jth period (start or end or middle by convention)

If the focus is on a parsimonious description of the overall behavior, then option 2 may be favored, as it is relatively nonparametric. However, if a specific linear or polynomial estimate of trend is required, then option 3 may be preferred. The individual frailty would usually be assumed to have a uncorrelated zero-mean Gaussian distribution:

$$v_i \sim N(0, \tau_v).$$

Incorporation of spatial contextual effects can follow as before by extending the model in (18.7) to include a individual contextual component:

$$\log(\mu_{ij}) = \mathbf{x}_i'\boldsymbol{\beta} + v_i + w_i + \gamma_j, \tag{18.8}$$

where $w_i = w(\underset{i \in l}{c_l})$ and $w(c_l)|w(c_{-l}) \sim N(\overline{w}(c_{\delta_l}), \tau_w/n_{\delta_l})$, where $w(c_l)$ is the value of w for the lth county and $\overline{w}(c_{\delta_l})$ is the average value of w for the neighborhood (δ_l) of c_l. The number of counties in this neighborhood is given by n_{δ_l}. Note that v_i is an individual frailty effect here. An alternative specification could also assume an uncorrelated county effect, as for w_i, except without correlation. Steele et al. (2004) describe essentially the model in (18.7), albeit with multiple events.

18.3.3.2 Multiple event types

In the case of multiple event types, assume a count of the form y_{ijl} where l denotes the event type. Steele et al. (2004) describe a competing risk model where a multinomial form is assumed for the vector of events within a time period. The multinomial probability ratio (relative to the zero-event case) was assumed to be defined by a logit link to a linear predictor with covariate random effect and trend components similar to (18.8), but without the spatial dependence. In general, one approach to these problems assumes that, conditionally on $N_{ij} = \sum_l y_{ijl}$,

$$\mathbf{y}_{ij} \sim Mult(\mathbf{p}_{ij}, N_{ij}).$$

If it is important to consider the relative preference for visit types, then this might be useful. Otherwise, without conditioning, it would be possible to consider

$$y_{ijl} \sim Pois(\mu_{ijl}).$$

A log-linear link could be assumed whereby

$$\log(\mu_{ijl}) = \mathbf{x}_i'\boldsymbol{\beta} + v_i + w_i + u_{il} + \gamma_{jl}, \tag{18.9}$$

where v_i, w_i are contextual effects, as before, u_{il} is an individual-level effect specific to the event type, and γ_{jl} is a temporal effect specific to the event type. As before, the contextual random effects can be defined to depend on small-area-level geographies, while the temporal effects could be regression based or random walk based. In the multinomial model of Steele et al. (2004), the temporal component γ_{jl} was assumed to be defined by a quadratic regression in time and the individual random effect u_{il} was assumed to have a multivariate normal distribution (between event types only).

Asthma comorbidity Medicaid example. Sutton (2005) provides an example of the analysis of individual-level outcomes with fixed time periods in a Bayesian setting. This work was based on Medicaid data on asthma (International Classification of Diseases, Ninth Revision [ICD-9] 493) and congestive heart failure (CHF) (ICD-9 428, 402, 518.4), comorbidities for recipients between age 50 and 64 in South Carolina for the period 1997–1999. Three groups were identified: asthma only, CHF only, and asthma and CHF. Of the 1857 individuals, 223 were in the comorbidity group. Recorded for each recipient were number of

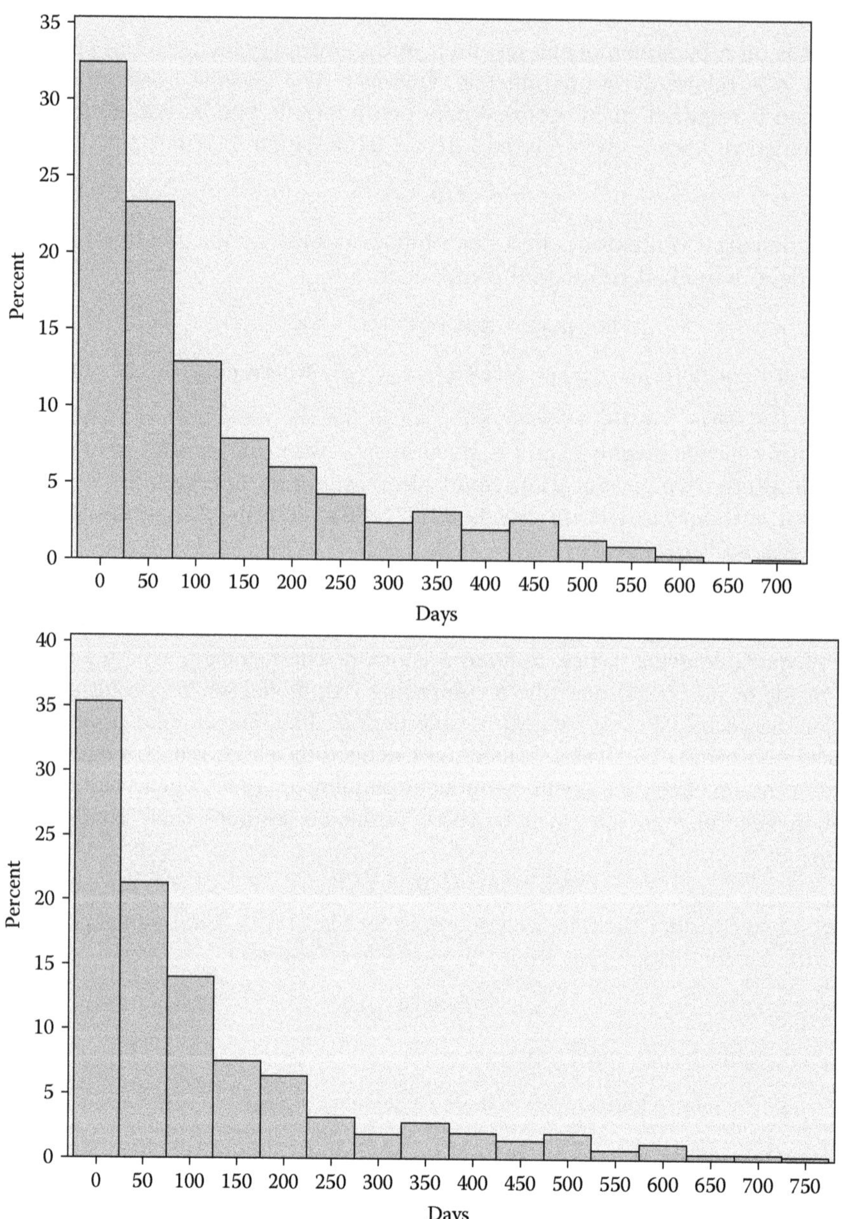

FIGURE 18.5
Distribution of number of days between first and second visits for recipients with (top panel) asthma and (bottom panel) CHF.

days between multiple dates of medical service; type of visit (inpatient, ER, outpatient, or doctor's office), and recipients, demographic information (age at first visit, gender, race, and county of residence). As a brief guide, Figure 18.5 displays the event profiles summarized by time to second visit for the asthma and CHF individuals separately.

An analysis was performed with the South Carolina Medicaid data mentioned above where asthma and CHD were analyzed together. Here, I only display the comorbidity group analysis. The analysis was carried out for 21 time periods (of 50 days each).

A model of the form

$$y_{ijl} \sim Poiss(\mu_{ijl})$$
$$\log(\mu_{ijl}) = \mathbf{x}_i'\boldsymbol{\beta} + v_i + w_i + \gamma_{jl}$$

was fitted to the data for the comorbidity group. Individual frailty was not required in this example, based on a variable selection criterion. Here, the fixed covariates were β_0 (common intercept), β_1(age), factors for race and gender, and two spatial contextual effects at the county level: $v_i = v(\underset{i \in l}{c_l})$ with $v_i \sim N(0, \tau_v)$ and $w_i = w(\underset{i \in l}{c_l})$ and $w(c_l)|w(c_{-l}) \sim N(\overline{w}(c_{\delta_l}), \tau_w/n_{\delta_l})$, where $w(c_l)$ is the value of w for the lth county and $\overline{w}(c_{\delta_l})$ is the average value of w for the neighborhood (δ_l) of c_l. All regression parameters had overdispersed zero-mean Gaussian prior distributions, while variances were assume to have $Ga(0.05, 0.0005)$ distributions. The temporal effects for each type were assumed to have independent random walk prior distributions

$$\gamma_{jl} \sim N(\gamma_{j-1,l}, \tau_\gamma) \; \forall l.$$

The converged model fit yields the following results:

1. The posterior expected estimates (standard deviations [sds]) of β_0, β_1, respectively, were -4.6 (0.9982) and -0.008553 (0.01118).
2. The posterior expected estimates (sds) of the race and gender effects were not significant for gender, but showed a significance for the white versus African American racial groups.
3. The temporal effects were estimated and are shown in Figure 18.6. The sequence of visit types is $l = 1$, $l = 2$, $l = 3$, $l = 4$.

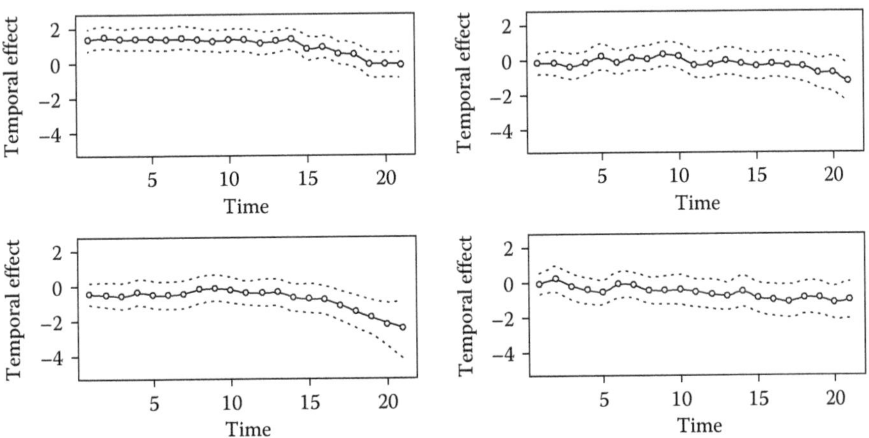

FIGURE 18.6
Posterior mean temporal profiles for the effect γ_{jl}, for $l = 1, \ldots, 4$. Mean estimates and 95% credible interval are shown. Top left, $l = 1$; top right, $l = 2$; bottom left, $l = 3$; bottom right, $l = 4$.

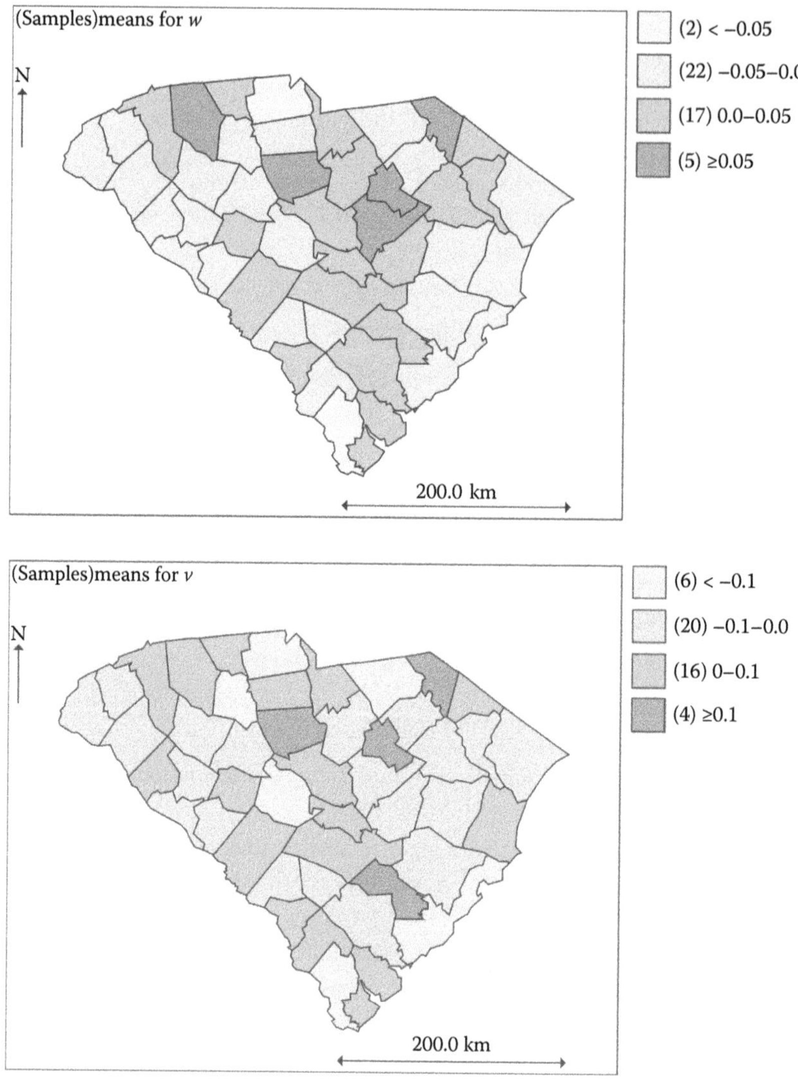

FIGURE 18.7
Posterior average maps of the county-level random effects (w_i, v_i) form the multiple-event model for the Medicaid data.

Finally, the posterior average maps of the county-specific random contextual effects are shown in Figure 18.7. It is clear that there is some spatial effect (w map) displayed in the northeast of the state within largely rural areas, whereas the uncorrelated heterogeneity (UH) effect (v map) seems to be largely random. It should be noted here, however, that analysis of discrete time events lacks a considerable amount of information due to the grouping within time periods and information about sequencing is lost; hence, there is an inevitable limitation to this form of analysis.

References

Berman, M. and T. R. Turner. (1992). Approximating point process likelihoods with GLIM. *Applied Statistics* 41, 31–38.

Clayton, D. G., L. Bernardinelli, and C. Montomoli. (1993). Spatial correlation in ecological analysis. *International Journal of Epidemiology* 22, 1193–1202.

Cook, R. and J. Lawless. (2007). *The Statistical Analysis of Recurrent Events.* New York: Springer.

Daniels, M. and J. Hogan. (2008). *Missing Data in Longitudinal Studies.* New York: CRC Press.

Diggle, P., P. Heagerty, K.-Y. Liang, and S. Zeger. (2002). *Analysis of Longitudinal Data* (2nd ed.). New York: Oxford University Press.

Diggle, P. and P. Ribeiro Jr. (2007). *Model-Based Geostatistics.* New York: Springer.

Fahrmeir, L. and T. Kneib. (2011). *Bayesian Smoothing for Longitudinal, Spatial and Event History Data.* New York: Oxford University Press.

Fitzmaurice, G., M. Davidian, G. Verbeke, and G. Molenberghs (eds.). (2008). *Longitudinal Data Analysis.* New York: CRC Press.

Guttorp, P. (1995). *Stochastic Modeling of Scientific Data.* New York: CRC Press.

Ibrahaim, J., M. Chen, and D. Sinha. (2000). *Bayesian Survival Analysis.* New York: Springer.

Kawachi, I. and L. Berkman (eds.). (2003). *Neighborhoods and Health.* New York: Oxford University Press.

Lachin, J. L. (2000). Statistical consdierations in the intent-to-treat principle. *Controlled Clinical Trials 21*, 167–189.

Lawson, A. B., R. Carroll, and M. Castro. (2014). Joint spatial Bayesian modeling for studies combining longitudinal and cross-sectional data. *Statistical Methods in Medical Research* 23, 611–624.

Lesaffre, E. and A. B. Lawson. (2012). *Bayesian Biostatistics.* New York: Wiley.

Maheu-Giroux, M., and M. Castro. (2013). Impact of community-based larviciding on the prevalence of malaria infection in Dar es Salaam, Tanzania. *PLoS One* 8, e71638.

Molenberghs, G. and G. Verbeke. (2005). *Models for Discrete Longitudinal Data.* New York: Springer.

Sibson, R. (1980). The Dirichlet tesselation as an aid in data analysis. *Scandinavian Journal of Statistics* 7, 14–20.

Steele, F., H. Goldstein, and W. Browne. (2004). A general multilevel multistate competing risks model for event history data, with an application to a study of contraceptive use dynamics. *Statistical Modelling* 4, 145–159.

Sutton, S. (2005). The modeling of spatially-referenced recurrent event data in a South Carolina population. PhD thesis, University of South Carolina.

Verbeke, G. and G. Molenberghs. (2000). *Linear Mixed Models for Longitudinal Data.* New York: Springer.

19

Spatiotemporal Disease Mapping

Andrew B. Lawson
Division of Biostatistics and Bioinformatics
Department of Public Health Sciences
Medical University of South Carolina
Charleston, South Carolina

Jungsoon Choi
Department of Mathematics
Hanyang University
Seoul, South Korea

CONTENTS

As in other application areas, it is possible to consider the analysis of disease maps that have an associated temporal dimension. The two most common formats for observations are

1. Georeferenced case events that have associated with them a time of diagnosis or registration or onset; that is, we observe within a fixed time period J and fixed spatial window W m cases at locations $\{s_i, t_i\}, i = 1, \ldots, m$.
2. Counts of cases of disease within tracts that are available for a sequence of J time periods; that is, we observe a binning of case events within $m \times J$ space–time units: $y_{ij}, i = 1, \ldots, m,\ j = 1, \ldots, J$.

   The analysis found for spatial data (see Chapter 6) can be extended into the time domain without significant difficulty.

19.1 Case Event Data

In the case event situation, few examples exist of mapping analysis. However, it is possible to specify a model to describe the first-order intensity of the space–time process (as in the spatial case). The intensity at time t can be specified as

$$\lambda(s,t) = \rho g(s,t) \cdot f_1(s;\theta_x) \cdot f_2(t;\theta_t) \cdot f_3(s,t;\theta_{xt}), \tag{19.1}$$

where ρ is a constant background rate (in space $\times$ time units); $g(s,t)$ is a modulation function describing the spatiotemporal at-risk population background in the study region;

f_k are appropriately defined functions of space, time, and space–time; and θ_x, θ_t, and θ_{xt} are parameter vectors relating to the spatial, temporal, and spatiotemporal components of the model.

Here, each component of the f_k can represent a *full* model for the component; that is, f_1 can include spatial trend, covariate, and covariance terms, and f_2 can contain similar terms for the temporal effects, while f_3 can contain *interaction* terms between the components in space and time. Note that this final term can include *separate* spatial structures relating to interactions that are not included in f_1 or f_2. The exact specification of each of these components will depend on the application, but the separation of these three components is helpful in the formulation of components.

The above intensity specification can be used as a basis for the development of Bayesian models for case events. If it can be assumed that the events form a modulated Poisson process (PP) in space–time, then a likelihood can be specified, as in the spatial case. For example, a parsimonious model could be proposed where a regression component and a random effect component is assumed:

$$\lambda(s,t) = \rho g(s,t)\exp\{\mathbf{P}(s,t)'\boldsymbol{\beta} + T(s,t)\}, \tag{19.2}$$

where $\mathbf{P}(s,t)$ is a covariate vector, $\boldsymbol{\beta}$ is a regression parameter vector, and $T(s,t)$ is a random component representing extra variation in risk. The term $T(s,t)$ could be decomposed in a number of ways. For example, it could represent a spatiotemporal Gaussian process (Brix and Diggle, 2001; Diggle, 2007). However, a simpler approach might be to consider $T(s,t) = a(s) + b(t) + c(s,t)$, where a discretized version of the random fields could be envisaged so that any realization of the field $\{s_i, t_i\}$ has separable correlation structure and

$$\begin{aligned} a(\mathbf{s}) &\sim MVN(\mathbf{0}, K_a(\tau_x, \phi)) \\ b(t) &\sim N(f(\Delta t), \tau_b) \\ c(s,t) &\sim N(0, \tau_c I), \end{aligned} \tag{19.3}$$

where $K_a(\tau_x, \phi)$ is a parameterized spatial covariance matrix and I is an identity matrix, with variances τ_b and τ_c, and Δt is a distance measured in time. In this approach, the likelihood remains that of a conditionally modulated Poisson process.

This type of model can be included within a likelihood specification, and a full Bayesian analysis can proceed using extensions to the analysis for purely spatial data. In these extensions, either the integrated intensity of the process,

$$\Lambda(\boldsymbol{\theta}) = \int_W \int_0^J \lambda(u,v)\,dv\,du,$$

where $\boldsymbol{\theta} = (\boldsymbol{\beta}, \tau_x, \phi, \tau_b, \tau_c)$ is the parameter vector, must be estimated or the background is concentrated out of the model by conditioning. In Lawson (2006), an example of application of this model to a well known space–time case event dataset was given (Burkitt's lymphoma in the Western Nile district of Uganda for the period 1960–1975). In that dataset, the location of cases and a diagnosis date (days from January 1, 1960) are known, as well as the age of the case. There was no background population information in this example. While it might be possible to consider an approach where a population effect in $g(s,t)$ was estimated from the case event data, this is particularly assumption dependent and was not pursued.

Conditioning on the realization of cases and controls allows the derivation of a simpler likelihood (see, e.g., Lawson, 2012). Assuming that $\lambda(s,t) = g(s,t)\lambda_1(s,t)$, consider the cases governed by a PP with intensity $g(s,t)\lambda_1(s,t)$ and the controls governed by $g(s,t)$, then the superposition of the cases and controls has intensity $g(s,t)[1 + \lambda_1(s,t)]$, and so the probability of any given location being a case is just $\lambda_1(s,t)/[1+\lambda_1(s,t)]$. This can be used to form a logistic likelihood for m cases and n controls of the form

$$L = \prod_{i=1}^{m+n} \frac{[\lambda_1(s_i,t_i)]^{y_i}}{[1+\lambda_1(s_i,t_i)]},$$

where the indicator

$$y_i = \begin{cases} 1 \text{ if } i\text{th event is a case} \\ 0 \text{ otherwise} \end{cases}$$

and (s_i,t_i) are the location and time of the ith event. This is just a binary logistic regression with linear predictor $\log(\lambda_1(s,t))$. Hence, the formulation in (19.3) could be assumed with y_i as outcome and $\alpha + a(s) + b(t) + c(s,t)$ as linear predictor. Within a Bayesian formulation then, the component prior distributions could be

$$\begin{aligned} \alpha &\sim N(0,\tau_\alpha) \\ a(s_i) &\sim N(0,\Sigma) \\ b(t_i) &\sim N(\phi b(t_{i-1})/\Delta t_i, \tau_b) \\ c(s_i,t_i) &\sim N(0,\tau_c), \end{aligned} \tag{19.4}$$

where Σ is a spatial covariance matrix, and $\Delta t_i = t_i - t_{i-1}$. Note that due to the time differences between events, a form of the Orstein–Uhlenbeck process must be assumed for $b(t_i)$ (see, e.g., Lawson et al., 2014). The spatial component can be assumed to have a variety of forms. A simple example could be to assume a Markov random field prior for the effect, where $\phi = 1$. An intrinsic conditional autoregression (ICAR) prior distribution can be assumed for the $a(s_i)$ term if a neighborhood adjacency structure can be defined. This is possible if a tiling of the spatial point process is available. A Dirichlet tesselation of a spatial point process (Sibson, 1980) can provide "natural" neighbors based on its tiling, as can the Delauney triangulation. The R package DELDIR (Turner, 2012) provides this information. Once adjacencies are defined, then an ICAR prior distribution can be considered.

In the following example, we apply model (19.4) to a synthetic realization of a space–time case event process. This realization is synthetic, as it is based on a real distribution in space–time of cancer cases of a specific type in the state of South Carolina. The distribution is simulated to mimic the real case distribution, but the locations and times are random. This approach is adopted, as the exact locations of cancer cases are subject to medical confidentiality restrictions. Figure 19.1 displays the spatial distribution of cases and controls (early- and late-stage cancer). The marginal temporal distribution is displayed in Figure 19.2.

A period of 25 years is used as a time domain and the time unit is one week. There are 188 diagnoses, and of that number, 73 are late stage and 115 are early stage. The diagnosis times range from week 1 to week 298. We have initially applied a simple conditional logistic spatiotemporal model to the early-stage (control) and late-stage (case) cancer. Following the definition in model (19.4), we assume an ICAR model for the spatial effect, based on Delauney triangulation neighbors from DELDIR, and an Orstein–Uhlenbeck model for the temporal effect. An uncorrelated (residual) effect was also assumed. Other prior distributions assumed were as follows: Standard deviations of Gaussian distributions were assumed to have uniform prior distributions, that is, $\tau_*^{1/2} \sim U(0,C)$. The upper limit (C) was varied to check for informativeness. A value of $C = 10$ was found acceptable. This model was fitted using Markov chain Monte Carlo (MCMC) methods (OpenBUGS), and convergence of posterior sampling was achieved by 30,000 iterations. A total of 5000 from two chains was sampled. Convergence was based on the Brooks–Gelman–Rubin $\widehat{R}$ diagnostic. Goodness of fit was measured by the deviance information criterion (DIC) and, in this case, yielded 222.50 with pD = 35.34. Figure 19.3 displays four image-contour surface plots

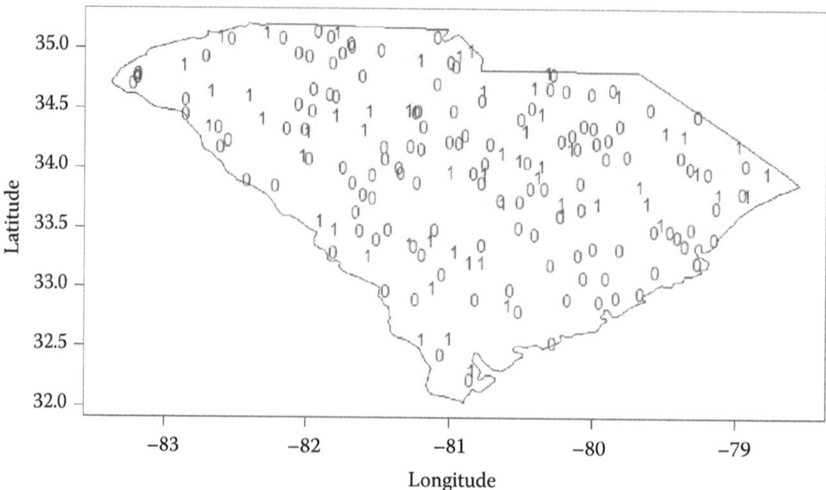

FIGURE 19.1
Cancer case distribution in South Carolina (synthetic example): early stage (0) and late stage (1) over 25 years.

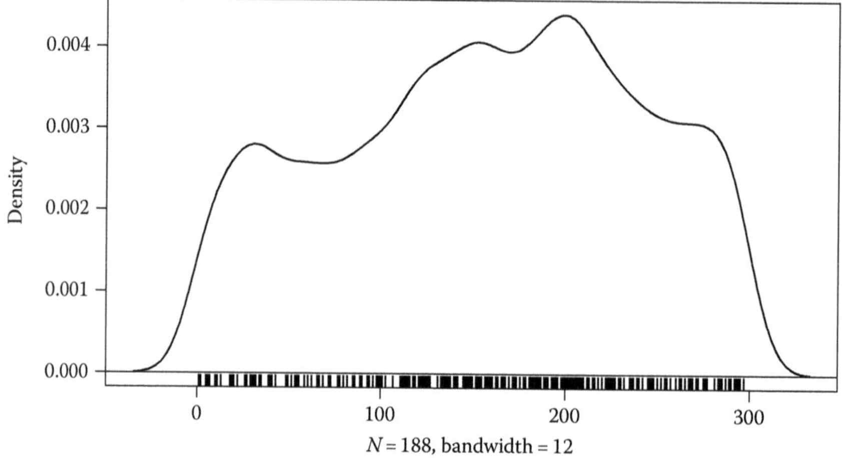

FIGURE 19.2
Marginal kernel density of the diagnosis times for the superposition of cases and controls over 298 weeks.

of the resulting posterior mean estimates of the case probability p_i, $c(s_i, t_i)$, $a(s_i)$, and $b(t_i)$ based on a logit link to the linear predictor of $\alpha + a(s) + b(t) + c(s, t)$.

As a comparison, we have also fitted a full Gaussian process prior distribution for $a(s_i)$. We assumed a power exponential covariance model with $\Sigma_{ij} = \tau_a \exp\{-(\phi_a d_{ij})^{\kappa}\}$ (Diggle and Ribeiro, 2007). We assumed that $\kappa = 1$ and $\phi_a \sim Unif(0.5, 4.5)$ based on the maximum and minimum distances in the sample. All the standard deviations of Gaussian prior distributions were assume to have uniform prior distributions, and all other settings were as per model (19.4). Convergence was achieved by 20,000 iterations, and a sample of 4000 from two chains was taken. The resulting DIC for this model was 229.0 with pD = 25.5. Hence, the use of a geostatistical model in place of the ICAR model was not supported based on

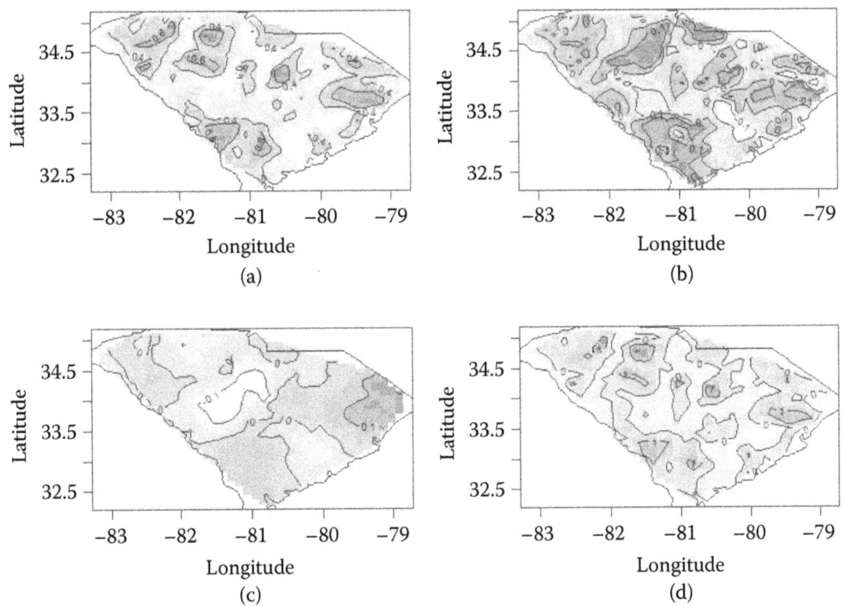

FIGURE 19.3
Multiple image-contour plots of the posterior mean estimates under an ICAR spatial prior distribution of p_i (a), $c(s_i, t_i)$ (b), $a(s_i)$ (c), and $b(t_i)$ (d).

the DIC criterion, although the pD was lower for the exponential covariance model. The posterior mean estimate of the ϕ_a was 2.885 with $sd = 0.762$. Figure 19.4 displays four image-contour surface plots of the resulting posterior mean estimates of p_i, $c(s_i, t_i)$, $a(s_i)$, and $b(t_i)$ based on a logit link with this geostatistical model.

A diagnostic that can be used to assess areas of excess risk in both spatial and spatiotemporal disease data is the exceedence probability. While this can be very sensitive to model assumptions (Lawson, 2013, chapter 6), it has achieved wide use as a local measure of adverse risk. A comparison of exceedence probabilities for the resulting relative risk estimators for the two different spatial models is provided in Figures 19.5 and 19.6.

It is noticeable that there are slight differences in the resulting peaks of excess risk (hot spots) in exceedence of the posterior marginal probability surfaces: $\widehat{Pr}(p_i > 0.5) = \sum_{g=1}^{G} I\left(p_i^{(g)} > 0.5\right)/G$, where G is the number of samples and $p_i^{(g)}$ is the sampled value of p_i. Overall, the same pattern emerges under either model, with the ICAR providing a slightly smoother reconstruction of $a(s_i)$.

Following Besag and Tantrum (2003), it would also be feasible to consider a variant of an autologistic model to allow for spatial correlation in the binary outcome in this case. Conditioning on the fixed neighborhood sum of binary indicators could provide a simpler approach to incorporation of spatial or spatiotemporal dependency in this conditional logistic model. Note that extension to all of the above models with predictors could straightforwardly be made by adding a linear or nonlinear function of covariates to the log-linear predictor.

Computation for Bayesian log Gaussian Cox process models has recently been enhanced by the implementation of Laplace approximation software in the R package INLA (Rue et al., 2009; Li et al., 2012; Blangiardo et al., 2013). It is now feasible to fit these more complex models without recourse to posterior sampling. Schrodle and Held (2011) and Blangiardo et al. (2013) provided a range of examples of the use of such software in space–time disease mapping applications.

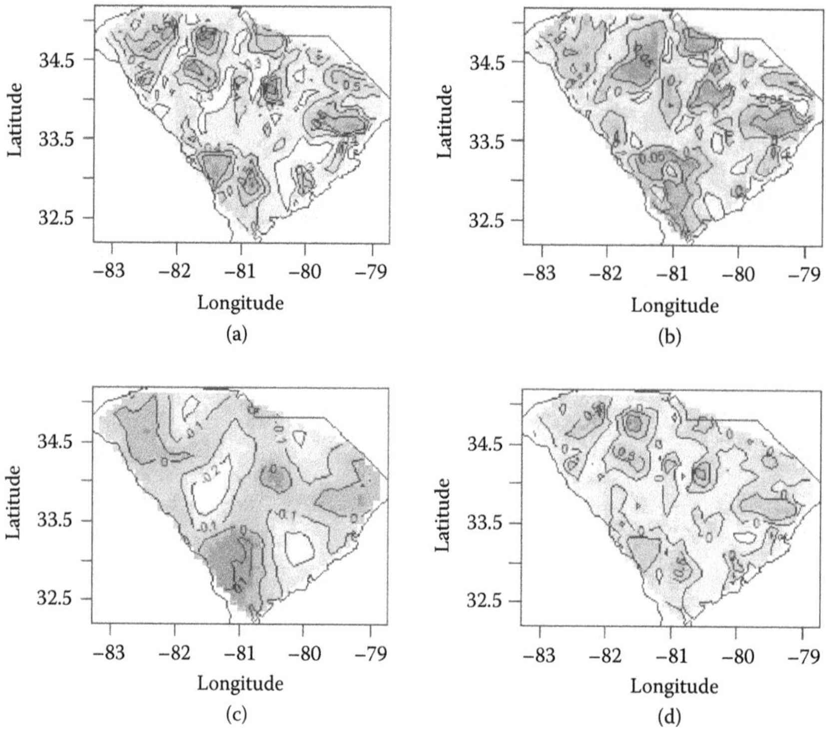

FIGURE 19.4
Multiple image-contour plots of the posterior mean estimates, under a spatial Gaussian process prior distribution with a power exponential covariance, of p_i (a), $c(s_i, t_i)$ (b), $a(s_i)$ (c), and $b(t_i)$ (d).

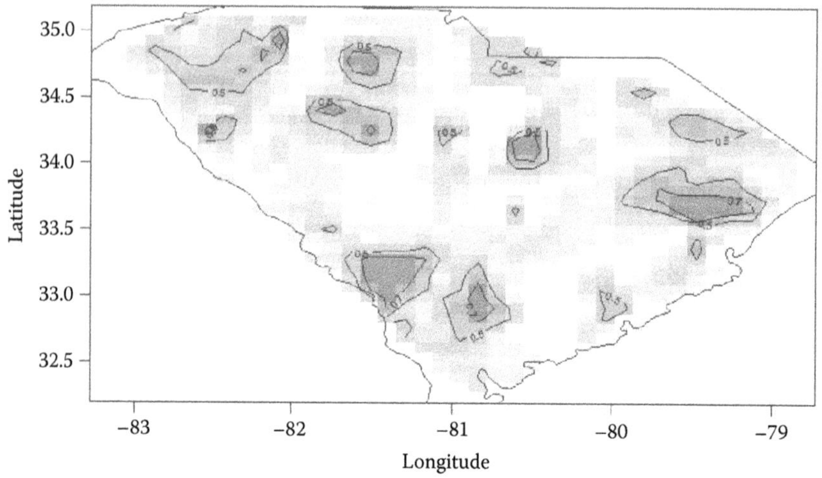

FIGURE 19.5
Posterior marginal exceedence probability estimates for the case probability (p_i) being greater than 0.5, under a model with an ICAR spatial prior distribution.

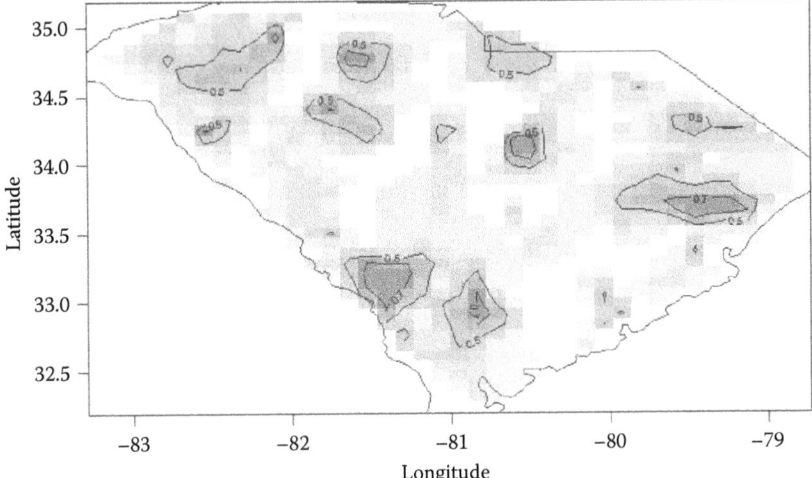

FIGURE 19.6
Posterior marginal exceedence probability estimates for the case probability (p_i) being greater than 0.5 under a model with a spatial Gaussian process prior distribution with a power exponential covariance.

19.2 Count Data

Modeling for spatiotemporal count data can be naturally extended from the space count modeling discussed in previous chapters. Suppose that y_{ij} for $i = 1, \ldots, m$ and $j = 1, \ldots, J$ is the observed count of disease in the ith space unit at the jth time unit. A general count data model might be a Poisson regression model as

$$y_{ij} \sim Pois(\lambda_{ij})$$
$$E(y_{ij}) = \lambda_{ij} = e_{ij}\theta_{ij},$$

where e_{ij} is the expected count for space unit i and time unit j and θ_{ij} is the relative risk. The expected count is usually assumed to be observed (fixed) given the space and time domain in the study. Modeling of the relative risk is the focus of interest.

A large literature on space–time disease mapping has been proposed based on the Bayesian framework by using conditional autoregressive (CAR) models.

Bernardinelli et al. (1995) first introduced a model for θ_{ij} where the log relative risk is defined as

$$\log(\theta_{ij}) = \mu + u_i + (\alpha + \tau_i)j,$$

where μ is an intercept describing the overall log relative risk rate and u_i is a spatial random effect that can follow a conditional autoregressive Gaussian distribution, $\mathbf{u} = (u_1, \ldots, u_m)^T \sim \text{CAR}(\sigma_u^2)$. The temporal trend $(\alpha + \tau_i)j$ is expressed as the sum of the linear temporal trend term αj and the space–time interaction term $\tau_i j$, where τ_i is a spatial random effect. Thus, temporal trends in disease risk may be varying with spatial locations. However, this model only includes a simple linear time trend and no temporal random effect. In this model, Assunção et al. (2001) considered a second-order polynomial temporal trend structure.

To allow for the temporally varying spatial patterns, Waller et al. (1997) assumed that the log relative risk for the kth subgroup (e.g., male and female subgroup in gender) is given by

$$\log(\theta_{ijk}) = \mathbf{X}_k^T \boldsymbol{\beta} + \mathbf{Z}_i^T \mathbf{w} + v_i^j + u_i^j, \tag{19.5}$$

where $\mathbf{X}_k$ and $\mathbf{Z}_i$ are the vectors of the kth group covariates and the ith spatial covariates, respectively. The spatial random effects v_i^j and u_i^j are spatially uncorrelated and correlated heterogeneity structures for each time point j. That is, $v_i^j \sim N(0, \sigma_v^2)$ and $\mathbf{u}^j = (u_1^j, \ldots, u_m^j)^T \sim \text{CAR}(\sigma_u^2)$. This model was applied to annual county-level Ohio lung cancer mortality data during the years 1968–1988. With the same dataset, Xia and Carlin (1998) further developed a different spatiotemporal model to allow for errors in the covariates. They assumed a model for the log relative risk of the form

$$\log(\theta_{ijk}) = \mu + \mathbf{X}_k^T \boldsymbol{\beta} + \rho p_i + \alpha j + \phi_{ij},$$

where μ is an intercept, p_i is a smoking variable for the ith county, αj is a temporal linear trend, and ϕ_{ij} is a spatial random effect that is nested within the time period. For each time point j, the spatial effect $\boldsymbol{\phi}_j = (\phi_{1j}, \ldots, \phi_{mj})^T$ follows a CAR distribution. The model considers a temporal linear trend term and a space–time interaction, but space and time random effects separately.

Knorr-Held and Besag (1998) considered a different model to analyze the same Ohio lung cancer mortality data. They assumed that the observed disease count for the kth subgroup, y_{ijk}, follows the binomial distribution with the number at risk n_{ijk} and probability π_{ijk}, and employed a logit link function to the linear predictor,

$$y_{ijk} \sim Bin(n_{ijk}, \pi_{ijk})$$
$$\log\left(\frac{\pi_{ijk}}{1-\pi_{ijk}}\right) = v_i + u_i + \beta z_i + \gamma_j + \alpha_{jk},$$

where v_i and u_i are the spatially uncorrelated and correlated random components over time, z_i is the ith covariate variable, γ_j is a temporal random component, and α_{jk} is a kth subgroup effect at the jth time point. The model combines spatial and temporal effects additively, has no space–time interactions, and has no spatial or temporal trend structures.

Recently, Knorr-Held (2000) proposed a more general space–time interaction structure in space–time modeling. The log relative risk is modeled as

$$\log(\theta_{ij}) = \mu + v_i + u_i + \delta_j + \gamma_j + \phi_{ij}, \tag{19.6}$$

where μ is an intercept quantifying the log of the overall risk rate, v_i and u_i are the uncorrelated and correlated spatial random components, δ_j and γ_j are the uncorrelated and correlated temporal random components, and ϕ_{ij} is the space–time interaction term. The temporal dependent component γ_j can be assumed to be an autoregressive (AR) distribution or a random walk distribution: for example, $\gamma_j \sim N(\tau\gamma_{j-1}, \sigma_\gamma^2)$ is an AR(1) with $|\tau| < 1$, and when $\tau = 1$, γ_j follows a random walk. Like the structure of the uncorrelated spatial component v_i, the parameter δ_j is assumed to be a normal distribution with mean 0 and variance σ_δ^2. The matrix of interaction terms $\{\phi_{ij}\}$, $i = 1, \ldots, m$ and $j = 1, \ldots, J$, defined by $\boldsymbol{\Phi}$ is assumed to be normally distributed as $\boldsymbol{\Phi} \sim N(\mathbf{0}, \sigma_\phi^2 \boldsymbol{\Sigma}_\phi)$, where σ_ϕ^2 is an overall variance and $\boldsymbol{\Sigma}_\phi$ is a structure matrix expressed as the Kronecker product of the structure matrices of those main effects. The author considered four types of space–time interaction structures in $\boldsymbol{\Phi}$. In type I interactions, all ϕ_{ij}'s are a prior independent and $\boldsymbol{\Sigma}_\phi$ is defined as $\mathbf{I}_S \otimes \mathbf{I}_T$, where $\mathbf{I}_S$ and $\mathbf{I}_T$ are identity matrices of space and time, respectively. This means that unobserved covariates for space i and time j do not have any structure in space $\times$ time. In type II interactions, $\boldsymbol{\phi}_i = (\phi_{i1}, \ldots, \phi_{iT})^T$ for $i = 1, \ldots, m$ is assumed to follow an independent random walk, so $\boldsymbol{\Sigma}_\phi = \mathbf{I}_S \otimes \boldsymbol{\Sigma}_T$ and $\boldsymbol{\Sigma}_T$ is a covariance matrix of a random walk distribution. This specification explains the spatially varying temporal trends but no temporally varying space trends. In type III interactions, $\boldsymbol{\phi}_j = (\phi_{11}, \ldots, \phi_{mj})^T$ for $j = 1, \ldots, J$ is assumed to follow a CAR distribution, independently of all other time points. The matrix $\boldsymbol{\Sigma}_\phi$ is specified as $\boldsymbol{\Sigma}_S \otimes \mathbf{I}_T$, and $\boldsymbol{\Sigma}_S$ is a covariance matrix of a CAR distribution. Thus, type III interactions will be suitable if the temporally varying spatial trends exist but no spatially varying temporal trends exist in relative risk. In type IV interactions, $\boldsymbol{\Phi}$ is completely dependent over space and time and $\boldsymbol{\Sigma}_\phi = \boldsymbol{\Sigma}_S \otimes \boldsymbol{\Sigma}_T$. Thus, this type of interaction can be interpreted as different spatial trends from time to time, as well as different temporal trends from space to space. The spatiotemporal model in (19.6) with these different types of interactions was applied to the Ohio lung cancer data for white males for the years 1968–1988. It was found that a model with type II interaction has the smallest deviance among the models with four different interactions. Thus, the type II interaction model provided the best model in terms of measures of model fit and complexity. Ugarte et al. (2014) fitted these types of interactions easily within INLA.

One topic of recent interest for space–time count data is multivariate spatiotemporal modeling since multiple-related diseases can have common spatial or temporal trends in risk. Richardson et al. (2006) proposed an extension of the shared component models in two spatial disease datasets introduced by Knorr-Held and Best (2001) to the space–time setting. Similarly, Tzala and Best (2008) employed a Bayesian factor analysis to examine the shared spatial and temporal patterns for mortality from multiple cancers. They analyzed male and female lung cancer mortality data jointly by using shared space–time interactions.

There are other developments when the effects of covariates on the space–time disease data are the main focus. For example, the effects of poverty rates on low birth weight may vary with space and time. Based on the model of Knorr-Held (2000), one can define a general model as

$$\log(\theta_{ij}) = \mathbf{X}_{ij}^T \boldsymbol{\beta}_{ij} + v_i + u_i + \delta_j + \gamma_j + \phi_{ij},$$

where $\mathbf{X}_{ij}^T$ is a vector of fixed covariates and $\boldsymbol{\beta}_{ij}$ is a spatiotemporally varying coefficient for space i at time j. Various structures on $\boldsymbol{\beta}_{ij}$ can be considered. Assunção et al. (2002) and Gamerman et al. (2003) developed space–time models with spatially dependent coefficients, and Dreassi et al. (2005) considered temporally dependent coefficients. More recently, Cai et al. (2012) and Choi et al. (2012) proposed hierarchical latent models with spatiotemporally varying coefficients to analyze Georgia low birth weight data.

19.2.1 Georgia chronic obstructive pulmonary disease example

County-level chronic obstructive pulmonary disease (COPD) data in Georgia during the years 1999–2007, as an example, are considered to compare a variety of the space–time models discussed. The observed data consist of yearly counts of COPD incidents in counties of Georgia for the years 1999–2007. These data are publicly obtained from the state health information system OASIS (Georgia Division of Public Health: http://oasis.state.ga.us/). The expected counts were computed by using the internal standardization method (Banerjee et al., 2004). Figure 19.7 displays the standardized incidence ratios for each year. The spatiotemporal variation of standardized incidence ratios can be clearly seen. In particular, southeast areas and some northeast areas in Georgia have high standardized incidence ratios of COPD over the years of study.

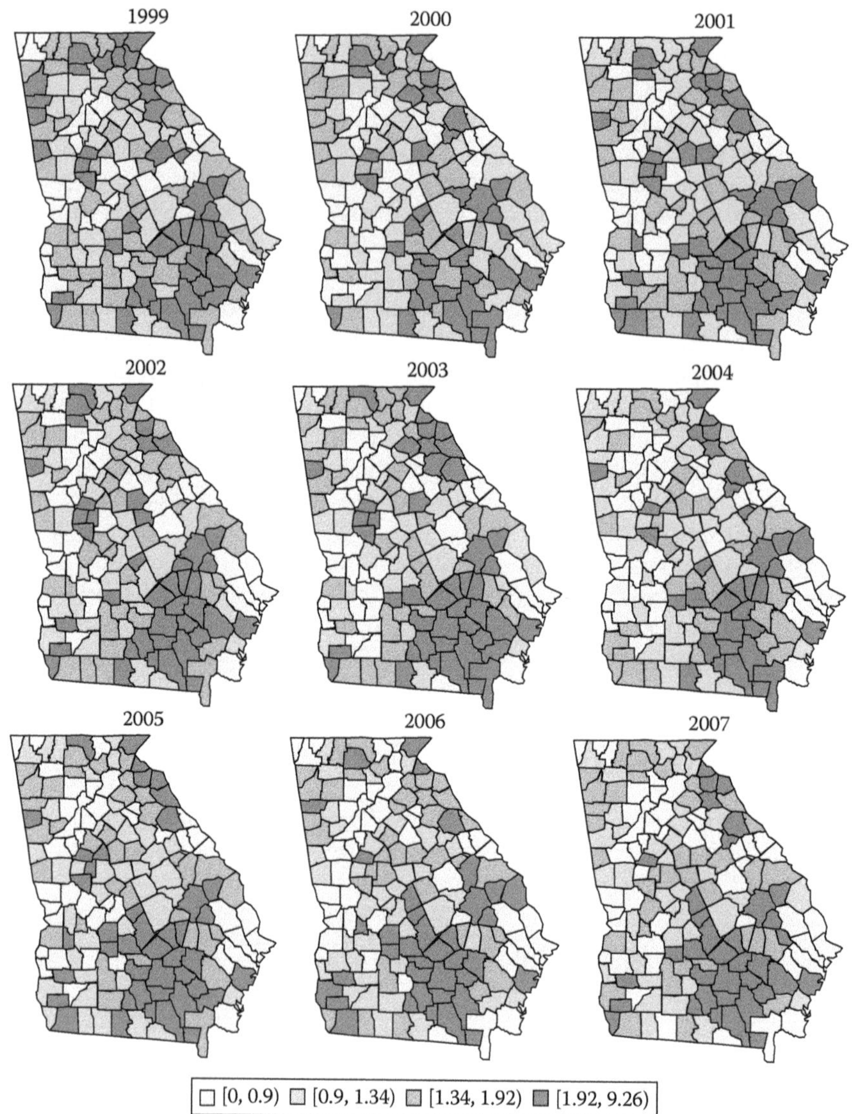

FIGURE 19.7
Georgia county-level standardized incidence ratios for COPD data during the years 1999–2007.

Space–time variation in disease risk should be considered when analyzing COPD data. To this end, I have examined a set of eight models based on the above discussion. This is an example of how to compare different space–time models, and readers can consider alternative models. We assume that the number of cases follows a Poisson distribution as $y_{ij} \sim Pois(\lambda_{ij} = e_{ij}\theta_{ij})$. The first model (Model 1) has spatially uncorrelated and temporally uncorrelated terms in relative risk, and the second model (Model 2) has spatially uncorrelated and temporal linear terms. These two models are simple separable models without space–time interaction. Model 3 is the original Bernardinelli et al. (1995) model, and Model 4 is extended by adding a spatial uncorrelated component in Model 3.

Space–time models with the four types of space–time interactions proposed by Knorr-Held (2000) are considered (Models 5–8).

$$\begin{aligned}
\text{Model 1: } \log\theta_{ij} &= \alpha + v_i + \delta_j \\
\text{Model 2: } \log\theta_{ij} &= \alpha + v_i + \beta j \\
\text{Model 3: } \log\theta_{ij} &= \alpha + u_i + (\beta + \tau_i)j \\
\text{Model 4: } \log\theta_{ij} &= \alpha + u_i + v_i + (\beta + \tau_i)j \\
\text{Models 5–8: } \log\theta_{ij} &= \alpha + u_i + v_i + \delta_j + \gamma_j + \phi_{ij},
\end{aligned}$$

where

$$\begin{aligned}
&\alpha \sim N(0, 10^6), \beta \sim N(0, 10^6) \\
&v_i \sim N(0, \sigma_v^2), \delta_j \sim N(0, \sigma_\delta^2) \\
&u_i \sim \text{CAR}(\sigma_u^2), \tau_i \sim \text{CAR}(\sigma_\tau^2), \gamma_j \sim N(\gamma_{j-1}, \sigma_\gamma^2).
\end{aligned}$$

Model 5 has $\phi_{ij} \sim N(0, \sigma_\phi^2)$, Model 6 has $\phi_{ij} \sim N(\phi_{ij-1}, \sigma_\phi^2)$, and Model 7 has $\phi_{ij} \sim \text{CAR}(\sigma_\phi^2)$. Model 8 has $\boldsymbol{\Phi} \sim N(\mathbf{0}, \sigma_\phi^2 \boldsymbol{\Sigma}_S \otimes \boldsymbol{\Sigma}_T)$, where $\boldsymbol{\Sigma}_S$ is the covariance of a CAR prior and $\boldsymbol{\Sigma}_T$ is the covariance of a random walk. All the standard deviation parameters in the models have uniform prior distributions (Gelman, 2006). The estimations of the parameters are obtained by MCMC algorithms. To ensure MCMC convergence, several diagnostics, such as autocorrelation functions, trace plots, and the Gelman–Rubin statistic (Gelman and Rubin, 1992), are used. WinBUGS (http://www.mrc-bsu.cam.ac.uk/software/bugs/the-bugs-project-winbugs/) is used to implement the models considered here.

Table 19.1 presents the DIC and the mean square prediction error (MSPE), defined as

$$\text{MSPE} = \frac{1}{mJ} \sum_{i=1}^{m} \sum_{j=1}^{J} (y_{ij} - \hat{y}_{ij})^2,$$

where y_{ij} is the observed value and $\hat{y}_{ij}$ is the predicted value of y_{ij}, that is, the posterior mean from the posterior predictive distribution. It is clear that Models 1–4 have quite large DIC values in comparing the models of Knorr-Held (2000). Especially models with no space–time interaction are far from the best model. Bernardinelli et al.'s (1995) models provide better model fit than Models 1–2. Overall, the models proposed by Knorr-Held (2000) have small DIC values among the models considered. In particular, the type II space–time interaction model (Model 6) provides the smallest DIC value. Models 5 and 6 also have similar small MSPE values compared to the other models. Thus, Model 6 is the best model in terms of goodness-of-fit measures. These results depend on prior distributions, and sensitivity analysis to prior distribution should be conducted in any application.

TABLE 19.1
Georgia COPD data: goodness of fit

Model	$\overline{D}$	pD	DIC	MSPE
1	12,712.5	164.9	12,877.4	540.6
2	12,811.8	158.0	12,969.8	548.7
3	11,301.6	288.1	11,589.7	328.5
4	11,296.2	284.9	11,581.1	328.4
5	9847.6	918.8	10,766.5	188.6
6	9880.9	730.6	10,611.5	188.9
7	9923.6	867.4	10,791.0	194.4
8	9949.0	700.5	10,649.5	195.4

There are various posterior summaries for the best model (Model 6). Figure 19.8 displays the maps of exceedence probabilities for Model 6, where estimates of these probabilities are obtained as

$$\widehat{Pr}(\theta_{ij} > 1) = \frac{1}{G}\sum_{g=1}^{G} I(\theta_{ij}^{(g)} > 1),$$

where $\theta_{ij}^{(g)}$ is the sampled value of θ_{ij} from the posterior distribution and G is the number of samples. It is seen that rural areas, mainly southeast areas and some northeast areas, in Georgia have high exceedences, while urban areas (Atlanta) have low exceedences. We also found that the number of counties with high exceedence probabilities tends to decrease over time.

FIGURE 19.8
Georgia county-level exceedence probability from a space–time interaction model with type II interaction.

References

Assunção, R. M., J. E. Potter, and S. M. Cavenaghi. (2002). A Bayesian space varying parameter model applied to estimating fertility schedules. *Statistics in Medicine* 21(14), 2057–2075.

Assunção, R. M., I. A. Reis, and C. D. L. Oliveira. (2001). Diffusion and prediction of leishmaniasis in a large metropolitan area in Brazil with a Bayesian space-time model. *Statistics in Medicine* 20(15), 2319–2335.

Banerjee, S., B. P. Carlin, and A. E. Gelfand. (2004). *Hierarchical Modeling and Analysis for Spatial Data.* Boca Raton, FL: Chapman & Hall/CRC.

Bernardinelli, L., D. G. Clayton, C. Pascutto, C. Montomoli, M. Ghislandi, and M. Songini. (1995). Bayesian analysis of space-time variation in disease risk. *Statistics in Medicine* 14, 2433–2444.

Besag, J. and J. Tantrum. (2003). Likelihood analysis of binary data in space and time. In Green P., N. Hjort, and S. Richardson (eds.), *Highly Structured Stochastic Systems*, pp. 289–295. Oxford: Oxford University Press.

Blangiardo, M., M. Cameletti, G. Baio, and H. Rue. (2013). Spatial and spatio-temporal models with R-INLA. *Spatial and Spatio-Temporal Epidemiology* 4, 33–49.

Brix, A. and P. Diggle. (2001). Spatio-temporal prediction for log-Gaussian Cox processes. *Journal of the Royal Statistical Society B* 63, 823–841.

Cai, B., A. B. Lawson, M. M. Hossain, and J. Choi. (2012). Bayesian latent structure models with space-time dependent covariates. *Statistical Modelling* 12(2), 145–164.

Choi, J., A. B. Lawson, B. Cai, M. M. Hossain, R. S. Kirby, and J. Liu. (2012). A Bayesian latent model with spatio-temporally varying coefficients in low birth weight incidence data. *Statistical Methods in Medical Research* 21(5), 445–456.

Diggle, P. (2007). Spatio-temporal point processes: methods and applications. In Finkenstadt B., L. Held, and V. Isham (eds.), *Statistical Methods for Spatio-Temporal Systems*, pp. 1–45. London: CRC Press.

Diggle, P. and P. Ribeiro Jr. (2007). *Model-Based Geostatistics.* New York: Springer.

Dreassi, E., A. Biggeri, and D. Catelan. (2005). Space-time models with time-dependent covariates for the analysis of the temporal lag between socioeconomic factors and lung cancer mortality. *Statistics in Medicine* 24(12), 1919–1932.

Gamerman, D., A. R. Moreira, and H. Rue. (2003). Space-varying regression models: specifications and simulation. *Computational Statistics and Data Analysis* 42(3), 513–533.

Gelman, A. (2006). Prior distributions for variance parameters in hierarchical models. *Bayesian Analysis* 1, 515–533.

Gelman, A. and D. B. Rubin. (1992). Inference from iterative simulation using multiple sequences. *Statistical Science* 7, 457–472.

Knorr-Held, L. (2000). Bayesian modelling of inseparable space-time variation in disease risk. *Statistics in Medicine* 19, 2555–2567.

Knorr-Held, L. and J. Besag. (1998). Modelling risk from a disease in time and space. *Statistics in Medicine* 17, 2045–2060.

Knorr-Held, L. and N. G. Best. (2001). A shared component model for detecting joint and selective clustering of two diseases. *Journal of the Royal Statistical Society A* 164(1), 73–85.

Lawson, A. B. (2006). *Statistical Methods in Spatial Epidemiology* (2nd ed.). New York: Wiley.

Lawson, A. B. (2012). Bayesian point event modeling in spatial and environmental epidemiology: a review. *Statistical Methods in Medical Research* 21, 509–529.

Lawson, A. B. (2013). *Bayesian Disease Mapping: Hierarchical Modeling in Spatial Epidemiology* (2nd ed.). New York: CRC Press.

Lawson, A. B., R. Carroll, and M. Castro. (2014). Joint spatial Bayesian modeling for studies combining longitudinal and cross-sectional data. *Statistical Methods in Medical Research* 23, 611–624.

Li, Y., P. Brown, D. Gesink, and H. Rue. (2012). Log Gaussian Cox processes and spatially aggregated disease incidence data. *Statistical Methods in Medical Research* 21(5), 479–507.

Richardson, S., J. J. Abellan, and N. Best. (2006). Bayesian spatio-temporal analysis of joint patterns of male and female lung cancer risks in Yorkshire (UK). *Statistical Methods in Medical Research* 15(4), 385–407.

Rue, H., S. Martino, and N. Chopin. (2009). Approximate Bayesian inference for latent Gaussian models by using integrated nested Laplace approximations. *Journal of the Royal Statistical Society B* 71, 319–392.

Schrodle, B. and L. Held. (2011). Spatio-temporal disease mapping using INLA. *Environmetrics* 22, 725–734.

Sibson, R. (1980). The Dirichlet tesselation as an aid in data analysis. *Scandinavian Journal of Statistics* 7, 14–20.

Turner, R. (2012). DELDIR: Delaunay triangulation and Dirichlet (Voronoi) tessellation. R package version 0.0-19. http://www.r-pkg.org/pkg/deldir.

Tzala, E. and N. Best. (2008). Bayesian latent variable modelling of multivariate spatio-temporal variation in cancer mortality. *Statistical Methods in Medical Research* 17(1), 97–118.

Ugarte, M. D., A. Adin, T. Goicoa, and A. F. Militino. (2014). On fitting spatio-temporal disease mapping models using approximate Bayesian inference. *Statistical Methods in Medical Research* 23(6), 507–530.

Waller, L. A., B. P. Carlin, H. Xia, and A. E. Gelfand. (1997). Hierarchical spatio-temporal mapping of disease rates. *Journal of the American Statistical Association* 92, 607–617.

Xia, H. and B. P. Carlin. (1998). Spatio-temporal models with errors in covariates: mapping Ohio lung cancer mortality. *Statistics in Medicine* 17(18), 2025–2043.

20

Mixtures and Latent Structure in Spatial Epidemiology

Md. Monir Hossain
Division of Biostatistics and Epidemiology
Cincinnati Children's Hospital and Medical Center
University of Cincinnati College of Medicine
Cincinnati, Ohio

Andrew B. Lawson
Division of Biostatistics and Bioinformatics
Department of Public Health Sciences
Medical University of South Carolina
Charleston, South Carolina

CONTENTS

20.1 Introduction

Latent structure models are regarded as an effective tool in disease mapping for modeling the heterogeneity or discontinuity in risk surface. They provide a convenient framework in which homogeneous areas can be grouped together into clusters for discrimination of risk levels. While the applications of these models in spatial epidemiology are emerging, the wider use is hampered mainly because of limited software implementations in commonly used statistical software such as SAS or STATA.

These models are often implemented in a Bayesian framework where the hierarchical parameters can flexibly be estimated. The Bayesian approach also provides a wide variety of model selection criteria, such as the deviance information criterion (DIC), mean square predictive error (MSPE), or Bayesian p-value. Among these criteria, some are specifically developed for mixture-type models (Celeux et al., 2006). The computational flexibility of latent models in generating results for epidemiological or public health focus such as relative risk estimation or finding areas where the relative risk estimates exceed certain threshold values varies widely, depending on distributional assumptions involved in these models and on whether the number of components is fixed or needs to be estimated.

When the component number is fixed, the model can easily be implemented in many freely available software packages, for example, WinBUGS (Spiegelhalter et al., 2003) or R-INLA (Rue et al., 2009). A simple example of a fixed-component latent model is the well-known BYM convolution model (see Chapter 7). With a variable-component dimension, the estimation process often requires specifically written software. In the simpler case of orthogonal components, which can be modeled via factor decomposition, efficient matrix computation is required. When mixtures are considered, however, alternatives must be sought. Two possible procedures, often followed in such cases, are transdimensional Markov chains (Sisson, 2005) in the parametric setting and Dirichlet process priors in the nonparametric setting. Although transdimensional models in the parametric setting are conceptually intriguing approaches, the lack of model diagnostic tools hampers wider adoption and the ability to ascertain the validity of the inference, and objectively quantifying the model uncertainty as the dimension varies within the model space remains challenging (Hastie and Green, 2012). A number of methods are proposed for the transdimensional approach; among them, three widely used approaches are product space Markov chain Monte Carlo (MCMC) (Dellaportas et al., 2002), reversible jump MCMC (RJMCMC) (Green, 1995; Richardson and Green, 1997), and birth-and-death MCMC (BDMCMC) (Stephens, 2000; Hurn et al., 2003). In a recent work (Cappe et al., 2003), RJMCMC and BDMCMC were formulated in a continuous-time MCMC (CTMCMC) framework.

Spatial mixture models are often constructed by defining mixtures in mean (Lawson and Clark, 2002) or mixtures in distribution (Green and Richardson, 2002; Fernandez and Green, 2002). Both of these approaches have a long pedigree of development. In Sections 20.3 and 20.5, we summarize these two approaches. The extensions of these spatial models to the spatiotemporal domain are summarized in Sections 20.4 and 20.6. In Section 20.7, we provide a short description for the Dirichlet process mixture (see Chapter 21 for more detail). Since in many epidemiologic studies outcomes are often recorded as counts of events aggregated at the spatial unit, for example, at the county level or census tract level, we focus mainly on the Poisson model that is frequently assumed for these kinds of data sets. Applications of other models, such as normal or binomial models, can also be found in relevant research.

20.2 Decomposition Methods: FA and PCA

A classical approach to the latent structure for multiple diseases is to consider a dimension reduction to allow a more parsimonious representation of risk. This can be achieved in a variety of ways, but the most common is to assume that latent factors underlie the risk. Wang and Wall (2003) define a spatial factor model for $i = 1, \ldots, n$ regions and q diseases, $j = 1, \ldots, q$, that allows the log relative risk to be decomposed as the product of a common spatial factor (f_i) and a disease loading (λ_j), that is, $\log(\theta_{ij}) = \log(e_{ij}) + \lambda_j f_i$.

Here, the expected count for the ith area and jth disease (e_{ij}) is included as a log (offset) as per usual, and the loading is disease dependent, while the common factor is area specific. Conventionally, when a Bayesian approach is assumed, prior distributions are used to provide for prior knowledge and also for identification. Here, for identification, it is often assumed that $f_i \sim N(0,1)$. However, the factors can be allowed to be spatially correlated, and so the vector $\mathbf{f} = (f_1, \ldots, f_n)^T$ is assumed to have zero-mean MVN with a spatially dependent covariance matrix with variance 1. A sum-to-zero constraint is also usually applied to this vector. For the loading vector, it is assumed that each element is independently distributed as $\lambda_j \sim N(0, \tau_\lambda)$ and the variance is assumed to be large for noninformativeness.

Note that this is different from the idea of a decomposition of a predictor design matrix, where a set of lower dimensional factors are sought that have loadings from the covariates and predictors in a model. That approach would lead to factorial or PCA (Poisson) regression in a disease mapping example.

Extension of the decomposition approach for spatiotemporal applications is straightforward if it is assumed that the loading vector has temporal dependence. Hence, for outcome and expected count y_{ijt}, e_{ijt} for the tth time, and relative risk θ_{ijt}, the log relative risk model could be specified as $\log(\theta_{ijt}) = \log(e_{ijt}) + \lambda_{jt} f_i$, and then a random walk or autoregressive lag-one prior distribution could be assumed. The random walk version could be specified thus: $\lambda_{jt} \sim N(\lambda_{j,t-1}, \tau_\lambda)$. A further extension would be to consider f_i as time dependent as well as spatially dependent (see, e.g., Lopes et al., 2008; Wang and Wall, 2003).

20.3 Simple Mean Mixtures

Another approach that has received attention is to consider the mean of the process to consist of combinations of effects. A simple example of this is the convolution model of Besag et al. (1991), where an additive model is assumed with two fixed (random) effects and an intercept: $\log(\theta_i) = \alpha + v_i + u_i$. This is in effect a mean mixture model with fixed (known) number of components. The components are uncorrelated (v_i) and correlated (u_i) spatially. The second component is usually assumed to have a conditional autoregressive (CAR) prior distribution (see Chapter 7 for further details). This was extended by Lawson and Clark (2002) to a three-component model with two spatially structured components.

In general, a mixture could have L components and, if additive, could form a sum of terms. For example, $\log(\theta_i) = \alpha + \sum_{l=1}^{L} \omega_l \mu_{li}$ could be specified where $E(y_i) = e_i \exp\left\{\alpha + \sum_{l=1}^{L} \omega_l \mu_{li}\right\}$. The term μ_{li} represents the lth latent effect, while we include a weight ω_l to allow differential weighting of effects. In a model such as this, the effects are unobserved and, in general, the number of effects or components (L) is unknown. Some prior distributional assumptions must be made to allow the effects and weights to be estimable or identified. If the components are levels of risk, then it is possible to order the levels to ensure identification (Stephens, 2000a). In general, it is possible to make further assumptions about the weights and the effects. First, the weights could sum to 1 and are essentially probabilities: $0 < \omega_l < 1; \sum_l \omega_l = 1$. In a Bayesian setting, this could be achieved by assuming a Dirichlet distribution for $\boldsymbol{\omega}$. Second, the effect levels could be assumed to be strictly positive and so may be assumed to have gamma or lognormal distributional constraints. However, there is no requirement for this, and so the effects could be assigned Gaussian prior distributions instead.

With unknown numbers of components as well as unknown component levels, there is an added computational burden. The dimension of the mixture is unknown and has to be estimated, as well as the effects and weights. A variety of proposals have been made to deal with this situation, such as transdimensional Markov chains (Sisson, 2005), or entry parameters (Kuo and Mallick, 1998). In the latter approach, the mixture is replaced by

$\log(\theta_i) = \alpha + \sum_{l=1}^{L} w_l \phi_l \mu_{li}$, with fixed large L and the ϕ_l being allowed to take the binary value of 0 or 1 (via Bernoulli prior distributions). Within a Bayesian posterior sampling algorithm, the entry parameters can allow the components to enter or leave as their importance dictates (Lawson et al., 2010).

20.4 Spatiotemporal Mean Mixtures

Extension of the mean mixture formulation has been explored in space–time also. In that case, the mixture can be assigned temporal as well as spatial labels. Define the model: $y_{it} \sim Poisson(e_{it}\theta_{it})$, and, $\log(\theta_{it}) = \alpha + R_{it}$, where the subscript t denotes the time label denoting a discrete time unit.

The terms in R_{it} could take a variety of forms. In the case of a mean mixture, one such specification could be $\log(\theta_{it}) = \alpha + \sum_{l=1}^{L} w_{ilt}\lambda_{tl}$, where L temporal components are convolved with weights that are indexed by region and time. A simpler and more identified version has weights indexed only by spatial region. These models have been used in a sequence of papers (Lawson et al., 2010, 2012; Choi et al., 2011). A variant based on model choice, rather than component selection, has also been proposed by Li et al. (2012).

Note that often the use of mixtures of the above kind can lead to improved goodness of fit overall. The additional reward in the use of these methods is that underlying components can also be estimated. While identification of components must remain a concern, the gains in predictive performance over conventional random effect convolution models support their use.

20.5 Simple Distribution Mixtures

In two contemporary seminal papers, Green and Richardson (2002) and Fernandez and Green (2002) introduced spatial mixture models to take into account the presence of discontinuity in the spatial structure of risk within hierarchical approaches. The model development and the parameter estimation followed a fully Bayesian setting using a reversible jump MCMC algorithm. The marginal distribution of y_i is assumed to follow a mixture of Poisson distributions and is given by $y_i \sim \sum_{l=1}^{L} \omega_{il} Poisson\left(\theta_l e_i \exp\left(\eta_i\right)\right)$, where the offset e_i is the expected incidence at county i, and L is the number of mixing components and is assumed to be unknown. The linear predictor has the form $\eta_i = \beta_0 + \sum_{j=1}^{p} \beta_j x_{ij}$.

The mixing weight ω_{il} is assumed to be space and component dependent. The weights satisfy two conditions, $0 < \omega_{il} < 1$ and $\sum_{l=1}^{L} \omega_{il} = 1$, and are determined by a hidden (or unobserved) allocation (or cluster) variable, $\{Z_i\}_{i=1,\ldots,n}$ as $\omega_{il} = \Pr\left(Z_i = l | L, \omega, \theta\right)$. Note that, by making the mixing weights vary with space, as opposed to the traditional mixing weights, where they depend only on the components, the model for the cluster variable, Z_i, allows borrowing information from neighboring areas by incorporating spatial dependence functions in the estimation of mixing weight ω_{il} so that similar outcomes close in space can be assigned a similar risk level. It can also include the covariates, depending on the study's purpose. Much of the work of Green and Richardson (2002) and Fernandez and Green (2002) is devoted to specifying the function for Z_i, that is, how to incorporate spatial dependence in the estimation of ω_{il}. Green and Richardson (2002) use a discrete state-space Markov random field model to account for the spatial dependencies, and the joint distribution for the $\mathbf{Z}$ is defined according to a Potts model, $p(\mathbf{Z}) = \exp(\psi U(\mathbf{Z}) - H(\psi))$, where $U(\mathbf{Z})$ is the potential function of like-labeled neighbor pairs and $H(\psi)$ is the normalizing constant. The Potts model is defined with a single interaction parameter, ψ, for the interaction between neighboring areas and may fail to capture variable risk patterns if there is reasonable discontinuity in the outcome. Forbes et al. (2013) and Alfó et al. (2009) propose a general form such as the Gibbs distribution, for which the Potts model is a specific case. The Gibbs distribution models the interactions by a $L \times L$ parameter matrix, allowing more flexibility in modeling the discontinuities, but the computation could be cumbersome.

Instead of specifying a prior distribution for the allocation variable, Fernandez and Green (2002) propose assigning a prior distribution for the mixing weights. The spatial dependence was incorporated through this prior specification. For example, one of their prior specifications, the logistic normal model, has the form $\omega_{il} = \exp(\tau_{il}/\phi)/\sum_{l=1}^{L} \exp(\tau_{il}/\phi)$; the spatial dependence was introduced by defining a Markov random field–type model for $\tau_l = (\tau_{1l}, \ldots, \tau_{nl})^T$. A conditional autoregressive–type model (Besag et al., 1991) can also be used to achieve better computational stability.

20.5.1 Label switching and identifiability constraint

The Bayesian implementation of mixture models is often subject to the *label switching* problem since the likelihood estimate is invariant to the permutation of the labels; that is, the estimates of mixture components in every iteration of an MCMC algorithm are insensitive to the estimates of the hidden allocation variable (Jasra et al., 2005). The presence of this problem can be verified by looking into the MCMC samples, that is, if there is any structured pattern or the presence of multiple modes in the posterior samples of model parameters.

Earlier work (e.g., Richardson and Green, 1997) to circumvent the problem of label switching focused on imposing an ordering restriction to the mean parameter, which they referred to as *identifiability constraints.* Stephens (2000b) demonstrated that imposing the identifiability constraint does not always produce the expected results, and that the estimates could be misleading (see also Marin et al., 2005). To resolve this issue, Stephens (2000b) proposed *relabeling algorithms* based on the postprocessing of posterior samples from a longer MCMC run generated from a model without restricting the parameter space. The labels are estimated from the postprocessing of posterior samples, for example, by minimizing the posterior expectation of some loss function. The loss function proposed by Stephens (2000b) for the relabeling algorithms depends on the posterior estimates of the mixing weights. Another way to implement the relabeling algorithm is to use the *posterior similarity matrix* as an approximation of true clusters and minimize Binder's loss function (Binder, 1978, 1981) for the estimated and true clusters, with the posterior similarity matrix defined by pairwise coincidences in MCMC samples. Lau and Green (2007) formulated the optimization of Binder's expected loss as a binary integer programming problem. Fritsch and Ickstadt (2009) extended the adjusted Rand index for the relabeling algorithm, named it the posterior expected adjusted Rand, and compared it with the Binder's loss function approach.

20.5.2 Covariate inclusion and variable selection

The covariates in mixture models can be allowed to have homogeneous or heterogeneous effects, allowing for the capturing of the effect of the discontinuity pattern in covariates on outcome. For example, the geographical distribution of the county-specific percentage of population living under the poverty line may show a different discontinuity pattern than the outcome, such as county-specific low birth rate. Following Viallefont et al. (2002), the covariate effects are homogeneous if they are the same for all the mixing components; otherwise, the effects are heterogeneous. In the presence of a heterogeneous covariate effect, the marginal distribution of outcome can be written as $y_i \sim \sum_{l=1}^{L} \omega_{il} Poisson\left(\theta_l e_i \exp\left(\eta_{il}\right)\right)$, where the linear predictor has the form $\eta_{il} = \beta_{0l} + \sum_{j=1}^{p} \beta_{jl} x_{ij}$.

Selecting significant covariates in a spatial mixture model is critical for the correct classification of risk labels. It is a complex issue, especially when the component number is not known and requires the simultaneous estimation of component number and significant covariates. The research is very limited in this area as far as we know. Two methods have recently been proposed for nonspatial data. One method combines the Dirichlet process mixture with the variable selection method (Kim et al., 2006), while the other is the reversible jump MCMC with the variable selection method (Tadesse et al., 2005). In both approaches, entry parameters are considered in the linear predictor model for the selection of covariates. It appears that within a fully Bayesian setting, these approaches can be extended for the spatial mixture model by incorporating spatial dependence through a logistic normal prior for the mixing weights or by specifying a Gibbs prior distribution for the allocation variable.

20.5.3 Multiple outcomes

Often, the presence of one disease in an area may correlate with the incidence of another disease over the same region; joint modeling of multiple diseases in a multivariate modeling framework allows us to gain a better understanding of underlying common risk factors, while accounting for spatial dependence between the adjacent areas. The underlying risk factor can also be viewed as an unobserved latent factor, and the models proposed in spatial latent class analysis (e.g., Wall et al., 2012; Wall and Liu, 2009) can be applied here. The spatial latent class models use the probit prior for the modeling of mixing weights and incorporate the spatial dependence by a linear combination of CAR components, that is, by a linear model of coregionalization (Wackernagel, 2003). It separates out the spatial dependence into within-site spatial correlation (i.e., between outcomes) and across-site spatial correlation components. Multivariate CAR models (Jin et al., 2007) can also be applied instead, which provides a better parameterization for the covariance structure and a general framework for dealing with multiple outcomes. In a fully Bayesian setting using the reversible jump algorithm, a multivariate extension of Poisson mixtures is also proposed by Meligkotsidou (2007). More recently, Alfó et al. (2009) proposed a multivariate spatial mixture model where the relative risk parameters are parameterized according to a shared component model (Knorr-Held and Best, 2001). The spatial dependence was assigned by a CAR prior model for the shared component.

20.6 Spatiotemporal Distribution Mixtures

Correlated outcomes can also arise from repeated measures, that is, for the outcomes measured at each time point for a period over the same area. This repeated time information introduces a further dimension that is often important because of changes in socio-demographic structure or other health factors during the time period of the study. It has also been observed that any significant changes close in space can be close in time too. This observation underscores the importance of space–time modeling and emphasizes the consideration of two potential dimensions of dependency: between spatially neighbor locations, and between successive time points. Including these dimensions in the analysis adjusts the parameter estimates for spatial, temporal, and spatial-temporal dependencies, and contributes to the better understanding of disease etiology.

We assume that the marginal distribution of y_{it} follows a mixture of Poisson distributions and is given by $y_{it} \sim \sum_{l=1}^{L} \omega_{itl} Poisson\left(\theta_l e_{it} \exp\left(\eta_{it}\right)\right)$, where the offset e_{it} is the expected incidence at county i and year t. The linear predictor with the homogeneous covariate effect and area-specific regression coefficients has the form $\eta_{it} = \beta_{i0} + \sum_{j=1}^{p} \beta_{ij} x_{itj}$. The prior distribution for the coefficients can be specified by a multivariate CAR model, which allows the correlation between covariates to be better controlled. Since the regression coefficients are spatially varying, a condition $\sum_{i=1}^{n} \beta_{ij} = C$, as mentioned in Assunção (2003), has to be imposed to ensure identifiability. Including the two intercept parameters in the above model, component-specific $\log\left(\theta_l\right)$ and area-specific β_{i0} offer greater flexibility, such as allowing mimicking if there are any inherent differences in the ratios of observed to expected that vary with the components or clusters, in addition to the area-level intercepts. A model with heterogeneous covariate effect can also be considered, but often for this model, the coefficient estimates are subject to lack of stability and the improvement in goodness of fit is not significant.

The mixing weight can assume a logistic transformed function such as $\omega_{itl} = \exp\left(h_{itl}/\phi\right)/\sum_{l=1}^{L} \exp\left(h_{itl}/\phi\right)$. As in the spatial model, by making the mixing weights vary with space and time, the model for the cluster variable, Z_{it}, allows borrowing information from neighboring areas and preceding times by incorporating spatial and temporal dependence functions in the specification of h_{itl}, so that similar outcomes close in space and time can be assigned with similar cluster labels. The function h_{itl}, in its simplest form, will include spatial and temporal random effects (Hossain and Lawson, 2010; Hossain et al., 2014), such as $h_{itl} = \kappa_{il} + \gamma_{tl}$, where the κ_l's are the structured spatial random effects and the γ_l's are the structured temporal random effects; a complicated structure could include the space–time nonseparable interaction effect (Knorr-Held, 2000).

20.7 Dirichlet Process Mixture

Dirichlet process (DP) prior distributions are also proposed (Gelfand et al., 2005; Griffin and Steel, 2006; Duan et al., 2007) for better risk estimation of spatially referenced health data. In these works, the spatial dependence was introduced by defining the spatial model for either the mixing weights or the mixing components. The dependent DP introduced by Griffin and Steel (2006) involves an ordered stick-breaking prior distribution for the mixing weights, and the ordering depends on the closeness of a covariate value; that is, distributions for similar covariate values will be assigned to similar orderings. Reich and Fuentes (2007) extended the kernel-based stick-breaking processes (Dunson and Park, 2008) to include spatial processes in order to model the spatial dependence where the weights are spatially indexed. The development of space–time-dependent DPM is very limited. Gelfand et al. (2005) used the temporal observations as the replications at each spatial location in their development of a spatial dependence DP. Instead of considering this temporal information as independent replications in spatial dependence DP, Kottas et al. (2007) viewed it as a temporally evolving spatial process. The proposal is based on the decomposition of mixing components with temporal index into a first-order autoregressive term and a random innovation term. Dependent DP prior distributions are assumed for the random innovations, and a zero-mean stationary Gaussian process is considered for the base distribution. Hossain et al. (2013) propose a space–time dependence DP mixture model and introduce the dependencies for spatial and temporal effects for the mixing weights by using space–time covariate-dependent kernel stick-breaking processes.

20.8 Case Study

We analyzed the county-specific South Carolina low-birth-weight (LBW) incidence data for the years 1997–2007 to show the implementation of some of the above models and for the discussion of relevant issues. The data are publicly available and can be acquired from the South Carolina Department of Health and Environmental Control (http://www.scdhec.gov/), and population, income, poverty, and unemployment data were obtained from the U.S. Census Bureau (http://www.census.gov). LBW is defined as a child weighing less than 2500 g at live birth. The county- and year-specific expected LBW incidence, e_{it}, is calculated by $n_{it}R$, where n_{it} is the total birth in county i and year t, and R is the South Carolina overall LBW incidence rate calculated by the ratio of total LBW to total birth over the entire spatial-temporal domain. The histogram and density plots of observed county-specific standardized incidence ratios (SIRs) for 1 year (1998) are presented in Figure 20.1. The figure shows the presence of clustering patterns in the observed SIRs and justifies why we are using mixture models for this data set. The number of breaks and the bandwidth in this figure are estimated by using a robust method available through the bw.SJ function in R (R Development Core Team, 2004), rather than fixing these to predetermined values. The county-specific SIRs for the years 1997, 2001, 2004, and 2007 are given in Figure 20.2 (top row). The observed SIRs show the presence of a small cluster of excess risks in the east in 2001. This cluster appears again in 2004. Although there is no spatial cluster that is

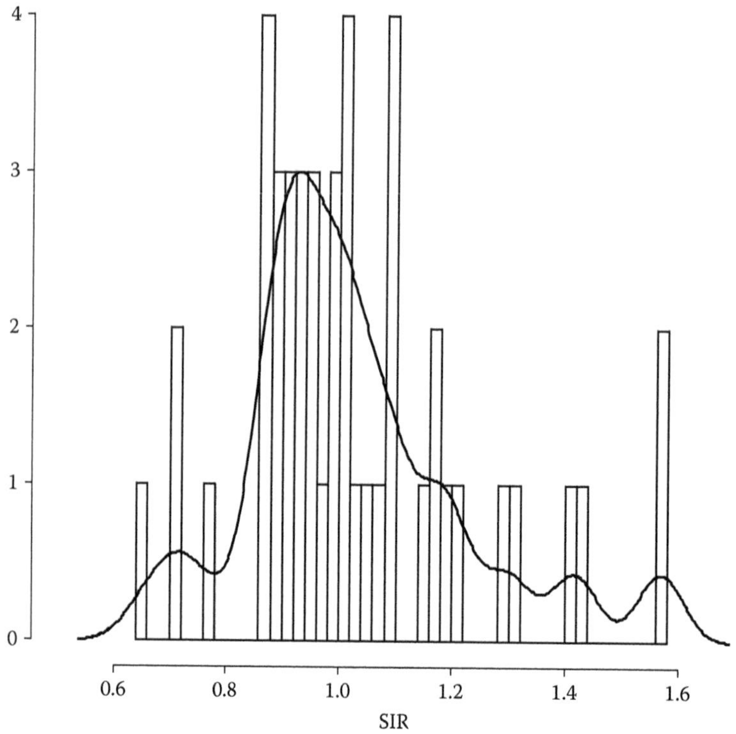

FIGURE 20.1
Histogram and density plots of observed standardized incidence ratio (SIR) of county-specific South Carolina LBW count data for the year 1998. The number of breaks and bandwidth are determined by using the bw.SJ function in R.

persistent over the study period, there are some sporadic regions of excess risks. In our first application, we applied the stick-breaking process space–time (SBPST) method (Hossain et al., 2013) to three models based on their specification of the function h_{itl}. Model 1 considers no covariates, specified only by the structured spatial and structured temporal random effects, that is, $h_{itl} = \kappa_{il} + \gamma_{tl}$. Model 2 considers covariates, and we consider the county- and year-specific covariates: population density (PD), proportion of African Americans (PAA), median household income (MHI), proportion of poverty (PP), and unemployment rate (UR). These covariates are considered by following previous studies by Fang et al. (1999) and Pearl et al. (2001). Thus, we use the form $h_{itl} = \beta_{0l} + \beta_{1l}PD_{it} + \beta_{2l}PAA_{it} + \beta_{3l}MHI_{it} + \beta_{4l}PR_{it} + \beta_{5l}UR_{it} + \kappa_{il} + \gamma_{tl}$. This model assumes that the regression parameters are component specific and has the flexibility to capture any effect of discontinuity or clustering pattern in covariates on LBW. Model 3 considers a general approach by defining a function $h_{itl} = \beta_{0il} + \beta_{1il}PD_{it} + \beta_{2il}PAA_{it} + \beta_{3il}MHI_{it} + \beta_{4il}PR_{it} + \beta_{5il}UR_{it} + \beta_{6il}t$. This model illustrates a simpler way to model a space–time interaction effect, and specifying a multivariate prior distribution for β_{il}'s ensures a nonseparable covariance for this effect. Using a multivariate prior distribution for the joint modeling of covariate effects is very practical in the sense that counties with a higher UR are likely to have a higher PP and a lower MHI. The deviance information criterion (DIC) and the mean square predictive error (MSPE) values for these models are reported in Table 20.1. We have observed similar trends in DIC and MSPE values for the three models specified above. The smallest values are obtained for Model 1, which includes only the structured spatial and structured temporal components. The second smallest values are obtained for Model 3. This model considers a multivariate conditional autoregressive model as a prior distribution for the spatially varying regression parameters, and also includes a space–time nonseparable effect. The DIC and

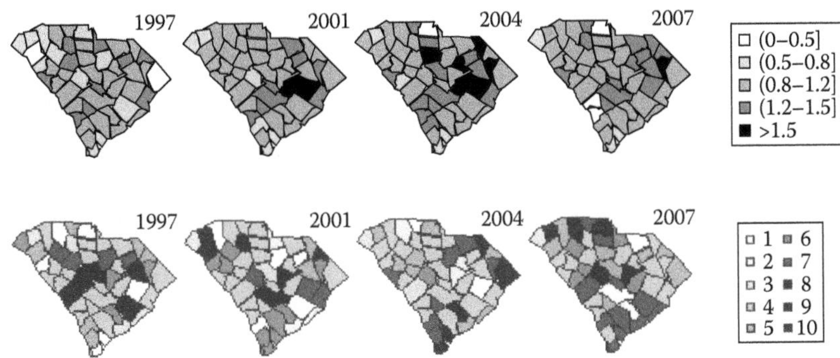

FIGURE 20.2
Thematic maps of standardized incidence ratios (top row) and the group membership labels of South Carolina LBW incidences from SBPST Model 1 (bottom row) for the selected years (1997, 2001, 2004, and 2007).

TABLE 20.1
DIC and MSPE values for the three competing SBPST models for the South Carolina LBW incidences for the years 1997–2007

Model	Mean of deviance	Variance of deviance	DIC	MSPE
Model 1	3636.0	27.71	3649.86	224.2
Model 2	3676.0	28.66	3690.33	256.0
Model 3	3664.0	20.76	3674.38	244.6

MSPE values for Model 3 are lower than those of Model 2, indicating the presence of spatial correlation within the covariates, although this evidence is not substantial, since Model 1 has the smallest DIC and MSPE values. In the following, we report the results for Model 1.

In estimating the latent cluster label, we intend to find the group in which observations are more similar to each other than to observations in other groups. One way of assigning the appropriate group label to each observation is by calculating the frequency distribution of the latent variable Z_{it} over all posterior samples, and then assigning the label for which the frequency is maximized. The frequency value for each l can be calculated by $n_{it}^{(l)} = \sum_{g=1}^{G} I(Z_{it}^{(g)} = l)$, where $Z_{it}^{(g)}$ is the gth posterior latent variable Z_{it}, which indicates the group membership; $l = 1, \ldots, K$; and G is the posterior sample size. The group label for the (i, t)th observation will be l', for which $n_{it}^{(l')} = \max_{l \in K}(n_{it}^{(l)})$. The membership labels are in the range of 1–10, since the number of mixing components L in the computation was set to 10. It appears that over the 11-year study period, the number of grouping for each county varies in the range of three to six levels. The maximum number of grouping occurs for Charleston and Greenwood Counties. The thematic maps of this membership label for the years 1997, 2001, 2004, and 2007 are presented in Figure 20.2 (bottom). In both maps, top and bottom of Figure 20.2, the darker colors may not bear the same meaning. In the top map, the darker colors indicate high-risk counties, and in the bottom, they distinguish the group membership.

Note that, in this case study, we did not consider models where the covariates are homogeneous with respect to the mixture components. Those models would be useful to explore in a fuller investigation.

References

Alfó M, Nieddu L, Vicari D. Finite mixture models for mapping spatially dependent disease counts. *Biometrical Journal* 2009; 51:84–97.

Assunção RM. Space varying coefficient models for small area data. *Environmetrics* 2003; 14:453–473.

Besag J, York J, Mollie A. Bayesian image restoration with two applications in spatial statistics [with discussion]. *Annals of the Institute of Statistical Mathematics* 1991; 43: 1–59.

Binder D. Bayesian cluster analysis. *Biometrika* 1978; 65:31–38.

Binder D. Approximations to Bayesian clustering rules. *Annals of Statistics* 1981; 68: 275–285.

Cappe O, Robert C, Ryden T. Reversible jump, birth-and-death and more general continuous time Markov chain Monte Carlo samplers. *Journal of the Royal Statistical Society: Series B* 2003; 65:670–700.

Celeux G, Forbes F, Robert C, Titterington M. Deviance information criteria for missing data models. *Bayesian Analysis* 2006; 1:651–674.

Choi J, Lawson A, Cai B, Hossain M. Evaluation of Bayesian spatial-temporal latent models in small area health data. *Environmetrics* 2011; 22:1008–1022.

Dellaportas P, Forster J, Ntzoufras I. On Bayesian model and variable selection using MCMC. *Statistics and Computing* 2002; 12:27–36.

Duan J, Guindani M, Gelfand A. Generalized spatial Dirichlet process models. *Biometrika* 2007; 94(4):809–825.

Dunson D, Park JH. Kernel stick-breaking processes. *Biometrika* 2008; 95(2):307–323.

Fang J, Madhavan S, Alderman M. Low birthweight: Race and maternal nativity-impact of community income. *Pediatrics* 1999; 103:E5.

Fernandez C, Green PJ. Modelling spatially correlated data via mixtures: A Bayesian approach. *Journal of the Royal Statistical Society: Series B* 2002; 64:805–826.

Forbes F, Charras-Garrido M, Azizi L, Doyle S, Abrial D. Spatial risk mapping for rare disease with hidden Markov fields and variational EM. *Annals of Applied Statistics* 2013; 7:1192–1216.

Fritsch A, Ickstadt K. Improved criteria for clustering based on the posterior similarity matrix. *Bayesian Analysis* 2009; 4(2):367–392.

Gelfand AE, Kottas A, MacEachern S. Bayesian nonparametric spatial modeling with Dirichlet process mixing. *Journal of the American Statistical Association* 2005; 100: 1021–1035.

Green J. Reversible jump MCMC computation and Bayesian model determination. *Biometrika* 1995; 82:711–732.

Green J, Richardson S. Hidden Markov models and disease mapping. *Journal of the American Statistical Association* 2002; 97:1055–1070.

Griffin J, Steel M. Order-based dependent Dirichlet processes. *Journal of the American Statistical Association* 2006; 101:179–194.

Hastie DI, Green PJ. Model choice using reversible jump Markov chain Monte Carlo. *Statistica Neerlandica* 2012; 66:309–338.

Hossain M, Lawson A, Cai B, Choi J, Liu J, Kirby R. Space-time areal mixture model: Relabeling algorithm and model selection issues. *Environmetrics* 2014; 25:84–96.

Hossain M, Lawson A, Cai B, Jungsoon C, Liu J, Kirby R. Space-time Dirichlet process mixture models for small area disease risk and cluster estimation. *Environmental and Ecological Statistics* 2013; 20:91–107.

Hossain MM, Lawson AB. Space-time Bayesian small area disease risk models: Development and evaluation with a focus on cluster detection. *Environmental and Ecological Statistics* 2010; (17):73–95.

Huang K. Space-time latent component modeling of geo-referenced health data. *Statistics in Medicine* 2010; 29:2012–2017.

Hurn M, Justel A, Robert C. Estimating mixtures of regressions. *Journal of Computational and Graphical Statistics* 2003; 12:55–79.

Jasra A, Holmes C, Stephans D. Markov chain Monte Carlo methods and the label switching problem in Bayesian mixture modeling. *Statistical Science* 2005; 20:50–67.

Jin X, Banerjee S, Carlin B. Order-free co-regionalized areal data models with application to multiple-disease mapping. *Journal of the Royal Statistical Society: Series B* 2007; 69:817–838.

Kim S, Tadesse M, Vannucci M. Variable selection in clustering via Dirichlet process mixture models. *Biometrika* 2006; 93:877–893.

Knorr-Held L. Bayesian modelling of inseparable space-time variation in disease risk. *Statistics in Medicine* 2000; 19:2555–2567.

Knorr-Held L, Best N. A shared component model for detecting joint and selective clustering of two diseases. *Journal of the Royal Statistical Society: Series A* 2001; 164:73–85.

Kottas A, Duan J, Gelfand AE. Modeling disease incidence data with spatial and spatio-temporal Dirichlet process mixtures. *Biometrical Journal* 2007; 49(5):1–14.

Kuo L, Mallick B. Variable selection for regression models. *Sankhya: Series B* 1998; 60: 65–81.

Lau J, Green P. Bayesian model based clustering procedure. *Journal of Computational and Graphical Statistics* 2007; 16:526–558.

Lawson A, Choi J, Cai B, Hossain M, Kirby R, Liu J. Bayesian 2-stage space-time mixture modeling with spatial misalignment of the exposure in small area health data. *Journal of Agricultural, Biological, and Environmental Statistics* 2012; 17:417–441.

Lawson AB, Clark AB. Spatial mixture relative risk models applied to disease mapping. *Statistics in Medicine* 2002; 21(3):359–370.

Lawson AB, Song HR, Cai B, Hossain MM, Huang, K. Space-time latent component modeling for geo-referenced health data. *Statistics in Medicine* 2010; 29:2012–2027.

Li G, Best N, Hansell A, Ahmed I, Richardson S. BaySTDetect: Detecting unusual temporal patterns in small area data via Bayesian model choice. *Biostatistics* 2012; 13:695–710.

Lopes H, Salazar E, Gamerman D. Spatial dynamic factor analysis. *Bayesian Analysis* 2008; 3:759–792.

Marin J-M, Mengersen K, Robert C. Bayesian modelling and inference on mixtures of distributions. In Rao C, Dey D (eds.), *Handbook of Statistics*. Springer-Verlag, New York, 2005.

Meligkotsidou L. Bayesian multivariate Poisson mixtures with an unknown number of components. *Statistics and Computing* 2007; 17:93–107.

Pearl M, Braveman P, Abrams B. The relationship of neighborhood socioeconomic characteristics to birthweight among 5 ethnic groups in California. *American Journal of Public Health* 2001; 91:1808–1814.

R Development Core Team. R: A language and environment for statistical computing. R Foundation for Statistical Computing, Vienna, Austria, 2004.

Reich B, Fuentes M. A multivariate semiparametric Bayesian spatial modeling framework for hurricane surface wind fields. *Annals of Applied Statistics* 2007; 1(1):249–264.

Richardson S, Green P. On Bayesian analysis of mixtures with unknown number of components. *Journal of the Royal Statistical Society: Series B* 1997; 59:731–792.

Rue H, Martino S, Chopin N. Approximate Bayesian inference for latent Gaussian models by using integrated nested Laplace approximations [with discussion]. *Journal of the Royal Statistical Society: Series B* 2009; 71:319–392.

Sisson S. Transdimensional Markov chains: A decade of progress and future perspectives. *Journal of the American Statistical Association* 2005; 100:1077–1089.

Spiegelhalter D, Thomas A, Best N, Lunn D. WinBUGS user manual, v1.4. MRC Biostatistics Unit, Institute of Public Health, Cambridge, UK, 2003.

Stephens M. Bayesian analysis of mixture models with an unknown number of components—an alternative to reversible jump methods. *Annals of Statistics* 2000a; 28:40–74.

Stephens M. Dealing with label switching in mixture models. *Journal of the Royal Statistical Society: Series B* 2000b; 62:795–809.

Tadesse M, Sha N, Vannucci M. Bayesian variable selection in clustering high-dimensional data. *Journal of the American Statistical Association* 2005; 100:602–617.

Viallefont V, Richardson S, Green P. Bayesian analysis of Poisson mixtures. *Journal of Nonparametric Statistics* 2002; 14:181–202.

Wackernagel H. *Multivariate Geostatistics: An Introduction with Applications.* Springer-Verlag, New York, 2003.

Wall M, Guo J. Mixture factor analysis for approximating a nonnormally distributed continuous latent factor with continuous and dichotomous observed variables. *Multivariate Behavioral Research* 2012; 47:276–313.

Wall M, Liu X. Spatial latent class analysis model for spatially distributed multivariate binary data. *Computational Statistics and Data Analysis* 2009; 53:3057–3069.

Wang F, Wall M. Generalized common spatial factor model. *Biostatistics* 2003; 4:569–582.

21

Bayesian Nonparametric Modeling for Disease Incidence Data

Athanasios Kottas
Department of Applied Mathematics and Statistics
University of California
Santa Cruz, California

CONTENTS

21.1 Introduction

Disease incidence or mortality data are routinely recorded as summary counts for contiguous geographical regions (e.g., census tracts, zip codes, districts, or counties) and collected over discrete time periods. The count responses are typically accompanied by covariate information associated with the region (e.g., median family income or percent with a specific type of education), and occasionally, by covariate information associated with each incidence case (e.g., sex, race, or age), even though we only know the region into which the case falls. A key inferential objective in the analysis of disease incidence data is identification and explanation of spatial and spatiotemporal patterns of disease risk (disease mapping). Also of interest is forecasting of disease risk.

The statistical literature of the past 25 years or so has witnessed a growing emphasis on fairly sophisticated methods to model heterogeneity in disease event rates. Most of the methodology has been developed within a hierarchical framework through introduction of spatial and spatiotemporal models tailored to the disease mapping inference goals. In this context, the Bayesian approach to modeling and inference is naturally attractive.

In this chapter, we review Bayesian nonparametric spatial and spatiotemporal modeling approaches for disease incidence data. Section 21.2 provides the necessary background on nonparametric priors, mainly the Dirichlet process prior and its extension to spatial Dirichlet process models. Bayesian nonparametric work has focused on modeling methods for the stochastic mechanism that generates the region-specific count responses, and this is where we place the emphasis in this review, providing only brief discussion on modeling the covariate information. We thus focus on distributional specifications for the disease incidence counts (number of observed cases of the disease). These count responses are denoted by y_{it}, where $i = 1, \ldots, m$ indexes the geographic regions A_i, and $t = 1, \ldots, T$ indexes the (discrete) time periods. Note that although cases occur at specific spatial point locations, the available responses are associated with entire subregions, $A_1, \ldots, A_m$, that form a partition of the study region. As a consequence, there exist two distinct perspectives to model formulation. The more straightforward approach is to develop the hierarchical spatiotemporal model building the spatial dependence through a finite set of spatial random effects, one for each region. An alternative prior specification approach emerges by modeling the underlying continuous-space disease risk surface, which yields an implied prior for the finite dimensional distribution of the spatial random effects through aggregation of the continuous surface. Sections 21.3 and 21.4 discuss Bayesian nonparametric methods under these two modeling frameworks. In the former case, nonparametric mixtures of Poisson distributions have been used to model directly the distribution for the disease incidence counts (e.g., Hossain et al., 2013). In the latter setting, nonparametric spatial or spatiotemporal prior models have been considered for the disease risk (or rate) surface (e.g., Kelsall and Wakefield, 2002; Kottas et al., 2008). Finally, Section 21.5 provides concluding remarks.

21.2 Background on Bayesian Nonparametrics

Bayesian nonparametric methods enable flexible modeling and inference for a wide range of problems, since they are built from prior probability models for entire spaces of distributions or functions, instead of scalar or vector parameters, as in traditional parametric Bayesian modeling. Such methods have led to substantive applications in several fields, since they free the data analyst from customary parametric modeling restrictions, yielding more accurate inference and more reliable predictions. Bayesian nonparametrics is at this point a burgeoning area of Bayesian statistics; we refer to Hjort et al. (2010) and Müller and Mitra (2013) for general reviews of related theory, methods, and applications.

Most of the Bayesian nonparametric methods for disease mapping discussed here are based on (extensions of) the Dirichlet process (DP) prior (Ferguson, 1973), the earliest example of a nonparametric prior for spaces of distributions. The DP can be defined in terms of two parameters: a parametric baseline distribution G_0, which defines the expectation of the process, and a scalar parameter $\alpha > 0$, which can be interpreted as a precision parameter, since larger α values result in DP realizations that are *closer* to G_0. We use $G \sim \mathrm{DP}(\alpha, G_0)$ to denote that a DP prior, with parameters α and G_0, is placed on random distribution G. The most widely used definition of the DP is the constructive definition given by Sethuraman (1994). According to this definition, a distribution G generated from a $\mathrm{DP}(\alpha, G_0)$ prior is (almost surely) of the form $G = \sum_{i=1}^{\infty} w_i \, \delta_{\vartheta_i}$, where δ_x denotes a point mass at x. Here, the ϑ_i are independent and identically distributed (i.i.d.) from G_0, and the weights are constructed through a *stick-breaking* procedure, specifically, $w_1 = \zeta_1$, $w_i = \zeta_i \prod_{k=1}^{i-1}(1 - \zeta_k), i = 2, 3, \ldots$, with the ζ_k i.i.d. $\mathrm{Beta}(1, \alpha)$; moreover, the sequences $\{\zeta_k : k = 1, 2, \ldots\}$ and $\{\vartheta_i : i = 1, 2, \ldots\}$ are independent. Hence, the DP generates discrete distributions that can be represented as countable mixtures of point masses, with locations drawn independently from G_0 and weights generated according to a stick-breaking mechanism based on i.i.d. draws from a $\mathrm{Beta}(1, \alpha)$ distribution.

A natural way to increase the applicability of DP-based modeling is by using the DP as a prior for the mixing distribution in a mixture model with a parametric kernel density function (or probability mass function) $k(\cdot \mid \boldsymbol{\theta})$. This approach yields the class of

DP mixture models, which can be generically expressed as $f(\cdot \mid G) = \int k(\cdot \mid \boldsymbol{\theta}) \, \mathrm{d}G(\boldsymbol{\theta})$, with $G \sim \mathrm{DP}(\alpha, G_0)$. The model is typically extended by adding hyperpriors to the DP precision parameter α or the parameters of G_0. Semiparametric versions are also possible by mixing on only a portion of the kernel parameters. The kernel can be chosen to be a (possibly multivariate) continuous distribution, thus overcoming the discreteness of the DP. In fact, the discreteness of G is an asset in this context, since, given the data, it enables ties among the corresponding mixing parameters. Thus, the class of DP mixture models offers an appealing choice for applications where clustering is anticipated, as in, for example, density estimation, classification, and regression, and is indeed the most widely used Bayesian nonparametric method in applications.

The DP constructive definition has motivated extensions in several directions. One such extension is the spatial DP (Gelfand et al., 2005), a nonparametric prior for the distribution of random fields $\boldsymbol{W}_D = \{W(\boldsymbol{s}) : \boldsymbol{s} \in D\}$ over a region $D \subseteq R^d$. To model the distribution of $\boldsymbol{W}_D$, the atoms in the DP stick-breaking representation, ϑ_i, are extended to realizations from a random field, $\boldsymbol{\vartheta}_{i,D} = \{\vartheta_i(\boldsymbol{s}) : \boldsymbol{s} \in D\}$. Thus, G_0 is extended to a spatial stochastic process G_{0D} over D. For instance, a Gaussian process (GP) can be used for G_{0D}. The resulting spatial DP provides a random distribution for $\boldsymbol{W}_D$, with realizations G_D given by $\sum_{i=1}^{\infty} w_i \delta_{\boldsymbol{\vartheta}_{i,D}}$. Consequently, for any (finite) set of spatial locations in D, $\boldsymbol{s} = (\boldsymbol{s}_1, \ldots, \boldsymbol{s}_M)$, G_D induces a random distribution $G_{\boldsymbol{s}}^{(M)}$ for $(W(\boldsymbol{s}_1), \ldots, W(\boldsymbol{s}_M))$. In fact, $G_{\boldsymbol{s}}^{(M)} \sim \mathrm{DP}(\alpha, G_0^{(M)})$, where $G_0^{(M)}$ is the M-variate normal distribution induced by the GP for G_{0D} at $(\boldsymbol{s}_1, \ldots, \boldsymbol{s}_M)$. It can be shown that the random process G_D yields non-Gaussian finite dimensional distributions, has nonconstant variance, and is nonstationary, even if it is centered around a stationary GP G_{0D}. Moreover, if G_{0D} has continuous sample paths, then as the distance between two spatial locations $\boldsymbol{s}$ and $\boldsymbol{s}'$ becomes smaller, the difference between distributions $G_{\boldsymbol{s}}$ and $G_{\boldsymbol{s}'}$ becomes *smaller*; formal details can be found in Gelfand et al. (2005) and Guindani and Gelfand (2006). For alternative constructions of nonparametric prior models for spatial random surfaces, we refer to Griffin and Steel (2006), Duan et al. (2007), Reich and Fuentes (2007), and Rodriguez and Dunson (2011). Bayesian nonparametric mixture modeling has also been explored for spatial (marked) point processes, including the work of Wolpert and Ickstadt (1998), Ishwaran and James (2004), Kottas and Sansó (2007), and Taddy and Kottas (2012).

21.3 Nonparametric Mixture Modeling for Incidence Counts

In order to model heterogeneity or discontinuity in disease event rates, one is naturally led to mixture models. Indeed, various types of parametric mixture model specifications have been explored in the disease mapping literature. Chapter 20 in this volume reviews several of the existing methods based on both spatial and spatiotemporal mixture models. As discussed in Section 21.2, a significant portion of methodological and applied work in Bayesian nonparametrics has built from (countable) nonparametric mixtures. It is therefore natural to consider nonparametric mixture prior models for the disease incidence count distribution in order to expand on the inferential power of corresponding parametric finite mixtures.

To arrive at a generic mixture model formulation, and to fix notation, recall the typical assumption for the disease incidence counts: $y_{it} \mid \theta_{it} \overset{ind.}{\sim} \mathrm{Poisson}(y_{it} \mid e_{it}\theta_{it})$; that is, conditionally on parameters θ_{it}, the y_{it} are independent Poisson distributed with mean $e_{it}\theta_{it}$. Here, e_{it} is the expected disease count for region i at time period t, and θ_{it} is the associated relative risk. The expected counts are typically computed through $e_{it} = R n_{it}$, where n_{it} is the specified number of individuals at risk in region i at time t, and R is an overall disease rate. The given data set can be used to obtain R, for example, $R = \sum_{i,t} y_{it} / \sum_{i,t} n_{it}$ (internal standardization), or R can be developed from reference tables (external standardization). Standard parametric hierarchical models explain the relative risk parameters through different types of random effects. For instance, a specification with random effects additive in space and time is

$$\log \theta_{it} = \mu_{it} + u_i + v_i + \delta_t, \tag{21.1}$$

where μ_{it} is a component for the regional covariates, u_i are regional random effects (typically, assumed i.i.d. from a zero-mean normal distribution), δ_t are temporal effects (say, with an autoregressive prior), and v_i are spatial random effects with a prior typically built from a conditional autoregressive (CAR) structure. For further details, we refer to Banerjee et al. (2015), as well as Chapters 7 and 15 in this volume. When spatiotemporal interaction is sought, $v_i + \delta_t$ in (21.1) may be replaced by space–time random effects v_{it}, which have been modeled using independent CAR structures over time, dynamically with independent CAR innovations, or as a CAR in space and time; see Chapter 19 in this volume for related references.

A general (finite) mixture model formulation arises by replacing the continuous mixing distribution for the v_i (or the v_{it}), implied by the CAR prior, with a discrete distribution taking K possible values. These values represent the relative risks for K underlying space or space–time clusters, and have corresponding mixing weights that form probability vectors on the $(K-1)$-dimensional simplex. The simplest form for the discrete mixing distribution involves values ϕ_j with corresponding probabilities ω_j, for $j = 1, \ldots, K$, which, upon marginalization over the random effects, results in the mixture $\sum_{j=1}^{K} \omega_j \text{Poisson}(y_{it} \mid e_{it}\phi_j)$ (e.g., Böhning et al., 2000). In the setting without a temporal component, related is the work of Knorr-Held and Rasser (2000), Denison and Holmes (2001), and Hegarty and Barry (2008) based on spatial partition structures, which divide the study region into a number of distinct clusters (sets of contiguous regions) with constant relative risk, assuming a priori random number, size, and location for the clusters. More flexible mixture model specifications arise through the use of spatially dependent vectors of weights, ω_{ij}, such that $\sum_{j=1}^{K} \omega_{ij} = 1$ for all regions i. For instance, for spatial-only incidence counts, Fernández and Green (2002) model ω_{ij} through a logistic transformation, $\omega_{ij} = \exp(\eta_{ij}/\psi) / \sum_{\ell=1}^{K} \exp(\eta_{i\ell}/\psi)$, where the vectors $(\eta_{1j}, \ldots, \eta_{mj})$, for $j = 1, \ldots, K$, arise conditionally independent from a Markov random field prior model, inducing spatial dependence to the collection of weights $(\omega_{1j}, \ldots, \omega_{mj})$ for each mixture component. Finally, as discussed in Chapter 20, for spatiotemporal disease incidence data, one can envision the further extension to space–time-dependent weights, ω_{itj}, where now $\sum_{j=1}^{K} \omega_{itj} = 1$ for any region i and any time period t.

As a nonparametric version of this last modeling scenario, Hossain et al. (2013) developed mixtures of Poisson distributions for the count responses with weights that depend on both space and time. In particular, Hossain et al. (2013) propose the model

$$y_{it} \mid \boldsymbol{\omega}, \boldsymbol{\phi} \sim \sum_{j=1}^{K} \omega_{itj} \text{Poisson}(y_{it} \mid e_{it}\phi_j), \tag{21.2}$$

where $\boldsymbol{\phi} = (\phi_1, \ldots, \phi_K)$ is the vector of mixing relative risk parameters that are common across both space and time, and $\boldsymbol{\omega} = \{(\omega_{it1}, \ldots, \omega_{itK}) : i = 1, \ldots, m;\, t = 1, \ldots, T\}$ collects the space–time-dependent mixing weights, which satisfy $\sum_{j=1}^{K} \omega_{itj} = 1$ for any i and t. The weights are defined through $\omega_{itj} = B_{itj} \prod_{k=1}^{j-1}(1 - B_{itk})$. Here, $B_{itj} = q_{itj}\zeta_j$, where q_{itj} is a space–time-dependent kernel function (that may also depend on regional covariates), and the ζ_j are i.i.d. Beta$(1, \alpha)$. This construction can be recognized as an extension of the stick-breaking representation for the DP weights and is an example of a kernel stick-breaking process (Dunson and Park, 2008). The kernel function is specified by an extension of the logistic form from Fernández and Green (2002). More specifically, $q_{itj} = \exp(\eta_{itj}/\psi)/\sum_{\ell=1}^{K}\exp(\eta_{it\ell}/\psi)$, where $\psi > 0$ plays the role of a smoothness parameter, and the η_{itj} may be modeled through different types of spatial and temporal random effects, and possibly also as a function of time-varying regional covariates. Hossain et al. (2013) implemented the model in (21.2) with a specified number of components K, which is a finite truncation approximation to the general kernel stick-breaking mixture model that involves a countable number of components in the prior. This approximation allows ready Markov chain Monte Carlo (MCMC) posterior simulation for the model through the use of routine techniques for discrete finite mixture models.

Structured nonparametric mixtures of Poisson distributions are also explored in Li et al. (2015) for spatial-only count data. In particular, the development of their "areal-referenced spatial stick-breaking process" prior model is along the lines of the finite mixture models discussed above. Using the previous notation, but excluding the time component, that model involves the first-stage specification: $y_i \mid \mu_i, v_i \overset{ind.}{\sim} \text{Poisson}(y_i \mid e_i \exp(\mu_i + v_i))$, for $i = 1, \ldots, m$, where the μ_i are defined through a linear regression on regional covariates. The prior for the spatial random effects is built from $v_i \mid G^{(i)} \overset{ind.}{\sim} G^{(i)}$, where $G^{(i)} = \sum_{j=1}^{K}\omega_{ij}\delta_{\varphi_j}$, with a normal prior for the mixing parameters φ_j, and a kernel stick-breaking structure for the weights. More specifically, $\omega_{ij} = q_{ij}\zeta_j\prod_{k=1}^{j-1}(1 - q_{ik}\zeta_k)$, where the ζ_j are i.i.d. Beta$(1, \alpha)$, and a CAR prior on the logit scale is used for $(q_{1j}, \ldots, q_{mj})$. Evidently, collapsing the two stages of the model by marginalizing over the v_i yields a finite mixture of Poisson distributions with spatially dependent vectors of weights, in the spirit of the model specification in Fernández and Green (2002). The focus in Li et al. (2015) is on detection of boundaries between disparate neighboring regions. This is formulated as a multiple-hypothesis testing problem, based on the posterior probabilities of the events $\{v_i = v_{i'}\}$ for each pair of adjacent regions A_i and $A_{i'}$, for which discreteness is a necessary property of the prior probability model for the spatial random effects.

Finally, we note that a different semiparametric extension of the hierarchical structure for the relative risks in (21.1) can be developed by replacing the normal distribution for the regional random effects, u_i, with a nonparametric prior. Malec and Müller (2008) provide an example in the context of small-area estimation with binary responses under a setting that involves multivariate regional random effects, modeled with a DP mixture of multivariate normal distributions.

21.4 Nonparametric Prior Models for Continuous-Space Risk Surfaces

Here, we review methods that build the hierarchical model for disease incidence counts (and related covariates) from prior models for the underlying continuous-space relative risk (or rate) surface, which is aggregated to provide the induced prior for the relative risk (or rate) spatial random effects. Although less commonly used in disease mapping, this approach offers a more coherent modeling framework, since it avoids the dependence of the prior model on the data collection procedure (e.g., the number, shapes, and sizes of the regions in the particular study), and it can more naturally accommodate data sources available at different levels of spatial aggregation. Focusing in all cases on the modeling for the spatial or space–time random effects, Section 21.4.1 provides an overview of two methods for spatial-only data (Best et al., 2000; Kelsall and Wakefield, 2002), and Section 21.4.2 discusses a spatiotemporal approach (Kottas et al., 2008).

21.4.1 Spatial models

For disease counts recorded over space only, Kelsall and Wakefield (2002) construct a hierarchical model, the first stage of which assumes $y_i \mid \theta_i \overset{ind.}{\sim} \text{Poisson}(y_i \mid e_i\theta_i)$, for $i = 1, \ldots, m$ (again, the notation is similar to the that in Section 21.3, excluding the temporal component). This familiar specification is derived through Poisson process continuous-space models for the area or stratum population at risk and corresponding cases. In particular, the population at risk within stratum k is assumed to follow a nonhomogeneous Poisson process (NHPP) with intensity $\lambda_k(\boldsymbol{s})$, and the cases are viewed as a spatial point pattern from a NHPP with intensity $\lambda_k(\boldsymbol{s})p_k(\boldsymbol{s})$, where $p_k(\boldsymbol{s})$ denotes the probability of disease for stratum k at location $\boldsymbol{s}$. It is further assumed that $p_k(\boldsymbol{s}) = p_k\theta_k(\boldsymbol{s})$, where $\theta_k(\boldsymbol{s})$ is the relative risk for stratum k at location $\boldsymbol{s}$, and p_k is the reference disease probability in stratum k. For locations $\boldsymbol{s}$ in region A_i, the intensity $\lambda_k(\boldsymbol{s})$ is conceptualized as $\lambda_k(\boldsymbol{s}) = N_{ik}f_{ik}(\boldsymbol{s})$, where N_{ik} is the area or stratum population count, and $f_{ik}(\boldsymbol{s})$ is the density function for the spatial distribution of the population in stratum k and region A_i. Then, using (for rare diseases) the Poisson approximation to the binomial distribution for the number of cases in stratum k and region A_i, and summing over k, the distribution for y_i arises as Poisson with mean $\sum_k N_{ik}p_k \int_{A_i} \theta_k(\boldsymbol{s})f_{ik}(\boldsymbol{s})\mathrm{d}\boldsymbol{s}$. The mean can be rewritten as $e_i \sum_k w_{ik}\theta_{ik}$, where $e_i = \sum_k N_{ik}p_k$ and $\theta_{ik} = \int_{A_i} \theta_k(\boldsymbol{s})f_{ik}(\boldsymbol{s})\mathrm{d}\boldsymbol{s}$, such that the relative risk for region A_i is given by $\theta_i = \sum_k w_{ik}\theta_{ik}$, a weighted average of stratum-specific relative risks θ_{ik}, with weights $w_{ik} = N_{ik}p_k/e_i$, the expected proportions of cases in region A_i that are in stratum k. Kelsall and Wakefield (2002) make two simplifying assumptions or approximations to arrive at the final form for the θ_i. First, the relative risk surface is assumed constant across strata, that is, $\theta_k(\boldsymbol{s}) = \theta(\boldsymbol{s})$, for all k and $\boldsymbol{s}$. This results in $\theta_i = \int_{A_i} \theta(\boldsymbol{s})f_i(\boldsymbol{s})\mathrm{d}\boldsymbol{s}$, where $f_i(\boldsymbol{s}) = \sum_k w_{ik}f_{ik}(\boldsymbol{s})$ is a weighted average of the stratum-specific population densities over region A_i. Finally, the population density is assumed uniform across regions (an assumption implicitly made in most disease mapping modeling approaches), such that $f_i(\boldsymbol{s}) = |A_i|^{-1}$ for all $\boldsymbol{s} \in A_i$.

The model is completed with a prior for the vector of log relative risk random effects, $(\log(\theta_1), \ldots, \log(\theta_m))$, where, based on the argument above, $\theta_i = |A_i|^{-1} \int_{A_i} \theta(\boldsymbol{s})\mathrm{d}\boldsymbol{s}$. This prior is given by an m-dimensional normal distribution with a structured covariance matrix H, induced by a GP prior for the underlying log relative risk surface, $\{\log(\theta(\boldsymbol{s})) : \boldsymbol{s} \in D\}$, where D is the region under study. Kelsall and Wakefield (2002) use an isotropic GP with cubic correlation function defined in terms of a single range of dependence parameter. For computational feasibility, in particular, to simplify the form of the elements for the covariance matrix H, an additional approximation is applied. More specifically, the distribution of $\log(\theta_i) = \log(|A_i|^{-1} \int_{A_i} \theta(\boldsymbol{s})\mathrm{d}\boldsymbol{s})$ is approximated by the distribution of $|A_i|^{-1} \int_{A_i} \log(\theta(\boldsymbol{s}))\mathrm{d}\boldsymbol{s}$. Posterior inference under the model is implemented using standard MCMC methods for GP-based models. A benefit of the continuous-space modeling approach is that, in addition to estimation for the relative risks, predictive inference for the underlying relative risk surface is also possible.

Similar in spirit is also the contribution of Best et al. (2000), although their data structure does not exactly fit within the standard disease mapping setting. This work develops a spatial regression model to study the effect of traffic pollution on respiratory disorders in children. The particular study encompasses data on the residential postcode for "severe wheezing" cases, case-specific individual attributes (e.g., age and gender) and home environment covariates (e.g., home dampness and maternal smoking), population density available at a district level, and traffic pollution levels available on a spatial grid. The modeling approach incorporates the incidence cases essentially as spatially referenced data, using the centroid locations of the home postcode for all 191 cases; five postcodes contain 2 cases each, with the remaining 181 cases corresponding to a unique postcode. Best et al. (2000) model these locations of severe wheezing cases along with the individual attributes as realizations from a marked NHPP process, using a semiparametric formulation for its intensity measure. The parametric component of the intensity incorporates the information from the covariates, population density, and risk factor (traffic pollution). The nonparametric component models the (continuous-space) spatial random effects through a kernel mixture with a gamma process for the mixing measure, using the approach in Wolpert and Ickstadt (1998). The specific model for the spatial random effects can also be represented as a DP mixture, using the direct connection between the gamma process and the Dirichlet process.

21.4.2 Space–time modeling

The model of Kelsall and Wakefield (2002) is extended in Kottas et al. (2008), where a spatial DP prior is used for the underlying disease rate surface under a dynamic setting that handles disease incidence data collected over both space and time. We first discuss the spatial component of the modeling approach, followed by the dynamic spatial extension.

The first stage of the hierarchical model involves $\text{Poisson}(y_{it} \mid n_{it}\exp(\gamma_{it}))$ distributions, where n_{it} is the number of individuals at risk for region A_i and time t, and $p_{it} = \exp(\gamma_{it})$ is the corresponding disease rate (with rare diseases, the logarithmic and logit transformations are practically equivalent). Kottas et al. (2008) argue for this form for the Poisson mean instead of $e_{it}\theta_{it}$, since it avoids the need to develop the e_{it} through standardization. The γ_{it} are viewed as log-rate spatial effects arising by aggregating log-rate surfaces $\boldsymbol{\gamma}_{t,D} = \{\gamma_t(\boldsymbol{s}) : \boldsymbol{s} \in D\}$ over the regions A_i. That is, $\gamma_{it} = |A_i|^{-1}\int_{A_i}\gamma_t(\boldsymbol{s})\mathrm{d}\boldsymbol{s}$ is the block average of the surface $\boldsymbol{\gamma}_{t,D}$ over region A_i. Similar to Kelsall and Wakefield (2002), the first-stage Poisson specification is derived through aggregation of an underlying NHPP under certain assumptions and approximations. For time period t, the disease incidence cases are assumed to follow a NHPP with intensity function $n_t(\boldsymbol{s})p_t(\boldsymbol{s})$, where $\{n_t(\boldsymbol{s}) : \boldsymbol{s} \in D\}$ is the population density surface and $p_t(\boldsymbol{s}) = \exp(\gamma_t(\boldsymbol{s}))$ is the disease rate at time t and location $\boldsymbol{s}$. Assuming a uniform population density over each region at each time period, $n_t(\boldsymbol{s}) = n_{it}|A_i|^{-1}$, for $\boldsymbol{s} \in A_i$. Hence, aggregating the NHPP over the regions A_i, each y_{it} follows a Poisson distribution with mean $\int_{A_i} n_t(\boldsymbol{s})p_t(\boldsymbol{s})\mathrm{d}\boldsymbol{s} = n_{it}p^*_{it}$, where $p^*_{it} = |A_i|^{-1}\int_{A_i} p_t(\boldsymbol{s})\mathrm{d}\boldsymbol{s}$. If the distribution of the p^*_{it} is approximated by the distribution of the $\exp(\gamma_{it})$, one obtains $y_{it} \mid \gamma_{it} \overset{ind.}{\sim} \text{Poisson}(y_{it} \mid n_{it}\exp(\gamma_{it}))$ for the first-stage distribution.

To develop the prior model for the spatial log-rate random effects, first, the log-rate surfaces $\boldsymbol{\gamma}_{t,D}$ are taken as realizations from a mean-zero isotropic GP with variance σ^2 and exponential correlation function $\exp(-\varphi\|\boldsymbol{s}-\boldsymbol{s}'\|)$. The induced distribution for $\boldsymbol{\gamma}_t = (\gamma_{1t}, \ldots, \gamma_{mt})$ is an m-dimensional normal with covariance matrix $\sigma^2 H(\phi)$, where the (i,j)th element of $H(\phi)$ is given by $|A_i|^{-1}|A_j|^{-1}\int_{A_i}\int_{A_j}\exp(-\varphi\|\boldsymbol{s}-\boldsymbol{s}'\|)\mathrm{d}\boldsymbol{s}\mathrm{d}\boldsymbol{s}'$. Next, a DP prior is assumed for the distribution of the $\boldsymbol{\gamma}_t$ with centering distribution given by the m-dimensional normal above, $\mathrm{N}_m(\boldsymbol{0}, \sigma^2 H(\phi))$. The choice of the DP in this context allows for data-driven deviations from the normality assumption for the spatial random effects.

Note that this structure implies for the vector of counts $\boldsymbol{y}_t = (y_{1t}, \ldots, y_{mt})$ a Poisson DP mixture model: $\int \prod_{i=1}^{m} \text{Poisson}(y_{it} \mid n_{it}\exp(\gamma_{it}))\mathrm{d}G(\boldsymbol{\gamma}_t)$, where $G \sim \mathrm{DP}(\alpha, \mathrm{N}_m(\boldsymbol{0}, \sigma^2 H(\phi)))$. To overcome the discreteness of the distribution for the log-rate vectors (induced by the discreteness of DP realizations), the DP prior for the $\boldsymbol{\gamma}_t$ can be replaced with a DP mixture prior,

$$\boldsymbol{\gamma}_t \mid \tau^2, G \sim \int \mathrm{N}_m(\boldsymbol{\gamma}_t \mid \boldsymbol{\gamma}^*_t, \tau^2 I_m)\mathrm{d}G(\boldsymbol{\gamma}^*_t), \quad G \sim \mathrm{DP}(\alpha, \mathrm{N}_m(\boldsymbol{0}, \sigma^2 H(\phi))),$$

where $\boldsymbol{\gamma}_t^* = (\gamma_{1t}^*, \ldots, \gamma_{mt}^*)$. This extension essentially involves the introduction of a heterogeneity effect, writing $\gamma_{it} = \gamma_{it}^* + u_{it}$, with u_{it} i.i.d. $\mathrm{N}(0, \tau^2)$. The mixture model for the $\boldsymbol{y}_t$ now becomes $f(\boldsymbol{y}_t \mid \tau^2, G) = \int \prod_{i=1}^m p(y_{it} \mid \tau^2, \gamma_{it}^*) dG(\boldsymbol{\gamma}_t^*)$, where $p(y_{it} \mid \tau^2, \gamma_{it}^*) = \int \mathrm{Poisson}(y_{it} \mid n_{it} \exp(\gamma_{it})) \mathrm{N}(\gamma_{it} \mid \gamma_{it}^*, \tau^2) \mathrm{d}\gamma_{it}$ is a Poisson–lognormal mixture. In full hierarchical form, the model is given by

$$\begin{aligned} y_{it} \mid \gamma_{it} &\overset{ind.}{\sim} \mathrm{Poisson}(y_{it} \mid n_{it} \exp(\gamma_{it})), \;\; i = 1, \ldots, m, \;\; t = 1, \ldots, T \\ \gamma_{it} \mid \gamma_{it}^*, \tau^2 &\overset{ind.}{\sim} \mathrm{N}(\gamma_{it} \mid \gamma_{it}^*, \tau^2), \;\; i = 1, \ldots, m, \;\; t = 1, \ldots, T \\ \boldsymbol{\gamma}_t^* \mid G &\overset{i.i.d.}{\sim} G, \;\; t = 1, \ldots, T \\ G \mid \alpha, \sigma^2, \phi &\sim \mathrm{DP}(\alpha, \mathrm{N}_m(\mathbf{0}, \sigma^2 H(\phi))). \end{aligned} \tag{21.3}$$

The model is completed with independent hyperpriors for τ^2 and for the DP prior parameters.

Both to establish the connection with a spatial DP prior and for MCMC posterior simulation, it is useful to marginalize the random mixing distribution G in (21.3) over its DP prior (Blackwell and MacQueen, 1973). The resulting joint prior distribution for the $\boldsymbol{\gamma}_t^*$ is given by

$$\mathrm{N}_m(\boldsymbol{\gamma}_1^* \mid \mathbf{0}, \sigma^2 H(\phi)) \prod_{t=2}^{T} \left\{ \frac{\alpha}{\alpha + t - 1} \mathrm{N}_m(\boldsymbol{\gamma}_t^* \mid \mathbf{0}, \sigma^2 H(\phi)) + \frac{1}{\alpha + t - 1} \sum_{j=1}^{t-1} \delta_{\boldsymbol{\gamma}_j^*}(\boldsymbol{\gamma}_t^*) \right\}. \tag{21.4}$$

Hence, the $\boldsymbol{\gamma}_t^*$ arise according to a Pólya urn scheme, which highlights the DP-induced clustering: $\boldsymbol{\gamma}_1^*$ is drawn from the centering distribution, and then for each $t = 2, \ldots, T$, $\boldsymbol{\gamma}_t^*$ is either set equal to $\boldsymbol{\gamma}_j^*$, $j = 1, \ldots, t-1$, with probability $(\alpha + t - 1)^{-1}$ or drawn from the centering distribution with the remaining probability.

The prior model for the spatial random effects $\boldsymbol{\gamma}_t^*$ discussed above is defined starting with a GP prior for the corresponding surfaces $\{\gamma_t^*(\boldsymbol{s}) : \boldsymbol{s} \in D\}$, block averaging the associated GP realizations over the regions to obtain the $\mathrm{N}_m(\mathbf{0}, \sigma^2 H(\phi))$ distribution, and finally centering a DP prior for the $\boldsymbol{\gamma}_t^*$ around this m-dimensional normal distribution. Kottas et al. (2008) show that the joint prior distribution for the $\boldsymbol{\gamma}_t^*$ is exactly as in (21.4) if one starts instead with a spatial DP prior for the distribution G_D of the $\{\gamma_t^*(\boldsymbol{s}) : \boldsymbol{s} \in D\}$ (centered around the same isotropic GP used above), marginalizes G_D over its spatial DP prior, and then block averages the (marginal) realizations from the spatial DP prior over the regions. Hence, the marginal version of model (21.3) (which is the one used for posterior predictive inference) is consistent with the marginal version of the corresponding (continuous-space) spatial DP mixture model, regardless of the number and geometry of the subregions chosen to partition the region under study.

Finally, to extend the spatial model described above to a spatiotemporal setting, Kottas et al. (2008) use a dynamic spatial modeling framework, viewing the log-rate process $\boldsymbol{\gamma}_{t,D}$ as a temporally evolving spatial process. In particular, the log-rate surface is modeled as $\gamma_t(\boldsymbol{s}) = \xi_t + \gamma_t^*(\boldsymbol{s})$, adding temporal structure through transition equations for the $\gamma_t^*(\boldsymbol{s})$. (Note that both here and in the spatial model discussed above, a mean structure $\mu_t(\boldsymbol{s})$ would typically be added to $\gamma_t(\boldsymbol{s})$ to incorporate covariate information.) For instance, $\gamma_t^*(\boldsymbol{s}) = \nu\gamma_{t-1}^*(\boldsymbol{s}) + \eta_t(\boldsymbol{s})$, where the innovations $\boldsymbol{\eta}_{t,D} = \{\eta_t(\boldsymbol{s}) : \boldsymbol{s} \in D\}$ are independent realizations from a spatial stochastic process with distribution G_D. A spatial DP prior is assigned to G_D, with parameters α and $G_{0D} = \mathrm{GP}(\mathbf{0}, \sigma^2 \exp(-\varphi||\boldsymbol{s} - \boldsymbol{s}'||))$. Marginalizing G_D over its prior, the induced joint prior $p(\boldsymbol{\eta}_1, \ldots, \boldsymbol{\eta}_T \mid \alpha, \sigma^2, \phi)$ for the block-averaged $\boldsymbol{\eta}_t = (\eta_{1t}, \ldots, \eta_{mt})$, where $\eta_{it} = |A_i|^{-1} \int_{A_i} \eta_t(\boldsymbol{s}) \mathrm{d}\boldsymbol{s}$, is given by (21.4) (with $\boldsymbol{\eta}_t$ replacing $\boldsymbol{\gamma}_t^*$). Block averaging the surfaces in the transition equations results in $\boldsymbol{\gamma}_t^* = \nu\boldsymbol{\gamma}_{t-1}^* + \boldsymbol{\eta}_t$. And, adding again the i.i.d. $N(0,\tau^2)$ terms to the γ_{it}, the following general form for the spatiotemporal model emerges:

$$\begin{aligned}
y_{it} \mid \gamma_{it} &\overset{ind.}{\sim} \text{Poisson}(y_{it} \mid n_{it}\exp(\gamma_{it})), \;\; i=1,\dots,m, \;\; t=1,\dots,T \\
\gamma_{it} \mid \xi_t, \gamma^*_{it}, \tau^2 &\overset{ind.}{\sim} \text{N}(\gamma_{it} \mid \xi_t + \gamma^*_{it}, \tau^2), \;\; i=1,\dots,m, \;\; t=1,\dots,T \\
\boldsymbol{\gamma}^*_t &= \nu\boldsymbol{\gamma}^*_{t-1} + \boldsymbol{\eta}_t \\
\boldsymbol{\eta}_1,\dots,\boldsymbol{\eta}_T \mid \alpha,\sigma^2,\phi &\sim p(\boldsymbol{\eta}_1,\dots,\boldsymbol{\eta}_T \mid \alpha,\sigma^2,\phi).
\end{aligned}$$

The ξ_t could be i.i.d. $N(0,\sigma^2_\xi)$, modeled with a parametric autoregressive structure, or explained through a parametric trend.

21.5 Conclusions

Although Bayesian nonparametric methodology is now an integral component of applied Bayesian modeling, applications in spatial epidemiology, and more specifically in disease mapping, are relatively limited. We have reviewed Bayesian nonparametric spatial and spatiotemporal models for disease incidence data, categorizing the modeling approaches according to whether the nonparametric prior is placed on the finite dimensional distribution of the region-specific spatial effects or, more generally, on the latent disease risk or rate surface.

From a practical point of view, more work is needed on empirical comparison between the existing methods, as well as with more standard parametric hierarchical models. In terms of new methodological developments, it is arguably of interest to expand the scope of existing methods that build from modeling the underlying temporally evolving continuous-space disease risk (or rate) surfaces. In this context, it is important to elaborate on the modeling framework to handle spatial misalignment issues for data settings where the disease counts are observed for one set of areal units, while covariate information is supplied for a different set of units. Finally, it would be of practical and methodological interest to explore flexible nonparametric methodology for describing and forecasting patterns of joint incidence of multiple diseases.

Acknowledgments

The author wishes to thank Andrew Lawson and Sudipto Banerjee for the invitation to write this book chapter, as well as a reviewer for useful feedback. This work was supported in part by the National Science Foundation under award DMS 1310438.

References

Banerjee, S., B. P. Carlin, and A. E. Gelfand. (2015). *Hierarchical Modeling and Analysis for Spatial Data* (2nd ed.). Boca Raton, FL: Chapman & Hall/CRC.

Best, N. G., K. Ickstadt, and R. L. Wolpert. (2000). Spatial Poisson regression for health and exposure data measured at disparate resolutions. *Journal of the American Statistical Association* 95, 1076–1088.

Blackwell, D. and J. MacQueen. (1973). Ferguson distributions via Pólya urn schemes. *Annals of Statistics* 1, 353–355.

Böhning, D., E. Dietz, and P. Schlattmann. (2000). Space-time mixture modelling of public health data. *Statistics in Medicine* 19, 2333–2344.

Denison, D. G. T. and C. C. Holmes. (2001). Bayesian partitioning for estimating disease risk. *Biometrics* 57, 143–149.

Duan, J. A., M. Guindani, and A. E. Gelfand. (2007). Generalized spatial Dirichlet process models. *Biometrika* 94, 809–825.

Dunson, D. and J. Park. (2008). Kernel stick-breaking processes. *Biometrika* 95, 307–323.

Ferguson, T. S. (1973). A Bayesian analysis of some nonparametric problems. *Annals of Statistics* 1, 209–230.

Fernández, C. and P. J. Green. (2002). Modelling spatially correlated data via mixtures: a Bayesian approach. *Journal of the Royal Statistical Society: Series B* 64, 805–826.

Gelfand, A. E., A. Kottas, and S. MacEachern. (2005). Bayesian nonparametric spatial modeling with Dirichlet process mixing. *Journal of the American Statistical Association* 100, 1021–1035.

Griffin, J. and M. Steel. (2006). Order-based dependent Dirichlet processes. *Journal of the American Statistical Association* 101, 179–194.

Guindani, M. and A. E. Gelfand. (2006). Smoothness properties and gradient analysis under spatial Dirichlet process models. *Methodology and Computing in Applied Probability* 8, 159–189.

Hegarty, A. and D. Barry. (2008). Bayesian disease mapping using product partition models. *Statistics in Medicine* 27, 3868–3893.

Hjort, N. L., C. Holmes, P. Müller, and S. G. Walker. (2010). *Bayesian Nonparametrics*. Cambridge: Cambridge University Press.

Hossain, M., A. B. Lawson, B. Cai, J. Choi, J. Liu, and R. S. Kirby. (2013). Space-time stick-breaking processes for small area disease cluster estimation. *Environmental and Ecological Statistics* 20, 91–107.

Ishwaran, H. and L. F. James. (2004). Computational methods for multiplicative intensity models using weighted gamma processes: proportional hazards, marked point processes, and panel count data. *Journal of the American Statistical Association* 99, 175–190.

Kelsall, J. and J. Wakefield. (2002). Modeling spatial variation in disease risk: a geostatistical approach. *Journal of the American Statistical Association* 97, 692–701.

Knorr-Held, L. and G. Rasser. (2000). Bayesian detection of clusters and discontinuities in disease maps. *Biometrics* 56, 13–21.

Kottas, A., J. Duan, and A. E. Gelfand. (2008). Modeling disease incidence data with spatial and spatio-temporal Dirichlet process mixtures. *Biometrical Journal* 50, 29–42.

Kottas, A. and B. Sansó. (2007). Bayesian mixture modeling for spatial Poisson process intensities, with applications to extreme value analysis. *Journal of Statistical Planning and Inference* 137, 3151–3163.

Li, P., S. Banerjee, T. A. Hanson, and A. M. McBean. (2015). Bayesian models for detecting difference boundaries in areal data. *Statistica Sinica* 25, 385–402.

Malec, D. and P. Müller. (2008). A Bayesian semi-parametric model for small area estimation. In B. Clarke and S. Ghosal (eds.), *Pushing the Limits of Contemporary Statistics: Contributions in Honor of Jayanta K. Ghosh*, pp. 223–236. Beachwood, OH: Institute of Mathematical Statistics.

Müller, P. and R. Mitra. (2013). Bayesian nonparametric inference—why and how [with discussion]. *Bayesian Analysis* 8, 269–360.

Reich, B. and M. Fuentes. (2007). A multivariate semiparametric Bayesian spatial modeling framework for hurricane surface wind fields. *Annals of Applied Statistics* 1, 249–264.

Rodriguez, A. and D. B. Dunson. (2011). Nonparametric Bayesian models through probit stick-breaking processes. *Bayesian Analysis* 6, 145–178.

Sethuraman, J. (1994). A constructive definition of Dirichlet priors. *Statistica Sinica* 4, 639–650.

Taddy, M. A. and A. Kottas. (2012). Mixture modeling for marked Poisson processes. *Bayesian Analysis* 7, 335–362.

Wolpert, R. L. and K. Ickstadt. (1998). Poisson/gamma random field models for spatial statistics. *Biometrika* 85, 251–267.

22

Multivariate Spatial Models

Sudipto Banerjee
Department of Biostatistics
University of California
Los Angeles, California

CONTENTS

This chapter deals with multivariate spatial modeling. Spatial data are often *multivariate* in the sense that multiple (i.e., more than one) outcomes are measured at each spatial unit. As in the univariate case, the spatial units can be referenced by points or by areal units. Multivariate areal data are conspicuous in public health, where each county or administrative unit supplies counts or rates for a number of diseases. For example, epidemiologists often encounter measurements on several diseases from each spatial unit (e.g., county) and wish to account for dependence among the different diseases, as well as the spatial dependence between sites. The natural and environmental sciences are teeming with examples of multivariate point-referenced data. For example, in ambient air quality assessment, we seek to jointly model multiple contaminants (e.g., ozone, nitric oxide, carbon monoxide, and $PM_{2.5}$) at a fixed set of monitoring sites. Inference focuses upon three major aspects: (1) estimating associations among the contaminants, (2) estimating the strength of spatial association for each contaminant, and (3) predicting the contaminants at arbitrary locations. We will first treat multivariate areal data and then attend to point-referenced data.

22.1 Modeling Multivariate Areal Data

Multivariate areal models introduce multiple, dependent spatial random effects associated with areal units and find applications in multivariate disease mapping problems, where we account for spatial dependence as well as dependence between diseases. Therefore, we seek multivariate extensions of areal models, such as the conditionally autoregressive (CAR) models. In this regard, Mardia [1] proposed a separable multivariate CAR (MCAR) model whose dispersion structure was the Kronecker product of two positive definite matrices—one capturing spatial covariances and the other capturing the dependence among diseases. Separable models, while simpler to construct, are limited by the assumption that the dependence among diseases remains invariant across the areal units. Much of the subsequent research on statistical models has focused on nonseparable models. Kim et al. [2] presented a nonseparable "twofold CAR" model to model counts for two different types of disease over each areal unit [3] and developed a "shared-component" model. Knorr-Held and Rue [4] illustrate sophisticated Markov chain Monte Carlo (MCMC) blocking approaches in a model placing three conditionally independent CAR priors on three sets of spatial random effects in a shared-component model setting.

MCAR models can also provide coefficients in a multiple regression setting that are dependent and spatially varying at the areal unit level. For example, Gamerman et al. [5] investigate a Gaussian Markov random field (GMRF) model (a multivariate generalization of the pairwise difference intrinsic autoregressive [IAR] model) and compare various MCMC blocking schemes for sampling from the posterior that results under a Gaussian multiple linear regression likelihood. They also investigate a "pinned down" version of this model that resolves the impropriety problem by centering the $\boldsymbol{\phi}_i$ vectors around some mean location. These authors also place the spatial structure on the spatial regression coefficients themselves, instead of on extra intercept terms. Assunção et al. [6] refer to these models as *space-varying coefficient* models, and illustrate in the case of estimating fertility schedules. Assunção [7] offers a nice review of the work to that time in this area. Also working with areal units, Sain and Cressie [8] offer multivariate GMRF models, proposing a generalization that permits asymmetry in the spatial conditional cross-correlation matrix. They use this approach to jointly model the counts of white and minority persons residing in the census block groups of St. James Parish, Louisiana, a region containing several hazardous waste sites.

Multivariate CAR models are not the only available option for analyzing multivariate areal data. Zhang et al. [9] develop an arguably much simpler alternative approach building upon the techniques of smoothed ANOVA (SANOVA) by Hodges et al. [10] to smooth spatial random effects by exploiting the spatial structure. The underlying idea is to extend SANOVA to cases in which one factor is a spatial lattice, which is smoothed using a CAR model, and a second factor is, for example, type of disease. Data sets routinely lack enough information to identify the additional structure of MCAR. SANOVA offers a simpler and more intelligible structure than the MCAR while performing as well. Also see Chapter 32 of this handbook. Nevertheless, the MCAR and more general CAR-based models offer a rich inferential framework for capturing complex spatial associations. We focus on MCAR models and their variants within the disease mapping context in the remainder of this chapter.

22.1.1 Spatial modeling of a single disease: a brief review

Disease incidence or mortality data are often reported as counts or rates at a regional level (county, census tract, zip code, etc.) and are called areal (or lattice) data. Markov random field (MRF) models for lattice data are based on the Markov property, where the conditional distribution of a site's response given the responses of all the other sites depends only on the observations in the neighborhood of this site. Here, we define the neighborhood by area adjacency, although other definitions are sometimes used (e.g., regions with centroids within a given fixed distance).

Let Y_i be the observed number of cases of a certain disease in region i, $i = 1, \ldots, n$, and E_i be the expected number of cases in this same region. A popular likelihood for mapping a single disease is

$$Y_i \overset{ind}{\sim} Poisson(E_i e^{\mu_i}), \quad i = 1, \ldots, n, \tag{22.1}$$

where $\mu_i = \mathbf{x}_i^T \boldsymbol{\beta} + \phi_i$ represents the log relative risk, estimates of which are often based on the departures of observed from expected counts; the $\mathbf{x}_i$'s are explanatory variables or covariates associated with region i having parameter coefficients $\boldsymbol{\beta}$; and the ϕ_i's are spatially correlated random effects. We place a form of the Gaussian MRF model, commonly referred to as the conditionally autoregressive (CAR) prior, on the random effects $\boldsymbol{\phi} = (\phi_1, \ldots, \phi_n)^T$, that is,

$$\boldsymbol{\phi} \sim N_n\left(\mathbf{0}, [\tau(D - \alpha W)]^{-1}\right), \tag{22.2}$$

where N_n denotes the n-dimensional normal distribution, D is an $n \times n$ diagonal matrix with diagonal elements m_i that denote the number of neighbors of region i, and W is the adjacency matrix of the map (i.e., $W_{ii} = 0$ and $W_{ii'} = 1$ if i' is adjacent to i and 0 otherwise). In the joint distribution (22.2), τ^{-1} is the spatial dispersion parameter, and α is the spatial autocorrelation parameter. The CAR prior corresponds to the following conditional distribution of ϕ_i:

$$\phi_i \mid \phi_j, j \neq i, \sim N\left(\frac{\alpha}{m_i}\sum_{i\sim j}\phi_j, \frac{1}{\tau m_i}\right), \quad i, j = 1, \ldots, n, \tag{22.3}$$

where $i \sim j$ denotes that region j is a neighbor of region i. The CAR structure (22.2) reduces to the well-known intrinsic conditionally autoregressive (ICAR) model (described in Besag et al. [11]) if $\alpha = 1$ or an independence model if $\alpha = 0$. The ICAR model induces "local" smoothing by borrowing strength from the neighbors, while the independence model assumes independence of spatial rates and induces "global" smoothing. The smoothing parameter α in the CAR prior (22.2) controls the strength of spatial dependence among regions, though it has long been appreciated that a fairly large α may be required to deliver significant spatial correlation; see Wall [12] for details on this.

A similar approach proposes a Gaussian convolution prior for the modeling of the random effects $\boldsymbol{\phi}$. The random effects $\boldsymbol{\phi}$ are assumed to be the sum of the two independent components, with one having a Gaussian independence prior and the other a Gaussian ICAR prior (as in Besag et al. [11], Leroux et al. [13], and Dean et al. [14]). With such a convolution prior, we may capture both the relative contributions of regionwide heterogeneity and local clustering.

22.1.2 Spatial modeling of multiple diseases

Turning to multiple diseases, let Y_{ij} be the observed number of cases of disease j in region i, $i = 1, \ldots, n$, $j = 1, \ldots, p$, and let E_{ij} be the expected number of cases for the same disease in this same region. As in Section 22.1.1, the Y_{ij} are thought of as random variables, while the E_{ij} are thought of as fixed and known. For the first level of the hierarchical model, conditional on the random effects ϕ_{ij}, we assume the Y_{ij} are independent of each other such that

$$Y_{ij} \overset{ind}{\sim} Poisson(E_{ij} e^{\mathbf{x}_{ij}^T \boldsymbol{\beta}_j + \phi_{ij}}), \quad i = 1, \ldots, n, \ j = 1, \ldots, p, \tag{22.4}$$

where the $\mathbf{x}_{ij}$ are explanatory, region-level spatial covariates for disease j having (possibly region-specific) parameter coefficients $\boldsymbol{\beta}_j$.

Carlin and Banerjee [15] and Gelfand and Vounatsou [16] generalized the univariate CAR (22.2) to a joint model for the random effects ϕ_{ij} under a separability assumption, which permits modeling of correlation among the p diseases while maintaining spatial dependence across space. Separability assumes that the association structure separates into a nonspatial and a spatial component. More precisely, the joint distribution of $\boldsymbol{\phi}$ is assumed to be

$$\boldsymbol{\phi} \sim N_{np}\left(\mathbf{0}, [\Lambda \otimes (D - \alpha W)]^{-1}\right), \tag{22.5}$$

where $\boldsymbol{\phi} = (\boldsymbol{\phi}_1^T, \ldots, \boldsymbol{\phi}_p^T)^T$, $\boldsymbol{\phi}_j = (\phi_{1j}, \ldots, \phi_{nj})^T$, Λ is a $p \times p$ positive definite matrix that is interpreted as the nonspatial precision matrix (inverse of the dispersion matrix) between cancers, and $\otimes$ denotes the Kronecker product. We denote (22.5) by $MCAR(\alpha, \Lambda)$.

As a specific example, suppose $p = 2$ (e.g., two cancers in each county), and define $\boldsymbol{\phi}_1^T = (\phi_{11}, \ldots, \phi_{n1})$ and $\boldsymbol{\phi}_2^T = (\phi_{12}, \ldots, \phi_{n2})$. Then the MCAR formulation in (22.5) can be written as

$$\begin{pmatrix} \boldsymbol{\phi}_1 \\ \boldsymbol{\phi}_2 \end{pmatrix} \sim N\left(\begin{pmatrix} \mathbf{0} \\ \mathbf{0} \end{pmatrix}, \begin{pmatrix} (D-\alpha W)\Lambda_{11} & (D-\alpha W)\Lambda_{12} \\ (D-\alpha W)\Lambda_{12} & (D-\alpha W)\Lambda_{22} \end{pmatrix}^{-1}\right), \tag{22.6}$$

where $\Lambda_{ij}, i = 1, 2, j = 1, 2$ are the elements of Λ. More generally, we may need three different α_i parameters in (22.6) to explain the correlation between the two types of cancer and across the counties that neighbor each other [2]. The covariance matrix Σ would then be revised to

$$\Sigma = \begin{pmatrix} (D-\alpha_1 W)\Lambda_{11} & (D-\alpha_3 W)\Lambda_{12} \\ (D-\alpha_3 W)\Lambda_{12} & (D-\alpha_2 W)\Lambda_{22} \end{pmatrix}^{-1}, \tag{22.7}$$

where α_1 and α_2 are the smoothing parameters for the two cancer types, and α_3 is the "bridging" or "linking" parameter associating ϕ_{i1} with ϕ_{j2}, $i \neq j$. Unfortunately, with this general covariance matrix, it is difficult to check the conditions guaranteeing positive definiteness, since they depend on the unknown Λ matrix.

We can extend (22.5) by introducing different smoothing parameters for each disease, that is,

$$\boldsymbol{\phi} \sim N_{np}\left(\mathbf{0}, \left[Diag(R_1, \ldots, R_p)(\Lambda \otimes I_{n\times n})Diag(R_1, \ldots, R_p)^T\right]^{-1}\right), \tag{22.8}$$

where $R_j R_j^T = D - \alpha_j W$, $j = 1, \ldots, p$. We denote (22.8) by $MCAR(\alpha_1, \ldots, \alpha_p, \Lambda)$. Note that the off-diagonal block matrices (the R_i's) in the precision matrix in (22.8) are completely determined by the diagonal blocks. Thus, there is no separate modeling of the cross-covariance matrices; they are induced by the spatial precision matrices for each disease.

Recently, Jin et al. [17] developed a more flexible generalized multivariate CAR (GMCAR) model for the random effects $\boldsymbol{\phi}$. For example, in the bivariate case ($p = 2$), they specify the conditional distribution $\boldsymbol{\phi}_1|\boldsymbol{\phi}_2$ as $N\left((\eta_0 I + \eta_1 W)\boldsymbol{\phi}_2, [\tau_1(D-\alpha_1 W)]^{-1}\right)$ and the marginal distribution of $\boldsymbol{\phi}_2$ as $N\left(\mathbf{0}, [\tau_2(D-\alpha_2 W)]^{-1}\right)$, both of which are univariate CAR, as in (22.2). This formulation yields the models of Kim et al. [2] as a special case and recognizes explicit smoothing parameters (η_0 and η_1) for the cross-covariances, unlike the MCAR models in (22.8), where the cross-covariances are not smoothed explicitly.

Kim et al. [2] and Jin et al. [17] demonstrate that explicit smoothing of the cross-covariances deliver improved model fits to areally referenced bivariate data. However, to model the random effects $\boldsymbol{\phi}$ with the GMCAR model, we need to specify the conditioning order, since different conditioning orders will result in different marginal distributions for $\boldsymbol{\phi}_1$ and $\boldsymbol{\phi}_2$, and hence different joint distributions for $\boldsymbol{\phi}$. In disease mapping contexts, a natural conditioning order is often not evident—a problem that is exacerbated when we have more than two diseases. What we seek, therefore, are models that avoid this dependence on conditional ordering, yet are computationally feasible with sufficiently rich spatial structures.

22.1.3 Conditional generalized MCAR

Jin et al. [17] expand upon this idea by building the joint distribution for a multivariate Markov random field (MRF) through specifications of simpler conditional and marginal models. The approach can be regarded as the analogue of the conditioning approach of Royle and Berliner [18] for areal models.

Assume a bivariate setting with the joint distribution of $\boldsymbol{\phi}_1$ and $\boldsymbol{\phi}_2$ as

$$\begin{pmatrix} \boldsymbol{\phi}_1 \\ \boldsymbol{\phi}_2 \end{pmatrix} \sim N\left(\begin{pmatrix} \mathbf{0} \\ \mathbf{0} \end{pmatrix}, \begin{pmatrix} \Sigma_{11} & \Sigma_{12} \\ \Sigma_{12}^T & \Sigma_{22} \end{pmatrix} \right),$$

where the Σ_{kl}, $k,l = 1,2$, are $n \times n$ covariance matrices. From standard multivariate normal theory, we obtain $\mathrm{E}(\boldsymbol{\phi}_1|\boldsymbol{\phi}_2) = \Sigma_{12}\Sigma_{22}^{-1}\boldsymbol{\phi}_2$ and $\mathrm{Var}(\boldsymbol{\phi}_1|\boldsymbol{\phi}_2) = \Sigma_{11\cdot 2} = \Sigma_{11} - \Sigma_{12}\Sigma_{22}^{-1}\Sigma_{12}^T$. Now writing $A = \Sigma_{12}\Sigma_{22}^{-1}$, we can rewrite the joint distribution of $\boldsymbol{\phi}_1$ and $\boldsymbol{\phi}_2$ as

$$\begin{pmatrix} \boldsymbol{\phi}_1 \\ \boldsymbol{\phi}_2 \end{pmatrix} \sim N\left(\begin{pmatrix} \mathbf{0} \\ \mathbf{0} \end{pmatrix}, \begin{pmatrix} \Sigma_{11\cdot 2} + A\Sigma_{22}A^T & A\Sigma_{22} \\ (A\Sigma_{22})^T & \Sigma_{22} \end{pmatrix} \right). \tag{22.9}$$

If Σ_{22} and $\Sigma_{11\cdot 2}$ are positive definite, then (22.9) is proper. Since $\boldsymbol{\phi}_1|\boldsymbol{\phi}_2 \sim N(A\boldsymbol{\phi}_2, \Sigma_{11\cdot 2})$ and $\boldsymbol{\phi}_2 \sim N(0, \Sigma_{22})$, we can construct $p(\boldsymbol{\phi}) = p(\boldsymbol{\phi}_1|\boldsymbol{\phi}_2)p(\boldsymbol{\phi}_2)$, where $\boldsymbol{\phi}^T = (\boldsymbol{\phi}_1^T, \boldsymbol{\phi}_2^T)$. For the joint distribution of $\boldsymbol{\phi}$, then, we need to specify the matrices $\Sigma_{11\cdot 2}$, Σ_{22}, and A.

Jin et al. [17] propose specifying the conditional distribution for $\boldsymbol{\phi}_1 \,|\, \boldsymbol{\phi}_2$ as $\boldsymbol{\phi}_1|\boldsymbol{\phi}_2 \sim N\left(A\boldsymbol{\phi}_2, [(D - \rho_1 W)\tau_1]^{-1}\right)$ and the marginal distribution $\boldsymbol{\phi}_2 \sim N\left(\mathbf{0}, [(D - \rho_2 W)\tau_2]^{-1}\right)$, where ρ_1 and ρ_2 are the smoothing parameters associated with the conditional distribution of $\boldsymbol{\phi}_1|\boldsymbol{\phi}_2$ and the marginal distribution of $\boldsymbol{\phi}_2$, respectively, and τ_1 and τ_2 scale the precision of $\boldsymbol{\phi}_1|\boldsymbol{\phi}_2$ and $\boldsymbol{\phi}_2$, respectively. The induced joint distribution will always be proper as long as these two CAR distributions are valid, so the positive definiteness of the covariance matrix in (22.9) is easily verified. If $D = Diag(m_i)$ and W is the adjacency matrix, then the positive definiteness conditions require only that $|\rho_1| < 1$ and $|\rho_2| < 1$. Further restricting these parameters between 0 and 1, $0 < \rho_1 < 1$, avoids negative spatial autocorrelation.

Regarding the A matrix, since $\mathrm{E}(\boldsymbol{\phi}_1 \,|\, \boldsymbol{\phi}_2) = A\boldsymbol{\phi}_2$, we assume its elements are of the form

$$a_{ij} = \begin{cases} \eta_0 & \text{if } j = i \\ \eta_1 & \text{if } j \in N_i \text{ (i.e., if region } j \text{ is a neighbor of region } i) \\ 0 & \text{otherwise} \end{cases}.$$

Thus, $A = \eta_0 I + \eta_1 W$ and $E(\boldsymbol{\phi}_1|\boldsymbol{\phi}_2) = (\eta_0 I + \eta_1 W)\boldsymbol{\phi}_2$. Here, η_0 and η_1 are the bridging parameters associating ϕ_{i1} with ϕ_{i2} and ϕ_{j2}, $j \neq i$. One could easily augment A with another bridging parameter η_2 associated with the *second-order* neighbors (neighbors of neighbors) in each region, but we do not pursue this generalization here. Under these assumptions, the covariance matrix in the joint distribution (22.9) can be written as $\Sigma = \begin{pmatrix} \Sigma_{11} & \Sigma_{12} \\ \Sigma_{12}^T & \Sigma_{22} \end{pmatrix}$, where

$$\begin{aligned} \Sigma_{11} &= [\tau_1(D - \rho_1 W)]^{-1} + (\eta_0 I + \eta_1 W)[\tau_2(D - \rho_2 W)]^{-1}(\eta_0 I + \eta_1 W) \\ \Sigma_{12} &= (\eta_0 I + \eta_1 W)[\tau_2(D - \rho_2 W)]^{-1} \\ \Sigma_{22} &= [\tau_2(D - \rho_2 W)]^{-1}. \end{aligned} \tag{22.10}$$

Jin et al. [17] denote this new model by $GMCAR(\rho_1,\ \rho_2,\ \eta_1,\ \eta_2,\ \tau_1,\ \tau_2)$. This bivariate GMCAR model has the same number of parameters as the twofold CAR model of Kim et al. [2], and has one more parameter than the $MCAR(\rho_1, \rho_2, \Lambda)$ model in (22.8).

Setting $\rho_1 = \rho_2 = \rho$ and $\eta_1 = 0$ in (22.10), and using a standard result from matrix theory, produces the separable precision matrix $\Sigma^{-1} = \Lambda \otimes (D - \rho W)$, where $\tau_1 = \Lambda_{11}$, $\tau_2 = \Lambda_{22} - \frac{\Lambda_{12}^2}{\Lambda_{11}}$, and $\eta_0 = -\frac{\Lambda_{12}}{\Lambda_{11}}$. Further assuming $\rho = 1$ produces an improper multivariate intrinsic autoregressive (MIAR) model. If we assume $\rho_1 \neq \rho_2$ and $\eta_0 = \eta_1 = 0$, then we ignore dependence between the multivariate components, and the model turns out to be equivalent to fitting two separate univariate CAR models. Finally, if we instead assume $\rho_1 = \rho_2 = 0$, $\eta_0 \neq 0$, and $\eta_1 = 0$, the model becomes an i.i.d. bivariate normal model.

The MCAR model in (22.6) has $\mathrm{E}(\boldsymbol{\phi}_1 \mid \boldsymbol{\phi}_2) = -\frac{\Lambda_{12}}{\Lambda_{11}}\boldsymbol{\phi}_2$, which reveals that the conditional mean is merely a scale multiple of $\boldsymbol{\phi}_2$. Since $Var(\boldsymbol{\phi}_1|\boldsymbol{\phi}_2) = [\Lambda_{11}(D - \rho_1 W)]^{-1}$, which is free of $\boldsymbol{\phi}_2$, the distribution of the random variable at a particular site in one field is independent of neighbor variables in another field *given* the value of the related variable at the same area. The extended MCAR model (22.8) has

$$\mathrm{E}(\boldsymbol{\phi}_1 \mid \boldsymbol{\phi}_2) = -\frac{\Lambda_{12}}{\Lambda_{11}}(D - \rho_1 W)^{-\frac{1}{2}}(D - \rho_2 W)^{\frac{1}{2}}\boldsymbol{\phi}_2$$

and $\mathrm{Var}(\boldsymbol{\phi}_1|\boldsymbol{\phi}_2)$ identical to that of model (22.6). Therefore, the distribution of the random variable at a particular site in one field is no longer conditionally independent of neighboring variables in another field. However, this dependence is determined implicitly by ρ_1 and ρ_2 and is difficult to interpret.

By contrast, the GMCAR model has $\mathrm{E}(\boldsymbol{\phi}_1 \mid \boldsymbol{\phi}_2) = (\eta_0 I + \eta_1 W)\boldsymbol{\phi}_2$ and $\mathrm{Var}(\boldsymbol{\phi}_1 \mid \boldsymbol{\phi}_2) = [\tau_1(D - \rho_1 W)]^{-1}$. Thus, while the conditional variance remains free of $\boldsymbol{\phi}_2$, the GMCAR allows spatial information (via the W matrix) to enter the conditional mean in an intuitive way, with a free parameter (η_1) to model the weights. That is, the GMCAR models the conditional mean of $\boldsymbol{\phi}_1$ for a given region as a sensible weighted average of the values of $\boldsymbol{\phi}_2$ for that region *and* a neighborhood of that region.

The GMCAR also allows us to incorporate different weighted adjacency matrices in the $MCAR(\rho, \boldsymbol{\Lambda})$ distribution. Suppose we wish to extend the precision matrix in model (22.6) to

$$\Sigma^{-1} = \begin{pmatrix} (D_1 - \rho W^{(1)})\Lambda_{11} & (D_3 - \rho W^{(3)})\Lambda_{12} \\ (D_3 - \rho W^{(3)})\Lambda_{12} & (D_2 - \rho W^{(2)})\Lambda_{22} \end{pmatrix}, \tag{22.11}$$

where $D_k = Diag\left(\sum_{j=1}^n W_{1j}^{(k)}, \ldots, \sum_{j=1}^n W_{nj}^{(k)}\right)$ and $W^{(k)}$ is the weighted adjacency matrix with ij element $W_{ij}^{(k)}$, $k = 1, 2, 3$, and $i, j = 1, \ldots, n$. The conditions for the precision matrix in (22.11) to be positive definite are less obvious. But in our GMCAR case, we obtain

$$\begin{aligned} \boldsymbol{\phi}_1|\boldsymbol{\phi}_2 &\sim N\left((\eta_0 I + \eta_1 W^{(3)})\boldsymbol{\phi}_2,\, [\tau_1(D_1 - \rho_1 W^{(1)})]^{-1}\right), \\ \text{and } \boldsymbol{\phi}_2 &\sim N\left(0,\ [\tau_2(D_2 - \rho_2 W^{(2)})]^{-1}\right). \end{aligned}$$

The conditions for positive definiteness can be easily seen to be $|\alpha_1| < 1$ and $|\alpha_2| < 1$ using the fact that diagonally dominant matrices are always positive definite.

Since we specify the joint distribution for a multivariate MRF directly through specification of simpler conditional and marginal distributions, an inherent problem with these methods is that their conditional specification imposes a potentially arbitrary order on the variables being modeled, as they lead to different marginal distributions, depending on the conditioning sequence (i.e., whether to model $p(\boldsymbol{\phi}_1|\boldsymbol{\phi}_2)$ and then $p(\boldsymbol{\phi}_2)$, or $p(\boldsymbol{\phi}_2|\boldsymbol{\phi}_1)$ and then $p(\boldsymbol{\phi}_1)$). This problem is somewhat mitigated in certain (e.g., medical and environmental) contexts where a *natural* order is reasonable, but in many disease mapping contexts, this is not the case. Although Jin et al. [17] suggest using model comparison techniques to decide on the proper modeling order, since all possible permutations of the variables would need to be considered, this seems feasible only with relatively few variables. In any case, the principle of choosing among conditioning sequences using model comparison metrics is perhaps not uncontroversial.

22.2 Joint MCAR Distributions

We mentioned in Section 22.1.3 that the GMCAR is afflicted by the dependence of the joint distribution on the sequence of ordering in the hieraracby. To obviate this issue, Jin et al. [19] developed an order-free, joint framework for multivariate areal modeling that allows versatile spatial structures, yet is computationally feasible for many outcomes. In particular, suppose we assume a common proximity specification for each component of the random effects vector, $\boldsymbol{\phi}$. Then, we could write $\boldsymbol{\phi} = A\boldsymbol{\psi}$, where $\boldsymbol{\psi}_j$, the jth component of $\boldsymbol{\psi}$, is a univariate intrinsic CAR with precision parameter τ_j^2, and each of the component CAR models are independent. We can take A to be lower triangular without loss of generality. The resulting multivariate CAR model is, of course, improper, which is fine as a prior for random effects. Moreover, it is easy to work with since we will only fit the model in the space of the independent CAR models. If we wanted, we could make the prior proper by making each of the components of $\boldsymbol{\psi}$ proper CARs. We offer some details below.

The essential idea is to develop richer spatial association models using linear transformations of much simpler spatial distributions. The objective is to allow explicit smoothing of cross-covariances, while at the same time not being hampered by conditional ordering. The most natural model here would parametrize the cross-covariances themselves as $D - \gamma_{ij}W$, instead of using the U_k's, as in (22.6). Unfortunately, except in the separable model with only one smoothing parameter ρ, constructing such dispersion structures is not trivial and leads to identifiability issues on the γ's (see, e.g., Gelfand and Vonatsou [16]). Kim et al. [2] resolve these identifiability issues in the bivariate setting using diagonal dominance, but recognize the difficulty in extending this to the multivariate setting.

We address this problem using a *linear model of coregionalization* (LMC), as described in Gelfand et al. [20]. Here, we build desired association structures by linearly transforming independent spatial effects. However, to explicitly smooth the cross-covariances with identifiable parameters, we will relax the independence of latent effects. Still, in our ensuing parametrization, we are able to derive conditions that produce valid joint distributions. To be precise, let $\boldsymbol{\phi} = (\boldsymbol{\phi}_1^T, \ldots, \boldsymbol{\phi}_p^T)^T$ be an $np \times 1$ vector, where each $\boldsymbol{\phi}_j = (\phi_{1j}, \ldots, \phi_{nj})^T$ is $n \times 1$, representing the spatial effects corresponding to disease j. We can write $\boldsymbol{\phi} = (A \otimes I_{n\times n})\mathbf{u}$, where $\mathbf{u} = (\mathbf{u}_1^T, \ldots, \mathbf{u}_p^T)^T$ is $np \times 1$, with each $\mathbf{u}_j$ being an $n \times 1$ areal process. Indeed, a proper distribution for $\mathbf{u}$ ensures a proper distribution for $\boldsymbol{\phi}$ subject only to the nonsingularity of A. The flexibility of this approach is apparent: we obtain different multivariate lattice models with rich spatial covariance structures by making different assumptions about the p spatial processes $\mathbf{u}_j$.

22.2.1 Case 1: Independent and identical latent CAR variables

First, we will assume that the random spatial processes $\mathbf{u}_j$, $j = 1, \ldots, p$, are independent and identical. Since each spatial process $\mathbf{u}_j$ is a univariate process over areal units, we might adopt a CAR structure for each of them, that is,

$$\mathbf{u}_j \sim N_n\left(\mathbf{0}, (D - \alpha W)^{-1}\right), \quad j = 1, \ldots, p. \tag{22.12}$$

Since the $\mathbf{u}_j$'s are independent of each other, the joint distribution of $\mathbf{u} = (\mathbf{u}_1^T, \ldots, \mathbf{u}_p^T)^T$ is $\mathbf{u} \sim N_{np}\left(\mathbf{0}, I_{p\times p} \otimes (D - \alpha W)^{-1}\right)$. The joint distribution of $\boldsymbol{\phi} = (A \otimes I_{n\times n})\mathbf{u}$ is

$$\boldsymbol{\phi} \sim N_{np}\left(\mathbf{0}, \Sigma \otimes (D - \alpha W)^{-1}\right), \tag{22.13}$$

defining $\Sigma = AA^T$. We denote the distribution in (22.13) by $MCAR(\alpha, \Sigma)$. Note that the joint distribution of (22.13) is identifiable up to $\Sigma = AA^T$, and is independent of the choice of A. Thus, without loss of generality, we can specify the matrix A as the upper-triangular Cholesky decomposition of Σ.

Since $\boldsymbol{\phi} = (A \otimes I_{n\times n})\mathbf{u}$, a valid joint distribution of $\boldsymbol{\phi}$ requires valid joint distributions of the $\mathbf{u}_j$, which happens if and only if $\frac{1}{\xi_{min}} < \alpha < \frac{1}{\xi_{max}}$, where ξ_{min} and ξ_{max} are the minimum and maximum eigenvalues of $D^{-\frac{1}{2}} W D^{-\frac{1}{2}}$. Note if $\alpha = 1$ in CAR structure (22.12), which is an ICAR, the joint distribution of $\boldsymbol{\phi}$ in (22.13) becomes the multivariate intrinsic CAR (see, e.g., Gelfand and Vonatsou [16]).

Currently, `BUGS` offers an implementation of the $MCAR(\alpha = 1,\ \Sigma)$ distribution (using its `mv.car` distribution), but not the $MCAR(\alpha,\ \Sigma)$. However, we still can fit the $MCAR(\alpha,\ \Sigma)$ in `BUGS` by writing $\boldsymbol{\phi} = (A \otimes I_{n\times n})\mathbf{u}$ and assigning proper CAR priors (via the `car.proper` distribution) for each $\mathbf{u}_j$, $j = 1, \ldots, p$, with a common smoothing parameter α. Regarding the prior on A, note that since $AA^T = \Sigma$ and A is the Cholesky decomposition of Σ, there is a one-to-one relationship between the elements of Σ and A. In Section 22.3, we argue that assigning a prior to Σ is computationally preferable.

22.2.2 Case 2: Independent but not identical latent CAR variables

Here, we continue to assume that the $\mathbf{u}_j$ are independent, but relax them being identically distributed. Adopting the CAR structure, we assume

$$\mathbf{u}_j \sim N_n\left(\mathbf{0},\ (D - \alpha_j W)^{-1}\right), \quad j = 1, \ldots, p, \tag{22.14}$$

where α_j is the smoothing parameter for the jth spatial process. Since the $\mathbf{u}_j$'s are independent of each other and $\boldsymbol{\phi} = (A \otimes I_{n\times n})\mathbf{u}$, the joint distribution of $\boldsymbol{\phi}$ is

$$\boldsymbol{\phi} \sim N_{np}\left(\mathbf{0}, (A \otimes I_{n\times n})\Gamma^{-1}(A \otimes I_{n\times n})^T\right), \tag{22.15}$$

where $\Sigma = AA^T$ and Γ is an $np \times np$ block-diagonal matrix with $n \times n$ diagonal entries $\Gamma_j = D - \alpha_j W, j = 1, \ldots, p$. We denote the distribution in (22.15) by $MCAR(\alpha_1, \ldots, \alpha_p,\ \Sigma)$.

It follows from (22.15) that different joint distributions of $\boldsymbol{\phi}$ having different covariance matrices emerge under different linear transformation matrices A. To ensure A is identifiable, we could again specify it to be the upper-triangular Cholesky decomposition of Σ, although this might not be the best choice computationally. Through the LMC approach in this case, the distribution in (22.15) is similar to the $MCAR(\alpha_1, \ldots, \alpha_p,\ \Lambda)$ structure (22.8), developed in Carlin and Banerjee [15] and Gelfand and Vonatsou [16]. All of these have the same number of parameters, and there is no unique joint distribution for $\boldsymbol{\phi}$ with the $MCAR(\alpha_1, \ldots, \alpha_p,\ \Lambda)$ structure, since there is not a unique R_j matrix such that $R_j R_j^T = R_j P P^T R_j^T = D - \alpha_j W$ (P being an arbitrary orthogonal matrix).

Again, a valid joint distribution in (22.15) requires p valid distributions for $\mathbf{u}_j$, that is, $\frac{1}{\xi_{min}} < \alpha_j < \frac{1}{\xi_{max}}$, $j = 1, \ldots, p$. Through the LMC approach, we can also fit the data with the $MCAR(\alpha_1, \ldots, \alpha_p,\ \Sigma)$ prior distribution (22.15) on $\boldsymbol{\phi}$ in `WinBUGS`, as in the previous subsection, by writing $\boldsymbol{\phi} = (A \otimes I_{n\times n})\mathbf{u}$ and assigning proper CAR priors (via the `car.proper` distribution) with a distinct smoothing parameter α_j for each $\mathbf{u}_j$, $j = 1, \ldots, p$.

As mentioned in the preceding section, we assign a prior to $AA^T = \Sigma$ (e.g., an inverse Wishart) and determine A from the one-to-one relationship between the elements of Σ and A; Section 22.3 provides details.

22.2.3 Case 3: Dependent and not identical latent CAR variables

Finally, in this case, we will assume that the random spatial processes $\mathbf{u}_j = (u_{1j}, \ldots, u_{nj})^T$, $j = 1, \ldots, p$, are neither independent nor identically distributed. We now assume that u_{ij} and $u_{i,l\neq j}$ are independent given $u_{k\neq i,j}$ and $u_{k\neq i,l\neq j}$, where $l, j = 1, \ldots, p$ and $i, k = 1, \ldots, n$, implying that latent effects for different diseases in the same region are conditionally independent given those for diseases in the neighboring regions. Based on the Markov property and similar to the conditional distribution in the univariate case, we specify the ij^{th} conditional distribution as Gaussian with mean

$$E(u_{ij} \mid u_{k\neq i,j}, u_{i,l\neq j}, u_{k\neq i,l\neq j}) = b_{jj}\left(\sum_{k\sim i} u_{kj}/m_i\right) + \sum_{l\neq j}\left[b_{jl}\left(\sum_{k\sim i} u_{kl}/m_i\right)\right]$$

and conditional variance $Var(u_{ij} \mid u_{k\neq i,j},\, u_{i,l\neq j},\, u_{k\neq i,l\neq j}) \propto 1/m_i$, where b_{jj} denotes the spatial autocorrelation for the random spatial process $\mathbf{u}_j$, while b_{jl} $(l \neq j,\ l, j = 1, \ldots, p)$ denotes the cross-spatial correlation between the random spatial processes $\mathbf{u}_j$ and $\mathbf{u}_l$. Putting these conditional distributions together reveals the joint distribution of $\mathbf{u} = (\mathbf{u}_1^T, \ldots, \mathbf{u}_p^T)^T$ to be

$$\mathbf{u} \sim N_{np}\left(\mathbf{0}, (I_{p\times p} \otimes D - B \otimes W)^{-1}\right), \tag{22.16}$$

where I is a $p \times p$ identity matrix and B is a $p \times p$ symmetric matrix with the elements b_{jl}, $j, l = 1, \ldots, p$. As long as the dispersion matrix in (22.16) is positive definite, which boils down to $(I_{p\times p} \otimes D - B \otimes W)$ being positive definite, (22.16) is itself a valid model. To assess nonsingularity, note $I_{p\times p} \otimes D - B \otimes W = (I_{p\times p} \otimes D)^{\frac{1}{2}}\left(I_{pn\times pn} - B \otimes D^{-\frac{1}{2}} W D^{-\frac{1}{2}}\right)$ $(I_{p\times p} \otimes D)^{\frac{1}{2}}$. Denoting the eigenvalues for $D^{-\frac{1}{2}} W D^{-\frac{1}{2}}$ as ξ_i, $i = 1, \ldots, n$, and the eigenvalues for B as ζ_j, $j = 1, \ldots, p$, one finds the eigenvalues for $B \otimes (D^{-\frac{1}{2}} W D^{-\frac{1}{2}})$ as $\xi_i \times \zeta_j$, $i = 1, \ldots, n$, $j = 1, \ldots, p$. Hence, the conditions for $I_{p\times p} \otimes D - B \otimes W$ being positive definite become $\xi_i \zeta_j < 1$, that is, $\frac{1}{\xi_{min}} < \zeta_j < \frac{1}{\xi_{max}}$, $i = 1, \ldots, n$, $j = 1, \ldots, p$, where ξ_{min} and ξ_{max} are the minimum and maximum eigenvalues of $D^{-\frac{1}{2}} W D^{-\frac{1}{2}}$. Thus, $\frac{1}{\xi_{min}} < \zeta_j < 1$, $j = 1, \ldots, p$, ensures the positive definiteness of the matrix $I_{p\times p} \otimes D - B \otimes W$, and hence the validity of the distribution of $\mathbf{u}$ given in (22.16). In fact, $\xi_{max} = 1$ and $\xi_{min} < 0$, which makes this formulation easier to work with in practice (e.g., in choosing priors; see Section 22.3) than the alternative parametrization $\frac{1}{\zeta_{min}} < \xi_j < \frac{1}{\zeta_{max}}$.

The model in (22.16) introduces smoothing parameters in the cross-covariance structure through the matrix B, but unlike the $MCAR$ models in Sections 22.2.1 and 22.2.2, it does not have the Σ matrix to capture nonspatial variances. To remedy this, we model $\boldsymbol{\phi} = (A \otimes I_{n\times n})\mathbf{u}$ so that the joint distribution for the random effects $\boldsymbol{\phi}$ is

$$\boldsymbol{\phi} \sim N_{np}\left(\mathbf{0},\, (A \otimes I_{n\times n})\,(I_{p\times p} \otimes D - B \otimes W)^{-1}\,(A \otimes I_{n\times n})^T\right). \tag{22.17}$$

Since $\boldsymbol{\phi} = (A \otimes I_{n\times n})\mathbf{u}$, it is immediate that the validity of (22.16) ensures a valid joint distribution for (22.17). We denote distribution (22.17) by $MCAR(B,\, \Sigma)$, where $\Sigma = AA^T$. Again, A identifies with the upper-triangular Cholesky square root of Σ. Note that with $\Sigma = I$, we recover (22.16), which we henceforth denote as $MCAR(B, I)$.

To see the generality of (22.17), we find that $\boldsymbol{\phi} \sim MCAR(\alpha_1, \ldots, \alpha_p,\, \Sigma)$ distribution (22.15) if $b_{jl} = 0$ and $b_{jj} = \alpha_j$, or the $MCAR(\alpha,\, \Sigma)$ distribution (22.13) if $b_{jl} = 0$ and

$b_{jj} = \alpha$, in both cases for $j,l = 1,\ldots,p$. Also note that the distribution in (22.17) is invariant to orthogonal transformations (up to a reparametrization of B) in the following sense: let $T = AP$, with P being a $p \times p$ orthogonal matrix such that $TT^T = APP^TA^T = \Sigma$. Then the covariance matrix in (22.17) can be expressed as $(A \otimes I_{n\times n})(I_{p\times p} \otimes D - B \otimes W)^{-1}(A \otimes I_{n\times n})^T = (T \otimes I_{n\times n})(I_{p\times p} \otimes D - C \otimes W)^{-1}(T \otimes I_{n\times n})^T$, where $C = P^TBP$. Without loss of generality, then, we can choose the matrix A as the upper-triangular Cholesky decomposition of Σ.

To understand the features of the $MCAR(B,\ \Sigma)$ distribution (22.17), we illustrate in the bivariate case ($p = 2$). Define

$$(AA^T)^{-1} = \Sigma^{-1} = \begin{pmatrix} \Lambda_{11} & \Lambda_{12} \\ \Lambda_{12} & \Lambda_{22} \end{pmatrix}$$

and $B = A^T \begin{pmatrix} \gamma_1\Lambda_{11} & \gamma_{12}\Lambda_{12} \\ \gamma_{12}\Lambda_{12} & \gamma_2\Lambda_{22} \end{pmatrix} A$, where $A = \begin{pmatrix} a_{11} & a_{12} \\ 0 & a_{22} \end{pmatrix}$. For convenience, we will denote the entries of B as b_{ij}. Note that the γ's are not identifiable from the matrix Λ and our reparametrization in terms of B must be used to conduct posterior inference on B and Λ (see Section 22.3), from which the cross-covariances may be recovered. The above expression does allow the $MCAR(B,\ \Sigma)$ distribution (22.17) to be rewritten as

$$\boldsymbol{\phi} \sim N_{2n}\left(\mathbf{0}, \begin{pmatrix} (D-\gamma_1W)\Lambda_{11} & (D-\gamma_{12}W)\Lambda_{12} \\ (D-\gamma_{12}W)\Lambda_{12} & (D-\gamma_2W)\Lambda_{22} \end{pmatrix}^{-1}\right), \tag{22.18}$$

which is precisely the general dispersion structure we set out to achieve. Jin et al. [19] provide explicit expressions for the conditional means and variances, which offer further insight into how the parameters in (22.18) affect smoothing.

22.3 Modeling with Joint or Coregionalized MCARs

The $MCAR(B,\ \Sigma)$ model is straightforwardly implemented in a Bayesian framework using MCMC methods. As in Section 22.2.3, we write $\boldsymbol{\phi} = (A \otimes I_{n\times n})\mathbf{u}$, where $\mathbf{u} = (\mathbf{u}_1^T, \mathbf{u}_2^T)$ and $\mathbf{u}_j = (u_{1j},\ldots,u_{nj})^T$. The joint posterior distribution is $p\left(\boldsymbol{\beta},\sigma^2,\mathbf{u},A,B \mid \mathbf{Y}_1,\mathbf{Y}_2\right)$, which is proportional to

$$L\left(\mathbf{Y}_1,\mathbf{Y}_2 \mid \mathbf{u},\sigma^2,A\right)p\left(\mathbf{u} \mid B\right)p(B)p(\boldsymbol{\beta})p(A)p(\sigma^2), \tag{22.19}$$

where $\mathbf{Y}_1 = (Y_{11},\ldots,Y_{n1})^T$ and $\mathbf{Y}_2 = (Y_{12},\ldots,Y_{n2})^T$, $L\left(\mathbf{Y}_1,\mathbf{Y}_2 \mid \mathbf{u},\sigma^2,A\right)$ is the data likelihood, and $p(\mathbf{u} \mid B) = N_{np}\left(\mathbf{0},\ (I_{p\times p} \otimes D - B \otimes W)^{-1}\right)$. As mentioned in Section 22.2.3, propriety of this distribution requires the eigenvalues ζ_j of B to satisfy $\frac{1}{\xi_{min}} < \zeta_j < 1$ ($j = 1,\ldots,p$). When p is large, it is hard to determine the intervals over the elements of B that result in $\frac{1}{\xi_{min}} < \zeta_j < 1$, and thus designing priors for B that guarantee this condition is awkward. In principle, one might impose the constraint numerically by assigning a flat prior or a normal prior with a large variance for the elements of B, and then simply check whether the eigenvalues of the corresponding B matrix are in that range during a random-walk Metropolis–Hastings (MH) update. If the resulting eigenvalues are out of range, the values are thrown out since they correspond to prior probability 0; otherwise, we perform the standard MH comparison step. In our experience, however, this does not work well, especially when p is large.

Instead, here we outline a different strategy to update the matrix B. Our approach is to represent B using the spectral decomposition, which we write as $B = P\Delta P^T$, where P is the corresponding orthogonal matrix of eigenvectors and Δ is a diagonal matrix of ordered eigenvalues, $\zeta_1, \ldots, \zeta_p$. We parametrize the $p \times p$ orthogonal matrix P in terms of the $p(p-1)/2$ Givens angles θ_{ij} for $i = 1, \ldots, p-1$ and $j = i+1, \ldots, p$ [21]. The matrix P is written as the product of $p(p-1)/2$ matrices, each one associated with a Givens angle. Specifically, $P = G_{12}G_{13} \ldots G_{1p} \ldots G_{(p-1)p}$, where i and j are distinct and G_{ij} is the $p \times p$ identity matrix with the ith and jth diagonal elements replaced by $\cos(\theta_{ij})$, and the (i, j)th and (j, i)th elements replaced by $\pm \sin(\theta_{ij})$, respectively. Since the Givens angles θ_{ij} are unique with a domain $(-\pi/2, \pi/2)$ and the eigenvalues ζ_j of B are in the range $(\frac{1}{\xi_{min}}, 1)$, we then put a Uniform$(-\pi/2, \pi/2)$ prior on the θ_{ij} and a Uniform$(\frac{1}{\xi_{min}}, 1)$ prior on the ζ_j. To update θ_{ij}'s or ζ_j's using random-walk Metropolis–Hastings steps with Gaussian proposals, we need to transform them to have support equal to the whole real line. A straightforward solution here is to use $g(\theta_{ij}) = \log(\frac{\pi/2+\theta_{ij}}{\pi/2-\theta_{ij}})$, a transformation having Jacobian $\prod_{i=1}^{p-1}\prod_{j=i+1}^{p}(\pi/2+\theta_{ij})(\pi/2-\theta_{ij})$. In practice, the ζ_j must be bounded away from 1 (say, by insisting $\frac{1}{\xi_{min}} < \zeta_j < 0.999$, $j = 1, \ldots, p$) to maintain identifiability and hence computational stability. In fact, with our approach, it is also easy to calculate the determinant of the precision matrix, that is, $|I_{p\times p} \otimes D - B \otimes W| \propto \prod_{i=1}^{n}\prod_{j=1}^{p}(1 - \xi_i\zeta_j)$, where ξ_i are the eigenvalues of $D^{-\frac{1}{2}}WD^{-\frac{1}{2}}$, which can be calculated prior to any MCMC iteration. For the special case of the $MCAR(\alpha_1, \ldots, \alpha_p, \Sigma)$ models, one could assign each $\alpha_i \sim U(0, 1)$, which would be sufficient to ensure a valid model (see, e.g., Jin et al. [19]). However, we also investigated with more informative priors on the α_i's, such as the $Beta(2, 18)$ that centers the smoothing parameters closer to 1 and leads to greater smoothing.

With respect to the prior distribution $p(A)$ on the right-hand side of (22.19), we can put independent priors on the individual elements of A, such as inverse gamma for the square of the diagonal elements of A and normal for the off-diagonal elements. In practice, we cannot assign noninformative priors here, since then MCMC convergence is poor. In our experience, it is easier to assign a vague (i.e., weakly informative) prior on Σ than to put such priors on the elements of A in terms of letting the data drive the inference and obtaining good convergence. Since Σ is a positive definite covariance matrix, the inverse Wishart prior distribution renders itself a natural choice, that is, $\Sigma^{-1} \sim Wishart\left(\nu, (\nu R)^{-1}\right)$. Hence, we instead place a prior directly on Σ, and then use the one-to-one relationship between the elements of Σ and the Cholesky factor A. Then, the prior distribution $p(A)$ becomes

$$p(A) \propto |AA^T|^{-\frac{\nu+4}{2}} exp\left\{-\frac{1}{2}tr[\nu D(AA^T)^{-1}]\right\}\left|\frac{\partial\Sigma}{a_{ij}}\right|,$$

where $\left|\frac{\partial\Sigma}{a_{ij}}\right|$ is the Jacobian $2^p\prod_{i=1}^{p}a_{ii}^{p-i+1}$. For example, when $p = 2$, the Jacobian is $4a_{22}^2a_{11}$. Rather than updating Σ as a block using a Wishart proposal, updating the elements a_{ij} of A offers better control. These are updated via a random-walk Metropolis, using lognormal proposals for the diagonal elements and normal proposals for the off-diagonal elements. With regard to choosing ν and R in the $Wishart\left(\nu, (\nu R)^{-1}\right)$, since $E(\Sigma^{-1}) = R^{-1}$, if there is no information about the prior mean structure of Σ, a diagonal matrix R can be chosen, with the scale of the diagonal elements being judged using ordinary least-squares estimates based on independent models for each response variable. While this leads to a data-dependent prior, typically the Wishart prior lets the data drive the results, leading to robust posterior inference. In this study, we adopt $\nu = 2$ (i.e., the smallest value for which this Wishart prior is proper) and $R = Diag(0.1, 0.1)$. Finally, for the remaining terms on the right-hand side of (22.19), flat priors are chosen for β_1 and β_2, while σ^2 is assigned a vague inverse gamma prior, that is, a $IG(1, 0.01)$ parametrized so that $E(\sigma^2) = b/(a-1)$. In this study, $\boldsymbol{\beta}$ and σ^2 have closed-form full conditionals, and so can be directly updated using Gibbs sampling.

22.4 Illustrating MCAR Models with Three Cancers

We present a part of the analysis conducted by Jin et al. [19] to estimate different coregionalized MCAR models with a data set consisting of the numbers of deaths due to cancers of the lung, larynx, and esophagus in the years 1990–2000 at the county level in Minnesota. The larynx and esophagus are sites of the upper aerodigestive tract, so they are closely related anatomically. Epidemiological evidence shows a strong and consistent relationship between exposure to alcohol and tobacco and the risk of cancer at these two sites. Meanwhile, lung cancer is the leading cause of cancer death for both men and women. An estimated 160,440 Americans died in 2004 from lung cancer, accounting for 28% of all cancer deaths. It has long been established that tobacco, and particularly cigarette smoking, is the major cause of lung cancer. More than 87% of lung cancers are smoking related (http://www.lungcancer.org).

Following Jin et al. [19], we estimate the model

$$Y_{ij} \overset{ind}{\sim} Po(E_{ij}e^{\mu_{ij}}),\ i = 1, \ldots, n,\ j = 1, 2, 3, \tag{22.20}$$

where Y_{ij} is the observed number of cases of cancer type j (one of three types) in region i, $\log \mu_{ij} = \beta_j + \phi_{ij}$, with the β_j's being cancer-specific intercepts and the ϕ_{ij}'s being spatial random effects that are distributed according to some version of the coregionalized MCARs we discussed earlier. To calculate the expected counts E_{ij}, we have to take each county's age distribution (over the 18 age groups) into account. To do so, we calculate the expected *age-adjusted* number of deaths due to cancer j in county i as $E_{ij} = \sum_{k=1}^{m} \omega_j^k N_i^k$, $i = 1, \ldots, 87$, $j = 1, 2, 3$, $k = 1, \ldots, 18$, where $\omega_j^k = (\sum_{i=1}^{87} D_{ij}^k)/(\sum_{i=1}^{87} N_i^k)$ is the age-specific death rate due to cancer j for age group k over all Minnesota counties, D_{ij}^k is the number of deaths in age group k of county i due to cancer j, and N_i^k is the total population at risk in county i, age group k, which we assume to be the same for each type of cancer.

The county-level maps of the raw age-adjusted standardized mortality ratios (i.e., $\text{SMR}_{ij} = Y_{ij}/E_{ij}$) shown in Figure 22.1 exhibit evidence of correlation both across space and among the cancers, motivating use of our proposed multivariate lattice model. Using the likelihood in (22.20), we model the random effects ϕ_{ij} using our proposed $MCAR(B, \Sigma)$ model (22.17). In what follows, we compare it with other MCAR models, including the $MCAR(\alpha, \Sigma)$ and $MCAR(1, \Sigma)$ from Section 22.2.1; a "three separate CARs" model ignoring correlation between cancers; and a trivariate i.i.d. model ignoring correlations of any kind. We also compare one of the $MCAR(\alpha_1, \alpha_2, \alpha_3, \Sigma)$ models given in (22.15) of Section 22.2.2 by choosing the matrix A as the upper-triangular Cholesky decomposition of Σ. Note that we do not consider the order-specific GMCAR model (Section 22.1.3), since with no natural causal order for these three cancers, it is hard to choose among the six possible conditioning orders.

For priors, we follow the guidelines outlined earlier and use the same specifications as in Jin et al. [19]. Since $p = 3$ in this example, we choose the inverse Wishart distribution with $\nu = 3$ and $R = Diag(0.1, 0.1, 0.1)$ for Σ. For a model comparison, using the deviance information criterion (DIC), we retain the same "focus" parameters and likelihood across the models. We used 20,000 pre-convergence burn-in iterations, followed by a further 20,000 production iterations for posterior summarization. To see the relative performance of these models, we use DIC. As in the previous section, the deviance is the same for the models we wish to compare since they differ only in their random effects distributions $p(\phi|B, \Sigma)$.

In what follows, Models 1–6 are multivariate lattice models with different assumptions about the smoothing parameters. Model 1 is the full model $MCAR(B, \Sigma)$ (with a 3×3 matrix B whose elements are the six smoothing parameters), while Model 2 is the

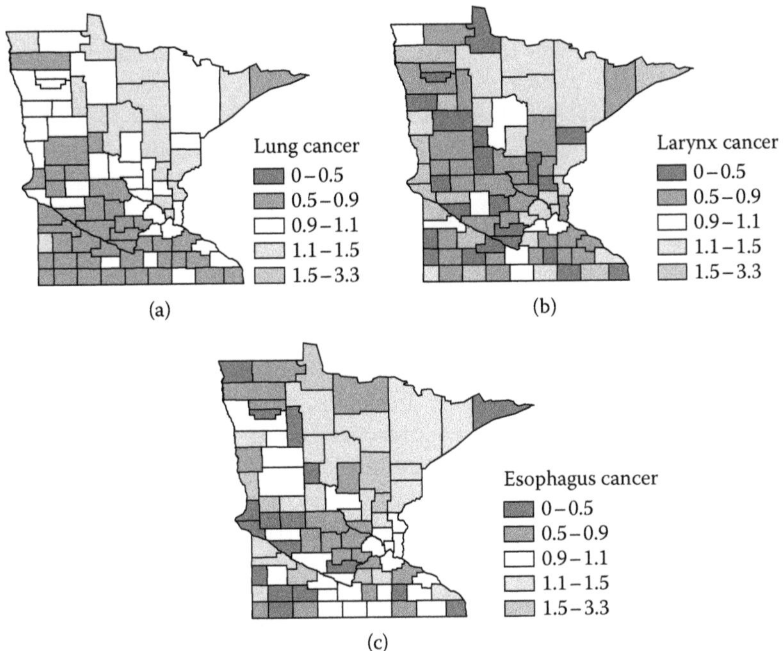

FIGURE 22.1
Maps of raw standardized mortality ratios (SMRs) of (a) lung, (b) larynx, and (c) esophagus cancers in the years 1990–2000 in Minnesota.

$MCAR(B, I)$ model. Model 2 is the $MCAR(\alpha_1, \alpha_2, \alpha_3; \Sigma)$ model (22.15) with a different smoothing parameter for each cancer. Model 3 assumes a common smoothing parameter α, and Model 4 fits the three separate univariate CAR models, while Model 6 is the trivariate i.i.d. model. Fit measures $\overline{D}$, effective numbers of parameters p_D, and DIC scores for each model are seen in Table 22.1. We find that the $MCAR(B, \Sigma)$ model has the smallest $\overline{D}$ and DIC values for this data set. The $MCAR(B, I)$ model again disappoints, excelling over the nonspatial model and the separate CAR models only (very marginally over the latter). The $MCAR(\alpha, \Sigma)$ and $MCAR(\alpha_1, \alpha_2, \alpha_3, \Sigma)$ models perform slightly worse

TABLE 22.1
Model comparison using DIC statistics, Minnesota cancer data analysis

	Model	$\overline{D}$	p_D	**DIC**
1	$MCAR(B, \Sigma)$	138.8	82.5	221.3
2	$MCAR(B, I)$	147.6	81.4	229.0
3	$MCAR(\alpha_1, \alpha_2, \alpha_3, \Sigma)$	139.6	86.4	226.0
4	$MCAR(\alpha, \Sigma)$	143.4	81.9	225.3
5	Separate CAR	147.6	82.8	230.4
6	Trivariate i.i.d.	146.8	91.3	238.1
7	$MCAR(B, \Sigma)$ + trivariate i.i.d.	129.6	137.6	267.2
8	$MCAR(B, I)$ + trivariate i.i.d.	139.5	155.2	294.7
9	$MCAR(\alpha_1, \alpha_2, \alpha_3, \Sigma)$ + trivariate i.i.d.	137.4	155.0	292.4
10	$MCAR(\alpha, \Sigma)$ + trivariate i.i.d.	138.2	151.0	289.2
11	Separate CAR + trivariate i.i.d.	139.2	162.8	302.0

than the $MCAR(B, \Sigma)$ model, suggesting the need for different spatial autocorrelation and cross-spatial correlation parameters for this data set. Note that the effective numbers of parameters p_D in Model 3 are a little larger than in Model 1, even though the latter has three extra parameters. Finally, the MCAR models do better than the separate CAR model or the i.i.d. trivariate model, suggesting that it is worth taking account of the correlations both across counties and among cancers. Model 6 exhibits a large p_D score, suggesting it does not seem to allow sufficient smoothing of the random effects. This is what we might have expected, since the spatial correlations are missed by this model.

Models 7–11 are the convolution prior models corresponding to Models 1–5 formed by adding i.i.d. effects (following $N(0, \tau^2)$) to the ϕ_{ij}'s. Here the distinctions between the models are somewhat more pronounced due to the added variability in the models caused by the i.i.d. effects. The relative performances of the models remain the same with the $MCAR(B, \Sigma)+$i.i.d. model emerging as best. Interestingly, none of the convolution models perform better than their purely spatial counterparts, as the improvements in $\bar{D}$ in the former are insignificant compared to the increase in the effective dimensions brought about. This is indicative of the dominance of the spatial effects over the i.i.d. effects, from where the convolution models seem to be rendering overparametrized models.

We summarize our results from the $MCAR(B, \Sigma)$, which is Model 1 in Table 22.1. Table 22.2 provides posterior means and associated standard deviations for the parameters $\boldsymbol{\beta}$, Σ, and b_{ij} in this model, where b_{ij} is the element of the symmetric matrix B. Instead of reporting Σ_{12}, Σ_{13}, and Σ_{23}, we provide the mean and associated standard deviations for the correlation parameters ρ_{12}, ρ_{13}, and ρ_{23}, which are calculated as $\rho_{ij} = \Sigma_{ij}/\sqrt{\Sigma_{ii}\Sigma_{jj}}$. We also plot histograms of the posterior samples ρ_{ij} in Figure 22.2 and histograms of the posterior samples b_{ij} in Figure 22.3.

Table 22.2 and Figure 22.2 reveal correlations between cancers, in particular a strong correlation between lung and esophagus (ρ_{13}). This might explain why the DIC scores for Models 1–4 in Table 22.2 are smaller than those under the separate CAR model. The b_{ij} in Table 22.2 are spatial autocorrelation and cross-spatial correlation parameters for the latent spatial processes $\mathbf{u}_j$, $j = 1, 2, 3$. Figure 22.3 shows that most of the b_{12} and b_{13} posterior samples are positive; the means of these two parameters are 0.323 and 0.389, respectively. Consistent with the DIC results in Table 22.1, these suggest it is worth fitting our proposed $MCAR(B, \Sigma)$ model to these data.

Turning to geographical summaries, Figure 22.4 maps the posterior means of the fitted standard mortality ratios (SMRs) of lung, larynx, and esophagus cancers from our $MCAR(B, \Sigma)$ model. From Figure 22.4, the correlation among the cancers is apparent, with higher fitted ratios extending from the Twin Cities metro area to the north and northeast

TABLE 22.2
Posterior summaries of parameters in $MCAR(B, \Sigma)$ model for Minnesota cancer data

	Lung median (2.5%, 97.5%)	**Larynx median (2.5%, 97.5%)**	**Esophagus mean (2.5%, 97.5%)**
$\beta_1, \beta_2, \beta_3$	−0.093 (−0.179,−0.006)	−0.128 (−0.316, 0.027)	−0.080 (−0.194, 0.025)
$\Sigma_{11}, \Sigma_{22}, \Sigma_{33}$	0.048 (0.030, 0.073)	0.173 (0.054, 0.395)	0.107 (0.044, 0.212)
ρ_{12}, ρ_{13}		0.277 (−0.112, 0.643)	0.378 (−0.022, 0.716)
ρ_{23}			0.337 (−0.311, 0.776)
b_{11}, b_{22}, b_{33}	0.442 (−0.302, 0.921)	0.036 (−0.830, 0.857)	0.312 (−0.526, 0.901)
b_{12}, b_{13}		0.323 (−0.156, 0.842)	0.389 (−0.028, 0.837)
b_{23}			0.006 (−0.519, 0.513)

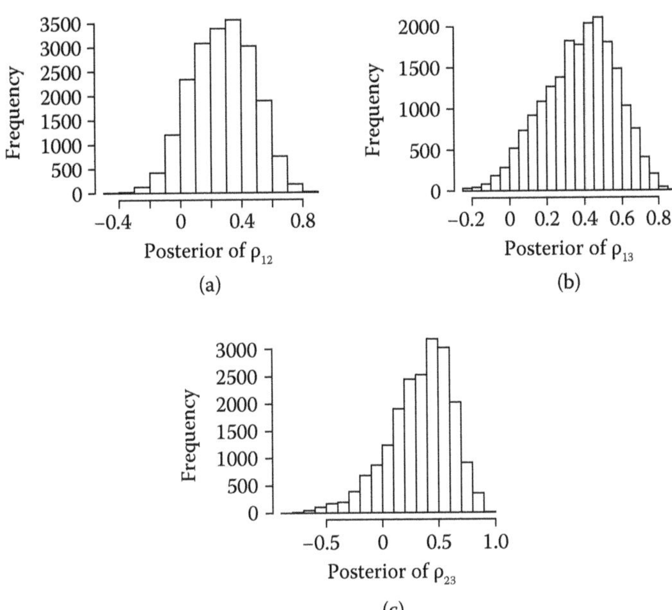

FIGURE 22.2
Posterior samples of ρ_{12}, ρ_{13}, and ρ_{23} in the Minnesota cancer data analysis using the $MCAR(B, \Sigma)$ model: (a) estimated posterior for correlation ρ_{12} between lung and larynx, (b) estimated posterior for correlation ρ_{13} between lung and esophagus, and (c) estimated posterior for correlation ρ_{23} between larynx and esophagus.

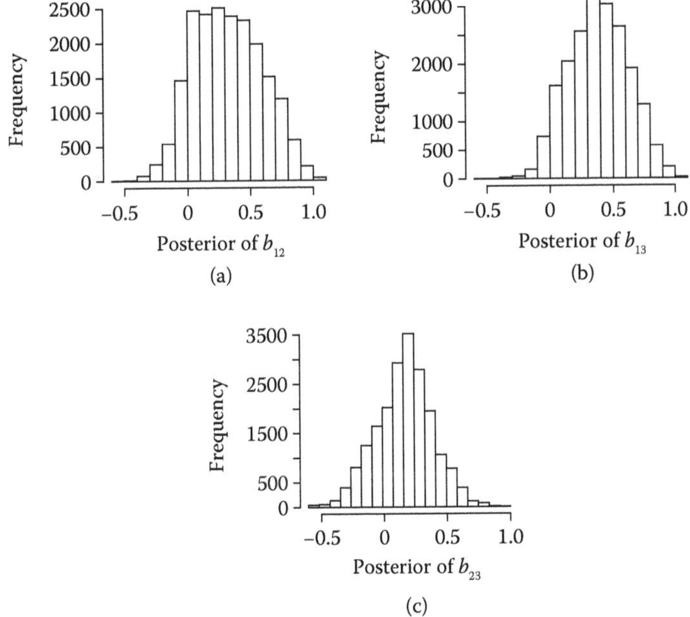

FIGURE 22.3
Posterior samples of b_{12}, b_{13}, and b_{23} in the Minnesota cancer data analysis using the $MCAR(B, \Sigma)$ model: (a) estimated posterior for b_{12}, (b) estimated posterior for b_{13}, and (c) estimated posterior for b_{23}.

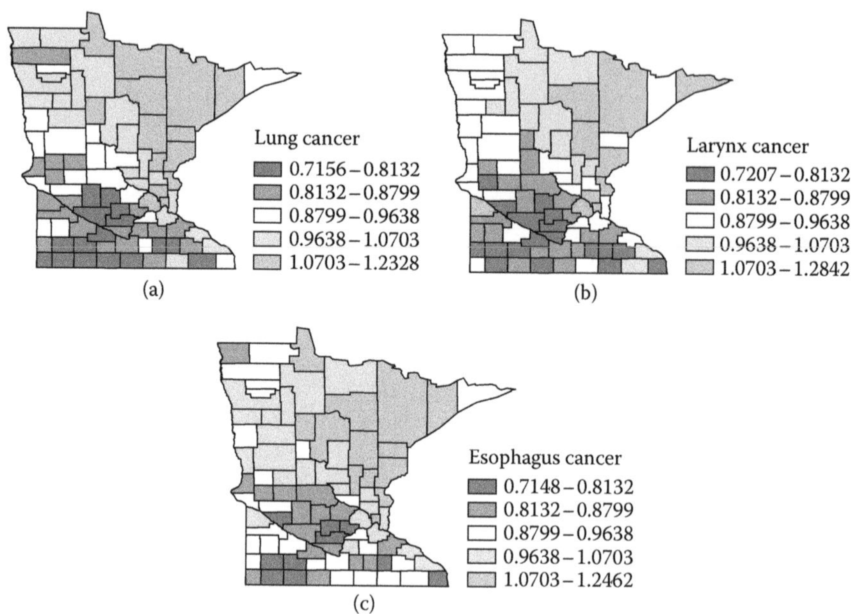

FIGURE 22.4
Maps of posterior means of the fitted standard mortality ratios (SMRs) of lung, larynx, and esophagus cancers in the years 1990–2000 in Minnesota from $MCAR(B, \Sigma)$ model.

(an area where previous studies have suggested smoking may be more common). In 22.1, the range of the raw SMRs is seen to be from 0 to 3.3, while in Figure 22.4, the range of the fitted SMRs is from 0.7 to 1.3, due to spatial shrinkage in the random effects.

22.5 Multivariate Point-Referenced Models

Statistical modeling and analysis for multivariate point-referenced data constitute the subject of multivariate geostatistics, which is too extensive to be comprehensively reviewed here; however, we will briefly discuss multivariate spatial process models for inference about parameters or for interpolation, which require specification of a valid *cross-covariance* function. A cross-covariance function yields the covariance between two variables at any pair of spatial locations. These are not routinely specified since they demand that for any number of locations and any choice of these locations, the resulting covariance matrix for the associated data be positive definite. Various constructions are possible. Constructing multivariate spatial process models essentially comes down to arriving at a legitimate cross-covariance function. We consider a conditional approach and a joint approach, but only skim the surface of this vast field. We refer the reader to Wackernagel [22], Cressie and Wikle [23], and Gelfand and Banerjee [24] for details.

22.5.1 Conditional models for multivariate geostatistics

Classical multivariate geostatistics begins, as with much of geostatistics, with the early work of Matheron [25,26]. The basic ideas here include cross-variograms and cross-covariance functions, intrinsic coregionalization, and co-kriging. The emphasis is on prediction. A thorough discussion of the work in this area is provided by Wackernagel [22].

Analysis of multivariate point-referenced data can proceed using either a *conditioning approach*, for example, X followed by $Y \mid X$, or a *joint approach* that directly models the joint distribution of the outcome variables. Both these approaches derive from approaches in multivariate geostatistics, or multivariate kriging, where the conditional approach is also referred to as *kriging with external drift* (see, e.g., Royle and Berliner [18]), while the joint approach is often called *co-kriging* (see, e.g., Wackernagel [22]).

Let $\mathbf{Y}(\mathbf{s}) = (Y_1(\mathbf{s}), Y_2(\mathbf{s}), \ldots, Y_p(\mathbf{s}))^T$ be a $p \times 1$ vector, where each $Y_i(\mathbf{s})$ represents an outcome of interest referenced by $\mathbf{s} \in \mathcal{D}$. We seek to capture the association both within components of $\mathbf{Y}(\mathbf{s})$ and across $\mathbf{s}$. A *stationary* cross-covariance function is defined as

$$C_{ij}(\mathbf{h}) = E[(Y_i(\mathbf{s}+\mathbf{h}) - \mu_i)(Y_j(\mathbf{s}) - \mu_j)], \tag{22.21}$$

where, for each i, a constant mean μ_i is assumed for $Y_i(\mathbf{s})$. The associated $p \times p$ matrix $C(\mathbf{h})$ is called the cross-covariance matrix. Note that it need not be symmetric (think of a setting where you might expect $C_{ij}(\mathbf{h}) \neq C_{ji}(\mathbf{h})$). Of course, $C(\mathbf{h})$ need not be positive definite, but in the limit, as $\|h\| \to 0$, it becomes positive definite.

We will turn to some properties of (22.21) and approaches for constructing them in Section 22.5.2, but before that, it is worth pointing out that one can obviate the problems of specifying a valid cross-covariance function by building hierarchical dependence structures, where we first model one variable and then, conditional upon the first, model the second, and then, conditional upon the first two, model the third, and so on. The matter is best explained with two variables. Consider two spatial variables $Y(\mathbf{s})$ and $X(\mathbf{s})$ observed over a finite set of locations $\mathcal{S} = \{\mathbf{s}_1, \mathbf{s}_2, \ldots, \mathbf{s}_n\}$. Let $\mathbf{Y}$ and $\mathbf{X}$ represent $n \times 1$ vectors of the observed $Y(\mathbf{s}_i)$'s and $X(\mathbf{s}_i)$'s, respectively. The conditional approach specifies the distribution of $\mathbf{X}$ and, subsequently, the conditional distribution of $\mathbf{Y}$ given $\mathbf{X}$. This is attractive in that valid specification of these two distributions yields a legitimate joint distribution. Cressie and Wikle [23] discuss this approach in detail and in a framework suitable for spatiotemporal extensions. Banerjee et al. [27] also offer a brief discussion. We provide a brief overview of this approach.

Recall that our inferential objectives are to interpolate or predict both $Y(\mathbf{s})$ and $X(\mathbf{s})$ at any arbitrary spatial location, while accounting for spatial dependence between the two variables. Assume that $X(\mathbf{s})$ is a univariate Gaussian spatial process with mean $\mu_X(\mathbf{s})$ and covariance function $C_X(\cdot;\boldsymbol{\theta}_X)$ indexed by process parameters $\boldsymbol{\theta}_X$. Therefore, $\mathbf{X} \sim N(\boldsymbol{\mu}_X\ \Sigma_X(\boldsymbol{\theta}_X))$, where $\boldsymbol{\mu}_X$ is $n \times 1$, with $\mu_X(\mathbf{s}_i)$ as its ith entry, and $\Sigma_X(\boldsymbol{\theta}_X)$ is an $n \times n$ spatial covariance matrix with entries $C_X(\mathbf{s}_i, \mathbf{s}_j; \boldsymbol{\theta}_X)$. If $\mathcal{S} = \{\mathbf{s}_1, \mathbf{s}_2, \ldots, \mathbf{s}_n\}$ is the set of locations where both $Y(\mathbf{s})$ and $X(\mathbf{s})$ have been observed, we can construct a conditional distribution for the $Y(\mathbf{s}_i)$'s given the $X(\mathbf{s}_i)$'s to yield

$$Y(\mathbf{s}_i) = \mathbf{z}_Y(\mathbf{s}_i)^T \boldsymbol{\gamma}_Y + \sum_{j=1}^{n} b_{ij} X(\mathbf{s}_j) + e(\mathbf{s}_i), \quad \text{for } i = 1, 2, \ldots, n. \tag{22.22}$$

Here, $\mathbf{z}_Y(\mathbf{s})$ is a known $p_2 \times 1$ vector of explanatory variables associated with $Y(\mathbf{s})$ with corresponding slope vector $\boldsymbol{\gamma}_Y$, b_{ij}'s are scalars, and the $e(\mathbf{s}_i)$'s are realizations of a spatial Gaussian process with zero mean and covariance function $C_e(\cdot;\boldsymbol{\theta}_e)$. The process $e(\mathbf{s})$ is independent of $X(\mathbf{s})$. Letting $\mathbf{Y}$ and $\mathbf{X}$ be the vectors with elements $Y(\mathbf{s}_i)$ and $X(\mathbf{s}_i)$, respectively, we can write (22.22) more compactly as

$$\mathbf{Y} \mid \mathbf{X} \sim N(Z_Y \boldsymbol{\gamma}_Y + B\mathbf{X},\, \Sigma_e(\boldsymbol{\theta}_e)),$$

where Z_Y is $n \times p_2$ with $\mathbf{z}_Y(\mathbf{s}_i)^T$ as its rows, B is $n \times n$ with entries b_{ij}, and Σ_e is $n \times n$ with entries $C_e(\mathbf{s}_i - \mathbf{s}_j; \boldsymbol{\theta}_e)$. The joint distribution of $\mathbf{Y}$ and $\mathbf{X}$ is

$$\begin{pmatrix} \mathbf{X} \\ \mathbf{Y} \end{pmatrix} \sim N\left(\begin{pmatrix} \boldsymbol{\mu}_X \\ \boldsymbol{\mu}_Y \end{pmatrix}, \begin{pmatrix} \Sigma_X(\boldsymbol{\theta}_X) & \Sigma_X(\boldsymbol{\theta}_X)B \\ B^T\Sigma_X(\boldsymbol{\theta}_X) & \Sigma_e(\boldsymbol{\theta}_e) + B\Sigma_X(\boldsymbol{\theta}_X)B^T \end{pmatrix} \right), \tag{22.23}$$

where $\boldsymbol{\mu}_X = Z_X\boldsymbol{\gamma}_X$ and $\boldsymbol{\mu}_Y = Z_Y\boldsymbol{\gamma}_Y + B\boldsymbol{\mu}_X$.

We remark that for (22.23) to be a valid distribution, it is not necessary that $\mathbf{X}$ and $\mathbf{Y}$ have the same dimensions. In that case, B will not be square—it will have as many rows as the dimension of $\mathbf{Y}$ and as many columns as the dimension of $\mathbf{X}$. In fact, the two outcomes need not be observed at the same set of locations—a situation referred to as *spatial misalignment*. Nevertheless, the joint distribution in (22.23) is well defined as long as $X(\mathbf{s})$ and $e(\mathbf{s})$ are legitimate spatial processes.

Royle and Berliner [18] consider several specifications for B that produce specific spatial regression models, including simple linear regression or kriging with external drift (KED), spatially varying regressions where the entries in B are functions of the locations, and nearest-neighbor regressions arising from sparse structures for B. However, in general, (22.22) does not correspond to a bivariate Gaussian process. In other words, it is not necessary for a variance–covariance matrix in the joint distribution in (22.22) to be constructed from a cross-covariance function that is well defined over the entire spatial domain, that is, for any two arbitrary locations. Banerjee et al. [27], however, point out that a special case of the conditional approach does produce a valid bivariate process model. Then, for any finite collection of n locations, suppose that

$$Y(\mathbf{s}_i) = \beta_0 + \beta_1 X(\mathbf{s}_i) + e(\mathbf{s}_i), \quad \text{for } i = 1, 2, \ldots, n. \tag{22.24}$$

The joint distribution of $\mathbf{X}$ and $\mathbf{Y}$ is

$$\begin{pmatrix} \mathbf{X} \\ \mathbf{Y} \end{pmatrix} \sim N\left(\begin{pmatrix} \boldsymbol{\mu}_X \\ \boldsymbol{\mu}_Y \end{pmatrix}, \begin{pmatrix} \Sigma_X(\boldsymbol{\theta}_X) & \beta_1\Sigma_X(\boldsymbol{\theta}_X) \\ \beta_1\Sigma_X(\boldsymbol{\theta}_X) & \Sigma_e(\boldsymbol{\theta}_e) + \beta_1^2\Sigma_X(\boldsymbol{\theta}_X) \end{pmatrix} \right), \tag{22.25}$$

where $\boldsymbol{\mu}_Y = \beta_0\mathbf{1} + \beta_1\boldsymbol{\mu}_X$ and $\Sigma_e(\boldsymbol{\theta}_e)$ is the $n \times n$ variance–covariance matrix for the $e(\mathbf{s}_i)$'s. The above joint distribution arises from a legitimate bivariate spatial process $\mathbf{W}(\mathbf{s}) = (X(\mathbf{s}), Y(\mathbf{s}))^T$, with mean $\boldsymbol{\mu}_{\mathbf{W}}(\mathbf{s}) = \begin{pmatrix} \mu_X(\mathbf{s}) \\ \beta_0 + \beta_1\mu_X(\mathbf{s}) \end{pmatrix}$ and cross-covariance

$$C_{\mathbf{W}}(\mathbf{s}, \mathbf{s}') = \begin{pmatrix} C_X(\mathbf{s}, \mathbf{s}') & \beta_1 C_X(\mathbf{s}, \mathbf{s}') \\ \beta_1 C_X(\mathbf{s}, \mathbf{s}') & \beta_1^2 C_X(\mathbf{s}, \mathbf{s}') + C_e(\mathbf{s}, \mathbf{s}') \end{pmatrix}, \tag{22.26}$$

where we have suppressed the dependence of $C_X(\mathbf{s}, \mathbf{s}')$ and $C_e(\mathbf{s}, \mathbf{s}')$ on $\boldsymbol{\theta}_X$ and $\boldsymbol{\theta}_e$, respectively. Equation 22.24 implies that $\mathrm{E}[Y(\mathbf{s}) \mid X(\mathbf{s})] = \beta_0 + \beta_1 X(\mathbf{s})$ for any arbitrary location $\mathbf{s}$, thereby specifying a well-defined spatial regression model for an arbitrary $\mathbf{s}$. A recent article by Cressie and Zammit-Mangion [28] demonstrates optimal spatial prediction (i.e., kriging) of $Y(\mathbf{s}_0)$ using the conditional approach.

The conditional approach is not bereft of problems. Consider predicting $X(\mathbf{s}_0)$ and $Y(\mathbf{s}_0)$ at a new location $\mathbf{s}_0$. It is clear that $X(\mathbf{s}_0)$ is conditionally independent of $\mathbf{Y}$ given $\mathbf{X}$ since $\mathbf{s}_0$ does not feature in (22.22). So $p(X(\mathbf{s}_0) \mid \mathbf{X}, \mathbf{Y}) = p(X(\mathbf{s}_0) \mid \mathbf{X})$ is the posterior predictive distribution of $X(\mathbf{s}_0)$ given the observed data. Therefore, prediction of $X(\mathbf{s})$ at a new location is carried out by borrowing information only from the observed $X(\mathbf{s}_i)$'s; information from $Y(\mathbf{s}_i)$'s is not utilized in any manner, no matter how strong the association between $Y(\mathbf{s})$ and $X(\mathbf{s})$. In this sense, predicting $X(\mathbf{s})$ is less than ideal. On the other hand, once $X(\mathbf{s}_0)$ has been obtained, one can predict $Y(\mathbf{s}_0)$ using the conditional posterior predictive distribution $p(Y(\mathbf{s}_0) \mid \mathbf{Y}, \mathbf{X})$, which is easily derived from the joint distribution in (22.23).

22.5.2 Multivariate process models

In multivariate geostatistical regression, dependence is incorporated using multivariate spatial processes. We extend the definition in (22.21) to general cross-covariances for arbitrary sets of random variables. Using the generic notation $\mathbf{w}(s)$ to denote a $k \times 1$ vector of random variables, with $w_i(\mathbf{s})$ as its ith entry, at location $\mathbf{s}$, we seek flexible, interpretable, and computationally tractable models to describe the process $\{\mathbf{w}(\mathbf{s}) : \mathbf{s} \in D\}$. A $k \times 1$ multivariate Gaussian process, written as $\mathbf{w}(\mathbf{s}) \sim GP_k(\mathbf{0}, C_w(\cdot))$, is specified by its matrix-valued *cross-covariance* function $C_w(\mathbf{s}, \mathbf{s}')$, whose (i,j)th element is given by $\text{Cov}\{w_i(\mathbf{s}), w_j(\mathbf{s}')\}$. For any integer n and any collection of sites $\mathbf{s}_1, \ldots, \mathbf{s}_n$, we write the multivariate realizations as a $kn \times 1$ vector $\mathbf{w} \sim N(\mathbf{0}, \Sigma_{\mathbf{w}})$, where $\Sigma_{\mathbf{w}}$ is the $kn \times kn$ variance–covariance matrix for $\mathbf{w}$. A *valid* cross-covariance function ensures that $\Sigma_{\mathbf{w}}$ is symmetric and positive definite. Therefore, the cross-covariance function must satisfy the following two conditions:

$$C_w(\mathbf{s}, \mathbf{s}') = C_w^T(\mathbf{s}', \mathbf{s}) \tag{22.27}$$

$$\sum_{i=1}^{n} \sum_{j=1}^{n} \mathbf{x}_i^T C_w(\mathbf{s}_i, \mathbf{s}_j) \mathbf{x}_j > 0 \ \forall \ \mathbf{x}_i, \mathbf{x}_j \in \Re^p \setminus \{\mathbf{0}\}. \tag{22.28}$$

The first condition ensures that while the cross–covariance function itself need not be symmetric, $\Sigma_{\mathbf{w}}$ is. The second condition ensures the positive definiteness of $\Sigma_{\mathbf{w}}$ and is in fact quite stringent; it must hold for all integers n and any arbitrary collection of sites $\mathcal{S} = \{\mathbf{s}_1, \ldots, \mathbf{s}_n\}$. Conditions (22.27) and (22.28) imply that $C_w(\mathbf{s}, \mathbf{s})$ is a symmetric and positive definite function. In fact, it is precisely the variance–covariance matrix for the elements of $\mathbf{w}(\mathbf{s})$ within site $\mathbf{s}$. When $\mathbf{s} = \mathbf{s}'$, $C_w(\mathbf{s}, \mathbf{s})$ is precisely the variance–covariance matrix for the elements of $\mathbf{w}(\mathbf{s})$.

Now, suppose each location $\mathbf{s}$ yields observations on q dependent variables given by a $q \times 1$ vector $\mathbf{Y}(\mathbf{s})$. For each $Y_l(\mathbf{s})$, we also observe a $p_l \times 1$ vector of regressors $\mathbf{x}_l(\mathbf{s})$. Therefore, for each location, we have q univariate spatial regression equations,

$$Y_l(\mathbf{s}) = \mathbf{x}_l^T(\mathbf{s})\boldsymbol{\beta}_l + \mathbf{z}_l^T(\mathbf{s})\mathbf{w}(\mathbf{s}) + \epsilon_l(\mathbf{s}), \quad l = 1, \ldots, q. \tag{22.29}$$

Here, each $\boldsymbol{\beta}_l$ is a $p_l \times 1$ regression coefficient associated with $\mathbf{x}_l^T(\mathbf{s})$, $\mathbf{z}_l(\mathbf{s})$ is a $k \times 1$ coefficient vector for the $k \times 1$–dimensional spatial process $\mathbf{w}(\mathbf{s})$, and $\epsilon_l(\mathbf{s})$ is the measurement error or nugget associated with $Y_l(\mathbf{s})$. These can be combined into a multivariate regression model written as

$$\mathbf{Y}(\mathbf{s}) = \mathbf{X}^T(\mathbf{s})\boldsymbol{\beta} + \mathbf{Z}^T(\mathbf{s})\mathbf{w}(\mathbf{s}) + \boldsymbol{\epsilon}(\mathbf{s}), \tag{22.30}$$

where $\mathbf{X}^T(\mathbf{s})$ is a $q \times p$ matrix ($p = \sum_{l=1}^{q} p_l$) having a block-diagonal structure, with its lth diagonal being the $1 \times p_l$ vector $\mathbf{x}_l^T(\mathbf{s})$, and $\boldsymbol{\beta} = (\boldsymbol{\beta}_1, \ldots, \boldsymbol{\beta}_p)^T$ is a $p \times 1$ vector of regression coefficients, with $\boldsymbol{\beta}_l$ being the $p_l \times 1$ vector of regression coefficients corresponding to $\mathbf{x}_l^T(\mathbf{s})$. The spatial effects $\mathbf{w}(\mathbf{s})$ form a $k \times 1$ coefficient vector of the $q \times k$ design matrix $\mathbf{Z}^T(\mathbf{s})$, with $\mathbf{z}_l(\mathbf{s})^T$ as its rows. The $q \times 1$ vector $\boldsymbol{\epsilon}(\mathbf{s})$ follows an $MVN(\mathbf{0}, \Psi)$, modeling the measurement error effect with dispersion matrix Ψ. Model (22.30) acts as a general framework admitting several spatial models. For instance, letting $k = q$ and $\mathbf{Z}^T(\mathbf{s}) = I_q$ leads to the multivariate analogue of univariate geostatistical regression models, where $\mathbf{w}(\mathbf{s})$ acts as a *spatially varying intercept.* On the other hand, we could envision all coefficients to be spatially varying and set $k = p$ with $\mathbf{Z}^T(\mathbf{s}) = \mathbf{X}^T(\mathbf{s})$. This yields *multivariate spatially varying coefficients*, multivariate analogues of those discussed in Gelfand et al. [29].

The precise definition of $\mathbf{Z}^T(\mathbf{s})$ will depend on the specific application, perhaps comprising a subset of the regressors in $\mathbf{X}^T(\mathbf{s})$ or some other design matrix, as, for instance, in many field experiments (see, e.g., Banerjee and Johnson [30]). Further connections between joint modeling and conditional specifications are outlined in Gelfand et al. [20] and also in

Chapter 9 of Banerjee et al. [27]. These references also contain details on MCMC algorithms for estimating models that can be cast within the framework of (22.30).

References

[1] Mardia KV. Multi-dimensional multivariate Gaussian Markov random fields with application to image processing. *Journal of Multivariate Analysis*, 24:265–284, 1988.

[2] Kim H, Sun D, and Tsutakawa RK. A bivariate Bayes method for improving the estimates of mortality rates with a twofold conditional autoregressive model. *Journal of the American Statistical Association*, 96:1506–1521, 2001.

[3] Knorr-Held L and Best NG. A shared component model for detecting joint and selective clustering of two diseases. *Journal of the Royal Statistical Society: Series A*, 164:73–85, 2001.

[4] Knorr-Held L and Rue H. On block updating in Markov random field models for disease mapping. *Scandinavian Journal of Statistics*, 29:597–614, 2002.

[5] Gamerman D, Moreira ARB, and Rue H. Space-varying regression models: Specifications and simulation. *Computational Statistics and Data Analysis*, 42:513–533, 2003.

[6] Assunção RM, Potter JE, and Cavenaghi SM. A Bayesian space varying parameter model applied to estimating fertility schedules. *Statistics in Medicine*, 21:2057–2075, 2002.

[7] Assunção RM. Space-varying coefficient models for small area data. *Environmetrics*, 14:453–473, 2003.

[8] Sain SR and Cressie N. Multivariate lattice models for spatial environmental data. In *Proceedings of the ASA Section on Statistics and the Environment*, pp. 2820–2825, New York City, NY, 2002.

[9] Zhang Y, Hodges JS, and Banerjee S. Smoothed ANOVA with spatial effects as a competitor to MCAR in multivariate spatial smoothing. *Annals of Applied Statistics*, 3:1805–1830, 2009.

[10] Hodges JS, Cui Y, Sargent DJ, and Carlin BP. Smoothing balanced single-error-term analysis of variance. *Technometrics*, 49:12–25, 2007.

[11] Besag J, York JC, and Mollié A. Bayesian image restoration, with two applications in spatial statistics [with discussion]. *Annals of the Institute of Statistical Mathematics*, 43:1–59, 1991.

[12] Wall MM. A close look at the spatial structure implied by the CAR and SAR models. *Journal of Statistical Planning and Inference*, 121:311–324, 2004.

[13] Leroux BG, Lei X, and Breslow N. Estimation of disease rates in small areas: A new mixed model for spatial dependence. In ME Halloran and D Berry, eds., *Statistical Models in Epidemiology, the Environment, and Clinical Trials*, pp. 135–178. Springer, New York, 1999.

[14] Dean CB, Ugarte MD, and Militino AF. Detecting interaction between random region and fixed age effects in disease mapping. *Biometrics*, 57:197–202, 2001.

[15] Carlin BP and Banerjee S. Hierarchical multivariate car models for spatio-temporally correlated survival data. In JM Bernardo, MJ Bayarri, JO Berger, AP Dawid, D Heckerman, AFM Smith, and M West, eds., *Bayesian Statistics 7*, pp. 45–64. Oxford University Press, Oxford, 2003.

[16] Gelfand AE and Vounatsou P. Proper multivariate conditional autoregressive models for spatial data analysis. *Biostatistics*, 4:11–25, 2003.

[17] Jin X, Carlin BP, and Banerjee S. Generalized hierarchical multivariate CAR models for areal data. *Biometrics*, 61:950–961, 2005.

[18] Royle JA and Berliner LM. A hierarchical approach to multivariate spatial modeling and prediction. *Journal of Agricultural, Biological and Environmental Statistics*, 4: 29–56, 1999.

[19] Jin X, Banerjee S, and Carlin BP. Order-free coregionalized lattice models with application to multiple disease mapping. *Journal of the Royal Statistical Society: Series B*, 69:817–838, 2007.

[20] Gelfand AE, Schmidt AM, Banerjee S, and Sirmans CF. Nonstationary multivariate process modelling through spatially varying coregionalization [with discussion]. *Test*, 13, 2004.

[21] Daniels MJ and Kass RE. Nonconjugate Bayesian estimation of covariance matrices and its use in hierarchical models. *Journal of the American Statistical Association*, 94:1254–1263, 1999.

[22] Wackernagel H. *Multivariate Geostatistics: An Introduction with Applications*. 3rd ed. Springer, New York, 2003.

[23] Cressie N and Wikle CK. *Statistics for Spatio-Temporal Data*. 1st ed. Wiley, New York, 2011.

[24] Gelfand AE and Banerjee S. Multivariate spatial process models. In AE Gelfand, P Diggle, P Guttorp, and M Fuentes, eds., *Handbook of Spatial Statistics*, pp. 495–516. Taylor & Francis/CRC, Boca Raton, FL, 2010.

[25] Matheron G. The intrinsic random functions and their applications. *Advances in Applied Probability*, 5:437–468, 1973.

[26] Matheron G. Recherche de simplification dans un probleme de cokrigeage. N-698. Centre de Géostatistique, Fountainebleau, France, 1979.

[27] Banerjee S, Carlin BP, and Gelfand AE. *Hierarchical Modeling and Analysis for Spatial Data*. 2nd ed. Chapman & Hall/CRC Press, Boca Raton, FL, 2014.

[28] Cressie N and Zammit-Mangion A. Multivariate spatial covariance models: A conditional approach. arXiv, 1504.01865v1, 2015.

[29] Gelfand AE, Kim H-J, Sirmans CF, and Banerjee S. Spatial modeling with spatially varying coefficient processes. *Journal of the American Statistical Association*, 98: 387–396, 2003.

[30] Banerjee S and Johnson GA. Coregionalized single- and multi-resolution spatially-varying growth curve modelling with applications to weed growth. *Biometrics*, 61: 617–625, 2006.

Part IV

Special Problems and Applications

23

Bayesian Variable Selection in Semiparametric and Nonstationary Geostatistical Models: An Application to Mapping Malaria Risk in Mali

Federica Giardina
Swiss Tropical and Public Health Institute (Swiss TPH)
and
University of Basel
Basel, Switzerland

Nafomon Sogoba
Université des Sciences
des Techniques et des Technologies de Bamako
Bamako, Mali

Penelope Vounatsou
Swiss Tropical and Public Health Institute (Swiss TPH)
and
University of Basel
Basel, Switzerland

CONTENTS

23.1 Introduction

Geostatistical models are used to analyze data collected at a discrete set of locations (georeferenced data) within a continuous domain (see Chapter 11 in this volume). They have been widely applied to problems ranging from geology and ecology to epidemiology and public health (Gelfand et al., 2004). Applications in epidemiology are mainly concerned with relating disease data to a set of predictors (i.e., environmental or climatic variables) with the aim of determining the main risk factors and predicting disease outcome measures (e.g., risk, incidence, and mortality) at unobserved locations (Lawson, 2013). The Bayesian formulation of linear and generalized linear geostatistical models has been introduced by Diggle et al. (1998).

Bayesian geostatistical models have been widely used in mapping parasitic diseases such as malaria risk (Gemperli and Vounatsou, 2006; Gosoniu et al., 2006, 2010; Hay et al., 2009; Giardina et al., 2012; Noor et al., 2014), schistosomiasis risk (Raso et al., 2005; Clements et al., 2008; Wang et al., 2008), filarial worm risk infection (Diggle et al., 2007; Crainiceanu et al., 2008), hookworm infection (Raso et al., 2006b; Chammartin et al., 2013), and helminths co-infections (Raso et al., 2006a; Pullan et al., 2008; Schur et al., 2011). Disease maps can be used to identify possible clusters, define and monitor epidemics, or provide baseline risk estimates at high spatial resolution. They represent an essential tool to guide disease control programs in planning targeted interventions and in evaluating their effectiveness. Recent developments in satellite-based remote sensing (RS) for environmental monitoring and geographical information systems (GIS) have further boosted research in this area (Bauwens et al., 2012).

In most epidemiological applications, the impact of the predictors is modeled as a linear effect, constant throughout the study area. However, more flexible functional forms are often more suitable to capture the relationships between the covariates and the response (see Chapter 12 in this volume). Large study areas can often be partitioned in different ecological zones that influence the effect of the predictors on the disease outcome. In large areas, the underlying spatial structure that models the geographical dependence among neighboring locations may vary also according to the geographic position. Therefore, a flexible model specification is required to enable choosing different predictors as well as different functional forms in each zone, while modeling a nonstationary spatial process.

Bayesian statistical methods for choosing an appropriate subset of covariates among many potential predictors have received increasing attention in recent years. A comprehensive review of the most commonly used methods can be found in O'Hara and Sillanpää (2009). Chen and Dunson (2003) and Kinney and Dunson (2007) studied both fixed and random effects selection in linear and logistic models. Tüchler (2008) and Wagner and Duller (2012) proposed an approach that links Bayesian variable selection methods to random effects variance selection by a reparametrization of the random effects. Less work has been done on functional form selection methods, including nonlinear effects: only recently Scheipl et al. (2012) proposed a stochastic search-based approach employing a modified spike-and-slab mixture prior for the coefficients, and Bové et al. (2012) developed an extension of the classical Zellner's g-prior (hyper-g-priors) to identify the presence of a variable and its spline transformation in generalized additive models. Curtis et al. (2014) provides a review of variable selection methods for additive models.

The literature on variable selection for spatial data is limited. Typically, spatial correlation is ignored in the selection of explanatory variables, influencing model selection as well as parameter estimation (Hoeting et al., 2006). In the work by Smith and Fahrmeir (2007), an Ising prior is used to allow dependence among variable inclusion probabilities at neighboring locations for linear regression models defined on a regular lattice. A similar approach is adopted by Scheel et al. (2013) studying the effect of climate change on the insurance industry at local geographic scale (municipalities). Reich et al. (2010) proposed a stochastic search approach to select covariates with constant or spatially varying effects (Gelfand et al., 2003).

A review of methods used for constructing nonstationary spatial processes can be found in Sampson (2010). These methods range from spatial deformation models (Sampson and Guttorp, 1992) to spatial process decomposition in terms of empirical orthogonal functions

(Nychka et al., 2002) and process convolution models (Higdon, 1998). Smoothing and kernel-based methods (Fuentes, 2001) model nonstationarity as spatially weighted combinations of stationary spatial covariance functions. This approach was applied by Banerjee et al. (2004) to model house prices in California and by Gosoniu et al. (2009) in malaria risk mapping in West Africa. In the latter, the relation between climate factors and malaria risk was modeled separately in each ecological zone by penalized B-splines. In this chapter, we extend the work of Gosoniu et al. (2009) by developing Bayesian nonstationary geostatistical models that choose among different functional forms, allowing variations across partitions of the area of interest. Furthermore, spatially varying weights are proposed to take into account irregularly shaped partitions of the study area. The application that motivated the work comes from the area of malaria epidemiology. Over the last few years, national malaria surveys have been carried out routinely in several countries in Sub-Saharan Africa with the aim of monitoring and evaluating progress in disease control.

The chapter is structured as follows: Section 23.2 describes the problem and the data used. Section 23.3 introduces variable selection methods for functional forms in nonstationary geostatistical models. Section 23.4 presents the results of the proposed methodology applied to a national malaria prevalence survey in Mali. Validation results compare predictions of the model determined by the developed methods to those obtained with a full B-spline model (Gosoniu et al., 2009) and with the same model with stationary covariance matrix. Section 23.5 provides concluding remarks and suggests further lines of research and areas of application.

23.2 Material and Methods

Malaria transmission is strongly influenced by climatic conditions that determine the abundance and seasonal dynamics of the *Anopheles* mosquito vector. The amount and duration of malaria transmission is influenced by the ability of a parasite and a mosquito vector to coexist long enough to enable transmission to occur. The distribution and abundance of the parasite and mosquito population are sensitive to environmental factors such as temperature, rainfall, humidity, presence of water, and vegetation. Environmental factors affect the biological cycle of both the vector and parasite, allowing or interrupting the different development stages, and therefore favoring or inhibiting transmission. Usually, *Anopheles* do not fly more than 2 km, but in certain circumstances, they can fly up to 5 km. The distance mosquitoes fly is determined largely by the environment: if suitable hosts and breeding places are nearby, mosquitoes do not disperse far, but if one or more are more distant, greater dispersal may be necessary (Schlagenhauf-Lawlor, 2008).

Mali is divided into five ecological zones based on Food and Agriculture Organization (FAO) methodology (FAO, 2000): the Sahara desert, the South Saharan zone, the Sahelian zone, the central delta of the Niger River, and the west Sudanian region, as shown in Figure 23.1. The northern part of Mali is occupied by the Sahara desert, which is a hyperarid zone with scarce water and precipitation; the first Sub-Saharan zone presents steppe and woodlands, and it is also well an arid and desertic zone. These two regions do not represent a favorable environment for the malaria vector. The Sahelian zone is mainly characterized by acacia savanna; it is arid with rainfall between 250 and 550 mm. The central region of the Niger delta presents similar characteristics in terms of rainfall, but it is mainly constituted by flooded savanna. The Sudan zone, in the southwest of the country, is a semiarid to subhumid region with abundant rainfall (between 550 and 1100 mm).

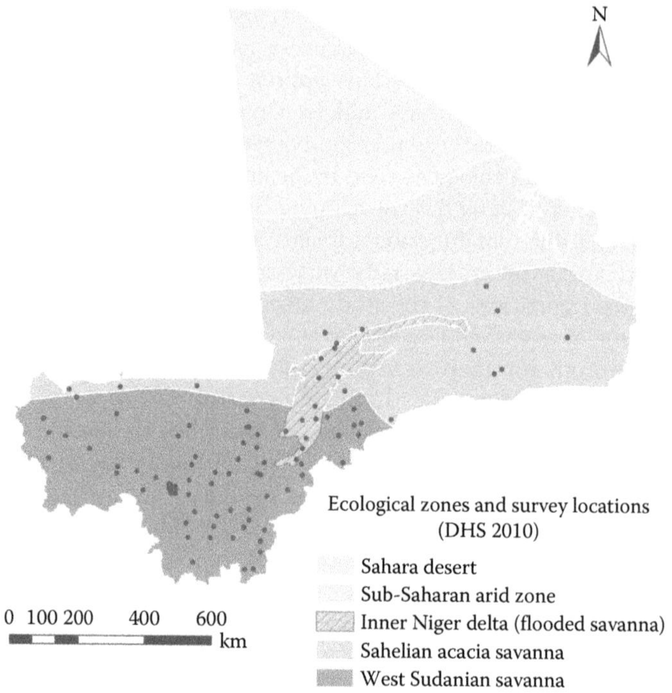

FIGURE 23.1
Ecological zones in Mali. (From FAO, Global forest resources assessment, 2000, www.fao.org/forestry/fra/2000/report/en/.)

23.2.1 National malaria survey in Mali

A Demographic and Health Survey (DHS) was carried out in Mali between August and October 2010 by the National Malaria Control Program in collaboration with Macro International and the Malaria Research and Training Center in Bamako. The information collected in the survey consists of georeferenced data with parasitemia measurements (malaria test positivity) among 1788 children below the age of 5 years (Ministère de la Santé, 2010).

23.2.2 Environmental predictors

RS and GIS have emerged as methods for exploring environmental factors potentially associated with malaria outcomes. With the purpose of deriving explanatory variables for our application, we have collated environmental and climatic data provided by satellite images. Vegetation measures such as the Normalized Difference Vegetation Index (NDVI) and Enhanced Vegetation Index (EVI), as well as temperatures proxies (night and day land surface temperature), were obtained by the Moderate Resolution Imaging Spectroradiometer (MODIS) at 1 km spatial resolution for the year 2010. Decadal rainfall data were extracted at 8 km resolution via the Africa Data Dissemination Service (ADDS) and aggregated over a year previous to the survey time. Water bodies were identified using the world water bodies layer provided by the ArcGIS website. The shortest Euclidean distance between the locations and the water bodies was calculated in ArcGIS version 10.0 (ESRI, 2011). Altitude data were obtained from an interpolated digital elevation model by the U.S. Geological Survey Earth Resources Observation and Science Data Center at a spatial resolution of 1 km. Information on area type (rural/urban) was provided by the

Global Rural-Urban Mapping Project (GRUMP) website, and population density data by the Afripop project (Tatem et al., 2007).

All environmental and climatic data have been associated with observed locations with the shortest Euclidean distance from the layers.

23.3 Models

Let $N(s)$ be the number of individuals screened for parasitemia at location s, $s = 1, \ldots, m$, $Y(s)$ be the number of those tested positive, and $\mathbf{x}(s) = (x_1(s), x_2(s), \ldots, x_p(s))^T$ be the vector of p potential predictors observed at location s. We assume that $Y(s)$ arises from a binomial distribution:

$$Y(s)|\pi(s), N(s) \sim Binomial(\pi(s), N(s)) \qquad \forall s = 1, \ldots, m \text{ sites}$$

and the probability $\pi(s)$ of being infected at location s is modeled through an additive logistic regression,

$$log\left(\frac{\pi(s)}{1-\pi(s)}\right) = \mu(s) + \omega(s), \tag{23.1}$$

where $\mu(\cdot)$ represents the mean structure and $\omega(\cdot)$ models the spatial correlation through Gaussian processes. The mean structure takes the general form

$$\mu(s) = \beta_0 + \sum_{i=1}^{p}\sum_{j=1}^{J}\sum_{k=1}^{K} f_{ijk}(x_i(s), \beta_{ijk}),$$

where $f_{ijk}(\cdot)$ indicates each one of the J possible functional forms that relate the observed variable X_i to the disease risk $\pi(s)$ in ecological zone k via the coefficients β_{ijk}, and β_0 is a common intercept term. We model nonstationarity in the spatial process through a mixture of stationary spatial processes smoothing at the borders between the zones through the definition of distance-dependent weights, as in Gosoniu et al. (2009). A stationary spatial process $\boldsymbol{\phi}_k$ is defined as $\boldsymbol{\phi}_k \sim N(0, \boldsymbol{\Sigma}_k)$ $\forall k = 1, \ldots, K$ ecological zone, where $(\Sigma_k)_{ss'} = \sigma_k^2 corr(||s - s'||; \rho_k, \nu)$ and *corr* is a parametric function of the Euclidean distance $||s - s'||$ between sites s and s'. The Matérn family describes most of the correlation function used in geostatistical models:

$$corr(||s-s'||; \rho_k, \nu) = \frac{1}{2^{\nu-1}\Gamma(\nu)}(\rho_k||s-s'||)^{\nu} K_{\nu}(\rho_k||s-s'||),$$

where K_ν is a modified Bessel function with smoothing parameter ν, while $\rho_k > 0$ controls the rate of correlation decay between observations as distance increases. The choice $\nu = 1/2$ leads to the commonly used exponential correlation function, that is, $corr(||s - s'||) = \exp(-\rho_k||s - s'||)$.

A nonstationary spatial process $\boldsymbol{\omega}$ is generated as a weighted sum of the above defined spatial processes as follows: $\boldsymbol{\omega} \sim N(0, \sum_{k=1}^{K} A_k \Sigma_k A_k)$, where A_k is a diagonal matrix with $(A_k)_{ss} = a_{sk}$. The weights a_{sk} are chosen as decreasing functions of the Euclidean distance between location s and "knots" of the subregion k. The knots are selected over a grid covering the entire region in order to take into account the irregularly shaped subregions. Further details on the choice of the weights are given in Section 23.3.2.

23.3.1 Mean structure selection

We describe a Bayesian variable selection procedure to choose an appropriate subset of potential covariates for malaria risk and determine whether a linear, piecewise constant or a smoother functional form is required to model the effect of the respective covariates, allowing them to vary across ecological zones. For each variable X_i in ecological zone k, we consider the following four scenarios: (1) there is *no relationship* between X_i and the infection probability π, or there is a relationship that can be described by (2) *linear*, (3) *piecewise constant*, or (4) *smooth* functions.

The variable selection approach is defined by the following hierarchy:

$$\boldsymbol{\alpha} = (\alpha_1, \ldots, \alpha_4)^T$$

$$\mathbf{p}_k|\boldsymbol{\alpha} = (p_{1k}, \ldots, p_{4k})^T \sim Dir(4, \boldsymbol{\alpha}) \quad \forall k = 1, \ldots, K,$$

where $\mathbf{p}_k$ follows a Dirichlet distribution of concentration hyperparameters $\boldsymbol{\alpha}$. Each element $p_{jk}, j = 1, \ldots, 4$ corresponds to the selection probability of the different functional forms (1 = linear, 2 = piecewise constant, 3 = smooth, 4 = none) in region k. For each predictor i in region k, a categorical variable c_{ik} can be defined to indicate the different functional forms, with probability mass function $p(c_{ik}|\boldsymbol{p}_k) = \prod_{j=1}^{4} p_{jk}^{\delta_j(c_{ik})}$, where $\delta_j(\cdot)$ denotes the Dirac delta function evaluated at j. We build the auxiliary variables γ_{ijk} to indicate presence or absence of the j functional form of covariate X_i in region k

$$\gamma_{ijk} = \delta_j(c_{ik}) + \epsilon_0(1 - \delta_j(c_{ik})) \; \forall j = 1, \ldots, 3,$$

and we assign a normal prior to the coefficients β_{ijk}, that is,

$$\beta_{ijk}|\gamma_{ijk}, \tau_{ij}^2 \sim N(0, \gamma_{ijk}\tau_{ij}^2)$$

$$\tau_{ij}^2|a, b \sim IG(a, b),$$

where ϵ_0 is some very small positive constant and the variance τ_{ij}^2 is sampled from an inverse gamma (IG) with shape parameter a and scale b.

This prior specification defines a continuous bimodal distribution on the hypervariance of β_{ijk} with a spike at ϵ_0, which shrinks the coefficients that are not relevant for the model, and a right continuous tail (slab) to identify nonzero parameters. In particular, if $\gamma_{ijk} = 1$, the covariate effect β_{ijk} is estimated by assuming a normal prior distribution with mean 0 and variance τ_{ij}^2; otherwise, $\gamma_{ijk} = \epsilon_0$ and β_{ijk} is shrunk toward 0. Therefore, the predictor X_i is not included in the model for region k in functional form j. The Dirichlet prior on the selection probability p allows flexibility in estimating model sizes by introducing another level of hierarchy in the model specification. If $\gamma_{i1k} = 1$, the relationship between the predictor X_i and the disease risk π is *linear* in region k and $f_k(x_i) = \beta_{i1k}x_i$. If $\gamma_{i2k} = 1$, the relationship between the predictor x_i and the disease risk is *piecewise constant*, where x_i has been categorized into Q quantiles and $f_k(x_i) = \sum_{q=1}^{Q} \beta'_{iqk}x'_{iq}$. Spike-and-slab priors perform poorly in identifying nonlinear forms of variables, which include groups of coefficients (Scheipl et al., 2012). In particular, switching status (i.e., inclusion or exclusion of the coefficient's vector) becomes very unlikely, resulting in very poor mixing of the indicator variables. Parameter expansion (Gelman et al., 2008) offers a method to improve mixing in the Markov chain Monte Carlo (MCMC) while selecting simultaneously a batch of coefficients. More specifically, we define $\boldsymbol{\beta}'_{ik} = \beta_{i2k}\eta_{ik}$, where

$$\eta_{iqk}|m_{iqk} \sim N(m_{iqk}, 1) \text{ and } m_{iqk} \sim 1/2N(-1, 1) + 1/2N(1, 1) \quad \forall q = 1, \ldots, Q$$

quantiles. The two parameters η_{iqk} and β_{iqk} are not identifiable, but inference can be obtained about their product β'_{ik}. If $\gamma_{i3k} = 1$, the relationship between the predictor X_i and the disease risk π includes nonlinear terms in region k, expressed in the form of a penalized B-spline, that is, $f_k(x_i) = bx_i + \sum_{l=1}^{L} u_{ilk} z_l(x_i)$, where $z_l, \forall l = 1, \ldots, L$ is an appropriate spline basis for covariate x_i, that is, radial cubic basis function.

Following Ruppert et al. (2003), a quadratic penalty is placed on **u**, which translates into the constraint $\boldsymbol{u}_{ik}^T \boldsymbol{u}_{ik} \leq \lambda$, where λ is the smoothing parameter. The above functional form can be written in a mixed-model representation (Zhao et al., 2006) as follows:

$$f_k(x_i) = \beta_{i3k} x_i + \mathbf{Z}_{x_i} \boldsymbol{u}_{ik},$$

where

$$\mathbf{Z}_{x_i} = \left[|x_i - \kappa_l|^3\right] \left[|\kappa_l - \kappa_{l'}|^3\right]^{-1/2}$$

and $\boldsymbol{u}_{ik} \sim N(0, \sigma^2_{u_{ik}} \mathbf{I})$. The knots κ_l are defined as the sample quintiles specific to each covariate X_i. To ensure identifiability of the model, we do not include a constant term in the spline representation. We apply random effect selection methods to choose whether a smooth term has to be included in the models. We follow the approach suggested by Wagner and Duller (2012) reparametrizing the variance component and perform variable selection on the standard deviation, treating it as a covariate effect. In particular, we rewrite the random effects associated with the spline terms $\boldsymbol{u}_{ik}$ as $\boldsymbol{u}_{ik} = \pm\sigma_{u_{ik}} \boldsymbol{\theta}_{ik}$, where $\boldsymbol{\theta}_{ik} \sim N(0, \mathbf{I})$, and assign $\sigma_{u_{ik}}$ the same spike-and-slab prior as for the parameter β_{i3k}. The sign of both $\sigma_{u_{ik}}$ and $\boldsymbol{\theta}_{ik}$ is not identifiable, but the product $\pm\sigma_{u_{ik}} \boldsymbol{\theta}_{ik}$, as well as the associated indicator γ_{i3k}, can be estimated. In fact, as in the case of batches of coefficients, for the selection of the *piecewise constant* functional form, this redundant parametrization has computational advantages in the MCMC implementation.

The procedure described above can be adopted only for continuous predictors. Categorical predictors, such as area type, which is a dummy variable, were modeled using piecewise constant functional forms.

It is not realistic to assume independence across the predictors selected in each ecological zone. To take into account spatial dependence in the mean structure and to obtain smooth prediction maps at the zone borders, we introduce spatially varying weights $\psi_k(s)$ in the regression coefficients and define $\beta^*_{ij}(s) = \psi_k(s)\beta_{ijk}$. Details on the specification of the weights can be found in Section 23.3.2. Equation 23.1 now takes the form $logit(\pi(s)) = \sum_{i=1}^{P} \sum_{j=1}^{J} \sum_{k=1}^{K} f_{ik}(x_i(s), \beta^*_{ij}(s)) + \omega(s)$. In most applications, model fit and prediction are performed using the model with the highest posterior probability. However, Barbieri and Berger (2004) show that for normal linear models, the one with the best predictive ability, that is, that minimizes the squared error loss, is the so-called median-probability model. The latter is defined as the model consisting of those variables that have an overall posterior probability greater than or equal to 1/2 of being included in a model. The median-probability model may differ from the highest-probability model.

23.3.2 Spatially varying weights

Spatially varying weights a_{sk} and $\psi_k(s)$ have been introduced to model the variance structure of the nonstationary Gaussian spatial process and the mean structure, respectively. While a_{sk} smooths the values of the spatial process at the border of the zones, $\psi_k(s)$ takes into account that the risk in neighboring points across the borders of the zones should be affected similarly by covariates, although the zones may have different predictors. Therefore, $\psi_k(s)$ smooths the mean structure at the borders. For convenience, we chose $a_{sk} = \psi_k(s)$. We define $d_k(s)$ as the Euclidean distance between a given location s and the closest of the knots belonging to ecological zone k. The knots are equally spaced points over a grid covering the study area. We obtain weights that are decreasing functions of the shared area between two circles of radius r, the first one centered in s and the other one in the point at distance $d_k(s)$. Following the definition of spherical correlation function in two dimensions (circular correlation function), we construct the spatially varying weights $\psi_k(s)$ as follows:

$$\psi_k(s) = \begin{cases} \frac{2}{\pi}\left(\arccos d_k(s)/r - (d_k(s)/r)\sqrt{1-(d_k(s)/r)^2}\right) & \text{if } d_k(s) < r \\ 0 & \text{otherwise} \end{cases}.$$

Therefore, the weights allow covariate effects of neighboring zones to be considered for all locations within a radius r from the border. Furthermore, each zone has a separate spatial process and the weights allow a mixing of neighboring processes only for locations close to the borders (in proximity defined by the radius r).

The weights were normalized (divided by their length) so that $\sum_{k=1}^{K} \psi_k(s)^2 = 1$. However, the choice of the grid spacing g and the radius r might influence posterior inference. Therefore, the main analysis has been carried out defining g and fixing $r = g$. Nevertheless, a sensitivity analysis has been conducted to assess the robustness of the results under different values of the radius and keeping the spacing of the grid knots constant.

23.4 Results

The model presented in Section 23.3 was applied on the national malaria survey data from Mali to identify the most important climatic predictors by ecological zone and perform spatial risk prediction over the study area at 2 km resolution. Three different malaria endemic ecological zones and eight predictors with three functional forms were considered in the analysis. All continuous covariates have been centered and standardized to obtain a better mixing of the Markov chains arising from simulations. The spatially varying weights defined for a specific observed point and for the whole study area are shown in Figure 23.2.

Posterior analysis was performed by MCMC (JAGS) (Plummer et al., 2003) using samples collected over 100,000 iterations after a burn-in of 10,000. Convergence was monitored by examining trace plots and autocorrelation plots for several representative parameters. Results of the variable selection procedure are given in Table 23.1 and Figure 23.3. Table 23.1 shows the models selected with the highest posterior probabilities (only the first three are listed). Figure 23.3 shows the posterior inclusion probability of the environmental variables for each zone and functional form, that is, the overall posterior probability that each variable is in the model.

The model selected with the highest posterior probability (Model 1 in Table 23.1) coincided with the median-probability model, and it was used for posterior inference on risk factors and spatial structure, as well as for predictions. Model 1 includes the variable rainfall in linear form in the Sahelian zone, the day temperature in linear form, the area type in the flooded zone of the Niger delta, the NDVI as smooth term, and the area type in the Sudanian zone. The functional forms of the selected predictors are shown in Figure 23.4. Living in rural areas is associated with a reduction in the odds of being infected with malaria by 23%, 95% BCI (19%–41%) in the flooded zone and by 52%, 95% BCI (41%–63%) in the Sudanian zone. Rainfall was associated with a significant increase of malaria risk in the Sahelian zone (OR = 1.22, 95% BCI, 1.11–1.41). Day temperature was found to be the main risk factor in the Niger delta (OR = 1.66, 95% BCI, 1.49–1.86). Nonlinearity was detected in the relationship between NDVI and the malaria risk in the Sudanian zone.

To study the predictive ability of the model, we divided the data into a training set used to fit the model and a test set for evaluating predictions. The training set consists

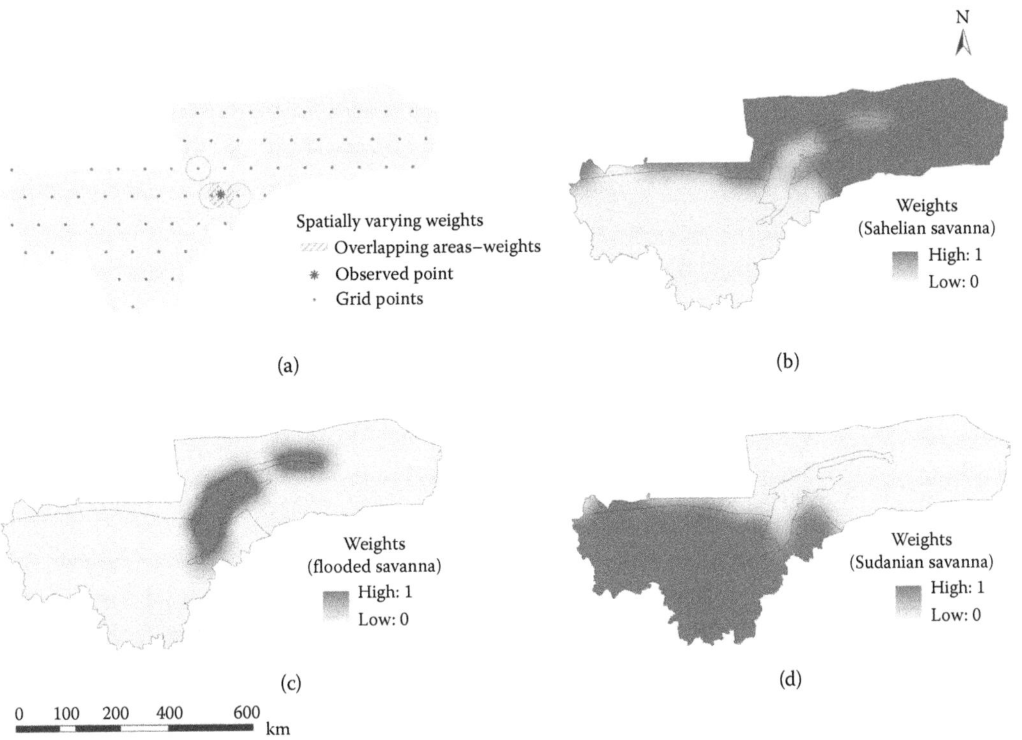

FIGURE 23.2
Spatially varying weights for an observed location and the three closest knots in each zone (a). Spatially varying weights for each prediction location: Sahelian zone (b), flooded zone (c), and Sudanian zone (d).

TABLE 23.1
Posterior model probabilities

Model	Mean structure	*p*
1	Rainfall (Sahelian zone, *linear*) + day temperature (flooded zone, *linear*) + area type (flooded zone, *piecewise constant*) + NDVI (Sudanian zone, *spline*) + area type (Sudanian zone, *piecewise constant*)	0.54
2	Rainfall (Sahelian zone, *linear*) + night temperature (flooded zone, *spline*) + area type (flooded zone, *piecewise constant*) + NDVI (Sudanian zone, *spline*) + area type (Sudanian zone, *piecewise constant*)	0.12
3	Rainfall (Sahelian zone, *spline*) + day temperature (flooded zone, *linear*) + area type (flooded zone, *piecewise constant*) + NDVI (Sudanian zone, *spline*) + area type (Sudanian zone, *piecewise constant*)	0.11

of 80% of survey locations randomly sampled for each ecological zone, and the test set includes the remaining points. The procedure was repeated five times (with five different training or testing sets), and the model predictive ability was assessed using a log-score criterion, defined as the negative log-likelihood evaluated at the testing locations. For the

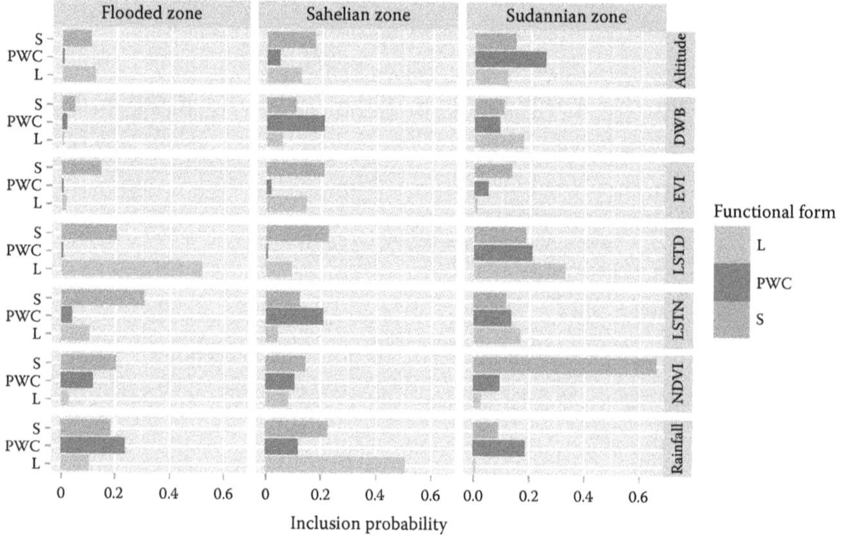

FIGURE 23.3
Inclusion probabilities per ecological zone and functional form (L = linear, PWC = piecewise constant, S = spline). DWB, LSTD and LSTN indicate distance to water bodies, day and night temperature, respectively.

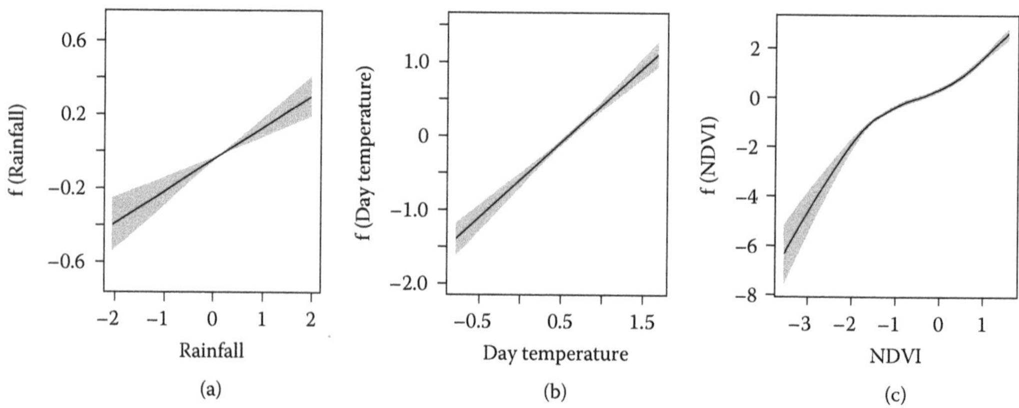

FIGURE 23.4
Estimated relationship between predictors and malaria risk in the three different ecological zones: (a) Sahelian zone, (b) flooded zone, and (c) Sudanian zone.

purpose of comparison, the model proposed by Gosoniu et al. (2009) (full B-spline model), as well as the same model with a stationary covariance matrix, was used for fitting and prediction on the same sets. Figure 23.5 compares the averaged log-likelihood between the three models. The plot indicates that Model 1 had a lower median log score and smaller variability.

The three models were used to perform spatial prediction throughout the study area at 2 km resolution. The predicted parasitemia risk in Figure 23.6 obtained using Model 1 suggests an overall trend of increasing risk from the north to the south. The regions with the highest risk are Sikasso and Segou in the south of the country and Kayes at the border

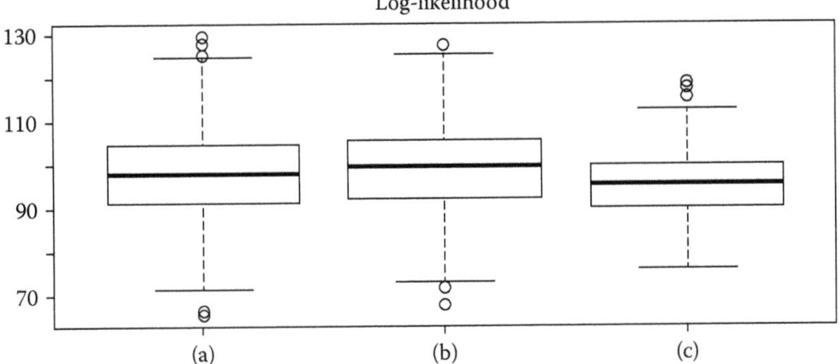

FIGURE 23.5
Log-score comparison between the full B-spline model, as in Gosoniu et al. (2009) (a), the full B-spline with a stationary covariance structure (b), and Model 1 (c).

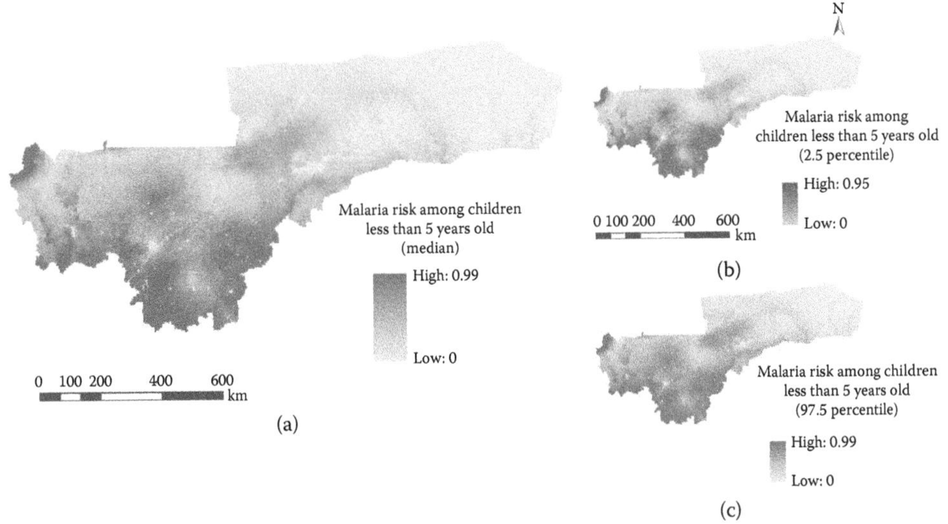

FIGURE 23.6
Predicted parasitemia risk in children under 5 years. Map produced using the nonstationary model (Model 1) with different predictors in each ecological zone and spatially varying weights. Median (a) and credible intervals (b and c).

with Senegal and Mauritania. Figures 23.7 and 23.8 show a similar geographical pattern but larger uncertainties. Moreover, the different mean structures in each ecological zone produce discontinuities at the borders in absence of spatially varying weights.

A sensitivity analysis was performed to study the effect of the spatially varying weights in the selection of covariates running MCMC under different specifications (different values of r, keeping fixed the spacing between the grid points g). Results are shown in Table 23.2 and expressed in terms of the ratio between the radius and the grid spacing. Under the three settings (radius smaller than the spacing, equal, or bigger), the model with the highest posterior probability remains the same, but the probabilities are different. In our analysis, we have defined the radius equal to the grid spacing. When the radius was smaller than the spacing, Model 1 was selected with a posterior probability of 0.51; very few points were affected by the covariates selected in the neighboring zones, and this resulted in discontinuities in

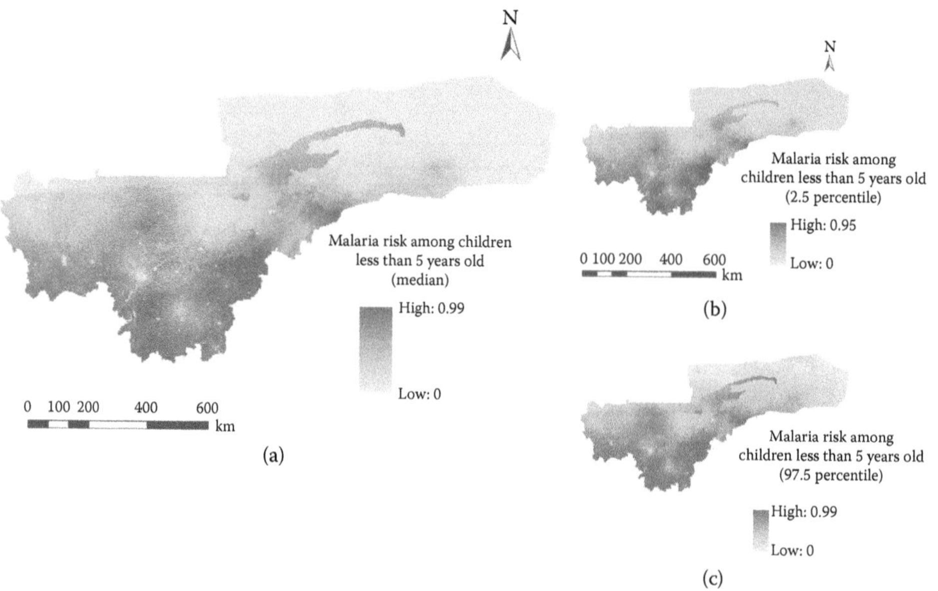

FIGURE 23.7
Predicted parasitemia risk in children under 5 years. Map produced using a nonstationary full B-spline model, as in Gosoniu et al. (2009). Median (a) and credible intervals (b and c).

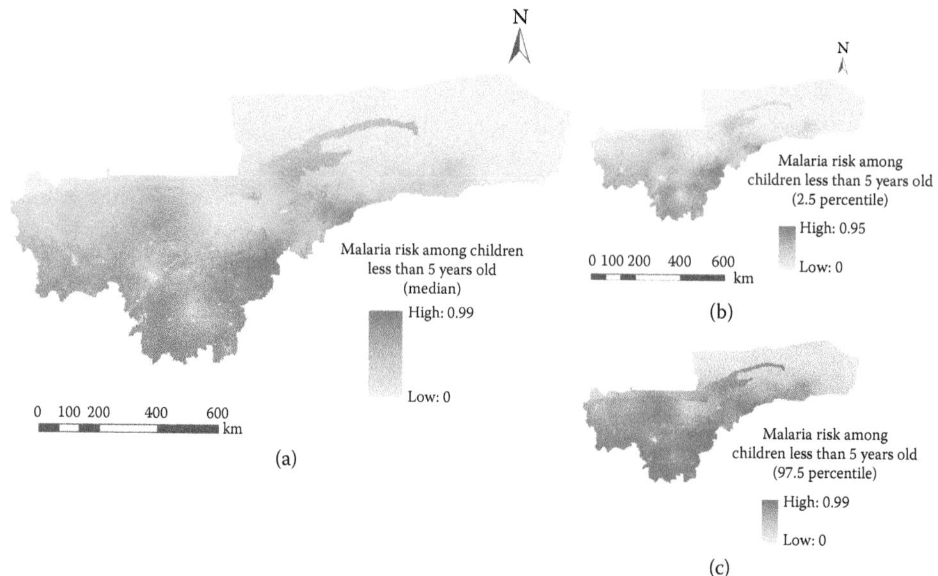

FIGURE 23.8
Predicted parasitemia risk in children under 5 years. Map produced using a full B-spline model with stationary covariance structure. Median (a) and credible intervals (b and c).

the prediction map. When the radius was higher than the spacing, Model 1 was selected with a low posterior probability (0.31), and several other competing models appeared to be selected with a probability of around 10%. This specification produced an oversmoothed prediction map.

TABLE 23.2
Posterior model probabilities of the first selected model with different values of the ratio between the radius and the grid spacing

Ratio	Sahelian zone	Flooded zone	Sudanian zone	p
r/g = 1	Rainfall (*linear*)	Day temperature (*linear*) Area type (*piecewise constant*)	NDVI (*spline*) Area type (*piecewise constant*)	0.54
r/g = 0.5	Rainfall (*linear*)	Day temperature (*linear*) Area type (*piecewise constant*)	NDVI (*spline*) Area type (*piecewise constant*)	0.51
r/g = 1.5	Rainfall (*linear*)	Day temperature (*linear*) Area type (*piecewise constant*)	NDVI (*spline*) Area type (*piecewise constant*)	0.31

23.5 Discussion

We have developed a Bayesian methodology to model nonstationary geostatistical data when the study area consists of irregularly shaped zones with different characteristics. The methods described allow the choice of covariates and their corresponding functional forms by zone via a Bayesian variable selection procedure. Spatially varying weights were used in the regression model to take into account the dependence of the covariates affecting the disease outcome at a given location not only on the zone associated with the location, but also on the neighboring regions within a certain radius. The weights introduced into the model smooth the predicted surface at the borders of the zones. Modeling a nonstationary spatial process enables the incorporation of the heterogeneity generated by effects of covariates, as well as unmeasured factors that vary geographically in the study area.

The choice of the radius might influence posterior inference even though the weights were normalized. In particular, a large radius could lead to oversmoothing, while a small one may introduce discontinuities. In our model formulation, the radius was fixed during the estimation process; alternatively, it could be considered a parameter estimated by the data.

Our modeling approach shares similarities with other approaches to model nonstationarity, such as spatially varying coefficient (SVC) models (Gelfand et al., 2003) and the geographically weighted regression (GWR) (Fotheringham et al., 2003). GWR is commonly seen as a descriptive approach that uses spatial weights to estimate spatially adaptive coefficients, whereas SVC places either a univariate or multivariate spatial process on those regression coefficients that are thought to vary spatially (Finley, 2011). GWR has recently been shown to produce biased estimates, and its application is not straightforward for generalized models. SVC offers a richer inferential framework at the cost of being computationally demanding. Moreover, identifiability issues may arise from the estimation of several spatially structured covariates.

Our model can be easily implemented in standard software for Bayesian inference (e.g., BUGS) and allows a parsimonious model definition, leading to best predictive performance. The model in its current formulation does not take into account potential interaction terms between the covariates. The variable selection procedure can be easily extended to identify important interactions.

A natural field of application of the proposed methods is that of spatial epidemiology of environmentally driven diseases, where the study area is often large, contains different ecological zones, and the effects of predictors may depend on the zone. Our example is focused on a study of malaria risk in Mali. The malaria endemic area in the country is divided into three different ecological zones. Malaria transmission is influenced by suitable rainfall and temperature that affect mosquito survival and longevity, and therefore contribute to abundance of the mosquito population. The Bayesian variable selection procedure identified the most important environmental predictors of parasitemia risk in each ecological zone. These predictors had meaningful biological interpretation. In particular, our analysis showed that in the Sahel, where the amount of precipitation is very low, an increase in the amount of rainfall was associated with an increase in malaria risk. In the flooded region in the center of the country, temperature was the most important predictor. In the Sudanian ecological zone, vegetation index, which is a proxy of humidity, was identified as the main factor affecting the disease risk. Malaria in Africa is present in both rural and urban areas (Machault et al., 2011), but as confirmed by our analysis, levels of transmission in urban areas are usually lower than those in peri-urban and rural places. The estimated malaria prevalence map identified high-risk areas in the center (Sigou region) and south of the country (Sikasso region).

Earlier mapping efforts of malaria risk in Mali are based on compilation of historical survey data. Our results are consistent with the ones obtained by Gemperli et al. (2006a, 2006b). A similar pattern is also observed comparing our map with the one produced by Gosoniu et al. (2009), with the exception of the parts of the country at the border with Burkina Faso and Côte d'Ivoire. The map of Mali produced by the Malaria Atlas Project (Hay et al., 2009) shows similar values of predicted risk in the areas of Sikasso and Sigou, but much lower values in Kayes and in the southeast region.

The proposed methodology improves disease risk prediction over large areas compared to commonly used stationary geostatistical models. The described models can be used to address the current needs of international agencies (e.g., World Health Organization and the Global Fund) that are interested in global atlases of infectious disease burden and estimates of the required amount of preventive and curative treatments.

Acknowledgments

The work was supported by ERC grant 323180-IMCCA and the Swiss National Science Foundation (SNSF) Swiss Programme for Research on Global Issues for Development (R4D) project no. 1Z01Z0-147286. The authors thank Measure DHS for providing the malaria data. Thanks also to Dr. Seydou Doumbia and Dr. Mahamadou Diakité for their helpful comments.

References

Banerjee, S., Gelfand, A., Knight, J., Sirmans, C. 2004. Spatial modeling of house prices using normalized distance-weighted sums of stationary processes. *Journal of Business and Economic Statistics* 22, 206–213.

Barbieri, M., Berger, J. 2004. Optimal predictive model selection. *Annals of Statistics* 32, 870–897.

Bauwens, I., Franke, J., Gebreslasie, M. 2012. Malareo—earth observation to support malaria control in Southern Africa. In *IEEE International Geoscience and Remote Sensing Symposium*, Munich, Germany, pp. 3–6.

Bové, D. S., Held, L., Kauermann, G. 2012. Mixtures of g-priors for generalised additive model selection with penalised splines. http://arxiv.org/pdf/1108.3520.pdf.

Chammartin, F., Scholte, R. G., Guimarães, L. H., Tanner, M., Utzinger, J., Vounatsou, P. 2013. Soil-transmitted helminth infection in South America: a systematic review and geostatistical meta-analysis. *Lancet Infectious Diseases* 13 (6), 507–518.

Chen, Z., Dunson, D. B. 2003. Random effects selection in linear mixed models. *Biometrics* 59, 762–769.

Clements, A., Garba, A., Sacko, M., Touré, S., Dembelé, R., Landouré, A., Bosque-Oliva, E., Gabrielli, A. F., Fenwick, A. 2008. Mapping the probability of schistosomiasis and associated uncertainty, West Africa. *Emerging Infectious Diseases* 14, 1629–1632.

Crainiceanu, C., Diggle, P., Rowlingson, B. 2008. Bivariate binomial spatial modeling of loa loa prevalence in tropical Africa. *Journal of the American Statistical Association* 103, 21–37.

Curtis, S., Banerjee, S., Ghosal, S. 2014. Fast Bayesian model assessment for nonparametric additive regression. *Computational Statistics and Data Analysis* 71, 347–358.

Diggle, P., Tawn, J., Moyeed, R. 1998. Model-based geostatistics. *Applied Statistics* 47, 299–350.

Diggle, P., Thomson, M., Christensen, O., Rowlingson, B., Obsomer, V., Gardon, J., Wanji, S., et al. 2007. Spatial modelling and the prediction of loa loa risk: decision making under uncertainty. *Annals of Tropical Medicine and Parasitology* 101, 499–509.

ESRI. 2011. ArcGIS Desktop: Release 10. Redlands, CA: Environmental Systems Research Institute.

FAO (Food and Agriculture Organization). 2000. Global forest resources assessment. http://www.fao.org/forestry/fra/2000/report/en/.

Finley, A. O. 2011. Comparing spatially-varying coefficients models for analysis of ecological data with non-stationary and anisotropic residual dependence. *Methods in Ecology and Evolution* 2 (2), 143–154.

Fotheringham, A. S., Brunsdon, C., Charlton, M. 2003. *Geographically Weighted Regression: The Analysis of Spatially Varying Relationships*. New York: John Wiley & Sons.

Fuentes, M. 2001. A new high frequency kriging approach for nonstationarity environmental processes. *Environmetrics* 12, 469–483.

Gelfand, A. E., Banerjee, S., Carlin, B. P. 2004. *Hierarchical Modeling and Analysis for Spatial Data*. Boca Raton, FL: Chapman & Hall.

Gelfand, A. E., Sirmans, H. K. K. C. F., Banerjee, S. 2003. Spatial modelling with spatially varying coefficient processes. *Journal of the American Statistical Association* 98, 387–396.

Gelman, A., Dyk, D. A. V., Huang, Z., Boscardin, W. J. 2008. Using redundant parameterizations to fit hierarchical models. *Journal of Computational and Graphical Statistics* 17, 95–122.

Gemperli, A., Sogoba, N., Fondjo, E., Mabaso, M., Bagayoko, M., Briët, O. J. T., Anderegg, D., Liebe, J., Smith, T., Vounatsou, P. 2006a. Mapping malaria transmission in West and Central Africa. *Tropical Medicine and International Health* 11, 1032–1046.

Gemperli, A., Vounatsou, P. 2006. Strategies for fitting large, geostatistical data in MCMC simulation. *Communications in Statistics—Simulation and Computation* 35, 331–345.

Gemperli, A., Vounatsou, P., Sogoba, N., Smith, T. 2006b. Malaria mapping using transmission models: application to survey data from Mali. *American Journal of Epidemiology* 163, 289–297.

Giardina, F., Gosoniu, L., Konate, L., Diouf, M. B., Perry, R., Gaye, O., Faye, O., Vounatsou, P. 2012. Estimating the burden of malaria in Senegal: Bayesian zero-inflated binomial geostatistical modeling of the MIS 2008 data. *PLoS One* 7, e32625.

Gosoniu, L., Veta, A. M., Vounatsou, P. 2010. Bayesian geostatistical modeling of Angola malaria indicator survey data. *PLoS One* 5, e9322.

Gosoniu, L., Vounatsou, P., Sogoba, N., Maire, N., Smith, T. 2009. Mapping malaria risk in West Africa using a Bayesian nonparametric non-stationary model. *Computational Statistics and Data Analysis* 53, 3358–3371.

Gosoniu, L., Vounatsou, P., Sogoba, N., Smith, T. 2006. Bayesian modelling of geostatistical malaria risk data. *Geospatial Health* 1, 127–139.

Hay, S., Guerra, C., Gething, P. W., Patil, A. P., Tatem, A. J., Noor, A. M., Kabaria, C. W., et al. 2009. A world malaria map: *Plasmodium falciparum* endemicity in 2007. *PLoS One* 6, e1000048.

Higdon, D. 1998. A process-convolution approach to modeling temperatures in the north Atlantic Ocean. *Journal of Environmental and Ecological Statistics* 5, 173–190.

Hoeting, J. A., Davis, R., Merton, A., Thomspon, S. 2006. Model selection for geostatistical models. *Ecological Applications* 16, 87–98.

Kinney, S. K., Dunson, D. B. 2007. Fixed and random effects selection in linear and logistic models. *Biometrics* 63, 690–698.

Lawson, A. 2013. Bayesian Disease Mapping: Hierarchical Modeling in Spatial Epidemiology, 2nd ed. Boca Raton, FL: Chapman & Hall/CRC Interdisciplinary Statistics, Taylor & Francis Group.

Machault, V., Vignolles, C., Borchi, F., Vounatsou, P., Pages, F., Briolant, S., Lacaux, J., Rogier, C. 2011. The use of remotely sensed environmental data in the study of malaria. *Geospatial Health* 5, 151–168.

Ministère de la Santé, Programme National de Lutte contre le Paludisme, INFO-STAT, ICF Macro. 2010. Enquête sur la prévalence de l'anémie et de la parasitémie palustre chez les enfants au Mali. Technical report, Bamako, Mali, and Calverton, MD: Ministère de la Santé.

Noor, A. M., Kinyoki, D. K., Mundia, C. W., Kabaria, C. W., Mutua, J. W., Alegana, V. A., Fall, I. S., Snow, R. W. 2014. The changing risk of *Plasmodium falciparum* malaria infection in Africa: 2000–10: a spatial and temporal analysis of transmission intensity. *Lancet* 383, 1739–1747.

Nychka, D., Wikle, C., Royle, J. 2002. Multiresolution models for nonstationary spatial covariance functions. *Statistical Modelling* 2, 315–331.

O'Hara, R., Sillanpää, M. J. 2009. A review of Bayesian variable selection methods: what, how and which. *Bayesian Analysis* 4, 85–118.

Plummer, M. 2003. JAGS: A program for analysis of Bayesian graphical models using Gibbs sampling. In *Proceedings of the 3rd International Workshop on Distributed Statistical Computing* (eds. Hornik K, Leisch F, and Zeileis A), pp. 20–22, Vienna, Austria.

Pullan, R. L., Bethony, J. M., Geiger, S., Cundill, B., Correa-Oliveira, R., Quinnell, R. J., Brooker, S. 2008. Human helminth co-infection: analysis of spatial patterns and risk factors in a Brazilian community. *PLoS Neglected Tropical Disease* 6, e352.

Raso, G., Matthys, B., N'Goran, E. K., Tanner, M., Vounatsou, P., Utzinger, J. 2005. Spatial risk mapping and prediction of *Schistosoma mansoni* infections among schoolchildren living in western Côte d'Ivoire. *Parasitology* 131, 97–108.

Raso, G., Vounatsou, P., Gosoniou, L., Tanner, M., N'Goran, E. K., Utzinger, J. 2006a. Risk factors and spatial patterns of hookworm infection among schoolchildren in a rural area of western Cote d'Ivoire. *International Journal for Parasitology* 36, 201–210.

Raso, G., Vounatsou, P., Singer, B. H., N'Goran, E. K., Tanner, M., Utzinger, J. 2006b. An integrated approach for risk assessment and spatial prediction for *Schistosoma mansoni*-hookworm co-infection. *Proceedings of the National Academy of Sciences of the United States of America* 103, 6934–6939.

Reich, B. J., Fuentes, M., Herring, A. H., Evenson, K. R. 2010. Bayesian variable selection for multivariate spatially varying coefficient regression. *Biometrics* 66, 772–782.

Ruppert, D., Wand, M., Carroll, R. 2003. *Semiparametric Regression.* Cambridge: Cambridge University Press.

Sampson, P., Guttorp, P. 1992. Nonparametric estimation of nonstationary spatial covariance structure. *Journal of the American Statistical Association* 87, 108–119.

Sampson, P. D. 2010. Constructions for nonstationary spatial processes. In A. E. Gelfand, P. Diggle, P. Guttorp, M. Fuentes (eds.), *Handbook of Spatial Statistics.* Boca Raton, FL: CRC Press, pp. 119–130.

Scheel, I., Ferkingstad, E., Frigessi, A., Haug, O., Hinnerichsen, M., Meze-Hausken, E. 2013. A Bayesian hierarchical model with spatial variable selection: the effect of weather on insurance claims. *Journal of the Royal Statistical Society: Series C (Applied Statistics)* 62, 85–100.

Scheipl, F., Fahrmeir, L., Kneib, T. 2012. Spike-and-slab priors for function selection. *Journal of the American Statistical Association* 107, 1518–1532.

Schlagenhauf-Lawlor, P. 2008. *Travelers' Malaria.* Shelton, CT: PMPH-USA.

Schur, N., Gosoniu, L., Raso, G., Utzinger, J., Vounatsou, P. 2011. Modelling the geographical distribution of co-infection risk from single-disease surveys. *Statistics in Medicine* 30, 1761–1776.

Smith, M., Fahrmeir, L. 2007. Spatial Bayesian variable selection with application to functional magnetic resonance imaging. *Journal of the American Statistical Association* 102, 417–431.

Tatem, A. J., Noor, A. M., Von Hagen, C., Di Gregorio, A., Hay, S. I. 2007. High resolution population maps for low income nations: combining land cover and census in East Africa. *PLoS One* 2, e1298.

Tüchler, R. 2008. Bayesian variable selection for logistic models using auxiliary mixture sampling. *Journal of Computational and Graphical Statistics* 17, 76–94.

Wagner, H., Duller, C. 2012. Bayesian model selection for logistic regression models with random intercept. *Computational Statistics and Data Analysis* 56, 1256–1274.

Wang, X. H., Zhou, X. N., Vounatsou, P., Chen, Z., Utzinger, J., Yang, K., Steinmann, P., Wu, X. H. 2008. Bayesian spatio-temporal modeling of *Schistosoma japonicum* prevalence data in the absence of a diagnostic gold standard. *PLoS Neglected Tropical Diseases* 2, e250.

Zhao, Y., Staudenmayer, J., Coull, B. A., Wand, M. P. 2006. General design Bayesian generalized linear mixed models. *Statistical Science* 21, 35–51.

24

Computational Issues and R *Packages for Spatial Data Analysis*

Marta Blangiardo
MRC-PHE Centre for Environment and Health
Department of Epidemiology and Biostatistics
Imperial College London
London, UK

Michela Cameletti
Department of Management, Economics and Quantitative Methods
University of Bergamo
Bergamo, Italy

CONTENTS

As seen in previous chapters, several types of models are used with spatial data, depending on the aim of the study. If we are interested in summarizing spatial variation between areas using risks or probabilities, then we could rely on statistical methods such as disease mapping to compare maps and identify clusters (see Ugarte et al. 2005 and Chapters 3 and 28 in this book). Moran's Index is extensively used to check for spatial autocorrelation (Moran 1950), while the scan statistics, implemented in SaTScan (Kulldorf 1997), performs cluster detection and geographical surveillance. The same type of tools can also be adopted in studies where there is an aetiological aim to assess the potential effect of risk factors on outcomes (see Chapter 5).

A different type of study considers the quantification of the risk of experiencing an outcome as the distance from a certain source increases. This typically is framed in an environmental context, so that the source could be a point (e.g., waste site or radio transmitter) or a line (e.g., power line or road). In this case, the methods commonly used vary from nonparametric tests, proposed by Stone (1988), to the parametric approach introduced by Diggle et al. (1998). See Chapter 14 for a detailed description of such studies.

In a different context, when the interest lies in mapping continuous spatial variables that are measured only at a finite set of specific points in a given region, and in predicting their values at unobserved locations, geostatistical methods—such as kriging—are employed (Cressie 1991; Stein 1999; see also Chapter 11). This may play a significant role in environmental risk assessment in order to identify areas characterized by high risk of exceeding potentially harmful thresholds.

This chapter aims to present the computational resources available for the analysis of spatial data. We focus on the `R` environment, which includes several packages for spatial analysis, and briefly recall the theory behind these, as well as present an example for the most used ones.

24.1 `R`

`R` is a statistical computer program that provides a rich environment for data analysis and graphics. It can be freely downloaded from the website http://www.r-project.org/ under the GNU general public license. Its installation is straightforward, and we will assume that the reader is familiar with its basics features, which can be found in Crawley (2007), Dalgaard (2008), Chambers (2008), Everitt and Hothorn (2009), and Zuur et al. (2012). Moreover, the literature is rich with books devoted to applications with `R`, such as, for example, time series analysis and econometrics (Kleiber and Zeileis 2008; Cowpertwait and Metcalfe 2009; Shumway and Stoffer 2011), ecology (Stevens 2009; Zuur et al. 2009), Bayesian inference (Albert 2009), geostatistics (Bivand et al. 2013), epidemiology and biostatistics (Peng and Dominici 2008; Logan 2010).

On `CRAN` (Comprehensive `R` Archive Network), a `task view` page regarding spatial statistical analysis is maintained; see http://cran.r-project.org/web/views/Spatial.html. It contains a list and descriptions of packages to manage spatial data, to import and export data from other software (e.g., geographic information systems [GIS]), and to implement models in the context of geostatistics, point pattern analysis, and disease mapping.

In the following section, we describe the main `R` packages for visualizing spatial data.

24.1.1 Reading and visualizing spatial data

The package `sp` (Pebesma and Bivand 2005; Bivand et al. 2013) was created in order to set a new reference set of classes and functions able to deal with all possible kinds of spatial data (i.e., points, lines, polygons, and grids). Type `demo(gallery)` in the R console for a list of all the possible plots that can be obtained with `sp`.

24.1.1.1 Point-referenced data

Now we present how to visualize point-referenced data using as an example the `meuse` dataset contained in the `sp` package:

```
> library(sp)
> data(meuse)
> colnames(meuse)

 [1] "x"       "y"       "cadmium" "copper"  "lead"    "zinc"    "elev"
 [8] "dist"    "om"      "ffreq"   "soil"    "lime"    "landuse" "dist.m"
```

The `meuse` object is a `data.frame` including coordinates (`x` and `y`) and some variables regarding metal concentrations in soil and landscape features observed at 155 locations near the Meuse River in the Netherlands. In order to plot the data, we need to change the `data.frame` to a `SpatialPointsDataFrame` object (one of the classes of `sp`), setting the spatial coordinates as follows:

```
> coordinates(meuse) <- c("x", "y")
> class(meuse)

[1] "SpatialPointsDataFrame"
attr(,"package")
[1] "sp"
```

The command `summary(meuse)` can be used to retrieve the spatial information of the `meuse` object and obtain summaries of the variables included in it.

The `spplot` function is the main method for plotting spatial data, and we use it to produce a colored points plot of the `zinc` variable, split in four groups using cutoffs given by quartiles and specified in the `cuts` option:

```
> q.cuts <- quantile(meuse$zinc)
> spplot(meuse, zcol="zinc", cuts=q.cuts,
        scales=list(draw=T), #to draw axes scale
        key.space=list(x=0.2,y=0.9), #to set the key position
        col.regions=gray(seq(0.8,0,length=4))) #change colour scale
```

The resulting plot is displayed in Figure 24.1a. Note that in the code, the argument `col.regions` is used to change the default point colors (here we adopt a gray color scale with four levels). It is also possible to specify more variables at the same time (e.g., `spplot(meuse, zcol=c("cadmium", "copper", "lead", "zinc"),...)`): in this case, the output is a multipanel plot with four maps and a shared legend.

For the same spatial domain, spatial coordinates regarding the Meuse River are available as a matrix named `meuse.riv`. To include it in the plot (see Figure 24.1b), we need to define

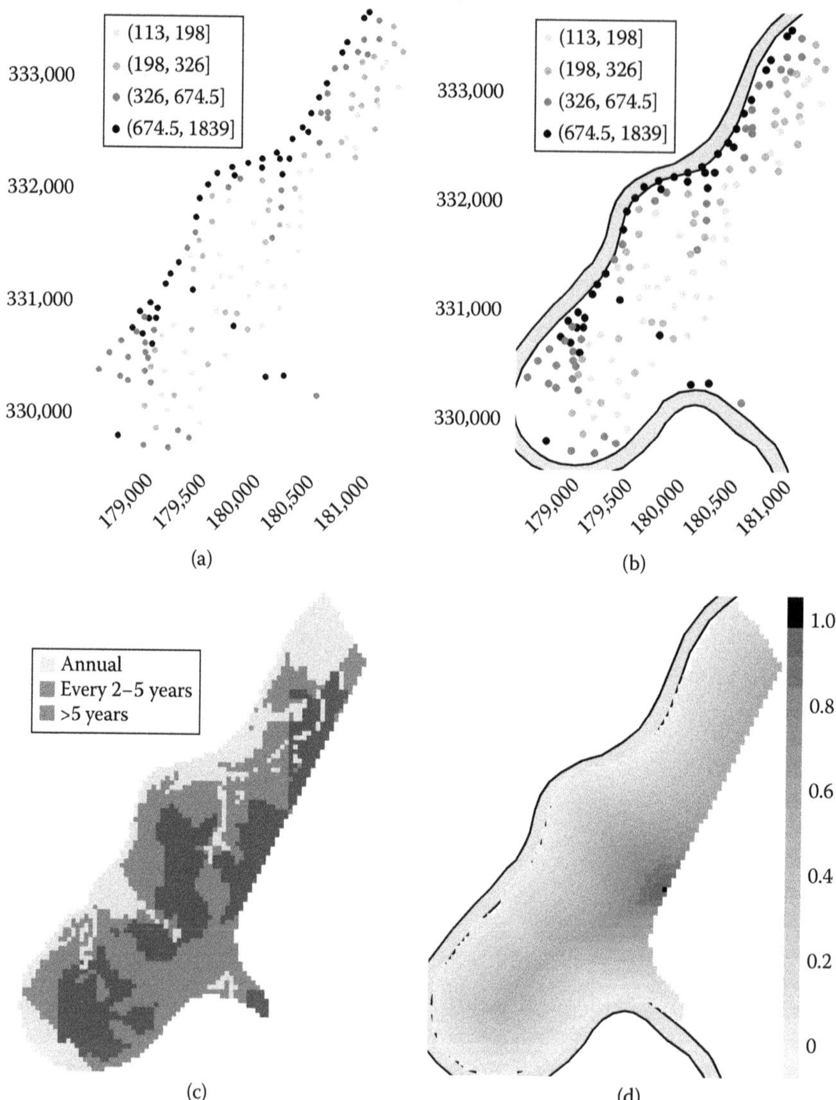

FIGURE 24.1
Maps for the Meuse case study: zinc concentration (a), zinc concentration with the inclusion of the Meuse River (b), flooding frequency class (c), and distance to the Meuse River (d).

an object of class `SpatialPolygons` using the following code:

```
> data(meuse.riv)
> meuse.sr <- SpatialPolygons(list(Polygons(list(Polygon(meuse.riv)),
                                              ID="meuse.riv")))
> river <- list("sp.polygons", meuse.sr, fill="grey")
> spplot(meuse, zcol="zinc", cuts=q.cuts, scales=list(draw=T),
        key.space=list(x=0.2,y=0.9), sp.layout = river,
        col.regions=gray(seq(0.8,0,l=4)))
```

where the `sp.layout` argument includes a list with the additional layout required for the river.

24.1.1.2 Lattice data

Alternatively, we might have spatial data available for a regular grid of points. For example, meuse.grid is a dataframe referring to a grid of 40×40 m that can be loaded using the following command:

```
> data(meuse.grid)
```

and which can be changed to a sp object of class SpatialPixelsDataFrame through

```
> coordinates(meuse.grid) <- ~x+y # or c("x","y")
> gridded(meuse.grid) <- TRUE
> class(meuse.grid)

[1] "SpatialPixelsDataFrame"
attr(,"package")
[1] "sp"

> names(meuse.grid)

[1] "part.a" "part.b" "dist"   "soil"   "ffreq"
```

The following code is used to plot the flooding frequency class (available as a three-level factor variable named ffreq) using the image function of the graphics package as an alternative to the spplot function:

```
> library(graphics)
> meuse.grid$ffreq.num <- as.numeric(meuse.grid$ffreq)
> image(meuse.grid,"ffreq.num",col=gray(seq(0.8,0,l=3)), axes=F)
> labels = c("annual", "every 2-5 years", "> 5 years")
> legend(x=179000,y=332900,legend=labels,fill=gray(seq(0.8,0,l=3)),bty="n")
```

and the resulting plot is reported in Figure 24.1c.

It is also interesting to plot a continuous variable such as distance to the Meuse River (dist in meuse.grid) using the spplot function:

```
> spplot(meuse.grid,zcol="dist",col.regions=gray(seq(0.8,0,l=64)),
         sp.layout = river)
```

Note that the resulting plot, which is displayed in Figure 24.1d, is characterized by a continuous scale of colors.

24.1.1.3 Area data

When the data are available at irregular grids (e.g., administrative areas), typically they are stored as polygons in a shapefile (a format used in GIS software) and can be imported in R using the readShapePoly[1] function of the maptools library, which depends on sp. As an example, we consider lung cancer mortality in Ohio counties for 1988. The data

[1]This function can be used to import shapefiles for point-referenced and lattice data as well, but we did not show this here, as we used the Meuse dataset already included in the sp package.

were originally presented in Lawson (2001) and are stored in the DataLung.csv file, which consists of a matrix with 88 rows (as the number of counties) and 6 columns with the name and the ID of the county (NAME and ID), the number of deaths (y), the number of exposed individuals (n), and the expected number of deaths (e). The data and the shapefile can be loaded in R as follows:

```
> dataDM <- read.csv("DataLung.csv",sep="")
> library(maptools)
> ohio <- readShapePoly("DataLung.shp")
```

The object ohio belongs to the SpatialPolygonsDataFrame class and includes the following slots:

```
> slotNames(ohio)

[1] "data"          "polygons"    "plotOrder"    "bbox"    "proj4string"
```

where the data can be accessed using ohio@data or attr(ohio,"data"). Figure 24.2 displays the rate of mortality for 100,000 people (computed as $p = y/n \times 100{,}000$) and classified in four classes using the following code:

```
> p <- dataDM$y/dataDM$n*100000 #mortality rate
> p.cutoff <- c(0.0, 30, 40, 50, 130) #cutoffs
> p.factor <- cut(p,breaks=p.cutoff,include.lowest=TRUE)
```

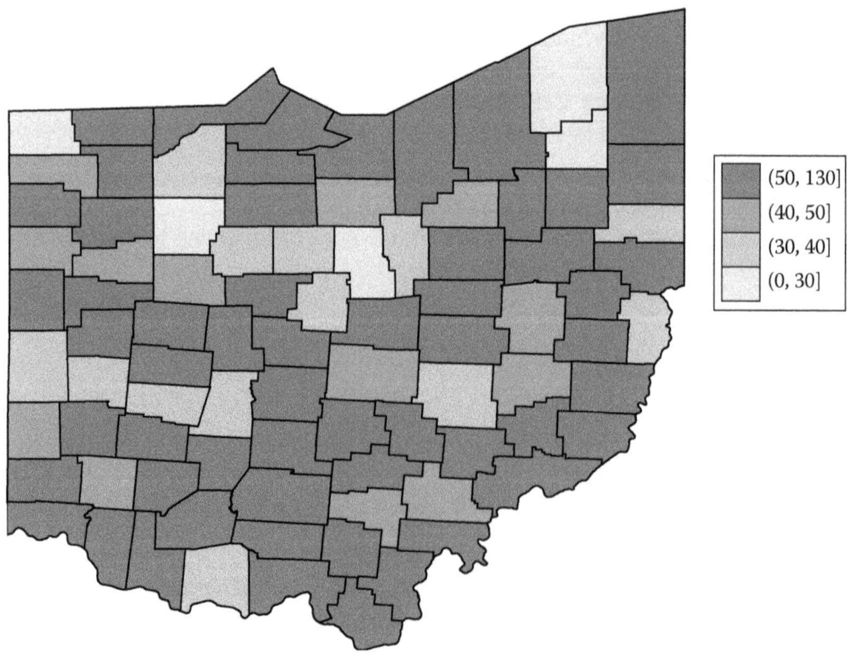

FIGURE 24.2
Distribution of mortality rates of lung cancer for 100,000 people in 88 counties in Ohio in 1988.

The discretized rate of mortality (`p.factor`) is added to the original `data.frame` (`ohio@data`) with

```
> ohio@data <- data.frame(ohio@data, p=p.factor)
```

and finally, the `spplot` function is run to draw the map depicted in Figure 24.2, which represents the mortality rates for the 88 counties in Ohio.

```
> spplot(obj=ohio, zcol= "p",col.regions=gray(3.5:0.5/4), main="")
```

24.2 Classical and Likelihood-Based Approach for Spatial Analyses

Spatial data are defined as realizations of a stochastic process indexed by space

$$Y(\boldsymbol{s}) \equiv \{y(\boldsymbol{s}), \boldsymbol{s} \in \mathcal{D}\},$$

where $\mathcal{D}$ is a (fixed) subset of $\mathbb{R}^d$ (here we consider $d = 2$). The actual data can then be represented by a collection of observations $\boldsymbol{y} = \{y(\boldsymbol{s}_1), \ldots, y(\boldsymbol{s}_m)\}$, where the set $(\boldsymbol{s}_1, \ldots, \boldsymbol{s}_m)$ indicates the spatial locations at which the measurements are taken. Depending on $\mathcal{D}$ being a continuous surface or a countable collection of d-dimensional spatial units, the problem can be specified as a spatially continuous or discrete random process, respectively (Gelfand et al. 2010).

For example, we can consider a collection of air pollutant measurements obtained by monitors located in the set $(\boldsymbol{s}_1, \ldots, \boldsymbol{s}_m)$ of m *points*. In this case, $\boldsymbol{y}$ is a realization of the air pollution process that changes continuously in space, and we usually refer to it as geostatistical or point-referenced data. Alternatively, we may be interested in studying the spatial pattern of a certain health condition observed in a set $(\boldsymbol{s}_1, \ldots, \boldsymbol{s}_m)$ of m *areas* (rather than points), defined, for example, by census tracts or counties; in this case, $\boldsymbol{y}$ may represent a suitable summary, for example, the number of cases observed in each area.

24.2.1 Geostatistical data

When the variable of interest $y(\boldsymbol{s})$ is observed at m locations $(\boldsymbol{s}_1, \ldots, \boldsymbol{s}_m)$ in the considered domain, the following simple model with a latent continuous process may be assumed

$$y(\boldsymbol{s}) = \mu(\boldsymbol{s}) + \xi(\boldsymbol{s}) + \epsilon(\boldsymbol{s}), \tag{24.1}$$

where $\mu(\boldsymbol{s})$ is the so-called large-scale component, defined by a polynomial trend surface or by some covariates, and $\xi(\boldsymbol{s})$ is a zero-mean Gaussian spatial process commonly assumed to be stationary with covariance function $Cov(\xi(\boldsymbol{s}), \xi(\boldsymbol{s}'))$, which depends only on the distance between the locations, for each $\boldsymbol{s}$ and $\boldsymbol{s}'$. Finally, $\epsilon(\boldsymbol{s})$ represents the measurement error, and its variance is usually known as a nugget effect (see Chapter 11 for more details).

The analysis of point-referenced data can be performed using the so-called classical geostatistical approach or, if independent Gaussian distributions are assumed for $\xi(\boldsymbol{s})$ and $\epsilon(\boldsymbol{s})$, a likelihood-based perspective. In the classical geostatistical approach, the empirical variogram is used as an exploratory tool and the mean and the covariance structure parameters (basically, the spatial variance and the decay parameter) are estimated through least-squares methods, usually adopting a two-step procedure (first, the mean is estimated, and then residuals are used to make inference on the spatial parameters). Then the estimated parameters are plugged in (thus ignoring parameter uncertainty) and kriging is performed; that is, prediction is computed in new spatial locations, basically as a linear combination of the observed values. The geostatistical approach is very simple in principle, but it provides nonoptimal estimators. From this point of view, the likelihood principle is preferable, even if it requires extensive computations, especially for inverting iteratively the spatial covariance matrix defined by the covariance function $Cov(\xi(\boldsymbol{s}), \xi(\boldsymbol{s}'))$.

The most suited R packages, for these tasks are geoR (Ribeiro and Diggle 2001; Diggle and Ribeiro 2007) and gstat (Pebesma 2004). The package geoR is based on its own class of spatial data, named geodata, which are defined as a list with coords and data as obligatory arguments. The main functions for variogram estimation and kriging are variog, variofit, likfit, and krige.conv. The geoRglm package extends geoR for implementing Binomial and Poisson generalized linear geostatistical models. The drawback of geoR is that it is completely written in R, and thus it becomes infeasible when moderately large spatial datasets are analyzed.

The package gstat (Bivand et al. 2013) is an interface to the original stand-alone C program (Pebesma and Wesseling 1998), which performs geostatistical modeling, prediction, and simulation. In particular, gstat can calculate sample variograms (function variogram), estimate the variogram parameters using the least-squares method (fit.variogram), and perform kriging (krige) or cokriging for multiple correlated variables (function predict).

24.2.2 Area-level data

When the data are counts at the area level, it is possible to define the variable of interest as

$$y(\boldsymbol{s}) \mid \lambda(\boldsymbol{s}) \sim \text{Poisson}(\lambda(\boldsymbol{s})) \tag{24.2}$$

$$\log(\lambda(\boldsymbol{s})) = \alpha + \log(e(\boldsymbol{s})) + \nu(\boldsymbol{s}) + \upsilon(\boldsymbol{s}), \tag{24.3}$$

where α is the global intercept; $\log(e(\boldsymbol{s}))$ is the logarithm of the expected counts, which typically is used to adjust for the population structure of the area (e.g., age and sex) and is obtained using a standardization method (Elliott et al. 2000); $\nu(\boldsymbol{s})$ is the unstructured area effect and $\upsilon(\boldsymbol{s})$ is the area effect modeled through a spatial structure (loosely corresponds to the $\xi(\boldsymbol{s})$ in Equation 24.1). Here, the spatial similarity is obtained through a neighborhood structure that could be based on shared boundaries or distance, instead of the covariance function as seen above (see Chapter 6 for a detailed description).

Full maximum likelihood inference for this model requires numerical integration methods. A computationally effective alternative is given by the penalized quasi-likelihood (PQL), a method that simplifies the maximum likelihood approach by using a Laplace approximation to the quasi-likelihood (Breslow and Clayton 1993). As described in Dean et al. (2004) and Ugarte et al. (2008), the PQL method provides good point estimates for the Poisson models with spatially correlated components, even if it underestimates the variability of the random effects because the uncertainty associated with the variance estimation is not taken into account.

When the aim of the analysis of area-level data is to discover a potential cluster of diseases (or deaths), for example neighboring areas with a risk of experiencing an outcome higher than the average, the R package DCluster can be used. It includes a collection of tests to assess the presence of clusters at different levels: if we are interested in evaluating the homogeneity of the relative risks across the study region, without specifying a spatial structure, then Pearson chi-square (implemented through the R function achisq) or Potthoff–Whittinghill's test (Potthoff and Whittinghill 1966) (pottwhitt) can be used.

To assess the degree of spatial autocorrelation of the data, indices such as Moran's I (Moran 1948) or Geary's (Geary 1954), can be computed through the `moranI` and `gearyc` functions, respectively. Scan statistics such as `besagnewell` (Besag and Newell 1991) and `kullnagar` (Kulldorff and Nagarwalla 1995) are used to evaluate the presence of small-area clusters, while Stone's test (`stone.test`) is used to identify a single region and test the presence of clustering around it at several distances (Stone 1988). Finally, for each statistic described, bootstrap can be used to obtain its sampling distributions so that p-values can be derived.

24.2.3 Limitations

The main disadvantage of these classical frequentist approaches arises when hierarchical models are required to account for complex processes, multiple sources of data, and different sources of uncertainty. In this case, the Bayesian perspective, which is described in the next section, is more suitable, as it easily allows us to specify a hierarchical structure on the data or parameters, with the added benefit of making prediction for new observations and missing data imputation relatively straightforward. Moreover, in the Bayesian framework, the specification of prior distributions allows the formal inclusion of information that can be obtained through previous studies or from expert opinion.

24.3 Bayesian Approach

In the Bayesian approach inference is based on Bayes' theorem:

$$p(\boldsymbol{\theta} \mid \boldsymbol{y}) = \frac{p(\boldsymbol{y} \mid \boldsymbol{\theta}) \times p(\boldsymbol{\theta})}{p(\boldsymbol{y})}, \tag{24.4}$$

where the quantity of interest is $p(\boldsymbol{\theta} \mid \boldsymbol{y})$, the posterior distribution of the parameters $\boldsymbol{\theta}$, which is a combination of the distribution of the data $p(\boldsymbol{y} \mid \boldsymbol{\theta})$ and the prior distribution for the parameters, $p(\boldsymbol{\theta})$. In interpretative terms, $p(\boldsymbol{\theta} \mid \boldsymbol{y})$ represents the uncertainty about the parameters of interest $\boldsymbol{\theta}$ after having observed the data, thus the conditioning on $\boldsymbol{y}$. Notice that $p(\boldsymbol{y})$, at the denominator of Equation 24.4, is the marginal distribution of the data, and it is considered a normalization constant as it does not depend on $\boldsymbol{\theta}$, so that Bayes' theorem is often reported as

$$p(\boldsymbol{\theta} \mid \boldsymbol{y}) \propto p(\boldsymbol{y} \mid \boldsymbol{\theta}) \times p(\boldsymbol{\theta}),$$

where the *equal to* sign ($=$) is replaced by the *proportional to* sign ($\propto$).

Typically, it is not possible to manipulate the posterior distribution analytically and we need to resort to simulations or approximative methods in order to explore it. Statistical simulations generate random values from a given density function by a computer and can become a computationally intensive procedure when the involved distributions are complex or the amount of data is large. Here, we show two alternative ways of obtaining the posterior distribution, the first based on Markov chain Monte Carlo simulations (through the `R2WinBUGS` and `spBayes` `R` packages) and the second based on Integrated Nested Laplace Approximations (INLA), a deterministic approach implemented through the `R-INLA` package.

24.3.1 Markov chain Monte Carlo

The essence of Markov chain Monte Carlo (MCMC) is to draw samples from the required distribution running a convenient Markov chain for a long time.

A Markov chain is a sequence of random variables $X_0, X_1, X_2, \ldots$, that satisfies the following relation: at each time $t \geq 0$, sampling from the distribution $p(X_{t+1} \mid X_t)$ depends only on the current state X_t and not on the previous state of the chain, $X_0, X_1, \ldots, X_{t-1}$. Under regularity conditions, the chain will forget the initial state, and after a sufficiently long burn-in of n iterations, the chain will converge to a unique stationary distribution, which does not depend on t or X_0. When the chain has reached the stationary distribution, we can obtain statistics through Monte Carlo integration. We do not present the way to obtain a Markov chain, but refer readers to Metropolis et al. (1953) and Hastings (1970) for the general framework from which all the other methods derive. Out of them, Gibbs sampler (Geman and Geman 1984) is the method implemented in BUGS (Bayesian Using Gibbs Sampling), the leading software for Bayesian analysis.

The steps needed to perform a MCMC simulation are the following:

1. Define initial values to be arbitrarily assigned to the parameters of interest. The sampling procedure starts from those values.
2. Perform a set of simulations, named the *burn-in*, during which the Markov chain converges to the stationary distribution. Convergence can be monitored by suitable statistics (Gelman et al. 2013).
3. Draw a sample of values from the estimated target distribution, once convergence is reached. Using this sample, all the inferences of interest can be performed. For instance, the whole distribution might be analyzed (i.e., by means of graphical methods, such as histograms or kernel density estimations). Moreover, point estimations such as the mean or the median can be computed.

For the theoretical basics of MCMC refer to Robert and Casella (2004) or Brooks et al. (2011), a handbook also including many applications and case studies involving MCMC.

24.3.2 `R2WinBUGS`

`R2WinBUGS` is the `R` interface of WinBUGS, the Windows version of BUGS, which includes `GeoBUGS`, a collection of specific functions for dealing with spatial data using MCMC simulation methods.

In this section, we illustrate the use of `R2WinBUGS` through the example on Ohio lung cancer mortality presented in Section 24.1.1 to assess the spatial pattern of lung cancer deaths in Ohio counties. We want to specify a disease mapping model, as presented in Section 24.2.2, so that for the ith area, the number of cases y_i is modeled as

$$y_i \mid \lambda_i \sim \text{Poisson}(\lambda_i); \qquad \log(\lambda_i) = \alpha + \log(e_i) + \upsilon_i + \nu_i, \tag{24.5}$$

where α is the intercept quantifying the average death rate in all 88 counties, e_i is the expected number of deaths so that the relative risk can be written as λ_i/e_i, and υ_i and ν_i are the two area-specific effects.

We assume a Besag–York–Mollie (BYM) specification (Besag et al. 1991), so υ_i is the spatially structured residual, modeled using an intrinsic conditional autoregressive structure (iCAR) specified as

$$\upsilon_i \mid \upsilon_{j\neq i} \sim \text{Normal}(\overline{\upsilon}_i, s_i^2)$$
$$\overline{\upsilon}_i = \frac{\sum_{j\in\mathcal{N}(i)} \upsilon_j}{n_i}$$

and

$$s_i^2 = \frac{\sigma_\upsilon^2}{n_i},$$

where n_i is the number of areas that share boundaries with the ith one (i.e., its neighbors denoted by $\mathcal{N}(i)$), as presented in Banerjee (2004). Note that due to the nonpositive definition of its covariance matrix, $\boldsymbol{\upsilon} = \{\upsilon_1, \ldots, \upsilon_{88}\}$ does not have a joint distribution, as it would be possible to add any constant to each υ_i without changing the distribution; this creates an issue of identifiability that can be rectified fixing a constraint such as $\sum_i \upsilon_i = 0$. Finally, the parameter ν_i represents the spatially unstructured effect, modeled as $\nu_i \sim \text{Normal}(0, \sigma_\nu^2)$.

To run this model in R2WinBUGS, we need three objects: (1) model file, (2) data object, and (3) initial values. We will briefly describe each of these.

24.3.2.1 Model file

The model needs to be specified in WinBUGS format, as follows:

```
model {
  for (i in 1 : 88) {
  # Data distribution
  y[i]  ~ dpois(lambda[i])
  log(lambda[i]) <- log(e[i]) + alpha + v[i] + u[i]

  # Area-specific residual relative risk (for maps)
  theta[i] <- exp(v[i] + u[i])

  # Probability that theta>1 (excess risk)
  ptheta[i] <- step(theta[i]-1)

  # Exchangeable prior on unstructured random effects v[i]
  v[i] ~ dnorm(0, tau.v)
  }

    # iCAR prior distribution for spatial random effects u[i]
    u[1:88] ~ car.normal(adj[], weights[], num[], tau.u)
    for(k in 1:sumNumNeigh) {
    weights[k] <- 1
    }

    # Other priors:
    alpha  ~ dflat()
    tau.v  ~ dgamma(0.5, 0.0005)
    sigma.v <- sqrt(1 / tau.v)
    tau.u  ~ dgamma(0.5, 0.0005)
    sigma.u <- sqrt(1 / tau.u)
}
```

Note that the model includes random nodes (identified in the code by the $\sim$ symbol, for example the distribution of the data, the prior distribution on the parameters[2]), as well as logical nodes identified in the code by the $<$ symbol, for example the residual relative risks theta. The model should be saved in a file with a .bug extension (e.g., modelDM.bug). The node theta is the *residual relative risk* calculated as the exponential of the sum of the spatially structured and unstructured effects (v[i] + u[i]), and it should be interpreted

[2]A peculiarity of the WinBUGS language is that the Normal distribution is specified in terms of mean and precision, instead of the usual mean and variance, where the precision is the inverse of the variance, and it is usually denoted as τ.

with respect to the average relative risk of the whole study region given by exp(alpha). An alternative risk measure can be calculated as exp(alpha + v[i] + u[i]) and should be interpreted compared to the expected number of cases in that area. We think that the former specification is easier to interpret, so in the rest of the chapter we will refer to it.

24.3.2.2 Data object

The data should be arranged in a list, containing the number of observed (y) and expected cases (e), as well as the neighborhood structure needed for modeling the spatial random effect $\boldsymbol{\upsilon} = \{\upsilon_1, \ldots, \upsilon_m\}$ as presented above, i.e., the number of neighbors (num), the identifier of the neighboring areas (adj), and the total number of neighbors for all the areas (sumNumNeigh) obtained from the shapefile as follows:

```
> library(maptools)
> library(spdep)
> temp <- poly2nb(ohio)
> neigh <- nb2WB(temp)
```

so that

```
> dataDM <- list(y=dataDM$y,e=dataDM$e, num=neigh$num,
                adj=neigh$adj,sumNumNeigh=sum(neigh$num))
```

24.3.2.3 Initial values

Lastly, we need to specify initial values for all the random nodes (i.e., the ones with $\sim$ in the model file). This is because we are running a MCMC simulation and we have to define the starting points, e.g., the first value that WinBUGS will extract from each distribution. In our case, we need initial values for α, $\boldsymbol{\upsilon}$, $\boldsymbol{\nu}$, τ_ν, and τ_υ, as these are all the random variables in our model. It is good practice to generate initial values from appropriate distributions, and we can use the ones defined as priors in the model. For instance, we can generate ν_i and υ_i from a Normal distribution centered around 0 and with a relatively small variance through the command rnorm in R. Similarly, for the precisions τ_ν and τ_u we can use a Gamma distribution through rgamma so that the list of our initial values looks like

```
> inits1 <- list(alpha = rnorm(n=1,mean=0,sd=1),
            v = rnorm(n=88,mean=0,sd=1), u = rnorm(n=88,mean=0,sd=1),
            tau.v = rgamma(n=1,shape=1,scale=1),
            tau.u = rgamma(n=1,shape=1,scale=1))
> inits2 <- list(alpha = rnorm(n=1,mean=0,sd=1),
            v = rnorm(n=88,mean=0,sd=1), u = rnorm(n=88,mean=0,sd=1),
            tau.v = rgamma(n=1,shape=1,scale=1),
            tau.u = rgamma(n=1,shape=1,scale=1))
```

Note that we have defined two sets of initial values (inits1 and inits2). This is because, as said before, an important part of the MCMC simulation consists of checking the convergence to the stationary distribution, and for this we need to use two (or more) chains, e.g., we run the simulation with different starting points. After a while all the chains should be mixed, i.e., they have converged to the stationary distribution.

24.3.2.4 Running the model and getting the output

Having defined all the objects that we need, we can now run the model typing

```
> library(R2WinBUGS)
> DMout <- bugs(data=dataDM, inits=list(inits1,inits2),
                parameters.to.save=c("sigma.u","sigma.v","theta","ptheta"),
                model.file="modelDM.bug",n.chains=2,n.iter=10000,
                n.burnin=5000, n.thin=10, codaPkg = TRUE)
```

where in addition to the data, model, and initial values, we need to define the parameters that we want to monitor (σ_u, σ_v, θ, and $p(\theta_i > 1)$), the number of iterations for the simulation, the number of chains, and the length of the burn-in (recall that this is the number of iterations discarded before reaching convergence).

To check that convergence is reached, we can use the tools within the `coda` package (automatically loaded using the option `codaPkg = TRUE` in the `bugs` function). The function `read.bugs` creates a list with a number of elements equal to the number of chains and, in each of these, includes all the values that each parameter assumes in the simulation:

```
> codaobject <- read.bugs(DMout)
```

and can be plotted using

```
> traceplot(codaobject)
> densplot(codaobject)
```

The trace and density plots presented in Figure 24.3 for θ_1 (first area) show good mixing and the absence of bimodality for the two chains; the same results is observed for all the other parameters (not shown), suggesting that convergence is obtained.

Summary information (e.g., the posterior mean and standard deviation, together with a 95% credibility interval) can be obtained through

```
> summary(codaobject)
```

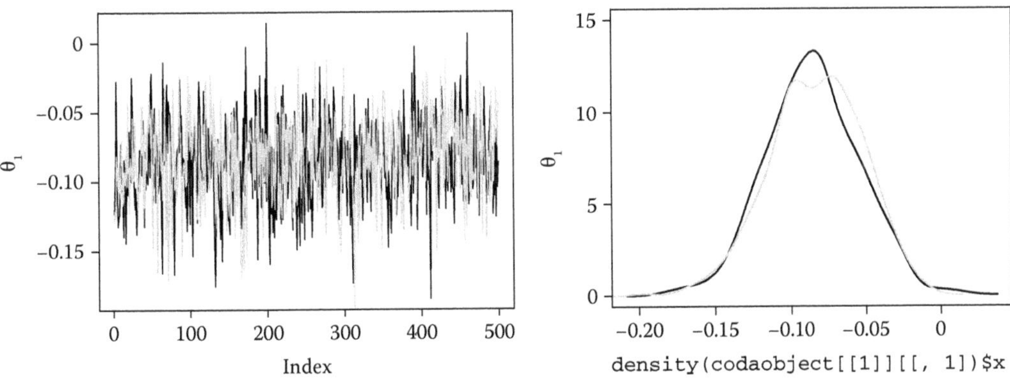

FIGURE 24.3
Trace plot (left) and density plot (right) for $\theta_1 = \exp(\upsilon_1 + \nu_1)$ for the two MCMC chains.

An important part of the interpretation of spatial data models consists of visualizing the results in a map. In our case, we are interested in the relative risk (`theta`), which will tell us the risk of dying from lung cancer in each of the 88 areas under study compared to the whole study region. In addition, the uncertainty associated with the posterior means can also be mapped and provide useful information (Richardson et al. 2004). In particular, as the interest lies in the excess risk, we can visualize $p(\theta_i > 1 \mid \boldsymbol{y})$. We can obtain the maps presented in Figure 24.4 through the commands below:

```
> #Define the cutoffs
> theta.cutoff <- c(0.5, 0.9, 1.0, 1.1, 1.5)
> ptheta.cutoff <- c(0,0.2,0.8,1)
> #Transform theta and ptheta in categorical variable
> cat.theta <- cut(summary(codaobject)[[1]][93:180,1],
                   breaks=theta.cutoff,include.lowest=TRUE)
> cat.ptheta <- cut(summary(codaobject)[[1]][3:90,1],
                    breaks=ptheta.cutoff,include.lowest=TRUE)
> #Create a dataframe with all the information needed for the map
> maps.cat.theta <- data.frame(NAME00=ohio$NAME00, cat.theta=cat.theta,
                               cat.ptheta=cat.ptheta)
> #Add the categorized zeta to the spatial polygon
> data.counties <- attr(ohio, "data")
> attr(ohio, "data") <- merge(data.counties, maps.cat.theta, by="NAME00")

> #Map RR
> spplot(obj=ohio, zcol= "cat.theta",
         col.regions=gray(seq(0.9,0.1,length=4)),asp=1)
> #Map pRR
> spplot(obj=ohio, zcol= "cat.ptheta",
         col.regions=gray(seq(0.9,0.1,length=3)),asp=1)
```

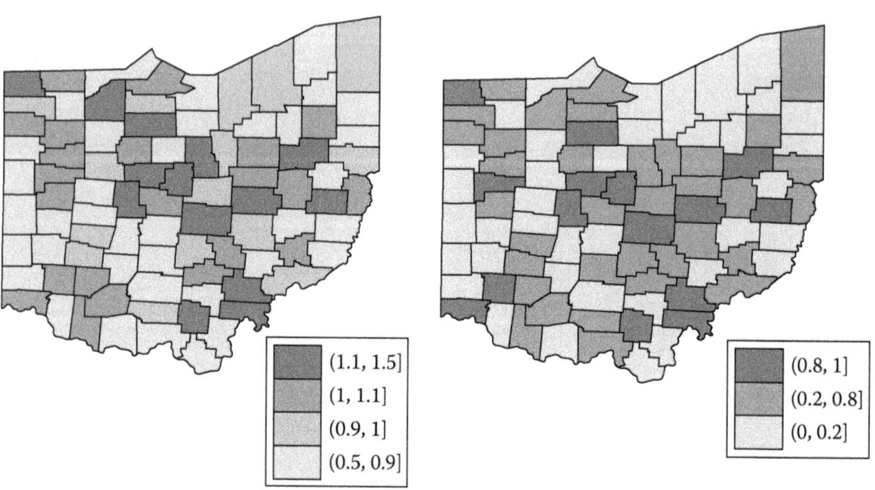

FIGURE 24.4
Posterior mean (left) for the county-specific relative risk $\theta_i = \upsilon_i + \nu_i$ compared to the whole study region and the posterior probability $p(\theta_i > 1 \mid \boldsymbol{y})$ (right).

24.3.3 spBayes

The spBayes package (Finley et al. 2007) represents a generalized and flexible template able to fit a wide range of models for univariate and multivariate spatial continuous processes. It is based on MCMC, written in C++ and uses LAPACK and BLAS libraries to efficiently perform matrix computations. The spBayes adopts the Gibbs sampler for parameter estimation and the Metropolis–Hastings algorithm when required (i.e., when the conditional distribution does not have a closed form).

In this section, we show how to use spBayes for analyzing point-referenced data through a simulated example available as a data.frame named SPDEtoy in the R-INLA package (which will be described extensively in Section 24.3.5). The data, which can be loaded with

```
> library(INLA)
> data(SPDEtoy)
> colnames(SPDEtoy)

[1] "s1" "s2" "y"

> coords <- as.matrix(SPDEtoy[,1:2])
> m <- dim(SPDEtoy)[1]
```

consist of $m = 200$ simulated values for the variable y, which refer to as many randomly sampled locations in the unit square area delimited by the points $(0, 0)$ and $(1, 1)$ and with coordinates given by s1 and s2 (saved separately in the coords object). The model used for simulating the SPDEtoy data assumes that the distribution of the observation y_i is

$$y_i \sim \text{Normal}(\eta_i, \sigma_e^2) \qquad i = 1, \ldots, 200, \tag{24.6}$$

where σ_e^2 is the variance of the zero-mean measurement error e_i, which is supposed to be normally distributed and independent of e_j for each $i \neq j$. Note that this is a different parameterization of Equation 24.1. The response mean, which coincides with the linear predictor, is defined as

$$\eta_i = \alpha + \xi_i \tag{24.7}$$

and includes the intercept α and a random effect represented by ξ_i, which is the realization of the latent GF $\xi(\boldsymbol{s}) \sim \text{MVNormal}(\boldsymbol{0}, \boldsymbol{\Sigma})$. The covariance matrix $\boldsymbol{\Sigma}$ is defined by the Matérn spatial covariance function given by

$$\mathcal{C}(\Delta_{ij}) = \frac{\sigma^2}{\Gamma(\lambda)2^{\lambda-1}} (\kappa \Delta_{ij})^{\lambda} K_{\lambda} (\kappa \Delta_{ij}), \tag{24.8}$$

where $\Delta_{ij} = ||\boldsymbol{s}_i - \boldsymbol{s}_j||$ is the Euclidean distance between locations. The term K_λ denotes the modified Bessel function of the second kind and order $\lambda > 0$, which measures the degree of smoothness of the process and is usually kept fixed. Conversely, $\kappa > 0$ is a scaling parameter related to the range r, i.e., the distance at which the spatial correlation becomes almost null. Typically, the empirically derived definition $r = \frac{\sqrt{8\lambda}}{\kappa}$ is used (see Section 2 in Lindgren et al. 2011). The parameter values chosen for simulating the data are $\alpha = 10$, $\sigma_e^2 = 0.3$, $\sigma^2 = 5$, $\kappa = 7$, and $\lambda = 1$. Note that spBayes uses different names for the model parameters: beta for the mean coefficients (in this case, just the intercept α), sigma.sq for the variance of the spatial process σ^2, tau.sq for the measurement error variance σ_e^2, phi for the decay parameter κ, and nu for the smoothness term λ. In this example, we assume that λ is fixed and equal to 1.

To use spBayes, we first define the matrix of regressors X, which coincides in this case with a vector of 1's, as we have only the intercept

```
> X <- as.matrix(rep(1,m))
```

Then, we specify the prior distributions in the following list using the parameter names adopted by spBayes:

```
> priors <- list("beta.Norm"=list(0, 1000),"phi.Unif"=c(3, 3/0.1),
           "sigma.sq.IG"=c(2, 2), "tau.sq.IG"=c(2, 2),"nu.Unif"=c(0.8,1.2))
```

In the model specification, we are assuming a Normal distribution for the intercept (beta.Norm), Uniform distribution for κ (phi.Unif) and λ (nu.Unif), and Inverse Gamma distribution for both variances (tau.sq.IG and sigma.sq.IG). Then we set the starting values for all the parameters in the following list simulating values from the corresponding prior distributions (with the exception of nu, which is kept fixed and equal to 1):

```
> starting_values <- list("beta"=rnorm(1,0,sqrt(1000)),"nu"=1,
                         "phi"=runif(1,3, 3/0.1), "sigma.sq"=1/rgamma(1,2,2),
                         "tau.sq"=1/rgamma(1,2,2))
```

Finally, we define a list containing the values of the variance of the proposal distribution (i.e., the tuning parameter) used for the Metropolis–Hastings algorithm (if we increase the variance, the proposed values will be sparser and the algorithm will tend to refuse too often; on the other hand, for small values, the procedure will accept too often, giving rise to a lower convergence speed). The following list defines the tuning values for all the parameters except the regression ones:

```
> tuning_values <- list("phi"=0.05, "sigma.sq"=0.05, "tau.sq"=0.05, "nu"=0)
```

where a tuning value equal to zero means that the parameter has to be kept fixed and is not estimated. The main function for fitting the univariate Gaussian model with spatial random effect is spLM, which can be run as follows:

```
> n.iter <- 5000
> formula_SPDEtoy <- SPDEtoy$y ~ X - 1 #explicit intercept
> m1 <- spLM(formula_SPDEtoy, coords=coords, starting=starting_values,
            tuning=tuning_values, priors=priors,
            cov.model="matern", n.samples=n.iter)
```

The spatial covariance function is specified through cov.model (other possibilities are exponential, spherical, and gaussian), and the number of MCMC iterations is defined by n.samples. To retrieve the simulated values for the regression coefficients (α in this case) and all the other parameters, the spRecover function is used, which also gives the possibility of defining a burn-in period or thinning factor:

```
> burnin <- 0.5*n.iter #exclude half of the MCMC iterations
> m1.output <- spRecover(m1, start=burnin, verbose=FALSE)
```

The object `m1.output$p.theta.recover.samples` contains the values simulated from the posterior distributions of σ^2, σ_e^2, and κ; similarly, `m1.output$p.beta.recover.samples` provides the output for the intercept α. The following code is used to compute the posterior 0.025, 0.5 and 0.975 quantiles for all the parameters and to check that the true values used for simulating the data are contained in the 95% credibility intervals:

```
> cbind(True=c(5,0.3, 7,1),
 round(summary(m1.output$p.theta.recover.samples)$quantiles[,c(1,3,5)],2))

         True 2.5%  50% 97.5%
sigma.sq  5.0 1.98 3.18  5.93
tau.sq    0.3 0.22 0.32  0.45
phi       7.0 5.49 8.50 12.27
nu        1.0 1.00 1.00  1.00

> c(True=10,
 round(summary(m1.output$p.beta.recover.samples)$quantiles[c(1,3,5)],2))

 True  2.5%   50% 97.5%
10.00  8.34  9.56 10.71
```

Given a `spLM` object (like `m1`), spatial prediction can be performed easily using the `spPRedict` function, which requires the coordinates of the new locations and the covariate values, if included in the model. The output is a matrix with samples from the posterior predictive distribution for the response variable.

The `spBayes` package is very flexible and can implement more complex univariate models (e.g., with site-specific parameters), as well as multivariate models through the `spMvLM` function. Moreover, univariate and multivariate logistic and Poisson regression are an option thanks to the `spGLM` and `spMvGLM` functions. Finally, it is worth citing the new functions `spMisalignLM` and `spMisalignGLM`, which can be employed in case of misaligned spatial data (Finley et al. 2014).

24.3.4 Limitations of MCMC methods

Independently of the software used to implement MCMC methods, particular attention should be paid to the analysis of the output. In fact, even though the MCMC theory states that the distribution of the simulated values converges to the target density (i.e., the posterior distribution) when the iteration number goes to infinity, it is not feasible to run a Markov chain infinitely. Thus, we run it long enough to achieve convergence, even if no stopping rule exists suggesting how many MCMC iterations are required for convergence or how long the burn-in period should be for eliminating the influence of starting values. The usual practice consists of employing convergence diagnostic tools on multiple chains (Cowles and Carlin 1996) in order to explore the MCMC output and detect a failure in convergence (unfortunately, even when no problems are observed, we are not guaranteed that convergence has successfully occurred). The convergence diagnostics comprise graphs, such as the trace and the density plot (presented in Section 24.3.2) and some test statistics (the most used are the Gelman–Rubin and Raftery–Lewis methods, which are also available in the `coda` package by Plummer et al. 2006).

The key point is that, when making inference with MCMC, great care (and a lot of time) is devoted to the tuning and monitoring convergence phases in order to find the best setting (in terms of parameterization, prior distributions, initial values, and Metropolis–Hastings proposal distributions) that produces the most reliable and accurate MCMC output (see Brooks et al. 2011 and references therein for recommendations about MCMC implementation). Another crucial aspect that cannot be ignored concerns the computational costs of MCMC methods. When models are complex (especially when designed in a hierarchical fashion) or we deal with large datasets, MCMC algorithms may be extremely slow and even become computationally infeasible. This computational burden occurs particularly in the case of spatial (and spatiotemporal) models and is usually known as the "big n problem" (Banerjee 2004, p. 387). A viable alternative to MCMC methods able to reduce the computational costs of Bayesian inference is the INLA algorithm described in the next section.

24.3.5 Integrated Nested Laplace Approximations

The Integrated Nested Laplace Approximations (INLA) method, developed by Rue et al. (2009), is designed for the class of latent Gaussian models where the response variable y_i observed for the ith unit ($i = 1, \dots, m$) is assumed to belong to a distribution family (not necessarily part of the exponential family) characterized by a parameter ϕ_i (usually the mean) which is linked to a structured additive predictor η_i through a link function $g(\cdot)$, such that $g(\phi_i) = \eta_i$. A general way for specifying the linear predictor η_i is

$$\eta_i = \alpha + \sum_{l=1}^{L} f_l(z_{li}), \tag{24.9}$$

where α is a scalar representing the intercept and $\boldsymbol{f} = \{f_1(\cdot), \dots, f_L(\cdot)\}$ is a collection of unknown functions defined in terms of a set of covariates $\boldsymbol{z} = (z_1, \dots, z_L)$. Upon varying the form of the functions $f_l(\cdot)$, this formulation can accommodate a wide range of models, from (generalized) linear mixed to spatial and spatiotemporal models (see Martins et al. 2013 and references therein). A Gaussian prior is assigned to α and $\boldsymbol{f}$, and then all the latent Gaussian components are collected in the vector of parameters (or latent field) $\boldsymbol{\theta} = \{\alpha, \boldsymbol{f}\}$, which is a function of some hyperparameters $\boldsymbol{\psi} = \{\psi_1, \dots, \psi_K\}$.

Latent Gaussian models can be formulated also by means of a hierarchical structure. At the first stage, the likelihood function for the set of observations $\boldsymbol{y} = y_1, \dots, y_m$ is defined by assuming the conditional independence property given the latent field $\boldsymbol{\theta}$:

$$\boldsymbol{y} \mid \boldsymbol{\theta}, \boldsymbol{\psi} \sim p(\boldsymbol{y} \mid \boldsymbol{\theta}, \boldsymbol{\psi}) = \prod_{i=1}^{m} p\,(y_i \mid \theta_i, \boldsymbol{\psi}).$$

At the second stage, we assume a multivariate Normal prior on $\boldsymbol{\theta}$ with mean $\boldsymbol{0}$ and precision matrix $\boldsymbol{Q}(\boldsymbol{\psi})$, that is,

$$\boldsymbol{\theta} \mid \boldsymbol{\psi} \sim p(\boldsymbol{\theta} \mid \boldsymbol{\psi}) = \text{Normal}\left(\boldsymbol{0}, \boldsymbol{Q}(\boldsymbol{\psi})^{-1}\right).$$

For a large range of models, we can assume that the components of the latent Gaussian field $\boldsymbol{\theta}$ admit conditional independence; hence, the precision matrix $\boldsymbol{Q}(\boldsymbol{\psi})$ is sparse and the latent field $\boldsymbol{\theta}$ is a Gaussian Markov random field (see Rue and Held 2005). Note that the sparsity of the precision matrix is crucial for computational benefits, as it allows us to use numerical methods for sparse matrices that are faster than the general algorithms for dense matrices. The hierarchical model specification is then completed with a prior for the hyperparameters $\boldsymbol{\psi} \sim p(\boldsymbol{\psi})$.

The objectives of Bayesian inference are the marginal posterior distributions for each element of the parameter vector,

$$p(\theta_i \mid \boldsymbol{y}) = \int p(\theta_i \mid \boldsymbol{\psi}, \boldsymbol{y}) p(\boldsymbol{\psi} \mid \boldsymbol{y}) \mathrm{d}\boldsymbol{\psi}, \tag{24.10}$$

and for each element of the hyperparameter vector

$$p(\psi_k \mid \boldsymbol{y}) = \int p(\boldsymbol{\psi} \mid \boldsymbol{y}) \mathrm{d}\boldsymbol{\psi}_{-k}. \tag{24.11}$$

The INLA algorithm substitutes MCMC simulations with accurate deterministic approximations to the distributions in Equations 24.10 and 24.11 obtained through nested Laplace approximations. In particular, the approximated posteriors returned by INLA are

$$\begin{aligned}
\tilde{p}(\psi_k \mid \boldsymbol{y}) &= \int \tilde{p}(\boldsymbol{\psi} \mid \boldsymbol{y}) \mathrm{d}\boldsymbol{\psi}_{-k} \\
\tilde{p}(\theta_i \mid \boldsymbol{y}) &= \sum_j \tilde{p}(\theta_i \mid \psi_j, \boldsymbol{y}) \tilde{p}(\psi_j \mid \boldsymbol{y}) \Delta_j,
\end{aligned}$$

where $\{\psi_j\}$ is a set of relevant points—with corresponding weights $\{\Delta_j\}$—obtained through a grid exploration on $\tilde{p}(\boldsymbol{\psi} \mid \boldsymbol{y})$ (see details in Rue et al. 2009). Moreover, as described in Martins et al. (2013), INLA can provide approximations to the posterior marginals of linear combinations of the latent field defined as $\boldsymbol{\nu} = \boldsymbol{B}\boldsymbol{\theta}$, where matrix $\boldsymbol{B}$ contains the weights defining the linear combinations.

The INLA approach described above is implemented in the `R` package `R-INLA`, which substitutes the stand-alone `INLA` program built upon the `GMRFLib` library (Martino and Rue 2010). `R-INLA` is available for Linux, Mac and Windows operating systems. Extensive documentation that comprises theory, examples, and `R` code can be found in Blangiardo and Cameletti (2015) and on the website http://www.r-inla.org/, which also includes a discussion forum. Here, we use `R-INLA` to run the same example used in the `R2WinBUGS` section.

Given the vector of observations $\boldsymbol{y}$, Equation 24.9 is reproduced in `R-INLA` through the command `formula`:

```
> formula <- y ~ 1 + f(ID, ...)
```

where `y` is the column names of the dataframe containing the data. The term `f(ID,...)` is used to specify the structure of the function $f(\cdot)$ on the area identifier `ID`, which assumes a sequential number between 1 and 88 (total number of areas in the study region), using the following notation:

```
> f(ID, model = "bym", graph=ohio.adj)
```

where the string associated with the option `model` specifies the type of random effect. The default choice is `model="iid"`, documented typing `inla.doc("iid")`, and it amounts to assuming exchangeable Normal distributions for all the areas (equivalent to include only α and v_i in Equation 24.5). This specification can be used to build standard hierarchical models. If we want to include a conditional autoregressive term and reproduce the model specification in Equation 24.5, we can just substitute `model="iid"` with `model="bym"`. The list of the other alternatives and options is available by typing `names(inla.models()$latent)`; in addition, a detailed description of all the possible choices is available at the web page http://www.r-inla.org/models/latent-models. Note that in R-INLA, the sum to zero constraint that we introduced in Section 24.3.2 on the iCAR specification for $\boldsymbol{\upsilon}$ is automatically defined on any intrinsic model.

Similarly to what we have done in the R2WinBUGS model, we need to define the adjacency matrix when we specify spatial models. We can do this by starting from the shapefile as follows:

```
> library(spdep)
> temp <- poly2nb(ohio)
> nb2INLA("ohio.graph", temp)
> ohio.adj <- paste(getwd(),"/ohio.graph",sep="")
```

Now, we have a file called ohio.graph stored in the current working directory that is in the right format to be read by R-INLA. The last command creates an object (ohio.adj) with the location of the graph.

Once the model has been specified, we can run the INLA algorithm using the inla function:

```
> mod <- inla(formula, family = "...", data)
```

where formula has been specified above, data is the dataframe containing all the variables in the formula, and family is a string that specifies the distribution of the data (likelihood). By typing names(inla.models()$likelihood), it is possible to retrieve all the available data distributions; moreover, complete descriptions with examples are provided at http://www.r-inla.org/models/likelihoods. The inla function includes many other options; see ?inla for a complete list.

Summary information (e.g., the posterior mean and standard deviation, together with a 95% credibility interval) can be obtained for the so-called fixed effects (α, in this case) and for the random effects (i.e., $\boldsymbol{\theta} = \boldsymbol{\upsilon} + \boldsymbol{\nu}$ and $\boldsymbol{\upsilon}$) by typing

```
> round(mod$summary.fixed,3)

              mean    sd 0.025quant 0.5quant 0.975quant   mode kld
(Intercept) -0.085 0.031     -0.148   -0.084     -0.024 -0.083   0

> round(head(mod$summary.random$ID),3) #partial output

  ID   mean    sd 0.025quant 0.5quant 0.975quant   mode kld
1  1  0.011 0.174     -0.338    0.013      0.348  0.017   0
2  2 -0.052 0.115     -0.283   -0.051      0.169 -0.047   0
3  3  0.083 0.147     -0.211    0.084      0.367  0.087   0
4  4  0.133 0.113     -0.094    0.134      0.352  0.136   0
5  5 -0.190 0.148     -0.490   -0.186      0.093 -0.180   0
6  6 -0.057 0.154     -0.368   -0.054      0.240 -0.049   0
```

The latter is a dataframe formed by $2m$ rows: the first m rows include information on the area-specific residual relative risk θ_i, which is the primary interest in a disease mapping study, while the remaining rows present information on the spatially structured residual υ_i only. Recall that all these parameters are on the logarithmic scale; for the sake of interpretability, it would be more convenient to transform them back to the natural scale.

To compute the posterior mean and 95% credibility interval for the fixed effect α on the original scale, we type

```
> exp.alpha.mean <- inla.emarginal(exp,mod$marginals.fixed[[1]])
> exp.alpha.mean

[1] 0.9192447

> exp.alpha.95CI <- inla.qmarginal(c(0.025,0.975),
                    inla.tmarginal(exp,mod$marginals.fixed[[1]]))
> exp.alpha.95CI

[1] 0.8631675 0.9753600
```

The computation of the posterior mean for the random effects $\boldsymbol{\theta}$ is performed in two steps, as we have more than one parameter:

```
> logtheta <- mod$marginals.random$ID[1:88]
> theta <- lapply(logtheta,function(x) inla.emarginal(exp,x))
```

First, we extract the marginal posterior distribution for each element of $\log(\boldsymbol{\theta})$, and then we apply the exponential transformation and calculate the posterior mean for each of them using the `lapply` function. The equivalent map presented in Figure 24.4 can be obtained using the same code described in that section and shows no differences (see also Figure 24.5).

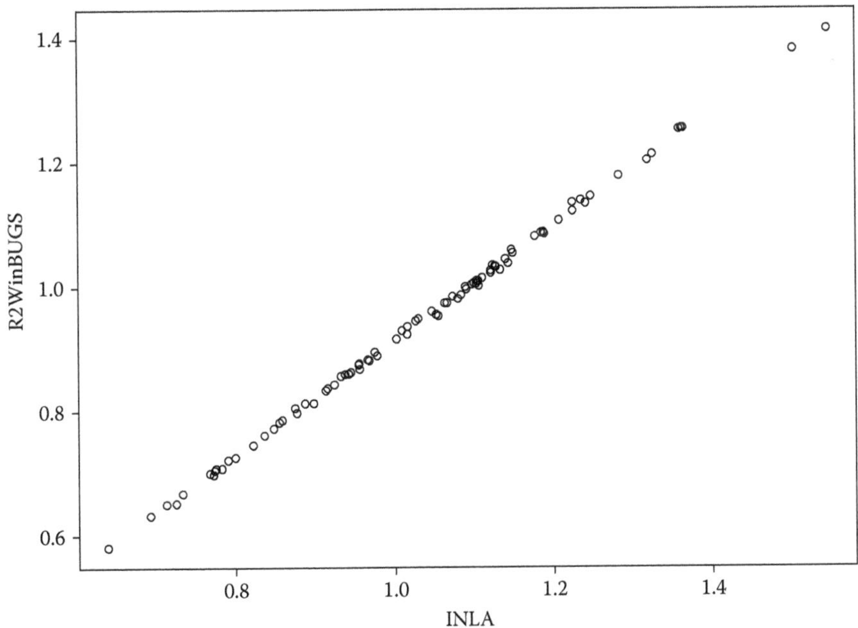

FIGURE 24.5
Comparison between R2WinBUGS and R-INLA: the plot reports the posterior mean for the county-specific relative risk $\theta_i = \exp(\upsilon_i + \nu_i)$ obtained using the simulative approach (MCMC) and the deterministic approximative approach (INLA).

Finally, the posterior probability of $\theta_i > 1$, evidentiating an excess risk, can be accessed using

```
> a <- 0
> prob.logtheta <- lapply(logtheta, function(x) {1 - inla.pmarginal(a, x)})
```

and mapped in the same way as presented in Section 24.3.2.

24.3.6 SPDE

A computationally effective alternative to MCMC to deal with the big n problem is given by the stochastic partial differential equation (SPDE) approach proposed by Lindgren et al. (2011). It represents a continuous spatial process $\xi(\boldsymbol{s})$ with Matérn covariance function using a discretely indexed spatial random process (i.e., a GMRF), which is characterized by a sparse precision matrix and enjoys computational benefits in terms of fast inference. This representation is based on the following finite combination of piecewise linear functions defined over a triangulation of the domain $\mathcal{D}$ with G vertexes:

$$\xi(\boldsymbol{s}) = \sum_{g=1}^{G} \varphi_g(\boldsymbol{s})\tilde{\xi}_g, \tag{24.12}$$

where $\{\varphi_g\}$ is the set of basis functions and $\{\tilde{\xi}_g\}$ are zero-mean Gaussian distributed weights. Lindgren et al. (2011) show that the vector of basis weights $\tilde{\boldsymbol{\xi}} = \{\tilde{\xi}_1, \ldots, \tilde{\xi}_G\}$ is a GMRF with sparse precision matrix $\boldsymbol{Q}_{\tilde{\xi}}(\boldsymbol{\psi})$ depending on the Matérn parameter $\boldsymbol{\psi} = \{\kappa, \sigma^2\}$ (in R-INLA the smoothness parameter λ is fixed equal to 1; see the description of the Matérn covariance function in Section 24.3.3 and Equation 24.8).

Here, we use the SPDEtoy simulated data, first introduced in Section 24.3.3, to illustrate the basic functions in R-INLA for the SPDE approach. This is based on a triangulation of the spatial domain, with the definition of the mesh being a trade-off between the accuracy of the GMRF representation and computational costs, both depending on the number of vertices used in the triangulation: the bigger the number of mesh triangles, the finer the GF approximation, but the higher the computational costs. In R-INLA, we make use of the helper function inla.mesh.2d to create the mesh. This function requires as input some information about the spatial domain, given either by some relevant spatial points (not necessarily the points where observations are available) or by the domain extent, to be specified using the loc or loc.domain arguments, respectively. Another nonoptional argument is max.edge, which represents the largest allowed triangle edge length. If a vector of two values is provided, the spatial domain is divided into an inner and an outer area whose triangle resolution is specified by max.edge (the higher the value for max.edge, the lower the resolution and the accuracy). This extension of the original domain can be useful in order to avoid the boundary effects related to the SPDE approach (the Neumann boundary conditions used in R-INLA have as a drawback an increase of the variance near the boundary). The optional argument offset can be used to define how much the domain should be extended in the inner and the outer part (the default values are offset = c(-0.05, -0.15)). Another optional argument of the inla.mesh.2d function is cutoff, which can be used to avoid building too many small triangles around clustered data locations (the default value is equal to 0). In the following, we define the domain for the SPDEtoy data and create a proper mesh:

```
> domain <- matrix(cbind(c(0,1,1,0.7,0), c(0,0,0.7,1,1)), ncol=2)
> mesh_toy <- inla.mesh.2d(loc.domain=domain,
   max.edge=c(0.04, 0.2), offset = c(0.1, 0.4), cutoff=0.05)
```

which can be plotted just typing plot(mesh_toy) (see Figure 24.6, left).

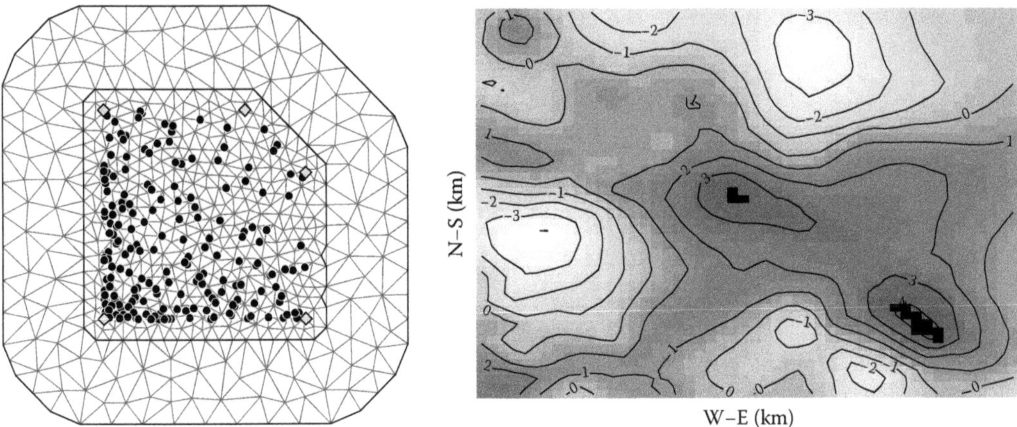

FIGURE 24.6
Triangulation for the SPDEtoy dataset (left); black points and gray diamonds denote the observation locations and the boundary points, respectively. Posterior mean (right) of the spatial latent field predicted at the grid level.

Given the basis function representation of Equation 24.12, the linear predictor in Equation 24.7 can be rewritten as

$$\eta_i = \alpha + \sum_{g=1}^{G} A_{ig}\tilde{\xi}_g,$$

with $A_{ig} = \varphi_g(\boldsymbol{s}_i)$ being the generic element of the sparse matrix $\boldsymbol{A}$, which maps the GMRF $\tilde{\boldsymbol{\xi}}$ from the G triangulation vertices to the m observation locations. The R-INLA function inla.spde.make.A creates the sparse weight matrix $\boldsymbol{A}$ by identifying the data locations in the mesh and organizing the corresponding values of the basis functions. The following code is used to create the $\boldsymbol{A}$ matrix for the mesh_toy triangulation characterized by 549 vertices:

```
> A.est <- inla.spde.make.A(mesh=mesh_toy, loc=coords)
> dim(A.est)
```

```
[1] 200 549
```

The dimension of the resulting $\boldsymbol{A}$ matrix is given by the number of data locations times the number of mesh nodes. For the considered case, it happens that each data spatial location is placed inside a triangle that is delimited by three vertices; for this reason, A.est is characterized by three nonzero elements for each row whose sum is equal to 1 (i.e., $\sum_{g=1}^{G} A_{ig} = 1$). In addition, there are some columns with zero values corresponding to vertices not connected to points.

In R-INLA, the vector of parameters is defined as $\boldsymbol{\theta} = \{\tilde{\boldsymbol{\xi}}, \alpha\}$ with hyperparameter vector $\boldsymbol{\psi} = \{\sigma_e^2, \kappa, \sigma^2\}$, where κ and σ^2 are the Matérn covariance function parameters. We show now how to estimate the model parameters using the mesh_toy triangulation and the projector matrix A.est created previously and choosing the default prior specifications for the SPDE parameters; see Blangiardo and Cameletti (2015) for more details about how to change priors. First, we need to create a Matérn SPDE object through

```
> spde <- inla.spde2.matern(mesh=mesh_toy, alpha=2)
```

from which it is possible to retrieve the number of mesh vertices with

```
> spde$n.spde
```

```
[1] 549
```

Then we define the linear predictor through the following `formula`:

```
> formula <- y ~ -1 + intercept + f(spatial.field, model=spde)
```

removing the default intercept (with the code specification `-1`) and including an explicit one named `intercept`. The spatial random effect is included with the `f()` term, where `spatial.field` is a proper index variable and `spde` is the model created previously with `inla.spde2.matern`. As usual, the model is fitted by means of the `inla` function:

```
> output <- inla(formula,
   data = list(y=SPDEtoy$y, intercept=rep(1,spde$n.spde),
               spatial.field=1:spde$n.spde),
   control.predictor=list(A=A.est,compute=TRUE))
```

Note that the `data` list also includes the `spatial.field` variable for the random effect, defined as the sequence of integers from 1 to the number of mesh vertices. Moreover, the projector matrix is provided through the `control.predictor` option; by setting `compute=TRUE`, it is possible to obtain the posterior marginals for the linear predictor. The posterior summaries of the intercept and of the precision for the Gaussian observations ($1/\sigma_e^2$) can be retrieved with the following commands:

```
> round(output$summary.fixed,3)
```

```
           mean    sd 0.025quant 0.5quant 0.975quant  mode kld
intercept 9.505 0.696      8.036    9.524     10.865 9.555   0
```

```
> round(output$summary.hyperpar[1,],3)
```

```
     mean          sd 0.025quant   0.5quant 0.975quant        mode
    2.860       0.467      2.041      2.827      3.870       2.765
```

If the interest is on the variance σ_e^2 (and not on the precision), it is possible to compute the summary of the transformed posterior marginal through

```
> inla.zmarginal(inla.tmarginal(function(x) 1/x,
  output$marginals.hyper[[1]]))
```

```
Mean            0.358993
Stdev           0.0584133
Quantile  0.025 0.258954
Quantile  0.25  0.317158
Quantile  0.5   0.35358
Quantile  0.75  0.394964
Quantile  0.975 0.488078
```

The posterior summaries of the spatial parameters can be extracted from the `inla` output by means of the `inla.spde2.result` function, which returns all the relevant bits of information from the list `output`. Moreover, the function provides the posterior distributions for the nominal variance σ^2 and the nominal range r. In the following code,

```
> output.field <- inla.spde2.result(inla=output, name="spatial.field",
                                    spde=spde, do.transf=TRUE)
```

`name` denotes the name of the SPDE effect used in the `formula` (in this case, `spatial.field`) and `spde` is the result of the call to `inla.spde2.matern`. The option `do.transf=TRUE` is used to compute marginals of the parameters transformed to user-scale. See `names(output.field)` for exploring the elements of the resulting list. The posterior mean of κ, σ^2 and the range r can be obtained by typing

```
> inla.emarginal(function(x) x, output.field$marginals.kappa[[1]])

[1] 8.276316

> inla.emarginal(function(x) x, output.field$marginals.variance.nominal[[1]])

[1] 4.026735

> inla.emarginal(function(x) x, output.field$marginals.range.nominal[[1]])

[1] 0.361024
```

The corresponding 95% highest probability density intervals are

```
> inla.hpdmarginal(0.95, output.field$marginals.kappa[[1]])

                low     high
level:0.95 4.69942 12.10467

> inla.hpdmarginal(0.95, output.field$marginals.variance.nominal[[1]])

                low     high
level:0.95 1.940859 6.497331

> inla.hpdmarginal(0.95, output.field$marginals.range.nominal[[1]])

                low     high
level:0.95 0.2089509 0.537118
```

which contain the true parameter values.

In geostatistical applications, the main interest resides in the spatial prediction of the spatial latent field (or of the response variable) in new locations belonging to $\mathcal{D}$. This is straightforward as SPDE provides the approximation of the entire spatial process $\xi(\boldsymbol{s})$ through the representation given in Equation 24.12. To describe how to perform spatial prediction with R-INLA, we create a regular grid of 50×50 points in the same spatial domain of the SPDEtoy dataset:

```
> x.res <- y.res <- 50
> seq.x.grid <- seq(0, 1, length.out = x.res)
> seq.y.grid <- seq(0, 1, length.out = y.res)
> pred.grid <- expand.grid(x=seq.x.grid, y=seq.y.grid)
```

The simplest and computationally cheap procedure for predicting the spatial random field $\xi(\boldsymbol{s})$ consists of projecting the latent field—estimated at the mesh vertices—onto the grid locations, once the parameters have been estimated. To follow this approach, a linkage between the mesh and the grid is created using the inla.mesh.projector command:

```
> proj.grid <- inla.mesh.projector(mesh_toy,
              xlim=range(pred.grid[,1]), ylim=range(pred.grid[,2]),
              dims=c(x.res,y.res))
```

Then, the posterior mean of the spatial random field is projected from the mesh to the grid using the inla.mesh.project function:

```
> post.mean.pred <- output$summary.random$spatial.field$mean
> post.mean.pred.grid <- inla.mesh.project(proj.grid, post.mean.pred)
```

The map of the posterior mean at the grid level (see Figure 24.6, right panel) is obtained through the following code based on the image.plot function of the fields package:

```
> library(fields)
> image.plot(seq.x.grid,seq.y.grid,post.mean.pred.grid,
  xlab="W-E (km)", ylab="N-S (km)", col=grey(20:0/20))
> contour(seq.x.grid,seq.y.grid,post.mean.pred.grid, add=T, lwd=2)
```

When the model complexity increases, for example, covariates or many random effects are included in the linear predictor or more likelihoods are involved, matrix and vector manipulation of SPDE objects can be tedious and lead to errors. To avoid this kind of complications for the users, a function named inla.stack has been introduced in R-INLA for optimal and easy management of the SPDE objects (see Lindgren and Rue 2015). This new function can also be used to perform spatial prediction, together with the parameter estimation process, by joining linear predictors (one for parameter estimation and one for prediction); see Blangiardo and Cameletti (2015) for some advanced examples.

24.3.7 Limitations of INLA

R-INLA is very powerful and allows us to greatly save computing time; thus it is a valid alternative to MCMC, but due to its recent inception it is less established. Its main limitation resides in the types of models that can be specified, as they need to be expressed as latent Gaussian models. Also, there are restrictions on the type of likelihood, link function, or prior that can be specified; for instance, the multinomial or Dirichlet distributions are not implemented as well as the probit function, so that a model with ordinal variables cannot be run. Finally, the number of hyperparameters must be relatively small (less than 20 is recommended) to avoid an exponential increase in the computational costs of the model.

Nevertheless, its development is still ongoing, particularly with respect to some more advanced features (e.g., the SPDE module described in Section 24.3.6), meaning that in the future, some of these issues might be resolved.

24.4 From Spatial to Spatiotemporal Models

The methods presented can be easily extended to accomodate a temporal dimension in addition to the spatial one. The data are then defined by a process

$$Y(\boldsymbol{s},t) \equiv \{y(\boldsymbol{s},t), (\boldsymbol{s},t) \in \mathcal{D} \in \mathbb{R}^2 \times \mathbb{R}\}$$

and are observed at n spatial locations or areas and at T time points. When spatiotemporal geostatistical data are considered (Gelfand et al. 2010, chapter 23), we need to define a valid spatiotemporal covariance function. If we assume stationarity in space and time, the space–time covariance function can be written as a function of the spatial Euclidean distance and the temporal lag; several examples of valid nonseparable space–time covariance functions are reported in Cressie and Huang (1999) and Gneiting (2002).

In practice, to overcome the computational complexity of nonseparable models, some simplifications are introduced. For example, under the separability hypothesis, the space–time covariance function is decomposed into the sum (or the product) of a purely spatial and a purely temporal term (Gneiting et al. 2006). Alternatively, it is possible to assume that the spatial correlation is constant in time, giving rise to a space–time covariance function that is purely spatial within each time point and is zero otherwise. In this case, the temporal evolution could be introduced assuming that the spatial process evolves in time following an autoregressive dynamics (Harvill 2010).

Similar reasoning can be applied to area-level data, where a neighborhood structure also needs to be included on the time dimension. Several papers have discussed this extension; see, for instance, Waller et al. (1999) for a description of MCMC methods in this context, Abellan et al. (2008) for implementation of spatiotemporal models in WinBUGS, and Blangiardo et al. (2013) and Ugarte et al. (2014) for a description of how to use Laplace approximations for spatiotemporal disease mapping. If a space–time interaction is included, its precision can be obtained through the Kronecker product of the precision matrices for the space and time effects interacting—see Clayton (1996) and Knorr-Held (2000) for a detailed description.

24.5 Closing Remarks

In this chapter, we presented the most used computational resources to import, visualize, and run statistical models on spatial data aiming at assessing and investigating spatial variability in a region, evaluating the disease risk from a given source, or predicting the values of a spatially varying factor. We focused on the `R` language and briefly showed the resources for the frequentist approach to spatial inference, and then presented the commonly used Bayesian framework and, in particular, the MCMC simulation approach (`R2WinBUGS`, `spBayes`) and the INLA deterministic approximate method (`R-INLA`). Obviously, there are other resources for spatial analysis that we could not include here: we encourage the reader to access the complete list of `R` packages, available at http://cran.r-project.org/web/views/Spatial.html, and look at some stand-alone software, such as the Rapid Inquiry Facility (RIF), developed for working with environmental and health data, freely available at http://www.sahsu.org/content/rapid-inquiry-facility, and the software developed by the GeoDA Center for Geospatial Analysis and Computation (http://geodacenter.asu.edu/).

References

Abellan, J., S. Richardson, and N. Best. (2008). Use of space-time models to investigate the stability of patterns of disease. *Environmental Health Perspectives* 116(8), 1111–1119.

Albert, J. (2009). *Bayesian Computation with R*. Springer, New York.

Banerjee, S. (2004). Revisiting spherical trigonometry with orthogonal projectors. *Mathematical Association of America's College Mathematics Journal* 35, 375–381.

Besag, J., and J. Newell. (1991). The detection of clusters in rare diseases. *Journal of the Royal Statistical Society: Series A* 154, 143–155.

Besag, J., J. York, and A. Mollié. (1991). Bayesian image restoration, with two applications in spatial statistics. *Annals of the Institute of Statistical Mathematics* 43, 1–59.

Bivand, R., E. Pebesma, and V. Gómez-Rubio. (2013). *Applied Spatial Data Analysis with R*, 2nd ed. Springer, New York.

Blangiardo, M., and M. Cameletti. (2015). *Spatial and Spatio-Temporal Bayesian Models with R-INLA*. Wiley, New York.

Blangiardo, M., M. Cameletti, G. Baio, and H. Rue. (2013). Spatial and spatio-temporal models with R-INLA. *Spatial and Spatio-Temporal Epidemiology* 4, 33–49.

Breslow, N. E., and D. G. Clayton. (1993). Approximate inference in generalized linear mixed models. *Journal of the American Statistical Association* 88(421), 9–25.

Brooks, S., A. Gelman, G. Jones, and X. Meng. (2011). *Handbook of Markov Chain Monte Carlo*. CRC Press, Taylor & Francis Group, Boca Raton, FL.

Chambers, J. (2008). *Software for Data Analysis: Programming with R*. Springer, New York.

Clayton, D. (1996). Generalised linear mixed models. In W. Gilks, S. Richardson, and D. Spiegelhalter (eds.), *Markov Chain Monte Carlo in Practice*, pp. 275–301. Chapman & Hall, London.

Cowles, M., and B. Carlin. (1996). Markov chain Monte Carlo convergence diagnostics: A comparative review. *Journal of the American Statistical Association* 91, 883–904.

Cowpertwait, P., and A. Metcalfe. (2009). *Introductory Time Series with R*. Springer, New York.

Crawley, M. (2007). *The R Book*. John Wiley & Sons.

Cressie, N. (1991). *Statistics for Spatial Data*. Wiley, New York.

Cressie, N., and H. Huang. (1999). Classes of nonseparable, spatio-temporal stationary covariance functions. *Journal of the American Statistical Association* 94(448), 1330–1340.

Dalgaard, P. (2008). *Introductory Statistics with R*. Springer, New York.

Dean, C., M. D. Ugarte, and A. F. Militino. (2004). Penalized quasi-likelihood with spatially correlated data. *Computational Statistics and Data Analysis* 45(2), 235–248.

Diggle, P., R. Moyeed, and J. Tawn. (1998). Model-based geostatistics. *Journal of the Royal Statistical Society: Series C* 47, 299–350.

Diggle, P., and J. Ribeiro. (2007). *Model-Based Geostatistics*. Springer, New York.

Elliott, P., J. Wakefield, N. Best, and D. Briggs. (2000). *Spatial Epidemiology: Methods and Applications*. Oxford University Press, Oxford.

Everitt, B., and T. Hothorn. (2009). *A Handbook of Statistical Analyses Using R*, 2nd ed. CRC Press, Boca Raton, FL.

Finley, A., S. Banerjee, and B. Carling. (2007). spBayes: An R package for univariate and multivariate hierarchical point-referenced spatial models. *Journal of Statistical Software* 19(4), 1–24.

Finley, A., S. Banerjee, and B. Cook. (2014). Bayesian hierarchical models for spatially misaligned data in R. *Methods in Ecology and Evolution* 5(6), 514–523.

Geary, R. (1954). The contiguity ratio and statistical mapping. *Incorporated Statistician* 5, 115–145.

Gelfand, A., P. Diggle, M. Fuentes, and P. Guttorp (eds.), (2010). *Handbook of Spatial Statistics.* Chapman & Hall, London.

Gelman, A., J. Carlin, H. Stern, D. Dunson, A. Vehtari, and D. Rubin. (2013). *Bayesian Data Analysis.* Chapman & Hall, London.

Geman, S., and D. Geman. (1984). Stochastic relaxation, Gibbs distributions and the Bayesian restoration of images. *IEEE Transactions on Pattern Analysis and Machine Intelligence* 6, 721–741.

Gneiting, T. (2002). Nonseparable, stationary covariance functions for space-time data. *Journal of the American Statistical Association* 97(458), 590–600.

Gneiting, T., M. Genton, and P. Guttorp. (2006). Statistical methods for spatio-temporal systems. In B. Finkenstädt, L. Held, and V. Isham (eds.), *Statistical Methods for Spatio-Temporal Systems*, pp. 151–175. CRC Press, Chapman & Hall, Boca Raton, FL.

Harvill, J. (2010). Spatio-temporal processes. *Wiley Interdisciplinary Reviews: Computational Statistics* 2(3), 375–382.

Hastings, W. (1970). Monte Carlo sampling methods using Markov chains and their application. *Biometrika* 57, 97–109.

Kleiber, C., and A. Zeileis. (2008). *Applied Econometrics with R.* Springer, New York.

Knorr-Held, L. (2000). Bayesian modelling of inseparable space-time variation in disease risk. *Statistics in Medicine* 17–18(19), 2555–2567.

Kulldorf, M. (1997). A spatial scan statistics. *Communications in Statistics: Theory and Methods* 26, 1481–1496.

Kulldorff, M., and N. Nagarwalla. (1995). Spatial disease clusters: Detection and inference. *Statistics in Medicine* 14, 799–810.

Lawson, A. (2001). *Statistical Methods in Spatial Epidemiology.* Wiley, New York.

Lindgren, F., and H. Rue. (2015). Bayesian spatial modelling with R-INLA. *Journal of Statistical Software* 63(29), 1–25.

Lindgren, F., H. Rue, and J. Lindström. (2011). An explicit link between Gaussian fields and Gaussian Markov random fields: The stochastic partial differential equation approach [with discussion]. *Journal of the Royal Statistical Society: Series B* 73(4), 423–498.

Logan, M. (2010). *Biostatistical Design and Analysis Using R: A Practical Guide.* Wiley-Blackwell, New York.

Martino, S., and H. Rue. (2010). *Implementing Approximate Bayesian Inference Using Integrated Nested Laplace Approximation: A Manual for the INLA Program.* Available at http://www.math.ntnu.no/~hrue/GMRFsim/manual.pdf

Martins, G., D. Simpson, F. Lindgren, and H. Rue. (2013). Bayesian computing with INLA: New features. *Computational Statistics and Data Analysis* 67, 68–83.

Metropolis, N., A. Rosenbluth, A. Teller, and E. Teller. (1953). Equation of state calculations by fast computing machines. *Journal of Chemical Physics* 21(6), 1087–1092.

Moran, P. (1948). The interpretation of statistical maps. *Journal of the Royal Statistical Society: Series B* 10, 243–251.

Moran, P. (1950). Notes on continuous stochastic phenomena. *Biometrika* 37(1), 17–23.

Pebesma, E., and R. Bivand. (2005). Classes and methods for spatial data in R. *R News* 5(2), 9–13.

Pebesma, E., and C. Wesseling. (1998). gstat: A program for geostatistical modelling, prediction and simulation. *Computers and Geosciences* 24(1), 17–31.

Pebesma, E. J. (2004). Multivariable geostatistics in S: The gstat package. *Computers and Geosciences* 30, 683–691.

Peng, R., and F. Dominici. (2008). *Statistical Methods for Environmental Epidemiology with R: A Case Study in Air Pollution and Health.* Springer, New York.

Plummer, M., N. Best, K. Cowles, and K. Vines. (2006). CODA: Convergence diagnosis and output analysis for MCMC. *R News* 6, 7–11.

Potthoff, R., and P. Whittinghill. (1966). The interpretation of statistical maps. *Biometrika* 53, 183–190.

Ribeiro, P., and P. Diggle. (2001). geoR: A package for geostatistical analysis. *R-NEWS* 1(2), 14–18.

Richardson, S., A. Thomson, N. Best, and P. Elliott. (2004). Interpreting posterior relative risk estimates in disease-mapping studies. *Environmental Health Perspectives* 112(9), 1016–1025.

Robert, C., and G. Casella. (2004). *Monte Carlo Statistical Methods.* Springer, New York.

Rue, H., and L. Held. (2005). *Gaussian Markov Random Fields: Theory and Applications.* Chapman & Hall, London.

Rue, H., S. Martino, and N. Chopin. (2009). Approximate Bayesian inference for latent Gaussian models by using integrated nested Laplace approximations. *Journal of the Royal Statistical Society: Series B* 2(71), 1–35.

Shumway, R., and D. Stoffer. (2011). *Time Series Analysis and Its Applications, with R Examples*, 3rd ed. Springer, New York.

Stein, M. (1999). *Interpolation of Spatial Data: Some Theory of Kriging.* Springer, New York.

Stevens, M. (2009). *A Primer of Ecology with R.* Springer, New York.

Stone, R. (1988). Investigating of excess environmental risks around putative sources: Statistical problems and a proposed test. *Statistics in Medicine* 7, 649–660.

Ugarte, M. D., A. Adin, T. Goicoa, and A. F. Militino. (2014). On fitting spatio-temporal disease mapping models using approximate Bayesian inference. *Statistical Methods in Medical Research* 23(6), 507–530.

Ugarte, M. D., B. Ibáñez, and A. F. Militino. (2005). Detection of spatial variation in risk when using CAR models for smoothing relative risks. *Stochastic Environmental Research and Risk Assessment* 19(1), 33–40.

Ugarte, M. D., A. F. Militino, and T. Goicoa. (2008). Prediction error estimators in empirical Bayes disease mapping. *Environmetrics* 19(3), 287–300.

Waller, L., P. Carlin, H. X., and A. Gelfand. (1999). Hierarchical spatio-temporal models of disease rates. *Journal of the American Statistical Association* 92(438), 607–617.

Zuur, A., E. Ieno, and E. Meesters. (2012). *A Beginner's Guide to R*. Springer, New York.

Zuur, A., E. Ieno, N. Walker, A. Saveliev, and G. Smith. (2009). *Mixed Effects Models and Extensions in Ecology with R*. Springer, New York.

25

The Role of Spatial Analysis in Risk-Based Animal Disease Management

Kim B. Stevens
Veterinary Epidemiology, Economics and Public Health
Department of Production and Population Health
Royal Veterinary College
London, UK

Dirk U. Pfeiffer
Veterinary Epidemiology, Economics and Public Health
Department of Production and Population Health
Royal Veterinary College
London, UK

CONTENTS

Effective detection and control of diseases in animals (and humans) by health authorities needs to account for spatial patterns in both disease occurrence and any associated risk factors. This requires efficient data collection, management, and analysis. The integration of geographic information system (GIS) functionality into most modern disease information systems reflects the increased recognition of the importance of spatial patterns in disease control. As a result of the emerging disease risk and economic and climate challenges facing health authorities in the twenty-first century, decision makers are now looking for tools that make more effective use of the wide range of available data sources in an attempt to increase our ability to allow early detection of unusual occurrences of disease and targeted surveillance and control efforts that account explicitly for spatial variation in risk.

25.1 Mapping Disease Distribution

Disease distribution maps range from simple dot maps showing the location of disease events to predictive risk maps created using statistical algorithms applied to disease occurrence data with environmental covariates. But no matter what form they take, visualizing the spatial pattern of disease—be it at a global, national, or local scale—is fundamental in guiding risk-based disease surveillance and control strategies.

25.1.1 Why map?

Specifically, maps can be used to show the location and extent of disease distribution, to estimate disease burden, direct control and elimination efforts, and to integrate data from varied sources.

25.1.1.1 Location and extent

The simplest visualizations allow for the extent of the disease or vector habitat to be delineated and disease frequency monitored, but such maps, while useful for informing risk-based decision making, do not exploit all available data and knowledge. However, when combined with maps of environmental factors, or those highlighting the spatially heterogeneous distribution of at-risk populations, disease burden can be estimated and target populations identified (Hay et al., 2010; Tatem et al., 2011).

25.1.1.2 Estimate disease burden

Efficient location of disease control resources requires information on the spatial distribution of disease risk, and yet the epidemiology of certain diseases, such as malaria, makes it difficult to estimate disease burden using traditional surveillance-based methods. However, disease burden can be calculated by combining predictive risk maps of disease distribution with relevant population indices (Robinson et al., 2002; Hay et al., 2010). This cartographic approach allowed Hay et al. (2010) to estimate that, in 2007, 1.38 billion people lived in areas of stable *Plasmodium falciparum* malaria transmission. Estimates of disease burden can be further refined through the inclusion of additional spatial information, such as the distribution of relevant, constraining environmental factors (Guerra et al., 2006, 2008, 2010).

25.1.1.3 Direct control and elimination efforts

Visualizing disease distribution can also be fundamental for directing control and elimination efforts. Clements et al. (2013) describe how measures to eliminate malaria from endemic countries have generally adopted a spatially progressive elimination approach referred to as *shrinking the malaria map*, in which eradication efforts focus initially on the geographical perimeter of endemic areas and work inwards, effectively reducing the spatial extent of the area with disease, which allows for more efficient treatment and control. In addition, maps illustrating how the general pattern of disease distribution has changed in response to control efforts are key in documenting the progress of control and elimination efforts.

25.1.1.4 Data integration and synthesis

Apart from the fundamental role that disease distribution maps play in informing risk-based decision making, they also facilitate integration and synthesis of data from a wide range of diverse sources, each often capturing disease information at different scales (Bergquist and Tanner, 2012; Bennema et al., 2014). As such, one of the key components of spatial data integration is identifying the most appropriate scale at which to present the amalgamated data for it to be most useful to decision makers. For example, data presented at administrative level 1 (province or region) inevitably cannot capture the fine-scale heterogeneity of most infection patterns, and so tend to provide inaccurate estimates of the number of individuals requiring treatment (Brooker et al., 2010).

25.1.2 Tools for mapping disease distribution

Compared with previous decades, when the production of paper-based disease atlases was limited by the expense and inefficiency associated with producing something that was effectively out of date almost before it was published, the development of powerful geographical information system (GIS) software, together with the advent of interactive maps and virtual globes such as Google Maps™ (https://maps.google.com) and Google Earth™ (https://earth.google.com), allows for easy visualization of disease data. In addition, these Internet-based tools are a valuable alternative for nonprofit organizations and third-world countries that may lack access to the funding or training that makes even open-source GIS software, such as QGIS (Quantum GIS Development Team, 2009), an unrealistic option for visualizing the distribution of disease data.

Two examples serve to highlight the value of Google Earth technology in creating effective information resources for decision makers. First, *Nature* used the platform to track the global spatio-temporal spread of avian influenza (H5N1) (Butler, 2006; http://www.nature.com/nature/multimedia/googleearth)—a project that won the Association of Online Publishers (AOP) Use of a New Digital Platform Award in 2006. Second, when visualizing unconventional forms of georeferenced data, Google Earth is a useful alternative to GIS software. In a modern-day reprise of John Snow's 1856 cholera investigation, Google Earth enabled Baker et al. (2011) to map the spread of a typhoid outbreak in Kathmandu—where street names are not used—and trace the cause of the epidemic to low-lying public water resources. A list of all current projects using Google Earth and Google Maps can be found on the Google Earth Outreach website (https://www.google.co.uk/earth/outreach/index.html).

And yet, despite the ease with which disease distribution maps can be generated, and the myriad ways in which they inform and direct disease control and surveillance efforts, our knowledge of the global distribution of the vast majority of human and animal infectious diseases remains poor (Murray, 2012; Wertheim et al., 2012); only seven (2%) of the 355 clinically significant infectious human diseases have been comprehensively mapped (Hay et al., 2013).

Disease distribution maps can therefore play an important role in twenty-first century risk-based disease management, as the provision of timely and reliable information supports a proactive approach to disease control while allowing for early warning and response to emergent zoonoses and high-impact animal diseases. And yet, they are only a starting point; of strategic importance for decision makers is being able to identify areas of higher-than-normal risk of disease occurrence, and this can be achieved initially through the application of cluster detection tests.

25.2 Risk-Based Surveillance and Control Efforts through Identification of Disease Clusters

If spatial proximity has no influence on risk of infection and important risk factors are randomly distributed, it follows that infection should be spatially randomly distributed. A clustered spatial arrangement, on the other hand, suggests the presence of a contagious process or localized risk factor. Cluster detection tests allow for the identification of clustering of disease, and if significant, a more detailed causal epidemiological investigation may be justified. In fact, the Centers for Disease Control and Prevention (CDC) has devised a formal framework for such investigations (Anonymous, 1990), but as with other epidemiological investigations, the issues of data quality, bias and statistical power, together with diagnostic test sensitivity and specificity, must be considered when using one of the many cluster detection tests.

Two broad types of tests for spatial autocorrelation exist; global indices of spatial association—tests for clustering—provide inference in relation to the whole study area, while local methods of analysis—cluster detection tests—allow the location and extent of significant disease clusters to be characterized. The latter in particular are highly relevant to hypothesis-driven epidemiological investigations and disease surveillance. However, before going any further, it is necessary to define what is meant by the term *disease cluster.*

25.2.1 Defining disease clusters

In general, disease clusters, or ***hotspots***, are defined as regions and/or time periods in which, compared with the background probability, disease is more likely to occur. However, spatial heterogeneity of disease risk occurs over a wide range of spatial and temporal scales, and as a result, the precise designation and extent of a cluster varies, depending on the disease being studied and the specific objectives of the epidemiological investigation. For example, definitions of malaria transmission clusters differ between the pre-elimination and elimination phases. While interventions may be targeted initially at entire villages or towns having a high malaria incidence, in the latter stages of disease control, individual episodes of malaria (e.g., households) become the focus of control efforts (Clements et al., 2013). As a result, malaria hotspots can vary in size from entire countries or islands (Singh et al., 2009; Toty et al., 2010) to much smaller geographical areas (Bautista et al., 2006; Ernst et al., 2006; Bejon et al., 2010; Bousema et al., 2010). For other diseases where risk factors operate at a highly localized level, spatial heterogeneity may only be evident at a microscale (Standley et al., 2013).

25.2.2 Why identify hotspots?

Given the spatial and temporal heterogeneity of disease risk, spatial targeting of interventions is more cost-effective than uniform resource allocation (Stark et al., 2006), and therefore identification of high-risk areas is essential for informing risk-based surveillance and disease control efforts. Identification of significant disease clusters can advance our understanding of a disease in several ways, including suggesting potential risk factors for further investigation, indicating likely disease transmission routes or informing surveillance and control efforts. Cluster research typically falls into one of two broad categories.

25.2.2.1 Suggest potential risk factors

Identification of significant clusters can serve to identify potential risk factors for disease occurrence in two main ways: identifying the location of high-risk areas can generate hypotheses regarding disease occurrence and suggest potential risk factors for further investigation (Le et al., 2012; Vander Kelen et al., 2012; Calistri et al., 2013; Nogareda et al., 2013), while analysis of model residuals can highlight when the modelled predictors do not fully explain the spatial heterogeneity in disease distribution (Borba et al., 2013; Méroc et al., 2014). For example, when modelling risk factors for the occurrence of brucellosis in Maranhão, Brazil, spatial analysis of the model residuals identified a central region where disease risk remained unexplained by the predictors included in the final logistic regression model. Further investigation of this anomalous area could lead specifically to the identification of important local risk factors for brucellosis in Maranhão (Borba et al., 2013).

25.2.2.2 Explore routes of disease transmission

In contrast to cluster location suggesting potential risk factors for disease occurrence or spread, defining the scale of disease clustering through the use of global indices, such as the K-function and space–time K-function (Picado et al., 2007; Abatih et al., 2009; Minh et al., 2010; Le et al., 2012; Métras et al., 2012; Xu et al., 2012), can help to indicate possible transmission mechanisms involved in disease spread (Poljak et al., 2010; Métras et al., 2012; Sinkala et al., 2014). For example, disease clustering over short distances may suggest local transmission, such as direct contact or vector- or airborne dispersal of infected droplets (Picado et al., 2007), while clustering over distances of hundreds of kilometres can suggest long-distance spread (Ahmed et al., 2010; Picado et al., 2011) as a result of movement of infected animals or wind-borne dispersal of vectors (up to 100 km for some *Aedes* and *Culex* species [Service, 1997]) or infectious isolates (up to 300 km for the *C Noville* foot-and-mouth disease [FMD] isolate [Alexandersen et al., 2003]).

Furthermore, transmission mechanisms for a specific disease can differ between outbreaks. In their study of the spatio-temporal patterns of Rift Valley fever (RVF) in South Africa, Métras et al. (2012) suggest that while mosquito bites appeared to be the main mechanism of RVF virus infection, the differing extents and intensities of the space–time interactions exhibited by the five epidemics occurring between 2008 and 2011 suggest that other, concomitant transmission mechanisms may be involved in long-distance spread of the disease. Determining the relative contributions of the types of spread predominantly involved in an epidemic can help to inform disease control measures.

A retrospective spatio-temporal analysis (Wilesmith et al., 2003) of the UK 2001 FMD outbreak illustrates how the level of spatio-temporal interaction of a disease can both differ between regions and change over time, thereby highlighting important potential changes in the epidemiology of the disease during an outbreak—essential information if the most effective control measures are to be implemented at any one time. This study also highlighted the problems associated with analysing large-scale epidemics at an aggregated level and serves to illustrate Woolhouse's (2003) belief that "an understanding of the dynamics of the global epidemic is not a substitute for local decision making." Consequently, Picado et al. (2007) suggest that large-scale monitoring of an epidemic should occur in parallel with monitoring of events at a smaller, regional scale, thereby fostering better management of the outbreak through informed adaptation of any broad-scale disease control policy to account effectively for possible local differences in disease transmission. However, if techniques such as the space–time K-function are to be incorporated into the management of livestock epidemics, they should be interpreted together with other indicators, such as local and global predictions from disease transmission models (Picado et al., 2007).

In addition to enabling researchers to identify possible mechanisms of disease transmission, and thereby gain a better understanding of the underlying causes of the disease process, cluster detection can also be used to highlight possible regional differences in disease transmission (Vander Kelen et al., 2012), identify areas where vectors and hosts coincide, resulting in potentially increased risk of disease transmission (Hennebelle et al., 2013; Swirski et al., 2013), or track the direction and geographical extent of disease spread (Wilesmith et al., 2003; Denzin et al., 2013).

Apart from those already mentioned, other ways in which identification of disease clusters can inform surveillance and disease control strategies include cost-effective allocation of resources through targeted interventions within high-risk clusters, suggesting effective sizes for buffer and surveillance zones for specific diseases, evaluating the effectiveness of prevention plans in reducing the occurrence of disease outbreaks, and providing baseline information against which the progress of control and eradication measures can be assessed. In addition, high-risk clusters can be prioritized for future surveillance, and knowledge of their location can play an important role in disease control efforts.

Although most spatio-temporal analyses are retrospective, prospective cluster detection can also inform disease control measures, not only through syndromic surveillance (Kracalik et al., 2011; Madouasse et al., 2013), but also by identifying new or emerging diseases, by acting as early indicators of a new pathogen or outbreaks of existing disease not yet diagnosed, or by highlighting issues with existing data. For example, Hyder et al. (2011) used a prospective spatial scan statistic to identify clusters of unusually high levels of sample submissions where a diagnosis could not be reached; all clusters, except one, were the result of false positives or misclassification of sample submissions.

25.3 Network Analysis to Support Risk-Based Disease Management and Modelling of Disease Spread

Movement of livestock between farms has long been implicated as one of the main mechanisms for the spread of infectious diseases in livestock populations, and when disease incursion remains undetected, even for a short period, the results can be far reaching. In the 2001 UK FMD outbreak, it was estimated that at least 57 premises from 16 counties were infected before the first case was reported (Gibbens and Wilesmith, 2002), while in 2007, equine influenza spread rapidly through two Australian states as a result of infected horses attending an equestrian event (Cowled et al., 2009); approximately 70,000 horses on more than 9000 premises were infected, with most of the geographic dissemination occurring within the first 10 days of the epidemic.

With a view to minimizing the extent and impact of disease outbreaks, network research has focused primarily on characterizing livestock network topologies in order to identify vulnerable (likely to be infected [Rautureau et al., 2012]) and infective (likely to infect others [Bell et al., 1999]) entities, and assessing the implications of their existence for disease spread and control. While this information can be used to inform targeted tracing, surveillance, and disease control strategies, knowledge of contact network topology and incorporation of relevant network metrics have the added benefit of enhancing the realism of simulation modelling of the spread and control of disease, leading to more realistic model outcomes (Harvey et al., 2007; Dubé et al., 2011b; Lurette et al., 2011).

25.3.1 Characterization of livestock networks

Knowledge of livestock network topologies can be invaluable for informing targeted surveillance and disease control strategies, and developing strategic interventions for use in future outbreaks (Natale et al., 2009; Firestone et al., 2012). However, there are problems inherent with producing generalized pictures of livestock contact structures. Between-country comparisons of animal movement networks can be problematic due to differences in population sizes, the production types included, details in data structure, the measures applied, and the length of the period studied (Ribbens et al., 2009; Nöremark et al., 2011). For instance, livestock markets are expected to have great influence on the transmission of diseases in the UK (Robinson and Christley, 2007), while this situation is reportedly rare in Sweden (Nöremark et al., 2009).

In addition, livestock networks frequently exhibit distinct seasonal trends that can be fundamental in the development of an epidemic (Aznar et al., 2011; Bajardi et al., 2011; Vernon, 2011). For example, it has been shown that a large, nationwide UK FMD epidemic is more likely if the disease is introduced into the cattle population in late summer or early autumn, when the majority of cattle movements occur (Green et al., 2006). Thus, for research into network topology to be able to reliably inform future outbreak interventions, analysis of successive years of movement data is essential in order to identify potentially crucial seasonal behaviours and recurrent patterns, rather than looking at just a single snapshot in time.

Research into the characterization of livestock networks has focused primarily on pigs (Bigras-Poulin et al., 2007; Martínez-López et al., 2009; Lentz et al., 2011; Nöremark et al., 2011; Rautureau et al., 2012; Büttner et al., 2015) and cattle (Bigras-Poulin et al., 2006; Natale et al., 2009; Rautureau et al., 2011), with fewer studies devoted to sheep (Webb, 2005, 2006; Kiss et al., 2006a; Volkova et al., 2010) and poultry (Lockhart et al., 2010; Nickbakhsh et al., 2011; Van Steenwinkel et al., 2011; Rasamoelina-Andriamanivo et al., 2014).

25.3.1.1 Pig trade networks

Irrespective of country, swine networks generally demonstrate small-world and scale-free topologies (Bigras-Poulin et al., 2007; Lentz et al., 2011; Nöremark et al., 2011; Rautureau et al., 2012; Büttner et al., 2013; Dorjee et al., 2013; Smith et al., 2013; Thakur et al., 2014). In addition, the pig industry has a hierarchical structure with a unidirectional flow of animals from one production type to another (Lindström et al., 2010; Rautureau et al., 2012), but within a production type premises are in close trade contact with each other, thereby creating large-scale communities separated by "trade borders," as contact between such communities is rare (Lindström et al., 2010; Lentz et al., 2011; Rautureau et al., 2012). Identification of such communities could be useful when implementing regionalization or compartmentalization approaches to disease control (Scott et al., 2006); regions or groups of farms behaving as a single epidemiological unit (compartment) in terms of disease transmission risk could be identified, together with risk-free communities having no links to infected groups. Such an approach could facilitate resumption or continuation of trade from disease-free regions or risk-free herds within an infected region. Similar trade communities are also apparent in the UK cattle industry (Green et al., 2006).

The different pig production types exhibit different susceptibilities to disease introduction and spread. In general, grower units are well positioned to both spread and receive disease (having high degree and betweenness), while finishing units have a potentially high risk for disease introduction as a result of their high in-degree and in-going infection chain centrality values (Rautureau et al., 2012). Unsurprisingly, production type of swine holdings has been shown to be the most influential factor influencing final epidemic size and dynamic behaviour of an outbreak within the industry (Lindström et al., 2010, 2011). Despite inclusion of such information improving risk assessments (Lindström et al., 2011), European Union law does not require that this crucial data be recorded. However, certain countries, such as Sweden, have chosen to include data on production type in their livestock movement databases (Lindström et al., 2010).

25.3.1.2 Cattle trade networks

Cattle trade networks do not exhibit the hierarchical topology of the pig industry and, as a result, appear to be more vulnerable to disease spread than pig networks (Rautureau et al., 2012). Instead, cattle networks are generally characterized by the presence of a giant strong component (GSC) (Rautureau et al., 2012; Dutta et al., 2014), short path lengths—particularly in western European countries (Natale et al., 2009; Nöremark et al., 2009; Vernon, 2011), and high clustering, corresponding to a scale-free network (Natale et al., 2009; Aznar et al., 2011; Dubé et al., 2011a) with small-world properties (Natale et al., 2009; Dubé et al., 2011a). However, within these broad commonalities, cattle trade networks exhibit a large degree of heterogeneity. Apart from distinct seasonal trends (Aznar et al., 2011; Bajardi et al., 2011, 2012; Vernon, 2011; Dutta et al., 2014), variables related to number of movements and distances covered are generally highly right-skewed (Bigras-Poulin et al., 2006), which has implications for the rapid long-distance spread of infection. In fact, it was this characteristic of cattle trade networks that contributed to the failure of the culling strategy imposed during the 2001 UK FMD epidemic (Shirley and Rushton, 2005b).

Cattle movement networks also frequently display spatial heterogeneity, as certain areas can be predominantly exporters, while other areas may be predominantly importers of animals (Aznar et al., 2011), which has implications in outbreak situations and when implementing targeted surveillance and disease control strategies.

25.3.2 Using livestock network characteristics for surveillance and disease control

Infectious diseases are seldom homogenously spread within the population, which is why applying a risk-based sampling approach is favoured to optimize use of surveillance and disease control resources (Stark et al., 2006; Cannon, 2009). In fact, selection of herds based on network parameters, rather than random sampling, has been shown to enhance sensitivity of surveillance programmes (Callaway et al., 2000; Frossling et al., 2012). Understanding how the different network characteristics facilitate disease spread is fundamental for designing surveillance and disease control measures that work with, rather than against, network structure.

25.3.2.1 Small-world and scale-free networks

Small-world and scale-free networks present two distinct topologies when viewed from the perspective of disease spread and targeted surveillance and control measures. Small-world networks are characterized by high clustering coefficients and short average path lengths, although these may not be geographically short (Watts and Strogatz, 1998). Infectious diseases spread more easily in small-world networks than in regular (Watts and Strogatz, 1998) or random (Christley et al., 2005b) lattices, and disease can disseminate rapidly within clusters and has the potential to spread rapidly to geographically distant farms via only a few links (Christley et al., 2005a). However, the high level of clustering lowers the threshold at which an infectious agent can infect all nodes, effectively increasing the reproductive ratio of the epidemic. However, this may lead to localized depletion of susceptible individuals within clusters, and fewer individuals infected overall (Newman, 2003). Within a small-world network, disease spread can be successfully constrained by targeting the links that connect clusters (Cohen et al., 2000).

Scale-free networks are characterized by many nodes with few connections, while a small proportion of nodes will have many connections, resulting in hubs that can act as "super-spreaders" (Barabási and Albert, 1999). The presence of well-connected hubs means that infection initially spreads rapidly, but dissemination gradually slows once the primary (farms directly connected to hubs) and secondary (farms not directly connected to hubs but connected to primary contacts) contacts have been infected (Shirley and Rushton, 2005a; Kiss et al., 2006b). In such networks, disease control strategies are most effective when targeting the hubs (Shirley and Rushton, 2005a; Dubé et al., 2011a).

25.3.2.2 Degree

Degree refers to the number of links a node has. Farms with a high out-degree (number of off-farm links) are generally central to the spread of disease, while farms with a high in-degree (number of on-farm links) are at greatest risk of disease introduction (Shirley and Rushton, 2005a) and can be targeted for high surveillance sensitivity and to demonstrate freedom from disease (Frossling et al., 2012). Farms with both high in- and out-degree act as hubs for disease spread, as, once infected, disease can spread rapidly in a scale-free network. Control measures applied strategically to such hubs will be more effective than if applied to a random sample of farms (Cameron, 2012). Farms with a low in- and high out-degree have a low probability of disease introduction but high risk of spreading disease and can be targeted to mitigate the impact of exotic disease introduction (Cameron, 2012).

However, some farms with few direct contacts may have many indirect contacts, and as there is an association between disease occurrence and number of both direct and indirect sequential contacts, sampling or targeting decisions based only on degree could lead to high-risk farms being overlooked (Nöremark et al., 2011). Infection chain, which takes account of both direct and indirect contacts, is therefore possibly a better metric upon which to base surveillance and disease control decisions.

25.3.2.3 Infection chain

During an outbreak, analysis of in- and out-going infection chains (IIC and OIC) is invaluable for contact tracing, as it takes account of a node's direct and indirect contacts, together with the temporal sequence of these contacts (Dubé et al., 2008; Nöremark et al., 2011). Prior to any outbreak, characterization of infection chains in a given network can be used to prepare for and mitigate an outbreak, as well as estimate its likely magnitude.

IIC can be used to identify the most likely place for disease occurrence (Cannon, 2009; Nöremark et al., 2011). One example of when this could be useful is when a previously absent infection is first detected in a region. In such situations, it is often necessary to conduct a rapid survey to detect infected herds and to assess the spread within a region to support decisions regarding strategies for control. In such a survey, targeted sampling of herds with many contacts can mirror a larger part of the population, and the same number of samples could then give higher survey sensitivity than a random sample (Nöremark et al., 2011).

As farms with high OIC—also termed accessible world (Webb, 2006)—have the potential to spread disease to a large population (Nöremark et al., 2011), this metric is possibly best for estimating potential epidemic size (Büttner et al., 2015; Thakur et al., 2014).

EpiContactTrace (Nöremark and Widgren, 2014), an R-based social network analysis tool, enables identification of farms with many contacts, either directly through degree measures or sequentially through infection chain. This can be useful for risk-based surveillance when identifying parts of the population where the consequences would be large if infection were introduced. Correspondingly, the tool can identify farms with many in-going contacts and high likelihood of introduction, which can be useful for selection of strata to target for sampling, either in an outbreak situation or during ongoing surveillance programmes, with an aim to increase the likelihood of early detection or estimate probability of freedom.

25.3.2.4 Giant strong component

The presence of a giant strong component (GSC) can be useful when implementing disease control measures, as sorting network nodes on the basis of centrality scores such as degree allows the size of the largest connected component to be drastically reduced by isolating only a small percent of the nodes. However, the highly dynamic nature of certain trade networks (e.g., cattle) means that a specific node may be in a strategic position for controlling disease flow only within a certain time frame, after which it may cease to play any significant role (Vernon and Keeling, 2009; Bajardi et al., 2011).

However, parameterizing contact networks using only movement data may not adequately capture disease transmission dynamics, as a significant amount of local spread can increase epidemic size considerably, but may have only a small impact on the spatial extent of the disease (Green et al., 2006). To differentiate between network-based and local spatial spread, a spatially explicit element could be incorporated into the analysis. Examples of such approaches include combining network analysis with spatio-temporal cluster detection (Martínez-López et al., 2009; Firestone et al., 2011; Bajardi et al., 2012; Sánchez-Matamoros et al., 2013), including data on the spatial relationship between farms (Álvarez et al., 2011; Firestone et al., 2012) or whether farms occur in high or low-risk areas (Green et al., 2008). Networks incorporating both spatial and underlying contact relationships are useful for informing the placement of control zones (Firestone et al., 2012). However, network studies with a view to inform risk-based disease control are rare. Vicente-Rubiano et al. (2014) used network analysis together with spatio-temporal cluster detection to assess the risk of reintroduction of Aujeszky's disease into disease-free areas in Spain, where analysis of pig movements revealed distinct north–south compartments that could be useful in guiding risk-based measures that reduce the risk of reintroduction of Aujeszky's disease and allow authorities to stop vaccinating.

Disease control efforts based on knowledge of livestock movement data are fundamental for reducing risk of disease spread and preventing large economic losses, and can range from implementation of buffer zones and movement restrictions to inclusion of spatially explicit kernels in mathematical models of disease spread (Buhnerkempe et al., 2014). However, to correctly evaluate such preventive and control measures, detailed knowledge and regulation of animal movements and contact network topology are essential, as failure to account for the complexity of social networks through which diseases may be transmitted may in turn be responsible for the failure of disease control strategies based simply on one-off population reduction (Bajardi et al., 2011).

Although disease containment is key, when using network metrics to inform disease control measures, it is important to also consider other factors. For example, when compared with a random selection of premises, targeted control measures based on degree are more efficient in terms of epidemic containment, but produce similar impact in terms of trade detriment. On the other hand, targeted preventive and control measures based on betweenness and eigenvector values are more efficient in terms of epidemic containment, but have a lower impact on trade flow (Natale et al., 2009).

25.4 Spatial Modelling of Disease to Support Risk-Based Surveillance and Control Efforts

Spatial variation in disease risk is determined by interactions between pathogens, vectors, and hosts, and between these agents and their environment. In addition, the increased trade and globalization that characterizes the twenty-first century, together with the effects of climate change, has necessitated a growing demand to assess the risk of disease spread and emergence of new pathogens. As a result, spatially explicit models of disease occurrence are valuable tools for furthering our understanding of disease dynamics, while the associated risk maps provide a geographical representation of disease risk for informing risk-based disease control and surveillance strategies. The inclusion of uncertainty in model outputs (Clements et al., 2006; Stevens et al., 2013) using, for example, Dempster–Shafer theory, allows for identification of areas for which the collection of additional information might be beneficial. Expressions of uncertainty are an important attribute of risk predictions for policy development, in that they will allow the policy makers to appreciate the relative strength of the scientific evidence.

25.4.1 Modelling methods

Spatial modelling methods can be broadly grouped into data- and knowledge-driven methods (Pfeiffer et al., 2008; Stevens and Pfeiffer, 2011). The former use a dataset comprising several risk factors together with an outcome variable (e.g., presence or absence of disease), and risk factor effect estimates and significance levels are usually obtained using regression methods. Data-driven approaches can be further subdivided, depending on whether they require both disease presence and absence data to calibrate the model (e.g., generalized linear mixed models and boosted regression trees) or presence-only data (e.g., maximum entropy [Maxent] [Phillips and Dudík, 2008; Phillips, 2012] or genetic algorithm for rule-set production [GARP] [Stockwell and Peters, 1999]). Knowledge-driven methods, on the other hand, such as multicriteria decision making (MCDA), require prior definition of the risk factor variables and of the relationship between individual risk factors and the outcome (Malczewski, 2006).

Among data-driven methods, Bayesian approaches used to be a major focus of development, but they have recently been complemented by machine learning methods such as random forest and boosted regression tree methods—approaches that are considered to be less affected by missing values, nonlinearity, autocorrelation, lack of independence, and distributional assumptions than parametric methods. They are also better at dealing with large datasets. In addition, several comparative reviews of the performance of the different modelling methods (Hirzel et al., 2006; Elith and Graham, 2009) suggest that, in general, tree-based regression methods tend to perform slightly better than other spatial regression approaches. As a result, boosted regression trees are being used with increasing frequency to predict vector distributions and probability of disease risk (Mullins et al., 2011; Van Boeckel et al., 2012; Pigott et al., 2014; Stevens and Pfeiffer, 2015).

However, a common problem with disease regression modelling is that while the outcome variable may consist of fairly reliable disease presence data, absence data may not be available (e.g., surveillance data) or absence of disease reporting may not reflect true absence of disease. This is also common in ecological species distribution modelling and has led to the development of different sampling approaches to generate pseudoabsence data that can be used with regression methods requiring both presence and absence data, as well as the development of specific modelling techniques requiring presence-only data and background data (e.g., Maxent or GARP).

This increasingly wide range of spatial modelling approaches has been used to model the spatial distribution of an array of pathogens and disease vectors, including highly pathogenic avian influenza (HPAI) (H5N1) (Williams et al., 2008; Williams and Peterson, 2009; Hogerwerf et al., 2010; Stevens et al., 2013; Stevens and Pfeiffer, 2015), *Bacillus anthracis*—the causative agent of anthrax (Joyner et al., 2010; Mullins et al., 2011; Chikerema et al., 2013), Leishmaniasis vector species (Peterson and Shaw, 2003; Peterson et al., 2004; Nieto et al., 2006), and tuberculosis occurrence in domestic and wild ungulates in south-central Spain (Rodriguez-Prieto et al., 2012), to name but a few. In addition, studies have modelled the spatial distribution of freshwater snail-hosting trematodes in Zimbabwe under current and future climate scenarios (Pedersen et al., 2014), defined the climatic niches of tick species involved in disease transmission in the Mediterranean region and forecast changes in their habitat suitability as a result of the effects of climate change (Estrada-Pena and Venzal, 2007), and predicted the global risk of spread of the mosquito *Aedes albopictus* (Benedict et al., 2007). However, pathogen notwithstanding, spatial modelling studies generally focus on a few main themes, including the identification of factors associated with disease, distribution of vectors, disease emergence as a result of climate change or globalization, and more recently, the field of phylogeography.

25.4.2 Why model?

25.4.2.1 Identification of risk factors

The majority of spatial modelling studies focus on the identification of factors associated with disease introduction, spread, persistence, or occurrence. Until relatively recently, these studies used traditional regression methods to obtain risk factor effect estimates that, while informative, do not always provide a complete picture. However, newer methods, such as boosted regression trees and Maxent, can provide additional information, such as percentage contribution and profiles of the effect of each individual predictor on the outcome over the range of its values. Such information can be useful for designing surveillance strategies that target regions with these characteristics. For example, although studies using traditional regression methods have frequently identified an increasing density of domestic waterfowl (Gilbert et al., 2006, 2008; Pfeiffer et al., 2007; Minh et al., 2009; Martin et al., 2011) and rice growing (Pfeiffer et al., 2007; Gilbert et al., 2008) as important risk factors for occurrence of HPAI H5N1 in Asia, Stevens and Pfeiffer (2015) used boosted regression tree and Maxent models to show that roughly three-quarters of the spatial variation in distribution of HPAI H5N1 in domestic poultry in Asia could be attributed to these two variables. Similarly, Van Boeckel et al.'s (2012) use of a boosted regression tree model, and the resulting variable profiles for each predictor variable, not only enabled the identification of intensively raised ducks as an important risk factor for H5N1 occurrence, but also showed the range of values over which their effect was most important.

Although spatial modelling methods are valuable for identifying factors significantly associated with disease occurrence, users need to be aware of the limitations associated with these methods. For example, sampling biases due to variation in surveillance intensity and reporting motivation have the potential to effect model predictions considerably, resulting in overemphasis on the ecological variables in areas of intense sampling. In addition, certain factors, such as human population density, may act as confounders, with high human population densities increasing the probability of disease detection and leading to artificial bias in the disease data.

25.4.2.2 Vector distribution

Modelling suitable habitats for disease vectors allows us, by proxy, to identify areas with a high risk of occurrence of vector-borne disease, especially if these areas coincide with areas of high livestock distribution. Models have been developed for a number of important vector-borne diseases, including RVF (Clements et al., 2006, 2007a, 2007b; Mweya et al., 2013; Sallam et al., 2013; Conley et al., 2014), bluetongue (Purse et al., 2004; Guis et al., 2007; Calvete et al., 2008; Hartemink et al., 2014), leishmaniasis (Chamaillé et al., 2010; Gonzalez et al., 2010), and trypanosomiasis (Selby et al., 2013; Dicko et al., 2014). Such models can help to identify priority sites for implementation of control measures to reduce exposure to a vector or a pathogen of a vector-borne disease. For example, spatial MCDA was used to combine knowledge on tsetse fly distribution (the vector for trypanosomiasis), cattle density, land cover, crop-use intensity, bird species richness and erosion data, in order to prioritize areas for control of the vector in Zambia (Symeonakis et al., 2007). However, removal of tsetse flies from a region opens up those areas for livestock and cropping with the concomitant risk of overgrazing and soil erosion. Using a series of hypothetical scenarios, the authors were able to identify areas most likely to benefit from tsetse fly eradication without the associated detrimental environmental effects.

25.4.2.3 Climate change and disease emergence

Intensified global connectivity through trade and human travel offers increasing opportunities for the emergence and re-emergence of vector-borne and zoonotic diseases, followed by widespread dissemination. In addition, global averages of combined land and ocean surface temperatures show an increasing trend (IPCC, 2014), and changes in the distribution of species in response to climate change have been observed worldwide, with the range limits of more than 1700 species having shifted northward an average of 6 km each decade (Parmesan and Yohe, 2003). A similar change in distribution, together with potential elongation of seasons, has been predicted for a range of pathogens and parasites (Rose and Wall, 2011; Wall et al., 2011; Guichard et al., 2014; Jore et al., 2014; Simon et al., 2014; Seo et al., 2015).

Species distribution models such as Maxent are powerful tools for modelling the potential distribution of disease species under future climate scenarios. However, risk maps designed to identify suitable habitats beyond those currently occupied by a pathogen might seem unreliable if apparently false positive predictions show up on new continents, unless previous long-term events are taken into account. For example, 30 years ago, few would have believed that bluetongue would spread to Europe, yet the virus found not only suitable habitats, but also competent indigenous vectors. Bluetongue has shown two distinct patterns of geographical emergence over the past decade: a steady northward spread of the virus, associated with range expansion of the midge *Culicoides imicola* vector, together with long-distance jumps into new territories. The risk of the former was correctly predicted using statistical risk mapping methods (Tatem et al., 2003), although the latter was unforeseen, as the areas of emergence were beyond the range of *C. imicola*. However, northern expansion of

C. imicola led to range overlap with indigenous northern *Culicoides* species, which proved to be competent vectors for the disease (Randolph and Rogers, 2010).

There are many parts of the world where conditions are suitable for the establishment and spread of a range of pathogens and vectors, should they be introduced, and active surveillance aimed at detecting introduction into those areas highlighted by risk maps as suitable for a specific pathogen or vector is the best first line of defence (Randolph and Rogers, 2010).

25.4.2.4 Phylogeography

Increased access to molecular information on hosts and pathogens has resulted in the emerging field of phylogeography, which integrates geospatial with genetic data (Liang et al., 2010; Chan et al., 2011; Faria et al., 2011; Pybus et al., 2012; Carrel and Emch, 2013; Alvarado-Serrano and Knowles, 2014). Combining species distribution models (SDMs) and phylogeography is particularly informative for studies of globally distributed pathogens, where environmental associations may be linked to genetic variation. For example, within the narrow confines of South Africa's Kruger National Park, the evolutionarily-distinct *B. anthracis* A and B branches display spatial and ecological differences, with type B isolates generally found on soils with significantly higher calcium and pH levels than type A isolates (Smith et al., 2000). In addition, phylogeographic modelling of *B. anthracis* populations in the United States, Italy, and Kazakhstan identified different ecological associations within each country, suggesting niche specialization within the different sublineages (Mullins et al., 2013). Regionally specific genetic variation has been described in other diseases, including HPAI H5N1 (Carrel et al., 2010), rabies (Horton et al., 2015), and Lyme disease (Mechai et al., 2015).

From a methodological perspective, knowing that certain lineages exhibit niche specialization and unique geographic distributions can improve model accuracy by dividing a large population into biologically meaningful subpopulations (Mullins et al., 2013). However, ignoring such genetic variations may result in SDMs that are biased toward a dominant strain in a particular region.

25.5 Conclusion

Spatial analysis tools have become a key component for the assessment, management, and communication of animal disease risks. The visualization of disease risk patterns is immensely powerful as a communication tool aimed at policy makers as well as other stakeholders, but is also very useful for stimulating epidemiological thinking during the analysis process. The analytical tools for identifying hotspots and producing risk maps make an essential contribution to the scientific evidence base for informing policy development, although it needs to be acknowledged that many policy makers still struggle with the "black box" nature, in particular, of model outputs and how to account for the uncertainty associated with model predictions. The communication between data analysts and those generating data or using analytical outputs needs to become more effective so that we are able to better exploit what spatial analysis has to offer for the management of animal health risks. The volume and variety of data have increased enormously in recent years and will continue to do so at an accelerated pace. However, data quality will remain an issue, and that aspect will have to receive a stronger emphasis in the development of analytical tools.

References

Abatih, E.N., Ersbøll, A.K., Lo Fo Wong, D.M.A., Emborg, H.D. (2009). Space–time clustering of ampicillin resistant *Escherichia coli* isolated from Danish pigs at slaughter between 1997 and 2005. *Preventive Veterinary Medicine* 89, 90–101.

Ahmed, S.S.U., Ersboll, A.K., Biswas, P.K., Christensen, J.P. (2010). The space-time clustering of highly pathogenic avian influenza (HPAI) H5N1 outbreaks in Bangladesh. *Epidemiology and Infection* 138, 843–852.

Alexandersen, S., Zhang, Z., Donaldson, A.I., Garland, A.J.M. (2003). The pathogenesis and diagnosis of foot-and-mouth disease. *Journal of Comparative Pathology* 129, 1–36.

Alvarado-Serrano, D.F., Knowles, L.L. (2014). Ecological niche models in phylogeographic studies: applications, advances and precautions. *Molecular Ecology Resources* 14, 233–248.

Álvarez, L.G., Webb, C.R., Holmes, M.A. (2011). A novel field-based approach to validate the use of network models for disease spread between dairy herds. *Epidemiology and Infection* 139, 1863–1874.

Anonymous. (1990). Guidelines for investigating clusters of health events. *Morbidity and Mortality Weekly Report* 39, 1–16.

Aznar, M.N., Stevenson, M.A., Zarich, L., León, E.A. (2011). Analysis of cattle movements in Argentina, 2005. *Preventive Veterinary Medicine* 98, 119–127.

Bajardi, P., Barrat, A., Natale, F., Savini, L., Colizza, V. (2011). Dynamical patterns of cattle trade movements. *PLoS One* 6, e19869.

Bajardi, P., Barrat, A., Savini, L., Colizza, V. (2012). Optimizing surveillance for livestock disease spreading through animal movements. *Journal of the Royal Society Interface* 9, 2814–2825.

Baker, S., Holt, K.E., Clements, A.C.A., Karkey, A., Arjyal, A., Boni, M.F., Dongol, S., et al. (2011). Combined high-resolution genotyping and geospatial analysis reveals modes of endemic urban typhoid fever transmission. *Open Biology* 1, 110008. DOI: 10.1098/rsob.110008.

Barabási, A.-L., Albert, R. (1999). Emergence of scaling in random networks. *Science* 286, 509–512.

Bautista, C.T., Chan, A.S.T., Ryan, J.R., Calampa, C., Roper, M.H., Hightower, A.W., Magill, A.J. (2006). Epidemiology and spatial analysis of malaria in the northern Peruvian Amazon. *American Journal of Tropical Medicine and Hygiene* 75, 1216–1222.

Bejon, P., Williams, T., Liljander, A., Noor, A., Wambua, J., Ogada, E., Olotu, A., Osier, F., Hay, S., Farnert, A., Marsh, K. (2010). Stable and unstable malaria hotspots in longitudinal cohort studies in Kenya. *PLoS Medicine* 7, e1000304.

Bell, D.C., Atkinson, J.S., Carlson, J.W. (1999). Centrality measures for disease transmission networks. *Social Networks* 21, 1–21.

Benedict, M., Levine, R., Hawley, W., Lounibos, L. (2007). Spread of the tiger: global risk of invasion by the mosquito *Aedes albopictus*. *Vector Borne Zoonotic Diseases* 7, 76–85.

Bennema, S.C., Scholte, R.G.C., Molento, M.B., Medeiros, C., Carvalho, O.d.S. (2014). *Fasciola hepatica* in bovines in Brazil: data availability and spatial distribution. *Revista do Instituto de Medicina Tropical de São Paulo* 56, 35–41.

Bergquist, N., Tanner, M. (2012). Visual approaches for strengthening research, science communication and public health impact. *Geospatial Health* 6, 155–156.

Bigras-Poulin, M., Barfod, K., Mortensen, S., Greiner, M. (2007). Relationship of trade patterns of the Danish swine industry animal movements network to potential disease spread. *Preventive Veterinary Medicine* 80, 143–165.

Bigras-Poulin, M., Thompson, R.A., Chriel, M., Mortensen, S., Greiner, M. (2006). Network analysis of Danish cattle industry trade patterns as an evaluation of risk potential for disease spread. *Preventive Veterinary Medicine* 76, 11–39.

Borba, M.R., Stevenson, M.A., Gonçalves, V.S.P., Neto, J.S.F., Ferreira, F., Amaku, M., Telles, E.O., et al. (2013). Prevalence and risk-mapping of bovine brucellosis in Maranhão State, Brazil. *Preventive Veterinary Medicine* 110, 169–176.

Bousema, T., Drakeley, C., Gesase, S., Hashim, R., Magesa, S., Mosha, F., Otieno, S., et al. (2010). Identification of hot spots of malaria transmission for targeted malaria control. *Journal of Infectious Diseases* 201, 1764–1774.

Brooker, S., Hotez, P.J., Bundy, D.A.P. (2010). The global atlas of helminth infection: mapping the way forward in neglected tropical disease control. *PLoS Neglected Tropical Diseases* 4, e779.

Buhnerkempe, M.G., Tildesley, M.J., Lindström, T., Grear, D.A., Portacci, K., Miller, R.S., Lombard, J.E., Werkman, M., Keeling, M.J., Wennergren, U., Webb, C.T. (2014). The impact of movements and animal density on continental scale cattle disease outbreaks in the United States. *PLoS One* 9, e91724.

Butler, D. (2006). Mashups mix data into global service. *Nature* 439, 6–7.

Büttner, K., Krieter, J., Traulsen, I. (2015). Characterization of contact structures for the spread of infectious diseases in a pork supply chain in northern Germany by dynamic network analysis of yearly and monthly networks. *Transboundary and Emerging Diseases* 62, 188–199.

Büttner, K., Krieter, J., Traulsen, A., Traulsen, I. (2013). Static network analysis of a pork supply chain in northern Germany—characterisation of the potential spread of infectious diseases via animal movements. *Preventive Veterinary Medicine* 110, 418–428.

Calistri, P., Iannetti, S., Atzeni, M., Di Bella, C., Schembri, P., Giovannini, A. (2013). Risk factors for the persistence of bovine brucellosis in Sicily from 2008 to 2010. *Preventive Veterinary Medicine* 110, 329–334.

Callaway, D.S., Newman, M.E.J., Strogatz, S.H., Watts, D.J. (2000). Network robustness and fragility: percolation on random graphs. *Physical Review Letters* 85, 5468–5471.

Calvete, C., Estrada, R., Miranda, M.A., Borrás, D., Calvo, J.H., Lucientes, J. (2008). Modelling the distributions and spatial coincidence of bluetongue vectors *Culicoides imicola* and the *Culicoides obsoletus* group throughout the Iberian peninsula. *Medical and Veterinary Entomology* 22, 124–134.

Cameron, A. (2012). The consequences of risk-based surveillance: developing output-based standards for surveillance to demonstrate freedom from disease. *Preventive Veterinary Medicine* 105, 280–286.

Cannon, R.M. (2009). Inspecting and monitoring on a restricted budget—where best to look? *Preventive Veterinary Medicine* 92, 163–174.

Carrel, M., Emch, M. (2013). Genetics: a new landscape for medical geography. *Annals of the Association of American Geographers* 103, 1452–1467.

Carrel, M.A., Emch, M., Jobe, R.T., Moody, A., Wan, X.-F. (2010). Spatiotemporal structure of molecular evolution of H5N1 highly pathogenic avian influenza viruses in Vietnam. *PLoS One* 5, e8631.

Chamaillé, L., Tran, A., Meunier, A., Bourdoiseau, G., Ready, P., Dedet, J.-P. (2010). Environmental risk mapping of canine leishmaniasis in France. *Parasites and Vectors* 3, 31.

Chan, L.M., Brown, J.L., Yoder, A.D. (2011). Integrating statistical genetic and geospatial methods brings new power to phylogeography. *Molecular Phylogenetics and Evolution* 59, 523–537.

Chikerema, S.M., Murwira, A., Matope, G., Pfukenyi, D.M. (2013). Spatial modelling of *Bacillus anthracis* ecological niche in Zimbabwe. *Preventive Veterinary Medicine* 111, 25–30.

Christley, R.M., Pinchbeck, G.L., Bowers, R.G., Clancy, D., French, N.P., Bennett, R., Turner, J. (2005a). Infection in social networks: using network analysis to identify high-risk individuals. *American Journal of Epidemiology* 162, 1024–1031.

Christley, R.M., Robinson, S.E., Lysons, R., French, N.P. (2005b). Network analysis of cattle movements in Great Britain. In *Proceedings of the Society for Veterinary Epidemiology and Preventive Medicine (SVEPM)*, pp. 234–244, Nairn, Inverness, Scotland.

Clements, A., Pfeiffer, D., Martin, V. (2006). Application of knowledge-driven spatial modelling approaches and uncertainty management to a study of Rift Valley fever in Africa. *International Journal of Health Geographics* 5, 57.

Clements, A., Pfeiffer, D., Martin, V., Otte, M. (2007a). A Rift Valley fever atlas for Africa. *Preventive Veterinary Medicine* 82, 72–82.

Clements, A.C.A., Pfeiffer, D.U., Martin, V., Pittliglio, C., Best, N., Thiongane, Y. (2007b). Spatial risk assessment of Rift Valley fever in Senegal. *Vector Borne Zoonotic Diseases* 7, 203–216.

Clements, A.C.A., Reid, H.L., Kelly, G.C., Hay, S.I. (2013). Further shrinking the malaria map: how can geospatial science help to achieve malaria elimination? *Lancet Infectious Diseases* 13, 709–718.

Cohen, R., Erez, K., Ben-Avraham, D., Havlin, S. (2000). Resilience of the Internet to random breakdowns. *Physical Review Letters* 85, 4626–4628.

Conley, A.K., Fuller, D.O., Haddad, N., Hassan, A.N., Gad, A.M., Beier, J.C. (2014). Modeling the distribution of the West Nile and Rift Valley fever vector *Culex pipiens* in arid and semi-arid regions of the Middle East and North Africa. *Parasites and Vectors* 7, 289–289.

Cowled, B., Ward, M.P., Hamilton, S., Garner, G. (2009). The equine influenza epidemic in Australia: spatial and temporal descriptive analyses of a large propagating epidemic. *Preventive Veterinary Medicine* 92, 60–70.

Denzin, N., Borgwardt, J., Freuling, C., Müller, T. (2013). Spatio-temporal analysis of the progression of Aujeszky's disease virus infection in wild boar of Saxony-Anhalt, Germany. *Geospatial Health* 8, 2013–2213.

Dicko, A.H., Lancelot, R., Seck, M.T., Guerrini, L., Sall, B., Lo, M., Vreysen, M.J.B., Lefrançois, T., Fonta, W.M., Peck, S.L., Bouyer, J. (2014). Using species distribution models to optimize vector control in the framework of the tsetse eradication campaign in Senegal. *Proceedings of the National Academy of Sciences of the United States of America* 111, 10149–10154.

Dorjee, S., Revie, C.W., Poljak, Z., McNab, W.B., Sanchez, J. (2013). Network analysis of swine shipments in Ontario, Canada, to support disease spread modelling and risk-based disease management. *Preventive Veterinary Medicine* 112, 118–127.

Dubé, C., Ribble, C., Kelton, D., McNab, B. (2008). Comparing network analysis measures to determine potential epidemic size of highly contagious exotic diseases in fragmented monthly networks of dairy cattle movements in Ontario, Canada. *Transboundary and Emerging Diseases* 55, 382–392.

Dubé, C., Ribble, C., Kelton, D., McNab, B. (2011a). Estimating potential epidemic size following introduction of a long-incubation disease in scale-free connected networks of milking-cow movements in Ontario, Canada. *Preventive Veterinary Medicine* 99, 102–111.

Dubé, C., Ribble, C., Kelton, D., McNab, B. (2011b). Introduction to network analysis and its implications for animal disease modelling. *Revue Scientifique et Technique* 30, 425–436.

Dutta, B.L., Ezanno, P., Vergu, E. (2014). Characteristics of the spatio-temporal network of cattle movements in France over a 5-year period. *Preventive Veterinary Medicine* 117, 79–94.

Elith, J., Graham, C. (2009). Do they? How do they? Why do they differ? On finding reasons for differing performances of species distribution models. *Ecography* 30, 129–151.

Ernst, K., Adoka, S., Kowuor, D., Wilson, M., John, C. (2006). Malaria hotspot areas in a highland Kenya site are consistent in epidemic and non-epidemic years and are associated with ecological factors. *Malaria Journal* 5, 78.

Estrada-Pena, A., Venzal, J.M. (2007). Climate niches of tick species in the Mediterranean region: modeling of occurrence data, distributional constraints and impact of climate change. *Journal of Medical Entomology* 44, 1130–1138.

Faria, N.R., Suchard, M.A., Rambaut, A., Lemey, P. (2011). Toward a quantitative understanding of viral phylogeography. *Current Opinion in Virology* 1, 423–429.

Firestone, S.M., Christley, R.M., Ward, M.P., Dhand, N.K. (2012). Adding the spatial dimension to the social network analysis of an epidemic: investigation of the 2007 outbreak of equine influenza in Australia. *Preventive Veterinary Medicine* 106, 123–135.

Firestone, S.M., Ward, M.P., Christley, R.M., Dhand, N.K. (2011). The importance of location in contact networks: describing early epidemic spread using spatial social network analysis. *Preventive Veterinary Medicine* 102, 185–195.

Frossling, J., Ohlson, A., Bjorkman, C., Hakansson, N., Noremark, M. (2012). Application of network analysis parameters in risk-based surveillance—examples based on cattle trade data and bovine infections in Sweden. *Preventive Veterinary Medicine* 105, 202–208.

Gibbens, J.C., Wilesmith, J.W. (2002). Temporal and geographical distribution of cases of foot-and-mouth disease during the early weeks of the 2001 epidemic in Great Britain. *Veterinary Record* 151, 407–412.

Gilbert, M., Chaitaweesub, P., Parakamawongsa, T., Premashthira, S., Tiensin, T., Kalpravidh, W., Wagner, H., Slingenbergh, J. (2006). Free-grazing ducks and highly pathogenic avian influenza, Thailand. *Emerging Infectious Diseases* 12, 227–234.

Gilbert, M., Xiao, X., Pfeiffer, D.U., Epprecht, M., Boles, S., Czarnecki, C., Chaitaweesub, P., et al. (2008). Mapping H5N1 highly pathogenic avian influenza risk in Southeast Asia. *Proceedings of the National Academy of Sciences of the United States of America* 105, 4769–4774.

Gonzalez, C., Wang, O., Strutz, S.E., Gonzalez-Salazar, C., Sanchez-Cordero, V., Sarkar, S. (2010). Climate change and risk of leishmaniasis in North America: predictions from ecological niche models of vector and reservoir species. *PLoS Neglected Tropical Diseases* 4, e585.

Green, D., Kiss, I., Kao, R. (2006). Modelling the initial spread of foot-and-mouth disease through animal movements. *Proceedings of the Biological Sciences* 273, 2729–2735.

Green, D.M., Kiss, I.Z., Mitchell, A.P., Kao, R.R. (2008). Estimates for local and movement-based transmission of bovine tuberculosis in British cattle. *Proceedings of the Biological Sciences* 275, 1001–1005.

Guerra, C., Gikandi, P., Tatem, A., Noor, A., Smith, D. (2008). The limits and intensity of *Plasmodium falciparum* transmission: implications for malaria control and elimination worldwide. *PLoS Medicine* 5, e38.

Guerra, C., Howes, R., Patil, A., Gething, P., Van Boeckel, T., Temperley, W., Kabaria, C., Tatem, A., Manh, B., Elyazar, I. (2010). The international limits and population at risk of *Plasmodium vivax* transmission in 2009. *PLoS Neglected Tropical Diseases* 4, e774.

Guerra, C., Snow, R., Hay, S. (2006). Determining the global spatial limits of malaria transmission in 2005. *Advances in Parasitology* 62, 157–179.

Guichard, S., Guis, H., Tran, A., Garros, C., Balenghien, T., Kriticos, D.J. (2014). World-wide niche and future potential distribution of *Culicoides imicola*, a major vector of bluetongue and African horse sickness viruses. *PLoS One* 9, e112491.

Guis, H., Tran, A., Rocque, S.d.L., Baldet, T., Gerbier, G., Barragué, B., Biteau-Coroller, F., Roger, F., Viel, J.-F., Mauny, F. (2007). Use of high spatial resolution satellite imagery to characterize landscapes at risk for bluetongue. *Veterinary Research* 38, 669–683.

Hartemink, N., Vanwambeke, S.O., Purse, B.V., Gilbert, M., Van Dyck, H. (2014). Towards a resource-based habitat approach for spatial modelling of vector-borne disease risks. *Biological Reviews.* DOI: 10.1111/brv.12149.

Harvey, N., Reeves, A., Schoenbaum, M.A., Zagmutt-Vergara, F.J., Dubé, C., Hill, A.E., Corso, B.A., McNab, W.B., Cartwright, C.I., Salman, M.D. (2007). The North American Animal Disease Spread Model: a simulation model to assist decision making in evaluating animal disease incursions. *Preventive Veterinary Medicine* 82, 176–197.

Hay, S., Okiro, E., Gething, P., Patil, A., Tatem, A., Guerra, C., Snow, R. (2010). Estimating the global clinical burden of *Plasmodium falciparum* malaria in 2007. *PLoS Medicine* 7, e100029.

Hay, S.I., Battle, K.E., Pigott, D.M., Smith, D.L., Moyes, C.L., Bhatt, S., Brownstein, J.S., Collier, N., Myers, M.F., George, D.B., Gething, P.W. (2013). Global mapping of infectious disease. *Philosophical Transactions of the Royal Society B: Biological Sciences* 368, 1614.

Hennebelle, J.H., Sykes, J.E., Carpenter, T.E., Foley, J. (2013). Spatial and temporal patterns of *Leptospira* infection in dogs from northern California: 67 cases (2001–2010). *Journal of the American Veterinary Medical Association* 242, 941–947.

Hirzel, A., Le Lay, G., Helfer, V., Randin, C., Guisan, A. (2006). Evaluating the ability of habitat suitability models to predict species presences. *Ecological Modelling* 199, 142–152.

Hogerwerf, L., Wallace, R., Ottaviani, D., Slingenbergh, J., Prosser, D., Bergmann, L., Gilbert, M. (2010). Persistence of highly pathogenic avian influenza H5N1 virus defined by agro-ecological niche. *EcoHealth* 7, 213–225.

Horton, D.L., McElhinney, L.M., Freuling, C.M., Marston, D.A., Banyard, A.C., Goharrriz, H., Wise, E., et al. (2015). Complex epidemiology of a zoonotic disease in a culturally diverse region: phylogeography of rabies virus in the Middle East. *PLoS Neglected Tropical Diseases* 9, e0003569.

Hyder, K., Vidal-Diez, A., Lawes, J., Sayers, A., Milnes, A., Hoinville, L., Cook, A. (2011). Use of spatiotemporal analysis of laboratory submission data to identify potential outbreaks of new or emerging diseases in cattle in Great Britain. *BMC Veterinary Research* 7, 14.

IPCC (Intergovernmental Panel on Climate Change). 2014. Climate change 2014: impacts, adaptation, and vulnerability. Part A. Global and sectoral aspects. Contribution of Working Group II to the Fifth Assessment Report of the Intergovernmental Panel on Climate Change. Summary for policymakers. IPCC, Cambridge University Press, Cambridge.

Jore, S., Vanwambeke, S.O., Viljugrein, H., Isaksen, K., Kristoffersen, A.B., Woldehiwet, Z., Johansen, B., et al. (2014). Climate and environmental change drives *Ixodes ricinus* geographical expansion at the northern range margin. *Parasites and Vectors* 7, 11.

Joyner, T.A., Lukhnova, L., Pazilov, Y., Temiralyeva, G., Hugh-Jones, M.E., Aikimbayev, A., Blackburn, J.K. (2010). Modeling the potential distribution of *Bacillus anthracis* under multiple climate change scenarios for Kazakhstan. *PLoS One* 5, e9596.

Kiss, I., Green, D., Kao, R. (2006a). The network of sheep movements within Great Britain: network properties and their implications for infectious disease spread. *Journal of the Royal Society Interface* 3, 669–677.

Kiss, I.Z., Green, D.M., Kao, R.R. (2006b). Infectious disease control using contact tracing in random and scale-free networks. *Journal of the Royal Society Interface* 3, 55–62.

Kracalik, I., Lukhnova, L., Aikimbayev, A., Pazilov, Y., Temiralyeva, G., Blackburn, J.K. (2011). Incorporating retrospective clustering into a prospective cusum methodology for anthrax: evaluating the effects of disease expectation. *Spatial and Spatio-Temporal Epidemiology* 2, 11–21.

Le, H., Poljak, Z., Deardon, R., Dewey, C.E. (2012). Clustering of and risk factors for the porcine high fever disease in a region of Vietnam. *Transboundary and Emerging Diseases* 59, 49–61.

Lentz, H.H.K., Konschake, M., Teske, K., Kasper, M., Rother, B., Carmanns, R., Petersen, B., Conraths, F.J., Selhorst, T. (2011). Trade communities and their spatial patterns in the German pork production network. *Preventive Veterinary Medicine* 98, 176–181.

Liang, L., Xu, B., Chen, Y., Liu, Y., Cao, W., Fang, L., Feng, L., Goodchild, M.F., Gong, P. (2010). Combining spatial-temporal and phylogenetic analysis approaches for improved understanding on global H5N1 transmission. *PLoS One* 5, e13575.

Lindström, T., Lewerin, S.S., Wennergren, U. (2011). Influence on disease spread dynamics of herd characteristics in a structured livestock industry. *Journal of the Royal Society Interface* 9, 1287–1294.

Lindström, T., Sisson, S.A., Lewerin, S.S., Wennergren, U. (2010). Estimating animal movement contacts between holdings of different production types. *Preventive Veterinary Medicine* 95, 23–31.

Lockhart, C.Y., Stevenson, M.A., Rawdon, T.G., Gerber, N., French, N.P. (2010). Patterns of contact within the New Zealand poultry industry. *Preventive Veterinary Medicine* 95, 258–266.

Lurette, A., Belloc, C., Keeling, M. (2011). Contact structure and *Salmonella* control in the network of pig movements in France. *Preventive Veterinary Medicine* 102, 30–40.

Madouasse, A., Marceau, A., Lehébel, A., Brouwer-Middelesch, H., van Schaik, G., Van der Stede, Y., Fourichon, C. (2013). Evaluation of a continuous indicator for syndromic surveillance through simulation. Application to vector borne disease emergence detection in cattle using milk yield. *PLoS One* 8, e73726.

Malczewski, J. (2006). GIS-based multicriteria decision analysis: a survey of the literature. *International Journal of Geographic Information Science* 20, 703–726.

Martin, V., Pfeiffer, D.U., Zhou, X., Xiao, X., Prosser, D.J., Guo, F., Gilbert, M. (2011). Spatial distribution and risk factors of highly pathogenic avian influenza (HPAI) H5N1 in China. *PLoS Pathogens* 7, e1001308.

Martínez-López, B., Perez, A.M., Sánchez-Vizcaíno, J.M. (2009). Combined application of social network and cluster detection analyses for temporal-spatial characterization of animal movements in Salamanca, Spain. *Preventive Veterinary Medicine* 91, 29–38.

Mechai, S., Margos, G., Feil, E.J., Lindsay, L.R., Ogden, N.H. (2015). Complex population structure of *Borrelia burgdorferi* in southeastern and south central Canada as revealed by phylogeographic analysis. *Applied and Environmental Microbiology* 81, 1309–1318.

Méroc, E., De Regge, N., Riocreux, F., Caij, A.B., van den Berg, T., van der Stede, Y. (2014). Distribution of Schmallenberg virus and seroprevalence in Belgian sheep and goats. *Transboundary and Emerging Diseases* 61, 425–431.

Métras, R., Porphyre, T., Pfeiffer, D.U., Kemp, A., Thompson, P.N., Collins, L.M., White, R.G. (2012). Exploratory space-time analyses of Rift Valley fever in South Africa in 2008–2011. *PLoS Neglected Tropical Diseases* 6, e1808.

Minh, P.Q., Morris, R.S., Schauer, B., Stevenson, M., Benschop, J., Nam, H.V., Jackson, R. (2009). Spatio-temporal epidemiology of highly pathogenic avian influenza outbreaks in the two deltas of Vietnam during 2003–2007. *Preventive Veterinary Medicine* 89, 16–24.

Minh, P.Q., Stevenson, M.A., Jewell, C., French, N., Schauer, B. (2010). Spatio-temporal analyses of highly pathogenic avian influenza H5N1 outbreaks in the Mekong River Delta, Vietnam, 2009. *Spatial and Spatio-Temporal Epidemiology* 2, 49–57.

Mullins, J., Lukhnova, L., Aikimbayev, A., Pazilov, Y., Van Ert, M., Blackburn, J. (2011). Ecological niche modelling of the *Bacillus anthracis* A1.a sub-lineage in Kazakhstan. *BMC Ecology* 11, 32.

Mullins, J.C., Garofolo, G., Van Ert, M., Fasanella, A., Lukhnova, L., Hugh-Jones, M.E., Blackburn, J.K. (2013). Ecological niche modeling of *Bacillus anthracis* on three continents: evidence for genetic-ecological divergence? *PLoS One* 8, e72451.

Murray, C.J.L. (2012). Disability-adjusted life years (DALYs) for 291 diseases and injuries in 21 regions, 1990–2010: a systematic analysis for the Global Burden of Disease Study 2010. *Lancet* 380, 2197–2223.

Mweya, C.N., Kimera, S.I., Kija, J.B., Mboera, L.E.G. (2013). Predicting distribution of *Aedes aegypti* and *Culex pipiens* complex, potential vectors of Rift Valley fever virus in relation to disease epidemics in East Africa. *Infection Ecology and Epidemiology* 3. DOI: 10.3402/iee.v3403i3400.21748.

Natale, F., Giovannini, A., Savini, L., Palma, D., Possenti, L., Fiore, G., Calistri, P. (2009). Network analysis of Italian cattle trade patterns and evaluation of risks for potential disease spread. *Preventive Veterinary Medicine* 92, 341–350.

Newman, M.E. (2003). Properties of highly clustered networks. *Physical Review E: Statistical, Nonlinear and Soft Matter Physics* 68, 026121.

Nickbakhsh, S., Matthews, L., Bessell, P., Reid, S., Kao, R. (2011). Generating social network data using partially described networks: an example informing avian influenza control in the British poultry industry. *BMC Veterinary Research* 7, 66.

Nieto, P., Malone, J.B., Bavia, M.E. (2006). Ecological niche modeling for visceral leishmaniasis in the state of Bahia, Brazil, using genetic algorithm for rule-set prediction and growing degree day-water budget analysis. *Geospatial Health* 1, 115–126.

Nogareda, C., Juberta, A., Kantzoura, V., Kouam, M.K., Feidas, H., Theodoropoulos, G. (2013). Geographical distribution modelling for *Neospora caninum* and *Coxiella burnetii* infections in dairy cattle farms in northeastern Spain. *Epidemiology and Infection* 141, 81–90.

Nöremark, M., Håkansson, N., Lewerin, S.S., Lindberg, A., Jonsson, A. (2011). Network analysis of cattle and pig movements in Sweden: measures relevant for disease control and risk based surveillance. *Preventive Veterinary Medicine* 99, 78–90.

Nöremark, M., Håkansson, N., Lindstrom, T., Wennergren, U., Lewerin, S. (2009). Spatial and temporal investigations of reported movements, births and deaths of cattle and pigs in Sweden. *Acta Veterinaria Scandinavica* 51, 37.

Nöremark, M., Widgren, S. (2014). EpiContactTrace: an R-package for contact tracing during livestock disease outbreaks and for risk-based surveillance. *BMC Veterinary Research* 10, 71.

Parmesan, C., Yohe, G. (2003). A globally coherent fingerprint of climate change impacts across natural systems. *Nature* 421, 37–42.

Pedersen, U., Stendel, M., Midzi, N., Mduluza, T., Soko, W., Stensgaard, A.-S., Vennervald, B., Mukaratirwa, S., Kristensen, T. (2014). Modelling climate change impact on the spatial distribution of fresh water snails hosting trematodes in Zimbabwe. *Parasites and Vectors* 7, 536.

Peterson, A., Shaw, J. (2003). Lutzomyia vectors for cutaneous leishmaniasis in southern Brazil: ecological niche models, predicted geographic distributions, and climate change effects. *International Journal of Parasitology* 33, 919–931.

Peterson, A.T., Pereira, R.S., Neves, V.F.d.C. (2004). Using epidemiological survey data to infer geographic distributions of leishmaniasis vector species. *Revista de Sociedade Brasileira de Medicina Tropical* 37, 10–14.

Pfeiffer, D.U., Minh, P.Q., Martin, V., Epprecht, M., Otte, M.J. (2007). An analysis of the spatial and temporal patterns of highly pathogenic avian influenza occurrence in Vietnam using national surveillance data. *Veterinary Journal* 174, 302–309.

Pfeiffer, D.U., Robinson, T.P., Stevenson, M., Stevens, K.B., Rogers, D.J., Clements, A.C.A. (2008). Spatial risk assessment and management of disease. In *Spatial Analysis in Epidemiology.* Oxford University Press, Oxford, pp. 110–119.

Phillips, S. (2012). A brief tutorial on Maxent. *Lessons in Conservation* 3, 107–135.

Phillips, S.J., Dudík, M. (2008). Modeling of species distributions with Maxent: new extensions and a comprehensive evaluation. *Ecography* 31, 161–175.

Picado, A., Guitian, F., Pfeiffer, D. (2007). Space-time interaction as an indicator of local spread during the 2001 FMD outbreak in the UK. *Preventive Veterinary Medicine* 79, 3–19.

Picado, A., Speybroeck, N., Kivaria, F., Mosha, R.M., Sumaye, R.D., Casal, J., Berkvens, D. (2011). Foot-and-mouth disease in Tanzania from 2001 to 2006. *Transboundary and Emerging Diseases* 58, 44–52.

Pigott, D.M., Golding, N., Mylne, A., Huang, Z., Henry, A.J., Weiss, D.J., Brady, O.J., et al. (2014). Mapping the zoonotic niche of Ebola virus disease in Africa. *eLife* 3, e04395.

Poljak, Z., Dewey, C., Rosendal, T., Friendship, R., Young, B., Berke, O. (2010). Spread of porcine circovirus associated disease (PCVAD) in Ontario (Canada) swine herds. Part I. Exploratory spatial analysis. *BMC Veterinary Research* 6, 59.

Purse, B.V., Tatem, A.J., Caracappa, S., Rogers, D.J., Mellor, P.S., Baylis, M., Torina, A. (2004). Modelling the distributions of *Culicoides* bluetongue virus vectors in Sicily in relation to satellite-derived climate variables. *Medical and Veterinary Entomology* 18, 90–101.

Pybus, O.G., Suchard, M.A., Lemey, P., Bernardin, F.J., Rambaut, A., Crawford, F.W., Gray, R.R., Arinaminpathy, N., Stramer, S.L., Busch, M.P., Delwart, E.L. (2012). Unifying the spatial epidemiology and molecular evolution of emerging epidemics. *Proceedings of the National Academy of Sciences of the United States of America* 109, 15066–15071.

Quantum GIS Development Team. (2009). Quantum GIS Geographic Information System. Open Source Geospatial Foundation Project. http://qgis.osgeo.org.

Randolph, S.E., Rogers, D.J. (2010). The arrival, establishment and spread of exotic diseases: patterns and predictions. *Nature Reviews Microbiology* 8, 361–371.

Rasamoelina-Andriamanivo, H., Duboz, R., Lancelot, R., Maminiaina, O.F., Jourdan, M., Rakotondramaro, T.M.C., Rakotonjanahary, S.N., de Almeida, R.S., Rakotondravao, Durand, B., Chevalier, V. (2014). Description and analysis of the poultry trading network in the Lake Alaotra region, Madagascar: implications for the surveillance and control of Newcastle disease. *Acta Tropica* 135, 10–18.

Rautureau, S., Dufour, B., Durand, B. (2011). Vulnerability of animal trade networks to the spread of infectious diseases: a methodological approach applied to evaluation and emergency control strategies in cattle, France, 2005. *Transboundary and Emerging Diseases* 58, 110–120.

Rautureau, S., Dufour, B., Durand, B. (2012). Structural vulnerability of the French swine industry trade network to the spread of infectious diseases. *Animal* 6, 1152–1162.

Ribbens, S., Dewulf, J., Koenen, F., Mintiens, K., de Kruif, A., Maes, D. (2009). Type and frequency of contacts between Belgian pig herds. *Preventive Veterinary Medicine* 88, 57–66.

Robinson, S., Christley, R. (2007). Exploring the role of auction markets in cattle movements within Great Britain. *Preventive Veterinary Medicine* 81, 21–37.

Robinson, T.P., Harris, R.S., Hopkins, J.S., Williams, B.G. (2002). An example of decision support for trypanosomiasis control using a geographical information system in eastern Zambia. *International Journal of Geographical Information Science* 16, 345–360.

Rodriguez-Prieto, V., Martinez-Lopez, B., Barasona, J., Acevedo, P., Romero, B., Rodriguez-Campos, S., Gortazar, C., Sanchez-Vizcaino, J., Vicente, J. (2012). A Bayesian approach to study the risk variables for tuberculosis occurrence in domestic and wild ungulates in south central Spain. *BMC Veterinary Research* 8, 148.

Rose, H., Wall, R. (2011). Modelling the impact of climate change on spatial patterns of disease risk: sheep blowfly strike by *Lucilia sericata* in Great Britain. *International Journal for Parasitology* 41, 739–746.

Sallam, M.F., Al Ahmed, A.M., Abdel-Dayem, M.S., Abdullah, M.A.R. (2013). Ecological niche modeling and land cover risk areas for Rift Valley fever vector, *Culex tritaeniorhynchus* Giles in Jazan, Saudi Arabia. *PLoS One* 8, e65786.

Sánchez-Matamoros, A., Martínez-López, B., Sánchez-Vizcaíno, F., Sánchez-Vizcaíno, J.M. (2013). Social network analysis of equidae movements and its application to risk-based surveillance and to control of spread of potential equidae diseases. *Transboundary and Emerging Diseases* 60, 448–459.

Scott, A., Zepeda, C., Garber, L., Smith, J., Swayne, D., Rhorer, A., Kellar, J., Shimshony, A., Batho, H., Caporale, V., Giovannini, A. (2006). The concept of compartmentalisation. *Revue Scientifique et Technique* 25, 873–879.

Selby, R., Bardosh, K., Picozzi, K., Waiswa, C., Welburn, S. (2013). Cattle movements and trypanosomes: restocking efforts and the spread of *Trypanosoma brucei rhodesiense* sleeping sickness in post-conflict Uganda. *Parasites and Vectors* 6, 281.

Seo, H.-J., Park, J.-Y., Cho, Y., Cho, I.-S., Yeh, J.-Y. (2015). First report of bluetongue virus isolation in the Republic of Korea and analysis of the complete coding sequence of the segment 2 gene. *Virus Genes* 50, 156–159.

Service, M.W. (1997). Mosquito (Diptera: Culicidae) dispersal—the long and short of it. *Journal of Medical Entomology* 34, 579–588.

Shirley, M.D.F., Rushton, S.P. (2005a). The impacts of network topology on disease spread. *Ecological Complexity* 2, 287–299.

Shirley, M.D.F., Rushton, S.P. (2005b). Where diseases and networks collide: lessons to be learnt from a study of the 2001 foot-and-mouth disease epidemic. *Epidemiology and Infection* 133, 1023–1032.

Simon, J.A., Marrotte, R.R., Desrosiers, N., Fiset, J., Gaitan, J., Gonzalez, A., Koffi, J.K., et al. (2014). Climate change and habitat fragmentation drive the occurrence of *Borrelia burgdorferi*, the agent of Lyme disease, at the northeastern limit of its distribution. *Evolutionary Applications* 7, 750–764.

Singh, V., Mishra, N., Awasthi, G., Dash, A.P., Das, A. (2009). Why is it important to study malaria epidemiology in India? *Trends in Parasitology* 25, 452–457.

Sinkala, Y., Simuunza, M., Muma, J.B., Pfeiffer, D.U., Kasanga, C.J., Mweene, A. (2014). Foot and mouth disease in Zambia: spatial and temporal distributions of outbreaks, assessment of clusters and implications for control. *Onderstepoort Journal of Veterinary Research* 81, E1–E6.

Smith, K.L., DeVos, V., Bryden, H., Price, L.B., Hugh-Jones, M.E., Keim, P. (2000). *Bacillus anthracis* diversity in Kruger National Park. *Journal of Clinical Microbiology* 38, 3780–3784.

Smith, R.P., Cook, A.J.C., Christley, R.M. (2013). Descriptive and social network analysis of pig transport data recorded by quality assured pig farms in the UK. *Preventive Veterinary Medicine* 108, 167–177.

Standley, C.J., Vounatsou, P., Gosoniu, L., McKeon, C., Adriko, M., Kabatereine, N.B., Stothard, J.R. (2013). Micro-scale investigation of intestinal schistosomiasis transmission on Ngamba and Kimi islands, Lake Victoria, Uganda. *Acta Tropica* 128, 353–364.

Stark, K., Regula, G., Hernandez, J., Knopf, L., Fuchs, K., Morris, R., Davies, P. (2006). Concepts for risk-based surveillance in the field of veterinary medicine and veterinary public health: review of current approaches. *BMC Health Services Research* 6, 20.

Stevens, K.B., Gilbert, M., Pfeiffer, D.U. (2013). Modeling habitat suitability for occurrence of highly pathogenic avian influenza virus H5N1 in domestic poultry in Asia: a spatial multicriteria decision analysis approach. *Spatial and Spatio-Temporal Epidemiology* 4, 1–14.

Stevens, K.B., Pfeiffer, D.U. (2011). Spatial modelling of disease using data- and knowledge-driven approaches. *Spatial and Spatio-Temporal Epidemiology* 2, 125–133.

Stevens, K.B., Pfeiffer, D.U. (2015). Enhanced decision-making of HPAI H5N1 in domestic poultry in Asia: a comparison of spatial-modelling methods. In *Proceedings of the Society for Veterinary Epidemiology and Preventive Medicine (SVEPM)*, pp. 241–225, Ghent, Belgium.

Stockwell, D., Peters, D. (1999). The GARP modelling system: problems and solutions to automated spatial prediction. *International Journal of Geographical Information Science* 13, 143–158.

Swirski, A.L., Pearl, D.L., Williams, M.L., Homan, H.J., Linz, G.M., Cernicchiaro, N., LeJeune, J.T. (2013). Spatial epidemiology of *Escherichia coli* O157:H7 in dairy cattle in relation to night roosts of *Sturnus vulgaris* (European starling) in Ohio, USA (2007–2009). *Zoonoses and Public Health* 61, 427–435.

Symeonakis, E., Robinson, T., Drake, N. (2007). GIS and multiple-criteria evaluation for the optimisation of tsetse fly eradication programmes. *Environmental Monitoring and Assessment* 124, 89–103.

Tatem, A.J., Baylis, M., Mellor, P.S., Purse, B.V., Capela, R., Pena, I., Rogers, D.J. (2003). Prediction of bluetongue vector distribution in Europe and North Africa using satellite imagery. *Veterinary Microbiology* 97, 13–29.

Tatem, A., Campiz, N., Gething, P., Snow, R., Linard, C. (2011). The effects of spatial population dataset choice on estimates of population at risk of disease. *Population Health Metrics* 9, 4.

Thakur, K.K., Revie, C.W., Hurnik, D., Poljak, Z., Sanchez, J. (2014). Analysis of swine movement in four Canadian regions: network structure and implications for disease spread. *Transboundary and Emerging Diseases*. DOI: 10.1111/tbed.12225.

Toty, C., Barre, H., Le Goff, G., Larget-Thiery, I., Rahola, N., Couret, D., Fontenille, D. (2010). Malaria risk in Corsica, former hot spot of malaria in France. *Malaria Journal* 9, 231.

Van Boeckel, T.P., Thanapongtharm, W., Robinson, T., Biradar, C.M., Xiao, X., Gilbert, M. (2012). Improving risk models for avian influenza: the role of intensive poultry farming and flooded land during the 2004 Thailand epidemic. *PLoS One* 7, e49528.

Vander Kelen, P., Downs, J., Stark, L., Loraamm, R., Anderson, J., Unnasch, T. (2012). Spatial epidemiology of eastern equine encephalitis in Florida. *International Journal of Health Geographics* 11, 47.

Van Steenwinkel, S., Ribbens, S., Ducheyne, E., Goossens, E., Dewulf, J. (2011). Assessing biosecurity practices, movements and densities of poultry sites across Belgium, resulting in different farm risk-groups for infectious disease introduction and spread. *Preventive Veterinary Medicine* 98, 259–270.

Vernon, M. (2011). Demographics of cattle movements in the United Kingdom. *BMC Veterinary Research* 7, 31.

Vernon, M., Keeling, M. (2009). Representing the UK's cattle herd as static and dynamic networks. *Proceedings of the Biological Sciences* 276, 469–476.

Vicente-Rubiano, M., Martínez-López, B., Sánchez-Vizcaíno, F., Sánchez-Vizcaíno, J.M. (2014). A new approach for rapidly assessing the risk of Aujeszky's disease reintroduction into a disease-free Spanish territory by analysing the movement of live pigs and potential contacts with wild boar. *Transboundary and Emerging Diseases* 61, 350–361.

Volkova, V.V., Howey, R., Savill, N.J., Woolhouse, M.E.J. (2010). Sheep movement networks and the transmission of infectious diseases. *PLoS One* 5, e11185.

Wall, R., Rose, H., Ellse, L., Morgan, E. (2011). Livestock ectoparasites: integrated management in a changing climate. *Veterinary Parasitology* 180, 82–89.

Watts, D.J., Strogatz, S.H. (1998). Collective dynamics of 'small-world' networks. *Nature* 393, 440–442.

Webb, C.R. (2005). Farm animal networks: unravelling the contact structure of the British sheep population. *Preventive Veterinary Medicine* 68, 3–17.

Webb, C. (2006). Investigating the potential spread of infectious diseases of sheep via agricultural shows in Great Britain. *Epidemiology and Infection* 134, 31–40.

Wertheim, H., Horby, P., Woodall, J. (2012). *Atlas of Human Infectious Diseases.* Wiley-Blackwell, Oxford.

Wilesmith, J.W., Stevenson, M.A., King, C.B., Morris, R.S. (2003). Spatio-temporal epidemiology of foot-and-mouth disease in two counties of Great Britain in 2001. *Preventive Veterinary Medicine* 61, 157–170.

Williams, R., Fasina, F., Peterson, A. (2008). Predictable ecology and geography of avian influenza (H5N1) transmission in Nigeria and West Africa. *Transactions of the Royal Society of Tropical Medicine and Hygiene* 102, 471–479.

Williams, R.A., Peterson, A.T. (2009). Ecology and geography of avian influenza (HPAI H5N1) transmission in the Middle East and northeastern Africa. *International Journal of Health Geographics* 8, 47.

Woolhouse, M.E. (2003). Foot-and-mouth disease in the UK: what should we do next time? *Journal of Applied Microbiology* 94 (Suppl), 126S–130S.

Xu, B., Madden, M., Stallknecht, D.E., Hodler, T.W., Parker, K.C. (2012). Spatial and spatial–temporal clustering analysis of hemorrhagic disease in white-tailed deer in the southeastern USA: 1980–2003. *Preventive Veterinary Medicine* 106, 339–347.

26

Infectious Disease Modelling

Michael Höhle
Department of Mathematics
Stockholm University
Stockholm, Sweden

CONTENTS

Infectious diseases impose a critical challenge to human, animal, and plant health. Emerging and reemerging pathogens—such as SARS, influenza, and hemorrhagic fever among humans, or foot-and-mouth disease and classical swine fever among animals—hit the news coverage with regular certainty. Zoonoses and host-transmitted diseases underline how significant the connection is between human and animal diseases. While plant epidemics receive less immediate attention, they can severely impact crop yield or wipe out entire species. Unifying for the above epidemics is that they all represent realization of temporal processes. Why does the spatial dimension then matter for the modelling of epidemics? It depends very much on the aims of the analysis: many relevant questions can be adequately answered by models considering the population as homogeneous. However, in other situations, heterogeneity is important, for example, induced by age or spatial structure of the population. Spatially varying demographic and environmental factors could influence the disease transmission. Furthermore, having a spatial resolution allows the model to express spatial heterogeneity in the manifestation of the disease over time. This becomes particularly important when investigating the probability of fade-out or short-term prediction of the location of new cases. This kind of analysis represents an important mathematical contribution aimed at understanding the dynamics of disease transmission and predicting the course of epidemics in order to, for example, assess control measures or determine the source of an epidemic. This chapter is about the *spatiotemporal* analysis of epidemic processes.

26.1 Role of Space in Infectious Disease Epidemiology

The focus in this chapter is on communicable microparasite infections (typically viral or bacterial diseases) among humans—though the application of mathematical modelling is equally immediate in animal (Dohoo et al., 2010) and plant (Madden et al., 2007) epidemics.

Following Giesecke (2002), epidemiology is about "the study of diseases and their determinants in populations". The concepts of incidence and prevalence, known from chronic disease epidemiology, also apply to infectious disease epidemiology. However, contrary to chronic disease modelling, it is important to realize that in infectious disease epidemiology, each case is also a risk factor, and that not everyone is necessarily susceptible to a disease (e.g., immunity due to previous infections or vaccination). As a consequence, the study of infectious diseases is very much concerned with the study of interacting populations. Still, as treated in, for example, Becker (1989) or Hens et al. (2012), the principle of regression by linear models (LMs), generalized linear models (GLMs), or survival models applies to a variety of problems in infectious disease epidemiology. Hence, spatial extensions of such regression models, for example, structured additive regression models (Fahrmeir and Kneib, 2011), also immediately apply to the context of infectious disease epidemiology. Examples are the use of spatially enhanced ecological Poisson regression models with conditional autoregression (CAR) random effects to investigate the influence of socioeconomic factors on the incidence of specific infectious diseases (Wilking et al., 2012) or the assessment of mumps outbreak risk based on serological data by using generalized additive models (GAMs) with two-dimensional (2D) splines adjusting for location of individuals (Abrams et al., 2014). Another application of classical methods from spatial epidemiology is the use of spatial point process models for investigating putative sources for a food-borne outbreak (Diggle and Rowlingson, 1994).

In this chapter, however, we restrict the focus to the use of spatiotemporal *transmission models*, that is, dynamic models for the person-to-person spread of a disease in a well-defined population. The use of such models has become increasingly popular in the epidemiological literature in order to assess risk of emerging pathogens or evaluate control measures. Even spatial aspects of such evaluations are now feasible due to the fact that data become spatially more refined and computer power allows for more complex models to be investigated. The modelling of measles and influenza is two examples where the impact of space has been especially thoroughly investigated (Grenfell et al., 1995, 2001; Viboud et al., 2006; Eggo et al., 2011; Gog et al., 2014). As an illustration, we look in detail at the spatiotemporal modelling of biweekly measles counts in 60 towns and cities in England and the UK, 1944–1966. Figure 26.1 shows the locations of the 60 cities, including both very large and very small cities, and forming a subset of the data for 954 communities used in Xia et al. (2004). An illustration of the time series for the three largest and three smallest cities in the data set can be found as part of Figure 26.5. The aim of such an analysis is to quantify the effect of demographics and seasonality on the dynamics, but also to investigate the role of spatial spread on extinction and reintroduction.

In spatial analyses of this kind, the movement of populations plays an important role. From a historic perspective, especially mobility has undergone dramatic changes within the

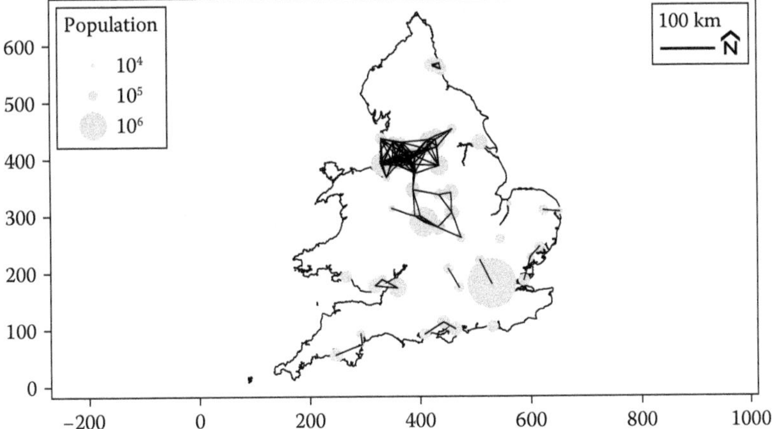

FIGURE 26.1
Location of the 60 cities in England and Wales (OSGB36 reference system) used in the measles modelling. The area of each circle is proportional to the city's population, and the connecting lines indicate immediate neighbourhood (distance $\leq$ 50 km).

last 200 years; for instance, Cliff and Haggett (2004) talk about this as the "collapse of geographical space": the time for travelling long distances has reduced immensely, which has resulted in populations mixing at increasingly higher rates. While there has been an abundance of papers and animations illustrating global spread of emerging diseases, for example, due to airline travelling, the core transmission of many "neglected" diseases still occurs at short range: in the household, in kindergarten, at work. The role of space in transmission models is to be studied within this dissonance between long-distance and short-distance spread. A number of chapters and articles have already surveyed this field, for example, Isham (2004), Deardon et al. (2015), Riley et al. (2015), and Held and Paul (2013). The emphasis in the present chapter is on metapopulation models and their likelihood-based inference. It is structured as follows: Section 26.2 consists of a primer on continuous-time and discrete-time epidemic modelling—first for a homogeneous population and then for spatially coupled populations. Section 26.3 then illustrates the application of discrete-time versions of such models to the 60 cities' measles data. A discussion in Section 26.4 ends the chapter.

26.2 Epidemic Modelling

Mathematical modelling of infectious diseases has become a key tool in order to understand, predict, and control the spread of infections. The fundamental difference to chronic disease epidemiology is that the temporal aspect is paramount. The aim of epidemic modelling is thus to model the spread of a disease in a population made up of a possibly large-integer number of individuals. To simplify the description of this population, it is common to use a compartmental approach to modelling; for instance, in its simplest form, the population is divided into classes of *susceptible*, *infective*, and *recovered* individuals. Disease dynamics can then be characterized by a mathematical description of each individual's transitions between compartments, subject to the state of the remaining individuals.

26.2.1 Continuous-time modelling

A number of books, for example, Anderson and May (1991), Diekmann and Heesterbeek (2000), and Keeling and Rohani (2008), give an introduction to epidemic modelling using primarily deterministic models based on ordinary differential equations (ODEs) in the setting of the susceptible–infective–recovered (SIR) model and its extensions. Let $S(t)$, $I(t)$, and $R(t)$ denote the number at time t of susceptible, infective, and recovered individuals, respectively. Then the dynamics in the basic *deterministic SIR model* in a population of fixed size can be expressed as in the seminal work by Kermack and McKendrick (1927):

$$\frac{dS(t)}{dt} = -\frac{\beta}{N}S(t)I(t), \qquad \frac{dI(t)}{dt} = \frac{\beta}{N}S(t)I(t) - \gamma I(t), \qquad \frac{dR(t)}{dt} = \gamma I(t),$$

where the parameter $\beta > 0$ is the *transmission rate* and $\gamma > 0$ describes the *removal rate*. The initial condition is given by $S(0)$, $I(0)$, which are known integers, and $R(0) = 0$. In a population of fixed size $N = S(0) + I(0)$, the expression for $dR(t)/dt$ in the above ODE system is redundant because $R(t)$ is implicitly given as $N - S(t) - I(t)$.

ODE modelling implies an approximation of the integer-sized population using continuous numbers, and that the stochastic behaviour of an epidemic is sacrificed by looking at a deterministic average behaviour. If the population under study is large enough, such approximations are reasonably valid to obtain a biological understanding. In small populations, however, stochasticity plays an important role for extinction, which cannot be ignored. Stochastic epidemic modelling is described, for example, in Bailey (1975), Becker (1989), Daley and Gani (1999), and Andersson and Britton (2000), who all rely heavily on the theory of stochastic processes. In its simplest form, the basic discrete-state *stochastic SIR model* can be described as a general birth and death process, where the event rates for infection and removal are given as follows:

Event	Rate
$(S(t), I(t)) \to (S(t)-1, I(t)+1)$	$\frac{\beta}{N}S(t)I(t)$
$(S(t), I(t)) \to (S(t), I(t)-1)$	$\gamma I(t)$

Again, the development of $R(t)$ is implicitly given, because a fixed population of size $S(0) + I(0)$ is assumed. Notice also how the integer size of the population is now taken into account: once $I(t) = 0$, the epidemic ceases. From a point process viewpoint, the above specification corresponds to an assumption of piecewise constant conditional intensities for the process of infection, while the length of the infective period is given by independent and identically distributed (i.i.d.) exponential random variables. An important point is that the deterministic SIR model is not just modelling the expectation of the stochastic SIR model. As an illustration, Renshaw (1991, chapter 10) shows, based on calculations in Bailey (1950), how the expected number of susceptibles $\mu(t) = \mathrm{E}(S(t))$ in the stochastic SI model differs from the solution of $S(t)$ in the deterministic SI model. Figure 26.2 shows the result for a population with $S(0) = 10$, $I(0) = 1$, $\beta = 11$, and $\gamma = 0$. One notices the differences between the deterministic and the stochastic model.

Finding an analytic expression for $\mu(t)$ or—even better—the probability mass function (PMF) of $S(t)$ at a specific time point t is less easy already for the SIR model and intractable for most models. Instead, a numerical approach is to formulate a first-order differential-difference equation describing the time evolution of $P((S(t), I(t)) = (x, y))$ for each possible (x, y). Such an approach is known as a master equation approach and corresponds to a discrete-state continuous-time Markov jump process with the solution of the master equation obeying the Chapman–Kolmogorov equation. These ODE equations are then solved numerically. For large populations, the problem can, however, become intractable to solve

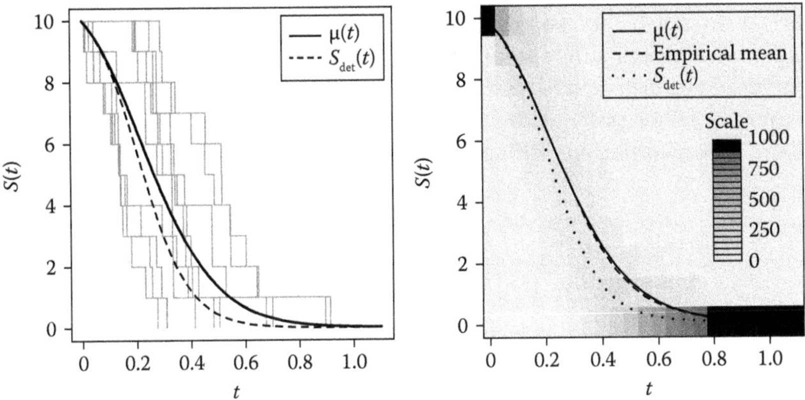

FIGURE 26.2
Left: $S(t)$ for 10 realizations of the SI model with parameters as described in the text. Also shown are $E(S(t))$ in the stochastic model and the solution $S(t)$ in the deterministic model. Right: Histograms of the $S(t)$ distribution based on 1000 realizations. The 23 histograms are computed for a grid between time 0.0 and 1.1 with a step size of 0.05. Also shown are the analytic mean $\mu(t)$ and the 23 empirical means.

even numerically. For further details on such stochastic population modelling see, for example, Renshaw (1991). An alternative to the above is to resort to Monte Carlo simulation of the stochastic epidemic process—a method that has become increasingly popular even for inferential purposes. For the basic SIR model, one needs to simulate a discrete-state continuous-time stochastic process with piecewise constant conditional intensity functions (CIFs). Several algorithms exist for doing this, see, for example, Wilkinson (2006) for an overview, of which the algorithm by Gillespie (1977) is the best known. As an example, Figure 26.2 shows 10 realizations of the previous SI model and the resulting empirical probability distribution of $S(t)$ computed from 1000 simulations. Besides simplicity, Monte Carlo simulations have the additional advantage of being very flexible. For example, it is easy to use the samples to compute pointwise 95% prediction intervals for $S(t)$ or to compute the final number of infected.

A further important difference between deterministic and stochastic modelling is the interpretation of the *basic reproduction number* R_0, which describes the reproductive potential of an infectious disease and which is defined as the average number of secondary cases directly caused by an infectious case in an entirely susceptible population. In deterministic models, a major outbreak can only occur if $R_0 > 1$. In stochastic models, if $R_0 > 1$, then a major outbreak will occur with a certain probability determined by the model parameters. See, for example, Andersson and Britton (2000) or Britton (2010) for further details.

Other variations of the basic SI and SIR model could be, for example, the SI-Susceptible, SIR-Susceptible, or S-Exposed-IR model. Further extensions consist of reflecting the protection due to maternal antibodies or vaccination by respective compartments in the models. Finally, the rates can be additionally modified to, for example, reflect the import of infected from outside the population or demographics such as birth and death of individuals. See, for example, Keeling and Rohani (2008, chapter 2) for additional details.

26.2.2 Spatial extension: the metapopulation model

If interest is in enhancing the homogeneous SIR model with heterogeneity due to spatial aspects, one common modelling approach is to divide the overall population into a number of subpopulations—a so-called *metapopulation model* (Keeling et al., 2004, chapter 17). Here, one assumes that within each subpopulation, the mixing is homogeneous, whereas coupling between the subpopulations occurs by letting the force of infection contain contributions of the infected from the other populations as well. Considering a total of K subpopulations, the deterministic metapopulation SIR model looks as follows (Keeling and Rohani, 2008, section 7.2):

$$\begin{aligned}\frac{dS_k(t)}{dt} &= -\frac{\beta_k}{N_k} S_k(t) \left(\sum_{l=1}^{K} w_{kl} I_l(t) \right), \quad k = 1, \ldots, K \\ \frac{dI_k(t)}{dt} &= \frac{\beta_k}{N_k} S_k(t) \left(\sum_{l=1}^{K} w_{kl} I_l(t) \right) - \gamma I_k(t),\end{aligned}$$

where the weights w_{kl} quantify the impact of one infected from population l on population k. Typically, these weights are scaled such that $w_{kk} = 1, k = 1, \ldots, K$. Consequently, for a susceptible in k, w_{kl} represents the influence of an infected in unit l relative to one in the same unit as the susceptible. The appropriate choice of weights depends on the modelled disease, its mode of transmission, and the questions to be answered. They can even be subject to parametric modelling: for plant or animal diseases, dispersal could, for example, be due to airborne spread, which makes distance kernels such as exponential $w_{kl} \propto \exp(-\rho \, \mathrm{dist}_{kl})$ or power-law $w_{kl} \propto \mathrm{dist}_{kl}^{-\rho}$ convenient choices. Here, dist_{kl} denotes the geographic distance between the populations k and l. When restricting attention to directly transmitted human diseases, the focus is instead on the movement of infectious individuals from population l to population k—not necessarily due to a permanent relocation of individuals, but more to temporary movement. This could be, for instance, in the case of seasonal influenza, the commuting of individuals or, in the case of emerging epidemics (e.g., SARS and swine influenza), long-distance airplane travelling. If commuter or airline data are available, these can be used to determine the weights. Movement exhibits a strong age dependence, though, which can make it difficult to extract the relevant information from such sources for childhood diseases. As a consequence, one might use distance-based kernels as a proxy for mobility—possibly augmented by population sizes as in the *gravity model* (Erlander and Stewart, 1990; Xia et al., 2004). If interest is in short-term prediction of an emerging pathogen, long-distance travelling is an important concept to capture. However, the main bulk of infections for an established disease typically happen at a much smaller geographical scale, as shown, for example, in recent reanalyses of the 1918 pandemic influenza (Eggo et al., 2011) and the 2009 swine influenza outbreak (Gog et al., 2014).

In analogy to the above deterministic metapopulation model, the system of rates of the continuous-time stochastic SIR model can be modified accordingly to obtain a stochastic metapopulation model with unit-specific (conditional) intensity function for infection events,

$$\lambda_k(t) = \frac{\beta_k}{N_k} S_k(t) \left(\sum_{l=1}^{K} w_{kl} I_l(t) \right), \tag{26.1}$$

while the unit-specific intensity function for recovery events is now $\gamma I_k(t)$ (assuming exponentially distributed infectious periods). One important insight is that the difference between deterministic and stochastic metapopulation models is increased as subpopulations become smaller. Studying the extinction and reintroduction of disease in such metapopulations is hence preferably conducted using stochastic metapopulation models, see, for example, Bjørnstad and Grenfell (2008). Two variants of the stochastic metapopulation model are of additional interest: The first variant is the so-called Levins-type metapopulation model (Keeling and Rohani, 2008, section 7.2.4), where one for each k is only interested in the probability of $I_k(t) > 0$ as a function of time. Such models have been used to study the arrival of the first disease cases in a city during a pandemic (Eggo et al., 2011; Gog et al., 2014). The second variant is the individual model, that is, when the number of considered subpopulations is equal to the population size $K = N$.

Both cases are instances of multivariate counting processes that can be consistently handled in the SIR-S framework using a so-called two-component SIRS model, which additionally allows for immigration of disease cases from external sources. This is particularly useful if the data contain multiple outbreaks. Following Höhle (2009), let $N_k(t)$, $k = 1, \ldots, K$, denote the counting process, which for unit k counts the number of changes from state susceptible to state infectious. The conditional intensity function given the history of all K processes up to, but not including, t is then given as

$$\begin{aligned}\lambda_k(t) &= \exp\left(h_0(t) + \boldsymbol{z}_k(t)^T \boldsymbol{\beta}\right) + \sum_{j \in I(t)} \left\{w(\text{dist}_{kj}) + \boldsymbol{v}_{kj}(t)^T \boldsymbol{\alpha}_e\right\} \\ &= Y_k(t) \cdot \left[\exp(h_0(t)) \exp\left(\boldsymbol{z}_k(t)^T \boldsymbol{\beta}\right) + \boldsymbol{x}_k(t)^T \boldsymbol{\alpha}\right],\end{aligned} \tag{26.2}$$

where $Y_k(t)$ is an indicator if unit k is susceptible, that is, $Y_k(t) = \mathbb{1}_{k \in S(t)})$, while $S(t)$ and $I(t)$ now denote the index set of all susceptibles and infectious, respectively. Furthermore, $w(\cdot)$ denotes a distance weighting kernel parametrized as a spline function, while $\boldsymbol{z}_k(t)$ and $\boldsymbol{v}_{kj}(t)$ denote possibly time-varying covariates affecting the introduction of new cases in the endemic and epidemic components, respectively, and $\boldsymbol{x}_k(t)$ denotes the combination of linear spline terms and epidemic covariates. Finally, $h_0(t)$ is the baseline rate of the endemic component. If import of new cases from external sources is not relevant, the first component in the above can be left out (i.e., set to zero).

26.2.3 Fitting SIR models to data

From a statistician's point of view, parameter inference in epidemic models appears to receive only marginal attention in the medical literature. One reason might be that little or no data are available, and hence parameters are "guesstimated" from literature studies or expert knowledge. Another reason is that inference often boils down to extracting information about parameter values from a *single* realization of the stochastic epidemic process. For deterministic models, estimation might also appear less of a statistical issue. Finally, estimation is complicated by the epidemic process only being partially observable: the number of susceptibles to begin with, as well as the time point of infection of each case, might be unknown. As a consequence, parameter estimation in ODE-based SIR models has, typically, been done using least-square-type estimation based on observable quantities. Only recently, the models have been extended to non-Gaussian observational components enabling count data likelihood-based statistical inference, see, for example, Pitzer et al. (2009),

$$\mathrm{E}(Y_{t,k}) = \int_t^{t+1} \left\{-\frac{dS_k(u; \boldsymbol{\theta})}{dt}\right\} du,$$

where $Y_{t,k}$ is the observed number of *new* infections within time interval t, which is assumed to follow a count data PMF f such as the Poisson or negative binomial distribution with expectation $\mathrm{E}(Y_{t,k})$. This setup allows the estimation of the model parameters $\boldsymbol{\psi}$ in a likelihood framework by the log-likelihood function $l(\boldsymbol{\psi}) = \sum_{t=1}^{T} \sum_{k=1}^{K} \log f(y_{t,k}; \boldsymbol{\psi})$, when assuming observations are independent given the model. However, a residual analysis often shows remaining autocorrelation, which means any quantification of estimation uncertainty based on asymptotic theory using the Hessian of the log-likelihood tends to be overly optimistic. As a consequence, the confidence intervals might be too narrow. A novel two-step approach to improve this shortcoming within the above framework is treated in Weidemann et al. (2014).

An advantage of stochastic continuous-time SIR modelling is that it allows for a quantification of the uncertainty of the estimates, even though the estimate is based on a single process realization. The work of Becker (1989) and the second part of Andersson and Britton (2000) are some of the few books dedicated to this task. If the epidemic process $(S(t), I(t))'$ is completely observed over the interval $(0, T]$, where T is the entire duration of the epidemic, the resulting data of the epidemic are given as $\{(t_i, y_i, k_i), i = 1, \ldots, n\}$, where n is the number of infections in the population during the epidemic, t_i denotes the time of infection of individual i, y_i is the length of the individuals' infectious period, and $k_i \in \{1, \ldots, K\}$ is the unit it belongs to. Further assuming that the probability distribution function (PDF) of the infectious period is exponential $f_{I \to R}(y) = \gamma \exp(-\gamma y)$, we obtain the log-likelihood for the parameter vector $\psi = (\beta, \gamma)'$ as

$$l(\psi) = \sum_{i=1}^{n} \log f_{I \to R}(y_i) + \sum_{i=1}^{n} \log \lambda_{k_i}(t_i) - \int_0^T \sum_{k=1}^{K} \lambda_k(u) du, \tag{26.3}$$

where $\lambda_k(t_i)$ is defined as in (26.1) and evaluated at the time just prior to t_i. Note that in (26.3), the CIFs have to be integrated over time; however, for the simple SIR model, the CIF is a piecewise constant function between events, and hence integration is tractable. Höhle (2009) develops these likelihood equations further using counting process notation for the two-component SIR model (26.2), whereas Lawson and Leimich (2000), Diggle (2006), and Scheel et al. (2007) contain accounts and examples of a partial-likelihood approach for spatial SIR-type model inference including covariates.

In applications, the times of infection t_i of infected individuals would typically be unknown. One way to make inference tractable is to assume that the duration of the infectious period is a constant, say, $\mu_{I \to R}$ (known or to be estimated). Furthermore, the initial number of susceptibles might also be unknown and require estimation. See Becker (1989) for details, which also covers a discrete-time approximation covered in the next section. More recently, Gibson and Renshaw (1998), O'Neill and Roberts (1999), O'Neill and Becker (2001), Neal and Roberts (2004), and Höhle et al. (2005) used a Bayesian data augmentation approach using Markov chain Monte Carlo (MCMC) to impute the missing infection times while simultaneously performing Bayesian parameter inference for the S(E)IR and metapopulation S(E)IR model. Model diagnostics can in the likelihood context be performed using a graphical assessment of residuals and forward simulation (Höhle, 2009); for the Bayesian models, one can use additional posterior predictive checks and latent residuals (Lau et al., 2014).

26.2.4 Discrete-time models

Up to now, focus has been on continuous-time epidemic modelling. However, data are usually only available at much coarser timescales: weekly or daily reporting is usual in public health surveillance. If individual data are available, this observational situation can be handled by considering the observed event times as being for the event times interval censored. However, when looking at large populations or routinely collected data, data are typically provided in aggregated form without access to the individual data. It is thus necessary to consider alternative ways of casting the continuous-time stochastic SIR approach into a discrete-time framework. Time series analysis is one such approach, providing a synthesis between complex stochastic modelling and available data.

In order to derive such a model, we consider a sufficiently small time interval $[t, t + \delta t)$. We could then—as an approximation—assume constant conditional intensity functions of our SIR model in $[t, t + \delta t)$. By definition, let these intensities be equal to the intensities at the left time point of the interval, that is, at time t. This implies that all individuals are independent for the duration of the interval. Looking at one susceptible individual, its probability of escaping infection during $[t, t + \delta t)$ is then equal to $\exp(-\beta I(t) \cdot \delta t)$. Denoting by $C_{[t,t+\delta t)}$ the number of newly infected and by $D_{[t,t+\delta t)}$ the number of recoveries in the interval $[t, t + \delta t)$, we obtain

$$C_{[t,t+\delta t)} \sim \text{Bin}\left(S(t), 1 - \exp(-\beta I(t) \cdot \delta t)\right) \tag{26.4}$$

$$D_{[t,t+\delta t)} \sim \text{Bin}\left(I(t), 1 - \exp(-\gamma \cdot \delta t)\right). \tag{26.5}$$

The state at time $t + \delta t$ is then given by $S(t + \delta t) = S(t) - C_{[t,t+\delta t)}$, $I(t + \delta t) = I(t) + C_{[t,t+\delta t)} - D_{[t,t+\delta t)}$. Now, changing notation to discrete time, with discrete-time subscript t denoting the time $t + \delta t$ and subscript $t - 1$ the time t, we write

$$S_t = S_{t-1} - C_t, \quad I_t = I_{t-1} + C_t - D_t, \tag{26.6}$$

for $t = 1, 2, \ldots$ and with $S_0 = S(0), I_0 = I(0)$. Consequently, the discrete quantities C_t, D_t, and so forth, now replace the continuous ones in (26.4) and (26.5). Such models are known as chain binomial models (Becker, 1989), if one assumes that $D_t = I_t$; that is, the timescale is chosen such that all infective individuals recover after one time step. In such models, one time step can be seen as one generation time (Daley and Gani, 1999, chapter 4). For large S_t and I_t, the binomial distributions can be further approximated by Poisson distributions: by a first-order Taylor expansion of the $1 - \exp(-x)$ terms, one obtains

$$C_t \sim \text{Po}\left(\beta S_{t-1} I_{t-1} \cdot \delta t\right) \tag{26.7}$$

$$D_t \sim \text{Po}\left(\gamma I_{t-1} \cdot \delta t\right). \tag{26.8}$$

Note that these approximations no longer ensure that $C_t \leq S_{t-1}$ and $D_t \leq I_{t-1}$. If this is a practical concern, one can instead use right-truncated Poisson distributions fulfilling the conditions. Altogether, we have transformed the continuous-time stochastic model into a discrete-time model. If $(S_t, I_t)'$ is known at each point in time $t = 1, \ldots, T$, estimates for β and γ can be found using maximum likelihood approaches based on (26.7) and (26.8). For example, Becker (1989) shows how the above equations can be used to fit a homogeneous SIR model to data using generalized linear model (GLM) software: (26.8) can be represented as a log-link Poisson GLM with offset $\log(S_{t-1} I_{t-1} \delta t)$ and intercept $\log(\beta)$ or an identity link GLM with covariate $S_{t-1} I_{t-1} \delta t$ and no intercept. For the stochastic metapopulation SIR model with known weights, the idea can similarly be extended by jointly modelling $C_{t,k}$ and $D_{t,k}$, $1 \leq t \leq T$, $1 \leq k \leq K$, by an appropriate Poisson GLM with either log-link or identity link with offset, intercept, and potential covariates derived from (26.1). Furthermore, Klinkenberg et al. (2002) uses the above equations to fit a spatial grid SIR model using numerical maximization of the binomial likelihood.

26.2.5 Time series SIR model

The model given by (26.7) and (26.8) corresponds to a simple version of the *time series SIR model* (TSIR model) initially proposed in Finkenstädt and Grenfell (2000) and since extended (Finkenstädt et al., 2002; Bjørnstad et al., 2002). For the TSIR model, it is assumed that $D_t = I_t$, that is, as in a chain binomial model, and $\delta t = 1$. As an example, one time unit corresponds to a biweekly scale when modelling measles. In addition, the TSIR model contains additional flexibility beyond the simple SIR model, for example, by taking population demographics into account, where births provide new susceptibles and immigration provides an influx of infectives allowing the reintroduction of cases into a population where the disease was already extinct. Further extensions are a time-varying transmission rate, for example, due to school closings, and a modification of the multiplication term $I(t)$ in the transmission rate to $I(t)^{\alpha}$, which allows for spatial substructures and other forms of heterogeneity in the population (Liu et al., 1987). Xia et al. (2004) extend the model to a multivariate (and hence spatial) time series model, where the transmission between populations is based on a gravity model.

Below, we present a slightly modified version of this *multivariate TSIR model* (mTSIR) given in Xia et al. (2004). Let $I_{t,k}$ and $S_{t,k}$ be the number of infected and susceptibles in region k at time t, where $1 \leq k \leq K$ and $1 \leq t \leq T$. For $t = 2, \ldots, T$, the model is now defined by

$$I_{t,k} | \lambda_{t,k}, \iota_{t,k} \sim \text{NB}(\lambda_{t,k}, I_{t-1,k}), \tag{26.9}$$

where

$$\begin{aligned}
\lambda_{t,k} | \iota_{t,k} &= \frac{1}{N_{t-1,k}} \beta(t) \, S_{t-1,k} \, (I_{t-1,k} + \iota_{t,k})^{\alpha} \\
\iota_{t,k} &\sim \text{Ga}(m_{t,k}, 1) \\
m_{t,k} &= \theta \, N_{t-1,k}^{\tau_1} \sum_{j \neq k} d_{j,k}^{-\rho} \, I_{t-1,j}^{\tau_2}.
\end{aligned}$$

Here, $\text{NB}(\mu, c)$ denotes the negative binomial distribution with expectation μ and clumping parameter c; that is, the variance is $\mu + \mu^2/c$. For example, Bailey (1964, chapter 8) shows that the number of offspring generated by a pure birth process with an according rate and starting with a population of c individuals has exactly this distribution. Note, however, that the model in (26.9) is a double-stochastic model, because the rate itself is a random variable due to the random influx of new infectives, which is gamma distributed. This implies that there is a small twist in case $I_{t-1,k} = 0$, because (26.9) would then imply that $I_{t,k} | (I_{t-1,k} = 0) \equiv 0$, which is not as intended. The reason is that the birth process motivation is not directly applicable in this situation—instead we will assume that $I_{t,k} | I_{t-1,k} = 0$ is just $\text{Po}(\lambda_t)$ distributed. Another approach is to assume that the clumping parameter in (26.9) is equal to $I_{t-1,k} + \iota_{t,k}$ (Morton and Finkenstädt, 2005; Bjørnstad and Grenfell, 2008). In practice, if $I_{t-1,k}$ is large, the negative binomial effectively reduces to the Poisson distribution. The additional complexity given by the negative binomial is thus somewhat artificial—it might be worthwhile to let the clumping parameter vary more freely instead.

Finally, $\beta(t) \geq 0$ is a periodic function with period 1 year and parametrized by the parameter vector $\boldsymbol{\beta}$. For biweekly measles data, this could, for example, be a 26-parameter function or a harmonic seasonal forcing function, that is,

$$\beta(t) = \beta_{t \bmod 26+1}$$

or

$$\beta(t) = \beta_1(1 + \beta_2 \cos(2\pi t/26)).$$

The number of susceptibles in the TSIR model is given by the recursion

$$S_{t,k} = S_{t-1,k} - I_{t,k} + B_{t,k} \quad \text{for } t = 2, \ldots, T, \tag{26.10}$$

where $B_{t,k}$ denotes the birth rate in region k at time t. One challenge when using the TSIR model is, however, that $S_{t,k}$ is only partially observable. Untangling the recursion yields

$$S_{t,k} = S_{1,k} - \sum_{u=2}^{t} I_{u,k} + \sum_{u=2}^{t} B_{u,k}, \quad t = 2, \dots, T,$$

where $S_{1,k}$ is usually unknown but needs to be such that $0 < S_{t,k} \le N_{t,k} - I_{t,k}$ for all $t = 2, \dots, T$. As a consequence, either all unknown $S_{1,k}$'s need to enter the analysis as K additional parameters or one has just a single extra parameter κ together with the coarse assumption that $S_{1,k} = \kappa N_{1,k}$. An alternative is to use a preprocessing step to determine it: Conditional on $S_{1,k}$ and the observed $I_{t,k}$, one can compute the resulting $S_{t,k}$'s. Inspired by the univariate procedure in Finkenstädt et al. (2002), one then considers $\iota_{t,k}$ within the unit to be a time series varying around its mean and uses a Taylor expansion up to order 3 to derive

$$\log I_{t,k} = \log \beta_t + \alpha \log(I_{t-1,k}) + c_1 I_{t-1,k}^{-1} + c_2 I_{t-1,k}^{-2} + c_3 I_{t-1,k}^{-3} + \log S_{t,k} + \epsilon_{t,k}, \qquad (26.11)$$

where the $\epsilon_{t,k}$'s are zero-mean random variables. Altogether, conditionally on $S_{1,k}$, expression (26.11) represents a linear regression model that can be fitted using maximum likelihood. The resulting profile log-likelihood in $S_{1,k}$ can then be optimized for each unit separately.

As already mentioned, one problem with the above model formulation is that the update given in (26.9) does not ensure that $I_{t,k} \le S_{t-1,k} + B_{t-1,k}$. In other words, $S_{t,k} \ge 0$ is not explicitly ensured by the model. When fitting the model to data, this is unproblematic, because the $S_{t,k}$'s are computed by a separate preprocessing step. However, when simulating from the model and hence computing $S_{t,k}$ in each step from (26.10), it becomes clear that these can become negative if $I_{t,k}$ is large enough. To ensure validity of the model, we thus right-truncate the $\text{NB}(\lambda_{t,k}, I_{t-1,k})$ at $S_{t-1,k}$.

26.2.6 Fitting the mTSIR model to data

In the work of Xia et al. (2004), a heuristic optimization criterion is applied, which consists both of a one-step-ahead squared prediction error assessment and a criterion aimed at minimizing the absolute difference between the observed and predicted cross-correlation at lag zero between the time series of a unit and the time series of London. The 26 seasonal parameters β_t are explicitly fixed at the values obtained in Finkenstädt and Grenfell (2000). However, from Xia et al. (2004), it is not entirely clear how the point prediction of $\hat{I}_{t,k}$ is defined. Instead, we proceed here using a marginal likelihood-oriented approach for inference in the mTSIR model. We do so by computing the marginal distribution of $I_{t,k}$ given $\boldsymbol{I}_{t-1} = \boldsymbol{i}_{t-1}$ as follows:

$$f_m(i_{t,k}|\boldsymbol{i}_{t-1}; \boldsymbol{\psi}) = \int_0^\infty f(i_{t,k}|\iota_{t,k}, \boldsymbol{i}_{t-1}; \boldsymbol{\psi}) f(\iota_{t,k}|\boldsymbol{i}_{t-1}; \boldsymbol{\psi}) d\iota_{t,k},$$

where the two integrated densities refer to the PMF of the truncated negative binomial and the PDF of the gamma distribution, respectively. The former is computed from the ordinary negative binomial PMF as $f_{\text{NB}}(i_{t,k})/F_{NB}(S_{t-1,k})$, with F denoting the cumulative distribution function. Based on the above, the marginal log-likelihood for the parameter vector $\boldsymbol{\psi} = (\boldsymbol{\beta}', \alpha, \theta, \tau_1, \tau_2, \rho)'$ is then given as

$$l(\boldsymbol{\psi}) = \sum_{t=2}^{T} \sum_{k=1}^{K} \log f_m(i_{t,k}|\boldsymbol{i}_{t-1}; \boldsymbol{\psi}). \qquad (26.12)$$

This expression is then optimized numerically for $\boldsymbol{\psi}$ in order to find the maximum likelihood estimator (MLE). In our R implementation (R Core Team, 2014) of the model, we use a Broyden–Fletcher–Goldfarb–Shanno (BFGS) method, as implemented in the function `optim`, for optimization, while using the function `gauss.quad.prob` to perform Gaussian quadrature based on orthogonal Laguerre polynomials for handling the integration over the gamma densities. In a recent work, Jandarov et al. (2014) presented a more complex Bayesian approach, including an evaluation of the required integrals by MCMC (Smyth, 1998). Note that the quadrature strategy can be numerically difficult in practice, because the the marginal density for some parameter configurations to be evaluated is so small that—with the given floating-point precision—it becomes zero. This then causes problems when taking the logarithm. A simplification to allow for estimation is to replace the stochastic $\iota_{t,k}$ component in (26.9) by its expectation $m_{t,k}$—as also done in Jandarov et al. (2014)—at the cost of reducing the variability of the model.

26.2.7 Endemic–epidemic modelling

Inspired by the SIR and mTSIR models, Held et al. (2005) presented a multivariate count data time series model for routine surveillance data, which does not require the number of susceptibles to be available. The formal inspiration for the model was the spatial branching process with immigration, which means that observation time and generation time have to correspond. In a series of successive papers, the modelling was subsequently extended such that it now constitutes a powerful and flexible regression approach for multivariate count data time series (Paul et al., 2008; Paul and Held, 2011; Held and Paul, 2012; Meyer and Held, 2014). From a spatial statistics perspective, this even includes the use of CAR-type random effects (Besag et al., 1991) (discussed in Chapter 7 of this book) in the time series modelling. The fundamental idea is to divide the infection dynamics into two components: an *endemic component* handles the influx of new infections from external sources, and an *epidemic component* covers the contagious nature by letting the expected number of transmissions be a function of the lag-one number of infections. The resulting so-called HHH model (named for authors L. Held, M. Höhle, and M. Hofmann) is given by

$$Y_{t,k} \sim \mathrm{NB}(\mu_{t,k}, \gamma_k), \quad t = 2, \ldots, T, \quad k = 1, \ldots, K \tag{26.13}$$

$$\mu_{t,k} = N_{t,k}\, \nu_{t,k} + \lambda_{t,k}\, Y_{t-1,k} + \phi_{t,k} \sum_{j \neq k} w_{jk}\, Y_{t-1,j}.$$

In Equation 26.13, $Y_{t,k}$ denotes the number of new cases in unit k at time t, which is assumed to follow a negative binomial distribution with expectation $\mu_{t,k}$ and region-specific clumping parameter γ_k. As before, $N_{t,k}$ denotes the corresponding population in region k at time t, and the w_{jk} are known weights describing the impact of cases in unit j on unit k. This can, for example, be population flux data such as airline passenger data (Paul et al., 2008) or neighbourhood indicators such as $w_{jk} \propto I(d_{jk} = 1)$ or a power-law weight $w_{jk} \propto d_{jk}^{-\rho}$, where in the last two examples, d_{jk} denotes the graph-based distance between units j and k in the neighbourhood graph, while ρ is a parameter to estimate (Meyer and Held, 2014). Furthermore, $\nu_{t,k}$, $\lambda_{t,k}$ and $\phi_{t,k}$ are linear predictors covering the endemic component, as well as the within- and between-unit autoregressive behaviour, respectively,

$$\log(\nu_{t,k}) = \alpha^{(\nu)} + b_k^{(\nu)} + \boldsymbol{z}_{t,k}^{(\nu)\prime} \boldsymbol{\beta}^{(\nu)}$$

$$\log(\lambda_{t,k}) = \alpha^{(\lambda)} + b_k^{(\lambda)} + \boldsymbol{z}_{t,k}^{(\lambda)\prime} \boldsymbol{\beta}^{(\lambda)}$$

$$\log(\phi_{t,k}) = \alpha^{(\phi)} + b_k^{(\phi)} + \boldsymbol{z}_{t,k}^{(\phi)\prime} \boldsymbol{\beta}^{(\phi)}.$$

In each case, $\alpha^{(\cdot)}$ denotes the overall intercept, $b_k^{(\cdot)}$ is a unit-specific random intercept, and $\boldsymbol{z}_{t,k}^{(\cdot)}$ represents a length $p^{(\cdot)}$ vector of possibly time-varying predictor-specific covariates with associated parameter vector $\boldsymbol{\beta}^{(\cdot)}$. The use of covariates allows for a flexible modelling of, for example, secular trends and concurrent processes influencing the disease dynamics (temperature, occurence of other diseases, vaccination, etc.). The vector of random effects per unit $(b_k^{(\nu)}, b_k^{(\lambda)}, b_k^{(\phi)})'$ can be i.i.d. normal with predictor-specific variance, trivariate normal with a correlation matrix (to be estimated), or of CAR type for each predictor. The HHH model allows the absence of one or several of the model components, for example, lack of endemic component, and also supports the use of offsets in the linear predictors, which is especially convenient to represent population influences, as we shall see subsequently. The HHH model has already been successfully used to describe influenza, measles, and meningococcal dynamics; see, for example, Herzog et al. (2011) or Geilhufe et al. (2014).

26.2.7.1 Linking the HHH and mTSIR model

As an example of the flexibility of HHH modelling, we consider it here as a framework for representing a simplified version of the mTSIR model, including a gravity model-based flux of infectives: replacing the stochastic $\iota_{t,k}$ component in (26.9) by its expectation and assuming $\alpha = \tau_2 = 1$ then yields

$$\lambda_{t,k}^{\text{mTSIR}} = \mu_{t,k}^{\text{HHH}} = \frac{1}{N_{t,k}}\beta(t)S_{t-1,k}\left(I_{t-1,k} + \theta\, N_{t-1,k}^{\tau_1} \sum_{j\neq k} d_{j,k}^{-\rho}\, I_{t-1,j}\right).$$

Ignoring the difference between $N_{t-1,k}$ and $N_{t,k}$, this corresponds to an HHH model without endemic component, that is, $\nu_{t,k} \equiv 0$, and

$$\begin{aligned}
\log(\lambda_{t,k}) &= \log(S_{t-1,k}/N_{t,k}) + \log\beta(t)\\
\log(\phi_{t,k}) &= \log(S_{t-1,k}) + (\tau_1 - 1)\log(N_{t,k}) + \log(\theta) + \log\beta(t),
\end{aligned}$$

together with weights $w_{jk} \propto d_{jk}^{-\rho}$, where d_{jk} is an integer-discretized version (e.g., in steps of 50 km) of the Euclidean distance between units j and i. Note that in both predictors, the first term is an offset and the number of susceptibles has been included as a time- and unit-specific covariate. Even if no data on the number of susceptibles are available, good proxies can be established. For example, Herzog et al. (2011) perform a spatiotemporal analysis using the county-specific vaccination rate against measles as a proxy for the availability of susceptibles, and hence, as the result of the modelling, obtain a quantification of the effectiveness of the vaccination. Note also that the clumping parameter of the negative binomial is different in the two models: in the HHH model, it is time constant but varies freely, whereas in the mTSIR model, it is time varying but fixed to be the previous number of infectives.

26.2.8 Fitting HHH models to data

Likelihood-based inference for the HHH model is performed by maximizing the corresponding (marginal) log-likelihood function, which is similar to (26.12), with f_m being the (marginalized) PMF of the negative binomial model given by (26.13). Altogether, a variant of the penalized quasi-likelihood approach discussed in Breslow and Clayton (1993) is used, if the predictors contain random effects. Model selection is performed by Akaike's Information Criterion (AIC) or—in the case of random effects—using proper scoring rules for count data based on one-step-ahead forecasts (Czado et al., 2009; Paul and Held, 2011); see the respective HHH papers for the inferential details.

26.3 Measles Dynamics in the Prevaccination Area

In this section, we analyse the 1944–1966 biweekly England and Wales measles data already presented in Section 26.1. These prevaccination data have been analysed univariately in Finkenstädt and Grenfell (2000) and Finkenstädt et al. (2002) and are available for download from the Internet (Grenfell, 2006). The present analysis is an attempt to analyse the data in spatiotemporal fashion. In a preprocessing step, we determine $S_{1,k}$ for each of the 60 series as follows: the linear model (26.11) is fitted for a grid of $S_{1,k}$ values, and the resulting log-likelihood is used to determine $\hat{S}_{1,k}$. Subsequently, the recursion in (26.10) is used to compute the time series of susceptibles in unit k given $\hat{S}_{1,k}$. Figure 26.3 shows this exemplarily for the time series of London.

26.3.1 Results of the mTSIR model

We use the individually reconstructed series of susceptibles for each city to fit the mTSIR model to the multivariate time series of the 60 cities. Fitting the model based on the likelihood proves to be a complicated matter: we therefore follow the strategy of Xia et al. (2004) by fixing $\beta(t)$ to the 26 parameters found in Finkenstädt and Grenfell (2000) and fixing $\alpha = 0.97$. Table 26.1 contains the ML results in a model where the $\iota_{t,k}$ variables are replaced with their expectation. The table also contains 95% Wald confidence intervals (CI) based on the observed inverse Fisher information.

Altogether, the results differ quite markedly from the results reported in Xia et al. (2004), where the estimates are $\hat{\tau}_1 \approx 1, \hat{\tau}_2 \approx 1.5$, $\hat{\rho} \approx 1$, and $\hat{\theta} \approx 4.6 \cdot 10^{-9}$. Several explanations for the differences exist: different estimation approaches and different data were used. Altogether, it appears that the full mTSIR model is too flexible for the available reduced data set containing only 60 cities. Still, the one-step-ahead 95% predictive distributions obtained by plug-in of the MLE show that the model enables too little variation: a substantial number

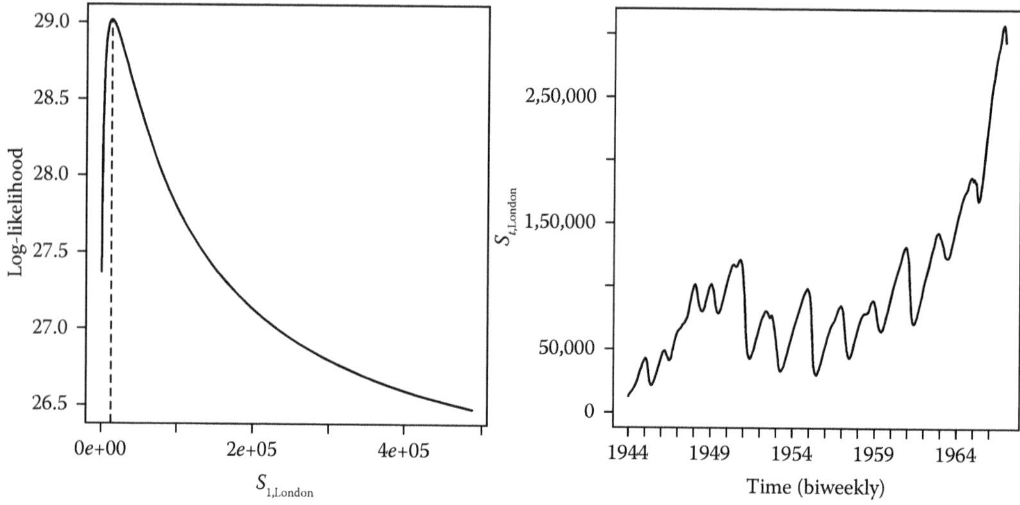

FIGURE 26.3
Left: Log-likelihood of the linear model as a function of $S_{1,k}$ for London. Right: Resulting time series of susceptibles obtained from the $S_{1,k}$ maximizing the log-likelihood.

TABLE 26.1
Parameter estimates and 95% CIs for the mTSIR model

	Estimate	95% CI	
τ_1	2.24	2.24	2.24
τ_2	1.10	1.10	1.10
ρ	$3.05 \cdot 10^{-2}$	$2.94 \cdot 10^{-2}$	$3.16 \cdot 10^{-2}$
θ	$1.79 \cdot 10^{-16}$	$1.77 \cdot 10^{-16}$	$1.81 \cdot 10^{-16}$

of observations lay outside the 95% predictive intervals obtained by computing the 2.5% and 97.5% quantiles of the truncated negative binomial in (26.9). Figure 26.4 shows this exemplarily for the London time series. One explanation might be the direct use of $I_{t-1,k}$ as clumping parameter for the negative binomial distribution in (26.9)—it appears more intuitive that the variance with increasing $I_{t-1,k}$ should increase instead of converging to the Poisson variance as implied by the model.

26.3.2 Results of the HHH modelling

Instead of trying to improve the mTSIR model manually, we do so by performing this investigation within the HHH model. As an initial HHH model capturing seasonality in both the endemic and epidemic components, we consider the following model:

$$\log(\nu_{t,k}) = \log(N_{t,k}) + \alpha^{(\nu)} + \beta_1^{(\nu)} \sin\left(\frac{2\pi t}{26}\right) + \beta_2^{(\nu)} \cos\left(\frac{2\pi t}{26}\right)$$

$$\log(\lambda_{t,k}) = \alpha^{(\lambda)} + \beta^{(\lambda)} \log(N_{t,k}) + \beta_1^{(\lambda)} \sin\left(\frac{2\pi t}{26}\right) + \beta_2^{(\lambda)} \cos\left(\frac{2\pi t}{26}\right)$$

$$\log(\phi_{t,k}) = \alpha^{(\phi)} + \beta^{(\phi)} \log(N_{t,k}) + \beta_1^{(\phi)} \sin\left(\frac{2\pi t}{26}\right) + \beta_2^{(\phi)} \cos\left(\frac{2\pi t}{26}\right),$$

and with weights $w_{jk} = I(0 < \text{dist}_{jk} \leq 50 \text{ km})$, where dist_{jk} denotes the geographic distance between cities j and k in kilometres. As a first step in our model selection strategy,

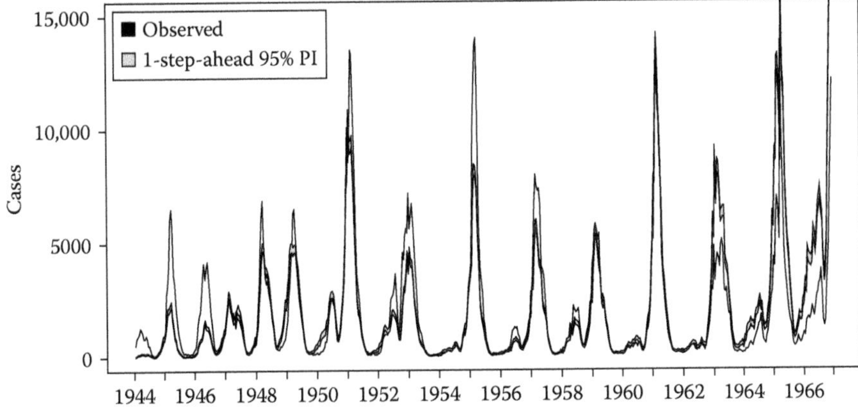

FIGURE 26.4
One-step-ahead 95% predictive intervals for the London time series obtained by plug-in of the MLE in the mTSIR model. Also shown is the actual observed time series.

we compare this model to one using weights $w_{jk} = d_{jk}^{-\rho}$, which corresponds to a power-law distance relationship with $d_{jk} = \lceil \text{dist}_{jk}/50 \text{ km} \rceil$. Based on AIC, such a power-law distance kernel is prefered (AIC $3.172 \cdot 10^5$ vs. $3.189 \cdot 10^5$), and we hence proceed analysing the power-law version. Figure 26.5 shows the model fit decomposed into the three components exemplarily for the three largest and three smallest cities. We observe that the endemic and neighbourhood components only play a small role in the measles transmission. However, neither excluding the endemic component (AIC $3.1981 \cdot 10^5$) nor excluding the neighbourhood component (AIC $3.1977 \cdot 10^5$) provides a better AIC. Furthermore, Figure 26.6 shows the weight-quantifying neighbourhood interaction as a function of distance (in steps of 50 km). The right panel of the figure contrasts the power-law model with a model containing individual coefficients for lags 1–4; this gives some indication that the distance influence might be stronger at short lags than implied by the power-law model. Also, the AIC of this more flexible model is slightly better (AIC $3.198 \cdot 10^5$). However, there is not enough information in the data to fit individual coefficients for lags 5 and higher.

The seasonal component in the two components is best illustrated graphically (Figure 26.7). We note that the seasonality of the dominating within-city transmission component has a shape very similar to what was found in previous work, that is, with a lower transmission during the summertime. On the other hand, during summertime, imports from external sources appear more likely. The interpretation of the neighbourhood transmission is slightly inconclusive, as it is shifted compared to the two other—this component only makes up a small part of the overall transmission, though.

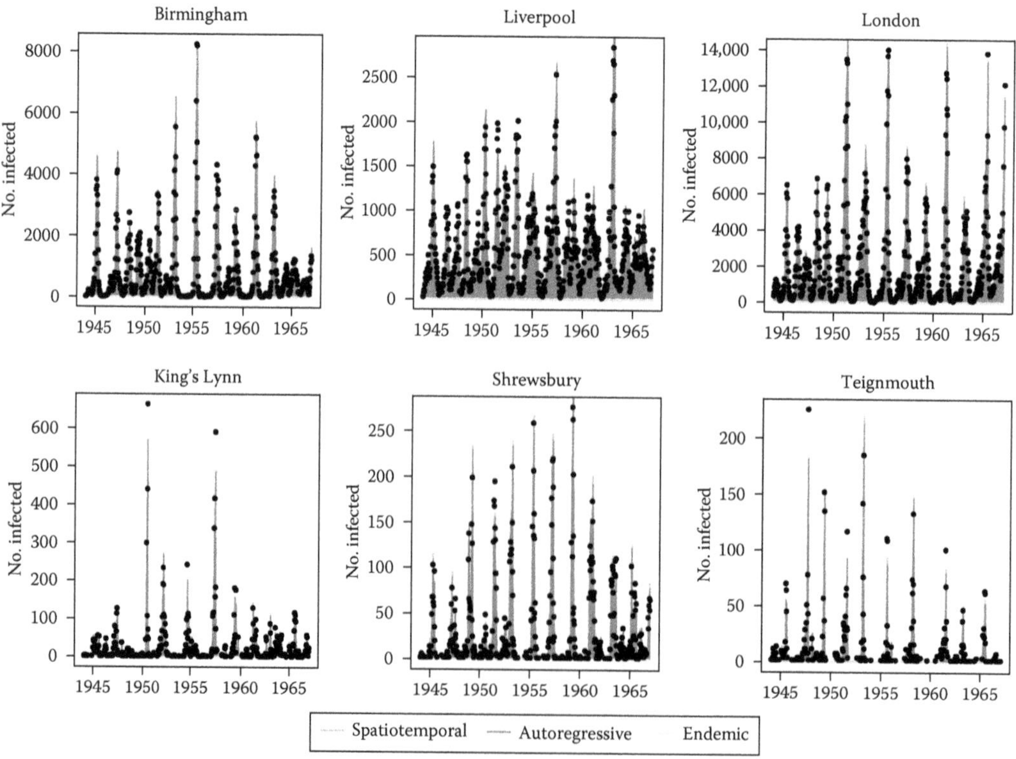

FIGURE 26.5
Time series of counts for the three largest (top row) and three smallest (bottom row) cities. Also shown are the model-predicted expectations decomposed into the three components: endemic, within city, and from outside cities.

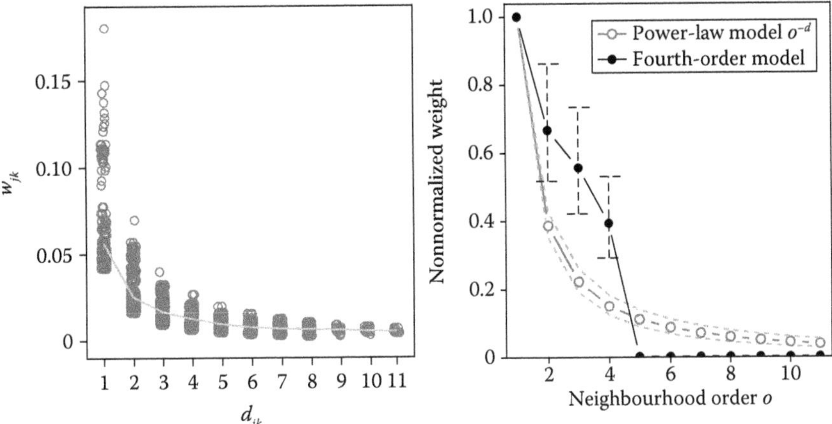

FIGURE 26.6
Left: Normalized weights as a function of d_{jk}. Right: Difference between the power-law model and a model containing individual coefficients for each neighbourhood order 1–4 and zero weight thereafter.

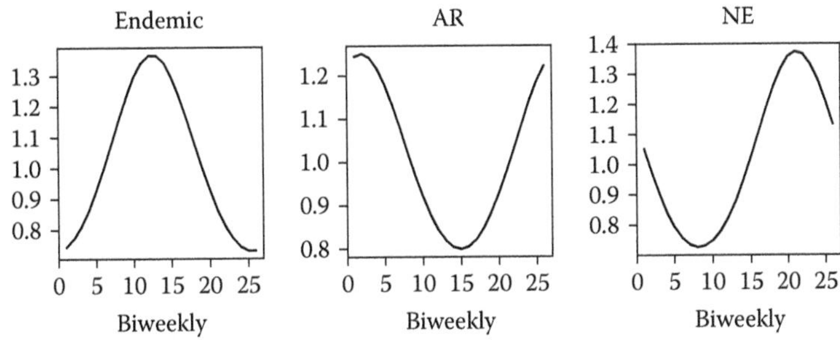

FIGURE 26.7
Illustration of the seasonal terms of the endemic, autoregressive (AR), and neighbourhood (NE) components of the power-law HHH model.

26.3.3 Results of the mTSIR mimicking HHH model

When $\alpha = 1$ and the autoregressive part is just $I_{t,k-1} + m_{t,k}$, we can, as described in Section 26.2.7, mimic the behaviour of the mTSIR model by using an HHH model with

$$\log(\lambda_{t,k}) = \beta_1^{(\lambda)} \sin\left(\frac{2\pi t}{26}\right) + \beta_2^{(\lambda)} \cos\left(\frac{2\pi t}{26}\right) + \log\left(\frac{S_{t-1,k}}{N_{t,k}}\right)$$

$$\log(\phi_{t,k}) = \log(\theta) + \beta_1^{(\phi)} \sin\left(\frac{2\pi t}{26}\right) + \beta_2^{(\phi)} \cos\left(\frac{2\pi t}{26}\right) + (\tau_1 - 1)\log(N_{t,k}) + \log(S_{t-1,k}),$$

and $w_{jk} = d_{jk}^{-\rho}$, which corresponds to a power-law distance relationship. In both these components, the last term of the predictor represents an offset—an alternative would have been to use terms of the type $\beta_3^{(\cdot)} \log(S_{t-1,k})$ in order to address additional nonlinearity in the susceptibles, as suggested in the original TSIR formulations (Finkenstädt and Grenfell, 2000; Finkenstädt et al., 2002). The results obtained from fitting this model to the 60 cities' data are given in Table 26.2.

TABLE 26.2
Parameter estimates and 95% CIs for the mTSIR-mimicking HHH model

	Estimate	95% CI	
ar.sin(2 * pi * t/26)	1.35	1.20	1.50
ar.cos(2 * pi * t/26)	4.12	4.05	4.18
ne.1	−10.58	−10.87	−10.29
ne.log(N)	0.10	0.08	0.13
ne.sin(2 * pi * t/26)	−0.20	−0.25	−0.15
ne.cos(2 * pi * t/26)	−1.31	−1.35	−1.27
neweights.d	0.84	0.75	0.93
overdisp	2.72	2.68	2.76

By transformation, we find the corresponding mTSIR model parameters to be $\hat{\theta} = 2.5 \cdot 10^{-5}$ (95% CI: $1.9 \cdot 10^{-5}$ to $3.4 \cdot 10^{-5}$), $\hat{\tau}_1 = 1.10$ (95% CI: 1.08–1.13), and $\hat{\rho} = 0.84$ (95% CI: 0.75–0.93). It also becomes clear that the clumping parameter of the negative binomial is estimated to be at a very different scale than suggested by the mTSIR model. As an additional sensitivity analysis, we investigate whether the addition of an endemic component of the form $\log(\nu_{t,k}) = \log(S_{t,k-1}) + \alpha^{(\nu)}$ is mandated. This would indicate that there is additional influx of measles cases from unspecified sources, for example, imports from cities other than the 60 included in the analysis. However, and as expected for measles, the AIC of the original model is not improved by this extension (difference in the log-likelihood almost immeasurable). Altogether, no subtantial improvement over the best-fitting model in Section 26.3.2 is seen, though.

26.4 Discussion

The focus in this chapter was on infectious disease data where disease location was known to occur at a fixed and known set of units (e.g., cities or county centroids), hence ultimately leading to the analysis of multivariate time series of counts. Underlying, however, the epidemic process may be continuous in space: for infectious diseases such continuous–space continuous-time models have been developed in, for example, Lawson and Leimich (2000), as well as Meyer et al. (2012). Other recent advances have been a more direct focus of the actual who-infected-who transmission chains manifesting as spatiotemporal point patters. This could, for example, be based on a covariate augmented network model of the underlying contact structure in the population (Groendyke et al., 2012). Another path is to supplement the reconstruction with available microbiological information about the disease strain in the cases, for example, by taking the difference in RNA sequences as an additional genetic distance metric supplementing spatial distance (Aldrin et al., 2011; Ypma et al., 2012). Others have taken the collapse of geographical space even further by visualizing and computing only effective distances based on, for example, mobility or population flux data (Brockmann and Helbing, 2013). Spatio and spatiotemporal methods have also been used for the detection of disease clusters (Kulldorff, 2001; Diggle et al., 2005; Lawson and Kleinman, 2005); see Chapter 27 of this book for further details. Another application of spatial methods in infectious disease epidemiology has been the source identification of large food-borne outbreaks. Here, point patterns of cases are aligned with knowledge about the spatial and spatiotemporal distribution of food items or appropriate proxies (Manitz et al., 2014; Kaufman et al., 2014).

In a world of commuting and travelling, location of a case is at best a "most likely" or "average" of the individuals' location over time. In the digital age, new data sources are emerging, allowing an even more exact spatiotemporal tracking of individuals. Hence, location is destined to become a more dynamic concept in future spatiotemporal epidemic models. In general and for stigmatizing diseases in particular, positional accuracy is destined to be in conflict with data protection of the individual. It is a matter of discussion and evaluation how the right balance is to be found between the two.

Addressing spatial heterogeneity can be one aspect to improving the usefulness of epidemic models, but it certainly also complicates the analysis. When such models are used to support the decision-making processes on intervention, control, and vaccination, it is important to remember how formidable a task the modelling of infectious diseases is and where the limitations lie. From a statistician's point of view, transmission modelling in practice misses a stronger emphasis on parameter estimation, uncertainty handling, and model selection. While the modelling community has published high-impact semi-mechanistic models, the concurrent advances in the field by the statistical community have gone somewhat unnoticed. In the last 10 years, though, mechanistic modelling, data science, and statistical inference for dynamic processes appear to have synthesized more. This means that Markov chain Monte Carlo, approximate Bayesian computation and plug-and-play simulation methods are now standard instruments for spatially enriched epidemic models (Mugglin et al., 2002; McKinley et al., 2009; He et al., 2010; Jandarov et al., 2014). Furthermore, flexible open-source R packages have become available for visualizing, analysing, and simulating such epidemic models. In particular, the R package `surveillance` (Höhle et al., 2015) implements both the two-component SIR model and the HHH model, as illustrated in Meyer et al. (2015). However, with powerful computational tools at hand, it is also worthwhile to discuss how complex models need to be in order to answer the questions they are designed for. Long computation times often imply little room for model diagnostics or model selection, which is equally important. The power of mathematical models lies in their abstraction, as May (2004) reminds us. What to include and what to abstract upon in order to make a model *relevant* requires an interdisciplinary approach.

Acknowledgement

Sebastian Meyer, University of Zurich, is thanked for his input and comments on the part of HHH modelling in this chapter.

References

Abrams, S., P. Beutels, and N. Hens. (2014). Assessing mumps outbreak risk in highly vaccinated populations using spatial seroprevalence data. *American Journal of Epidemiology* 179(8), 1006–1017.

Aldrin, M., T. M. Lyngstad, A. B. Kristoffersen, B. Storvik, Ø. Borgan, and P. A. Jansen. (2011). Modelling the spread of infectious salmon anaemia among salmon farms based on seaway distances between farms and genetic relationships between infectious salmon anaemia virus isolates. *Journal of the Royal Society Interface* 8(62), 1346–1356.

Anderson, R. M., and R. M. May. (1991). *Infectious Diseases of Humans: Dynamics and Control.* Oxford: Oxford University Press.

Andersson, H., and T. Britton. (2000). *Stochastic Epidemic Models and Their Statistical Analysis.* Vol. 151 of Lectures Notes in Statistics. Berlin: Springer-Verlag.

Bailey, N. T. J. (1950). A simple stochastic epidemic. *Biometrika* 37(3/4), 193–202.

Bailey, N. T. J. (1964). *The Elements of Stochastic Processes with Applications to the Natural Sciences.* New York: Wiley.

Bailey, N. T. J. (1975). *The Mathematical Theory of Infectious Diseases and Its Applications.* Spokane, WA: Griffin.

Becker, N. G. (1989). *Analysis of Infectious Disease Data.* London: Chapman & Hall/CRC.

Besag, J., J. York, and A. Mollié. (1991). Bayesian image restoration, with two applications in spatial statistics. *Annals of the Institute of Statistical Mathematics* 43, 1–59 [with discussion].

Bjørnstad, O. N., B. F. Finkenstädt, and B. T. Grenfell. (2002). Dynamics of measles epidemics: estimating scaling of transmission rates using a time series SIR model. *Ecological Monographs* 72(2), 169–184.

Bjørnstad, O. N., and B. T. Grenfell. (2008). Hazards, spatial transmission and timing of outbreaks in epidemic metapopulations. *Environmental and Ecological Statistics* 15, 265–277.

Breslow, N. E., and D. G. Clayton. (1993). Approximate inference in generalized linear mixed models. *Journal of the American Statistical Association* 88(421), 9–25.

Britton, T. (2010). Stochastic epidemic models: a survey. *Mathematical Biosciences* 225(1), 24–35.

Brockmann, D., and D. Helbing. (2013). The hidden geometry of complex, network-driven contagion phenomena. *Science* 342(6164), 1337–1342.

Cliff, A., and P. Haggett. (2004). Time, travel and infection. *British Medical Bulletin* 69, 87–99.

Czado, C., T. Gneiting, and L. Held. (2009). Predictive model assessment for count data. *Biometrics* 65, 1254–1261.

Daley, D. J., and J. Gani. (1999). *Epidemic Modelling.* Cambridge: Cambridge University Press.

Deardon, R., X. Fang, and G. Kwong. (2015). Statistical modelling of spatio-temporal infectious disease tranmission. In D. Chen, B. Moulin, and J. Wu (eds.), *Analyzing and Modeling Spatial and Temporal Dynamics of Infectious Diseases*, pp. 211–232, New York: John Wiley & Sons.

Diekmann, O., and J. A. P. Heesterbeek. (2000). *Mathematical Epidemiology of Infectious Diseases.* New York: Wiley.

Diggle, P. (2006). Spatio-temporal point processes, partial likelihood, foot and mouth disease. *Statistical Methods in Medical Research* 15(4), 325–336.

Diggle, P. J., B. Rowlingson, and T.-L. Su. (2005). Point process methodology for on-line spatio-temporal disease surveillance. *Environmetrics* 16, 423–34.

Diggle, P. J., and B. S. Rowlingson. (1994). A conditional approach to point process modelling of elevated risk. *Journal of the Royal Statistical Society: Series A* 157(3), 433–440.

Dohoo, I., W. Martin, and H. Stryhn. (2010). *Veterinary Epidemiologic Research* (2nd ed.). Charlottetown, Canada: VER Inc.

Eggo, R. M., S. Cauchemez, and N. M. Ferguson. (2011). Spatial dynamics of the 1918 influenza pandemic in England, Wales and the United States. *Journal of the Royal Society Interface* 8(55), 233–243.

Erlander, S., and N. F. Stewart. (1990). *The Gravity Model in Transportation Analysis: Theory and Extensions.* McCall, ID: VSP.

Fahrmeir, L., and T. Kneib. (2011). *Bayesian Smoothing and Regression for Longitudinal, Spatial and Event History Data.* Oxford: Oxford University Press.

Finkenstädt, B. F., O. N. Bjørnstad, and B. T. Grenfell. (2002). A stochastic model for extinction and recurrence of epidemics: estimation and inference for measles outbreaks. *Biostatistics* 3(4), 493–510.

Finkenstädt, B. F., and B. T. Grenfell. (2000). Time series modelling of infectious diseases: a dynamical systems approach. *Journal of the Royal Statistical Society: Series C* 49(2), 187–205.

Geilhufe, M., L. Held, S. O. Skrøvseth, G. S. Simonsen, and F. Godtliebsen. (2014). Power law approximations of movement network data for modeling infectious disease spread. *Biometrical Journal* 56(3), 363–382.

Gibson, G. J., and E. Renshaw. (1998). Estimating parameters in stochastic compartmental models using Markov chain methods. *IMA Journal of Mathematics Applied in Medicine and Biology* 15, 19–40.

Giesecke, J. (2002). *Modern Infectious Disease Epidemiology* (2nd ed.). London: Hodder Arnold.

Gillespie, D. T. (1977). Exact stochastic simulation of coupled chemical reactions. *Journal of Physical Chemistry* 81(25), 2340–2361.

Gog, J. R., S. Ballesteros, C. Viboud, L. Simonsen, O. N. Bjornstad, J. Shaman, D. L. Chao, F. Khan, and B. T. Grenfell. (2014). Spatial transmission of 2009 pandemic influenza in the US. *PLoS Compututational Biology* 10(6), e1003635.

Grenfell, B. T. (2006). TSIR analysis of measles in England and Wales. http://www.zoo.cam.ac.uk/zoostaff/grenfell/measles.htm. Available from the Internet Archive as http://web.archive.org/web/20060311051944; http://www.zoo.cam.ac.uk/zoostaff/grenfell/measles.htm.

Grenfell, B. T., O. N. Bjørnstad, and J. Kappey. (2001). Travelling waves and spatial hierarchies in measles epidemics. *Nature* 414(6865), 716–723.

Grenfell, B. T., A. Kleczkowski, C. A. Gilligan, and B. M. Bolker. (1995). Spatial heterogeneity, nonlinear dynamics and chaos in infectious diseases. *Statistical Methods in Medical Research* 4(2), 160–183.

Groendyke, C., D. Welch, and D. R. Hunter. (2012). A network-based analysis of the 1861 Hagelloch measles data. *Biometrics* 68(3), 755–765.

He, D., E. L. Ionides, and A. A. King. (2010). Plug-and-play inference for disease dynamics: measles in large and small populations as a case study. *Journal of the Royal Society Interface* 7(43), 271–283.

Held, L., M. Höhle, and M. Hofmann. (2005). A statistical framework for the analysis of multivariate infectious disease surveillance data. *Statistical Modelling* 5, 187–199.

Held, L., and M. Paul. (2012). Modeling seasonality in space-time infectious disease surveillance data. *Biometrical Journal* 54(6), 824–843.

Held, L., and M. Paul. (2013). Statistical modeling of infectious disease surveillance data. In N. M. M'ikanatha, R. Lynfield, C. A. Van Beneden, and H. de Valk (eds.), *Infectious Disease Surveillance* (2nd ed.), pp. 535–544, New York: Wiley-Blackwell.

Hens, N., Z. Shkedy, M. Aerts, C. Faes, P. Van Damme, and P. Beutels. (2012). *Modeling Infectious Disease Parameters Based on Serological and Social Contact Data.* Berlin: Springer.

Herzog, S. A., M. Paul, and L. Held. (2011). Heterogeneity in vaccination coverage explains the size and occurrence of measles epidemics in German surveillance data. *Epidemiology and Infection* 139(4), 505–515.

Höhle, M. (2009). Additive-multiplicative regression models for spatio-temporal epidemics. *Biometrical Journal* 51(6), 961–978.

Höhle, M., E. Jørgensen, and P. O'Neill. (2005). Inference in disease transmission experiments by using stochastic epidemic models. *Journal of the Royal Statistical Society: Series C* 54(2), 349–366.

Höhle, M., S. Meyer, and M. Paul. (2015). Surveillance: temporal and spatio-temporal modeling and monitoring of epidemic phenomena. R package version 1.9-1. https://cran.r-project.org/web/packages/surveillance/index.html.

Isham, V. (2004). Stochastic models for epidemics. Technical Report 263. Department of Statistical Science, University College London. https://www.ucl.ac.uk/statistics/research/pdfs/rr263.pdf.

Jandarov, R., M. Haran, O. Bjørnstad, and B. Grenfell. (2014). Emulating a gravity model to infer the spatiotemporal dynamics of an infectious disease. *Journal of the Royal Statistical Society: Series C* 63(3), 423–444.

Kaufman, J., J. Lessler, A. Harry, S. Edlund, K. Hu, J. Douglas, C. Thoens, B. Appel, A. Kasbohrer, and M. Filter. (2014). A likelihood-based approach to identifying contaminated food products using sales data: performance and challenges. *PLoS Computational Biology* 10(7), e1003692.

Keeling, M. J., O. N. Bjørnstad, and B. T. Grenfell. (2004). Metapopulation dynamics of infectious diseases. In I. Hanski and O. E. Gaggiotti (eds.), *Ecology, Genetics and Evolution of Metapopulations*, pp. 415–445. Burlington, MA: Academic Press.

Keeling, M. J., and P. Rohani. (2008). *Modeling Infectious Diseases in Humans and Animals.* Princeton, NJ: Princeton University Press.

Kermack, W. O., and A. G. McKendrick. (1927). A contribution to the mathematical theory of epidemics. *Proceedings of the Royal Society: Series A* 115, 700–721.

Klinkenberg, D., J. de Bree, H. Laevens, and M. C. M. de Jong. (2002). Within- and between-pen transmission of classical swine fever virus: a new method to estimate the basic reproduction ratio from transmission experiments. *Epidemiology and Infection* 128, 293–299.

Kulldorff, M. (2001). Prospective time periodic geographical disease surveillance using a scan statistic. *Journal of the Royal Statistical Society: Series A* 164, 61–72.

Lau, M. S. Y., G. Marion, G. Streftaris, and G. J. Gibson. (2014). New model diagnostics for spatio-temporal systems in epidemiology and ecology. *Journal of the Royal Society Interface* 11(93).

Lawson, A., and K. Kleinman (eds.). (2005). *Spatial and Syndromic Surveillance for Public Health.* New York: Wiley.

Lawson, A., and P. Leimich. (2000). Approaches to the space-time modelling of infectious disease behaviour. *IMA Journal of Mathematics Applied in Medicine and Biology* 17(1), 1–13.

Liu, W., H. W. Hethcote, and S. A. Levin. (1987). Dynamical behavior of epidemiological models with nonlinear incidence rates. *Journal of Mathematical Biology* 25, 359–380.

Madden, L. V., G. Hughes, and F. van den Bosch. (2007). *The Study of Plant Disease Epidemics.* St. Paul, MN: APS Press.

Manitz, J., T. Kneib, M. Schlather, D. Helbing, and D. Brockmann. (2014). Origin detection during food-borne disease outbreaks—a case study of the 2011 EHEC/HUS outbreak in Germany. *PLoS Currents* 6.

May, R. M. (2004). Uses and abuses of mathematics in biology. *Science* 303(5659), 790–793.

McKinley, T. J., A. R. Cook, and R. Deardon. (2009). Inference in epidemic models without likelihoods. *International Journal of Biostatistics* 5(1), article 24.

Meyer, S., J. Elias, and M. Höhle. (2012). A space-time conditional intensity model for invasive meningococcal disease occurrence. *Biometrics* 68(2), 607–616.

Meyer, S., and L. Held. (2014). Power-law models for infectious disease spread. *Annals of Applied Statistics* 8(3), 1612–1639.

Meyer, S., L. Held, and M. Höhle. (2015). Spatio-temporal analysis of epidemic phenomena using the R package surveillance. http://arxiv.org/pdf/1411.0416.

Morton, A., and B. Finkenstädt. (2005). Discrete time modelling of disease incidence time series by using Markov chain Monte Carlo methods. *Journal of the Royal Statistical Society: Series C* 54(3), 575–594.

Mugglin, A. S., N. Cressie, and I. Gemmell. (2002). Hierarchical statistical modelling of influenza epidemic dynamics in space and time. *Statistics in Medicine* 21(18), 2703–2721.

Neal, P., and G. O. Roberts. (2004). Statistical inference and model selection for the 1861 Hagelloch measles epidemic. *Biostatistics* 5(2), 249–261.

O'Neill, P. D., and N. G. Becker. (2001). Inference for an epidemic when susceptibility varies. *Biostatistics* 2(1), 99–108.

O'Neill, P. D., and G. O. Roberts. (1999). Bayesian inference for partially observed stochastic epidemics. *Journal of the Royal Statistal Society: Series A* 162, 121–129.

Paul, M., and L. Held. (2011). Predictive assessment of a non-linear random effects model for multivariate time series of infectious disease counts. *Statistics in Medicine* 30(10), 1118–1136.

Paul, M., L. Held, and A. M. Toschke. (2008). Multivariate modelling of infectious disease surveillance data. *Statistics in Medicine* 27, 6250–6267.

Pitzer, V. E., C. Viboud, L. Simonsen, C. Steiner, C. A. Panozzo, W. J. Alonso, M. A. Miller, R. I. Glass, J. W. Glasser, U. D. Parashar, and B. T. Grenfell. (2009). Demographic variability, vaccination, and the spatiotemporal dynamics of rotavirus epidemics. *Science* 325(5938), 290–294.

R Core Team. (2014). *R: A Language and Environment for Statistical Computing.* Vienna, Austria: R Foundation for Statistical Computing.

Renshaw, E. (1991). *Modelling Biological Populations in Space and Time.* Cambridge: Cambridge University Press.

Riley, S., K. Eames, V. Isham, D. Mollison, and P. Trapman. (2015). Five challenges for spatial epidemic models. *Epidemics* 10, 68–71.

Scheel, I., M. Aldrin, A. Frigessi, and P. Jansen. (2007). A stochastic model for infectious salmon anemia (ISA) in Atlantic salmon farming. *Journal of the Royal Society Interface* 4, 699–706.

Smyth, G. K. (1998). Numerical integration. In P. Armitage and T. Colton (eds.). *Encyclopedia of Biostatistics*, pp. 3088–3095. New York: Wiley.

Viboud, C., O. N. Bjørnstad, D. L. Smith, L. Simonsen, M. A. Miller, and B. T. Grenfell. (2006). Synchrony, waves, and spatial hierarchies in the spread of influenza. *Science* 312(5772), 447–451.

Weidemann, F., M. Dehnert, J. Koch, O. Wichmann, and M. Höhle. (2014). Bayesian parameter inference for dynamic infectious disease modelling: rotavirus in Germany. *Statistics in Medicine* 33(9), 1580–1599.

Wilking, H., M. Höhle, E. Velasco, M. Suckau, and T. Eckmanns. (2012). Ecological analysis of social risk factors for rotavirus infections in Berlin, Germany, 2007–2009. *International Journal of Health Geographics* 11(1), 37.

Wilkinson, D. J. (2006). *Stochastic Modelling for Systems Biology.* Boca Raton, FL: Chapman & Hall/CRC.

Xia, Y., O. N. Bjørnstad, and B. T. Grenfell. (2004). Measles metapopulation dynamics: a gravity model for epidemiological coupling and dynamics. *American Naturalist* 164(2), 267–281.

Ypma, R. J., A. M. Bataille, A. Stegeman, G. Koch, J. Wallinga, and W. M. van Ballegooijen. (2012). Unravelling transmission trees of infectious diseases by combining genetic and epidemiological data. *Proceedings of the Royal Society B: Biological Sciences* 279(1728), 444–450.

27

Spatial Health Surveillance

Ana Corberán-Vallet
Department of Statistics and Operations Research
University of Valencia
Valencia, Spain

Andrew B. Lawson
Division of Biostatistics and Bioinformatics
Department of Public Health Sciences
Medical University of South Carolina
Charleston, South Carolina

CONTENTS

27.1 Introduction

Spatial epidemiology focuses on the analysis of health data that are georeferenced, often with the objective of describing and analyzing geographic variations in disease in consideration of explanatory variables such as demographic, environmental, and socioeconomic factors. Consequently, the retrospective analysis of disease maps is a major focus. However, there are situations where real-time modeling and prediction play a crucial part. This is the case of public health surveillance, which is defined as (Thacker and Berkelman, 1992)

> the ongoing, systematic collection, analysis, and interpretation of health data essential to the planning, implementation, and evaluation of public health practice, closely integrated with the timely dissemination of these data to those who need to know. The final link of the surveillance chain is the application of these data to prevention and control.

This definition emphasizes that surveillance is essential in public health practice and that timeliness is a key concept to early detection of changes in disease incidence and, consequently, to facilitate timely public health response. Timeliness can be in terms of hours or days for diseases that spread quickly, or in terms of weeks, months, or even years for less infectious or noninfectious diseases. Regardless of the situation, sequential analyses of all the data collected so far are required to determine whether an outbreak is taking place as quickly as possible.

The most common form of disease surveillance is temporal surveillance, and so a wide range of statistical methods, including process control charts, temporal scan statistics, regression techniques, and time-series methods, have been proposed for surveillance of univariate time series of counts of disease (Sonesson and Bock, 2003; Le Strat, 2005; Unkel et al., 2012). Unlike testing methods, model-based surveillance techniques provide better insight into etiology, spread, prediction, and control of diseases, and so they are widely used for outbreak detection. For instance, the log-linear regression model of Farrington et al. (1996) is used routinely by the Health Protection Agency to detect aberrations in laboratory-based surveillance data in England and Wales. At each time point, the observed count of disease is declared aberrant if it lies above a threshold, which is computed from the estimated model using a set of recent observations with similar conditions.

Recent developments in temporal surveillance include time-series models of increasing complexity, such as hidden Markov models. These models assume that the probability density of the observations depends on the state of an underlying Markov chain, which segments the time series of disease counts into epidemic and nonepidemic phases (Le Strat and Carrat, 1999). Martínez-Beneito et al. (2008) used a Bayesian hidden Markov model to detect the onset of influenza epidemics. Unlike previous hidden Markov models, the authors modeled the series of differenced incidence rates rather than the series of incidence rates. Depending on whether the system is in an epidemic or a nonepidemic phase, the differenced series is modeled either with a first-order autoregressive process or with a Gaussian white noise process. More recently, Conesa et al. (2015) proposed an enhanced modeling framework that incorporates the magnitude of the incidence to better distinguish between epidemic and nonepidemic phases.

Currently, large amounts of surveillance data are routinely collected by laboratories, health care providers, and government agencies in an effort to prevent, control, and manage outbreaks of disease. Information on both the time and location of events is increasingly available, and as a consequence, the area of disease surveillance is experiencing a great enrichment. The use of spatial information enhances the ability to detect small localized outbreaks of disease relative to the surveillance of the overall count of disease cases across the entire study region, where increases in a relatively small number of regional counts may be diluted by the natural variation associated with overall counts. In addition, spatiotemporal surveillance facilitates public health interventions once an increased regional count has been identified.

At a fine level of spatial and temporal resolution, observation of disease usually consists of a sequence of disease occurrences $\{(s_1, t_1), (s_2, t_2), (s_3, t_3), \ldots\}$ indicating the exact location of events (s_i) and the date of diagnosis (t_i). For case event data, spatiotemporal point process models can help to develop prospective surveillance techniques (Lawson and Leimich, 2000; Diggle et al., 2005; Diggle, 2007). In practice, however, the study region is usually divided into m smaller, nonoverlapping tracts (small areas), and only partial information about the total count of disease cases in each small area and time period (y_{it}, where subindices i and t represent area i and time period t) is available. Consequently, practical statistical surveillance implies analyzing simultaneously multiple time series $\{y_{i1}, y_{i2}, y_{i3}, \ldots\}_{i=1,2,\ldots,m}$ that are spatially correlated. It is important to emphasize here that ignoring the spatial structure of the data will lead to insufficient and suboptimal methods because of a loss of information about the disease under study (Sonesson and Bock, 2003).

Some testing methods have been proposed in the literature for spatiotemporal disease surveillance (Kulldorff, 2001; Rogerson, 2005). Recent developments in the analysis of space–time disease surveillance data use a statistical model. Model-based approaches provide a flexible framework for the inclusion of space, time, space–time interaction, and possible covariates effects, and this is a very active area of statistical research (Robertson et al., 2010).

27.2 Modeling for Spatiotemporal Disease Surveillance

Model-based surveillance techniques use a statistical model to describe the behavior of disease over space and time during endemic periods, that is, when the disease occurs at its expected frequency of occurrence, and the emphasis is placed on detection of unusual departures from predictable patterns based on the estimated model. The correct modeling of this endemic behavior or background is a major issue, since detection of changes in disease incidence will depend on what is assumed to be the *normal* behavior of disease in the small areas under study. Usually, the background is estimated using historical data. This approach has some advantages, such as the ability to self-control area risk. However, the choice of a historical period that is stable could be misjudged. The model chosen to describe the background also plays an important role, and it must be able to describe the overall behavior of disease in space and time and be sensitive to temporal changes in the spatiotemporal structure. A relatively simple model capturing the normal historical variation without absorbing changes in the model fit is then advisable. To this end, it would be reasonable to include a spatial model to capture the spatial correlation in disease maps. However, the inclusion of adaptive time components is more problematic, since they are likely to track changes in risk. While seasonal effects may be important if the emphasis was on detection of unusual outbreaks of disease, it is less clear why temporal trend components or random effects should be included. If the time-series data were available over a relatively long period of time and an overall time trend was present, a temporal effect may be included in the model, so that changes in disease incidence beyond those expected based on the main spatial and temporal effects could be effectively detected. Otherwise, modeling the temporal variation is not advisable, since the objective of prospective surveillance is to monitor changes in disease risk over time (Corberán-Vallet and Lawson, 2011).

Let y_{it} be the count of disease in area i, $i = 1, 2, \ldots, m$, and time period t, $t = 1, 2, \ldots, T$. Usually, we assume that disease counts are described by a discrete probability model. For relatively rare diseases, the Poisson distribution is considered for modeling the within-area variability of the counts,

$$y_{it} \sim Po(e_{it}\theta_{it}), \tag{27.1}$$

where the expected count of disease e_{it} represents the background population effect, and θ_{it} is the unknown area-specific relative risk. For finite populations within small areas, we could assume a binomial likelihood instead. The estimator of e_{it} is usually based on a standard population rate (e.g., the whole study region). Once estimated, the e_{it} component is assumed fixed and known. A basic description of space–time risk variation, which has proved to be a robust and appropriate model, is (Knorr-Held, 2000; Lawson, 2013, chapter 12)

$$\log(\theta_{it}) = \alpha_0 + u_i + v_i + \gamma_t + \psi_{it}, \tag{27.2}$$

where the intercept α_0 represents the overall level of the relative risk, u_i and v_i represent, respectively, spatially correlated and uncorrelated extra variation, γ_t is the temporal random effect, and ψ_{it} represents the space–time interaction term. In terms of inferential paradigms, it is commonly found that a Bayesian approach is adopted to the formulation of the hierarchical model structure, and the ensuing estimation methods focus on posterior sampling via Markov chain Monte Carlo (McMC). The model hierarchy with suitable prior distributions could be

$$\begin{aligned} y_{it} &\sim Po(e_{it}\theta_{it}) \\ \theta_{it} &= \exp\{\alpha_0 + u_i + v_i + \gamma_t + \psi_{it}\} \\ \alpha_0 &\sim N(0, \sigma_\alpha^2) \\ u_i &\sim ICAR(\sigma_u^2/|n_i|) \\ v_i &\sim N(0, \sigma_v^2) \\ \gamma_t &\sim N(\gamma_{t-1}, \sigma_\gamma^2) \\ \psi_{it} &\sim N(0, \sigma_\psi^2), \end{aligned}$$

where $ICAR(\sigma_u^2/|n_i|)$ denotes the improper conditional autoregressive (CAR) model (Besag et al., 1991), which implies that the term u_i has a Markov random field specification: a conditional Gaussian distribution given its n_i neighboring region set $\left(u_i|u_{(i)} \sim N\left(\frac{1}{|n_i|}\sum_{j\in n_i} u_j, \frac{\sigma_u^2}{|n_i|}\right)\right)$.

As time progresses and new observations become available, detection of small areas of increased disease incidence in need of further investigation is the focus.

27.2.1 Residual-based approach

In order to detect changes in the relative risk pattern of disease, Vidal Rodeiro and Lawson (2006) introduced the surveillance residuals as the difference between the observed data for the new time period and the data predicted under the previously described spatiotemporal model when it is fitted using the data from previous periods, that is,

$$r_{it}^s = y_{it} - \frac{1}{J}\sum_{j=1}^{J} e_{it}\theta_{i,t-1}^{(j)}, \qquad (27.3)$$

$\{\theta_{i,t-1}^{(j)}\}_{j=1}^J$ being a set of relative risks sampled from the posterior distribution that corresponds to the previous time period. The absolute value of the surveillance residuals summed over space is monitored over time to detect changes in risk. Once a change is detected, the p-value surface is analyzed to determine the location and extension of possible outbreaks. For each small area and time point, the Bayesian p-value is defined as the probability that replicated data based on the posterior predictive distribution could be more extreme than the observed data,

$$p_{it} = \Pr(y_{it}^* > y_{it}|\text{data}),$$

extremely small p-values indicating that the observation for the new time period is not representative of the data expected under the fitted model. Note that the equivalent of a parametric bootstrap in the Bayesian setting is necessary to generate the set of simulated counts $\{y_{it}^*\}$.

27.2.2 EWMA approach

In an effort to overcome the estimation problem arising when Bayesian hierarchical Poisson models are used in a spatiotemporal surveillance context, Zhou and Lawson (2008) suggested using an approximated procedure where a spatial model is fitted to the data observed at each time period t. Changes in the risk pattern of disease can then be detected by comparing the estimated relative risk for each small area $\hat{\theta}_{it}$ with a baseline level $\tilde{\theta}_{it}$, $i = 1, 2, \ldots, m$, calculated as an exponentially weighted moving average (EWMA) of historical estimates,

$$\tilde{\theta}_{it} = \kappa\hat{\theta}_{i,t-1} + (1-\kappa)\tilde{\theta}_{i,t-1}. \tag{27.4}$$

In particular, the authors defined a sample-based Monte Carlo p-value as

$$\text{MCP}_{it} = \frac{1}{J}\sum_{j=1}^{J} I(\hat{\theta}_{it}^{(j)} < \tilde{\theta}_{it}^{(j)}), \tag{27.5}$$

extremely small p-values indicating that an increase in disease risk might have occurred. The estimated percentage of increase in disease relative risk, which is defined as

$$\text{PIR}_{it} = \frac{\frac{1}{J}\sum_{j=1}^{J}(\hat{\theta}_{it}^{(j)} - \tilde{\theta}_{it}^{(j)})}{\frac{1}{J}\sum_{j=1}^{J}\tilde{\theta}_{it}^{(j)}} \times 100, \tag{27.6}$$

can also be calculated to assess the magnitude of the change in disease risk. This is a computationally quick technique that has been shown to have a good performance in outbreak detection. However, a sliding window of length 1 cannot guarantee accurate model estimates. Also, an EWMA approach is only justified under Gaussian model assumptions.

27.2.3 Surveillance conditional predictive ordinate

More recently, a surveillance approach based on the conditional predictive ordinate (Geisser, 1980) was proposed (Corberán-Vallet and Lawson, 2011). Within this approach, a surveillance conditional predictive ordinate (SCPO) is sequentially computed to detect small areas of unusual disease aggregation. For each small area and time period, the SCPO is calculated as

$$\text{SCPO}_{it} = f(y_{it}|y_{1:t-1}) = \int\int f(y_{it}|\theta_{it})\, p(\theta_{it}|\theta_{i,t-1}, y_{1:t-1})\, p(\theta_{i,t-1}|y_{1:t-1})\, d\theta_{i,t-1}\, d\theta_{it}, \tag{27.7}$$

where $y_{1:t-1}$ represents the data observed up to time $t-1$, $f(y_{it}|\theta_{it})$ is Poisson, $p(\theta_{it}|\theta_{i,t-1}, y_{1:t-1})$ can be derived from the model describing the relative risk surface, and $p(\theta_{i,t-1}|y_{1:t-1})$ represents the marginal posterior distribution of parameter $\theta_{i,t-1}$ at time $t-1$.

As mentioned earlier, the inclusion of adaptive time components may hinder detection of changes in risk. In that paper, the authors found that removal of random temporal effects from the background aids the identification of temporal changes in risk, and consequently, the convolution model (defined as $\log(\theta_{it}) = \alpha_0 + u_i + v_i$) was used to model the behavior of nonseasonal disease data during endemic periods. In that case, the SCPO simplifies to

$$\text{SCPO}_{it} = f(y_{it}|y_{1:t-1}) = \int f(y_{it}|\theta_i)\, p(\theta_i|y_{1:t-1})\, d\theta_i. \tag{27.8}$$

A Monte Carlo estimate for the SCPO, which does not have a close form, can be obtained from a posterior sampling algorithm as

$$\text{SCPO}_{it} \approx \frac{1}{J}\sum_{j=1}^{J} Po(y_{it}|\, e_{it}\theta_i^{(j)}), \tag{27.9}$$

where $\{\theta_i^{(j)}\}_{j=1}^{J}$ is a set of relative risks sampled from the posterior distribution that corresponds to the previous time period. Hence, if there is no change in risk, y_{it} will be representative of the data expected under the previously fitted model. Otherwise, SCPO values close to zero will be obtained.

Corberán-Vallet and Lawson (2011) showed the application of the SCPO to salmonellosis cases in South Carolina from January 1995 to December 2003 (see Figure 27.1).

In order to detect occasional outbreaks beyond seasonal patterns, a generalization of the convolution model allowing for seasonal effects was used to model the regular behavior of disease. In particular, the relative risks were modeled as

$$\log(\theta_{it}) = \alpha_0 + u_i + v_i + \sum_{s=1}^{12} \alpha_s I_s(t), \tag{27.10}$$

where $\{\alpha_s\}_{s=1}^{12}$ are the seasonal effects and $I_s(t)$ is the indicator function that takes the value 1 if time t corresponds to month s and zero otherwise. Note that within each small area, different relative risks are defined to account for the seasonality of the data, but the risks so defined are constant over time. Figure 27.2 displays the spatial distribution of the SCPO for a selection of 6-month periods: September–October 1996, February–March 2001, and October–November 2002. As can be seen, values of the SCPO close to zero alert us to counties presenting unusually high counts of disease.

An important feature of the SCPO is that it can be easily extended to incorporate information from the spatial neighborhood, which facilitates outbreak detection capability when changes in risk affect neighboring areas simultaneously, and also information from multiple diseases observed within a predefined study region (see Section 27.3).

27.2.4 Other approaches

Alternative approaches to prospective surveillance of disease across space and time have been proposed. For example, Kleinman et al. (2004) proposed a method based on generalized linear mixed models to evaluate whether observed counts of disease are larger than would be expected on the basis of a history of naturally occurring disease. In that paper, the number of cases in area i and time t is assumed to follow a binomial distribution, $y_{it} \sim Bi(n_{it}, p_{it})$, n_{it} being the population and p_{it} the probability of an individual being a case. This probability is modeled as a function of covariate and spatial random effects; in particular, the authors model $logit(p_{it}) = x_{it}\beta + b_i$, where x_{it} is a set of covariates measured for area i and time t and b_i is a random effect for area i. Once the model is fitted using historical data observed under endemic conditions, the probability of seeing more cases than the current observed

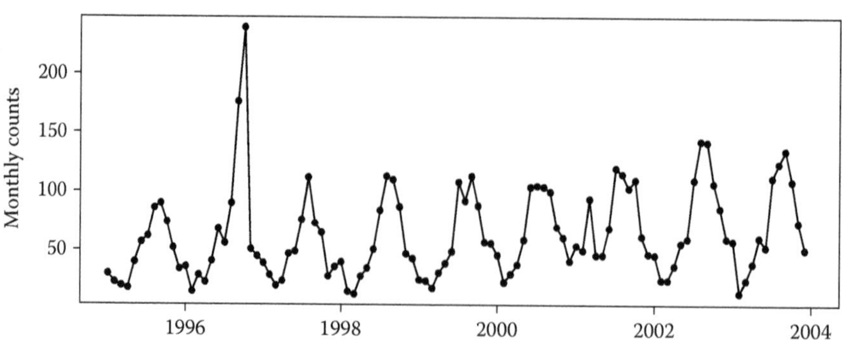

FIGURE 27.1
Monthly counts of reported salmonellosis cases in South Carolina for the period 1995–2003.

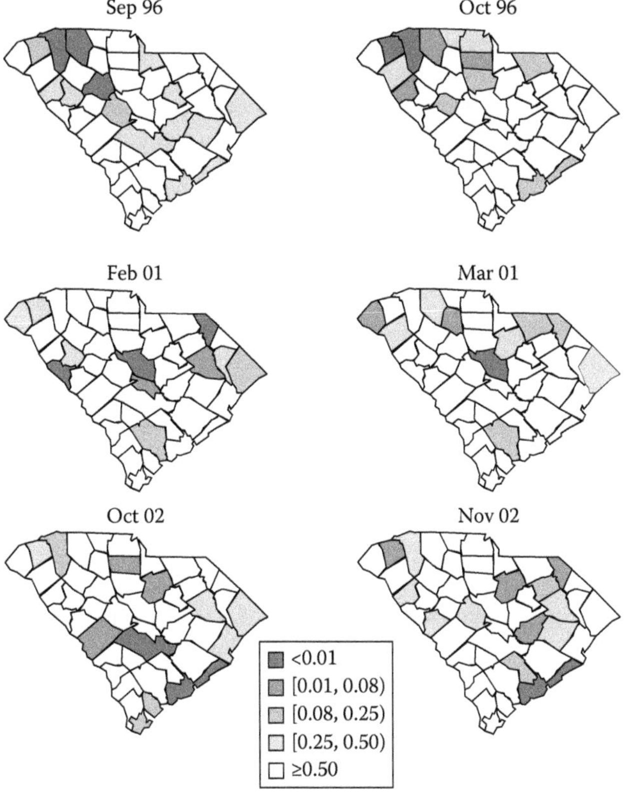

FIGURE 27.2
Spatial distribution of the scaled SCPO for the salmonellosis data at those months undergoing a possible outbreak of disease.

count of disease is calculated for each small area and time period to detect unusually high counts of disease.

Watkins et al. (2009) provided an extension of a purely temporal hidden Markov model (HMM) to incorporate spatially referenced data. In particular, the authors proposed a first-order HMM where the spatial structure of the data was incorporated by modeling the sum of reported cases for each small area and its nearest neighbors. More recently, Heaton et al. (2012) proposed a spatiotemporal conditional autoregressive hidden Markov model with an absorbing state. By considering the epidemic state to be absorbing, the authors avoid undesirable behavior, such as day-to-day switching between the epidemic and nonepidemic states. This feature, however, limits the application of the model to a single outbreak of disease at each location. A novelty of that model is that the nonabsorbing transition probabilities of the Markov chain are allowed to vary over space and time.

27.3 Surveillance of Multiple Diseases

Multivariate space–time surveillance data arise naturally in many public health applications. For instance, surveillance systems are often focused on more than one disease within a predefined area. Some examples are the monitoring of smoking-related cancers, respiratory diseases, or gastrointestinal illnesses. In a syndromic surveillance setting, different syndromes associated with disease are monitored simultaneously to detect outbreaks of disease at the earliest possible time (see Section 27.4). On those occasions when outbreaks of disease are likely to be correlated, the use of multivariate surveillance techniques integrating information from the different data sets is important to achieve higher detection power for events that are present simultaneously in more than one data set.

Kulldorff et al. (2007) proposed an extension of the space–time scan statistic to jointly monitor multiple data sets. The multivariate scan statistic is based on a combined log-likelihood that is defined as the sum of the individual log-likelihoods for those data sets with more counts than expected in the scanning window. A signal is generated if a cluster is detected in either one or a combination of data sets. Further extensions, such as the Bayesian multivariate scan statistic (Neill and Cooper, 2010), have been proposed. Banks et al. (2012) presented a model-based approach to surveillance of spatial data on multiple diseases. The proposed methodology, which is focused on syndromic surveillance, uses univariate Bayesian hierarchical models to describe counts of patients with specific symptoms indicative of the same disease in the absence of an outbreak. Indicator variables modeled as binary Markov random fields are then used to detect disease outbreaks. An increase in the number of cases is assumed for all the symptoms when the disease is present. An alternative model-based surveillance technique can be found in Corberán-Vallet (2012). Because the different diseases under study may be influenced by common confounding factors, and so are likely to be correlated, the author proposed an alternative shared-component model (Knorr-Held and Best, 2001; Held et al., 2005b) to describe the behavior of diseases under endemic periods. Assume that there are $K \geq 2$ diseases and a fixed study region common to all the diseases. Let $y_{it} = (y_{it1}, y_{it2}, \ldots, y_{itK})$ be the vector of observed counts of disease in area i and time t, $e_{it} = (e_{it1}, e_{it2}, \ldots, e_{itK})$ the vector of expected counts, and $\theta_i = (\theta_{i1}, \theta_{i2}, \ldots, \theta_{iK})$ the vector of relative risks (which are assumed to be constant during endemic periods to improve outbreak detection capability). The alternative shared-component model is defined as

$$\begin{aligned} y_{itk} &\sim Po(e_{itk}\theta_{ik}) \\ log(\theta_{ik}) &= \rho_k + \sum_{l=1}^{L} \phi_{l,k}\, \delta_{l,k}\, w_{l,i} + \psi_{ik}, \end{aligned} \tag{27.11}$$

where ρ_k is the disease-specific overall risk, L represents the number of spatial fields (CAR components) $w_l = (w_{l,1}, w_{l,2}, \ldots, w_{l,m})'$ needed to describe the correlation in the relative risks across both areas and diseases, $\phi_{l,k}$ is a binary indicator variable that takes the value 1 if the spatial random effect w_l has an influence on disease k and the value zero otherwise, $\delta_{l,k}$ is the scaling parameter that measures the contribution of w_l to disease k, and ψ_{ik} is the uncorrelated term, which is assumed to be zero-mean Gaussian distributed. Similar to the model proposed in Held et al. (2005b), this alternative formulation of the shared-component model assumes that there may be more than one latent spatial field that can be shared by some of the diseases or may be relevant only to one of them. However, by using indicator variables in the model formulation, it is not necessary to specify the structure of the multivariate model in advance.

A multivariate extension of the SCPO (see Section 27.2.3) integrating information from the K diseases was then proposed to detect disease outbreaks, which need not necessarily occur at the same time for all the diseases under surveillance or affect the same spatial units. For each area i and time t, let $\hat{\theta}_i = (\hat{\theta}_{i1}, \hat{\theta}_{i2}, \ldots, \hat{\theta}_{iK})$ be the vector of posterior relative risk estimates at the previous time point $(t-1)$ using the spatial-only shared-component model, and $y_{it}^h = (y_{itk_1}, y_{itk_2}, \ldots, y_{itk_n})$ the vector of observed counts higher than expected, that is, $y_{itk} > e_{itk}\hat{\theta}_{ik}$. The MSCPO for each small area and time period is defined as the conditional predictive distribution of those counts of disease higher than expected given the data observed up to the previous time period, that is,

$$\begin{aligned} \text{MSCPO}_{it} &= f(y_{itk_1}, y_{itk_2}, \ldots, y_{itk_n} | y_{1:t-1}) \\ &= \int\int \cdots \int f(y_{itk_1}, y_{itk_2}, \ldots, y_{itk_n} | \theta_{ik_1}, \theta_{ik_2}, \ldots, \theta_{ik_n}) \\ &\quad \times p(\theta_{ik_1}, \theta_{ik_2}, \ldots, \theta_{ik_n} | y_{1:t-1}) d\theta_{ik_1} d\theta_{ik_2} \ldots d\theta_{ik_n}, \end{aligned} \tag{27.12}$$

if y_{it}^h is not null, and MSCPO_{it} equal to 1 otherwise. Values of the MSCPO close to zero alert to both small areas of increased disease incidence and the diseases causing the alarm within each area. Note that when $y_{it}^h = \{y_{itk_1}\}$, the MSCPO_{it} corresponds to the SCPO_{it} for disease k_1. When $n \geq 2$, counts of disease higher than expected are looked at in conjunction to improve the outbreak detection capability.

Corberán-Vallet (2012) applied the MSCPO technique to emergency room discharges for a group of five respiratory diseases—acute upper respiratory infections (AURIs), influenza, acute bronchitis, asthma, and pneumonia—in South Carolina in 2009 (see Figure 27.3), and compared its performance with that of the univariate SCPO and the multivariate space–time scan statistic as implemented in the free SaTScan™ software (2009). Because the author was interested in detecting outbreak onsets, the analysis was confined to data collected from the week beginning June 28 (where all the diseases can be assumed to be in an endemic state) to the week beginning December 27 (weeks 26–52 in Figure 27.3).

A comparison with the univariate SCPO (see Figure 27.4) showed that when observed counts of disease are unusually high in comparison with the expected counts, the univariate and multivariate surveillance techniques signal an alarm at the same time. However, by borrowing information from different diseases, the MSCPO alerts us to unusual counts of disease that are not significant enough to cause an alert on their own. Hence, the MSCPO improves both detection time and recovery of the true outbreak behavior when changes in disease incidence happen simultaneously for two or more diseases.

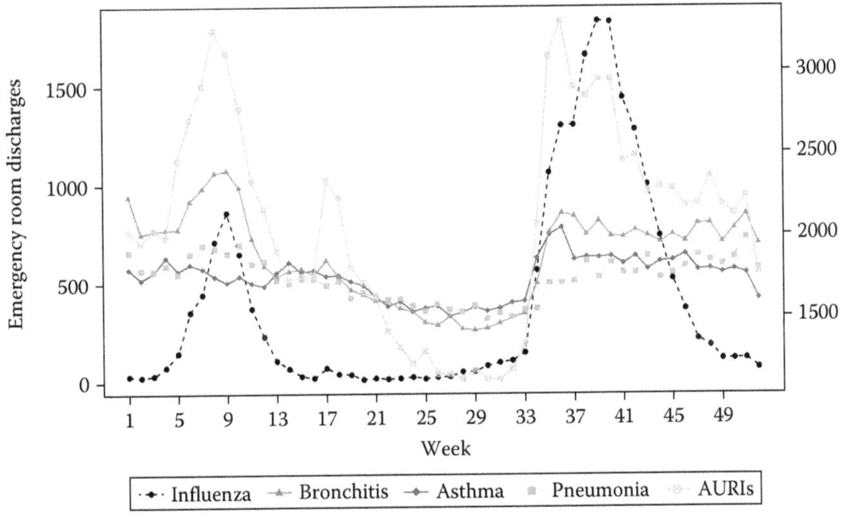

FIGURE 27.3
Weekly emergency room discharges for influenza, acute bronchitis, asthma, pneumonia, and acute upper respiratory infections (right y axis) in South Carolina in 2009.

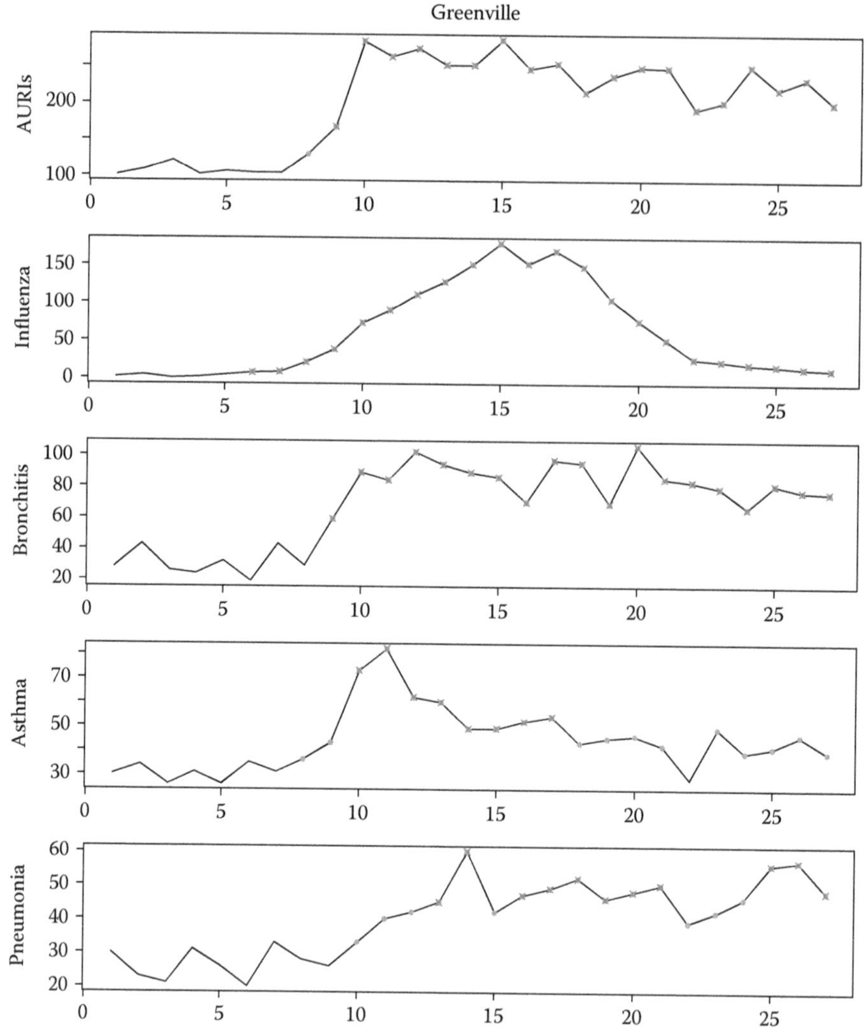

FIGURE 27.4
Temporal profile for Greenville County. Time points corresponding to outbreak periods as detected by the multivariate surveillance conditional predictive ordinate are represented by solid points. Unusual observations based on the univariate surveillance technique are represented by crosses.

Figure 27.5 compares the counties where an outbreak is declared based on the MSCPO and the multivariate scan statistic at six time periods. Several large clusters covering practically all the study region are reported by the multivariate scan statistic at each time point. These clusters usually include the five diseases. This is due to the fact that the multivariate space–time scan statistic pinpoints the general time and location of the most likely cluster (and possible significant secondary clusters), and so its exact boundaries remain uncertain. The MSCPO searches, at each time point, for counties of increased disease incidence and the diseases within each county with more counts than expected. Consequently, in this case study, the MSCPO identifies more accurately the areas affected by disease outbreaks at each time period, thus reducing the number of false alarms and enabling a more informed

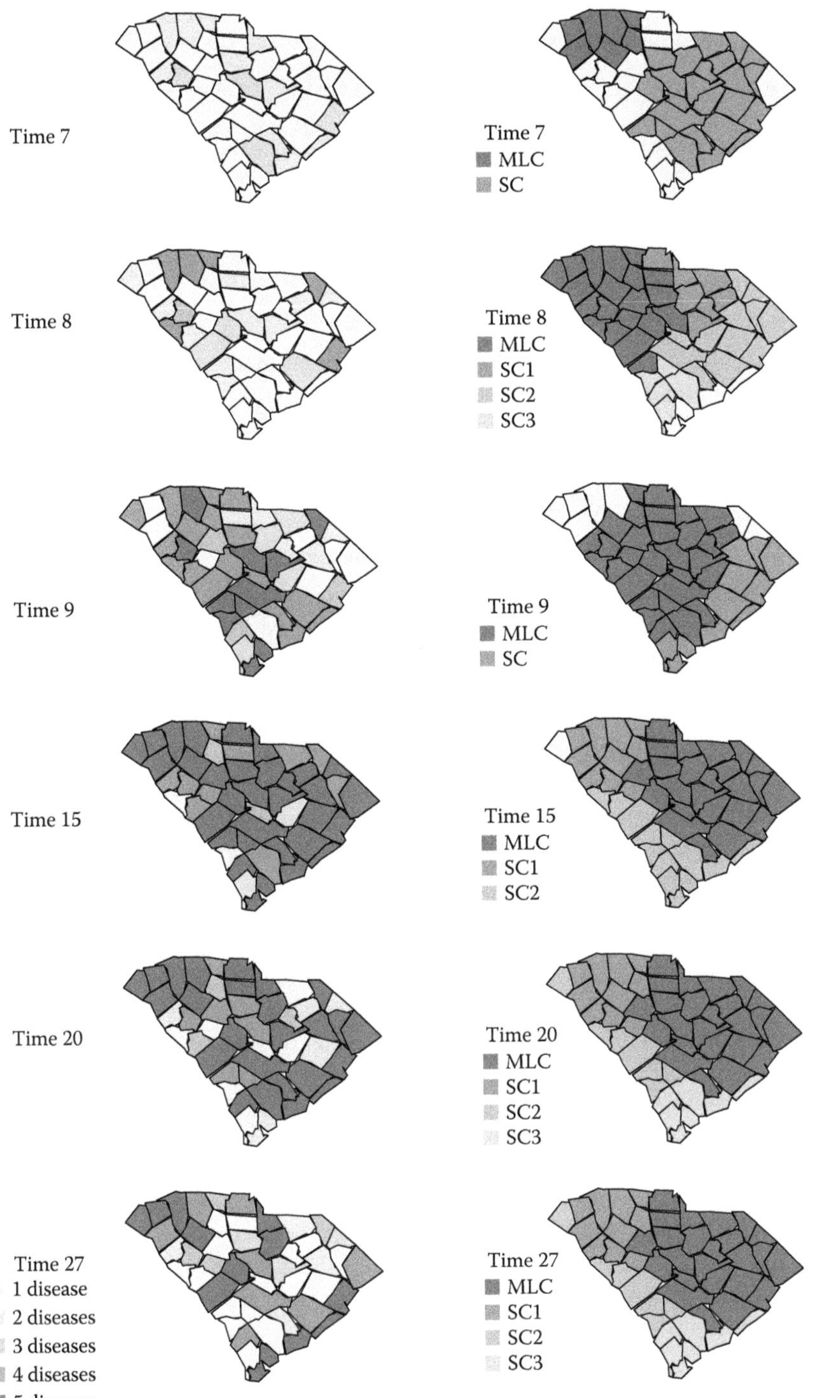

FIGURE 27.5
Counties where an outbreak is declared. Left: Areas signaling an alarm based on the multivariate surveillance conditional predictive ordinate. Darker shading indicates a higher number of diseases causing the alarm. Right: Most likely cluster (MLC) and secondary clusters (SCs) using the multivariate Poisson-based prospective space–time scan statistic.

response. It also allows reporting the end of an outbreak, which can be valuable for the planning and reduction of public health interventions.

27.4 Syndromic Surveillance

Much of the new literature in the area of prospective disease surveillance relates to syndromic surveillance, which can be defined as (CDC, 2004)

> an investigational approach where health department staff, assisted by automated data acquisition and generation of statistical signals, monitor disease indicators continually (real-time) or at least daily (near real-time) to detect outbreaks of diseases earlier and more completely than might otherwise be possible with traditional public health methods. The distinguishing characteristic of syndromic surveillance is the use of indicator data types.

So, unlike traditional surveillance that tends to focus on compulsory reporting of specific diseases, syndromic surveillance exploits more diverse sources of data that precede a clinical diagnosis, such as emergency department visits, hospital bed occupancy, physician telephone calls, school and work absenteeism, and over-the-counter medication sales.

Statistical challenges facing early detection of disease outbreaks have been reviewed recently by Shmueli and Burkom (2010). Katz et al. (2011) provide a recent review of the area, stressing the range of definitions found in the literature and also the usefulness of syndromic surveillance within the public health community. Without a doubt, syndromic surveillance is an efficient tool to detect outbreaks of disease at the earliest possible time, possibly even before definitive disease diagnoses are obtained. Some examples where syndromic surveillance provides faster detection can be found in Ginsberg et al. (2009) and Kavanagh et al. (2012). However, as Chan et al. (2010) emphasized, alerts based on syndrome aberrations surely contain uncertainty, and so they should be evaluated with a proper probabilistic measure. In that paper, the authors used a space–time Bayesian hierarchical model incorporating information from meteorological factors to model influenza-like illness visits. The risk of an outbreak was assessed using $Pr(y_{it} > \text{threshold}|y_{1:t-1})$.

Recently, Corberán-Vallet and Lawson (2014) have presented a flexible methodology to prospectively analyze infectious disease surveillance data using information from a syndromic disease, that is, a sentinel disease that precedes the disease of interest or other disease predisposing to the infection analyzed. Let y_{it} and y_{it}^s be the number of cases of both the disease of interest and the syndromic disease in area i, $i = 1, 2, \ldots, m$, and time period t, $t = 1, 2, \ldots, T$. At the first level of the model hierarchy, the Poisson distribution with two components is considered to model the within-area variability of the counts (Held et al., 2005a, 2006; Paul et al., 2008),

$$\begin{aligned} y_{it} &\sim Po(\mu_{it} + I_{it}) \\ y_{it}^s &\sim Po(\mu_{it}^s + I_{it}^s), \end{aligned} \tag{27.13}$$

where $\mu_{it} = e_{it}\theta_{it}$ and $\mu_{it}^s = e_{it}^s\theta_{it}^s$ are the nonepidemic components, and I_{it} and I_{it}^s are the epidemic components, which represent the expected additive increase in disease counts due to epidemics. During nonepidemic states, the two diseases may be influenced by common confounding factors, and so they are likely to be correlated. To account for risk correlations, the authors used the following generalized common spatial model (Wang and Wall, 2003; Ma and Carlin, 2007):

$$\begin{aligned} log(\theta_{it}) &= \rho + u_i + v_i + \delta_{it} \\ log(\theta^s_{it}) &= \rho^s + \psi u_i + v^s_i + \delta^s_{it}, \end{aligned} \tag{27.14}$$

where a single spatially correlated component (u_i) is used to model the correlation across diseases and locations. The scaling parameter $\psi \sim N(0,\sigma^2_\psi)$ allows for a different risk gradient for the syndromic disease. During an epidemic, counts of disease will trend upward over time until reaching the epidemic's peak, which will be followed by a gradual downward trend. In addition, epidemics of the syndromic disease will generally precede those of the infection of interest, and so this information can be used to provide a better description of the infection under study. In order to both capture the temporal and spatial dependence in disease counts and relate increases in the number of cases of the disease of interest to increases previously observed in the syndromic disease, the epidemic components were modeled as

$$I_{it} = \beta_{it}\left(y_{i,t-1} + \gamma_i \sum_{j \in n_i} y_{j,t-1}\right) + \phi_i I^s_{i,t-1} \tag{27.15}$$

$$I^s_{it} = \beta^s_{it}\left(y^s_{i,t-1} + \gamma^s_i \sum_{j \in n_i} y^s_{j,t-1}\right),$$

where β_{it} is the autoregressive parameter and γ_i measures, for the ith small area, the magnitude of the neighborhood effect. A beta prior distribution can be assumed for parameter γ_i. A key feature of the model formulation is that the autoregressive parameters are allowed to vary in both space and time to properly capture the pattern of infectious disease data in the different regions under study. Specifically, β_{it} was modeled as

$$\beta_{it} = \exp\left\{ b_{i,0} + \sum_{k=1}^{K}\left(b_{i,2k-1}\, sin\left(\frac{2k\pi t}{\omega}\right) + b_{i,2k}\, cos\left(\frac{2k\pi t}{\omega}\right)\right) + \epsilon_{it}\right\} \tag{27.16}$$

$$\epsilon_{it} \sim N(\epsilon_{i,t-1}, \sigma^2_\epsilon); \quad \epsilon_{i1} \sim N(0, \sigma^2_\epsilon),$$

where K is the number of sine–cosine waves necessary to model the seasonal variation, ω denotes the cycle (e.g., $\omega = 52$ for weekly data), each $b_{i,k}$ is assumed to follow a zero-mean Gaussian distribution, and ϵ_{it} allows for extra variation at each time period. Parameters β^s_{it} and γ^s_i were defined in a similar way. Finally, for each spatial unit, the parameter ϕ_i quantifies the effect that changes in the behavior of the syndromic disease have on the incidence of the infection of interest. A beta prior distribution can be assumed for this parameter. Hence, instead of analyzing syndromes associated with disease, the authors modeled the disease of interest and the syndromic disease jointly and related increases in the number of cases of the infection under study to increases previously observed in the syndromic disease.

An example of the application of this multivariate model to weekly cases of bronchitis in South Carolina in 2009, using as syndromic information the number of AURIs and the comparison with its univariate counterpart, where the infection of interest is analyzed individually via the model

$$\begin{aligned} y_{it} &\sim Po(e_{it}\theta_{it} + I_{it}) \\ log(\theta_{it}) &= \rho + u_i + v_i + \delta_{it} \\ I_{it} &= \beta_{it}\left(y_{i,t-1} + \gamma_i \sum_{j \in n_i} y_{j,t-1}\right), \end{aligned} \tag{27.17}$$

TABLE 27.1
DIC (pD) values for both the univariate and multivariate models

Data fitted	Univariate model	Multivariate model	Data fitted	Univariate model	Multivariate model
Weeks 1–30	6746.64 (363.27)	6538.69 (139.73)	Weeks 1–42	9347.27 (396.19)	9317.18 (322.82)
Weeks 1–33	7229.30 (274.11)	7117.22 (168.75)	Weeks 1–45	10018.80 (373.52)	9938.77 (262.01)
Weeks 1–36	7983.41 (392.86)	7763.08 (140.25)	Weeks 1–48	10692.20 (256.32)	10685.90 (343.46)
Weeks 1–39	8607.89 (316.69)	8398.41 (85.62)	Weeks 1–51	11445.30 (359.79)	11149.30 (66.69)

Note: Only the DIC value corresponding to bronchitis is shown in the multivariate scenario.

is demonstrated in Table 27.1 and Figure 27.6. Table 27.1 shows, for a selection of time periods, the deviance information criterion (DIC) values (together with the pD) obtained in the analysis of the bronchitis data with both the univariate and the multivariate model. Figure 27.6 compares the one-step-ahead forecasts for the number of cases of bronchitis in Charleston, Cherokee, Florence, Greenville, and Richland from week 31 to week 52 obtained with both the univariate and multivariate models. For illustrative purposes, the number of cases of AURIs, which are used to calculate forecasts in the multivariate scenario, are also shown.

As can be seen, the multivariate model provides a better description of the data. It is important to emphasize here that those are complex data that present a lot of variability. It may be the case that the dependence between increases in the incidence of the syndromic disease (AURI) and the disease of interest (bronchitis) changes during the forecast horizon and, as a consequence, forecasts obtained with the multivariate model are larger than the real values. Nevertheless, the forecasts obtained with the multivariate model are more accurate, especially at the epidemic onset if the syndromic disease outbreaks early enough. In addition, in a surveillance context where the main objective is to rapidly detect and follow outbreaks of disease, it is better to err on the side of caution and warn of possible increases in disease incidence that may need further investigation.

27.5 Evaluation of Surveillance Techniques

Sonesson and Bock (2003) discussed different ways of evaluating the effectiveness of surveillance techniques. A trade-off between false alarms and short delay times or detection probability must be struck, and so a single optimality criterion is not always enough, and evaluation by several measures might be necessary to characterize the behavior when the process is in and out of control. Let t_A represent the time of an alarm and τ the changepoint. One of the most common measures of in-control performance is the expected run length until the first false alarm, denoted $ARL^0 = E(t_A|\tau = \infty)$. A useful measure of evaluation with respect to a true change is the conditional expected delay, which is the average delay time for a motivated alarm when the change occurs at time point t: $CED(t) = E(t_A - \tau|t_A \geq \tau = t)$. Alternatively, when only a limited delay time d can be tolerated, we can consider the probability of successful detection defined as $PSD(d, t) = Pr(t_A - \tau \leq d|t_A \geq \tau = t)$.

An extensive discussion on measures of evaluation of surveillance systems can be found in the guidelines of the Centers for Disease Control and Prevention (CDC, 2001),

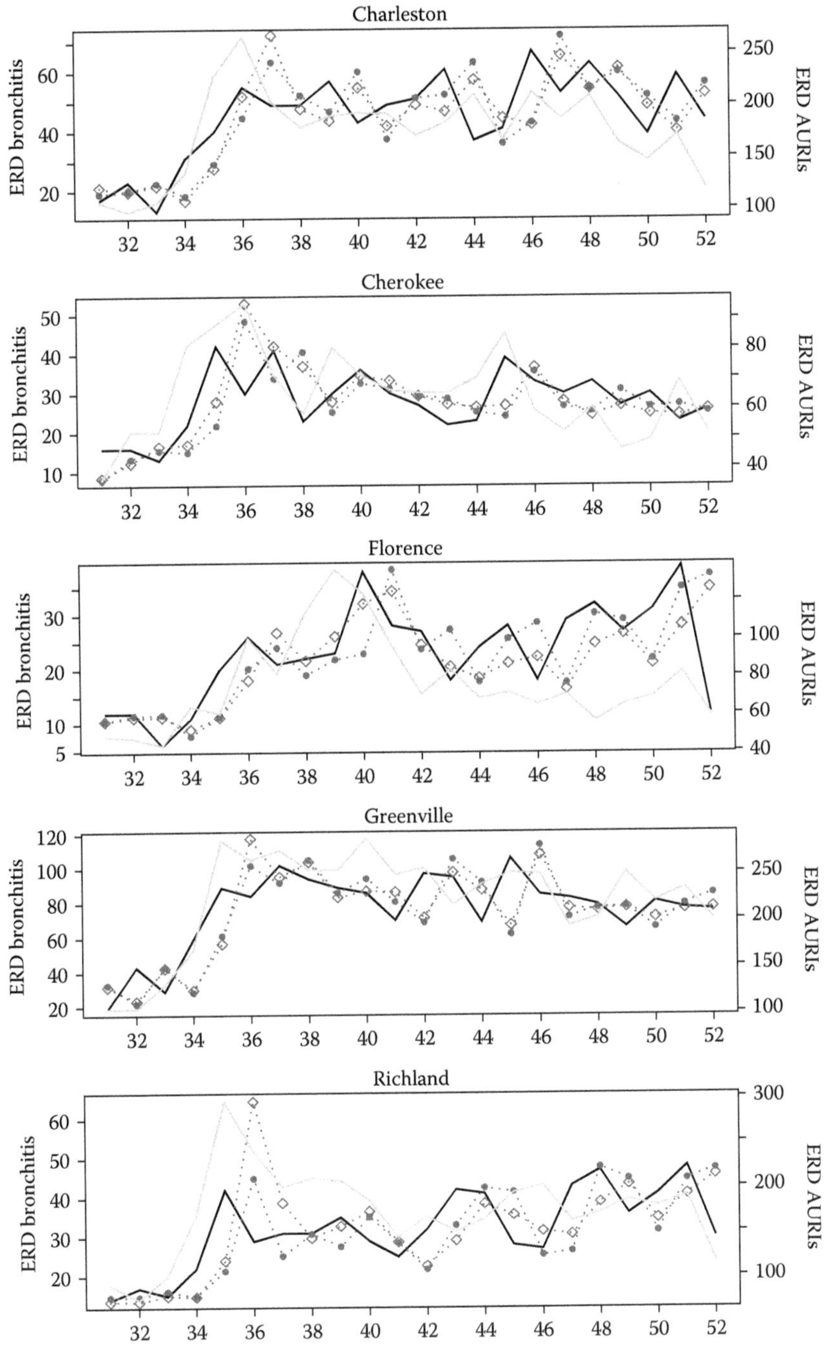

FIGURE 27.6
One-step-ahead forecasts for bronchitis infections from week beginning August 2 to week beginning December 27 (weeks 31–52) obtained with both the univariate model (points on dotted lines) and the multivariate model (diamonds on dotted lines) for a selection of five counties. Real values are represented by thick solid lines. Thin solid lines represent counts of AURIs.

where simplicity, flexibility, data quality, acceptability, sensitivity, predictive positive value (or specificity), representativeness, timeliness, and stability are mentioned as important attributes of public health surveillance systems. Additional reading can also be found in Unkel et al. (2012).

Different definitions of these measures can be found in a spatiotemporal surveillance setting. For instance, rather than just detecting a small area of unusual disease aggregation when a change in risk takes place, the primary concern in Corberán-Vallet and Lawson (2011) was to recover the true outbreak behavior. Let TA, FA, FNA, and TNA represent, respectively, the number of true alarms, false alarms, false no alarms, and true no alarms during the outbreak period, where $TA + FA + FNA + TNA = m$. In particular, the authors define the sensitivity as the proportion of all the areas undergoing a change in risk that signal an alarm at any time during the outbreak period, $TA/(TA + FNA)$. Similarly, the specificity is given by the proportion of in-control regions that are correctly identified as such, $TNA/(FA + TNA)$. The median time to outbreak detection (MTD) was used as an additional measure accounting for timeliness. For each small area undergoing an outbreak, the time to detection is defined as the number of time periods from the beginning of the outbreak until the first alarm is triggered, the outbreak taking place at any time during the surveillance period. If no alarm is sounded, that is, if the outbreak is missing, an infinite time to detection is assigned to that area. The MTD is then given by the median of the times to detection corresponding to those areas of increased incidence. Finally, in order to characterize the behavior of the surveillance mechanism when the process is in control, the authors considered the probability of a false alarm ($PrFA$), which can be estimated by the total number of time points associated with a false alarm divided by the total number of nonoutbreak time periods.

References

D. Banks, G. Datta, A. Karr, J. Lynch, J. Niemi, and F. Vera. Bayesian CAR models for syndromic surveillance on multiple data streams: Theory and practice. *Information Fusion*, vol. 13, pp. 105–116, 2012.

J. Besag, J. York, and A. Mollié. Bayesian image restoration, with two applications in spatial statistics. *Annals of the Institute of Statistical Mathematics*, vol. 43, pp. 1–59, 1991.

CDC. Updated guidelines for evaluating public health surveillance systems: Recommendations from the guidelines working group. *Morbidity and Mortality Weekly Report*, vol. 50(RR13), pp. 1–35, 2001.

CDC. Framework for evaluating public health surveillance systems for early detection of outbreaks: Recommendations from the CDC working group. *Morbidity and Mortality Weekly Report*, vol. 53(RR05), pp. 1–11, 2004.

T.-C. Chan, C.-C. King, M.-Y. Yen, P.-H. Chiang, C.-S. Huang, and C.K. Hsiao. Probabilistic daily ILI syndromic surveillance with a spatio-temporal Bayesian hierarchical model. *PLoS One*, vol. 5, e11626, 2010.

D. Conesa, M.A. Martínez-Beneito, R. Amorós, and A. López-Quílez. Bayesian hierarchical Poisson models with a hidden Markov structure for the detection of influenza epidemic outbreaks. *Statistical Methods in Medical Research*, vol. 24, pp. 206–223, 2015.

A. Corberán-Vallet. Prospective surveillance of multivariate spatial disease data. *Statistical Methods in Medical Research*, vol. 21, pp. 457–477, 2012.

A. Corberán-Vallet and A.B. Lawson. Conditional predictive inference for online surveillance of spatial disease incidence. *Statistics in Medicine*, vol. 30, pp. 3095–3116, 2011.

A. Corberán-Vallet and A.B. Lawson. Prospective analysis of infectious disease surveillance data using syndromic information. *Statistical Methods in Medical Research*, vol. 23, pp. 572–590, 2014.

P. Diggle. Spatio-temporal point processes: Methods and applications. In *Statistical Methods for Spatio-Temporal Systems.* B. Finkenstädt, L. Held, and V. Isham (eds.). London: Chapman & Hall, 2007, chapter 1.

P. Diggle, B. Rowlingson, and T. Su. Point process methodology for on-line spatio-temporal disease surveillace. *Environmetrics*, vol. 16, pp. 423–434, 2005.

C.P. Farrington, N.J. Andrews, A.D. Beale, and M.A. Catchpole. A statistical algorithm for the early detection of outbreaks of infectious disease. *Journal of the Royal Statistical Society: Series A*, vol. 159, pp. 547–563, 1996.

S. Geisser. Discussion on sampling and Bayes' inference in scientific modelling and robustness (by G.E.P. Box). *Journal of the Royal Statistical Society: Series A*, vol. 143, pp. 416-417, 1980.

J. Ginsberg, M.H. Mohebbi, R.S. Patel, L. Brammer, M.S. Smolinski, and L. Brilliant. Detecting influenza epidemics using search engine query data. *Nature*, vol. 457, pp. 1012–1014, 2009.

M.J. Heaton, D.L. Banks, J. Zou, A.F. Karr, G. Datta, J. Lynch, and F. Vera. A spatio-temporal absorbing state model for disease and syndromic surveillance. *Statistics in Medicine*, vol. 31, pp. 2123–2136, 2012.

L. Held, M. Hofmann, and M. Höhle. A two-component model for counts of infectious diseases. *Biostatistics*, vol. 7, pp. 422–437, 2006.

L. Held, M. Höhle, and M. Hofmann. A statistical framework for the analysis of multivariate infectious disease surveillance counts. *Statistical Modelling*, vol. 5, pp. 187–199, 2005a.

L. Held, I. Natário, S.E. Fenton, H. Rue, and N. Becker. Towards joint disease mapping. *Statistical Methods in Medical Research*, vol. 14, pp. 61–82, 2005b.

R. Katz, L. May, J. Baker, and E. Test. Redefining syndromic surveillance. *Journal of Epidemiology and Global Health*, vol. 1, pp. 21–31, 2011.

K. Kavanagh, C. Robertson, H. Murdoch, G. Crooks, and J. McMenamin. Syndromic surveillance of influenza-like illness in Scotland during the influenza A H1N1v pandemic and beyond. *Journal of the Royal Statistical Society: Series A*, vol. 175, pp. 939–958, 2012.

K. Kleinman, R. Lazarus, and R. Platt. A generalized linear mixed models approach for detecting incident clusters of disease in small areas, with an application to biological terrorism. *American Journal of Epidemiology*, vol. 159, pp. 217–224, 2004.

L. Knorr-Held. Bayesian modelling of inseparable space-time variation in disease risk. *Statistics in Medicine*, vol. 19, pp. 2555–2567, 2000.

L. Knorr-Held and N.G. Best. A shared component model for detecting joint and selective clustering of two diseases. *Journal of the Royal Statistical Society: Series A*, vol. 164, pp. 73–85, 2001.

M. Kulldorff. Prospective time periodic geographical disease surveillance using a scan statistic. *Journal of the Royal Statistical Society Series A*, vol. 164, pp. 61–72, 2001.

M. Kulldorff, F. Mostashari, L. Duczmal, W.K. Yih, K. Kleinman, and R. Patt. Multivariate scan statistics for disease surveillance. *Statistics in Medicine*, vol. 26, pp. 1824–1833, 2007.

A.B. Lawson. *Bayesian Disease Mapping: Hierarchical Modeling in Spatial Epidemiology* (2nd ed.). Boca Raton, FL: CRC Press, 2013.

A.B. Lawson and P. Leimich. Approaches to the space-time modelling of infectious disease behavior. *IMA Journal of Mathematics Applied in Medicine and Biology*, vol. 17, pp. 1–13, 2000.

Y. Le Strat. Overview of temporal surveillance. In *Spatial and Syndromic Surveillance for Public Health*. A.B. Lawson and K. Kleinman (eds.). New York: Wiley, 2005, chapter 2.

Y. Le Strat and F. Carrat. Monitoring epidemiologic surveillance data using hidden Markov models. *Statistics in Medicine*, vol. 18, pp. 3463–3478, 1999.

H. Ma and B.P. Carlin. Bayesian multivariate areal wombling for multiple disease boundary analysis. *Bayesian Analysis*, vol. 2, pp. 281–302, 2007.

M.A. Martínez-Beneito, D. Conesa, A. López-Quílez, and A. López-Maside. Bayesian Markov switching models for the early detection of influenza epidemics. *Statistics in Medicine*, vol. 27, pp. 4455–4468, 2008.

D.B. Neill and G.F. Cooper. A multivariate Bayesian scan statistic for early event detection and characterization. *Machine Learning*, vol. 79, pp. 261–282, 2010.

M. Paul, L. Held, and A.M. Toschke. Multivariate modeling of infectious disease surveillance data. *Statistics in Medicine*, vol. 27, pp. 6250–6267, 2008.

C. Robertson, T.A. Nelson, Y.C. MacNab, and A.B. Lawson. Review of methods for space-time disease surveillance. *Spatial and Spatio-Temporal Epidemiology*, vol. 1, pp. 105–116, 2010.

P.A. Rogerson. Spatial surveillance and cumulative sum methods. In *Spatial and Syndromic Surveillance for Public Health*. A.B. Lawson and K. Kleinman (eds.). New York: Wiley, 2005, chapter 6.

SaTScan™ v9.1.1: Software for the spatial and space-time scan statistcs. 2009. http://www.satscan.org/ (accessed October 26, 2011).

G. Shmueli and H. Burkom. Statistical challenges facing early outbreak detection in biosurveillance. *Technometrics*, vol. 52, pp. 39–51, 2010.

C. Sonesson and D. Bock. A review and discussion of prospective statistical surveillance in public health. *Journal of the Royal Statistical Society: Series A*, vol. 166, pp. 5–21, 2003.

S.B. Thacker and R.L. Berkelman. History of public health surveillance. In *Public Health Surveillance*. W. Halperin and E.L. Baker (eds.). New York: Van Nostrand Reinhold, 1992, chapter 1.

S. Unkel, C.P. Farrington, P.H. Garthwaite, C. Robertson, and N. Andrews. Statistical methods for the prospective detection of infectious diseases outbreaks: A review. *Journal of the Royal Statistical Society: Series A*, vol. 175, pp. 49–82, 2012.

C.L. Vidal Rodeiro and A.B. Lawson. Monitoring changes in spatio-temporal maps of disease. *Biometrical Journal*, vol. 48, pp. 463–480, 2006.

F. Wang and M.M. Wall. Generalized common spatial factor model. *Biostatistics*, vol. 4, pp. 569–582, 2003.

R.E. Watkins, S. Eagleson, B. Veenendaal, G. Wright, and A.J. Plant. Disease surveillance using a hidden Markov model. *BMC Medical Informatics and Decision Making*, vol. 9, pp. 39, 2009.

H. Zhou and A.B. Lawson. EWMA smoothing and Bayesian spatial modeling for health surveillance. *Statistics in Medicine*, vol. 27, pp. 5907–5928, 2008.

28

Cluster Modeling and Detection

Andrew B. Lawson
Division of Biostatistics and Bioinformatics
Department of Public Health Sciences
Medical University of South Carolina
Charleston, South Carolina

CONTENTS

28.1 Introduction

It is often appropriate to ask questions related to the local properties of the relative risk surface rather than models of relative risk per se. Local properties of the surface could include peaks of risk, sharp boundaries between areas of risk, or local heterogeneities in risk. These different features relate to surface properties, but not directly to a value at a specific location. Relative risk estimation (disease mapping; Chapters 6 and 7) concerns the "global" smoothing of risk and estimation of true underlying risk level (height of the risk surface), whereas cluster detection is focused on local features of the risk surface where elevations of risk or depressions of risk occur. Hence, it is clear that cluster detection is fundamentally different from relative risk estimation in its focus. However, the difference can become blurred, as methods that are used for risk estimation can be extended to allow certain types of cluster detection. This will be discussed in detail in later sections. While Chapters 3, 8, and 9 deal with clustering issues and related testing, this chapter focuses solely on cluster modeling and methods related to modeling. A general review of clustering in a nonspatial and a spatial context can be found in Xu and Wunsch (2009).

28.2 Cluster Definitions

Before discussing cluster detection and estimation methods, it is important to define the nature of the clusters or clustering to be studied. There are a variety of definitions of clusters and clustering. Different definitions of clusters or clustering will lead to differences in the ability of detection methods. First, it should be noted that sometimes the correlated heterogeneity term in relative risk models is called a clustering term (see, e.g., Clayton and Bernardinelli, 1992). This implies that the term captures aggregation in the risk, and indeed, this does lead to an effect where neighboring areas having similar risk levels. This is a global feature of the risk, however, and also induces a smoothing of risk. This begs the question of how we define clusters or clustering: Should it be a global feature or local in nature?

Global clustering basically assumes that the risk surface is clustered or has areas of like elevated (reduced) risk. An uncorrelated surface, on the other hand, should display random changes in risk with changes in location, and so should both be much more variable in risk level and have few contiguous areas of like risk. Figure 28.1 displays a comparison

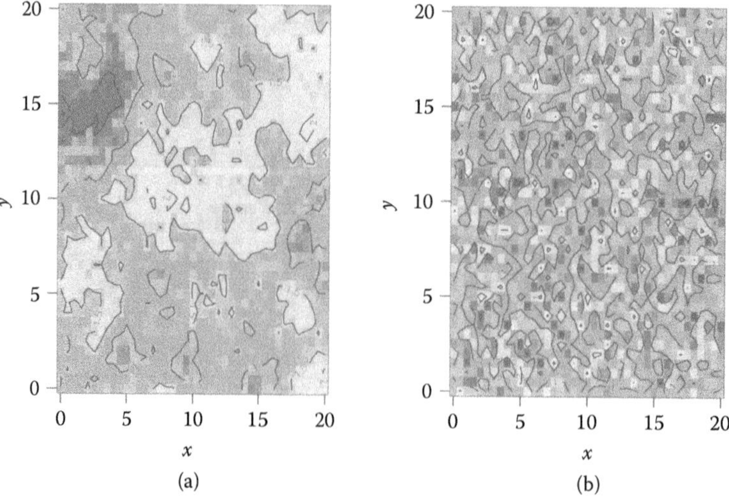

FIGURE 28.1
Simulated examples of spatially correlated (a) and uncorrelated (b) surfaces. Simulation using the R function GaussRF.

of correlated and uncorrelated surfaces. However, modeling the overall clustering does not address their locations specifically. Hence, this form of clustering does not address localized behavior or the location of clusters per se. This is often termed general clustering (see Chapter 8).

A general definition of a (spatial) cluster is

any spatially bounded area of significantly elevated (reduced) risk.

This is clearly very general and requires further definition. By "spatially bounded," I mean that the cluster must have some spatial integrity. This could be a neighborhood criterion such as *areas must be adjoining* or *at least two adjoining areas must meet a criterion*, or could be defined to have a certain type of external boundary (e.g., risk differences around the cluster must meet a criterion). A simple criterion that is often assumed is that known as *hot spot clustering*. In hot spot clustering, any area or region can be regarded as a cluster. This is due to the assumption of a zero-neighborhood criterion, that is, no insistence on adjacency of regions within clusters. This is a convenient and simple criterion and is often assumed to be the only criterion. It is commonly used in epidemiology (see, e.g., Richardson et al., 2004; Abellan et al., 2008). Without prior knowledge of the behavior of the disease, this criterion is appealing. It could be useful for preliminary screening of data, for example.

However, this hot spot definition ignores any contiguity that may be thought to be inherent in relevant clusters. For example, it might be important that clusters of a given threshold size be investigated. This threshold size could be defined as a minimum number of contiguous areas. Hence, only groups of contiguous regions of "unusual" risk could qualify as clusters. On the other hand, in the case of infectious diseases, it may be that a certain shape and size of cluster are important in understanding disease spread. Note that there are many tests now available to detect clustering of various types. In Chapters 3, 8, 9, and 10, these are discussed in detail. In this chapter, attention is confined to model-based assessment of clustering.

In this chapter, three different scenarios for clustering are considered:

1. Single region hot spot relative risk detection

2. Clusters as objects or groupings

3. Clusters defined as residuals

28.2.1 Hot spot clustering

Hot spot clustering is often the most intuitive form of clustering and may be that which most public health professionals consider as their definition. In hot spot clustering, any area or region can be regarded as a cluster. This is due to the assumption of a zero-neighborhood criterion, that is, no insistence on adjacency of regions within clusters.

28.2.2 Clusters as objects or groupings

Clustering might be considered to be apparent in a data set when a specific form of grouping is apparent. This grouping would usually be predefined. Usually, the criterion would also have a neighborhood or proximity condition. That is, only neighboring or proximal areas (which meet other criteria) can be considered to be in a cluster.

28.2.3 Clusters defined as residuals

Often it is convenient to consider clusters as a residual feature of data. For example, let's assume that y_i is the count of disease within the ith census tract within a study area. Let's also assume that our basic model for the average count μ_i (i.e., $E(y_i) = \mu_i$) is

$$\log \mu_i = a_i + e_i.$$

Here, a_i could consist of a linear or nonlinear predictor as a function of covariates and could also consist of random effects of different kinds. To simplify the idea, we assume that a_i is the smooth part of the model and e_i is the rough or residual part. The basic idea is that if we model a_i to include all relevant nonclustering confounder effects, then the residual component must contain residual clustering information. Hence, if we examine the estimated value of e_i, then this will contain information about any clusters unaccounted for in a_i. Of course, this does not account for any pure noise that might also be found in e_i. This means, of course, that an estimate of e_i could have at least two components: clustered and unclustered (or frailty). There could, of course, be additional components, depending on whether the confounding in a_i was adequately specified or estimated.

There are a number of approaches to isolating the residual clustering. First, it is possible to include a pure noise term within a_i and to consider e_i as a cluster term. For example, we could assume that $a_i = f(v_i; covariates)$, where $f(.)$ is a function of an uncorrelated noise at the observation level (v_i : frailty or random effect term) and a function of covariates. Second, a smoothed version of e_i, say $s(e_i)$, could be examined in the hope that the pure noise is smoothed out. Of course, this begs the question of which component should include the clustering: Should it be a model component or a residual component? If the clustering is likely to be irregular and we can be assured that no clustering confounding effects are to be found in the model component, then a residual or smoothed residual might be useful. On the other hand, if there is any prior knowledge of the form of clustering to be expected, then it may be more important to include some of that information within the model itself. The real underlying issue is the ability of models and estimation procedures to differentiate spatial scales of clustering.

28.3 Cluster Detection Using Residuals

First, assume that we observe disease outcome data within a spatial window.

28.3.1 Case event data

28.3.1.1 Unconditional analysis

For the case event scenario, we have $\{s_i\}, i = 1, \ldots, m$, events observed within the window T. Modeling here focuses on the first-order intensity and its parametrization. Assume that $\lambda(s|\psi) = \lambda_0(s|\psi_0).\lambda_1(s|\psi_1)$ as defined in Chapter 6. We focus first on the specification of a residual for a point process governed by $\lambda(s|\psi)$. First, in the spirit of classical residual analysis, it is clear that we can assume that we want to compare fitted values to observed values. This is not simple, as we have locations as observed data. One way to circumvent this problem is to consider a function of the observed data that can be compared with an intensity estimate at location s_i, say $\lambda(s_i|\widehat{\psi})$. One such function could be a saturated or nonparametric intensity estimate ($\widehat{\lambda}_{loc}(s_i)$ say, where loc denotes a local estimator). Essentially, this gives a slight aggregation of the data, but it allows for a direct comparison of model to data. Hence, we can define a residual as

$$r_i^{loc} = \widehat{\lambda}_{loc}(s_i) - \lambda(s_i|\widehat{\psi}),$$

or in the case of a saturated estimate (Lawson, 1993),

$$r_i^{sat} = \widehat{\lambda}_{sat}(s_i) - \lambda(s_i|\widehat{\psi}).$$

Baddeley et al. (2005) discuss more general cases applied to a range of processes. An example of a local estimate of intensity could be derived from a suitably edge-weighted density estimate (Diggle, 1985). An example of the use of the saturated estimator is as follows.

First, assume that an estimator is available for the background intensity $\lambda_0(s_i|\psi_0)$, say $\lambda_0(s_i|\widehat{\psi}_0) \equiv \lambda_{0i}$. Also, assume that it can be used as a plug-in estimator within $\lambda(s|\psi)$. If this is the case, then we can compute $r_i^{sat} = \lambda_{0i}[\widehat{\lambda_1}_{sat}(s_i) - \lambda_1(s_i|\widehat{\psi})]$. For a simple heterogeneous Poisson process model with intensity $\lambda_0(s|\psi_0).\lambda_1(s|\psi_1)$ and using an integral weighting scheme (as described in Chapter 6), the saturate estimate of the intensity at s_i is $1/(w_i\lambda_{0i})$. A simple weight (which provided a crude estimator of the local intensity) is $w_i = A_i$, where A_i is the Dirichlet tile area surrounding s_i, based on a tesselation of the case events. Hence, a simple residual could be based on

$$\begin{aligned} r_i^{sat} &= \lambda_{0i}[(w_i\lambda_{0i})^{-1} - \lambda_1(s_i|\widehat{\psi})] \\ &= w_i^{-1} - \lambda_{0i}\lambda_1(s_i|\widehat{\psi}). \end{aligned}$$

The use of such tile areas must be carefully considered, as edge effect distortion can occur with tesselation, and so boundary regions of the study window should be treated with caution. Of course, both the error in estimation of the background intensity is ignored here and a crude approximation to the saturated intensity is assumed. Note that r_i^{sat} or r_i^{loc} can be computed within a posterior sampler, and so a posterior expectation of the residuals can be estimated.

Figure 28.2 displays an example of the use of posterior expectation of r_i^{sat} for a model, for the well-known larynx cancer data set from Lancashire, UK, 1974–1983. This data set has been analyzed many times and consists of the residential address locations of cases of larynx cancer with the residential addresses of cases of respiratory cancer as a control disease (see, e.g., Diggle, 1990; Lawson, 2006b, chapter 1), with a distance decline component (variable d_i) around the fixed point (3.545, 4.140), an incinerator. The motivation for this type of analysis relates to assessment of health hazards around putative sources (putative source analysis). This is discussed more fully in Chapter 14. The model for the first-order intensity is defined to depend on this distance: $\lambda_1(s_i|\theta) = \beta_0[1+\exp(-\beta_1 d_i)]$. (WinBUGS code for this and other models can be found on the handbook website: https://www.crcpress.com/9781482253016.) The map displays the contours for the posterior sample estimate of $\Pr(r_i^{sat} > 0)$, the residual exceedence probability.

To allow for extra unobserved variation in this map, an uncorrelated random effect term can also be included in the model. Figure 28.3 displays the resulting posterior average residual exceedence probability map for the model with $\lambda_1(s_i|\theta) = \beta_0[1+\exp(-\beta_1 d_i)]\exp(v_i)$, where $v_i \sim N(0, \tau_v)$ and $\beta_* \sim N(0, \tau_{\beta_*})$. Both figures suggest that there is slight evidence for an excess of aggregation in the north of the study region (where there is a large area where $\Pr(r_i^{sat} > 0) > 0.9$). There is also weaker evidence of an excess in the area to the west of the putative source (3.545, 4.140), where $\Pr(r_i^{sat} > 0) > 0.8$ on average. There are also marked edge effects close to the study region corners due to the distortion of the tesselation suspension algorithm. Figure 28.3 displays a similar picture even after removal of extra noise.

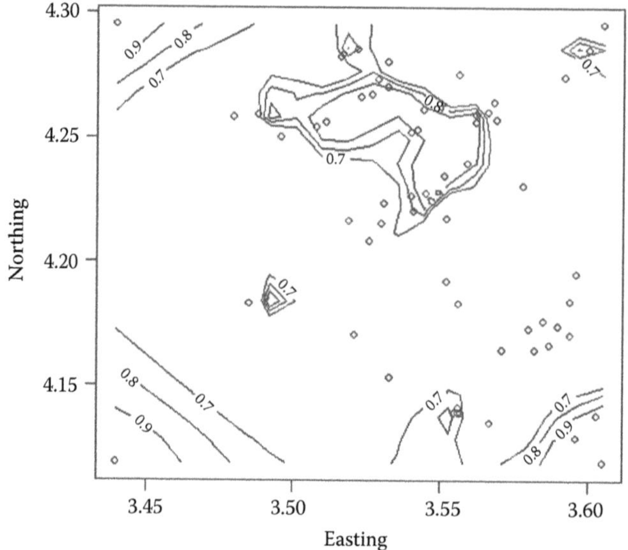

FIGURE 28.2
Map of Lancashire larynx cancer case distribution with, superimposed, a contour map of exceedence probability (0.7, 0.8, 0.9) for the residual (r_i^{sat}) from a Bayesian model assuming Berman–Turner Dirichlet tile integration weights and a nonparametric density estimate of background risk computed from the respiratory cancer control distribution.

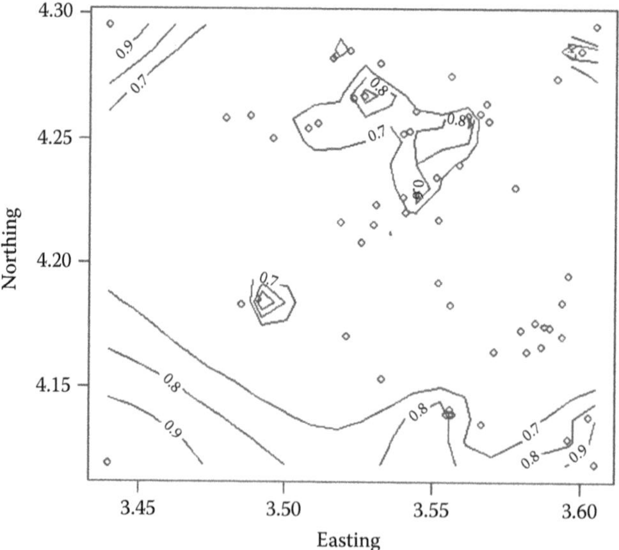

FIGURE 28.3
Same as Figure 28.2, but the model has included a random uncorrelated effect (v_i) to allow for extra variation in the risk: $\lambda_1(s_i|\theta) = \beta_0[1 + \exp(-\beta_1 d_i)] \exp(v_i)$.

28.3.1.2 Conditional logistic analysis

An alternative approach to the analysis of case event data is to consider the joint realization of cases and controls and to model the conditional probability of a case given an event has occurred at a location. This approach was discussed in Chapter 6, and has the advantage that the background effect factors out of the likelihood. Define the joint realization of m cases and n controls as $s_i : i = 1, \ldots, N$, with $N = m + n$. Also define a binary label variable $\{y_i\}$ that labels the event as either a case ($y_i = 1$) or a control ($y_i = 0$). The resulting conditional likelihood has a logistic form:

$$L(\psi_1|\mathbf{s}) = \prod_{i \in cases} p_i \prod_{i \in controls} 1 - p_i$$
$$= \prod_{i=1}^{N} \left[\frac{\{\exp(\eta_i)\}^{y_i}}{1 + \exp(\eta_i)} \right],$$

where $p_i = \frac{\exp(\eta_i)}{1+\exp(\eta_i)}$ and $\eta_i = x_i'\beta$, where x_i' is the ith row of the design matrix of covariates and β is the corresponding p-length parameter vector. Hence, in this form, a Bernoulli likelihood can be assumed for the data and a hierarchical model can be established for the linear predictor $\eta_i = x_i'\beta$. In general, it is straightforward to extend this formulation to the inclusion of random effects in a generalized linear mixed form. Bayesian residuals such as $r_i = y_i - \widehat{p}_i / \widehat{se}(y_i - \widehat{p}_i)$ (or directly standardized version: $r_i = (y_i - \widehat{p}_i)/\sqrt{\widehat{p}_i(1 - \widehat{p}_i)}$) are available, where $\widehat{p}_i$ is the average value of p_i from the posterior sample. Residuals from binary data models are often difficult to interpret due to the limited variation in the dependent variable (0/1), and the usual recommendation for their examination is to group or aggregate the results. A wide variety of aggregation methods could be used. Spatial aggregation methods might be considered here. Figure 28.4 displays the mapped surface of the standardized Bayesian residual using $r_i = (y_i - \widehat{p}_i)/\sqrt{\widehat{p}_i(1 - \widehat{p}_i)}$, where $\widehat{p}_i$ is computed

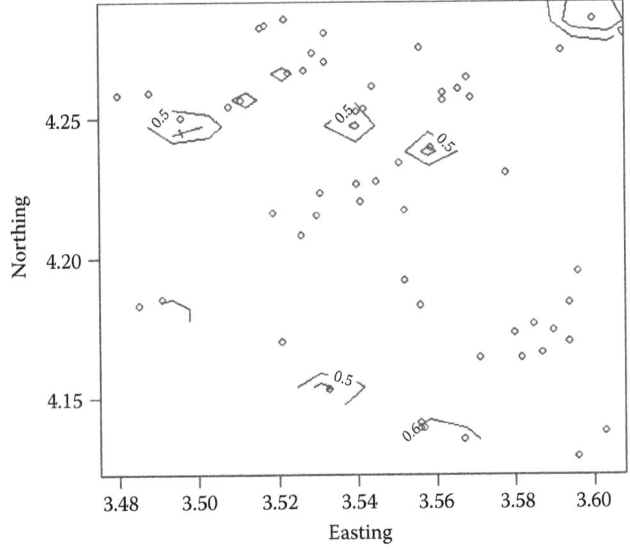

FIGURE 28.4
Contour map of the standardized Bayesian residual for the logistic case-control spatial model applied to the larynx cancer data from Lancashire, UK. The display shows the posterior average residual for a sample of 5000 after burn-in.

from the converged posterior sample via R2winBUGS. The model assumed for this example also has a additive distance effect and is specified by

$$p_i = \frac{\lambda_i}{1+\lambda_i}$$
$$\lambda_i = \exp\{\alpha_0 + v_i\}.\{1+\exp(-\alpha_1 d_i)\}.$$

Figure 28.5 displays the thresholded mapped surface of the $\Pr(r_i > 2)$ for values (0.05, 0.1, 0.2). This suggests some evidence of clustering or unusual aggregation in the north and also in the vicinity of the putative location in the south.

Note that all of these models assume that there is negligible clustering under the model and that any residual effects will include the clustering. These models do not explicitly model clustering, but only model long-range and uncorrelated variation. Hence, we make the tacit assumption that any remaining aggregation of cases will be found in the residual component. Of course, other effects that were excluded from the model could be present in the residuals.

28.3.2 Count data

For count data, it is assumed that either a Poisson data likelihood or a binomial likelihood is relevant. Note also that an autologistic model for binary outcomes could also be specified.

28.3.2.1 Poisson likelihood

In the case of a Poisson likelihood, assume that y_i, $i = 1,\ldots,m$, are counts of cases of disease and e_i, $i = 1,\ldots,m$, are expected rates of the disease in m small areas, and so $y_i \sim Poiss(e_i\theta_i)$ given θ_i. The log relative risk is usually modeled, and so $\log\theta_i$ is the modeling focus. Bayesian residuals for this likelihood are easily computed in standardized form as $r_i = (y_i - e_i\widehat{\theta}_i)/\sqrt{e_i\widehat{\theta}_i}$, where $\widehat{\theta}_i$ is the average value of the θ_i obtained from the

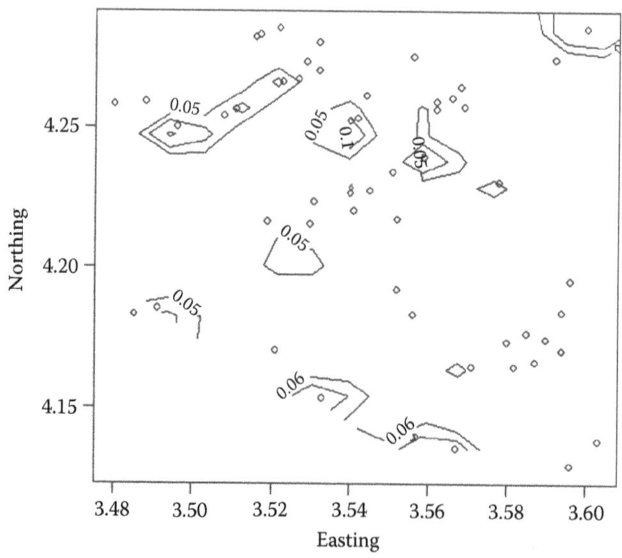

FIGURE 28.5
Map of the contoured surface of $\Pr(r_i > 2)$ estimated from the converged posterior sample for the standardized Bayesian residual in Figure 28.4.

converged posterior sample. An example of such count data is the U.S. state of Georgia's county-level oral cancer mortality data for 2004 (see Lawson, 2013, chapter 1, for more details), examined with a Poisson data likelihood and model $\log \theta_i = \alpha_0 + v_i$, where

$$\alpha_0 \sim U(a, b)$$
$$v_i \sim N(0, \tau_v),$$

with τ_v set large and (a, b) a large negative to positive range. No correlated random effect is included here, as it is assumed that clustering is to be found in residuals. Figure 28.6 displays the results from a converged sampler based on 10,000 burn-in and a sample size of 2000. The display shows the average estimate of $\Pr(r_i > 2)$ and $\Pr(r_i > 3)$ for r_i given above. The most extreme region appears to be in the far west of Georgia. It should be noted that there is considerable noise in these residuals, particularly for $\Pr(r_i > 2)$.

28.3.2.2 Binomial likelihood

In the case of a binomial likelihood, assume m small areas, and that in the ith area there is a finite population n_i out of which y_i disease cases occur. The probability of a case is p_i. The data model is thus $y_i \sim bin(p_i, n_i)$ given p_i, and the usual assumption is made that $\text{logit}(p_i) = f(\eta_i)$, where η_i is a linear or nonlinear predictor. Of course, various ingredients can be specified for $f(\eta_i)$, including the addition of random effects to yield a binomial generalized linear mixed model (GLMM). A Bayesian residual for this model is given in standardized form as $r_i = (y_i - n_i\widehat{p}_i)/\sqrt{n_i\widehat{p}_i(1 - \widehat{p}_i)}$, where $\widehat{p}_i$ is the average of p_i values found in the converged posterior sample.

While the above discussion has focused on simple residual diagnostics, albeit from posterior samples, there is also the possibility of examining *predictive* residuals for any given model. A predictive residual can be computed for each observation unit as

$$r_i^{pr} = y_i - y_i^{pr},$$

where $y_i^{pr} = \frac{1}{G}\sum_{g=1}^{G} f(y_i|\theta^g)$, and $f(y_i|\theta^g)$ is the likelihood given the current value of θ^g. Of course, this will usually be small compared to the standard Bayesian residual. Note that for a given data model, y_i^{pred} can be easily generated on WinBUGS. For the binomial example above, the code could be

```
y[i]~dbin(p[i],n[i])
ypred[i]~dbin(p[i],n[i])
rpred[i]<-y[i]-ypred[i]
```

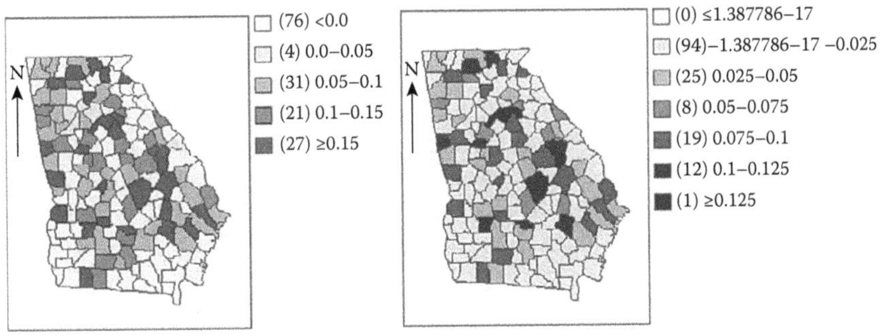

FIGURE 28.6
Georgia oral cancer: county maps of Bayesian residuals from a converged posterior sampler with an uncorrelated random effect term. From left to right, $\Pr(r_i > 2)$ and $\Pr(r_i > 3)$.

An alternative approach to residual analysis could be based in the construction of a residual envelope, based on the comparison of the Bayesian residual: $r_i = y_i - \widehat{y}_i$, with $r_i^* = y_i^{pred} - \widehat{y}_i$. Unusual residuals could be assessed by assessing the ranking of r_i among a series of B simulated $\{r_{ib}^*\}$ $b = 1, \ldots, B,$. Further, a p-value surface can be computed from a tally of exceedences:

$$P_{v_i} = \Pr(|r_i| > |r_i^*|) = \frac{1}{B}\sum_{b=1}^{B} I(|r_i| > |r_{ib}^*|).$$

The mapped surface of P_{v_i} could be examined for areas of unusually elevated values, and hence provide a tool for hot spot detection.

28.4 Cluster Detection Using Posterior Measures

Another approach to cluster detection is to consider measures of quantities monitored in the posterior that may contain clustering information. One such measure is related to estimates of first-order intensity (case event data) or relative risk (Poisson count data) or case probability (binomial count data). If we have captured the clustering tendency within our estimate of any of these quantities, then we could examine their posterior sample behavior. Perhaps the most commonly used example of this is the use of exceedence probability in relation to relative risk estimates for individual areas for count data (see, e.g., Richardson et al., 2004). Define the exceedence probability as the probability that the relative risk θ exceeds some threshold level (c): $\Pr(\theta > c)$. This is often estimated from posterior sample values $\{\theta_i^g\}_{g=1,\ldots,G}$ via

$$\widehat{\Pr}(\theta_i > c) = \sum_{g=1}^{G} I(\theta_i^g > c)/G,$$

where

$$I(a) = \begin{cases} 1 \text{ if } a \text{ true} \\ 0 \text{ otherwise} \end{cases}.$$

Of course, there are two choices that must be made when evaluating $\widehat{\Pr}(\theta_i > c)$. First, the value of c must be chosen. Second, the threshold for the probability must also be chosen, that is, $\widehat{\Pr}(\theta_i > c) > b$, where b might be set to some conventional level such as 0.95, 0.975, or 0.99. In fact, there is a trade-off between these two quantities, and usually one must be fixed before considering the value of the other. Figure 28.7 displays the posterior expected exceedence probability maps: $\widehat{\Pr}(\theta_i > c)$ for $c = 2$, and $c = 3$ for the Georgia oral cancer data when a relative risk model with an uncorrelated heterogeneity (UH) component is fitted.

One major concern with the use of exceedence probability for single regions is that it is designed only to detect *hot spot* clusters (i.e., single-region signaling) and does not consider any other information concerning possible forms of cluster or even neighborhood information. Some attempt has been made to enhance this post hoc measure by inclusion of neighborhoods by Hossain and Lawson (2006). For the neighborhood of the ith area defined as δ_i, and the number of neighbors as n_i,

$$\overline{q_i} = \sum_{j=0}^{n_i} q_{ij}/(n_i + 1),$$

where

$$q_{ij} = \Pr(\theta_j > c)\ \forall\ j \in \delta_i$$

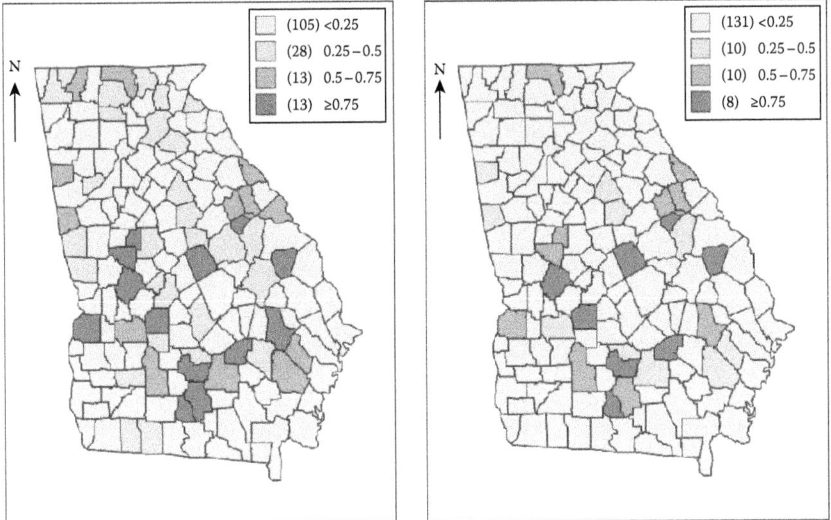

FIGURE 28.7
Georgia oral cancer: maps of $\widehat{\Pr}(\theta > c)$ for $c = 2$ (left panel) and $c = 3$ (right panel) for a model with an uncorrelated random effect (UH).

and

$$q_{i0} = \Pr(\theta_i > c).$$

This measures $\overline{q_i}$, and q_{i0} can be used to detect different forms of clustering. Other, more sophisticated measures have also been proposed (see, e.g., Hossain and Lawson, 2006, for details).

A second concern with the use of exceedence probabilities is of course that the usefulness of the measure depends on the model that has been fitted to the data. It is conceivable that a poorly fitting model will not demonstrate any exceedences relate to clustering and may leave the clustering of interest in the residual noise. An extreme example of this is displayed in Figure 28.8. In the figure, the same data set is examined with completely different models. The data set is South Carolina county-level congenital anomaly deaths for 1990 (see also Lawson et al., 2003, chapter 8). The expected rates were computed for an 8-year period. In the left panel, a Poisson log-linear trend model was assumed, and in the right panel, a convolution model. The trend model was $\log\theta_i = \alpha_0 + \alpha_1 x_i + \alpha_2 y_i$ with zero-mean Gaussian prior distributions for the regression parameters, whereas the right panel was $\log\theta_i = \alpha_0 + u_i + v_i$, where the u_i, v_i are correlated and uncorrelated heterogeneity terms with the usual conditional autoregression (CAR) and zero-mean Gaussian prior distributions. Without examination of the goodness of fit of these models, it is clear that there could be considerable latitude for misinterpretation if exceedence probabilities are used in isolation to assess (hot spot) clustering.

As in the count data situation, we can also examine exceedences for other data types and models. For example, in the case event example, intensity exceedence could be examined as $\Pr(\widehat{\lambda}_1(s_i) > 1)$, whereas for the binary or binomial data, the exceedence of the case probability could be used: $\Pr(\widehat{p}_i > 0.5)$. These can also be mapped, of course. However, the rider concerning the goodness of fit of the model, as highlighted in Figure 28.8, also applies here.

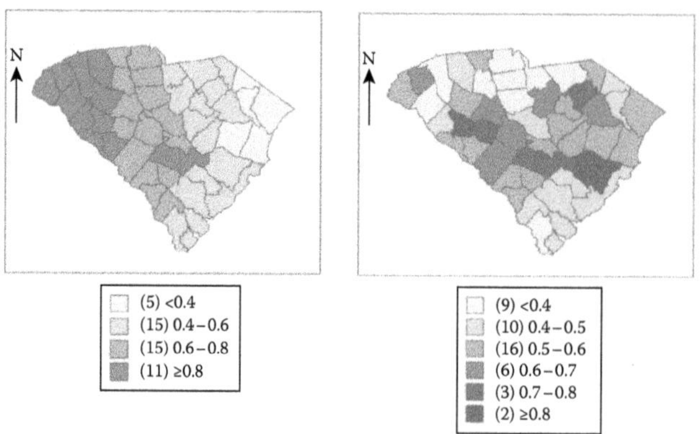

FIGURE 28.8
Display of exceedence probabilities for two models. Left: Simple first-order trend. Right: Convolution model with UH and CH only and no trend for the same data set: South Carolina county-level congenital mortality, 1990.

28.5 Cluster Models

It is also possible to design models that explicitly describe the clustering behavior of the data. In this way, parameters and functions can be defined that summarize this behavior. It should be noted that clustering behavior is often regarded as a second-order feature of the data. By second order, I mean relating to the mutual covariation of the data. Hence, it is often assumed that covariance modeling will capture clustering in data. This is often termed *general* clustering. However, as noted in Section 28.2, while general covariance modeling can capture the overall mutual covariation (as in Figure 28.1a), it does *not* lead to identification or detection of clusters per se. In the following, I focus on the detection of clusters, rather than general clustering.

28.5.1 Case event data

In the analysis of point processes (PPs), there is a set of models designed to describe clustering. For an introductory overview, which focuses mainly on general cluster testing, see Diggle (2003, chapter 9). Basic models often assumed for PPs, which allow clustering, are the Poisson cluster process and the Cox process. In the Poisson (Neyman–Scott) cluster process (PcP), an underlying process of parents (unobserved cluster centers) is assumed and offspring (observed points) are generated randomly in number and location. This generation is controlled by distributions. Clearly, this formulation is most appropriate in examples where parent generation occurs, such as seed dispersal in ecology.

An alternative to a PcP is found in the Cox process where a nonnegative stochastic process ($\Lambda(s)$) governs the intensity of a heterogeneous Poisson process (hPP). Conditional on the realization of the stochastic process, the events follow a hPP. In this case,

$$\lambda(s) = E[\Lambda(s)],$$

where the expectation is with respect to the process. Note that this formulation allows the inclusion of spatial correlation via a specification such as $\Lambda(s) = \exp\{S(s)\}$, where $S(s)$ is a spatial Gaussian process. This is sometimes known as a log-Gaussian Cox process (LGCP) (see, e.g., Møller et al., 1998; Brix and Diggle, 2001; Li et al., 2012). Instead, note also that an intensity process of the form

$$\Lambda(s) = \mu \sum_{j=1}^{\infty} h(s - c_j) \tag{28.1}$$

can be assumed, where $h(s - c_j)$ is a bivariate probability density function (pdf) and c_j are cluster centers. If the centers are assumed to have a homogeneous PP, then this is also a PcP. Of course, these models were derived mainly for ecological examples and not for disease case events. However, we can take as a starting point a model for case events that includes population modulation in the first-order intensity, and that also allows clustering via unobserved process of centers.

28.5.1.1 Object models

Define the first-order intensity as

$$\lambda(s|\psi) = \lambda_0(s|\psi_0).\lambda_1(s|\psi_1).$$

Assume that the case events form a hPP conditional on parameters in ψ_1. In the basic hPP likelihood, dependence on ψ_0 would also have to be considered. Often, $\lambda_0(s|\psi_0)$ is estimated nonparametrically and a profile likelihood is assumed. Alternatively, ψ_0 could be estimated within a posterior sampler. Here, focus is made on the specification of $\lambda_1(s|\psi_1)$. Following from the basic definitions of PcPs and Cox processes, it is possible to formulate a Bayesian cluster model that relies on underlying unobserved cluster center locations, but is not restricted to the restrictive assumptions of the classical PcP. Define the excess intensity at s_i as

$$\lambda_1(s_i|\psi_1) = \mu_0 \sum_{j=1}^{K} h(s_i - c_j; \tau), \tag{28.2}$$

where a finite number of centers is considered inside (or close to) the study window. For practical purposes, K is assumed to be relatively small (usually in the range of 1–20). The parameter τ controls the scale of the distribution. Note that in this formulation, we do not insist that $\{c_j\}$ follow a homogenous PP, nor is the cluster distribution function $h(s_i - c_j; \tau)$ restricted to a pdf, although it must be nonnegative. A simple extension of this allows for individual-level covariates within a predictor (η_i):

$$\lambda_1(s_i|\psi_1) = \exp(\rho_0 + \eta_i). \sum_{j=1}^{K} h(s_i - c_j; \tau), \tag{28.3}$$

where $\mu_0 = \exp(\rho_0)$.

In the following, intensity (28.2) will be examined. In general, intensity (28.2) can be regarded as a mixture intensity with an unknown number of components and component values (cluster center locations). A general Bayesian model formulation can be

$$[\{s_i\}|\psi_0, \mu_0, \tau, K, \mathbf{c}] \sim \prod_{i=1}^{m} \lambda(s_i|\%). \exp\left\{-\int_T \lambda(u|\%)du\right\},$$

where $\% \equiv \{\psi_0, \mu_0, K, \mathbf{c}\}$, with $\psi_1 \equiv \{\rho_0, \tau, K, \mathbf{c}\}$,

$$\begin{aligned}
\lambda(s_i|\psi_1) &= \lambda_0(s_i|\psi_0)\lambda_1(s_i|\psi_1)\\
\lambda_1(s_i|\psi_1) &= \exp(\rho_0).\sum_{j=1}^{K} h(s_i - c_j; \tau)\\
\rho_0 &\sim Ga(a, b)\\
K &\sim Pois(\gamma)\\
\{c_j\} &\sim U(A_T)\\
\tau &\sim Ga(c, d).
\end{aligned}$$

Here, the prior distributions reflect our beliefs concerning the nature of the parameter variation. As ρ_0 is the case event rate, we assume a positive distribution (in this case, a gamma distribution). The parameter γ essentially controls the parent rate (center rate), and in this case, the prior for the number of centers (K) is Poisson with rate γ. Other alternatives can be assumed for this distribution. A uniform distribution on a small positive range would be possible. Another possibility is to assume that the centers are mutually inhibited and to assume a distribution that will provide this inhibition. Such a distribution could be a Markov process form, such as a Strauss distribution (Móller and Waagepetersen, 2004, chapter 6). The τ parameter is assumed to appear as a precision term in the cluster distribution function: $h(s_i - c_j; \tau)$. A typical symmetric specification for this distribution is distance based:

$$h(s_i - c_j; \tau) = \frac{\tau}{2\pi} \exp\left\{-\tau d_{ij}^2/2\right\},$$

where

$$d_{ij} = \|s_i - c_j\|.$$

Other forms are of course possible, including allowing the precision to vary with location and asymmetry of the directional form. Many examples exist where variants of these specifications have been applied to cluster detection problems (e.g., Lawson, 1995, 2000; Cressie and Lawson, 2000; Clark and Lawson, 2002; Rotejanaprasert, 2014). One variant that has been assumed commonly is to change the link between the cluster term and the background risk. For example, there is some justification to assume that areas of maps could be little affected by clustering if far from a parent location. In these areas, the background rate ($\lambda_0(s_i|\psi_0)$) should remain. The multiplicative link, assumed in $\lambda_1(s_i|\psi_1)$, may be improved by assuming an additive–multiplicative link, as well as the introduction of linkage parameters (a, b):

$$\lambda_1(s_i|\psi_1) = \exp(\rho_0).\left\{a + b\sum_{j=1}^{K} h(s_i - c_j; \tau)\right\}. \qquad (28.4)$$

28.5.1.2 Estimation issues

The full posterior distribution for this model is proportional to

$$[\{s_i\}|\psi_0, \mu_0, a, b, \tau, K, \mathbf{c}].P_1(\psi_0, \mu_0, a, b, \tau).P_2(K, \mathbf{c}),$$

where $P_*(.)$ denotes the joint prior distribution. Given the mixture form of the likelihood, it is not straightforward to develop a simple posterior sampling algorithm. Both the number of centers (K) and their locations ($\mathbf{c}$) are unknown. Hence, it is not possible to use straightforward Gibbs sampling. In addition, we do not require there to be assignment of data to centers, and so no allocation variables are used, unlike other mixture problems (Marin and Robert, 2007, chapter 6). One simple approximate approach is to evaluate a range of fixed-component models with different fixed K. The model with the highest marginal posterior probability is chosen (K^*), and the sampler is rerun with fixed K^*. This two-stage method is not efficient, however. Another alternative would be to use the fixed-dimension metropolized Carlin–Chib algorithm (Godsill, 2001; Kuo and Mallick, 1998). Instead, for variable-dimension problems such as this, resort can be made to reversible jump Markov chain Monte Carlo (McMC) (Green, 1995). A special form of this algorithm can be used called a spatial birth–death McMC. In this algorithm, centers, at different iterations, are added, deleted, or moved based on proposal and acceptance criteria. In this way, the location and the number of centers can be sampled jointly. Details of these algorithms are given in van Lieshout and Baddeley (2002) and Lawson (2001, appendix C).

How do these models perform, and are they realistic for disease cluster detection? In general, the simplistic assumptions made by point process models are really inadequate to describe clustering in spatial disease data. First, clustering tends to occur not as a common spatial field, but often as isolated areas. Even when multiple clusters occur, it is unlikely they will be of similar size or shape. In addition, clusters do not form regular shapes, and any spatial time cross section may show different stages of cluster development. For instance, there may be an infectious agent that differentially affects different areas at different times. A time-slice spatial map will then show different cluster forms in different areas. Another factor is that scales of clustering can appear on spatial maps. This is not considered in simple cluster PP models.

Given the possibility that unobserved confounders are present, (1) the resulting clustering will be unlikely to be summarized by a common model with global clustering components, and (2) cluster distribution functions with regular forms may not fit the irregular variation found.

The use of birth–death McMC with cluster models is not as limited as it may at first seem, however. The disadvantages of this form of modeling are (1) tuning of reversible jump McMC is often needed, and so the method is not readily available, (2) interpretation of output is more difficult due to the sampling over a joint distribution of centers and number of centers, and (3) possible rigidity of the model specification.

However, there are a number of advantages. First, it can easily be modified to include variants such as spatially dependent cluster variances (thereby allowing different sizes of clusters in different areas) and even a semiparametric definition of $h(s_i - c_j; \tau)$, which would allow some adaptation to local conditions. Second, it is also important to realize that by posterior sampling and averaging over posterior samples, it is possible to gain flexibility: even with a rigid symmetric form such as $\frac{\tau}{2\pi} \exp\left\{-\tau d_{ij}^2/2\right\}$, it is easy to see that the resulting cluster density map does not reflect a common global form, and indeed highlights the irregularity in the data. This of course is quite unlike the rigidity found in commonly used cluster testing methods such as SatScan (http://www.satscan.org/) (see Chapter 9). In addition, there is a wealth of information provided from a posterior sampler that can even include additional clustering information. For instance, the posterior marginal distribution of number of centers can yield information about multiple scales of clustering (even when these are not included in the model specification). Finally, it is also possible to increase the flexibility of the model by introduction of extra noise in the cluster sum. For example, the introduction of a random effect parameter for each of the centers, $\sum_{j=1}^{K} \exp(\psi_j).h(s_i - c_j; \tau)$ with $\psi_j \sim N(0, \tau_\psi)$, can lead to improved estimation of the overall intensity of the process. Another option that could be exploited that allows the sampling of mixtures more nonparametrically is the use of Dirichlet process prior distributions for mixtures

(Ishwaran and James, 2001, 2002; Gelfand et al., 2005; Kim et al., 2006; Duan et al., 2007; Kottas et al., 2008). Chapters 20 and 21 deal with such processes in greater depth.

28.5.1.3 Data-dependent models

Another possible approach to modeling is to consider models that do not assume a hidden process of centers, but model the data interdependence directly. Such data-dependent models have various forms, depending on assumptions.

Partition models and regression trees. Partition models attempt to divide up the space of the point process into segments or partitions. Each partition has a parameter or parameters associated with it. The partitions are usually disjoint and provide complete coverage of the study domain (T). An example of disjoint partition (or tiling) is the Dirichlet tesselation, which is constructed around each point of the process. Each tile consists of allocations closer to the associated point than to any other. Figure 28.9 shows such a tesselation of the Lancashire larynx cancer data set. It is clear that small tiles are associated with aggregations of cases. The formal statistical properties of such a tesselation are known (Barndorff-Nielsen et al., 1999; Illian et al., 2008) for most processes (such as the marginal distribution of tile areas).

However, in partition modeling, the tesselation is used in a different manner. Byers and Raftery (2002) describe an approach where a Dirichlet tesselation is used to group events together. Hence, a tiling consisting of K tiles with areas a_k is superimposed on the points,

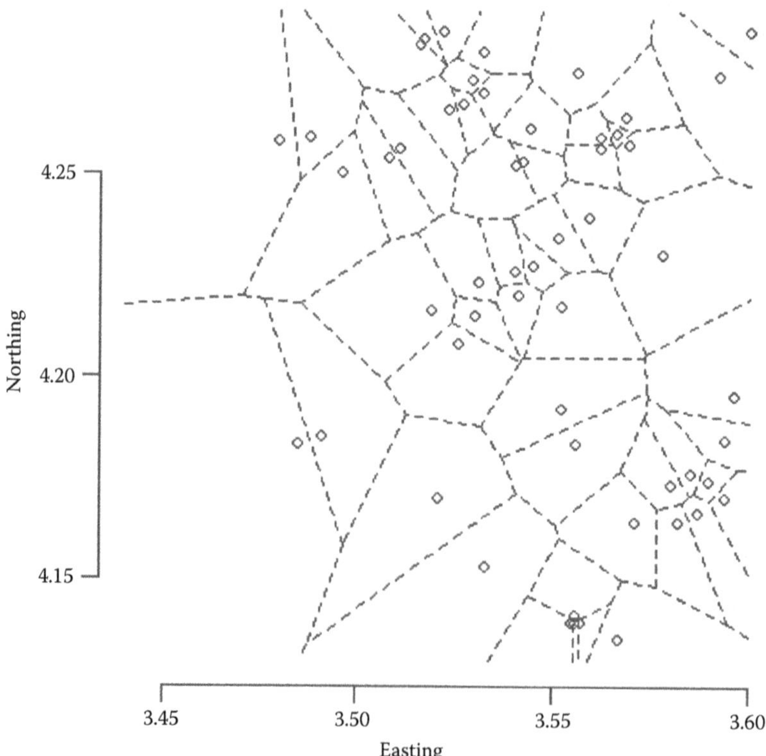

FIGURE 28.9
Lancashire larynx cancer data: Dirichlet tesselation produced with four external dummy points using the DELDIR package on R.

and the number of events within a tile (n_k) are recorded. The first-order intensity of the process is discretized to be constant within tiles (λ_k). The tile centers are defined to be $\{c_k\}$. Based on this definition, a posterior distribution can be defined, where

$$
\begin{aligned}
L(\mathbf{n}|\mathbf{l}) &\propto \prod_{k=1}^{K} \lambda_k^{n_k} \exp\{-\lambda_k a_k\} \\
K &\sim Poiss(\nu) \\
\{\lambda_k\}, k &= 1, \ldots, K|K \ iid \ Ga(a, b) \\
\{c_k\}, k &= 1, \ldots, K|K \ iid \ U(T).
\end{aligned}
$$

In this definition, the centers and areas are not given any stochastic dependency, whereas the areas are really dependent on the center locations. In addition, the number of centers is not fixed in general. This leads to a posterior distribution, within the general case, which does not have fixed dimension, but assuming ν, a, b fixed, is proportional to

$$
\frac{\nu^K}{K!} \prod_{k=1}^{K} \lambda_k^{(n_k+a-1)} \exp\{-\lambda_k(a_k + b)\}.
$$

In general, a reversible jump McMC algorithm or metropolized Carlin–Chib algorithm must be used to sample from this posterior distribution unless K is fixed. The focus of this work was the estimation of λ_k. Of course, in general, λ_k will vary over iterations of a converged posterior sample and will not be allocated to the same areas. Hence, any summarization of the output would have to overlay the realizations of λ_k for a predefined grid mesh of sites (possibly the data points), at which the average intensity would be estimated. Hence, a smoothly varying estimate of the intensity would result. In addition to simple intensity estimation, the authors also include a binary inclusion variable (d_k), which has a Bernoulli prior, and categorizes the tile as being in a high-intensity area ($d_k = 1$) or not. This allows a form of crude intensity segmentation (between areas of high and low intensity). In that sense, the method provides a clustering algorithm, albeit where only two states of intensity delineate the clusters. Mixing over the posterior allows for gradation of risk in the converged posterior sample (see, e.g., Byers and Raftery, 2002, figure 6.3b).

Note that in their application, Byers and Raftery (2002) have no background (population) effect, which would be needed in an epidemiological example. In application to disease cases, it may be possible to estimate a background effect using a control disease and to use this as a plug-in estimate (i.e., replace λ_k by $\widehat{\lambda}_{0k}\lambda_k$ in the likelihood, where $\widehat{\lambda}_{0k}$ is a background rate estimate in the kth tile). Alternatively, if counts of the control disease are available within tiles, then it would be possible to construct a joint model for both counts. Hegarty and Barry (2008) have also introduced a variant where product partitions are used to model risk. For a recent extension to case-control data, see Costain (2009).

Local likelihood. An alternative view considers the use of a grouping variable that relates to a sampling window. The sampling window is a subset of the study window. For example, a sampling window (lasso) is defined to be controlled by a parameter (δ). This parameter controls the size of the window. Usually (but not necessarily), the window is circular so that δ is a radius. First, consider cases of disease collected within a window of size δ and denote these as n_δ. Second, denote cases of a control disease as e_δ within the lasso. Now assume that the case disease and control disease are observed at a set of locations and denote these as $\{x_i\}$, $i = 1, \ldots, n$, and $\{x_i\}$, $i = n+1, \ldots, n+m$. The joint set of $\{x_i\}$ can be described jointly by a Bernoulli distribution with case probability $p(x_i) = \lambda(x_i)/(\lambda_0(x_i) + \lambda(x_i))$, conditional on $\lambda_0(x_i), \lambda(x_i)$ and their parameters. Further, assume that within the lasso there is a risk parameter θ_{δ_i} and that $\lambda(x_i) = \rho\theta_{\delta_i}\ \forall x_i \in \delta_i$. Now assume that within the lasso, the probability of a case or control is constant. In that case, we can write down a local likelihood of the form

$$\prod_{i=1}^{n+m}\left[\frac{\rho\theta_{\delta_i}}{1+\rho\theta_{\delta_i}}\right]^{n_{\delta_i}}\left[\frac{1}{1+\rho\theta_{\delta_i}}\right]^{e_{\delta_i}}.$$

Note that the lasso depends on δ_i, defined at the ith location, and different assumptions about these can be made. Attention focuses on the estimation of θ_{δ_i}, rather than δ_i, which can yield information about clustering behavior. Based on this local likelihood (Kauermann and Opsomer, 2003), it is possible to consider a posterior distribution with suitable prior distribution for parameters. For example, the δ can have a correlated prior distribution (either a fully specified Gaussian covariance model or a CAR model). Alternatively, it has been found that assuming an exchangeable gamma prior appears to work reasonably well. In addition, the dependence of θ_{δ_i} on δ_i across a range of δ_i's should be weak a priori, and so we assume a uniform distribution. Assuming a CAR specification, the prior distributions are then

$$\begin{aligned}[\theta_{\delta_i}|\delta_i] &\sim U(a,b)\ \forall i\\ \rho &\sim IGa(3,0.01)\\ [\delta_i|\boldsymbol{\delta}_{-i};\tau] &\sim N(\bar{\delta}_{\Delta_i},\tau/n_{\Delta_i})\\ \tau &\sim IGa(3,0.01),\end{aligned} \tag{28.5}$$

where Δ_i is a neighborhood of the ith point, $\bar{\delta}_{\Delta_i}$ is the mean of the neighborhood δs and n_{Δ_i} is the number of neighbors, $N(,)$ is an (improper) Gaussian distribution, τ is a variance parameter, and ρ is a rate with reasonably vague inverse gamma (IGa) distributions. An advantage of this approach is that a fixed-dimension posterior distribution can be specified, albeit with a local likelihood. Figure 28.10 displays a smoothed version of the posterior average value of the exceedence value of θ_{δ_i}: the function shown is $1-\widehat{\Pr}(\theta_{\delta_i}>1)$ for the converged sampler with the prior specification shown (28.5). The case-only map is shown with only the convex hull of the case distribution contoured. Further details of this model are given in Lawson (2006a). Note that it is clear that the southern area in the vicinity of $(3.55\times 10\text{e–}4, 4.15\times 10\text{e–}4)$ demonstrates a very high exceedence probability (<0.01).

The main advantages of this approach are the simplicity of programming (compared to birth–death McMC methods) and the ease with which interpretation of output can be made. The disadvantage is that in one sense, the methods do not directly model clustering, and may be regarded as smoothing methods, although their cluster detection performance in simulations is impressive (see, e.g., Hossain and Lawson, 2006).

28.5.2 Count data

28.5.2.1 Hidden process and object models

As in the situation where case events are analyzed, it is possible to specify object models for count data. Essentially, these assume a hidden process exists and this must be estimated. The hidden process could take a variety of forms. In the earliest examples, these forms followed those for case events where a hidden process of cluster centers was posited. The aggregation effects of accumulating case events into small areas (census tracts, zip codes, etc.) lead to integrals of the first-order case intensity in the expectation of the count. Hence, denoting the count of disease in the ith area (within area a_i) as n_i,

$$E(n_i)=\int_{a_i}\lambda(u)du,$$

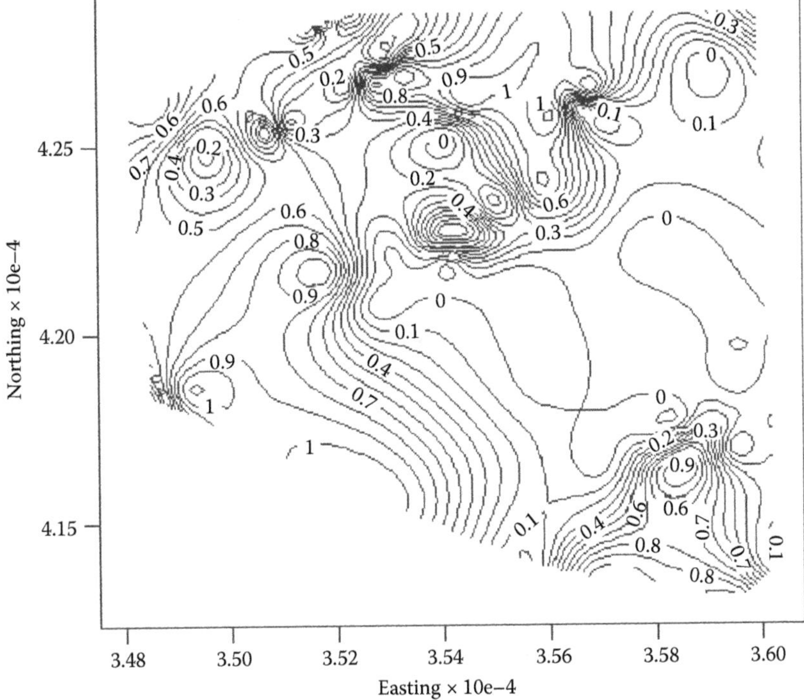

FIGURE 28.10
Lancashire larynx cancer data: exceedence probability map. Shown is the smoothed posterior average value, from a converged sampler, of $1 - \widehat{\Pr}(\theta_{\delta_i} > 1)$ for the case data only. The smoothing was done using the MBA (R) package.

and an "exact" hidden process model (HPM) would be defined as

$$n_i \sim Pois\left(\int_{a_i} \lambda(u)du\right) \tag{28.6}$$

$$\lambda(u) = \lambda_0(u|\psi_0)\lambda_1(u|\psi_1)$$

$$\lambda_1(u|\psi_1) = \sum_{j=1}^{K} \exp(\phi_j).h(u - c_j; \tau_h)$$

$$\phi_j \sim N(0, \tau_\psi).$$

Here, $h(u - c_j; \tau_h)$ is a cluster distribution function as before and $\{c_j\}$ is a set of hidden cluster centers, and the background population function $\lambda_0(u|\psi_0)$ must be estimated as before (Lawson and Clark, 1999). Often, a simplified (approximate) version of (28.6) is assumed where the expectation is simply a function of an expected rate (e_i) and relative risk parameter (θ_i):

$$n_i \sim Pois(e_i\theta_i),$$

$$\log\theta_i = \alpha_0 + \alpha_1 \sum_{j=1}^{K} \phi_j h(C_i - c_j; \tau_h),$$

where $\{C_i\}$ is a set of centroids of the small areas, α_0 is an intercept, and α_1 is a linking parameter. The $\{\phi_j\}$ play the same role as before as random effects, and the $\{c_j\}$ are hidden centers to be estimated. The expected rates are usually calculated based on an external standard population. An example of a variant of this model is given in Lawson (2006b, chapter 6), where $\theta_i = \exp(\phi_j).[1+\sum_{j=1}^{K} h(C_i - c_j; \tau_h)]$, with $\phi_j \sim N(0, \tau_\psi)$. Another variant of this model was specified by Gangnon (2006), where the log relative risk in a small area is assumed to be specified by $\log\theta_i = \alpha_0 + \Gamma_i + e_i$, where

$$\Gamma_i = \sum_{j=1}^{K} \phi_j h(C_i - c_j; \tau_h), \tag{28.7}$$

and ϕ_j is now a relative risk component associated with the jth cluster and $h(.)$ is a cluster membership function assumed to be uniform on a disc. The intercept term is α_0, while the e_i term is a random effect with zero-mean Gaussian prior distribution. The function $h(.)$ associates small areas to centers. The cluster centers are assumed to be centroids of the small areas (to avoid empty clusters that could arise due to the membership function definition). The main difference between these two variants is the inclusion of relative risk components in the latter model (28.7).

28.5.2.2 Data-dependent models

As in case event applications, it is possible for count data to be modeled without recourse to HP models. Two alternatives to such modeling are the use of partition models and splitting methods, and the local likelihood. A common feature of these methods is that spatial membership must be computed or present within an area. In either case, the methods require computation of distances within the spatial configuration of the data. Within McMC algorithms, these distances have to be recomputed as parameters are sampled.

Partition models and regression trees. One of the first examples of applying partition modeling to small-area count data was proposed by Knorr-Held and Rasser (2000). In their modeling approach, at the first level of the hierarchy, the small-area count y_i is assumed to be conditionally independent Poisson with expectation $e_i h_j$, where e_i is the usual standardized expected rate. The relative risk for a given area is chosen from a discrete set of risk levels (which are called clusters by the authors). These clusters are a set of contiguous regions $\{C_j\}$, $j = 1, \ldots, n$, which have associated constant risk $\{h_j\}$. The relative risk assigned to the ith area is just h_j if $i \in C_j$. Hence, $y_i \sim Pois(e_i h_j)$. The number of clusters (k) is treated as unknown: $k \in \{1, \ldots, n\}$. Hence, this method seeks a nonoverlapping partition of the map into areas of like risk. In this definition of clustering, clusters are areas of *constant* risk, and of course, this is a different definition of clustering from the more usual definitions using elevated risk. The essential difference is that the clusters are discrete partitions in this case. However, as in other cases, posterior averaging can lead to a more continuous relative risk estimate. Computational issues related to sampling partitions led the authors to consider reversible jump McMC. In the examples cited, the number of partitions were quite large (40–45) in the posterior realizations for male oral cavity cancer in 544 districts in Germany.

Partition models have since been extended (Denison and Holmes, 2001; Denison et al. 2002a; Ferreira et al., 2002). In application to count data in small areas, the usual Poisson assumption is made at the first level: $y_i \sim Pois(e_i \varepsilon(C_i))$. The partition is defined to be $\varepsilon(C_i) = \mu_{r(i)}$, the relative risk level associated with a region of a tesselation that the centroid C_i of the small area lies in. In this approach, a discrete tesselation is used as a partition. The prior distribution of the levels is assumed to be independent gamma: $[\mu_j|\gamma_1, \gamma_2] \sim Ga(\gamma_1, \gamma_2), j = 1, \ldots, k$. A prior model is also assumed for the location of tile centers:

$$p(\mathbf{c}) = \frac{1}{K}\frac{1}{\{Area(T)\}^k},$$

where K is the maximum number of centers or tiles and k is the current number. The marginal likelihood of the data given the centers is

$$\text{constant} + \sum_{j=1}^{k}\{\log\Gamma(\gamma_1 + n_j\overline{y}_j) - (\gamma_1 + n_j\overline{y}_j)\log(\gamma_2 + n_j\overline{N}_j)\},$$

where n_j is the number of points in the jth region, $\overline{y}_j$ is the mean of the observations in that tile, and $\overline{N}_j$ is the mean expected rate in the jth tile. An inclusion rule must be specified for the small areas being included within any given tile. The usual rule is to include if the centroid falls within the tile. Unlike the Knorr-Held–Rasser model, this allows there to be different partition shapes. Computation for the tesselation model is based on a birth–death McMC algorithm where the centers are added, deleted, or moved sequentially. An extension to these partition models has recently been proposed by Wakefield and Kim (2013), which assumes spatial contiguity for cluster membership. Anderson et al. (2014), on the other hand, postprocess the cluster membership and allow variation within groups. See also Charras-Garrido et al. (2012) for an expectation-maximization (EM) classifier and maximum aposteriori (MAP) estimation approaches.

Of course, there are considerable edge effect issues with partition models, and these apply to all partition models that use tesselations (rather than groupings): at the external study region, boundary tesselations will inevitably be distorted due to censoring adjacent regions *outside* the study window. Grouping algorithms (such as proposed by Knorr-Held and Rasser [2000]) could also be affected, as no information outside the boundaries has been reported, and this could distort the edge region allocations. These seem to be largely ignored in the literature on partition modeling.

Extensions to tree models could be imagined (Denison et al. 2002b). For example, all the clustering partition models proposed assume a flat level of partitioning. However, we could conceive of a form of splitting where a tree-based hierarchical cluster formation is conceived. It could be possible to construct a multilevel partition by allocation of higher levels of partition based on residuals from lower levels. This could be conceived as a multivariate partition function with L levels and K_l partitions at the lth level. There could be ordering constraints on the different levels. This could provide an approach to multiscale feature identification in the data.

Connections between partition models and mixture models for relative risk are also evident. For example, the model of Green and Richardson (2002) assumes that there are a small number of risk components $\{\theta_j, j = 1, 2, \ldots, k\}$ and the small areas are allocated one of these levels via the allocation variables $\{z_i, i = 1, \ldots, n\}$ (see Chapter 20 for more detail). Hence, $y_i \sim Pois(e_i\theta_{z_i})$, where θ_{z_i} is the allocated risk level for the ith area. Unlike other approaches, the allocation variables are given a spatially structured Potts prior distribution. This allows grouping at a higher level in the hierarchy. In contrast, Fernandez and Green (2002) describe a mixture model where the count is assumed to be governed by a weighted sum of Poisson distributions and the weights have spatial structure.

One general question concerning these methods is whether they really can be considered clustering methods. While all these methods could yield a variety of posterior information concerning risk, and some part of that information could be employed to detect clusters, they do not directly address the detection of clusters; rather, they seek to find underlying discrete risk levels that characterize the map. Often, the authors regard their methods as competitors with disease map smoothing models such as the convolution model of

Besag et al. (1991) (see, e.g., discussions in Knorr-Held and Rasser 2000; Ferreira et al. 2002; Fernandez and Green 2002), and they have been compared in simulations as such (Best et al., 2005). Essentially, mixture models are closely identified with latent structure models, and these are discussed more fully in Chapters 19 through 21.

Local likelihood. As in the case event situation, local likelihood models could also be applied. For small-area counts, within a lasso, centered at the centroid of the area, of dimension δ_i, with a Poisson assumption, the probability of y_{δ_i} counts with expected rate e_{δ_i} is

$$f(y_{\delta_i}|e_{\delta_i}, \theta_{\delta_i}; \delta_i) = [e_{\delta_i}\theta_{\delta_i}]^{y_{\delta_i}} \exp(-e_{\delta_i}\theta_{\delta_i})/y_{\delta_i}!.$$

This can be employed within a local likelihood as

$$L(\boldsymbol{\theta}|\mathbf{y}) = \prod_{i=1}^{m} f(y_{\delta_i}|e_{\delta_i}, \theta_{\delta_i}, \delta_i).$$

Note that in this case, the factorial term ($y_{\delta_i}!$) must be included in the likelihood as it varies with δ_i. In this case, the data and expected rate are accumulated within the lasso. A rule must be assumed for the inclusion of small areas. Centroids falling within the lasso are a common assumption. Then $y_{\delta_i} = \sum_{i \in \delta_i} y_i$ and $e_{\delta_i} = \sum_{i \in \delta_i} e_i$. In this definition, the counts and expected rates within neighboring lassos can overlap, and hence they can be correlated. Usually, the focus is on the estimation of θ_{δ_i} rather than δ_i. The linkage between the risk and the lasso is specified via

$$\log \theta_{\delta_i} = \delta_i + \varepsilon_i,$$

where $\varepsilon_i \sim N(0, \tau_\epsilon)$, an unstructured component, and δ_i is assumed to have a spatially structured prior distribution. In examples, a CAR prior distribution has been assumed for this purpose, that is,

$$[\delta_i|\delta_{-i}; \beta_\delta] \sim N(\overline{\delta}_i, \beta_\delta/d_i),$$

where $d_i = \sum_{j}^{m} I(j \in \delta_i)$ and $\overline{\delta}_i = \sum_{j \in \delta_i} \delta_j/d_i$. This allows the correlation inherent in the local likelihood to be modeled at a higher level of the hierarchy. Covariates can also be added to this formulation if required (see Hossain and Lawson, 2005, for discussion). Figure 28.11 displays an example of the application of the local likelihood model to a simulated realization where clusters of risk are introduced into the district geographies of the former East Germany. The counts are generated from a Poisson model with mean $e_i.\theta_i^{true}$, where the θ_i^{true} are the true risks. The left-hand panel depicts the true risks and the right-hand panel depicts the estimated risks under the local likelihood model with CAR prior distribution for the lasso parameters and inverse gamma distributions for the precision parameters (τ). In full simulation comparisons based on operating characteristic curves, the LL model performs well in recovering true risk, as well as some features of clusters (see, e.g., Hossain and Lawson, 2006). An application of local likelihood to clustering of intellectual delay can be found in Aelion et al. (2009).

28.5.3 Markov connected component field (MCCF) models

A number of alternative approaches to count data clustering have been proposed and should be mentioned as they relate to methods previously discussed. First, a cluster modeling

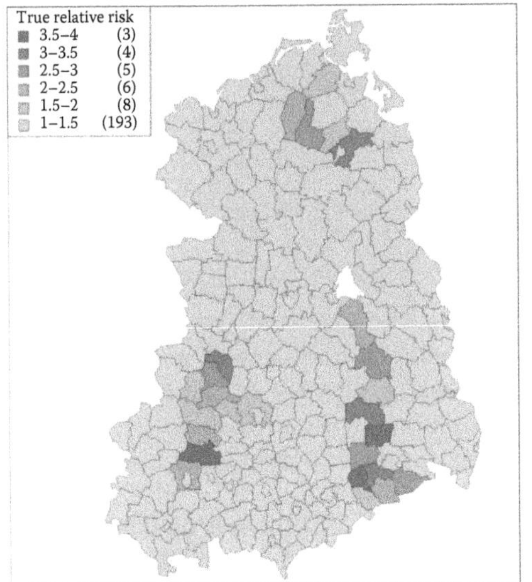

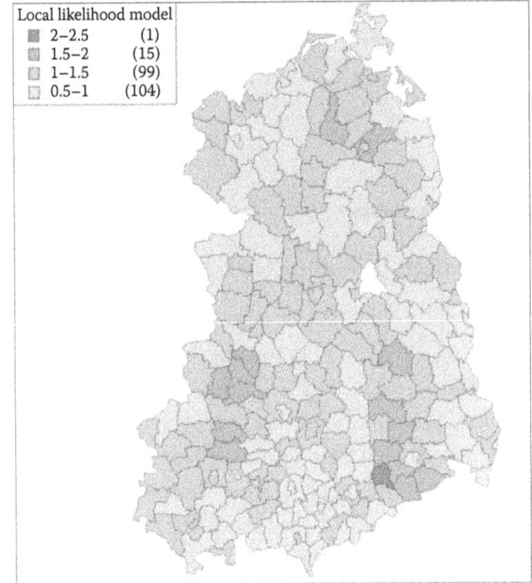

FIGURE 28.11
East German lip cancer mortality (1980–1989) data. Simulated true risk map (left) and local likelihood posterior mean relative risk estimates (right).

approach based on Markov connected component fields has been proposed (Gangnon and Clayton, 2000). This approach also seeks to find a grouping of small areas into clusters. The cluster set k clusters has associated a risk level λ_j, $j = 1, \ldots, k$. The membership of the clusters is defined for each of the N small areas ($i = 1, \ldots, N$) via a membership (allocation) variable: $\mathbf{z} = (z_1, \ldots, z_N)$. Hence, the counts in the N areas have distribution conditional on the cluster assignment ($\mathbf{z}$):

$$y_i \sim Pois(e_i \lambda_{z_i}),$$

where λ_{z_i} is the disease risk for the cluster associated with the ith area. Clusters are assumed to be aggregations of the basic small areas, and so the allocation assigns different disease risks to each area depending on prior distributional constraints. This setup was later used by Green and Richardson (2002), but they assumed a spatially structured prior distribution for the $\mathbf{z}$.

Here, the disease risks are assumed to have a $Ga(\alpha, \beta)$ prior distribution given the $\mathbf{z}$, where $Ga(\alpha, \beta)$ has mean α/β. However, the $\mathbf{z}$ are assumed to have a Markov connected component field prior governing their form (Móller and Waagepetersen, 1998). This construction allows there to be various cluster form specifications included in the prior structure. This construction allows a potential function to describe the clustering:

$$p(\mathbf{z}) \propto \exp\left\{ -\sum_{j=1}^{k} S_j \right\},$$

where the S_j is a score function for the jth cluster. The authors use properties of the clusters such as circularity (shape) and size to yield a composite score: $S_j = \alpha + S_{j1}(size) + S_{j2}(shape)$. Once specified, the authors use randomized model search criteria to evaluate different models. In a similar development, but different application, Móller and Skare (2001) further apply these priors to imaging problems.

Note that this MCCF approach has many advantages over purely spatially structured distributions for component assignment. It also allows the models to address specific features of the clusters that are not considered by other partitioning approaches.

28.6 Edge Detection and Wombling

A closely related area of concern to disease clustering is the idea that discontinuities in maps are the focus. These could be the edge of some uniform risk area or the edge of a variable risk cluster of some kind. In cluster modeling, usually there is some criteria defining the cluster and its boundary. In edge detection, the boundary is usually defined by a jump or discontinuity in risk, and so this in a sense takes the approach whereby the focus is the differences in risk, rather than finding the area of elevated risk.

Wombling is closely related to the edge effect problem in that the focus is on boundaries of regions. There is a growing interest in the ability to locate boundaries within a spatial domain. This has been developing within geography for some time (see, e.g., Oden et al., 1993, among others). These boundaries often have some natural context, such as catchments of health providers (Ma et al., 2010) or areas that vote predominantly for one party in elections (O'Loughlin, 2002). Basically, these methods are edge detection methods, and many such methods are commonly found in the image processing literature (see, e.g., the Canny operator, Mars–Hilldreth, and Gaussian derivative kernels) (Bankman, 2007). Often, they are intended for object recognition and use derivative-based methods for locating discontinuities.

The discovery of discontinuities in disease maps may be of interest in detection of natural boundaries for health provision (recent Bayesian examples are Ma et al. [2010] and Lu and Carlin [2005]). One simple approach to this problem can be examined within a Bayesian hierarchical model. Assume for small areas with observed tract counts y_i that $y_i \sim Poiss(e_i\theta_i)$.

Within a posterior sampling algorithm, θ_i^g is the estimate of θ_i at the gth iteration. It is possible to estimate the posterior expected value of the absolute difference between relative risks, $\Delta\theta_{ij}$, by simply computing

$$\Delta\theta_{ij} = \sum_{g\in d} |\theta_i^g - \theta_j^g|/n(d),$$

where d denotes the converged sample set and $n(d)$ is the number in that set. Hence, this estimator is available for all region boundaries. The parameters $\{\Delta\theta_{ij}\}$ really detect discontinuities between regions, and there is no information in this approach that informs the edge detection beyond the average difference between *modeled* estimates of adjoining area risks. Hence, cluster or object information that could "tie" the areas together is not used. Presmoothing the estimates of risk (e.g., via a convolution model) may lead to reduction in discontinuity, of course, and so the choice of model for the θ could be crucially important. More sophisticated approaches have also been proposed (see, e.g., Ma et al., 2010).

It is not clear how discontinuities are to be modeled for applications in disease incidence studies. Underlying most models of risk is the assumption of continuity of risk over space. The convolution smoothing models assume this, as do most other risk models. The fixed mixture model of Lawson and Clark (2002) and also the mixture models of Green and Richardson (2002) do attempt to honor discontinuities within a general framework of smooth risk and allow the smoothness to be selected differentially over space. It is a matter of debate whether pure discontinuity models have advantages in incidence studies.

References

Abellan, J., S. Richardson, and N. Best. (2008). Use of space-time models to investigate the stability of patterns of disease. *Environmental Health Perspectives* 116, 1111–1118. doi: 10.1289/ehp.10814.

Aelion, M., H. Davis, S. McDermott, and A. B. Lawson. (2009). Metal concentrations in rural topsoil in South Carolina: Potential for human health impact. *Science of the Total Environment* 407, 2216–2223.

Anderson, C., D. Lee, and N. Dean. (2014). Identifying clusters in Bayesian disease mapping. *Biostatistics* 15, 457–469.

Baddeley, A., R. Turner, J. Møller, and M. Hazelton. (2005). Residual analysis for spatial point processes. *Journal of the Royal Statistical Society B* 67, 617–666 [with discussion].

Bankman, I. (ed.). (2007). *Handbook of Medical Image Processing and Analysis* (2nd ed., vol. 1). New York: Academic Press.

Barndorff-Nielsen, O. E., W. S. Kendall, and M. N. M. van Lieshout. (1999). *Stochastic Geometry: Likelihood and Computation.* Boca Raton, FL: Chapman & Hall.

Besag, J., J. York and A. Mollié. (1991). Bayesian image restoration with two applications in spatial statistics. *Annals of the Institute of Statistical Mathematics* 43, 1–59.

Best, N., S. Richardson, and A. Thomson. (2005). A comparison of Bayesian spatial models for disease mapping. *Statistical Methods in Medical Research* 14, 35–59.

Brix, A. and P. Diggle. (2001). Spatio-temporal prediction for log-Gaussian Cox processes. *Journal of the Royal Statistical Society B* 63, 823–841.

Byers, S. and A. Raftery. (2002). Bayesian estimation and segmentation of spatial point processes using Voronoi tilings. In A. B. Lawson and D. Denison (eds.), *Spatial Cluster Modelling.* New York: CRC Press, chapter 6.

Charras-Garrido, M., D. Abrial, and J. de Goer. (2012). Classification method for disease risk mapping based on discrete hidden Markov random fields. *Biostatistics* 13, 241–255.

Clark, A. B., and A. B. Lawson. (2002). Spatio-temporal cluster modelling of small area health data. In A. B. Lawson and D. Denison (eds.), *Spatial Cluster Modelling,* pp. 235–258. New York: CRC Press.

Clayton, D. G. and L. Bernardinelli. (1992). Bayesian methods for mapping disease risk. In P. Elliott, J. Cuzick, D. English, and R. Stern (eds.), *Geographical and Environmental Epidemiology: Methods for Small-Area Studies.* Oxford: Oxford University Press.

Costain, D. (2009). Bayesian partitioning for modeling and mapping spatial case-control data. *Biometrics* 65, 1123–1132.

Cressie, N. C. and A. Lawson. (2000). Hierarchical probability models and Bayesian analysis of mine locations. *Advances in Applied Probability* 32(2), 315–330.

Denison, D., N. Adams, C. Holmes, and D. Hand. (2002a). Bayesian partition modelling. *Computational Statistics and Data Analysis* 38, 475–485.

Denison, D. and C. Holmes. (2001). Bayesian partitioning for estimating disease risk. *Biometrics* 57, 143–149.

Denison, D., C. Holmes, B. Mallick, and A. Smith. (2002b). *Bayesian Methods for Nonlinear Classification and Regression.* New York: Wiley.

Diggle, P. J. (1985). A kernel method for smoothing point process data. *Journal of the Royal Statistical Society C* 34, 138–147.

Diggle, P. J. (1990). A point process modelling approach to raised incidence of a rare phenomenon in the vicinity of a prespecified point. *Journal of the Royal Statistical Society A* 153, 349–362.

Diggle, P. J. (2003). *Statistical Analysis of Spatial Point Patterns* (2nd ed.). London: Arnold.

Duan, J., M. Guidani, and A. Gelfand. (2007). Generalized spatial Dirichlet process models. *Biometrika* 94, 809–825.

Fernandez, C. and P. Green. (2002). Modelling spatially correlated data via mixtures: A Bayesian approach. *Journal of the Royal Statistical Society B* 64, 805–826.

Ferreira, J., D. Denison, and C. Holmes. (2002). Partition modelling. In A. B. Lawson and D. Denison (eds.), *Spatial Cluster Modelling*, pp. 125–145. New York: CRC Press.

Gangnon, R. (2006). Impact of prior choice on local Bayes factors for cluster detection. *Statistics in Medicine* 25, 883–895.

Gangnon, R. and M. Clayton. (2000). Bayesian detection and modeling of spatial disease clustering. *Biometrics* 56, 922–935.

Gelfand, A., A. Kottas, and S. MacEachern. (2005). Bayesian nonparametric spatial modeling with Dirichlet process mixing. *Journal of the American Statistical Association* 100, 1021–1035.

Godsill, S. (2001). On the relationship between Markov chain Monte Carlo methods for model uncertainty. *Journal of Computational and Graphical Statistics* 10, 230–248.

Green, P. J. (1995). Reversible jump MCMC computation and Bayesian model determination. *Biometrika* 82, 711–732.

Green, P. J. and S. Richardson. (2002). Hidden Markov models and disease mapping. *Journal of the American Statistical Association* 97, 1055–1070.

Hegarty, A. and D. Barry. (2008). Bayesian disease mapping using product partition models. *Statistics in Medicine* 27, 3868–3893.

Hossain, M. and A. B. Lawson. (2005). Local likelihood disease clustering: Development and evaluation. *Environmental and Ecological Statistics* 12, 259–273.

Hossain, M. and A. B. Lawson. (2006). Cluster detection diagnostics for small area health data: With reference to evaluation of local likelihood models. *Statistics in Medicine* 25, 771–786.

Illian, J., A. Penttinen, H. Stoyan, and D. Stoyan. (2008). *Statistical Analysis and Modelling of Spatial Point Patterns.* New York: Wiley.

Ishwaran, H. and L. James. (2001). Gibbs sampling methods for stick-breaking priors. *Journal of the American Statistical Association* 96, 161–173.

Ishwaran, H. and L. James. (2002). Approximate Dirichlet process computing in finite normal mixtures: Smoothing and prior information. *Journal of Computational and Graphical Statistics* 11, 1–26.

Kauermann, G. and J. D. Opsomer. (2003). Local likelihood estimation in generalized additive models. *Scandinavian Journal of Statistics* 30, 317–337.

Kim, S., M. Tadesse, and M. Vanucci. (2006). Variable selection in clustering via Dirichlet process mixture models. *Biometrika* 93, 877–893.

Knorr-Held, L. and G. Rasser. (2000). Bayesian detection of clusters and discontinuities in disease maps. *Biometrics* 56, 13–21.

Kottas, A., J. Duan, and A. Gelfand. (2008). Modelling disease incidence data with spatial and spatio-temporal Dirichlet process mixtures. *Biometrical Journal* 50(2), 29–42.

Kuo, L. and B. Mallick. (1998). Variable selection for regression models. *Sankhya* 60, 65–81.

Lawson, A. B. (1993). A deviance residual for heterogeneous spatial Poisson processes. *Biometrics* 49, 889–897.

Lawson, A. B. (1995). Markov chain Monte Carlo methods for putative pollution source problems in environmental epidemiology. *Statistics in Medicine* 14, 2473–2486.

Lawson, A. B. (2000). Cluster modelling of disease incidence via RJMCMC methods: A comparative evaluation. *Statistics in Medicine* 19, 2361–2376.

Lawson, A. B. (2001). *Statistical Methods in Spatial Epidemiology*. New York: Wiley.

Lawson, A. B. (2006a). Disease cluster detection: A critique and a Bayesian proposal. *Statistics in Medicine* 25, 897–916.

Lawson, A. B. (2006b). *Statistical Methods in Spatial Epidemiology* (2nd ed.). New York: Wiley.

Lawson, A. B. (2013). *Bayesian Disease Mapping: Hierarchical Modeling in Spatial Epidemiology* (2nd ed.). New York: CRC Press.

Lawson, A. B., W. J. Browne, and C. L. Vidal-Rodiero. (2003). *Disease Mapping with WinBUGS and MLwiN*. New York: Wiley.

Lawson, A. B. and A. Clark. (1999). Markov chain Monte Carlo methods for clustering in case event and count data in spatial epidemiology. In M. E. Halloran and D. Berry (eds.), *Statistics and Epidemiology: Environment and Clinical Trials*, pp. 193–218. New York: Springer Verlag.

Lawson, A. B. and A. Clark. (2002). Spatial mixture relative risk models applied to disease mapping. *Statistics in Medicine* 21, 359–370.

Li, Y., P. Brown, D. Gesink, and H. Rue. (2012). Log Gaussian Cox processes and spatially aggregated disease incidence data. *Statistical Methods in Medical Research* 21, 479–507.

Lu, H. and B. Carlin. (2005). Bayesian areal wombling for geographical boundary analysis. *Geographical Analysis* 37, 265–285.

Ma, H., B. Carlin, and S. Banerjee. (2010). Hierarchical and joint site edge methods for Medicare hospice service region boundary analysis. *Biometrics* 66, 355–364.

Marin, J.-M. and C. Robert. (2007). *Bayesian Core: A Practical Approach to Computational Bayesian Statistics.* New York: Springer.

Móller, J. and O. Skare. (2001). Coloured Voronoi tesselations for Bayesian image analysis and reservoir modelling. *Statistical Modelling* 1, 213–232.

Møller, J., A. Syversveen, and R. P. Waagepetersen. (1998). Log Gaussian Cox processes. *Scandinavian Journal of Statistics* 25, 451–482.

Móller, J. and R. Waagepetersen. (1998). Markov connected component fields. *Advances in Applied Probability* 30, 1–35.

Móller, J. and R. Waagepetersen. (2004). *Statistical Inference and Simulation for Spatial Point Processes.* New York: CRC Press.

Oden, N., R. Sokal, M. Fortin, and H. Goebl. (1993). Categorical wombling: Detecting regions of significant change in spatially located categorical variables. *Geographical Analysis* 25(4), 315–336.

O'Loughlin, J. (2002). The electoral geography of Weimar Germany: Exploratory spatial data analyses (ESDA) of Protestant support for the Nazi Party. *Political Analysis* 10(3), 217–243.

Richardson, S., A. Thomson, N. Best, and P. Elliott. (2004). Interpreting posterior relative risk estimates in disease mapping studies. *Environmental Health Perspectives* 112, 1016–1025.

Rotejanaprasert, C. (2014). Evaluation of cluster recovery for small area relative risk models. *Statistical Methods in Medical Research* 23, 531–551.

van Lieshout, M. and A. Baddeley. (2002). Extrapolating and interpolating spatial patterns. In A. B. Lawson and D. Denison (eds.), *Spatial Cluster Modelling*, pp. 61–86. New York: CRC Press.

Wakefield, J. and A. Kim. (2013). A Bayesian model for cluster detection. *Biostatistics* 14, 752–765.

Xu, R. and D. C. Wunsch. (2009). *Clustering.* New York: Wiley.

29

Spatial Data Analysis for Health Services Research

Brian Neelon
Department of Public Health Sciences
Medical University of South Carolina
Charleston, South Carolina

CONTENTS

29.1 Introduction

Spatial data analysis is playing an increasingly important role in health services research, enabling policy makers to better understand the health needs of communities, to identify barriers to health care, and to allocate health resources in a cost-effective manner. To date, spatial models have been used to explore geographic variation in hospital referral rates (Congdon, 2006; Congdon and Best, 2000), emergency department visits (Congdon, 2006; Neelon et al., 2014, 2013), and access to primary care services (Mobley et al., 2006). Spatial methods have also been developed to examine geographic and temporal trends in health services costs and patient medical expenditures (Moscone et al., 2007; Moscone and Knapp, 2005; Neelon et al., 2015a, 2015b). These efforts continue to inform policy decisions and guide community-based efforts to improve access to health care.

The past two decades have seen a proliferation of spatial methods specifically designed to address problems arising in health services studies. One of the earliest approaches was the spatial gravity model proposed by Bailey and Gattrell (1995) to describe consumer "flow" between competing services. In the context of health services, gravity models are often used to evaluate patient referral rates from "origin" facilities, such as primary care clinics, to "destination" facilities (e.g., specialty clinics), while accounting for such factors as clinic capacity, patient demand, and distance between facilities. Congdon and Best (2000), for instance, proposed a gravity model to examine spatial variation in hospital emergency referrals. The approach was later refined by Congdon (2001) to accommodate health system changes—for example, the opening or closing of new facilities. More recent extensions include two-step floating catchment models for measuring spatial accessibility (Guagliardo, 2004; McGrail and Humphreys, 2009) and spatial optimization models that account for both availability of and accessibility to health services (Nobles et al., 2014).

Spatial analysis of zero-inflated count data is another active area in health services research. Zero inflation occurs when the number of observed zeros exceeds what is expected under a standard count model, such as a Poisson or negative binomial. In the presence of zero inflation, special two-part mixture models, such as hurdle models (Mullahy, 1986), are often needed to provide adequate fit to the data. Building on this approach, Neelon et al. (2013) proposed a spatial Poisson hurdle model to examine geographic variation in emergency department visits. They introduced spatial random effects via a bivariate conditionally autoregressive (CAR) prior (Besag et al., 1991; Mardia, 1988) that induces dependence between the model components and provides spatial smoothing across neighboring spatial regions. Neelon et al. (2014) later extended the method to the spatiotemporal setting by developing negative binomial and generalized Poisson (Consul and Jain, 1973) space–time hurdle models that are especially well suited for overdispersed count data.

Spatial methods have also been applied to the analysis of health service expenditures. Moscone and Knapp (2005) used spatial-lag models to examine spatial associations in mental health expenditures. Moscone et al. (2007) later developed a spatially lagged, seemingly unrelated regression (SUR) model to explore spatiotemporal patterns in mental health expenditures. More recently, Neelon et al. (2015b) introduced a set of Bayesian two-part models for the spatial analysis of semicontinuous medical expenditure data. Semicontinuous expenditure data are characterized by two distinct features: (1) a point mass at zero representing patients with no expenditures (or, equivalently, no health service use) and (2) a right-skewed distribution with positive support that describes the expenditure level conditional on health service use. Neelon et al. (2015b) discuss various models for spatially referenced semicontinuous data, including two-part lognormal models, log skew-elliptical models, and Dirichlet process mixture models. Taking a somewhat different approach, Neelon et al. (2015a) recently proposed a spatiotemporal quantile regression model for the analysis of emergency department expenditures. Their model yields distinct spatial patterns across time for each quantile of the response distribution, which is appealing in the spatial analysis of expenditures, as there is often little spatiotemporal variation in mean expenditures but more pronounced variation in the extremes.

As these recent developments suggest, spatial data analysis is occupying an increasingly important place in health services research. By discerning spatial patterns in health service outcomes, policy makers and health officials can design effective public health interventions, promote educational outreach programs, and develop cost-efficient outpatient health services. The remainder of this chapter presents three case studies that highlight the emerging role of spatial data analysis in health services studies.

29.2 Spatial Gravity Model for Hospital Referrals

For the first case study, we consider the spatial gravity model proposed by Congdon and Best (2000) to explore hospital referral rates among primary care practices in Greater London. Drawing on an analogy to Newton's laws of gravitation, gravity models were originally developed to describe market dynamics in terms of the attractiveness and proximity of various consumer goods. In the health setting, gravity models are often used to evaluate provider referral patterns while accounting for such features as patient demand, availability of services, and distance to services.

Congdon and Best begin by constructing primary care *catchment zones*, which are spatial cross-classifications of primary care practices and electoral wards. They define the catchment zone for practice j to include ward i if at least 5% of the practice patients reside in ward i. Similarly, each catchment zone is affiliated with a set of destination hospitals that account for at least 5% of patient referrals. The aim is to model the number of admissions to hospital k from catchment zone ij as a function of hospital capacity (here taken to be bed size), patient demand (catchment population size), and distance between zone ij and hospital k. Using these catchment definitions, Congdon and Best posit the following basic gravity model:

$$\begin{aligned} Y_{ijk} &\sim \text{Poisson}(\mu_{ijk}) \\ \mu_{ijk} &= AN_{ij}C_k d_{ijk}^{-\lambda}, \end{aligned} \tag{29.1}$$

where C_k denotes the capacity of the kth hospital, N_{ij} is the patient population size for zone ij, A is a balancing constant, and d_{ijk} is the distance from the centroid of zone ij to hospital k. The parameter $\lambda > 0$ controls the rate of decay in the number of referrals as the distance between hospital and catchment zone increases. They later extend the model to allow for separate capacity and demand effects:

$$\log(\mu_{ijk}) = \alpha + g_1 \log(C_k) + g_2 \log(N_{ij}) - \lambda \log(d_{ijk}), \tag{29.2}$$

where α is a constant and g_1 and g_2 are "calibration" parameters that quantify the effect of capacity and demand, respectively. The final gravity model includes additional hospital, practice, and ward effects, as well as ward × hospital interactions:

$$\log(\mu_{ijk}) = \alpha + g_1 \log(C_k) + g_2 \log(N_{ij}) - \lambda \log(d_{ijk}) + \omega_i + \pi_j + \eta_k + \psi_{ik}, \tag{29.3}$$

where ω_i is a random effect for the ith electoral ward, π_j is a practice-level random effect, η_k is a hospital effect, and ψ_{ik} is an interaction effect that accounts for unobserved factors such as travel time from ward i to hospital k.

The authors adopt a Bayesian modeling approach and assign prior distributions to the model parameters (for a general overview of Bayesian inference, see Chapter 7 of the handbook). For the ward effects, they first consider an intrinsic conditionally autoregressive (CAR) (Besag et al., 1991) prior, but after preliminary analysis, settle on a spatially independent normal prior. They assign normal priors to π_j and ψ_{ik}, and minimally informative Guassian priors to the fixed parameters α, g_1, g_2, $\log(\lambda)$, and η_k (with $\eta_1 \equiv 0$ for identifiability). Diffuse inverse-gamma priors are used to model the random effect variance components, σ^2_ω, σ^2_π, and σ^2_ψ.

Congdon and Best fit four separate gravity models to the data: (1) a model with capacity and demand terms, C_k and N_{ij}, as well as ward effects ω_i; (2) a model with additional practice effects, π_j; (3) a model that includes additional ward × hospital interaction terms, ψ_{ik}; and (4) a final full model that incorporates the distance term $\log(d_{ijk})$, as well as two ward-level covariates. The authors compared the four models using a posterior predictive loss criterion (Gelfand and Ghosh, 1998) and found that models 3 and 4 provided the best fit. Their results highlight the importance of adjusting for hospital and distance effects in such models. In particular, without these effects, wards located near hospitals appeared to have higher admission rates than less proximal wards. Once these factors were accounted for, this spatial artifact disappeared and more homogenous region-level patterns emerged.

Early formulations of the gravity model did not allow for health system changes or for nonspatial constraints on access to care. More recently, gravity models have been extended to accommodate system dynamics (Congdon, 2001) and to address additional barriers to health care, such as time constraints, financial burdens, and accessibility of transportation (Nobles et al., 2014).

29.3 Spatiotemporal Hurdle Models for Zero-Inflated Count Data

For the second case study, we review a recent analysis by Neelon et al. (2014) involving zero-inflated count data. Zero inflation arises commonly in health services research and occurs when the number of zeros is greater than what is expected under a standard count model. Failure to account for zero inflation can lead to poor model fit and invalid inferences.

As a motivating application, Neelon et al. examine spatial and temporal trends in emergency department (ED) visits among census block groups in Durham County, North Carolina, from 2007 to 2011. Figure 29.1 presents a partial histogram of the annual number of ED visits up to 10 visits. More than 70% of the counts were zero. The number of nonzero visits ranged from 1 to 91, with 5% of the patients having greater than six visits annually. Vuong's procedure (Vuong, 1989) indicated significant zero inflation relative to both the ordinary Poisson and negative binomial distributions. For more information on Vuong's test and its application to hurdle models, see Winkelmann (2008, p. 184).

Neelon et al. propose a spatiotemporal hurdle model for the analysis of the ED visits (for a detailed discussion of spatiotemporal models, see Chapter 19 in this volume). The hurdle

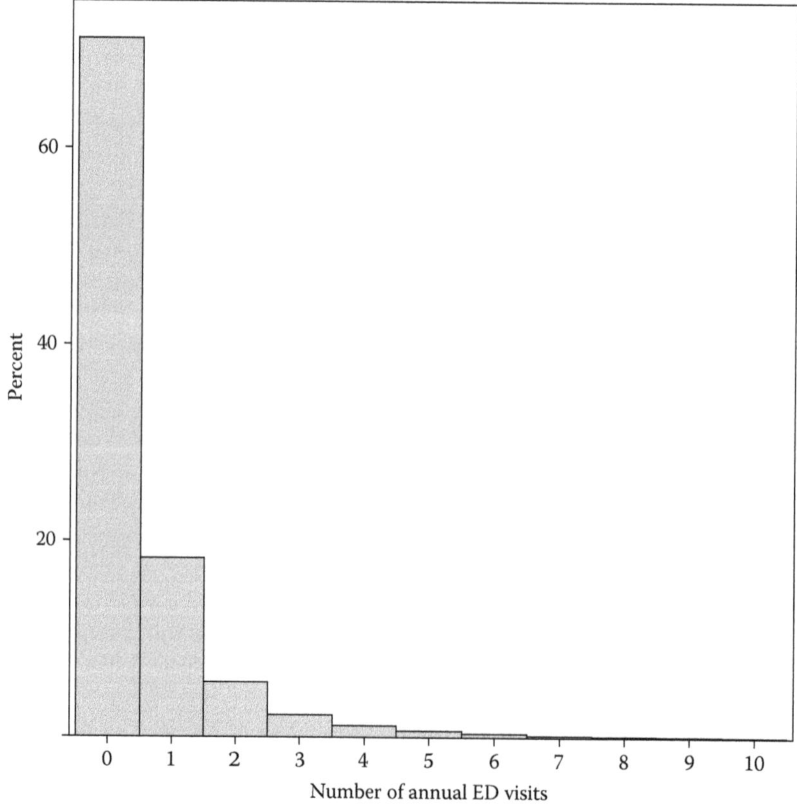

FIGURE 29.1
Partial histogram of annual ED visits from the Neelon et al. (2014) ED study.

model (Mullahy, 1986) is a two-part mixture consisting of a point mass at zero, followed by a zero-truncated count distribution for the positive observations. The spatiotemporal hurdle model introduced by Neelon et al. has a hierarchical structure incorporating both patient- and region-level covariates. Its basic form is expressed as

$$\begin{aligned} \Pr(Y_{ijk} = y_{ijk} | \boldsymbol{\phi}_i, \boldsymbol{\nu}_j, \boldsymbol{\delta}_{ij}) &= (1 - \pi_{ijk})\mathbb{I}(y_{ijk} = 0) + \frac{\pi_{ijk} p(y_{ijk}; \mu_{ijk}, \alpha)}{1 - p(0; \mu_{ijk}, \alpha)} \mathbb{I}(y_{ijk} > 0) \\ \text{logit}(\pi_{ijk}) &= \boldsymbol{x}_{ijk}^T \boldsymbol{\beta}_1 + \phi_{1i} + \nu_{1j} + \delta_{1ij} \\ \ln(\mu_{ijk}) &= \boldsymbol{x}_{ijk}^T \boldsymbol{\beta}_2 + \phi_{2i} + \nu_{2j} + \delta_{2ij}, \\ & i = 1, \ldots, m;\ j = 1, \ldots, J;\ k = 1, \ldots, n_{ij}, \end{aligned} \tag{29.4}$$

where Y_{ijk} denotes the number of ED visits for the kth patient in block group i and year j, $\mathbb{I}(\cdot)$ is the indicator function, $\pi_{ijk} = \Pr(Y_{ijk} > 0)$ is the probability of a positive (nonzero) count, $p\,(y_{ijk}; \mu_{ijk}, \alpha)$ is an untruncated, or *base*, count distribution with mean μ_{ijk} and dispersion parameter α, and $p(0; \mu_{ijk}, \alpha)$ is the base distribution evaluated at zero. The linear predictors include individual- and region-level covariates ($\boldsymbol{x}_{ijk}$), fixed-effect regression coefficients ($\boldsymbol{\beta}_1$ and $\boldsymbol{\beta}_2$), and spatiotemporal effects, $\boldsymbol{\phi}_i = (\phi_{1i}, \phi_{2i})'$, $\boldsymbol{\nu}_j = (\nu_{1j}, \nu_{2j})'$, and $\boldsymbol{\delta}_{ij} = (\delta_{1ij}, \delta_{2ij})'$. Here, $\boldsymbol{\phi}_i$ denotes the spatial "main effect" for the ith block group, $\boldsymbol{\nu}_j$ represents the temporal main effect for year j, and $\boldsymbol{\delta}_{ij}$ is a space–time interaction. Thus, the spatiotemporal effects are partitioned into three parts: a purely spatial component, a purely temporal component, and a residual interaction term.

Neelon et al. compare three versions of the hurdle model: a Poisson hurdle model, a negative binomial hurdle model, and a generalized Poisson (Consul and Jain, 1973) hurdle model. The negative binomial base distribution accounts for overdispersion, whereas the generalized Poisson allows for both over- and underdispersion and is ideally suited for highly skewed count data (Joe and Zhu, 2005).

The authors adopt a Bayesian inferential approach, and the models can be easily fit in standard Bayesian software such as `WinBUGS` (Lunn et al., 2000). For the spatial main effects, they assume a bivariate intrinsic CAR (BICAR) prior (Mardia, 1988):

$$\boldsymbol{\phi}_i | \boldsymbol{\phi}_{(-i)}, \boldsymbol{\Sigma}_\phi \sim \mathrm{N}_2 \left(\frac{1}{m_i} \sum_{l \in \partial_i} \boldsymbol{\phi}_l, \frac{1}{m_i} \boldsymbol{\Sigma}_\phi \right), \tag{29.5}$$

where ∂_i is the set of neighbors sharing a geographic border with block group i, m_i is the number of neighbors, and $\boldsymbol{\Sigma}_\phi$ denotes the 2×2 covariance of $\boldsymbol{\phi}_i$ conditional on the remaining spatial random effects, $\boldsymbol{\phi}_{(-i)}$ (further details on multivariate priors for spatial data can be found in Chapter 22 of this volume). To ensure an identifiable model, sum-to-zero constraints are applied to the random effects. The bivariate CAR prior is appealing because it incorporates dependence between the logistic and count components of the hurdle model, which has been shown in previous work to improve inferences (Neelon et al., 2013). For example, block groups with higher rates of ED use may tend to have higher mean counts among users. It is therefore desirable to build this association into the model.

The temporal effects are treated as fixed and assigned diffuse normal priors. The space–time interactions are assigned first-order dynamic BICAR priors in which $\boldsymbol{\delta}_{ij} = \rho \boldsymbol{\delta}_{i(j-1)} + \boldsymbol{\psi}_{ij}$, where $\boldsymbol{\psi}_{ij}$ is BICAR and ρ is a temporal smoothing parameter. For identifiability, $\boldsymbol{\delta}_{i1}$ is set to $\mathbf{0}$ for all i. To complete the prior specification, the authors assign diffuse normal priors to the remaining fixed effects, inverse-Wishart priors to covariance matrices, and a uniform$(0, 1)$ prior to ρ. For the negative binomial model, α is assigned a gamma prior, and for the generalized Poisson model, it is assumed to be uniform over a permissable range of values.

To choose between the Poisson, negative binomial, and generalized Poisson hurdle models, the authors suggest several model selection measures, including the deviance information criterion (DIC) (Spiegelhalter et al., 2002), defined as $\bar{D}+p_D$, where $\bar{D}$ is a measure of model fit and p_D is a penalty for model complexity. Additionally, they propose a series of posterior predictive assessments, whereby the observed data are compared to data replicated from the posterior predictive distribution (Gelman et al., 1996). If the model fits well, the replicated data should resemble the observed data. To quantify the degree of similarity, one typically chooses a "discrepancy statistic" that captures some important aspect of the data. For the ED analysis, Neelon et al. adopt three discrepancy measures: the sample proportion of zeros and the sample mean and variance among the positive observations. For each measure, they compute the posterior predictive mean and 95% credible interval. A 95% credible interval that includes the observed sample value suggests adequate model fit.

Table 29.1 presents the model comparison results from the analysis. The negative binomial and generalized Poisson hurdle models substantially outperformed the Poisson hurdle model with respect to DIC. Overall, the generalized Poisson hurdle model had the lowest DIC value. In terms of posterior predictions, all models accurately reproduced the observed proportion of zeros and the conditional mean among the positive counts. While none of the models fully captured the observed conditional variance, the Poisson hurdle model showed the poorest fit, further supporting the need to model overdispersion in the counts.

Figure 29.2 presents yearly maps of the predicted spatiotemporal effects, η_1 and η_2, from the generalized Poisson hurdle model, where $\eta_{1ij}=\nu_{1j}+\phi_{1i}+\delta_{1ij}$ and $\eta_{2ij}=\nu_{2j}+\phi_{2i}+\delta_{2ij}$. The majority of ED visits occurred among the central block groups, and the fewest among the block groups in the southwest corner of the county. Over time, the most significant change took place between 2007 and 2008, with several central block groups transitioning into the highest ED category (represented by the darkest shade), and the southwestern block groups transitioning into the lowest category (represented by lightest shade). The spatial pattern stabilized following 2008, with only minor fluctuations in select block groups.

The shift in spatial pattern from 2007 to 2008 is even more evident in Figure 29.3, which presents caterpillar plots of the spatiotemporal effects. Positive (negative) effects indicate above- (below-) average ED activity, adjusting for other factors. Overall, there was a notable change in the magnitude of the effects following 2007. However, this shift was not uniform across block groups, with several block groups showing large increases and others showing little or no change. This highlights the importance of modeling the space–time interactions in order to capture localized temporal trends. The authors go on to suggest that the changes observed following 2008 may be due in part to the economic downturn occurring during that time.

TABLE 29.1
Model comparison results for the Poisson, negative binomial, and generalized Poisson hurdle models in the Neelon et al. (2014) ED study

Base distribution	DIC	pD	Posterior predictive checks		
			Pr($Y=0$)	**E($Y\|Y>0$)**	**V($Y\|Y>0$)**
Poisson	232,158	566	0.709 (0.705, 0.714)*	1.93 (1.90, 1.95)	1.48 (1.43, 1.55)
Negative binomial	211,198	367	0.709 (0.705, 0.713)	1.93 (1.89, 1.96)	3.76 (3.42, 4.20)
Generalized Poisson	211,035	367	0.709 (0.706, 0.713)	1.93 (1.89, 1.97)	4.66 (4.10, 5.46)
			Observed: 0.709	Observed: 1.94	Observed: 5.89

*Posterior median and 95% credible interval.

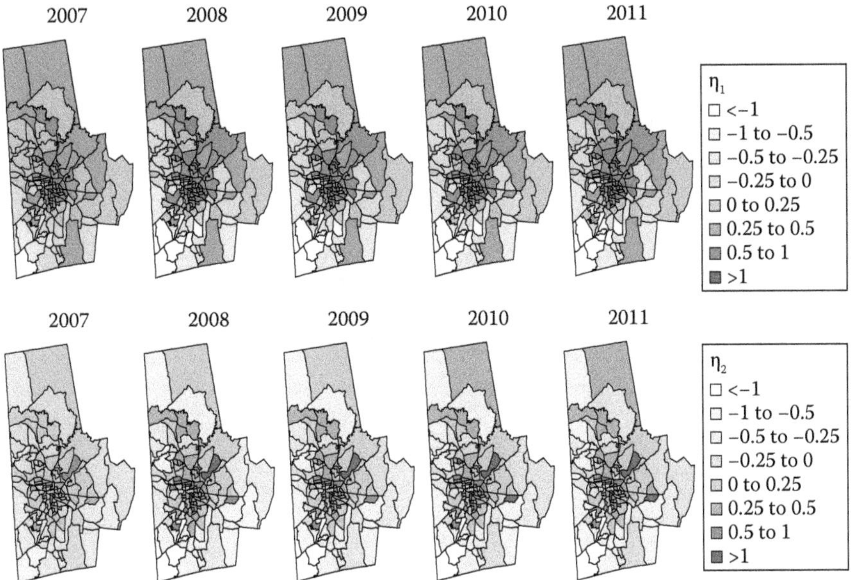

FIGURE 29.2
Spatiotemporal random effects for the binary (η_1) and count (η_2) components of the generalized Poisson hurdle model in the Neelon et al. (2014) ED study.

29.4 Spatiotemporal Quantile Regression of Health Service Expenditures

As a final case study, we consider an application of spatiotemporal quantile regression (Koenker and Bassett, 1978) to health service expenditures (for more on quantile regression, see Chapter 13 of the handbook). Unlike standard regression approaches that model the conditional mean of the response, quantile regression provides robust inference at conditional quantiles of the distribution, for example, at the median or 90th quantile. This is especially relevant in the context of health expenditures because there is often little spatial or temporal variation in average expenditures but noticeable variation in the extremes, that is, among high-cost patients who place the greatest burden on the health system.

Here, we review a recent analysis by Neelon et al. (2015a) of ED-related expenditures in Durham County, North Carolina. Neelon et al. propose the following quantile regression model:

$$\begin{aligned} y_{ijk} &= \eta_{ijk,\tau} + \epsilon_{ijk,\tau} \\ \eta_{ijk,\tau} &= \boldsymbol{x}_{ijk}^T \boldsymbol{\beta}_\tau + \phi_{i,\tau} + \psi_{j,\tau} + \theta_{ij,\tau}, \quad i = 1, \ldots, m;\ j = 1, \ldots, J;\ k = 1, \ldots, n_{ij}, \end{aligned} \tag{29.6}$$

where y_{ijk} denotes the annual expenditure for the kth patient in census block group i and year j; τ $(0 < \tau < 1)$ denotes the quantile level; $\epsilon_{ijk,\tau}$ is an error term with τth quantile equal to zero; $\boldsymbol{x}_{ijk}$ is a vector of individual- and region-level predictors with corresponding quantile-specific regression parameters, $\boldsymbol{\beta}_\tau$; $\phi_{i,\tau}$ and $\psi_{j,\tau}$ are the quantile-specific spatial and temporal main effects of block group i and year j, respectively; and $\theta_{ij,\tau}$ denotes a corresponding quantile-specific space–time interaction. The authors adopt a Bayesian approach,

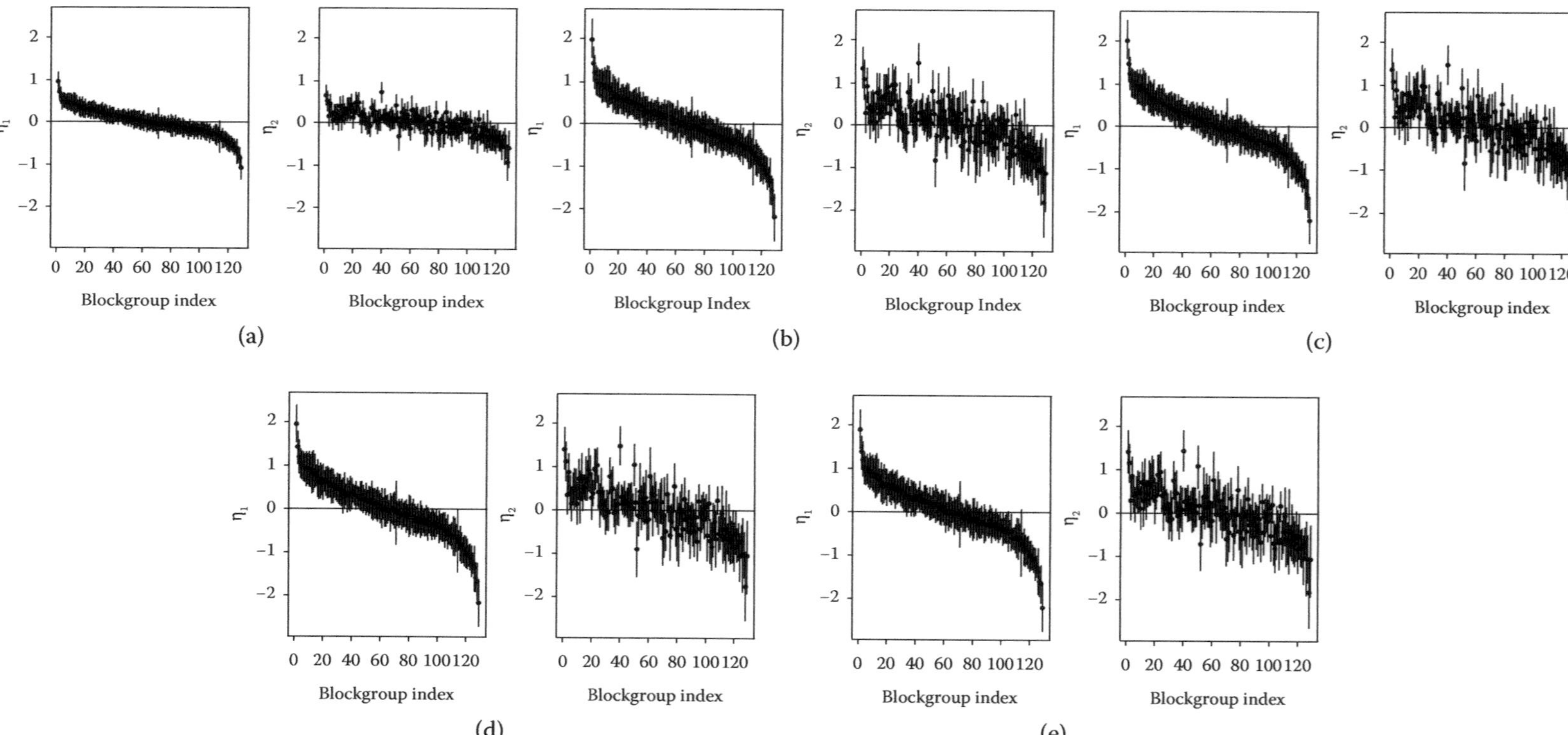

FIGURE 29.3
Spatiotemporal random effects for the binary ($\boldsymbol{\eta}_1$) and count ($\boldsymbol{\eta}_2$) components of the generalized Poisson hurdle model in the Neelon et al. (2014) ED study. Error bars denote 95% credible intervals.

and as a "working" likelihood, they assume an asymmetric Laplace distribution (ALD) for ϵ:

$$f(\epsilon) = \tau(1-\tau)\delta^2 \exp\left\{-\delta^2\epsilon\left[\tau - \mathbb{I}(\epsilon < 0)\right]\right\}, \tag{29.7}$$

where δ^2 and τ are rate and skewness parameters, respectively. The ALD has been shown to yield consistent posterior regression estimates in Bayesian quantile regression models (Sriram et al., 2013). However, posterior intervals can be biased for extreme quantiles (Neelon et al., 2015a; Waldmann et al., 2013). To address this issue, Sriram (2015) recently introduced a sandwich-likelihood correction for ALD-based posterior intervals. Additional approaches to quantile regression are discussed in Chapter 13 of the handbook.

To aid Bayesian computation, one can write ϵ as a linear combination of two random variables (Yue and Rue, 2011):

$$\epsilon = \frac{1-2\tau}{\tau(1-\tau)}W + \sqrt{\frac{2W}{\delta^2\tau(1-\tau)}}Z, \tag{29.8}$$

where W is an exponential variable with rate parameter δ^2 and Z is from a standard normal distribution. Marginally, ϵ maintains its ALD form. However, conditional on W, ϵ follows a normal distribution. Hence, posterior computation can be conveniently carried out using conventional Bayesian updates conditional on the exponential random variable W.

Prior specifications are similar to those presented in case study 2. Intrinsic CAR priors are assigned to spatial and temporal main effects, and a type IV spatiotemporal interaction Knorr-Held (2000) prior is used to model the space–time interaction terms. Prior specification is completed by assigning conjugate normal priors to fixed effects, a gamma prior to the ALD rate parameter, and inverse-gamma priors to the CAR variance parameters. Posterior computation is handled through straightforward Gibbs sampling, as all model parameters admit closed-form full conditional distributions.

Figures 29.4 and 29.5 display yearly maps of the predicted spatiotemporal random effects, represented by the expression $\phi_{i,\tau} + \psi_{j,\tau} + \theta_{ij,\tau}$ in Equation 29.6. Figure 29.4 presents maps for $\tau = 0.50$, and Figure 29.5 presents similar maps for $\tau = 0.90$. The upper panels

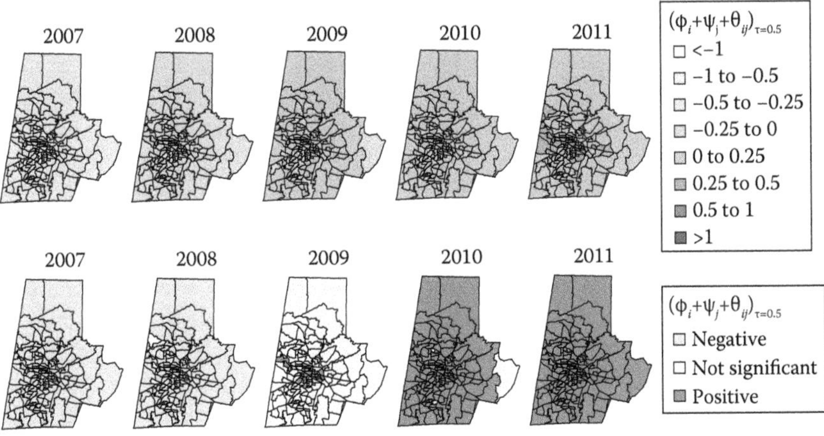

FIGURE 29.4
Predicted spatiotemporal random effects for the median model in the Neelon et al. (2015a) ED expenditure analysis. Upper panel: Predicted spatiotemporal effects ($1000s). Lower panel: Block groups with significant negative effects in light gray, significant positive effects in dark gray, and nonsignificant effects in white.

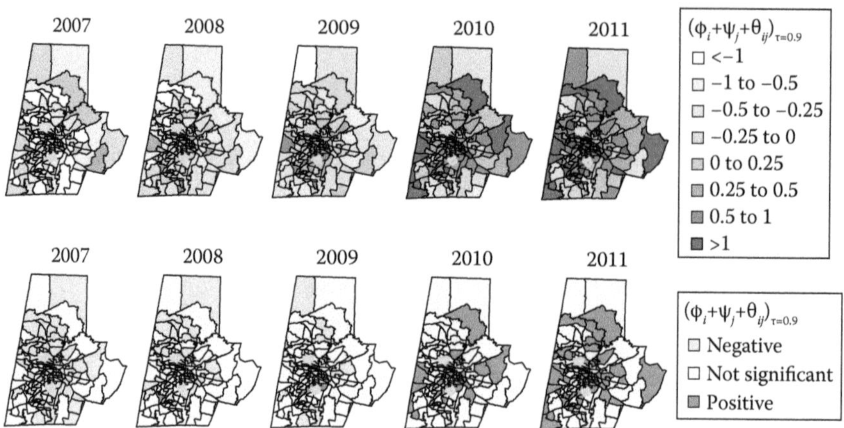

FIGURE 29.5
Predicted spatiotemporal random effects for the 90th quantile model in the Neelon et al. (2015a) ED expenditure analysis. Upper panel: Predicted spatiotemporal effects ($1000s). Lower panel: Block groups with significant negative effects in light gray, significant positive effects in dark gray, and nonsignificant effects in white.

show deviations (in $1000s) relative to an "average" block group with zero spatiotemporal effect. Thus, in Figure 29.4, the lightest shade represents block groups with median expenditures approximately $1000 lower than an average block group with zero spatiotemporal effect, and the darkest shade denotes block groups with approximately $1000 higher median expenditures than an average block group. In each figure, the lower panels display the posterior significance of the block group effects, as measured by whether the posterior 95% credible intervals (CrIs) overlap zero.

As Figure 29.4 indicates, there was a steady increase in median expenditures over time, ranging from $383 (95% CrI [289, 475]) below average in 2007 to $324 (95% CrI [240, 421]) above average in 2011. The temporal trend was fairly uniform across block groups, however, implying that there was little space–time interaction in median expenditures. The one notable exception occurred in 2009 among a small cluster of central block groups; these block groups appeared to increase median expenditures at a faster rate than surrounding areas.

There was considerably more spatiotemporal heterogeneity at $\tau = 0.90$ (Figure 29.5). As before, there was a general increase in expenditures over time, although this trend was not uniform across the county, with central and southwestern block groups showing larger increases in upper-tail expenditures than block groups in the southeast. This suggests that, over time, the expenditure distribution for the central and southwestern block groups grew more right-skewed than block groups in the southeast. As a concrete illustration, consider the two highlighted block groups in Figure 29.5. The more southern block group exhibited increasing 90th quantiles over time. In 2007, for example, this block group had 90th quantiles that were $1182 (95% CrI [677, 1620]) lower than one would expect for an average block group, after adjusting for patient- and census-level factors; in 2011, this differential increased to $1226 (95% CrI [388, 2308]) above average. Thus, this block group, along with its neighbors, might serve as an ideal candidate for local interventions to cap rising ED expenditures. In contrast, the northern block group had less extreme ED spending in all years, ranging from $882 (95% CrI [206, 1492]) below average in 2007 to $327 below average in 2011. This block group may therefore place less financial burden on the health system than the first block group.

29.5 Conclusion

These case studies highlight just a few of the recent applications of spatial statistics to health services research. As public attention turns toward rising health care costs and concerns over limited access to care, understanding spatial and temporal trends in health utilization and expenditures is becoming ever more important. As a result, spatial statistical methods continue to play a vital role in the design and implementation of community-based efforts to improve access while reducing costs. These efforts include establishing community health centers, developing medical education programs, and improving transportation services to and from primary care facilities. By identifying spatial patterns in health services outcomes, policy makers and health professionals can target resources in a cost-efficient manner to address the specific health needs of communities.

There are a number of potential areas for future work. For example, spatial latent class models could be developed to distinguish subpopulations of individuals with unique utilization or expenditure patterns. This could aid health officials in developing patient-centered interventions to promote health access. Additionally, joint modeling of health services utilization and chronic illness could provide insights into the spatial connections between availability of services and long-term patient health outcomes. And finally, spatial point process models (Diggle, 2013) could be used to study the regional distribution of health services and inform future allocation of resources.

Our hope is that this chapter has introduced readers—both statisticians and health services researchers alike—to a new set of analytic tools useful for their own research. Readers are encouraged to consult the references cited herein for further discussions of spatial methods and their ongoing application to health services research.

Supplementary Material

Color versions of Figures 29.2, 29.4, and 29.5 can be found at https://www.crcpress.com/9781482253016.

References

Bailey, T., and Gattrell, A. (1995). *Interactive Spatial Data Analysis.* Longman, London.

Besag, J., York, J., and Mollié, A. (1991). Bayesian image restoration, with two applications in spatial statistics. *Annals of the Institute of Statistical Mathematics*, 43(1):1–20.

Congdon, P. (2001). The development of gravity models for hospital patient flows under system change: A Bayesian modelling approach. *Health Care Management Science*, 4(4):289–304.

Congdon, P. (2006). Modelling multiple hospital outcomes: The impact of small area and primary care practice variation. *International Journal of Health Geographics*, 5:50.

Congdon, P., and Best, N. (2000). Small area variation in hospital admission rates: Bayesian adjustment for primary care and hospital factors. *Journal of the Royal Statistical Society: Series C*, 49(2):207–226.

Consul, P. C., and Jain, G. C. (1973). A generalization of the Poisson distribution. *Technometrics*, 15(4):791–799.

Diggle, J. P. (2013). *Statistical Analysis of Spatial Point Patterns*, 3rd ed. CRC Press, Boca Raton, FL.

Gelfand, A. E., and Ghosh, S. K. (1998). Model choice: A minimum posterior predictive loss approach. *Biometrika*, 85(1):1–11.

Gelman, A., Meng, X.-L., and Stern, H. (1996). Posterior predictive assessment of model fitness via realized discrepancies. *Statistica Sinica*, 6:733–807.

Guagliardo, M. (2004). Spatial accessibility of primary care: Concepts, methods and challenges. *International Journal of Health Geographics*, 3(1):3.

Joe, H., and Zhu, R. (2005). Generalized Poisson distribution: The property of mixture of Poisson and comparison with negative binomial distribution. *Biometrical Journal*, 47(2):219–229.

Knorr-Held, L. (2000). Bayesian modelling of inseparable space-time variation in disease risk. *Statistics in Medicine*, 19(17–18):2555–2567.

Koenker, R., and Bassett, G. (1978). Regression quantiles. *Econometrica*, 46(1):33–50.

Lunn, D. J., Thomas, A., Best, N., and Spiegelhalter, D. (2000). WinBUGS—a Bayesian modelling framework: Concepts, structure, and extensibility. *Statistics and Computing*, 10:325–337.

Mardia, K. (1988). Multi-dimensional multivariate Gaussian Markov random fields with application to image processing. *Journal of Multivariate Analysis*, 24:265–284.

McGrail, M. R., and Humphreys, J. S. (2009). The index of rural access: An innovative integrated approach for measuring primary care access. *BMC Health Services Research*, 9(1).

Mobley, L. R., Root, E., Anselin, L., Lozano-Gracia, N., and Koschinsky, J. (2006). Spatial analysis of elderly access to primary care services. *International Journal of Health Geographics*, 5(1).

Moscone, F., Knapp, M., and Tosetti, E. (2007). Mental health expenditure in England: A spatial panel approach. *Journal of Health Economics*, 26(4):842–864.

Moscone, L., and Knapp, M. (2005). Exploring the spatial pattern of mental health expenditure. *Journal of Mental Health Policy and Economics*, 8(4):205–217.

Mullahy, J. (1986). Specification and testing of some modified count data models. *Journal of Econometrics*, 33:341–365.

Neelon, B., Chang, H. H., Ling, Q., and Hastings, N. S. (2014). Spatiotemporal hurdle models for zero-inflated count data: Exploring trends in emergency department visits. *Statistical Methods in Medical Research*. DOI: 10.1177/0962280214527079.

Neelon, B., Ghosh, P., and Loebs, P. F. (2013). A spatial Poisson hurdle model for exploring geographic variation in emergency department visits. *Journal of the Royal Statistical Society: Series A*, 176:389–413.

Neelon, B., Li, F., Burgette, L. F., and Benjamin Neelon, S. E. (2015a). A spatiotemporal quantile regression model for emergency department expenditures. *Statistics in Medicine*, 34(17):2559–2575.

Neelon, B., Zhu, L., and Benjamin Neelon, S. E. (2015b). Bayesian two-part spatial models for semicontinuous data with application to emergency department expenditures. *Biostatistics*, 16(3):465–479.

Nobles, M., Serban, N., and Swann, J. (2014). Spatial accessibility of pediatric primary healthcare: Measurement and inference. *Annals of Applied Statistics*, 8(4):1922–1946.

Spiegelhalter, D., Best, N., Carlin, B. P., and Van Der Linde, A. (2002). Bayesian measures of model complexity and fit. *Journal of the Royal Statistical Society: Series B*, 64:583–639.

Sriram K. (2015). A sandwich likelihood correction for Bayesian quantile regression based on the misspecified asymmetric Laplace density. *Statistics & Probability Letters*, 107:18–26.

Sriram, K., Ramamoorthi, R., and Ghosh, P. (2013). Posterior consistency of Bayesian quantile regression based on the misspecified asymmetric Laplace density. *Bayesian Analysis*, 8(2):479–504.

Vuong, Q. H. (1989). Likelihood ratio tests for model selection and non-nested hypotheses. *Econometrica*, 57(2):307–333.

Waldmann, E., Kneib, T., Yue, Y. R., Lang, S., and Flexeder, C. (2013). Bayesian semiparametric additive quantile regression. *Statistical Modelling*, 13(3):223–252.

Winkelmann, R. (2008). *Econometric Analysis of Count Data*, 5th ed. Springer, Berlin.

Yue, Y. R., and Rue, H. (2011). Bayesian inference for additive mixed quantile regression models. *Computational Statistics and Data Analysis*, 55(1):84–96.

30

Spatial Health Survey Data

Christel Faes
Interuniversity Institute for Biostatistics and Statistical Bioinformatics
Hasselt University
Hasselt, Belgium

Yannick Vandendijck
Interuniversity Institute for Biostatistics and Statistical Bioinformatics
Hasselt University
Hasselt, Belgium

Andrew B. Lawson
Division of Biostatistics and Bioinformatics
Department of Public Health Sciences
Medical University of South Carolina
Charleston, South Carolina

CONTENTS

Public health is of major concern for society, public and private organizations, communities, and individuals. Often the question is raised whether there are disparities of illness and health problems across areas (see, e.g., Chapters 2, 3, 14, and 15). An increasing amount of information on individuals is collected in this respect. Bayesian methods in disease mapping based on census or population registry data are well developed and are used in a fairly standard manner (see, e.g., Lawson et al. [2013], Elliott et al. [2001], and Waller and Gotway [2004] for reviews of the methods, and Chapters 6, 7, and 20 in this book). Note that such population registries or census data record information pertaining to each member of the population of an area. Historically, focus was on the construction of cancer atlases and on mapping of other rare diseases based on registry data (see, e.g., Kemp et al. [1985] and Devesa et al. [1999]).

Since it is nearly always impossible to measure the outcome of interest in every individual of the population, one often relies on health surveys to record information from a representative sample of individuals from the population (Cochran, 1977). Such surveys are often characterized by a complex design, with stratification, clustering, and unequal sampling weights as common features. The simplest design is the stratified sampling design, where respondents are sampled randomly within different strata. Another complexity with surveys is that individuals can refuse to answer certain questions, leading to missing data. Since area-specific characteristics such as the total number or the prevalence of diseased cases per area are of interest to policy makers, the question is how to estimate these characteristics from survey data. This is well recognized in small-area estimation (Arora and Lahiri, 1997; Rao, 2003). The use of design-based inference plays a dominant role in sample surveys. This approach assumes that the observed health outcomes are fixed, but that the sampling indicators are random. A popular design-based estimator is the Horvitz–Tompson estimator (see Section 30.2), but it might fail in this context, because the number of samples per area could be very small or even zero, making the design-based area-specific direct estimation either unreliable or not feasible (Rao, 2011). In addition, special attention is needed to understand the geographical distribution of the health outcome in the population. Model-based approaches will therefore be of interest in the handling of spatially correlated health survey data. In contrast to the design-based approach, model-based approaches assume that the observed health outcomes are random and conditioned on the selected sample. The Fay–Herriot model is a commonly used model-based approach in small-area estimation (see Section 30.3.1).

Several methods to estimate the spatial pattern of a health outcome and that acknowledge and take into account the survey sampling design will be presented in this chapter. The focus is on hierarchical Bayesian modeling tools for the analysis of the spatial distribution of health outcomes, with the health outcome collected as a binary variable in a survey with a complex survey design. A first approach is to ignore the survey design features when modeling the survey outcome. This, however, can lead to biased inference. A second approach is to assume that the design is ignorable, in the sense that we can include all design variables that define the selection process as a covariate in the model, and then ignore the design. This might become problematic when many design variables are used or when design variables are not available. A third approach is to use a method that acknowledges and takes into account the survey sampling design. Several such methods will be discussed in this chapter.

One possibility is to use a direct estimator for each area based on the available data in each area and the sampling weight for each respondent. This is discussed in Section 30.2. Alternatively, one can use the direct estimators as a basis of a Bayesian model, taking into account the spatial variability of the direct estimators, as described in Section 30.3.1. In Section 30.3.2, a weighted likelihood method, also called pseudolikelihood, is used to deal with the complex design. These methods dealing with the spatial heterogeneity of survey data are described by Mercer et al. (2014) and Chen et al. (2014). Section 30.3.3 presents a weight-smoothing method. This method was proposed by Vandendijck et al. (2015). Methods are illustrated in the Health Interview Survey conducted in Belgium in Section 30.4.

30.1 Notation

Let Y_{ik} denote a binary outcome variable for individual k from area i ($k = 1, \ldots, N_i$, $i = 1, \ldots, I$), and N_i is the population size in area i. Interest is in the population count $T_i = \sum_{k=1}^{N_i} Y_{ik}$ in area i or in the population prevalence $P_i = T_i/N_i$ in area i. A sampling design is used to select the set of individuals, with s_i the set of sampled individuals in area i and n_i the sample size in area i. The set of binary survey variables for sampled individuals in area i is denoted as $\{y_{ik}; k \in s_i\}$. The total number of cases in the sample in area i is equal to $y_i = \sum_{k=1}^{n_i} y_{ik}$.

Individuals s_i are selected from the total population of individuals in area i according to some complex sampling design. Here, we will assume a probability sampling design, in the sense that each person in the sample represents, besides himself or herself, an entire slice of the population. For example, in a simple random sample in which $2\% = 2/100$ of a population is sampled, each person in the sample represents $100/2 = 50$ persons in the population. It is said that each person has a weight of 50. Depending on the sampling design, different individuals may have differing selection probabilities and, as a consequence, different weights. In most survey designs, samples are selected with unequal probability, leading to individual-specific sampling weights. Let w_{ik} be the design weight for individual k in area i, representing a certain number of population units attached to this respondent. A common choice is to take $w_{ik} = 1/\pi_{ik}$, where π_{ik} is the inclusion probability for individual k in area i, which depend on the sampling design. Note that, in general, the design weight can reflect a combination of the complex survey design and poststratification adjustments, and can also be adjusted for nonresponse in the survey.

30.2 Design-Based Estimator

A direct estimator for an area i uses the observed response values from the sampled individuals within this area only. This is in contrast with the indirect estimators, in that those methods "borrow strength" by also using values of the study variable from sampled individuals outside the area of interest.

The design weights play an important role in the construction of design-based direct estimators (Rao, 2003). A popular estimator of the total number of cases T_i in the population is given by

$$\widehat{T}_i = \sum_{k=1}^{n_i} w_{ik} y_{ik},$$

in which the design weights are used to adjust for the sampling design. Similarly, a commonly used estimator for the area-specific prevalence P_i is given as

$$\widehat{P}_i = \frac{1}{\widehat{N}_i} \sum_{k=1}^{n_i} w_{ik} y_{ik},$$

where $\widehat{N}_i = \sum_{k=1}^{n_i} w_{ik}$. These estimators are known as the Horvitz–Thompson (HT) estimators (Horvitz and Thompson, 1952; Cochran, 1977).

The HT estimator is a design-unbiased estimator and provides reliable inference in large samples without the need to make model assumptions. However, the method has the disadvantage that it may have unacceptable high variability if the sample size in the area is small and the estimators are potentially inefficient (Gelman, 2007; Rao, 2011). Note that most surveys are not designed to yield appropriate direct estimates for all the areas, as this would require large sample sizes for all the areas. Also, variance estimation requires knowledge of second-order inclusion probabilities, which can become cumbersome. In addition, in case some of the areas do not contain any sampled individuals, we cannot obtain an estimator for these areas, making the approach not feasible in this setting. In order to adequately estimate the total number of cases or prevalence in small geographical areas, the use of model-based indirect estimators, such as described in Section 30.3, is proposed.

30.3 Model-Based Estimators

Model-based methods assume a model for the sampled data and use the optimal predictor from this model as an estimate of the area characteristics of interest. Model-based methods have the advantage over design-based methods that they generally lead to smaller precision estimates (Pfeffermann, 2013). Small-area estimation methods can be used to reduce the variance of the estimators by borrowing strength across geographical areas, as will be discussed in the following sections. Attention is restricted to parametric Bayesian methods to model the prevalence of a certain health outcome measured in the survey. Ignoring the sampling design is often done in practice (see examples in Rao, 2003), but might lead to biased estimates when the design is informative, and will therefore be not considered here.

30.3.1 Fay–Herriot model and its extensions

A model that has been widely used in the literature of small-area estimation is the basic Fay–Herriot model. The model is composed of two stages (Fay and Herriot, 1979):

$$\text{Sampling model:} \quad \widehat{P}_i = \pi_i + \epsilon_i \tag{30.1}$$
$$\text{Linking model:} \quad \pi_i = \beta_0 + x_i'\beta + v_i, \tag{30.2}$$

where the outcome $\widehat{P}_i$ is a design-unbiased direct estimate for area i, $\epsilon_i \sim N(0, \sigma_i^2)$, and $v_i \sim N(0, \sigma_v^2)$, with all errors mutually independent. The sampling model takes into account the sampling variability of the direct survey estimates $\widehat{P}_i$. The term ϵ_i reflects the variation of the direct estimator around the true area prevalence π_i. It is customary to assume that the sampling variances σ_i^2 are known, equal to the variance of the direct estimator, though this might be restrictive in some cases (Rao, 2003). The linking model assumes that the area-specific true area prevalence π_i can be linearly modeled as a function of some known auxiliary variables x_i and random effects v_i. The parameters β_0, β, and σ_u^2 of the linking model are unknown and are estimated from the available data. In this model, the design is taken into account both in the use of the direct estimates and in the sampling variance.

Several authors have extended the Fay–Herriot model to incorporate spatial dependence (see, e.g., Cressie et al., 2006; Marhuenda et al., 2013; Molina and Rao, 2009; Pratesi and Salvati, 2009; Singh et al., 2005). A possible extension to the linking model in (30.2) is to assume a convolution model (Besag et al., 1991),

$$\pi_i = \beta_0 + x_i'\beta + u_i + v_i,$$

where the random effects terms u_i follow an intrinsic conditional autoregressive (ICAR) distribution, such as introduced by Besag and Kooperberg (1995):

$$\begin{aligned} u_i | u_{l,i\neq l} &\sim N(\bar{\pi}_i, \tau_i^2) \\ \bar{\pi}_i &= \frac{1}{\sum_l \nu_{il}} \sum_l \nu_{il} u_l \\ \tau_i^2 &= \frac{\sigma_u^2}{\sum_l \nu_{il}}, \end{aligned}$$

with $\nu_{il} = 1$ if areas i and l are adjacent and 0 otherwise.

A possible drawback of the Fay–Herriot model for binary spatial health outcomes is that the prevalences are not constrained to the $(0, 1)$ interval, which could cause difficulties in small areas (Mercer et al., 2014). The following extension of the Fay–Herriot model can be considered in such a situation, where $g(.)$ is an appropriate link function:

$$\text{Sampling model:} \quad g(\widehat{P}_i) = g(\pi_i) + \epsilon_i \tag{30.3}$$

$$\text{Linking model:} \quad g(\pi_i) = \beta_0 + x_i'\beta + u_i + v_i, \tag{30.4}$$

with ϵ_i, u_i, and v_i as specified before. Mercer et al. (2014) propose the use of a logit-link function in this respect, that is, $g(z) = \log(z/(1-z))$, with the variance σ_i^2 of the sampling model set equal to $\text{var}(\widehat{P}_i)/\widehat{P}_i(1-\widehat{P}_i)$. Alternatively, Efron and Morris (1975) and Raghunathan et al. (2007) propose the use of an arcsine square-root transformation, that is, $g(z) = \sin^{-1}(\sqrt{z})$, which stabilizes the variance approximately. The sampling variance σ_i^2 is approximately equal to $(4n_i^*)^{-1}$, where n_i^* denotes the effective sample size for area i. The effective sample size is obtained by dividing the actual sample size n_i by the design effect d_i, which is the ratio of the estimated design-based variance for $\widehat{P}_i$ to the estimated binomial variance $\widehat{P}_i(1-\widehat{P}_i)/n_i$. Raghunathan et al. (2007) extend this method to combine information from two surveys to estimate county-level prevalence rates of cancer risk factors and screening. This method is also applied by Davis et al. (2014) to obtain state-specific estimates of mammography screening.

30.3.2 Weighted likelihood method

An alternative strategy to take into account the complex sampling design is the use of a weighted likelihood, also referred to as pseudolikelihood (Skinner et al., 1989). Congdon and Lloyd (2010) and Mercer et al. (2014) extended this method with the inclusion of a spatial component.

For a binary health outcome, one can assume a weighted Bernouilli likelihood of the form

$$\prod_{i=1}^{I}\prod_{k=1}^{n_i} \pi_i^{w_{ik}^* y_{ik}} (1-\pi_i)^{w_{ik}^*(1-y_{ik})},$$

in which individual contributions to the likelihood are weighted by w_{ik}^* to reflect the complex survey design. We can further model the prevalences using any spatial model. In line with previous models, we assume that the spatial structure is modeled via a spatially structured ICAR model:

$$\text{logit}(\pi_i) = \beta_0 + x_i'\beta + u_i + v_i.$$

As recommended by Pfefferman et al. (1998) and Asparouhov (2006), Congdon and Lloyd (2010) define the weights

$$w_{ik}^* = \frac{w_{ik}}{\sum_k w_{ik}} n_i,$$

such that the weights sum to the sample size n_i. Mercer et al. (2014) note that this is equivalent to assuming that $y_i^P = \sum_k w_{ik}^* y_{ik}$ follows a binomial likelihood with n_i number of trials.

Chen et al. (2014), on the other hand, assume that $y_i^E = n_i^* \widehat{P}_i$, with n_i^*, being the effective sample size and $\widehat{P}_i$ the direct estimator of the area-specific prevalences, follow a binomial likelihood with n_i^* number of trials. This is equivalent to the use of a weighted likelihood with weights defined as

$$w_{ik}^* = \frac{w_{ik}}{\sum_k w_{ik}} n_i^*,$$

such that the weights sum up to the effective sample size. Mercer et al. (2014) studied the performance of this proposal with respect to other existing method.

30.3.3 Weight smoothing method

The model-based approach in survey sampling assumes a model for the health outcome of interest, which is then used to predict the outcome for the nonsampled individuals in the population (Little, 2011). Such estimators are consistent and efficient, but are prone to model misspecification. Several authors have argued that a model as a function of the weights (or probabilities of inclusion) could protect agains the effects of model misspecification (Little, 1983, 1991). Direct estimates are again obtained if the stratum means of the health outcomes are treated as fixed effects, but smoothing of the weights similar to weight trimming is achieved when the stratum means are treated as random effects (Holt and Smith, 1979; Lazzeroni and Little, 1998). These random effects models are often called weight smoothing models. Flexible models are further used in literature since these are more robust against model misspecification. Zheng and Little (2003) estimate the population totals using a nonparametric regression as a function of the inclusion probabilities. Chen et al. (2010) propose the use of a Bayesian P-spline for the inclusion probabilities.

The model-based estimator given by Chen et al. (2010) is based on the following probit truncated polynomial P-spline regression model (Ruppert et al., 2003):

$$\begin{aligned} y_{ik} &\sim \text{Bernoulli}(\tilde{p}_{ik}) \\ \Phi^{-1}(\tilde{p}_{ik}) &= \beta_0 + \sum_{l=1}^{p} \beta_l w_{ik}^l + \sum_{l=1}^{m} b_l (w_{ik} - \kappa_l)_+^p, \end{aligned}$$

where κ_l for $l = 1, \ldots, K$ is a set of prespecified knots, $\Phi^{-1}(.)$ is the inverse cumulative distribution function (CDF) of a standard normal distribution, and p is the degree of the truncated polynomial spline base function. The function $(x)_+ = x$ if $x > 0$ and 0 otherwise. When using a large set of knots, this model has much flexibility. However, when fitting the model using ordinary least squares, this method might overfit the data, following random fluctuations in the data as well as the main features. To overcome this problem, the spline function can be fitted as a mixed model (Ruppert et al., 2003). By specifying a normal distribution for b, $b \sim N(0, \sigma_b^2)$, the influence of the m knots is constrained. Bayesian specification of the model proceeds by assuming weak prior and hyperprior distributions. Vandendijck et al. (2015) extended this idea to small-area estimation, taking into account both spatially independent (v_i) and spatially dependent (u_i) random effects:

$$\Phi^{-1}\left(E(y_{ik}|\beta, b, w_{ik})\right) = \beta_0 + \sum_{l=1}^{p} \beta_l w_{ik}^l + \sum_{l=1}^{m} b_l (w_{ik} - \kappa_l)_+^p + u_i + v_i.$$

Other flexible functional forms for the weight component are considered in Vandendijck et al. (2015), including a random walk model of order 1.

Once estimates are obtained for the weight smoothing model, predictions of the health outcome $\widehat{y_{ik}}$ for the nonsampled individuals are calculated. An estimator of P_i is then given by (Royall, 1970)

$$\widehat{p}_i = N_i^{-1} \left(\sum_{k \in s_i} y_{ik} + \sum_{k \notin s_s} \widehat{y_{ik}} \right),$$

where the first term sums the outcome values of all the sampled individuals in area i and the second term sums the predicted values of all the nonsampled individuals in area i. Chen et al. (2010) simulate the posterior distribution of the population proportion by generating a large number D of draws of the posterior predictive distribution of the kth nonsampled individual from area i ($\widehat{y_{ik}}^{(d)}$) and calculate the predictive estimator

$$\widehat{p_i}^{(d)} = N_i^{-1}\left(\sum_{k\in s_i} y_{ik} + \sum_{k\notin s_s} \widehat{y_{ik}}^{(d)}\right).$$

The average of these draws is called the Bayesian P-spline predictive estimator of the finite population proportion.

Note that, in general, for nonsampled individuals, no information on the sampling weights w_{ik} is given, and thus it is not possible to obtain predictions of $\widehat{y_{ik}}$ for nonsampled individuals based on the weight smoothing model. However, Vandendijck et al. (2015) describe a method to estimate these weights for nonsampled individuals in a sampled area, as well as a method to estimate P_i for nonsampled areas.

30.4 Belgian Health Interview Survey

The Belgian Health Interview Survey (HIS) uses a complex multistage probability sampling scheme to select individuals for interviewing. In order to reach respondents, one first selects towns (municipalities), then a number of households within towns, and finally a number of household members within a household. A consequence of such a sampling scheme is that a number of respondents stem from the same household and the same town. We focus attention on the 2001 survey and investigate the prevalence of asthma in adults per district. Belgium is subdivided into 43 districts, with an adult population size of, on average, 200,000 per district. In total, 9921 adults responded to the question "Have you experienced asthma in the previous year?" The number of respondents per district varies between 41 and 2415, and four districts were not selected in the survey.

From the 9921 respondents, 511 responded positive to the question. Figure 30.1 (top left panel) shows the observed proportion of respondents that responded positive per district. The observed proportions range from 0 to 0.10. The white areas are those districts that do not have any respondents, and thus from which no information is available. Poststratification is performed on the age–gender strata. The weights range between 717 and 1212 across districts. Direct estimates of the prevalence are calculated based on these weights and presented in Figure 30.1 (top right panel). For areas with no information, direct estimates cannot be calculated.

We then applied the Bayesian hierarchical models as described in Sections 30.3.1, 30.3.2, and 30.3.3 to the HIS survey: the Fay–Herriot model with logit-link (Logit), the Fay–Herriot model with arcsine square-root transformation (Inv Sin), the weighted likelihood approach with weights summing up to sample size n_i (PL ss), and the weighted likelihood approach with weights summing up to the effective sample size n_i^* (PL ess). Also, a binomial likelihood approach on the observed number of cases is applied (Unadj). For all these models, both an independence model (with random effect v_i) and a spatially structured model (with both random effects u_i and v_i) are assumed. Models are estimates using the Bayesian inference framework, and computations are performed with integrated nested Laplace approximation (INLA). Uninformative priors are taken on the precision parameters and regression parameters. A box plot of the prevalence estimates based on these models is presented in Figure 30.2. Also, box plots of the observed proportions of asthma cases (Obs) and design-based direct estimates (Direct) are given.

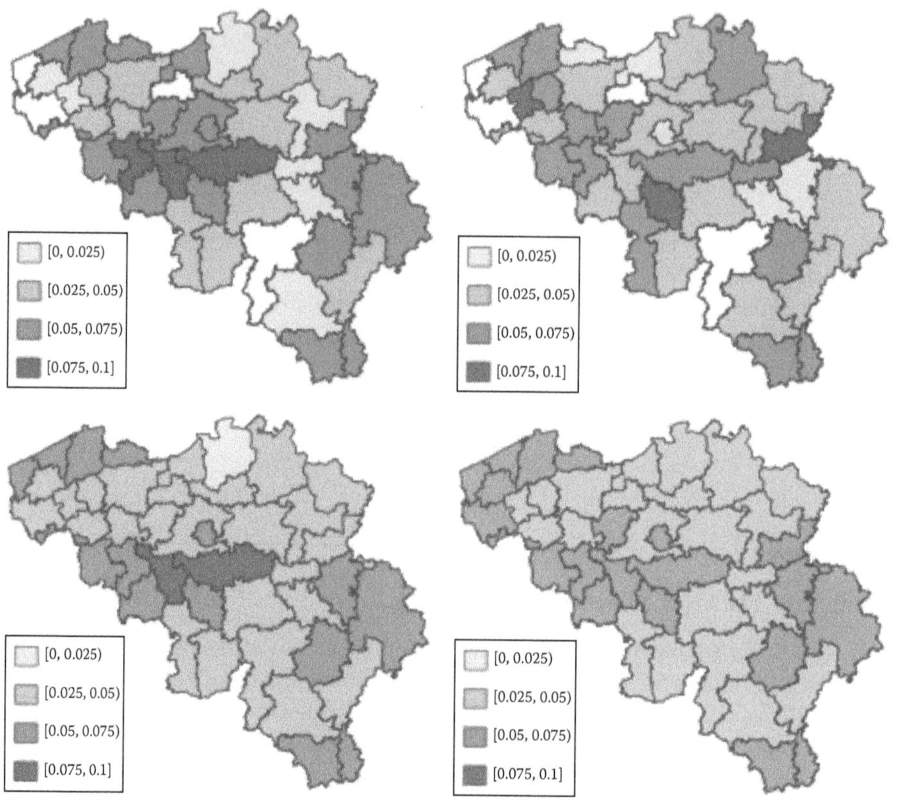

FIGURE 30.1
Predicted prevalence by district in Belgium in 2001. Top left: Observed prevalence. Top right: Direct estimated prevalence. Bottom left: Predicted under spatial logic transformed model. Bottom right: Predicted under spatial inverse sine transformed model.

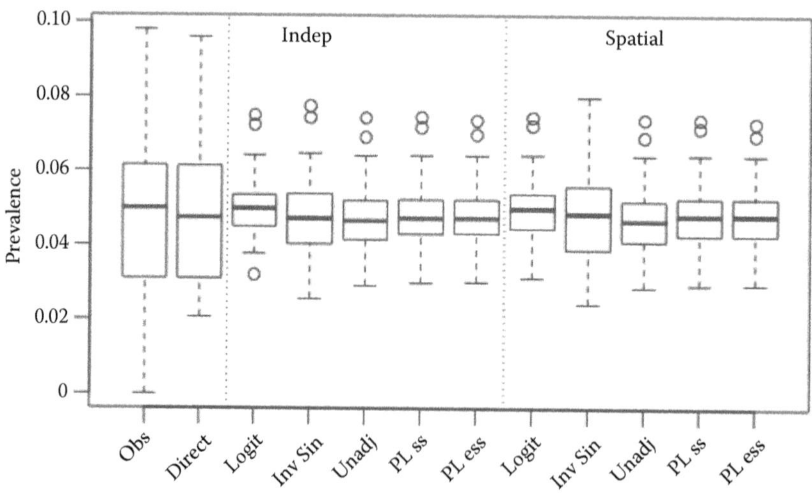

FIGURE 30.2
Predicted prevalence estimates across districts in Belgium in 2001, using various approaches.

As can be seen in Figure 30.2, the direct estimates have high variability, as a result of the small sample size in some of the districts. The indirect estimators share information from other areas, resulting in smaller variation between the areas, as seen in all Bayesian models. The median of the prevalence estimates of the binomial likelihood model of the observed number of cases is slightly smaller than that of the other models, reflecting possible bias due to the sampling design. In this example, we see that the arcsine transformed model has slightly higher variability than the other Bayesian models. It is important to note that all of the Bayesian models also allow prediction of the prevalence at non-observed areas. For illustration, Figure 30.1 shows the predicted estimates based on the logit-transformed spatial model (bottom left panel) and the arcsine transformed spatial model (bottom right panel), again showing that less smoothing is done in the arcsine-transformed model. The maps show higher prevalences of asthma in the center of the country, as well as in the northwest and northeast of the country.

30.5 Conclusions

Recently, methods have been developed to model the spatial structure of health outcomes based on survey data. Different approaches to deal with the complex design exist. We have considered a set of methods to deal with a binary response outcome measured in a health survey with a complex design, and where the design is reflected by design weights. The focus was on aggregate-level models. While each of the discussed methods has is own merits, further comparisons between the existing methods would be useful. Choosing an appropriate model is difficult based on a single data set, and needs further investigation.

Other important issues that need further attention in the context of spatial survey data include preferential sampling, missing data, and prediction for areas without observations. While missing data were handled in this chapter via the design weights, informative missingness could bias the results. This has received a lot of attention in the biostatistics literature, but less in the spatial statistics literature. Inference under preferential sampling arises when the sampling process and the process being modeled are stochastically dependent. Ignoring this association can bias the prediction. This issue has been recently investigated in the context of spatial geostatistical data and also needs further attention in the survey context.

References

Arora, V. and Lahiri, P. (1997). On the superiority of the Bayesian method over the BLUP in small area estimation problems. *Statistics Sinica*, 7, 1053–1063.

Asparouhov, T. (2006). General multi-level modeling with sampling weights. *Communications in Statistics—Theory and Methods*, 35, 439–460.

Besag, J. and Kooperberg, C. (1995). On conditional and intrinsic auro-regressions. *Biometrika*, 82, 733–746.

Besag, J., York, J., and Mollié, A. (1991). Bayesian image restoration with two applications in spatial statistics. *Annals of the Institute of Statistical Mathematics*, 43, 1–59.

Chen, Q., Elliott, M.R., and Little, R.J.A. (2010). Bayesian penalized spline model-based inference for finite population proportion in unequal probability sampling. *Survey Methodology*, 36, 23–34.

Chen, C., Wakefield, J., and Lumley, T. (2014). The use of sample weights in Bayesian hierarchical models for small area estimation. *Spatial and Spatio-Temporal Epidemiology.* doi: 10.1016/j.sste.2014.07.002.

Cochran, W.G. (1977). *Sampling Techniques.* Wiley, New York.

Congdon, P. and Lloyd, P. (2010). Estimating small area diabetes prevalence in the US using the Behavioral Risk Factor Surveillance System. *Journal of Data Science*, 8, 235–252.

Cressie, N., Frey, J., Harch, B., and Smith, M. (2006). Spatial prediction on a river network. *Journal of Agricultural, Biological, and Environmental Statistics*, 11, 127–150.

Davis, W.W., Parsons, V.L., Xie, D., Schenker, N., Twosn, M., Raghunathan, T.E., and Feuer, E.J. (2014). State-based estimates of mammography screening rates based on information from two health surveys. *Public Health Reports*, 125, 567–578.

Devesa, S.S., Grauman, D.G., Blot, W.J., Pennello, G., Hoover, R.N., and Fraumeni, J.F. (1999). Atlas of cancer mortality in the United States, 1950–94. U.S. Government Print Office, Washington, DC.

Efron, B.F. and Morris, C.N. (1975). Data analysis using Stein's estimator and its generalisations. *Journal of the American Statistical Association*, 70, 311–319.

Elliott, P., Wakefield, J., Best, N., and Briggs, D. (2001). *Spatial Epidemiology: Methods and Applications.* Oxford University Press, New York.

Fay, R.E. and Herriot, R.A. (1979). Estimates of income for small places: An application of James-Stein procedures to census data. *Journal of the American Statistical Association*, 74, 269–277.

Gelman, A. (2007). Struggles with survey weighting and regression modeling. *Statistical Science*, 22, 153–164.

Holt, D. and Smith, T.M.F. (1979). Poststratification. *Journal of the Royal Statistical Society: Series A*, 142, 33–36.

Horvitz, D. and Thompson, D. (1952). A generalization of sampling without replacement from a finite universe. *Journal of the American Statistical Association*, 47, 663–685.

Kemp, I., Boyle, P., Smans, M., and Muir, C.S. (1985). *Atlas of Cancer in Scotland, 1975–1980: Incidence and Epidemiological Perspective.* IARC publication no. 72. International Agency for Research on Cancer, Lyon, France.

Lawson, A.B., et al. (2013). *Bayesian Disease Mapping: Hierarchical Modeling in Spatial Epidemiology.* Chapman & Hall/CRC, Boca Raton, FL.

Lazzeroni, L.C. and Little, R.J.A. (1998). Random-effects models for smoothing post-stratification weights. *Journal of Official Statistics*, 14, 61–78.

Little, R.J. (2011). To model or not to model? Competing modes of inference for finite population sampling. *Journal of the American Statistical Association*, 99, 546–556.

Little, R.J.A. (1983). Estimating a finite population mean from unequal probability samples. *Journal of the American Statistical Sinica*, 78, 797–799.

Little, R.J.A. (1991). Inference with survey weights. *Journal of Official Statistics*, 7, 405–424.

Marhuenda, Y., Molina, I., and Morales, D. (2013). Small area estimation with spatio-temporal Fay-Herriot models. *Computational Statistics and Data Analysis*, 58, 308–325.

Mercer, L., Wakefield, J., Chen, C., and Lumley, T. (2014). A comparison of spatial smoothing methods for small area estimation with sampling weights. *Spatial Statistics*, 8, 69–85.

Molina, I. and Rao, J.N.K. (2009). Small area estimation of poverty indicators. *Canadian Journal of Statistics*, 38, 369–385.

Pfeffermann, D. (2013). New important developments in small area estimation. *Statistical Science*, 28, 40–68.

Pfefferman, D., Skinner, C., Holmes, D., Goldstein, H., and Rasbash, J. (1998). Weighting for unequal selection probabilities in multilevel models. *Journal of the Royal Statistical Society: Series B*, 60, 23–40.

Pratesi, M. and Salvati, N. (2009). Small area estimation: The EBLUP estimator based on spatially correlated random area effects. *Statistical Methods and Applications*, 17, 113–141.

Raghunathan, T.E., Xie, D., Schenker, N., Parsons, V.L., Davis, W.W., Dodd, K.W., and Feuer, E.J. (2007). Combining information from two surveys to estimate county-level prevalence rates of cancer risk factors and screening. *Journal of the American Statistical Association*, 102, 474–486.

Rao, J.N.K. (2003). *Small Area Estimation.* Wiley, Hoboken, NJ.

Rao, J.N.K. (2011). Impact of frequentist and Bayesian methods on survey sampling practice: A selective appraisal. *Statistical Science*, 26, 240–256.

Royall, R.M. (1970). On finite population sampling theory under certain linear regression modes. *Biometrika*, 57, 377–387.

Ruppert, D., Wand, M.P., and Carroll, R.J. (2003). *Semiprametric Regression.* Cambridge University Press, Cambridge.

Singh, B.B., Shukla, G.K., and Kundu, D. (2005). Spatio-temporal models in small area estimation. *Survey Methodology*, 31, 183–195.

Skinner, C., Hold, D., and Smidt, T. (1989). *Analysis of Complex Surveys.* Wiley, Chichester, UK.

Vandendijck, Y., Faes, C., Kirby, R., Lawson, A., and Hens, N. (2015). Model-based inference for small area estimation with sampling weights. Submitted for publication.

Waller, L.A. and Gotway, C.A. (2004). *Applied Spatial Statistics for Public Health Data.* Wiley, Hoboken, NJ.

Zheng, H. and Little, R.J.A. (2003). Penalized spline model-based estimation of finite population total from probability-proportional-to-size samples. *Journal of Official Statistics*, 19, 99–117.

31

Graphical Modeling of Spatial Health Data

Adrian Dobra
Department of Statistics
Department of Biobehavioral Nursing and Health Systems
and
Center for Statistics and the Social Sciences
University of Washington
Seattle, Washington

CONTENTS

Graphical models (Whittaker, 1990; Lauritzen, 1996) that encode multivariate independence and conditional independence relationships among observed variables $\mathbf{X} = (X_1, \ldots, X_p)$ have a widespread use in major scientific areas (e.g., biomedical and social sciences). In particular, a Gaussian graphical model (GGM) is obtained by setting off-diagonal elements of the precision matrix $\mathbf{K} = \mathbf{\Sigma}^{-1}$ to zero of a p-dimensional multivariate normal model (Dempster, 1972). Employing a GGM instead of a multivariate normal model leads to a significant reduction in the number of parameters that need to be estimated if most elements of $\mathbf{K}$ are constrained to be zero and p is large. A pattern of zero constraints in $\mathbf{K}$ can be recorded as an undirected graph $G = (V, E)$, where the set of vertices $V = \{1, 2, \ldots, p\}$ represent observed variables, while the set of edges $E \subset V \times V$ link all the pairs of vertices that correspond to off-diagonal elements of $\mathbf{K}$ that have not been set to zero. The absence of an edge between X_{v_1} and X_{v_2} corresponds with the conditional independence of these two random variables given the rest and is denoted by $X_{v_1} \perp\!\!\!\perp X_{v_2} \mid \mathbf{X}_{V \setminus \{v_1, v_2\}}$ (Wermuth, 1976). This is called the pairwise Markov property relative to G, which in turn implies the local as well as the global Markov properties relative to G (Lauritzen, 1996). The local Markov property plays a key role since it gives the regression model induced by G on each variable X_v. More explicitly, consider the neighbors of v in G, that is, the set of vertices $v' \in V$ such that $(v, v') \in E$. We denote this set by $\mathrm{bd}_G(v)$. The local Markov property relative to G says that $X_v \perp\!\!\!\perp \mathbf{X}_{V \setminus \{\{v\} \cup \mathrm{bd}_G(v)\}} \mid \mathbf{X}_{\mathrm{bd}_G(v)}$. This is precisely the statement we make when we drop the variables $\{X_{v'} : v' \in V \setminus \mathrm{bd}_G(v)\}$ from the regression of X_v on $\{X_{v'} : v' \in V \setminus \{v\}\}$.

The literature on GGMs contains two equally rich and significant domains of research. The first research domain relates to the problem of graph determination. That is, the underlying graph is unknown and needs to be inferred from the data. Frequentist methods estimate $\mathbf{K}$ and $\mathbf{\Sigma}$ given one graph that is best supported by the data in the presence of sparsity constraints that penalize for increased model complexity (i.e., for the addition of extra edges in the graph). Among numerous notable contributions, we mention the regularization methods of Meinshausen and Bühlmann (2006), Yuan and Lin (2007), Bickel and Levina (2008), and Friedman et al. (2008) as well as the simultaneous confidence intervals of Drton and Perlman (2004). Bayesian methods proceed by imposing suitable prior distributions for $\mathbf{K}$ or $\mathbf{\Sigma}$ (Leonard and Hsu, 1992; Yang and Berger, 1994; Daniels and Kass, 1999; Barnard et al., 2000; Smith and Kohn, 2002; Liechty et al., 2004; Rajaratnam et al., 2008). Inference can be performed based on the best model, that is, the graph having the highest posterior probability, or by Bayesian model averaging (Kass and Raftery, 1995) over all $2^{p(p-1)/2}$ possible graphs using Markov chain Monte Carlo (MCMC) approaches (Giudici and Green, 1999; Dellaportas et al., 2003; Wong et al., 2003). As the number of graphs grows, MCMC methods are likely to visit only subsets of graphs that have high posterior probabilities. To this end, various papers (Jones et al., 2005; Scott and Carvalho, 2008; Lenkoski and Dobra, 2011) have proposed stochastic search methods for fast identification of these high posterior probability graphs.

The second research domain on GGMs dominates the applications in spatial epidemiology. In this context, GGMs are referred to as Gaussian Markov random fields (GMRFs) (Besag, 1974, 1975; Besag and Kooperberg, 1995). The underlying graph G is assumed to be known: the vertices correspond to geographical areas, while the edges are associated with areas that are considered to be neighbors of each other (e.g., if they share a border). A GMRF is specified through the conditional distributions of each variable given the rest,

$$\left\{\mathsf{p}\left(X_v \mid \mathbf{X}_{V\setminus\{v\}}\right) : v \in V\right\}, \tag{31.1}$$

which are assumed to be normal. The local Markov property leads to a further reduction in the set of full conditionals:

$$\mathsf{p}\left(X_v \mid \mathbf{X}_{V\setminus\{v\}}\right) = \mathsf{p}\left(X_v \mid \mathbf{X}_{\mathrm{bd}_G(v)}\right). \tag{31.2}$$

Since it is typically assumed that phenomena (e.g., the occurrence of a disease) taking place in one area influence corresponding phenomena taking place in the remaining areas only through neighbor areas, the set of reduced conditionals (31.2) are employed to describe a full joint distribution of random spatial effects. GMRFs are conditional autoregression (CAR) models that have a subclass called simultaneous autoregressions (SARs). For a comprehensive account of inference in CAR, SAR, and GMRFs, see Cressie (1973), Rue and Knorr-Held (2005), Jin et al. (2005), Jin et al. (2007), Wakefield (2007), Zhang et al. (2009), and Gelfand et al. (2010). In particular, Zhang et al. (2009) discuss practical difficulties related to the interpretation and estimation of GMRFs and multivariate CAR (MCAR) models. Other key difficulties in the application of GMRFs arise from the conditions in which a joint distribution determined by (31.1) actually exists and, if it does, whether it is multivariate normal. This leads to particular parametric specifications for the set of conditionals (31.1) and (31.2) that are more restrictive than the general parametric specification of a GGM.

In this chapter, we examine the theoretical differences between GGMs and GMRFs. Dobra et al. (2011) developed efficient MCMC methods for inference in univariate and matrix-variate GGMs, and subsequently employed these methods to construct Bayesian hierarchical spatial models for mapping multiple diseases. We extend their results to multiway GGMs that can capture temporal dependencies in addition to several other relevant dimensions. We exemplify the construction of a Bayesian hierarchical spatiotemporal model based on three-way GGMs, and also present a related theoretical extension of multiway GGMs to dynamic multiway GGMs for array-variate time series. We end with a discussion section that includes links to relevant recent publications.

31.1 GGMs versus GMRFs

We consider a GGM defined by a graph $G = (V, E)$ for the multivariate normal distribution $\mathsf{N}_p(0, \mathbf{K}^{-1})$ of a vector $\mathbf{X} = (X_1, \ldots, X_p)$. The precision matrix $\mathbf{K}$ is constrained to belong to the cone P_G of positive definite matrices such that $K_{ij} = 0$ for all $(i, j) \notin E$. The full conditionals (31.1) associated with each X_v are expressed as a function of the elements of $\mathbf{K}$ as follows:

$$X_v \mid \mathbf{X}_{V\setminus\{v\}} = x_{V\setminus\{v\}} \sim \mathsf{N}\left(- \sum_{v' \in \mathrm{bd}_G(v)} (K_{vv'}/K_{vv}) x_{v'}, 1/K_{vv} \right). \tag{31.3}$$

Note that the variables $X_{v'}$ that are not linked by an edge with X_v are dropped from the full conditional (31.3) because $K_{vv'} = 0$. A GMRF with graph G is parametrized through the full conditionals

$$X_v \mid \mathbf{X}_{V\setminus\{v\}} = \mathbf{x}_{V\setminus\{v\}} \sim \mathsf{N}\left(\sum_{v' \in \mathrm{bd}_G(v)} \beta_{vv'} x_{v'}, \sigma_v^2 \right), \text{ for } v \in V. \tag{31.4}$$

The symmetry condition $\beta_{vv'}\sigma_{v'}^2 = \beta_{v'v}\sigma_v^2$ for all $v \neq v'$ is necessary for the conditionals (31.4) to define a proper GGM (Lauritzen, 1996). Additional constraints under which the set of regression parameters $\{\beta_{vv'}, \sigma_v^2\}$ induce a proper precision matrix $\mathbf{K} \in \mathsf{P}_G$ with $K_{vv} = \sigma_v^{-2}$ and $K_{vv'} = -\beta_{vv'}\sigma_v^{-2}$ are given in Besag and Kooperberg (1995). Besag et al. (1991) make use of a symmetric proximity matrix $\mathbf{W}$ with $w_{vv} = 0$, $w_{vv'} > 0$ if $v' \in \mathrm{bd}_G(v)$, and $w_{vv'} = 0$ if $v' \notin \mathrm{bd}_G(v)$. They define $\beta_{vv'} = \rho w_{vv'}/w_{v+}$ and $\sigma_v^2 = \sigma^2/w_{v+}$, where $w_{v+} = \sum_{v'} w_{vv'}$. Here, ρ is referred to as a spatial autocorrelation parameter. With this choice, for each $\rho \in (-1, 1)$ and $\sigma^2 > 0$, the GMRF specified by the full conditionals (31.4) has a precision $\mathbf{K} = \sigma^2(\mathbf{E_W} - \rho\mathbf{W})^{-1} \in \mathsf{P}_G$, where $\mathbf{E_W} = \mathrm{diag}\{w_{1+}, \ldots, w_{p+}\}$. As such, this widely used parametrization of GMRFs is quite restrictive since no matrix in the cone P_G can be represented through the two parameters ρ and σ^2 given a particular choice of $\mathbf{W}$. This difficulty originates from the parametrization (31.4) of a GGM. Instead, by imposing proper prior distributions for the precision matrix $\mathbf{K}$, we avoid the unnecessary representation of a GGM as the set of full conditionals (31.4). In particular, we use the G-Wishart prior $\mathsf{Wis}_G(\delta, \mathbf{D})$ with density

$$\mathsf{p}(\mathbf{K} \mid G, \delta, \mathbf{D}) = \frac{1}{I_G(\delta, \mathbf{D})} (\det \mathbf{K})^{(\delta-2)/2} \exp\left\{ -\frac{1}{2} \langle \mathbf{K}, \mathbf{D} \rangle \right\}, \tag{31.5}$$

with respect to the Lebesgue measure on P_G (Roverato, 2002; Atay-Kayis and Massam, 2005; Letac and Massam, 2007). Here, $\langle \mathbf{A}, \mathbf{B} \rangle = \mathrm{tr}(\mathbf{A}^T\mathbf{B})$ denotes the trace inner product. The normalizing constant $I_G(\delta, \mathbf{D})$ is finite provided $\delta > 2$ and $\mathbf{D}$ is positive definite (Diaconnis and Ylvisaker, 1979). The G-Wishart prior $\mathsf{Wis}_G(\delta, \mathbf{D})$ is conjugate to the normal likelihood. For a thorough account of its numerical properties, see Lenkoski and Dobra (2011) and the references therein.

For applications in hierarchical spatial models, Dobra et al. (2011) set $\mathbf{D} = (\delta - 2)\sigma^2(\mathbf{E_W} - \rho\mathbf{W})^{-1}$ because, with this choice, the prior mode for $\mathbf{K}$ is precisely

$\sigma^2(\mathbf{E_W} - \rho\mathbf{W})^{-1}$—the precision matrix of a GMRF. The prior specification for the precision matrix can therefore be completed in a manner similar to the current work from the existent literature on GMRFs. We note that the G-Wishart prior for $\mathbf{K}$ induces compatible prior distributions for the regression parameters (31.4) (Dobra et al., 2004). The advantage of this representation of GGMs is a more flexible framework for GMRFs that allows their regression coefficients to be determined from the data rather than being fixed or allowed to vary as a function of only two parameters.

31.2 Multiway Gaussian Graphical Models

We develop a framework for analyzing datasets that are associated with a random L-dimensional array $\mathbf{X}$. Such datasets are quite common in social and biomedical sciences. In particular, spatial epidemiology involves datasets recording standardized incidence rates (SIRs) of several diseases observed under different conditions at multiple time points. The notations, definitions, and operators related to tensors that appear throughout are introduced in De Lathauwer et al. (2000) and Kolda (2006). The elements of the observed multiway array are indexed by $\{(i_1, i_2, \ldots, i_L) : 1 \leq i_l \leq m_l\}$. The total number of elements of $\mathbf{X}$ is $m = \prod_{l=1}^{L} m_l$. We assume that $\mathbf{X}$ follows an array normal distribution

$$\mathrm{vec}(\mathbf{X}) \mid \mathbf{K} \sim \mathsf{N}_m\left(0, \mathbf{K}^{-1}\right),$$

whose $m \times m$ precision matrix $\mathbf{K}$ is separable across each dimension, that is,

$$\mathbf{K} = \mathbf{K}_L \otimes \mathbf{K}_{L-1} \otimes \ldots \otimes \mathbf{K}_1. \tag{31.6}$$

The $m_l \times m_l$ precision matrix $\mathbf{K}_l$ is associated with dimension l, while $\mathrm{vec}(\mathbf{X})$ is the vectorized version of $\mathbf{X}$. The separability assumption might seem restrictive in the sense that it captures only dependencies across each dimension of the data without directly taking into account the interactions that might exist among two, three, or more dimensions. However, this assumption reduces the number of parameters of the distribution of $\mathrm{vec}(\mathbf{X})$ from $2^{-1}m(m+1)$ to $2^{-L}\prod_{l=1}^{L} m_l(m_l+1)$, which constitutes a substantial advantage when a sample size is small. The probability density of $\mathbf{X}$ as an array is (Hoff, 2010)

$$\mathsf{p}(\mathbf{X} \mid \mathbf{K}_1, \ldots, \mathbf{K}_L) = (2\pi)^{-\frac{m}{2}} \left[\prod_{l=1}^{L} (\det \mathbf{K}_l)^{\frac{1}{m_l}}\right]^{\frac{m}{2}} \exp\left\{-\frac{1}{2}\langle \mathbf{X}, \mathbf{X} \times \{\mathbf{K}_1, \ldots, \mathbf{K}_L\}\rangle\right\}, \tag{31.7}$$

where

$$\mathbf{X} \times \{\mathbf{K}_1, \ldots, \mathbf{K}_L\} = \mathbf{X} \times_1 \mathbf{K}_1 \times_2 \ldots \times_L \mathbf{K}_L$$

is the Tucker product. Here $\mathbf{X} \times_l \mathbf{K}_l$ is the l-mode product of the tensor $\mathbf{X}$ and matrix $\mathbf{K}_l$. We refer to (31.7) as the mean-zero L-dimensional array normal distribution $\mathsf{AN}_L(\mathbf{0}; \{m_1, \mathbf{K}_1\}, \ldots, \{m_L, \mathbf{K}_L\})$.

Most of the existent literature has focused on two-dimensional (or matrix-variate) arrays; see, for example, Allen and Tibshirani (2010), Olshen and Rajaratnam (2010), and the references therein. Galecki (1994) studies the separable normal model (31.6) for $L = 3$, while Dawid (1981) presents theoretical results for matrix-variate distributions that include (31.6) with $L = 2$ as a particular case. Hoff (2010) has proposed a Bayesian inference framework for model (31.6) for an arbitrary number L of dimensions by assigning independent inverse-Wishart priors for the covariance matrices $\mathbf{\Sigma}_l = \mathbf{K}_l^{-1}$ associated with each dimension. Despite its flexibility and generality, the framework of Hoff (2010) does not allow any further reduction in the number of parameters of model (31.6). To this end, we propose a framework in which each precision matrix $\mathbf{K}_l$ is constrained to belong to a cone P_{G_l} associated with a GGM with graph $G_l \in \mathcal{G}_{m_l}$. We denote by $\mathcal{G}_{m_l}$ the set of undirected graphs with m_l vertices. Sparse graphs associated with each dimension lead to sparse precision matrices; hence, the number of parameters that need to be estimated could be significantly smaller than $2^{-L}\prod_{l=1}^{L} m_l(m_l+1)$.. A similar framework has been proposed in Wang and West (2009) for matrix-variate data ($L = 2$) and for row and column graphs restricted to the class of decomposable graphs. Our framework is applicable for any number of dimensions and allows arbitrary graphs (decomposable and nondecomposable) to be associated with each precision matrix $\mathbf{K}_l$.

The prior specification for $\{\mathbf{K}_l\}_{l=1}^{L}$ must take into account the fact that two precision matrices are not uniquely identified from their Kronecker product, which means that for any $z > 0$ and $l_1 \neq l_2$,

$$\mathbf{K}_L \otimes \cdots \otimes (z^{-1}\mathbf{K}_{l_1}) \otimes \cdots \otimes (z\mathbf{K}_{l_2}) \otimes \cdots \otimes \mathbf{K}_1 = \mathbf{K}_L \otimes \cdots \otimes \mathbf{K}_{l_1} \otimes \cdots \otimes \mathbf{K}_{l_2} \otimes \cdots \otimes \mathbf{K}_1$$

represents the same precision matrix for vec($\mathbf{X}$). We follow the idea laid out in Wang and West (2009) and impose the constraints

$$(\mathbf{K}_l)_{11} = 1, \text{ for } l = 2, \ldots, L. \tag{31.8}$$

Furthermore, we define a prior for $\mathbf{K}_l$, $l \geq 2$, through parameter expansion by assuming a G-Wishart prior $\mathsf{Wis}_{G_l}(\delta_l, \mathbf{D}_l)$ for the matrix $z_l\mathbf{K}_l$ with $z_l > 0$, $\delta_l > 2$, and $\mathbf{D}_l \in \mathsf{P}_{G_l}$. We denote $G_l = (V_l, E_l)$, where $V_l = \{1, 2, \ldots, m_l\}$ are vertices and $E_l \subset V_l \times V_l$ are edges. We consider the Cholesky decompositions of the precision matrices from (31.6),

$$\mathbf{K}_l = \mathbf{\Phi}_l^T \mathbf{\Phi}_l, \tag{31.9}$$

where $\mathbf{\Phi}_l$ is an upper triangular matrix with $(\mathbf{\Phi}_l)_{ii} > 0$, $1 \leq i \leq m_l$. Roverato (2002) proves that the set $\nu(G_l)$ of the free elements of $\mathbf{\Phi}_l$ consists of the diagonal elements together with the elements that correspond with the edges of G_l, that is,

$$\nu(G_l) = \{(i,i) : i \in V_l\} \cup \{(i,j) : i < j \text{ and } (i,j) \in E_l\}.$$

Once the free elements of $\mathbf{\Phi}_l$ are known, the remaining elements are also known. Specifically, $(\mathbf{\Phi}_l)_{1j} = 0$ if $j \geq 2$ and $(1,j) \notin E_l$. We also have

$$(\mathbf{\Phi}_l)_{ij} = -\frac{1}{(\mathbf{\Phi}_l)_{ii}} \sum_{k=1}^{i-1} (\mathbf{\Phi}_l)_{ki} (\mathbf{\Phi}_l)_{kj}$$

for $2 \leq i < j$ and $(i,j) \notin E_l$. The determination of the elements of $\mathbf{\Phi}_l$ that are not free based on the elements of $\mathbf{\Phi}_l$ that are free is called the completion of $\mathbf{\Phi}_l$ with respect to G_l (Roverato, 2002; Atay-Kayis and Massam, 2005). It is useful to remark that the free elements of $\mathbf{\Phi}_l$ fully determine the matrix K_l. The development of our framework involves the Jacobian of the transformation that maps $\mathbf{K}_l \in \mathsf{P}_{G_l}$ to the free elements of $\mathbf{\Phi}_l$ (Roverato, 2002):

$$J(\mathbf{K}_l \to \mathbf{\Phi}_l) = 2^{m_l} \prod_{i=1}^{m_l} (\mathbf{\Phi}_l)_{ii}^{d_i^{G_l}+1},$$

where $d_i^{G_l}$ is the number of elements in $\mathsf{bd}_{G_l}(i) \cap \{i+1, \ldots, m_l\}$ and $\mathsf{bd}_{G_l}(i) = \{j : (i,j) \in E_l\}$ is the boundary of vertex i in G_l.

Our proposed prior specification for the separable normal model (31.7) is

$$\mathbf{K}_1 \mid \delta_1, \mathbf{D}_1 \sim \mathsf{Wis}_{G_1}(\delta_1, \mathbf{D}_1),\ (z_l\mathbf{K}_l) \mid \delta_l, \mathbf{D}_l \sim \mathsf{Wis}_{G_l}(\delta_l, \mathbf{D}_l),\ \text{for } l = 2, \ldots, L. \quad (31.10)$$

The prior for $\mathbf{K}_1$ is

$$\mathsf{p}\left(\mathbf{K}_1 \mid G_1\right) = \frac{1}{I_{G_1}\left(\delta_1, \mathbf{D}_1\right)} \left(\det \mathbf{K}_1\right)^{\frac{\delta_1 - 2}{2}} \exp\left\{-\frac{1}{2}\langle \mathbf{K}_1, \mathbf{D}_1\rangle\right\}, \quad (31.11)$$

while the joint prior for $(z_l, \mathbf{K}_l)$ is

$$\mathsf{p}\left(z_l, \mathbf{K}_l \mid G_l\right) = \frac{1}{I_{G_l}\left(\delta_l, \mathbf{D}_l\right)} \left(\det \mathbf{K}_l\right)^{\frac{\delta_l - 2}{2}} \exp\left\{-\frac{1}{2}\langle \mathbf{K}_l, z_l\mathbf{D}_l\rangle\right\} z_l^{\frac{m_l(\delta_l - 2)}{2} + |\nu(G_l)| - 1} \quad (31.12)$$

for $l = 2, \ldots, L$.

31.3 Inference in Multiway GGMs

We assume that the observed samples $\mathcal{D} = \{\mathbf{x}^{(1)}, \ldots, \mathbf{x}^{(n)}\}$ are independently generated from the mean-zero array normal distribution $\mathsf{AN}_L(\mathbf{0}; \{m_1, \mathbf{K}_1\}, \ldots, \{m_L, \mathbf{K}_L\})$. The resulting likelihood is expressed by introducing an additional dimension $m_{L+1} = n$ with precision matrix $\mathbf{K}_{L+1} = \mathbf{I}_n$, where $\mathbf{I}_n$ is the $n \times n$ identity matrix. We see $\mathcal{D}$ as an $m_1 \times \ldots \times m_{L+1}$ array that follows an array normal distribution $\mathsf{AN}_{L+1}(\mathbf{0}; \{m_1, \mathbf{K}_1\}, \ldots, \{m_{L+1}, \mathbf{K}_{L+1}\})$. Furthermore, we define $\mathbf{\Phi}_{L+1} = \mathbf{I}_n$. The Cholesky decompositions (31.9) of the precision matrices $\mathbf{K}_l$ give the following form of the likelihood of $\mathcal{D}$:

$$\mathsf{p}(\mathcal{D} \mid \mathbf{K}_1, \ldots, \mathbf{K}_L) \propto \left[\prod_{l=1}^{L} (\det \mathbf{K}_l)^{\frac{1}{m_l}}\right]^{\frac{mn}{2}} \exp\left\{-\frac{1}{2}\|\mathcal{D} \times \{\mathbf{\Phi}_1, \ldots, \mathbf{\Phi}_{L+1}\}\|\right\}, \quad (31.13)$$

where $\|\mathbf{Y}\| = \langle \mathbf{Y}, \mathbf{Y}\rangle$ is the array norm (Kolda, 2006). Simple calculations show that the part of the likelihood (31.13) that depends on the precision matrix $\mathbf{K}_l$ is written as

$$\mathsf{p}(\mathcal{D} \mid \mathbf{K}_1, \ldots, \mathbf{K}_L) \propto (\det \mathbf{K}_l)^{\frac{mn}{2m_l}} \exp\left\{-\frac{1}{2}\langle \mathbf{K}_l, \mathbf{S}_l\rangle\right\}, \quad (31.14)$$

where $\mathbf{S}_l = \mathcal{D}^{[l]}_{(l)} \left(\mathcal{D}^{[l]}_{(l)}\right)^T$ and

$$\mathcal{D}^{[l]} = \mathcal{D} \times_1 \mathbf{\Phi}_1 \times_2 \ldots \times_{l-1} \mathbf{\Phi}_{l-1} \times_l \mathbf{I}_l \times_{l+1} \mathbf{\Phi}_{l+1} \times_{l+2} \ldots \times_{L+1} \mathbf{\Phi}_{L+1}.$$

$\mathbf{Y}_{(l)}$ is the l-mode matricization of an array $\mathbf{Y}$ (Kolda, 2006). We develop a Markov chain Monte Carlo sampler from the posterior distribution of precision matrices $\mathbf{K}_l \in \mathsf{P}_{G_l}$, graphs $G_l \in \mathcal{G}_{m_l}$, and auxiliary variables z_l for $1 \leq l \leq L$:

$$\begin{aligned} &\mathsf{p}\left(\mathbf{K}_1, G_1, (\mathbf{K}_l, G_l, z_l)_{l=2}^{L} \mid \mathcal{D}\right) \\ &\propto \mathsf{p}(\mathcal{D} \mid \mathbf{K}_1, \ldots, \mathbf{K}_L)\mathsf{p}(\mathbf{K}_1 \mid G_1) \prod_{l=2}^{L} \mathsf{p}(z_l, \mathbf{K}_l \mid G_l) \prod_{l=1}^{L} \pi_{m_l}(G_l). \end{aligned} \quad (31.15)$$

Here, $\pi_{m_l}(G_l)$ are prior probabilities on the set of graphs $\mathcal{G}_{m_l}$. The full conditionals of $\mathbf{K}_l$, $1 \leq l \leq L$, and z_l, $2 \leq l \leq L$, are G-Wishart and gamma, respectively:

$$\mathsf{p}(\mathbf{K}_l \mid \text{rest}) = \mathsf{Wis}_{G_l}\left(\frac{mn}{m_l} + \delta_l, \mathbf{S}_l + z_l D_l\right)$$

$$\mathsf{p}(z_l \mid \text{rest}) = \text{Gamma}\left(\frac{m_l(\delta_l - 2)}{2} + |\nu(G_l)|, \frac{1}{2}\langle \mathbf{K}_l, \mathbf{D}_l \rangle\right),$$

where Gamma(α, β) has mean α/β. We use the approach for updating $\mathbf{K}_l$ $(1 \leq l \leq L)$ described in Dobra and Lenkoski (2011). Their method sequentially perturbs each free element in the Cholesky decomposition of each precision matrix. The constraint (31.8) is imposed by not updating the free element $(\mathbf{\Phi}_l)_{11} = \sqrt{(\mathbf{K}_l)_{11}} = 1$.

The updates of the graphs G_l are based on the full joint conditionals of $\mathbf{K}_l$ and G_l, $1 \leq l \leq L$:

$$\mathsf{p}(\mathbf{K}_l, G_l \mid \text{rest})$$
$$\propto \frac{1}{I_{G_l}(\delta_l, \mathbf{D}_l)} (\det \mathbf{K}_l)^{\frac{1}{2}\left(\frac{mn}{m_l} + \delta_l - 2\right)} z_l^{\frac{m_l(\delta_l - 2)}{2} + |\nu(G_l)| - 1} \exp\left\{-\frac{1}{2}\langle \mathbf{K}_l, \mathbf{S}_l + z_l \mathbf{D}_l \rangle\right\},$$

since, once an edge in G_l is added or deleted, the corresponding set of free elements of $\mathbf{K}_l$ together with the remaining bound elements must also be updated.

We denote by $\mathsf{nbd}^+_{m_l}(G_l)$ the graphs that can be obtained by adding an edge to a graph $G_l \in \mathcal{G}_{m_l}$ and by $\mathsf{nbd}^-_{m_l}(G_l)$ the graphs that are obtained by deleting an edge from G_l. We call the one-edge-way set of graphs $\mathsf{nbd}_{m_l}(G_l) = \mathsf{nbd}^+_{m_l}(G_l) \cup \mathsf{nbd}^-_{m_l}(G_l)$ in the neighborhood of G_l in $\mathcal{G}_{m_l}$. These neighborhoods connect any two graphs in $\mathcal{G}_{m_l}$ through a sequence of graphs such that two consecutive graphs in this sequence are each other's neighbors. We sample a candidate graph $G'_l \in \mathsf{nbd}_{m_l}(G_l)$ from the proposal distribution:

$$q(G'_l \mid G_l, z_l) = \frac{1}{2} \frac{z_l^{|\nu(G'_l)|}}{\sum\limits_{G''_l \in \mathsf{nbd}^+_{m_l}(G_l)} z_l^{|\nu(G''_l)|}} \delta_{\{G'_l \in \mathsf{nbd}^+_{m_l}(G_l)\}} + \frac{1}{2} \frac{z_l^{|\nu(G'_l)|}}{\sum\limits_{G''_l \in \mathsf{nbd}^-_{m_l}(G_l)} z_l^{|\nu(G''_l)|}} \delta_{\{G'_l \in \mathsf{nbd}^-_{m_l}(G_l)\}}, \quad (31.16)$$

where δ_A is equal to 1 if A is true and is 0 otherwise. The proposal (31.16) gives an equal probability that the candidate graph is obtained by adding or deleting an edge from the current graph G_l.

We assume that the candidate graph G'_l is obtained by adding an edge (v_1, v_2), $v_1 < v_2$, to G_l. We have $\nu(G'_l) = \nu(G_l) \cup \{(v_1, v_2)\}$, $\mathsf{bd}_{G'_l}(v_1) = \mathsf{bd}_{G_l}(v_1) \cup \{v_2\}$, and $d_{v_1}^{G'_l} = d_{v_1}^{G_{sl}} + 1$. We define an upper diagonal matrix $\mathbf{\Phi}'_l$ such that $(\mathbf{\Phi}'_l)_{v'_1, v'_2} = (\mathbf{\Phi}_l)_{v'_1, v'_2}$ for all $(v'_1, v'_2) \in \nu(G_l)$. The value of $(\mathbf{\Phi}'_l)_{v_1, v_2}$ is sampled from an $\mathsf{N}\left((\mathbf{\Phi}_l)_{v_1, v_2}, \sigma_g^2\right)$ distribution. The bound elements of $\mathbf{\Phi}'_l$ are determined through completion with respect to G'_l. We form the candidate matrix $\mathbf{K}'_l = (\mathbf{\Phi}'_l)^T \mathbf{\Phi}'_l \in \mathsf{P}_{G'_l}$. Since the dimensionality of the parameter space increases by one, we must make use of the reversible jump method of Green (1995). We accept the update of $(\mathbf{K}_l, G_l)$ to $(\mathbf{K}_l', G_l')$ with probability $\min\{R_g, 1\}$, where

$$R_g = \frac{\mathsf{p}(\mathbf{K}_l', G_l' \mid \text{rest})}{\mathsf{p}(\mathbf{K}_l, G_l \mid \text{rest})} \frac{q(G_l \mid G_l', z_l)}{q(G_l' \mid G_l, z_l)} \frac{J(\mathbf{K}_l' \to \mathbf{\Phi}_l')}{J(\mathbf{K}_l \to \phi_l)} \frac{\pi_{m_l}(G_l')}{\pi_{m_l}(G_l)} \times \frac{J(\mathbf{\Phi}_l \to \mathbf{\Phi}_l')}{\frac{1}{\sigma_g\sqrt{2\pi}} \exp\left(-\frac{\left((\mathbf{\Phi}_l')_{v_1,v_2} - (\mathbf{\Phi}_l)_{v_1,v_2}\right)^2}{2\sigma_g^2} \right)}.$$

Since the free elements of $\mathbf{\Phi}_l'$ are the free elements of $\mathbf{\Phi}_l$ and $(\mathbf{\Phi}_l')_{v_1,v_2}$, the Jacobian of the transformation from $\mathbf{\Phi}_l$ to $\mathbf{\Phi}_l'$ is equal to 1. Moreover, $\det \mathbf{K}_l' = \det \mathbf{K}_l$ and

$$J(\mathbf{K}_l' \to \mathbf{\Phi}_l') = (\mathbf{\Phi}_l)_{v_1,v_1} J(\mathbf{K}_l \to \mathbf{\Phi}_l).$$

It follows that

$$R_g = \sigma_g\sqrt{2\pi}(\mathbf{\Phi}_l)_{v_1,v_1} z_l \frac{I_{G_l}(\delta_l, \mathbf{D}_l)}{I_{G_l'}(\delta_l, \mathbf{D}_l)} \frac{q(G_l \mid G_l', z_l)}{q(G_l' \mid G_l, z_l)} \frac{\pi_{m_l}(G_l')}{\pi_{m_l}(G_l)} \times \exp\left\{ -\frac{1}{2}\langle \mathbf{K}_l' - \mathbf{K}_l, \mathbf{S}_l + z_l\mathbf{D}_l\rangle + \frac{\left((\mathbf{\Phi}_l')_{v_1,v_2} - (\mathbf{\Phi}_l)_{v_1,v_2}\right)^2}{2\sigma_g^2} \right\}.$$

Next, we assume that G_l' is obtained by deleting the edge (v_1, v_2) from G_l. We have $\nu(G_l') = \nu(G_l) \setminus \{(v_1, v_2)\}$, $\mathsf{bd}_{G_l'}(v_1) = \mathsf{bd}_{G_l}(v_1) \setminus \{v_2\}$, and $d_{v_1}^{G_l'} = d_{v_1}^{G_l} - 1$. We define an upper diagonal matrix $\mathbf{\Phi}_l'$ such that $(\mathbf{\Phi}_l')_{v_1',v_2'} = (\mathbf{\Phi}_l)_{v_1',v_2'}$ for all $(v_1', v_2') \in \nu(G_l')$. The bound elements of $\mathbf{\Phi}_l'$ are obtained by completion with respect to G_l'. The candidate precision matrix is $\mathbf{K}_l' = (\mathbf{\Phi}_l')^T \mathbf{\Phi}_l' \in \mathsf{P}_{G_l'}$. Since the dimensionality of the parameter space decreases by 1, the acceptance probability of the update of $(\mathbf{K}_l, G_l)$ to $(\mathbf{K}_l', G_l')$ is $\min\{R_g', 1\}$, where

$$R_g' = \left(\sigma_g\sqrt{2\pi}(\mathbf{\Phi}_l)_{v_1,v_1} z_l\right)^{-1} \frac{I_{G_l}(\delta_l, \mathbf{D}_l)}{I_{G_l'}(\delta_l, \mathbf{D}_l)} \frac{q(G_l \mid G_l', z_l)}{q(G_l' \mid G_l, z_l)} \frac{\pi_{m_l}(G_l')}{\pi_{m_l}(G_l)} \times \exp\left\{ -\frac{1}{2}\langle \mathbf{K}_l' - \mathbf{K}_l, \mathbf{S}_l + z_l\mathbf{D}_l\rangle - \frac{\left((\mathbf{\Phi}_l')_{v_1,v_2} - (\mathbf{\Phi}_l)_{v_1,v_2}\right)^2}{2\sigma_g^2} \right\}.$$

31.4 Multiway GGMs with Separable Mean Parameters

So far we have discussed multiway GGMs associated with array normal distributions with an $m_1 \times \ldots \times m_L$ array mean parameter $\mathbf{M}$ assumed to be zero. In some practical applications, this assumption is too restrictive and $\mathbf{M}$ needs to be explicitly accounted for. The observed samples $\mathcal{D} = \{\mathbf{x}^{(1)}, \ldots, \mathbf{x}^{(n)}\}$ grouped as an $m_1 \times \ldots \times m_L \times n$ array are modeled as

$$\mathcal{D} = \mathbf{M} \circ \mathbf{1}_n + \mathbf{X},\ \ \mathbf{X} \sim \mathsf{AN}_{L+1}(\mathbf{0}; \{m_1, \mathbf{K}_1\}, \ldots, \{m_L, \mathbf{K}_L\}, \{n, \mathbf{K}_{L+1}\}),\ \ \mathbf{K}_{L+1} = \mathbf{I}_n. \tag{31.17}$$

If the sample size n is small or if the observed samples are not independent and their dependence structure is represented by removing the constraint $\mathbf{K}_{L+1} = \mathbf{I}_n$, estimating $m = \prod_{l=1}^{L} m_l$ mean parameters is unrealistic. The matrix-variate normal models of Allen and Tibshirani (2010) have separable mean parameters defined by row and column means, while Allen (2011) extends separable means for general array data. The L-dimensional mean array $\mathbf{M}$ is written as a sum of distinct mean arrays associated with each dimension:

$$\mathbf{M} = \sum_{l=1}^{L} \mathbf{M}_l, \text{ where } \mathbf{M}_l = \mathbf{1}_{m_1} \circ \ldots \circ \mathbf{1}_{m_{l-1}} \circ \boldsymbol{\mu}_l \circ \mathbf{1}_{m_{l+1}} \circ \ldots \circ \mathbf{1}_{m_L}, 1 \leq l \leq L.$$

Here, $\boldsymbol{\mu}_l \in \mathbb{R}^{m_l}$ represents the mean vector associated with dimension l of $\mathbf{X}$. This particular structure of the mean array $\mathbf{M}$ implies the following marginal distribution for each element of the array of random effects $\mathbf{X}$:

$$X_{i_1 \ldots i_L i_{L+1}} \sim \mathsf{N}\left(\sum_{l=1}^{L} (\boldsymbol{\mu}_l)_{i_l}, \prod_{l=1}^{L} (\mathbf{K}_l^{-1})_{i_l i_l} \right).$$

Thus, $\boldsymbol{\mu}_l$ can be interpreted as fixed effects associated with dimension l. The dependency between two different elements of $\mathbf{X}$ is represented by their covariance:

$$\mathsf{Cov}(X_{i_1 \ldots i_L i_{L+1}}, X_{i'_1 \ldots i'_L i'_{L+1}}) = \prod_{l=1}^{L} (\mathbf{K}_l^{-1})_{i_l i'_l}.$$

We remark that the individual mean arrays $\mathbf{M}_l$ are not identified, but their sum $\mathbf{M}$ is identified.

Bayesian estimation of $\mathbf{M}$ proceeds by specifying independent priors $\boldsymbol{\mu}_l \sim \mathsf{N}_{m_l}(\boldsymbol{\mu}_l^0, \boldsymbol{\Omega}_l^{-1})$. To simplify the notations, we take $n = 1$; hence, the arrays in Equation 31.17 have only L dimensions. We develop a Gibbs sampler in which each vector $\boldsymbol{\mu}_l$ is updated as follows. Denote $m_{-l} = \prod_{l' \neq l} m_{l'}$ and consider the l-mode matricizations $\mathcal{D}_{(l)}$, $\mathbf{M}_{(l)}$, and $\mathbf{X}_{(l)}$ of the arrays $\mathcal{D}$, $\mathbf{M}$, and $\mathbf{X}$. From Equation 31.17, it follows that the $m_l \times m_{-l}$ random matrix

$$\widetilde{\mathbf{X}}_{(l)} = \mathbf{X}_{(l)} - \sum_{l' \neq l} (\mathbf{M}_{l'})_{(l)}$$

follows a matrix-variate normal distribution with mean $(\mathbf{M}_l)_{(l)} = \boldsymbol{\mu}_l \mathbf{1}_{m_{-l}}^T$, row precision matrix $\mathbf{K}_l$, and column precision matrix $\mathbf{K}_{-l} = \mathbf{K}_L \otimes \ldots \otimes \mathbf{K}_{l+1} \otimes \mathbf{K}_{l-1} \otimes \ldots \otimes \mathbf{K}_1$. It follows that $\boldsymbol{\mu}_l$ is updated by direct sampling from the multivariate normal $\mathsf{N}_{m_l}(\mathbf{m}_{\boldsymbol{\mu}_l}, K_{\boldsymbol{\mu}_l}^{-1})$, where

$$\mathbf{K}_{\boldsymbol{\mu}_l} = \left(\mathbf{1}_{m_{-l}}^T \mathbf{K}_{-l} \mathbf{1}_{m_{-l}} \right) \mathbf{K}_l + m_{-l} \boldsymbol{\Omega}_l, \quad \mathbf{m}_{\boldsymbol{\mu}_l} = \mathbf{K}_{\boldsymbol{\mu}_l}^{-1} \left[\mathbf{K}_l \widetilde{\mathbf{X}}_{(l)} \mathbf{K}_{-l} \mathbf{1}_{m_{-l}} + m_{-l} \boldsymbol{\Omega}_l \boldsymbol{\mu}_l^0 \right].$$

31.5 Application to Spatiotemporal Cancer Mortality Surveillance

In this section, we construct a spatial hierarchical model for spatiotemporal cancer mortality surveillance based on the multiway GGMs we just developed. A relevant dataset could comprise counts $y_{i,j,t}$ for the number of deaths from cancer i in area j on year t, and can be seen as a three-dimensional array of size $m_C \times m_S \times m_T$. Our proposed model accounts for temporal and spatial dependence in mortality counts, as well as dependence across cancer types:

$$y_{i_C, i_S, i_T} \mid \theta_{i_C, i_S, i_T} \sim \mathsf{Poi}\left(\exp\{\mu_{i_C} + \log(h_{i_S, i_T}) + \theta_{i_C, i_S, i_T}\}\right).$$

Here, $i_C = 1, \ldots, m_C$, $i_S = 1, \ldots, m_S$, $i_T = 1, \ldots, m_T$, h_{i_S, i_T} denotes the population in area i_S during year i_T, μ_{i_C} is the mean number of deaths due to cancer i_C over the whole period and all locations, and θ_{i_C,i_S,i_T} is a zero-mean random effect, which is assigned the prior

$$\boldsymbol{\Theta} = (\theta_{i_C,i_S,i_T}) \sim \mathsf{AN}_3(\mathbf{0}; \{m_C, \mathbf{K}_C\}, \{m_S, \mathbf{K}_S\}, \{m_T, \mathbf{K}_T\}).$$

The matrix $\mathbf{K}_C$ models dependence across cancer types, $\mathbf{K}_S$ accounts for spatial dependence across neighboring areas, and $\mathbf{K}_T$ accounts for temporal dependence. Since we do not have prior information about dependence across cancer types, the prior for $\mathbf{K}_C$ is specified hierarchically by setting

$$\mathbf{K}_C \mid G_C \sim \mathsf{Wis}_{G_C}(\delta_C, \mathbf{I}_{m_C}), \qquad \mathsf{p}(G_C) \propto 1.$$

Thus, $G_C \in \mathcal{G}_{m_C}$ defines the unknown graphical model of cancer types. For the spatial component, we follow the approach of Dobra et al. (2011) and use a GGM to specify a conditionally autoregressive prior,

$$(z_S \mathbf{K}_S) \mid G_S \sim \mathsf{Wis}_{G_S}(\delta_S, (\delta_S - 2)\mathbf{D}_S),$$

where $\mathbf{D}_S = (\mathbf{E_W} - \rho\mathbf{W})^{-1}$ and W is the adjacency matrix for the m_S areas, so that $W_{i_S^1,i_S^2} = 1$ if areas i_S^1 and i_S^2 share a common border, and $W_{i_S^1,i_S^2} = 0$ otherwise, and $\mathbf{E_W} = \operatorname{diag}\{\mathbf{1}_{m_S}^T \mathbf{W}\}$. The graph G_S is fixed and given by the adjacency matrix $\mathbf{W}$. Furthermore, we assume that, *a priori*, there is a strong degree of positive spatial association, and choose a prior for spatial autocorrelation parameter ρ that gives higher probabilities to values close to 1 (Gelfand and Vounatsou, 2003):

$$\rho \sim \mathsf{Uni}(\{0, 0.05, 0.1, \ldots, 0.8, 0.82, \ldots, 0.90, 0.91, \ldots, 0.99\}).$$

For the temporal component, the prior for $\mathbf{K}_T$ is set to

$$(z_T \mathbf{K}_T) \mid G_T \sim \mathsf{Wis}_{G_T}(\delta_T, \mathbf{I}_{m_T}).$$

The graph G_T gives the temporal pattern of dependence and could be modeled in a manner similar to the graph G_C for cancer types. Instead of allowing G_T to be any graph with m_T vertices, we can constrain it to belong to a restricted set of graphs, for example, the graphs $G_T^{(1)}$, $G_T^{(2)}$, $G_T^{(3)}$, and $G_T^{(4)}$ with vertices $\{1, 2, \ldots, m_T\}$ and edges $E_T^{(1)}$, $E_T^{(2)}$, $E_T^{(3)}$, and $E_T^{(4)}$, where

$$\begin{aligned}
E_T^{(1)} &= \{(i_T - 1, i_T) : 2 \le i_T \le m_T\} \\
E_T^{(2)} &= E_T^{(1)} \cup \{(i_T - 2, i_T) : 3 \le i_T \le m_T\} \\
E_T^{(3)} &= E_T^{(2)} \cup \{(i_T - 3, i_T) : 4 \le i_T \le m_T\} \\
E_T^{(4)} &= E_T^{(3)} \cup \{(i_T - 4, i_T) : 5 \le i_T \le m_T\}.
\end{aligned}$$

These four graphs define AR(1), AR(2), AR(3), and AR(4) models. We set $\delta_C = \delta_S = \delta_T = 3$. We use a multivariate normal prior for the mean rates vector $\boldsymbol{\mu} = (\mu_1, \ldots, \mu_{m_C})^T \sim \mathsf{N}_{m_C}(\boldsymbol{\mu}^0, \boldsymbol{\Omega}^{-1})$, where $\boldsymbol{\mu}^0 = \mu_0 \mathbf{1}_{m_C}$ and $\boldsymbol{\Omega} = \omega^{-2}\mathbf{I}_{m_C}$. We set μ_0 to be the median log incidence rate across all cancers, all areas, and all time points, and ω to be twice the interquartile range of the raw log incidence rates.

The MCMC algorithm for this sparse multivariate spatiotemporal model involves iterative updates of the precision matrices $\mathbf{K}_C$, $\mathbf{K}_S$, and $\mathbf{K}_T$, as well as of the graph G_C, as described in Section 31.3. The mean rates $\boldsymbol{\mu}$ are sampled as described in Section 31.4. Here, the three-dimensional mean parameter array $\mathbf{M}$ is equal to the mean array associated with the first dimension (cancers), while the mean arrays associated with the other two dimensions (space and time) are set to zero:

$$\mathbf{M} = \boldsymbol{\mu} \circ \mathbf{1}_{m_S} \circ \mathbf{1}_{m_T}.$$

We consider the centered random effects $\widetilde{\boldsymbol{\Theta}} = \mathbf{M} + \boldsymbol{\Theta}$, which follows an array normal distribution with mean $\mathbf{M}$ and precision matrices $\mathbf{K}_C$, $\mathbf{K}_S$, and $\mathbf{K}_T$. We form $\widetilde{\boldsymbol{\Theta}}_{(1)}$—the one-mode matricization of $\widetilde{\boldsymbol{\Theta}}$. It follows that $\bar{\boldsymbol{\Theta}} = \widetilde{\boldsymbol{\Theta}}_{(1)}^T$ is a $(m_S m_T) \times m_C$ matrix that follows a matrix-variate normal distribution with mean $\mathbf{1}_{m_S m_T} \boldsymbol{\mu}^T$, row precision matrix $\bar{\mathbf{K}}_R = \mathbf{K}_T \otimes \mathbf{K}_S$, and column precision matrix $\mathbf{K}_C$. We resample $\widetilde{\boldsymbol{\Theta}}$ by sequentially updating each row vector $\bar{\boldsymbol{\Theta}}_{i\Lambda}$, $i = 1, \ldots, m_S m_T$. Conditional on the other rows of $\bar{\boldsymbol{\Theta}}$, the distribution of $\left(\bar{\boldsymbol{\Theta}}_{i\Lambda}\right)^T$ with $i = i_S i_T$ $(1 \le i_S \le m_S, 1 \le i_T \le m_T)$ is multivariate normal with mean $\mathbf{M}_i$ and precision matrix $\mathbf{V}_i$, where

$$\mathbf{M}_i = \boldsymbol{\mu} - \sum_{i'=1}^{m_S m_T} \frac{(\bar{\mathbf{K}}_R)_{ii'}}{(\bar{\mathbf{K}}_R)_{ii}} \left[\left(\bar{\boldsymbol{\Theta}}_{i'\Lambda}\right)^T - \boldsymbol{\mu} \right], \quad \mathbf{V}_i = \left(\bar{\mathbf{K}}_R\right)_{ii} \mathbf{K}_C.$$

Thus, the full conditional distribution of $\bar{\boldsymbol{\Theta}}_{i\Lambda}$ is proportional to

$$\prod_{i_C=1}^{m_C} \exp\left\{ y_{i_C,i_S,i_T} \left(\mu_{i_C} + \log(h_{i_S,i_T}) + \theta_{i_C,i_S,i_T} \right) - h_{i_S,i_T} \exp\left(\mu_{i_C} + \theta_{i_C,i_S,i_T} \right) \right\}$$
$$\times \exp\left\{ -\frac{1}{2} \left[\bar{\boldsymbol{\Theta}}_{i\Lambda} - (\mathbf{M}_i)^T \right] \mathbf{V}_i \left[\left(\bar{\boldsymbol{\Theta}}_{i\Lambda}\right)^T - \mathbf{M}_i \right] \right\}. \tag{31.18}$$

We make use of a Metropolis–Hastings step to sample from (31.18). We consider a strictly positive precision parameter $\widetilde{\sigma}$. For each $i_C = 1, \ldots, m_C$, we update the i_Cth element of $\bar{\boldsymbol{\Theta}}_{i\Lambda}$ by sampling $\gamma \sim \mathsf{N}\left(\bar{\Theta}_{i,i_C}, \widetilde{\sigma}^2\right)$. We define a candidate row vector $\bar{\boldsymbol{\Theta}}_{i\Lambda}^{new}$ by replacing $\bar{\Theta}_{i,i_C}$ with γ in $\bar{\boldsymbol{\Theta}}_{i\Lambda}$. We update the current ith row of $\bar{\boldsymbol{\Theta}}$ with $\bar{\boldsymbol{\Theta}}_{i\Lambda}^{new}$ with the Metropolis–Hastings acceptance probability corresponding with (31.18). Otherwise, the ith row of $\bar{\boldsymbol{\Theta}}$ remains unchanged.

31.6 Dynamic Multiway GGMs for Array-Variate Time Series

The cancer mortality surveillance application from Section 31.5 represented the time component as one of the dimensions of the three-dimensional array of observed counts. We give an extension of multiway GGMs to array-variate time series $\mathbf{Y}_t$, $t = 1, 2, \ldots, T$, where $\mathbf{Y}_t \in \mathbb{R}^{m_1 \times \ldots \times m_L}$. Our framework generalizes the results from Carvalho and West (2007a, 2007b) and Wang and West (2009), which assume vector ($L = 1$) or matrix-variate ($L = 2$) time series. We build on the standard specification of Bayesian dynamic linear models (West and Harrison, 1997) and assume that $\mathbf{Y}_t$ is modeled over time by

$$\mathbf{Y}_t = \boldsymbol{\Theta}_t \times_{L+1} \mathbf{F}_t^T + \boldsymbol{\Psi}_t, \quad \boldsymbol{\Psi}_t \sim \mathsf{AN}_L(\mathbf{0}; \{m_1, v_t^{-1}\mathbf{K}_1\}, \{m_2, \mathbf{K}_2\}, \ldots, \{m_L, \mathbf{K}_L\}) \tag{31.19}$$

$$\boldsymbol{\Theta}_t = \boldsymbol{\Theta}_{t-1} \times_{L+1} \mathbf{H}_t + \boldsymbol{\Gamma}_t, \quad \boldsymbol{\Gamma}_t \sim \mathsf{AN}_{L+1}(\mathbf{0}; \{m_1, \mathbf{K}_1\}, \ldots, \{m_L, \mathbf{K}_L\}, \{s, \mathbf{W}_t^{-1}\}), \tag{31.20}$$

where (1) $\boldsymbol{\Theta}_t \in \mathbb{R}^{m_1 \times \ldots \times m_L \times s}$ is the state array at time t, (2) $\mathbf{F}_t \in \mathbb{R}^s$ is a vector of known regressors at time t, (3) $\boldsymbol{\Psi}_t \in \mathbb{R}^{m_1 \times \ldots \times m_L}$ is the array of observational errors at time t, (4) $\mathbf{H}_t$

is a known $s \times s$ state evolution matrix at time t, (5) $\mathbf{\Gamma}_t \in \mathbb{R}^{m_1 \times \ldots \times m_L \times s}$ is the array of state evolution innovations at time t, (6) $\mathbf{W}_t$ is the $s \times s$ innovation covariance matrix at time t, and (7) $v_t > 0$ is a known scale factor at time t. Furthermore, the observational errors $\mathbf{\Psi}_t$ and the state evolution errors $\mathbf{\Gamma}_t$ follow zero-mean array normal distributions defined by $\mathbf{K}_1, \ldots, \mathbf{K}_L$ and $\mathbf{W}_t$, and are assumed to be both independent over time element-wise and mutually independent as sequences of arrays.

The observation equation (31.19) and the evolution equation (31.20) translate into the following dynamic linear model for the univariate time series $(\mathbf{Y}_t)_{i_1 \ldots i_L}$, $t = 1, 2, \ldots, T$:

$$(\mathbf{Y}_t)_{i_1 \ldots i_L} = \mathbf{F}_t^T (\mathbf{\Theta}_t)_{i_1 \ldots i_L \star} + (\mathbf{\Psi}_t)_{i_1 \ldots i_L}, \quad (\mathbf{\Psi}_t)_{i_1 \ldots i_L} \sim \mathsf{N}\left(0, v_t \prod_{l=1}^{L} (\mathbf{K}_l^{-1})_{i_l i_l}\right)$$

$$(\mathbf{\Theta}_t)_{i_1 \ldots i_L, \star} = \mathbf{H}_t (\mathbf{\Theta}_{t-1})_{i_1 \ldots i_L, \star} + (\mathbf{\Gamma}_t)_{i_1 \ldots i_L, \star}, \quad (\mathbf{\Gamma}_t)_{i_1 \ldots i_L, \star} \sim \mathsf{N}_s\left(\mathbf{0}, \prod_{l=1}^{L} (\mathbf{K}_l^{-1})_{i_l i_l} \mathbf{W}_t\right),$$

where $(\mathbf{\Theta}_t)_{i_1 \ldots i_L, \star} = ((\mathbf{\Theta}_t)_{i_1 \ldots i_L 1}, \ldots, (\mathbf{\Theta}_t)_{i_1 \ldots i_L s})^T$, while $(\mathbf{\Theta}_{t-1})_{i_1 \ldots i_L, \star}$ and $(\mathbf{\Gamma}_t)_{i_1 \ldots i_L, \star}$ are defined in a similar manner. The components $\mathbf{F}_t$, $\mathbf{H}_t$, and $\mathbf{W}_t$ are the same for all univariate time series, but the state parameters $(\mathbf{\Theta}_t)_{i_1 \ldots, i_L \star}$, as well as their scales of measurement defined by $\prod_{l=1}^{L} (\mathbf{K}_l^{-1})_{i_l i_l}$, could be different across series. The cross-sectional dependence structure across individual time series at time t is induced by $\mathbf{K}_1, \ldots, \mathbf{K}_L$ and $\mathbf{W}_t$:

$$\mathsf{Cov}\left((\boldsymbol{\nu}_t)_{i_1 \ldots i_L}, (\boldsymbol{\nu}_t)_{i_1' \ldots i_L'}\right) = v_t \prod_{l=1}^{L} (\mathbf{K}_l^{-1})_{i_l i_l'},$$

$$\mathsf{Cov}\left((\mathbf{\Gamma}_t)_{i_1 \ldots i_L, \star}, (\mathbf{\Gamma}_t)_{i_1' \ldots i_L', \star}\right) = \prod_{l=1}^{L} (\mathbf{K}_l^{-1})_{i_l i_l'} \mathbf{W}_t.$$

For example, if $\prod_{l=1}^{L} (\mathbf{K}_l^{-1})_{i_l i_l'}$ is large in absolute value, the univariate time series $(\mathbf{Y}_t)_{i_1 \ldots i_L}$ and $(\mathbf{Y}_t)_{i_1' \ldots i_L'}$ exhibit significant dependence on the variation of their observational errors and state vectors. Appropriate choices for the matrix sequence $\mathbf{W}_t$, $t = 1, 2, \ldots, T$, arise from the discount factors discussed in West and Harrison (1997), as exemplified, among others, in Wang and West (2009). The scale factors v_t can be set to 1, but other suitable values can be employed as needed.

The following result extends Theorem 1 of Wang and West (2009) to array-variate time series.

Theorem 31.1. *Let $\mathcal{D}_0$ be the prior information and denote by $\mathcal{D}_t = \{\mathbf{Y}_t, \mathcal{D}_{t-1}\}$ the information available at time $t = 1, 2, \ldots, T$. We assume to have specified precision matrices $\mathbf{K}_1, \ldots, \mathbf{K}_L$, the matrix sequence $\mathbf{W}_t$, $t = 1, 2, \ldots, T$, as well as an initial prior for the state array at time 0,*

$$(\mathbf{\Theta}_0 \mid \mathcal{D}_0) \sim \mathsf{AN}_{L+1}(\mathbf{M}_0; \{m_1, \mathbf{K}_1\}, \ldots, \{m_L, \mathbf{K}_L\}, \{s, \mathbf{C}_0^{-1}\}),$$

where $\mathbf{M}_0 \in \mathbb{R}^{m_1 \times \ldots \times m_L \times s}$ and $\mathbf{C}_0$ is an $s \times s$ covariance matrix. For every $t = 1, 2, \ldots, T$, the following distributional results hold:

i. Posterior at $t-1$:

$$(\mathbf{\Theta}_{t-1} \mid \mathcal{D}_{t-1}) \sim \mathsf{AN}_{L+1}(\mathbf{M}_{t-1}; \{m_1, \mathbf{K}_1\}, \ldots, \{m_L, \mathbf{K}_L\}, \{s, \mathbf{C}_{t-1}^{-1}\})$$

ii. Prior at t:

$$(\mathbf{\Theta}_t \mid \mathcal{D}_{t-1}) \sim \mathsf{AN}_{L+1}(\mathbf{a}_t; \{m_1, \mathbf{K}_1\}, \ldots, \{m_L, \mathbf{K}_L\}, \{s, \mathbf{R}_t^{-1}\}),$$

where $\mathbf{a}_t = \mathbf{M}_{t-1} \times_{L+1} \mathbf{H}_t$ and $\mathbf{R}_t = \mathbf{H}_t \mathbf{C}_{t-1} \mathbf{H}_t^T + \mathbf{W}_t$

iii. *One-step forecast at $t-1$:*

$$(\mathbf{Y}_t \mid \mathcal{D}_{t-1}) \sim \mathsf{AN}_L(\mathbf{f}_t; \{m_1, q_t^{-1}\mathbf{K}_1\}, \{m_2, \mathbf{K}_2\}, \ldots, \{m_L, \mathbf{K}_L\}),$$

where $\mathbf{f}_t = \mathbf{M}_{t-1} \times_{L+1} (\mathbf{F}_t^T \mathbf{H}_t) = \mathbf{a}_t \times_{L+1} \mathbf{F}_t^T$ *and* $q_t = \mathbf{F}_t^T \mathbf{R}_t \mathbf{F}_t + v_t$

iv. *Posterior at t:*

$$(\mathbf{\Theta}_t \mid \mathcal{D}_t) \sim \mathsf{AN}_{L+1}(\mathbf{M}_t; \{m_1, \mathbf{K}_1\}, \ldots, \{m_L, \mathbf{K}_L\}, \{s, \mathbf{C}_t^{-1}\}),$$

where $\mathbf{M}_t = \mathbf{a}_t + \mathbf{e}_t \times_{L+1} \mathbf{A}_t$, $\mathbf{C}_t = \mathbf{R}_t - \mathbf{A}_t \mathbf{A}_t^T q_t$. *Here* $\mathbf{A}_t = q_t^{-1} \mathbf{R}_t \mathbf{F}_t$ *and* $\mathbf{e}_t = \mathbf{Y}_t - \mathbf{f}_t$

The proof of Theorem 31.1 is straightforward. We write Equations 31.19 and 31.20 in matrix form:

$$(\mathbf{Y}_t)_{(L+1)} = \mathbf{F}_t^T (\mathbf{\Theta}_t)_{(L+1)} + (\mathbf{\Psi}_t)_{(L+1)}, \quad (\mathbf{\Psi}_t)_{(L+1)} \sim \mathsf{N}_m \left(\mathbf{0}, v_t \mathbf{K}^{-1}\right) \tag{31.21}$$

$$(\mathbf{\Theta}_t)_{(L+1)} = \mathbf{H}_t (\mathbf{\Theta}_{t-1})_{(L+1)} + (\mathbf{\Psi}_t)_{(L+1)}, \quad (\mathbf{\Psi}_t)_{(L+1)} \sim \mathsf{AN}_2 \left(\mathbf{0}; \{s, \mathbf{W}_t^{-1}\}, \{m, \mathbf{K}\}\right), \tag{31.22}$$

where $m = \prod_{l=1}^{L} m_l$ and $\mathbf{K}$ is given in Equation 31.6. The normal theory results laid out in West and Harrison (1997) apply directly to the dynamic linear model specified by Equations 31.21 and 31.22. The predictive distributions relevant for forecasting and retrospective sampling for array-variate time series can be derived from the corresponding predictive distributions for vector data.

We complete the definition and prior specification for the dynamic multiway GGMs with independent G-Wishart priors from Equation 31.10 for the precision matrices $\mathbf{K}_1, \ldots, \mathbf{K}_L$ and their corresponding auxiliary variables $z_2, \ldots, z_L$. The graphs $G_1, \ldots, G_L$ associated with the G-Wishart priors receive independent priors $\pi_{m_l}(G_l)$ on $\mathcal{G}_{m_l}$, $l = 1, \ldots, L$. Posterior inference in this framework can be achieved with the following MCMC algorithm that sequentially performs the following steps:

1. *Resampling the precision matrices, graphs, and auxiliary variables.* By marginalizing over the state arrays $\mathbf{\Theta}_1, \ldots, \mathbf{\Theta}_L$, we obtain the marginal likelihood (Carvalho and West, 2007a, 2007b):

$$\mathsf{p}\left(\mathbf{Y}_1, \ldots, \mathbf{Y}_T \mid \mathbf{K}_1, G_1, (\mathbf{K}_l, G_l, z_l)_{l=2}^{L}\right) = \prod_{t=1}^{T} \mathsf{p}\left(\mathbf{Y}_t \mid \mathcal{D}_{t-1}, \mathbf{K}_1, G_1, (\mathbf{K}_l, G_l, z_l)_{l=2}^{L}\right).$$

The one-step forecast distribution (iii) from Theorem 31.1 implies that

$$\left(q_t^{-1/2}(\mathbf{Y}_t - \mathbf{f}_t) \mid \mathcal{D}_{t-1}\right) \sim \mathsf{AN}_L(\mathbf{0}; \{m_1, \mathbf{K}_1\}, \{m_2, \mathbf{K}_2\}, \ldots, \{m_L, \mathbf{K}_L\}).$$

We use the filtering equations from Theorem 31.1 to produce the centered and scaled array data $\bar{\mathcal{D}} = \{q_t^{-1/2}(\mathbf{Y}_t - \mathbf{f}_t) : t = 1, \ldots, T\}$. Since the elements of $\bar{\mathcal{D}}$ are independent and identically distributed, we update each precision matrix $\mathbf{K}_l$, graph G_l, and auxiliary variable z_l as described in Section 31.3 based on $\bar{\mathcal{D}}$.

2. *Resampling the state arrays.* We employ the forward filtering backward algorithm (FFBS) proposed by Carter and Kohn (1994) and Frühwirth-Schnatter (1994). Given the current sampled precision matrices, we start by sampling $\mathbf{\Theta}_T$ given $\mathcal{D}_T$ from the posterior distribution given in (iv) of Theorem 31.1.

Then, for $t = T-1, T-2, \ldots, 0$, we sample $\mathbf{\Theta}_t$ given $\mathcal{D}_T$ and $\mathbf{\Theta}_{t+1}$ from the array normal distribution

$$\mathsf{AN}_{L+1}\left(\mathbf{M}_t^*; \{m_1, \mathbf{K}_1\}, \ldots, \{m_L, \mathbf{K}_L\}, \{s, (\mathbf{C}_t^*)^{-1}\}\right),$$

where

$$\mathbf{M}_t^* = \mathbf{M}_t + (\mathbf{\Theta}_{t+1} - \mathbf{a}_{t+1}) \times_{L+1} \left(\mathbf{C}_t \mathbf{G}_{t+1}^T \mathbf{R}_{t+1}^{-1}\right).$$

31.7 Discussion

Recent advances in data collection techniques have allowed the creation of high-dimensional public health datasets that monitor the incidence of many diseases across several areas, time points, and additional ecological sociodemographic groupings (Elliott et al., 2001). Jointly modeling the disease risk associated with each resulting cell count (i.e., a particular disease at a particular time point in a particular region given a particular combination of risk factors) is desirable since it takes into consideration interaction patterns that arise within each dimension or across dimensions. By aggregating data across time, key epidemiological issues related to the evolution of the risk patterns across time might not be given an appropriate answer (Abellan et al., 2008). The spatial structure of geographical regions must also be properly accounted for (Besag, 1974; Besag et al., 1991). Furthermore, since diseases are potentially related and share risk factors, it is critical that individual models should not be developed for each disease (Gelfand and Vounatsou, 2003; Wang and Wall, 2003). Rich, flexible classes of models that capture the joint variation of disease risk in the actual observed data without requiring the aggregation across one or more dimensions will be the fundamental aim of our proposed work related to disease mapping. Multiway GGMs can be used in Bayesian hierarchical models that produce estimates of disease risk by borrowing strength across time, areas, and the other dimensions. Due to the likely presence of small counts in many cells, the degree of smoothing will be controlled through a wide range of parameters that could be constrained to zero according to predefined interaction structures (e.g., the neighborhood structure of the areas) or by graphs that received the most support given the data.

We generalize the models from Section 31.5, and let $\mathbf{Y}$ be the L-dimensional array of observed disease counts $Y_{i_1 \ldots i_L}$ indexed by cells $\{(i_1, \ldots, i_L) : 1 \leq i_l \leq m_l\}$. We assume that the count random variable $Y_{i_1 \ldots i_L}$ associated with cell $(i_1, \ldots, i_L)$ follows a distribution from an exponential family (e.g., Poisson or binomial) with mean parameter $\theta_{i_1 \ldots i_L}$, that is,

$$Y_{i_1 \ldots i_L} \mid \theta_{i_1 \ldots i_L} \overset{iid}{\sim} H(\theta_{i_1 \ldots i_L}), \text{ for } 1 \leq i_l \leq m_l, 1 \leq l \leq L. \tag{31.23}$$

We assume that the cell counts $\mathbf{Y}$ are conditionally independent given the L-dimensional array of parameters $\mathbf{\Theta} = \{\theta_{i_1 \ldots i_L} : 1 \leq i_l \leq m_l\}$. Furthermore, given a certain link function $g(\cdot)$ (e.g., $\log(\cdot)$), the parameters $\mathbf{\Theta}$ follow a joint model

$$g(\theta_{i_1 \ldots i_L}) = \nu_{i_1 \ldots i_L} + X_{i_1 \ldots i_L}, \tag{31.24}$$

where $\nu_{i_1 \ldots i_L}$ is a known offset, while $\mathbf{X} = \{X_{i_1 \ldots i_L} : 1 \leq i_l \leq m_l\}$ is an array of zero-centered random effects. Equation 31.24 can subsequently include explanatory ecological covariates as needed. The multiway GGMs are employed in the context of non-Gaussian data as joint distribution for the array of random effects $\mathbf{X}$. Thus, $\mathbf{X}$ is assumed to follow the flexible joint distributions, and each dimension of the data is represented as a GGM in a particular dimension of the random effects $\mathbf{X}$.

This framework accommodates many types of interactions by restricting the set of graphs that are allowed to represent the dependency patterns of the corresponding dimensions. For example, if dimension l' of $\mathbf{Y}$ represents time, then the graphs associated with this dimension could be constrained to represent an autoregressive model AR(q), where $q = 1, 2, 3, \ldots$; see Section 31.5. Temporal dependence can also be modeled with the dynamic multiway GGMs from Section 31.6. If dimension l'' of $\mathbf{Y}$ represents spatial dependence, one could constrain the space of graphs for dimension l'' to consist of only one graph with edges defined by areas that are neighbors of each other in the spirit of Besag (1974) and Clayton and Kaldor (1987). As opposed to a modeling framework based on GMRFs, we can allow uncertainty around this neighborhood graph, in which case we let the space of graphs for dimension l'' include graphs that are obtained by adding or deleting one, two, or more edges from the neighborhood graph. This expansion of the set of spatial graphs is consistent with the hypothesis that interaction occurs not only between areas that are close to each other or share a border, but also between more distant areas. We can also allow all possible graphs to be associated with dimension l'' and examine the graphs that receive the highest posterior probabilities. Such graphs can be further compared with the neighborhood graph to see whether the spatial dependency patterns in observed data are actually consistent with the geographical neighborhoods.

To gain further insight on the flexibility of our modeling approach, we examine the case in which a two-dimensional array $\mathbf{Y} = \{Y_{i_1 i_2}\}$ is observed with the first dimension associated with m_1 diseases and the second dimension associated with m_2 areas. Under the framework of Gelfand and Vounatsou (2003) and Carlin and Banerjee (2003), the matrix of counts $\mathbf{Y}$ is modeled with a hierarchical Poisson model with random effects distributed as a multivariate CAR (MCAR) model (Mardia, 1988):

$$\text{vec}(\mathbf{X}') \sim \mathsf{N}_{m_1 m_2}(0, [\mathbf{K}_1 \otimes (\mathbf{E}_{\mathbf{W}} - \rho \mathbf{W})]^{-1}). \tag{31.25}$$

This structure of the random effects assumes separability of the association structure among diseases from the spatial structure (Waller and Carlin, 2010). The spatial autocorrelation parameter ρ is the only parameter that controls the strength of spatial dependencies, while the precision matrix $\mathbf{K}_1$ is not subject to any additional constraints on its elements. In our framework, the random effects $\mathbf{X}'$ follow a matrix-variate GGM prior obtained by taking $L = 2$ in Equation 31.6. The same separability of the association structure is assumed, but the precision matrices $\mathbf{K}_1$ and $\mathbf{K}_2$ follow G-Wishart hyperpriors as in Equation 31.10. The GGMs associated with the diseases are allowed to vary across all possible graphs with m_1 vertices, while the GGMs for the spatial structure can be modeled as we described earlier in this section.

The multivariate associations among the diseases are accounted for through Bayesian model averaging across the most relevant graphs supported by the available data. Since datasets typically provide scarce information to properly quantify the strength of associations between diseases, it is very likely to find not one, but many graphs that differ by only one, two, or a reduced set of edges that receive the highest posterior probabilities. These graphs will likely be sparse, thereby reducing the number of parameters used to model the associations between m_1 diseases from $m_1(m_1+1)/2$ (the diagonal and the unique off-diagonal elements of $\mathbf{K}_1$) to a much smaller number equal to m_1 plus the number of edges of a graph. Another approach for reducing the number of parameters that define the multivariate dependence between the diseases has been proposed by Zhang et al. (2009), and it is called smoothed analysis of variance (SANOVA). This class of models is also discussed, together with its extensions, in Chapter 32 in this volume. It is very flexible, as it can accommodate different specifications of design matrices among diseases. A design matrix gives m_1 linear combinations of the diseases that are represented by m_1 fixed effects. Therefore, the between-disease associations are modeled in SANOVAs with m_1 parameters, which leads to a parsimonious, interpretable, and computationally efficient framework.

Until recently, the application of GGMs with a G-Wishart prior for the precision matrix in large-scale Bayesian hierarchical models has been hindered by computational difficulties. For decomposable graphs, the normalizing constant of the G-Wishart distribution is calculated with formulas (Roverato, 2002; Atay-Kayis and Massam, 2005), and a direct sampler from this distribution existed for several years (Carvalho et al., 2007). But similar results did not exist for nondecomposable graphs. Fortunately, new methodological developments give formulas for the calculation of the G-Wishart distribution for arbitrary graphs (Uhler et al., 2014), and also a direct sampler for arbitrary graphs (Lenkoski, 2013). With these key results, the MCMC sampler developed in Section 31.3 can be significantly improved in its efficiency. The reversible jump algorithm that allows updates in the structure of the graphs associated with the dimensions of a multiway GGM can be subsequently refined to another transdimensional graph updating algorithm that bypasses the calculation of any normalizing constants of the G-Wishart distribution based on the double reversible jump algorithm of Wang and Li (2012) and Lenkoski (2013), or the birth–death MCMC sampling algorithm of Mohammadi and Wit (2015). Moreover, the G-Wishart distribution can be replaced altogether in the specification of priors for Bayesian hierarchical spatial models with the graphical lasso prior of Wang (2012). The application of these new theoretical results to spatial health data is a very intense area of research. In one of the latest contributions, Smith et al. (2015) developed a new distribution, which they call the negative G-Wishart distribution, that leads to positive associations between the random effects of neighboring regions while preserving the conditional independence of nonneighboring regions.

References

J. J. Abellan, S. Richardson, and N. Best. Use of space-time models to investigate the stability of patterns of disease. *Environmental Health Perspectives*, 116:1111–1119, 2008.

G. I. Allen. Comment on article by Hoff. *Bayesian Analysis*, 6:197–202, 2011.

G. I. Allen and R. Tibshirani. Inference with transposable data: Modeling the effects of row and column correlations. arXiv:1004.0209v1 [stat.ME], 2010.

A. Atay-Kayis and H. Massam. A Monte Carlo method for computing the marginal likelihood in nondecomposable Gaussian graphical models. *Biometrika*, 92:317–335, 2005.

J. Barnard, R. McCulloch, and X. Meng. Modeling covariance matrices in terms of standard deviations and correlations, with application to shrinkage. *Statistica Sinica*, 10:1281–311, 2000.

J. Besag. Spatial interaction and the statistical analysis of lattice systems [with discussion]. *Journal of the Royal Statistical Society: Series B*, 36:192–236, 1974.

J. Besag. Statistical analysis of non-lattice data. *The Statistician*, 24:179–195, 1975.

J. Besag and C. Kooperberg. On conditional and intrinsic autoregressions. *Biometrika*, 82:733–746, 1995.

J. Besag, J. York, and A. Mollié. Bayesian image restoration, with two applications in spatial statistics. *Annals of the Institute of Statistical Mathematics*, 43:1–59, 1991.

P. J. Bickel and E. Levina. Regularized estimation of large covariance matrices. *Annals of Statistics*, 36:199–227, 2008.

B. P. Carlin and S. Banerjee. Hierarchical multivariate CAR models for spatio-temporally correlated survival data [with discussion]. In J. M. Bernardo, M. J. Bayarri, J. O. Berger, A. P. Dawid, D. Heckerman, A. F. M. Smith, and M. West, eds. *Bayesian Statistics 7*, pp. 45–63. Oxford: Oxford University Press, 2003.

C. K. Carter and R. Kohn. Gibbs sampling for state space models. *Biometrika*, 81:541–553, 1994.

C. M. Carvalho, H. Massam, and M. West. Simulation of hyper-inverse Wishart distributions in graphical models. *Biometrika*, 94:647–659, 2007.

C. M. Carvalho and M. West. Dynamic matrix-variate graphical models. *Bayesian Analysis*, 2:69–98, 2007a.

C. M. Carvalho and M. West. Dynamic matrix-variate graphical models—a synopsis. In J. M. Bernardo, M. J. Bayarri, J. O. Berger, A. P. Dawid, D. Heckerman, A. F. M. Smith, and M. West, eds. *Bayesian Statistics VIII*, pp. 585–590. Oxford: Oxford University Press, 2007b.

D. G. Clayton and J. M. Kaldor. Empirical Bayes estimates of age-standardized relative risks for use in disease mapping. *Biometrics*, 43:671–681, 1987.

N. A. C. Cressie. *Statistics for Spatial Data*. New York: Wiley, 1973.

M. Daniels and R. Kass. Nonconjugate Bayesian estimation of covariance matrices. *Journal of the American Statistical Association*, 94:1254–1263, 1999.

A. P. Dawid. Some matrix-variate distribution theory: Notational considerations and a Bayesian application. *Biometrika*, 68:265–274, 1981.

L. De Lathauwer, B. De Moor, and J. Vandewalle. A multilinear singular value decomposition. *SIAM Journal on Matrix Analysis and Applications*, 21: 1253–1278, 2000.

P. Dellaportas, P. Giudici, and G. Roberts. Bayesian inference for nondecomposable graphical Gaussian models. *Sankhyā*, 65:43–55, 2003.

A. P. Dempster. Covariance selection. *Biometrics*, 28:157–175, 1972.

P. Diaconnis and D. Ylvisaker. Conjugate priors for exponential families. *Annals of Statistics*, 7:269–281, 1979.

A. Dobra, C. Hans, B. Jones, J. R. Nevins, G. Yao, and M. West. Sparse graphical models for exploring gene expression data. *Journal of Multivariate Analysis*, 90:196–212, 2004.

A. Dobra and A. Lenkoski. Copula Gaussian graphical models and their application to modeling functional disability data. *Annals of Applied Statistics*, 5:969–993, 2011.

A. Dobra, A. Lenkoski, and A. Rodriguez. Bayesian inference for non-decomposable general Gaussian graphical models with application to multivariate lattice data. *Journal of the American Statistical Association*, 106:1418–1433, 2011.

M. Drton and M. D. Perlman. Model selection for Gaussian concentration graphs. *Biometrika*, 91:591–602, 2004.

P. Elliott, J. Wakefield, N. Best, and D. Briggs. *Spatial Epidemiology: Methods and Applications.* Oxford: Oxford University Press, 2001.

J. Friedman, T. Hastie, and R. Tibshirani. Sparse inverse covariance estimation with the graphical lasso. *Biostatistics*, 3:432–441, 2008.

S. Frühwirth-Schnatter. Data augmentation and dynamic linear models. *Journal of Time Series Analysis*, 15:183–202, 1994.

A. T. Galecki. General class of covariance structures for two or more repeated factors in longitudinal data analysis. *Communications in Statistics—Theory and Methods*, 23:3105–3119, 1994.

A. E. Gelfand, P. J. Diggle, M. Fuentes, and P. Guttorp. *Handbook of Spatial Statistics.* Boca Raton, FL: CRC Press, Taylor & Francis Group, 2010.

A. E. Gelfand and P. Vounatsou. Proper multivariate conditional autoregressive models for spatial data analysis. *Biostatistics*, 4:11–25, 2003.

P. Giudici and P. J. Green. Decomposable graphical Gaussian model determination. *Biometrika*, 86:785–801, 1999.

P. J. Green. Reversible jump Markov chain Monte Carlo computation and Bayesian model determination. *Biometrika*, 82:711–732, 1995.

P. D. Hoff. Separable covariance arrays via the Tucker product, with applications to multivariate relational data. Technical report. Department of Statistics, University of Washington, 2010.

X. Jin, S. Banerjee, and B. P. Carlin. Order-free co-regionalized areal data models with application to multiple-disease mapping. *Journal of the Royal Statistical Society: Series B*, 69:817–838, 2007.

X. Jin, B. P. Carlin, and S. Banerjee. Generalized hierarchical multivariate CAR models for areal data. *Biometrics*, 61:950–961, 2005.

B. Jones, C. Carvalho, A. Dobra, C. Hans, C. Carter, and M. West. Experiments in stochastic computation for high-dimensional graphical models. *Statistical Science*, 20:388–400, 2005.

R. Kass and A. E. Raftery. Bayes factors. *Journal of the American Statistical Association*, 90:773–795, 1995.

T. G. Kolda. Multilinear operators for higher-order decompositions. Technical report. Albuquerque, NM: Sandia National Laboratories, 2006.

S. L. Lauritzen. *Graphical Models.* Oxford: Oxford University Press, 1996.

A. Lenkoski. A direct sampler for G-Wishart variates. *Stat*, 2:119–128, 2013.

A. Lenkoski and A. Dobra. Computational aspects related to inference in Gaussian graphical models with the G-Wishart prior. *Journal of Computational and Graphical Statistics*, 20:140–157, 2011.

T. Leonard and J. S. J. Hsu. Bayesian inference for a covariance matrix. *Annals of Statistics*, 20:1669–1696, 1992.

G. Letac and H. Massam. Wishart distributions for decomposable graphs. *Annals of Statistics*, 35:1278–1323, 2007.

J. C. Liechty, M. W. Liechty, and P. Müller. Bayesian correlation estimation. *Biometrika*, 91:1–14, 2004.

K. V. Mardia. Multi-dimensional multivariate Gaussian Markov random fields with application to image processing. *Journal of Multivariate Analysis*, 24:265–284, 1988.

N. Meinshausen and P. Bühlmann. High-dimensional graphs with the lasso. *Annals of Statistics*, 34:1436–1462, 2006.

A. Mohammadi and E. C. Wit. Bayesian structure learning in sparse Gaussian graphical models. *Bayesian Analysis*, 10:109–138, 2015.

R. A. Olshen and B. Rajaratnam. Successive normalization of rectangular arrays. *Annals of Statistics*, 38:1638–1664, 2010.

B. Rajaratnam, H. Massam, and C. M. Carvalho. Flexible covariance estimation in graphical Gaussian models. *Annals of Statistics*, 36:2818–2849, 2008.

A. Roverato. Hyper inverse Wishart distribution for non-decomposable graphs and its application to Bayesian inference for Gaussian graphical models. *Scandinavian Journal of Statistics*, 29:391–411, 2002.

H. Rue and L. Knorr-Held. *Gaussian Markov Random Fields: Theory and Applications.* Boca Raton, FL: Chapman & Hall/CRC, 2005.

J. G. Scott and C. M. Carvalho. Feature-inclusion stochastic search for Gaussian graphical models. *Journal of Computational and Graphical Statistics*, 17:790–808, 2008.

M. Smith and R. Kohn. Bayesian parsimonious covariance matrix estimation for longitudinal data. *Journal of the American Statistical Association*, 87:1141–1153, 2002.

T. R. Smith, J. Wakefield, and A. Dobra. Restricted covariance priors with applications in spatial statistics. *Bayesian Analysis*, 10: 965–990, 2015.

C. Uhler, A. Lenkoski, and D. Richards. Exact formulas for the normalizing constants of Wishart distributions for graphical models. arXiv:1406.4901 [math.ST], 2014.

J. Wakefield. Disease mapping and spatial regression with count data. *Biostatistics*, 8:158–183, 2007.

L. Waller and B. P. Carlin. Disease mapping. In A. E. Gelfand, P. J. Diggle, M. Fuentes, and P. Guttorp, eds. *Handbook of Spatial Statistics*, pp. 217–243. Boca Raton, FL: CRC Press, Taylor & Francis Group, 2010.

F. Wang and M. M. Wall. Generalized common factor spatial model. *Biostatistics*, 4:569–582, 2003.

H. Wang. The Bayesian graphical lasso and efficient posterior computation. *Bayesian Analysis*, 7:771–790, 2012.

H. Wang and S. Z. Li. Efficient Gaussian graphical model determination under G-Wishart prior distributions. *Electronic Journal of Statistics*, 6:168–198, 2012.

H. Wang and M. West. Bayesian analysis of matrix normal graphical models. *Biometrika*, 96:821–834, 2009.

N. Wermuth. Analogies between multiplicative models in contingency tables and covariance selection. *Biometrics*, 32:95–108, 1976.

M. West and J. Harrison. *Bayesian Forecasting and Dynamic Models*, 2nd ed. New York: Springer-Verlag, 1997.

J. Whittaker. *Graphical Models in Applied Multivariate Statistics.* New York: John Wiley & Sons, 1990.

F. Wong, C. K. Carter, and R. Kohn. Efficient estimation of covariance selection models. *Biometrika*, 90:809–830, 2003.

R. Yang and J. O. Berger. Estimation of a covariance matrix using the reference prior. *Annals of Statistics*, 22:1195–1211, 1994.

M. Yuan and Y. Lin. Model selection and estimation in the Gaussian graphical model. *Biometrika*, 94:19–35, 2007.

Y. Zhang, J. S. Hodges, and S. Banerjee. Smoothed ANOVA with spatial effects as a competitor to MCAR in multivariate spatial smoothing. *Annals of Applied Statistics*, 3:1805–1830, 2009.

32

Smoothed ANOVA Modeling

Miguel A. Martinez-Beneito
Fundación para el Fomento de la Investigación Sanitaria y Biomédica de la Comunidad Valenciana (FISABIO)
Valencia, Spain
and
CIBER de Epidemiología y Salud Publica (CIBERESP)
Madrid, Spain

James S. Hodges
Division of Biostatistics
School of Public Health
University of Minnesota
Minneapolis, Minnesota

Marc Marí-Dell'Olmo
CIBER de Epidemiología y Salud Pública (CIBERESP)
Madrid, Spain
and
Agència de Salut Pública de Barcelona
and
Institut d'Investigació Biomèdica Sant Pau (IIB Sant Pau)
Barcelona, Spain

CONTENTS

Smoothed analysis of variance, usually known as SANOVA, was proposed in different forms with different goals by Nobile and Green [1], Gelman [2], and Hodges et al. [3]. This chapter builds on the latter, which proposed a method for smoothing effects in balanced ANOVAs having a single error term, that is, without random effects as understood by, for example, Scheffé [4]. Zhang et al. [5] applied this approach to multivariate disease mapping as a simpler alternative to the intrinsic multivariate conditional autoregressive (MCAR) distribution, often used to analyze multivariate areal data (Section 1 of Zhang et al. [5] gives citations to pertinent MCAR literature). This application of SANOVA used specific known linear combinations of the diseases under study, presumably with particular meanings, to structure the covariance among diseases, which in most multivariate analyses is usually assumed to be unknown and unstructured [6, 7]. More recently Marí-Dell'Olmo et al. [8] proposed a reformulation of SANOVA for disease mapping that is simpler to implement and allows extensions such as multivariate ecological regression and spatiotemporal modeling.

This chapter reviews the SANOVA approach and shows some modeling possibilities it allows. The chapter is organized as follows: Section 32.1 introduces the original formulation of SANOVA for multivariate disease mapping and the advantageous reformulation. Section 32.2 discusses some settings where this approach can be applied, beyond its original use for multivariate modeling. Finally, Section 32.3 shows a multivariate ecological regression of mortality data in Barcelona, Spain, illustrating one use of SANOVA and the powerful epidemiological conclusions that can be drawn from it.

32.1 Smoothed ANOVA

For now, we consider the following multivariate disease mapping problem. Let O_{ij} and E_{ij} denote, respectively, the number of observed and expected health events in the ith geographical unit ($i = 1, \ldots, I$) for the jth outcome under study ($j = 1, \ldots, J$). From now on, without loss of generality, we refer to counties when talking of areal geographical units and to diseases when talking about outcomes. We assume

$$O_{ij} \sim Poisson(E_{ij} \exp(\mu_{ij})).$$

The multivariate disease mapping problem is mostly concerned with how to model μ, the matrix of log standardized mortality ratios (SMRs), to represent dependence both within diseases (spatial dependence) and between diseases.

32.1.1 Zhang et al.'s SANOVA proposal

SANOVA for multivariate disease mapping was proposed by Zhang et al. [5], using as an example the incidence of $J = 3$ cancers in the $I = 87$ counties of Minnesota. The idea was to model $vec(\mu) = (\mu'_{\cdot 1}, \ldots, \mu'_{\cdot J})'$, where each $\mu'_{\cdot j}$ is an I-vector, using a two-way ANOVA without replication, with factors disease and county. Because the number of diseases is usually much smaller than the number of counties, the disease main effect was modeled as a set of fixed effects. The proposed model did not include one fixed effect (indicator variable) for each disease, but rather one fixed effect for each of J specified linear combinations of the diseases. The coefficients of those linear combinations were arranged as the columns of a matrix $\boldsymbol{H}$. Zhang et al. proposed to set $\boldsymbol{H}_{\cdot 1}$ to $J^{-1/2}\boldsymbol{1}_J$, so the first linear combination corresponds to the ANOVA's grand mean. The remaining columns of $\boldsymbol{H}$ were called $\boldsymbol{H}^{(-)}$, so $\boldsymbol{H}$ may be written as $[\boldsymbol{H}_{\cdot 1} : \boldsymbol{H}^{(-)}]$. $\boldsymbol{H}^{(-)}$ was specified so that $(\boldsymbol{H})'\boldsymbol{H} = \boldsymbol{I}_{J-1}$; that is, the columns of $\boldsymbol{H}^{(-)}$ are orthogonal contrasts describing specific features of the diseases. Obviously, $\boldsymbol{H}^{(-)}$ could be defined infinitely many ways, yielding different SANOVA models. The choice of a specific $\boldsymbol{H}^{(-)}$ would depend on the questions of interest to the modeler. This is similar to a traditional ANOVA in which the selection of a specific set of contrasts usually depends on the questions to be answered or the statistical design used to answer them.

Thus, the disease main effect's contribution to the model for $vec(\boldsymbol{\mu})$ has this form:

$$\left(\boldsymbol{H} \otimes (I^{-1/2}\mathbf{1}_{\boldsymbol{I}})\right) \boldsymbol{\Theta}_{\boldsymbol{Dis}} = \left(\boldsymbol{H}_{\cdot 1} \otimes (I^{-1/2}\mathbf{1}_{\boldsymbol{I}})\right) \Theta_{GM} + \left(\boldsymbol{H}^{(-)} \otimes (I^{-1/2}\mathbf{1}_{\boldsymbol{I}})\right) \boldsymbol{\Theta}_{\boldsymbol{Contrast}}, \tag{32.1}$$

where Θ_{GM} denotes the first component of $\boldsymbol{\Theta}_{\boldsymbol{Dis}}$, used to model the grand mean, and $\boldsymbol{\Theta}_{\boldsymbol{Contrast}}$ is a $(J-1)$-vector modeling the effects of the contrasts. The vector $I^{-1/2}\mathbf{1}_{\boldsymbol{I}}$ applies the J disease effects to each of the I spatial regions of $vec(\boldsymbol{\mu})$; $I^{-1/2}$ is a normalizing constant.

Conversely, the number of counties is usually large in this kind of setting, which precludes modeling them as fixed effects. Moreover, it is convenient to use the counties' geographical arrangement to define dependence among their respective risks, especially given that counties are small areas. Thus, counties were modeled as a set of spatially correlated random effects. Zhang et al. proposed an intrinsic CAR distribution for modeling counties, with precision matrix $\tau\boldsymbol{Q}$, where $Q_{ii} = m_i$, the number of county i's neighbors, and $Q_{ii'} = -1$ if counties i and i' are neighbors and 0 otherwise. Let $\boldsymbol{Q}$ have spectral decomposition $\boldsymbol{Q} = \boldsymbol{V}\boldsymbol{D}\boldsymbol{V}'$, where $\boldsymbol{V}$ is an orthogonal matrix and $\boldsymbol{D}$ is diagonal. In the sequel, we assume the region of study defines a connected map (i.e., it consists of a single connected island), so $\boldsymbol{D}$ has exactly one diagonal element equal to 0 [9], which we assume to be the first diagonal element, contrary to the usual convention of sorting $\boldsymbol{D}$'s diagonal elements in decreasing order. Note that the eigenvector corresponding to that zero eigenvalue is $\boldsymbol{V}_{\cdot 1} = I^{-1/2}\mathbf{1}_{\boldsymbol{I}}$. We denote as $\boldsymbol{V}^{(-)}$ the $I \times (I-1)$ submatrix of $\boldsymbol{V}$ containing the columns with nonzero diagonal elements in $\boldsymbol{D}$, so $\boldsymbol{V}$ may be written as $[\boldsymbol{V}_{\cdot 1} : \boldsymbol{V}^{(-)}]$. Similarly, $\boldsymbol{D}^{(-)}$ denotes the submatrix of $\boldsymbol{D}$ with the first row and column removed. Zhang et al. proposed to model the county main effect as $\boldsymbol{V}^{(-)}\boldsymbol{\Theta}_{\boldsymbol{County}}$, where $\boldsymbol{\Theta}_{\boldsymbol{County}} \sim N_{I-1}(\mathbf{0}, (\tau\boldsymbol{D}^{(-)})^{-1})$, which yields the precision matrix

$$\left(\boldsymbol{V}^{(-)}(\tau\boldsymbol{D}^{(-)})^{-1}(\boldsymbol{V}^{(-)})'\right)^{-1} = \tau(\boldsymbol{V}^{(-)}\boldsymbol{D}^{(-)}(\boldsymbol{V}^{(-)})') = \tau\boldsymbol{Q}.$$

This model is equivalent to an intrinsic CAR distribution on the county main effect, which is a random effect (though not in the sense used by, e.g., Scheffé [4]). The contribution of the county random effect to the model for $vec(\boldsymbol{\mu})$ therefore has the following form:

$$\left(J^{-1/2}\mathbf{1}_{\boldsymbol{J}} \otimes \boldsymbol{V}^{(-)}\right) \boldsymbol{\Theta}_{\boldsymbol{County}} = \left(\boldsymbol{H}_{\cdot 1} \otimes \boldsymbol{V}^{(-)}\right) \boldsymbol{\Theta}_{\boldsymbol{County}}, \tag{32.2}$$

where the term $J^{-1/2}\mathbf{1}_{\boldsymbol{J}}$ applies the I county effects to each of the J diseases considered.

If the model included no more effects, the risks of all J diseases would have the same geographical pattern except for differences in their intercepts arising from the disease main effect. An interaction between disease and county is needed to allow deviation from this additive structure. The design matrices of the disease and county effects in Equations 32.1 and 32.2 are built using the components of the matrix modeling the between-disease structure $\boldsymbol{H} = [\boldsymbol{H}_{\cdot 1} : \boldsymbol{H}^{(-)}]$ and the components of the matrix modeling the spatial structure $\boldsymbol{V} = [\boldsymbol{V}_{\cdot 1} : \boldsymbol{V}^{(-)}]$. Thus, the design matrix for the grand mean is just $\boldsymbol{H}_{\cdot 1} \otimes \boldsymbol{V}_{\cdot 1}$; for the contrasts in the columns of $\boldsymbol{H}^{(-)}$, that is, the disease main effect, the design matrix is $\boldsymbol{H}^{(-)} \otimes \boldsymbol{V}_{\cdot 1}$; and for the county main effect, the design matrix is $\boldsymbol{H}_{\cdot 1} \otimes \boldsymbol{V}^{(-)}$. It seems natural therefore for the disease-by-county interaction to have design matrix $\boldsymbol{H}^{(-)} \otimes \boldsymbol{V}^{(-)}$, combining the dependence between diseases defined by $\boldsymbol{H}^{(-)}$ with the spatial dependence structure in $\boldsymbol{V}^{(-)}$. Thus, if $\boldsymbol{\Theta}_{\boldsymbol{inter}} \sim N_{(I-1)(J-1)}(\mathbf{0}, diag(\tau_1, \ldots, \tau_{J-1}) \otimes \boldsymbol{D}^{(-)})$ and the disease–county interaction is defined as $(\boldsymbol{H}^{(-)} \otimes \boldsymbol{V}^{(-)})\boldsymbol{\Theta}_{\boldsymbol{inter}}$, this version of SANOVA models the log SMRs as

$$\begin{aligned}
vec(\boldsymbol{\mu}) &= (\boldsymbol{H} \otimes \boldsymbol{V})\boldsymbol{\Theta} = ([\boldsymbol{H}_{.1} : \boldsymbol{H}^{(-)}] \otimes [\boldsymbol{V}_{.1} : \boldsymbol{V}^{(-)}])(\Theta_{GM}, \boldsymbol{\Theta}'_{\boldsymbol{County}}, \boldsymbol{\Theta}'_{\boldsymbol{Contrast}}, \boldsymbol{\Theta}'_{\boldsymbol{Inter}})' \\
&= [\boldsymbol{H}_{.1} \otimes \boldsymbol{V}_{.1} : \boldsymbol{H}_{.1} \otimes \boldsymbol{V}^{(-)} : \boldsymbol{H}^{(-)} \otimes \boldsymbol{V}_{.1} : \boldsymbol{H}^{(-)} \otimes \boldsymbol{V}^{(-)}] \\
&\quad \times (\Theta_{GM}, \boldsymbol{\Theta}'_{\boldsymbol{County}}, \boldsymbol{\Theta}'_{\boldsymbol{Contrast}}, \boldsymbol{\Theta}'_{\boldsymbol{Inter}})' \\
&= (\boldsymbol{H}_{.1} \otimes \boldsymbol{V}_{.1})\Theta_{GM} + (\boldsymbol{H}_{.1} \otimes \boldsymbol{V}^{(-)})\boldsymbol{\Theta}_{\boldsymbol{County}} + (\boldsymbol{H}^{(-)} \otimes \boldsymbol{V}_{.1})\boldsymbol{\Theta}_{\boldsymbol{Contrast}} \\
&\quad + (\boldsymbol{H}^{(-)} \otimes \boldsymbol{V}^{(-)})\boldsymbol{\Theta}_{\boldsymbol{Inter}}. \qquad (32.3)
\end{aligned}$$

This model implies $vec(\boldsymbol{\mu})$ has precision matrix $\boldsymbol{Q} \otimes (\boldsymbol{H} diag(\tau, \tau_1, \ldots, \tau_{J-1})\boldsymbol{H}')$, with known $\boldsymbol{H}$ [5]. By contrast, the multivariate intrinsic CAR (MCAR) distribution has a precision matrix of the form $\boldsymbol{Q} \otimes \boldsymbol{\Omega}$ for an unknown symmetric, positive definite $\boldsymbol{\Omega}$, the between-disease precision matrix. The fixed, known contrasts of SANOVA's $\boldsymbol{H}$ play the role of the eigenvectors of the MCAR's $\boldsymbol{\Omega}$, and the more they resemble $\boldsymbol{\Omega}$'s true eigenvectors, the better will be the fit of SANOVA. The drawback is that it is very difficult to have prior intuition about the eigenvectors of $\boldsymbol{\Omega}$ to help in specifying $\boldsymbol{H}$, although Zhang et al. presented a modest simulation experiment suggesting that in practice, this creates little or no disadvantage, most likely because the data provide weak information about $\boldsymbol{\Omega}$'s eigenvectors. Zhang et al. viewed this as a weakness of the proposed model, but Marí-Dell'Olmo et al. [8] saw it as an opportunity: if $\boldsymbol{H}$'s columns are chosen to focus on substantive questions of interest to the modeler, SANOVA becomes a way to simplify multivariate modeling of several diseases. From this viewpoint, SANOVA-based smoothing uses just J parameters $(\tau, \tau_1, \ldots, \tau_{J-1})$ to define the multivariate dependence between diseases, in contrast to MCAR, which uses $\boldsymbol{\Omega}$'s $J(J+1)/2$ parameters. In this sense, SANOVA can be considered a simpler and more convenient way to induce multivariate dependence between diseases.

32.1.2 Marí-Dell'Olmo et al.'s SANOVA proposal

The starting point of Marí-Dell'Olmo et al.'s [8] proposal is Equation 32.3. There, the log SMRs are modeled as the product $(\boldsymbol{H} \otimes \boldsymbol{V})\boldsymbol{\Theta}$, which can be expressed as

$$vec(\boldsymbol{\mu}) = (\boldsymbol{H} \otimes \boldsymbol{V})\boldsymbol{\Theta} = (\boldsymbol{H} \otimes \boldsymbol{I_I})(\boldsymbol{I_J} \otimes \boldsymbol{V})\boldsymbol{\Theta} = (\boldsymbol{H} \otimes \boldsymbol{I_I})vec(\boldsymbol{\Psi}), \qquad (32.4)$$

where the random effects in the $(I \cdot J)$-vector $vec(\boldsymbol{\Psi}) = (\boldsymbol{I_J} \otimes \boldsymbol{V})\boldsymbol{\Theta}$ follow an intrinsic CAR distribution. If $\boldsymbol{\Psi} = (\boldsymbol{\Psi}'_{.1}, \ldots, \boldsymbol{\Psi}'_{.J})'$ for I-vectors $\boldsymbol{\Psi}'_{.j}$, then Equation 32.4 can be written as

$$\begin{aligned}
(\boldsymbol{H} \otimes \boldsymbol{I_I})vec(\boldsymbol{\Psi}) &= \begin{pmatrix} H_{11}\boldsymbol{I_I} & \cdots & H_{1J}\boldsymbol{I_I} \\ \vdots & \ddots & \vdots \\ H_{J1}\boldsymbol{I_I} & \cdots & H_{JJ}\boldsymbol{I_I} \end{pmatrix} \begin{pmatrix} \boldsymbol{\Psi}_{.1} \\ \vdots \\ \boldsymbol{\Psi}_{.J} \end{pmatrix} = \begin{pmatrix} H_{11}\boldsymbol{\Psi}_{.1} + \ldots + H_{1J}\boldsymbol{\Psi}_{.J} \\ \vdots \\ H_{J1}\boldsymbol{\Psi}_{.1} + \ldots + H_{JJ}\boldsymbol{\Psi}_{.J} \end{pmatrix} \\
&= \boldsymbol{H}_{.1} \otimes \boldsymbol{\Psi}_{.1} + \ldots + \boldsymbol{H}_{.J} \otimes \boldsymbol{\Psi}_{.J}. \qquad (32.5)
\end{aligned}$$

Therefore, Zhang et al.'s proposal can be seen as the sum of J Kronecker products of disease contrasts and the spatial patterns. Because $\boldsymbol{H}_{.1}$ is simply $J^{-1/2}\boldsymbol{1_J}$, $\boldsymbol{\Psi}_{.1}$ contributes to the fit exactly the same way for every disease; that is, it models the component common to all the diseases, which we previously called the county main effect. $\boldsymbol{\Psi}_{.2}$ contributes to the fit in one way for diseases for which the corresponding element in $\boldsymbol{H}_{.2}$ is positive, and in the opposite way for diseases with negative elements in $\boldsymbol{H}_{.2}$. In general, then, for $j = 2, \ldots, J$, $\boldsymbol{\Psi}_{.j}$ models the spatial pattern associated with the jth contrast in diseases, identifying regions where this contrast takes higher or lower values. With this reformulation, SANOVA allows exploration of each contrast in which the modeler has an interest.

This reformulation of Zhang et al.'s proposal also has computational advantages. First, Zhang et al.'s approach requires that the matrix $\boldsymbol{Q}$ in the intrinsic CAR's precision matrix has no unknown parameters, so it does not extend to other spatial distributions, such as the proper CAR distribution, for which the analogous matrix and its spectral decomposition depend on unknown parameters. In that case, if MCMC was used to sample from the posterior distribution, this would require a new spectral decomposition of $\boldsymbol{Q}$ at every MCMC iteration, which could be prohibitive. Marí-Dell'Olmo et al.'s reformulation does not have this problem because computationally, it makes little difference if $\boldsymbol{\Psi}_{\cdot 1}, \ldots, \boldsymbol{\Psi}_{\cdot J}$ follow an intrinsic CAR distribution or any other spatially structured distribution. Moreover, even the graphical modeling approach in Chapter 31 could also be implemented within the SANOVA framework just introduced in order to ascertain an appropriate geographical dependence structure for the available data.

Marí-Dell'Olmo et al.'s reformulation can be used to extend the original SANOVA formulation to nonseparable multivariate dependence structures by putting different distributions on the $\boldsymbol{\Psi}_{\cdot 1}, \ldots, \boldsymbol{\Psi}_{\cdot J}$, in which case the resulting covariance structure cannot be the Kronecker product of a disease covariance matrix and a single spatial covariance. In this sense, the reformulated SANOVA generalizes the original because it can reproduce nonseparable covariance models. Moreover, Marí-Dell'Olmo et al.'s reformulation has a second advantage: it can be implemented in standard Bayesian software such as WinBUGS, OpenBUGS, or INLA. Equation 32.5 defines a SANOVA model as the sum of several Kronecker products of predefined contrasts and vectors of spatial random effects. For the jth disease, this sum of Kronecker products is

$$\boldsymbol{\mu}_j = H_{j1}\boldsymbol{\Psi}_{\cdot 1} + \ldots + H_{jJ}\boldsymbol{\Psi}_{\cdot J},$$

that is, a known linear combination of the spatial random effects. This simple expression of the log SMRs for any disease avoids Kronecker products and is therefore easily implemented in the aforementioned packages.

32.2 Some Specific Applications of Smoothed ANOVA

Although Zhang et al. [5] proposed SANOVA as a tool for traditional multivariate modeling in disease mapping studies, it can be used in a wider collection of settings. The contrasts in $\boldsymbol{H}$ are defined by the modeler, and this could be seen as a drawback. But these contrasts provide room for modeling; if properly used, they permit a great variety of models. For example, although $\boldsymbol{H}$ was described above as representing contrasts among levels of a single factor (diseases), with no change to the preceding theory, $\boldsymbol{H}$ can represent contrasts defining a balanced design with any number of factors, for example, a three-factor design with factors diseases, sex, and time periods. With this in mind, we now describe some settings where SANOVA can be applied for purposes somewhat different from its original conception.

32.2.1 Design-based studies in disease mapping

From their beginning, disease mapping studies have had mainly an observational aim, that is, obtaining reasonably reliable estimates for small areas to describe the geographical pattern underlying some diseases. At most, such studies may suggest the presence of a risk factor influencing the disease pattern, and this hypothesis could be tested in a confirmatory ecological regression study. Such a confirmatory study would ideally be done with new data to avoid post hoc analyses, possibly leading to the "Texas sharpshooter fallacy" [10].

Sometimes research questions involve comparing the geographical patterns of different diseases, different population groups, or different time periods. Unfortunately, traditional disease mapping methods do not address these questions; they were not conceived to do so. For example, suppose we have data for males and females for a disease and we want to explore the common geographical pattern of both sexes, as well as the geographical pattern of differences between sexes, that is, places with higher occurrence of the disease for one sex than for the other. These questions could be addressed only informally with traditional univariate disease mapping studies. Multivariate models such as MCAR include the correlation structure of the diseases and sexes, but that correlation does not necessarily address the questions of interest. Therefore, traditional disease mapping methods are not helpful. SANOVA, however, can incorporate those questions into the study's design through the matrix of contrasts $\boldsymbol{H}$. In this sense, smoothed ANOVA enables new multivariate disease mapping analyses, going beyond disease mapping's traditional descriptive purpose. Design-based studies to confirm or explore the hypothesis of interest become possible; indeed, this may require changing the descriptive conception of most disease mapping professionals.

32.2.2 Variance decomposition

The design matrices arising from the SANOVA approaches outlined in the previous section are orthonormal. Thus, if we use $J-1$ contrasts in diseases or groups in a SANOVA study, in addition to the linear combination modeling the grand mean, the design matrix's orthonormality allows us to decompose the variance of $vec(\boldsymbol{\mu})$ into these J components [8]. This decomposition can be a valuable epidemiological component in this kind of study, allowing us to see which elements of the decomposition explain most (or least) of the variance in the original data patterns.

This variance decomposition could be used, for example, in the study of lung cancer mortality in two periods and both sexes. In that case, we would define $\mathbf{H}$ with four columns: the common geographical pattern underlying all four maps (i.e., one map for each of the two periods $\times$ two sexes), the geographical pattern of differences between mortality in the two periods, the geographical pattern of differences between the mortality of the sexes, and the geographical pattern of deviations from these time and sex main effects, that is, the interaction of period and sex. If the four original sex-by-period maps are labeled so that the first two maps correspond to the first period and the first and third maps correspond to males, the $\mathbf{H}$ matrix arising from this design would be

$$\mathbf{H} = \frac{1}{2}\begin{pmatrix} 1 & 1 & 1 & 1 \\ 1 & 1 & -1 & -1 \\ 1 & -1 & 1 & -1 \\ 1 & -1 & -1 & 1 \end{pmatrix}. \tag{32.6}$$

Several obvious epidemiological questions arise here. Which of the four geographical patterns explains the most variance? Are geographical differences between sexes more important than those between periods, in terms of the variance explained? Is the sex-by-period interaction—the change between periods in the difference between sexes—important for explaining the original data pattern, after accounting for the effects of sex and period? Answers to these questions can provide important clues about the epidemiology of lung cancer. This kind of result is clearly beyond the scope of traditional univariate and even multivariate disease mapping studies.

32.2.3 Multivariate ecological regression

One more application of SANOVA is ecological regression [8, 11]. Following Marí-Dell'Olmo et al.'s approach, the original patterns in the data can be modeled or decomposed as a function of $\boldsymbol{\Psi}_{\cdot 1}, \ldots, \boldsymbol{\Psi}_{\cdot J}$, where each of these vectors contains the geographical pattern of a contrast in a column of $\boldsymbol{H}$, or the common underlying pattern in the case of $\boldsymbol{\Psi}_{\cdot 1}$. But these vectors $\boldsymbol{\Psi}_{\cdot j}$ could themselves be modeled by means of an ecological regression linking them to covariates of interest. In that case, we could determine the relationship between the covariate and the original data patterns through the estimated relationship between the covariate and the patterns corresponding to the contrasts in $\boldsymbol{H}$'s columns. In this sense, we use the contrasts to do a multivariate ecological regression study: we do not separately model the covariates' contribution to the original data patterns; we model these contributions entirely through the contrasts. Section 32.3 discusses an example in detail.

A second use of ecological regression in this context is linked to the variance decomposition described just above. As described so far, SANOVA splits the original variance into as many components as $\boldsymbol{H}$ has columns. But in ecological regression, we split these components further, into a part explained by the covariate and a second part attributed to other, possibly unknown factors, typically modeled using a spatial random effect. If the spatial random effect is defined to be orthogonal to the covariate [12, 13], we can split the variance explained by each contrast into at least two parts, the variance explained by the covariate and the "residual" unexplained variance [8]. Besides permitting the variance decomposition, placing an orthogonality restriction on the random effect avoids so-called spatial confounding, that is, confounding between the covariate of interest and the residual spatial pattern, a common problem in ecological regression problems [12].

32.2.4 Spatiotemporal modeling

Spatiotemporal problems [14] are just a kind of multivariate study with an order relationship (i.e., time sequence) on the geographical patterns being modeled, so SANOVA can be used for spatiotemporal studies [15]. In this case, it suffices to specify the columns of $\boldsymbol{H}$ as the elements of a basis of functions used to model the time trends for the geographical units composing the region of study. If an orthogonal basis is used, the variance decomposition mentioned above is retained, allowing us to explore which elements of the basis explain more and less of the variance in the spatiotemporal dataset.

The time trend for the ith geographical unit is modeled as

$$\boldsymbol{\mu}_{i\cdot} = (\boldsymbol{H}_{\cdot 1}\Psi_{i1} + \ldots + \boldsymbol{H}_{\cdot J}\Psi_{iJ})',$$

where $\boldsymbol{H}_{\cdot j}$ is the jth element of the basis for functions of time, evaluated at all the time units of the period of study. Since $\boldsymbol{\Psi}_{\cdot j}$ will typically have some spatial structure for each j, the parameters defining the time trend for the geographical units will be correlated: the time trends for nearby regions will be similar because they are similar combinations of the same basis elements.

The basis functions used to model time trends can be tailored to the data at hand. If no cyclic time trend is expected, a polynomial basis could be used. But if, as often happens, a cyclic trend is present, a Fourier basis could be used and will yield a much better fit. Therefore, SANOVA provides a powerful, versatile tool for modeling spatiotemporal disease mapping datasets.

Note also that other factors, such as sex or multiple diseases, can be modeled simultaneously with time, as indicated in this section's introduction.

32.3 Multivariate Ecological Regression Study Using SANOVA

We now illustrate the potential of SANOVA using chronic obstructive pulmonary disease (COPD) and lung cancer mortality data for the city of Barcelona, Spain. These two diseases have common risk factors, mainly tobacco consumption, so it seems reasonable to do a multivariate study of them. We have mortality data for both diseases and sexes, that is, observed and expected counts for all four combinations of these two factors on Barcelona's 1491 census tracts (each with about 1000–2000 inhabitants). Since tobacco consumption can be heavily influenced by deprivation, we also have this variable for every census tract so we can control for its effect, if possible.

Given these mortality data, researchers might be interested in several epidemiological questions, such as:

- Which census tracts show more mortality for all four combinations of disease and sex (common component)?
- Which census tracts show more mortality for one of the diseases regardless of sex (disease component)?
- Which census tracts show more mortality for one of the sexes regardless of disease (sex component)?
- Given the geographical distribution of the common, disease, and sex components, does the interaction between disease and sex make an important contribution to the variance of disease incidence?
- Is it possible to quantify the variability of the factors above with respect to the total variability of all four geographical patterns?
- What part of the variability of the common, disease, and sex components can be explained by deprivation?
- What is the geographical distribution of the common, disease, and sex components that cannot be attributed to deprivation?

These epidemiological questions cannot be addressed by traditional disease mapping techniques, but they can be addressed with SANOVA, as we will illustrate. The analysis we suggest is an example of what Section 32.2 called a *design-based study*, because both the design of the data to be studied and the questions to be answered make it convenient to consider specific relationships among all four geographical patterns in the study. This goal can be achieved easily using SANOVA.

From now on, we will label as geographical patterns 1 and 2 those corresponding to COPD deaths, for men and women, respectively, and label as 3 and 4 those corresponding to lung cancer deaths, also for men and women, respectively. We use expression (32.6) as the $\boldsymbol{H}$ matrix, so $\boldsymbol{\Psi}_{\cdot 1}$ represents the common component for all four geographical patterns, having higher risks for all four disease-by-sex groups than those regions i with $\Psi_{i1} > 0$. Similarly, $\boldsymbol{\Psi}_{\cdot 2}$ represents the disease-specific component, taking values higher than 0 for regions with a ratio of COPD versus lung cancer mortality higher than that for Barcelona in aggregate. $\boldsymbol{\Psi}_{\cdot 3}$ represents the sex-specific component, taking values higher than 0 for regions with a ratio of male versus female mortality higher than that for Barcelona in aggregate. Finally, $\boldsymbol{\Psi}_{\cdot 4}$ represents the disease-by-sex interaction, taking values higher than 0 for regions with particularly high mortality for COPD in men and lung cancer in women.

All four components in matrix $\boldsymbol{\Psi}$ are modeled in the same manner, as

$$\boldsymbol{\Psi}_{\cdot j} = \mu_j \cdot \mathbf{1} + f_j(\boldsymbol{D}) + \boldsymbol{S}_{\cdot j},\ j = 1, \ldots, 4,$$

where μ_j is the intercept, modeling the mean of $\boldsymbol{\Psi}_{\cdot j}$ for the whole city. As proposed in Marí-Dell'Olmo et al. [8], the expression $f_j(\cdot)$ is a step function of $\boldsymbol{D}$, the deprivation index. Values of $\boldsymbol{D}$ are split into groups at specific quantiles, and $f_j(\cdot)$ assigns the same value to all census tracts in the same group. In our study, we used 40 groups to define $f_j(\cdot)$ and modeled $(f_j(1), f_j(2), \ldots, f_j(40))$ with an intrinsic CAR distribution, considering consecutive quantile groups as neighbors. The vector $\boldsymbol{S}_{\cdot j}$, modeled as the usual sum of heterogeneous and intrinsic CAR random effects [16], models the residual variability in each component that cannot be explained by deprivation. We impose the condition that $\boldsymbol{S}_{\cdot j}$ sums to 0 for every group defined by quantiles of the deprivation index to guarantee orthogonality of $f_j(\cdot)$ and $\boldsymbol{S}_{\cdot j}$. This has two benefits: first, the variance of the original dataset can be decomposed as a function of all the terms in the model, and second, this avoids potential spatial confounding of $f_j(\cdot)$ and $\boldsymbol{S}_{\cdot j}$, which otherwise could compete to explain the same variation in the data. All computations were made using INLA [17]. Further modeling and computational details are in Marí-Dell'Olmo et al. [8], which used a very similar model.

Table 32.1 shows the variance explained by each component in the model. The variance attributable to the intercept of each component has not been included in Table 32.1 because it is 0 for all components. This is because expected cases have been calculated by internal standardization for each disease–sex combination, so the mean relative risk for each combination is 1, and this term does not induce any variability in the model. Among the four components considered, the common component explains the largest proportion of variance, followed by the sex and disease main effects, in that order. The disease-by-sex interaction explains hardly any variance, so henceforth we will ignore it. As a consequence, the maps for the four sex–disease combinations will be similar because of the common component's large fraction of the total variance, while maps for the two sexes will be less similar than those for the two diseases. Finally, the effect on the map of changing sex will be the same regardless of the disease, and analogously for the effect of changing disease.

Regarding the effect of deprivation, most of the variance in the data (77.5%) is associated with this factor. Nevertheless, deprivation is not equally associated with all four components considered. For instance, deprivation explains almost 99% of the variance of the difference between maps for males and females, with negligible variance attributable to other factors. On the other hand, deprivation accounts for only 68% of the variance of the difference between diseases, with, presumably, other factors underlying the remaining differences.

Figure 32.1 shows the estimated association between deprivation and the common, disease, sex, and interaction components, $\exp(f_j(\cdot)), j = 1, \ldots, 4$, respectively. All four plots show the posterior mean and 95% posterior credible interval for the 40 deprivation groups.

TABLE 32.1
Percentage of variance explained by each component in the model

Component	Variance deprivation (%)	Variance random effect (%)	Total (%)
Common	34.5	15.2	49.7
Disease	13.2	6.2	19.4
Sex	29.2	0.3	29.5
Interaction disease–sex	0.6	0.8	1.4
Total	77.5	22.5	

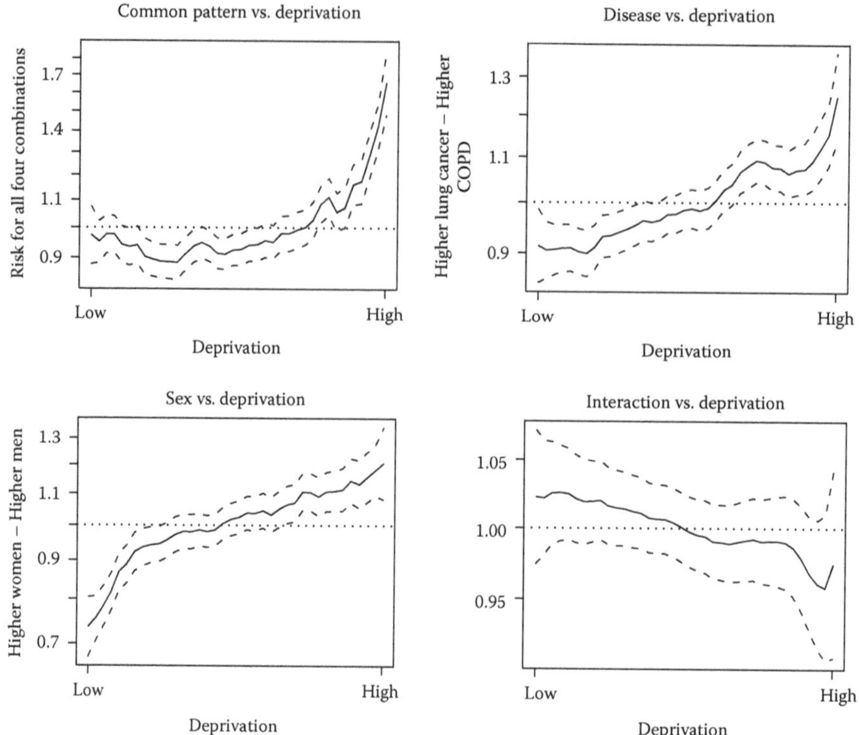

FIGURE 32.1
Relationship between deprivation and all four components included in the model.

Many deprivation levels have posterior credible intervals completely above or below 1, providing evidence that those census tracts have particular features linked to deprivation, making them different from the city's mean level. Specifically, the most deprived regions have, in general, higher mortality (for all four combinations of disease and sex), with risk up to 70% higher than the city's mean. Also, the most deprived groups show particularly high COPD mortality compared to lung cancer mortality, in contrast to the most affluent census tracts, which show the opposite trend. The most deprived regions show a higher mortality for men than for women, while the most affluent regions show higher mortality for women. This effect is especially pronounced for census tracts with the lowest deprivation, where the fitted curve has its steepest slope; this is consistent with the historically high prevalence of tobacco consumption by women in Spain's most affluent social groups [18]. As expected from Table 32.1, deprivation is not associated with the disease–sex interaction component.

Figures 32.2 and 32.3 show choropleth maps of the parts of the common and disease specific terms that are not related to deprivation, $\exp(\boldsymbol{S}_{\cdot 1})$ and $\exp(\boldsymbol{S}_{\cdot 2})$, respectively. The analogous maps for the sex and interaction components are not shown because they explain very little variance and neither has any census tract with significant excess risk compared to the city's mean risk (i.e., with 95% credible interval excluding 1). Ellipses in these figures indicate regions with significant deviations from the level of the city as a whole. Full-color versions of these figures can be found as annex material at https://www.crcpress.com/9781482253016.

The patterns in Figures 32.2 and 32.3 are uncorrelated with deprivation by construction, so they reflect the presence of other risk factors. Both components have regions with significant departures from Barcelona's mean mortality. Thus, for all four combinations of

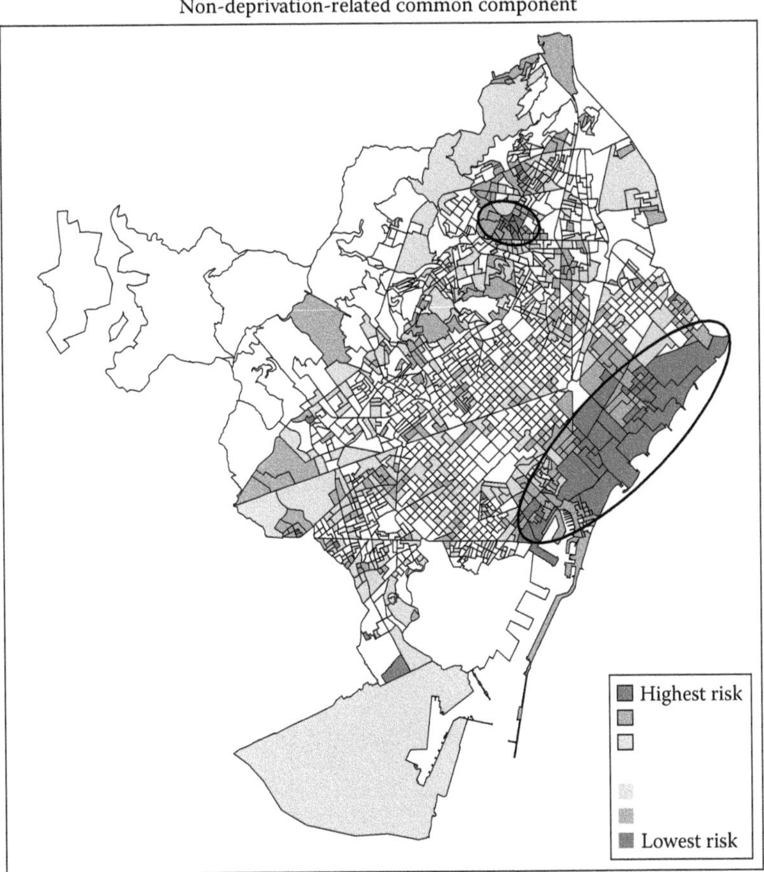

FIGURE 32.2
Geographical distribution of the part of the common component that is not related to deprivation. Ellipses indicate regions with large risk deviations compared to the whole city. Darker regions stand for larger deviations showing either higher (regions with thick border) or lower (regions with thin border) risk than the mean of the city.

disease and sex (Figure 32.2), a large region along the city's northern shoreline (the lower ellipse) has a significant risk excess. This risk excess cannot be explained by deprivation; indeed, that region includes census tracts with both the highest and lowest deprivation levels. Moreover, that region also includes both relatively new and old neighborhoods with very different demographic and social groups, suggesting an environmental risk factor as a possible explanation. Figure 32.3 shows regions with a risk excess for just one of the diseases, which cannot be explained by deprivation, pointing to the presence of risk factors for just one disease. Risk excesses have been found for both COPD (both upper-side ellipses) and lung cancer (lower-side ellipse).

This example shows that SANOVA is a powerful tool, making it possible to address questions that traditional disease mapping methods cannot. Indeed, all the questions posed at the beginning of this section have been answered using SANOVA. In this sense, SANOVA as a data analysis technique is particularly fitted to discerning mechanisms underlying diseases, beyond the exploratory aim of most disease mapping methods. This can make SANOVA a particularly appropriate tool to push disease mapping toward more analytical purposes.

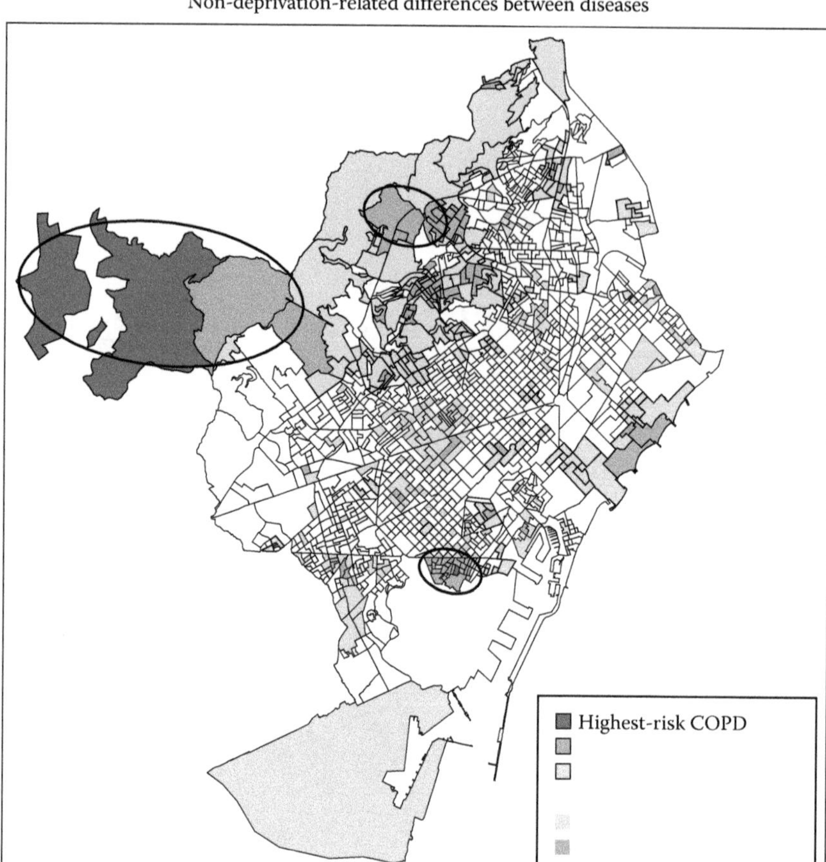

FIGURE 32.3
Geographical distribution of the part of the disease-specific component that is not related to deprivation. Ellipses indicate regions with large risk deviations compared to the whole city. Darker regions stand for larger deviations showing either higher (regions with thick border) or lower (regions with thin border) risk than the mean of the city.

References

[1] Agostino Nobile and Peter J. Green. Bayesian analysis of factorial experiments by mixture modelling. *Biometrika*, 87:15–35, 2000.

[2] Andrew Gelman. Analysis of variance—why it is more important than ever [with discussion]. *Annals of Statistics*, 33:1–53, 2005.

[3] James S. Hodges, Yue Cui, Daniel J. Sargent, and Bradley P. Carlin. Smoothing balanced single-error-term analysis of variance. *Technometrics*, 49:12–25, 2007.

[4] Henry Scheffé. *The Analysis of Variance*. Wiley, New York, 1959.

[5] Yufen Zhang, James S. Hodges, and Sudipto Banerjee. Smoothed ANOVA with spatial effects as a competitor to MCAR in multivariate spatial smoothing. *Annals of Applied Statistics*, 3(4):1805–1830, 2009.

[6] Paloma Botella-Rocamora, Miguel A. Martinez-Beneito, and Sudipto Banerjee. A unifying modeling framework for highly multivariate disease mapping. *Statistics in Medicine*, 34(9):1548–1559, 2015.

[7] Miguel A. Martinez-Beneito. A general modelling framework for multivariate disease mapping. *Biometrika*, 100(3):539–553, 2013.

[8] Marc Marí-Dell'Olmo, Miguel A. Martinez-Beneito, Mercè Gotséns, and Laia Palència. A smoothed ANOVA model for multivariate ecological regression. *Stochastic Environmental Research and Risk Assessment*, 28(3):695–706, 2014.

[9] James S. Hodges, Bradley P. Carlin, and Qiao Fan. On the precision of the conditionally autoregressive prior in spatial models. *Biometrics*, 59:317–322, 2003.

[10] Atul Gawande. The cancer-cluster myth. *The New Yorker*, pp. 34–37, 1998.

[11] Marc Marí-Dell'Olmo, Mercè Gotséns, Carme Borrell, Miguel A. Martinez-Beneito, Laia Palència, Glòria Pérez, Lluís Cirera, et al. Trends in socioeconomic inequalities in ischemic heart disease mortality in small areas of nine Spanish cities from 1996 to 2007 using smoothed ANOVA. *Journal of Urban Health*, 91:46–61, 2014.

[12] James S. Hodges and Brian J. Reich. Adding spatially-correlated errors can mess up the fixed effect you love. *American Statistician*, 64(4):325–334, 2010.

[13] John Hughes and Murali Haran. Dimension reduction and alleviation of confounding for spatial generalized linear mixed models. *Journal of the Royal Statistical Society: Series B*, 75(1):139–159, 2013.

[14] Miguel A. Martinez-Beneito, Antonio López-Quílez, and Paloma Botella-Rocamora. An autoregressive approach to spatio-temporal disease mapping. *Statistics in Medicine*, 27:2874–2889, 2008.

[15] Francisco Torres-Avilés and Miguel A. Martinez-Beneito. STANOVA: A smooth-ANOVA-based model for spatio-temporal disease mapping. *Stochastic Environmental Research and Risk Assessment*, 29:131–141, 2014. doi: 10.1007/s00477-014-0888-1.

[16] Julian Besag, Jeremy York, and Annie Mollié. Bayesian image restoration, with two applications in spatial statistics. *Annals of the Institute of Statistical Mathemathics*, 43:1–21, 1991.

[17] Håvard Rue, Sara Martino, and Nicolas Chopin. Approximate Bayesian inference for latent Gaussian models by using integrated nested Laplace approximations. *Journal of the Royal Statistical Society: Series B*, 71(2):319–392, 2009.

[18] Anna Schiaffino, Esteve Fernandez, Carme Borrell, Esteve Salto, Montse Garcia, and Josep Maria Borras. Gender and educational differences in smoking initiation rates in Spain from 1948 to 1992. *European Journal of Public Health*, 13(1):56–60, 2003.

33

Sociospatial Epidemiology: Segregation

Sue C. Grady
Department of Geography
Michigan State University
East Lansing, Michigan

CONTENTS

33.1 Introduction

Sociospatial epidemiological research focuses on understanding why certain populations experience poorer health outcomes or premature mortality than other populations within and across geographic areas. Populations comprise individuals of similar characteristics, for example, biological (sex, age, and genetic predisposition), behavioral (exercise, diet, smoking, alcohol, and illicit drug use), and social (race, education, occupation, and income). In the United States, *race* is defined by the Office of Management and Budget (2014) for use in the decennial census as a means by which to stratify the U.S. population. In the 2010 census, five racial groups were defined: white, black or African American, American Indian or Alaska Native, Asian, and Native Hawaiian or Pacific Islander. There was also a multiple-race category. People interviewed or surveyed selected a race based on their social identity, color of their skin, phenotype, and ancestry. The two most common reported racial groups were white (72.4%) and black (12.6%) (Census Briefs, 2010). Over the last several decades, the National Center for Health Statistics (NCHS, 2014) and state health departments have used the racial categories to monitor racial disparities in population health. In states across the United States, it is reported that black men, women, and children have a two- to threefold increase in many diseases, conditions, and premature mortality compared to whites despite their smaller population size (CDC WONDER, 2014). Over the last several decades, substantial resources from many branches of government have been allocated to understand and reduce these black–white disparities in health (CDC Healthy People, 2014).

Some medical experts surmise that blacks have a genetic predisposition to certain diseases because of their African ancestry (Kumar et al., 2012; Boone et al., 2014) and the generational transfer of stress over the last century (Williams and Mohammed, 2013). Social scientists recognize race as a social construct, focusing more pointedly on the historical practices of social oppression and the persistence of prejudices and discrimination that blacks experience today. Of particular concern is the continued residential confinement of blacks in black-segregated urban neighborhoods. Residential segregation is defined as the physical sorting of two or more racial groups into different neighborhoods (Massey and Denton, 1988). The origin of black segregation in the United States is multifaceted but can be summarized by the historical policies of restrictive covenants (i.e., neighborhood association contractual agreements that whites do not sell or lease their homes to blacks), blockbusting (i.e., real estate agents instilling fear in white residents to capitalize on the sale of their homes during flight), and redlining (i.e., the refusal of banks to provide housing loans to blacks wanting to move into white neighborhoods). Although these policies were outlawed in the 1950s (Massey and Denton, 1988), urban renewal programs subsequently exacerbated segregation by relocating urban poor blacks into neighborhoods on the periphery of city centers, such as high-density housing, which led to additional racial turnover and the expansion of segregation into those areas. In the 1970s, structural adjustment forced industries to move from the U.S. Northeast and Midwest regions to the South and overseas, thereby reducing employment opportunities in urban areas in these regions. Subsequently, white residents relocated to the suburbs, where new employment opportunities arose. Blacks, however, experienced real and perceived discriminatory resistance to their movement into the suburbs, and were therefore confined to the city centers in urban areas. With limited social mobility and high unemployment, city tax bases declined, reducing the availability and quality of services, resources, and amenities to residents in these areas. As neighborhood quality deteriorated, local retail disinvested, leading to higher crime and violence, further exacerbating economic and social decline. Today, living in poverty in black-segregated neighborhoods, also referred to as concentrated poverty, is more detrimental to health and well-being than living in poverty in white-segregated neighborhoods (Massey and Fischer, 2000), largely because of limited social mobility. Without social mobility, poor blacks have fewer incentives to travel outside of segregated neighborhoods for work or higher-quality education that could lead to future employment, higher incomes, and wealth. Importantly, many medium- and high-income blacks now reside on the periphery of concentrated poverty, also limiting their availability and access to resources and amenities that whites of similar economic standing enjoy. Less is known about the social pressures experienced by low-, medium-, and high-income blacks residing in white-segregated or racially mixed neighborhoods and their impacts on health and well-being.

33.1.1 Purpose of Study

The purpose of this study is to estimate the effect of maternal exposure to racial residential segregation and concentrated poverty on black singleton preterm births, that is, infants born before 37 weeks gestation. Infants born prematurely have less developed organs and organ systems than infants born at full gestation and are therefore at increased risk of neonatal mortality, that is, infant death within 30 days of birth. This study takes place in neighborhoods in lower Michigan's 16 urban areas—all of which are highly segregated by race. Between 2008 and 2011, the incidence of premature birth for black mothers residing in these urban areas was 12.4 per 1000 live births compared to 7.9 for white mothers (rate ratio [RR] = 1.6) and 7.8 for mothers of other racial groups (RR = 1.6) (MDCH, 2014), demonstrating the need to study the contribution of segregation and concentrated poverty on the incidence of preterm birth among black mothers in these areas. This study investigates preterm birth among Medicaid mothers because Medicaid black mothers are assumed to be more susceptible to the untoward effects of segregation and concentrated poverty because of their limited social mobility. Segregation indices are calculated and multilevel models are estimated to answer the following questions:

1. Is there significant variation in Medicaid preterm births across neighborhoods and urban areas in lower Michigan?
2. Are Medicaid black mothers at higher odds of preterm birth than Medicaid white mothers and Medicaid mothers of other racial groups? How much of the racial disparity in preterm birth is attributed to residing in high and medium levels of segregation and concentrated poverty?

33.1.2 Definition of Segregation

There are five aspatial global dimensions of segregation: evenness, exposure, clustering, centralization, and concentration (Massey and Denton, 1988). *Evenness* represents the proportion of blacks that would have to change neighborhoods to achieve an even distribution of racial groups across an urban area. *Exposure* represents the probability that blacks will share a neighborhood with whites and other racial groups in an urban area. In contrast, racial isolation is the probability that blacks will share a neighborhood with other blacks in an urban area (higher isolation signifying greater segregation). *Clustering* is the extent to which black neighborhoods adjoin each other, resulting in one large contiguous black area. *Centralization* is the degree to which blacks reside around the central business district. *Concentration* is the relative amount of physical space within an urban area that is occupied by blacks, and as segregation increases, blacks become increasingly concentrated. The methods by which each of these aspatial global dimensions of segregation is measured can be found in Massey and Denton (1988). In their research, Massey and Denton (1988) also report that urban areas segregated on three or more of these dimensions are referred to as hypersegregated. In 2004, blacks were hypersegregated in 29 metropolitan areas in the United States, with Chicago, Cleveland, Detroit, Milwaukee, Newark, and Philadelphia hypersegregated on all five dimensions (Wilkes and Iceland, 2004).

Reardon and O'Sullivan (2004) reduced these five aspatial dimensions of segregation to two primary spatial dimensions: spatial evenness and clustering. The centralization and concentration dimensions became subdimensions of spatial unevenness, and exposure or isolation become a subdimension of clustering. These spatial dimensions addressed two important limitations in the aspatial dimensions, specifically the checkerboard problem (i.e., the measurement of racial composition in a study area without consideration of the spatial proximity of racial groups [Morrill, 1991; Wong, 1993, 2002]) and the modifiable areal unit problem (MAUP) (i.e., using the same method but obtaining different estimates of segregation at different geographic scales or zone design schemes [Fotheringham and Wong, 1991]). Wong (2002) and Reardon and O'Sullivan (2004) later developed local spatial indices of segregation on both of these dimensions.

Many studies that estimate the effect of segregation on health utilize the index of dissimilarity (Duncan and Duncan, 1955), an aspatial measure of the dimension evenness (LaVeist, 1993; Polednak, 1996; Ellen, 2000; Cooper et al., 2001; Robert and Ruel 2006), or the aspatial index of racial isolation (Chang, 2006; Bell et al., 2006). A few studies have utilized the local spatial isolation index developed by David Wong (2002) (Grady, 2006; Grady and

Ramírez, 2008) or the local G-statistic (Getis and Ord, 1992) as a measure of racial clustering (Grady and Darden, 2012). This study utilizes the generalized spatial dissimilarity index $\bar{D}(m)$ (Feitosa et al., 2007) that considers the potential movement of residents across census tract boundaries (social mobility).

33.2 Data and Methods

33.2.1 Data

All Medicaid singleton live births ($n = 127{,}116$) that occurred in 2008–2011 in lower Michigan's 16 urbanized areas are used in this analysis. The birth data were obtained from the Vital Statistics Birth Registry, Division of Vital Records and Health Statistics, Michigan Department of Community Health (MDCH), at the individual level. The birth records were geocoded to Michigan's street files and spatially joined to 2010 census tract boundaries (U.S. Census Bureau TIGER Products, 2014) in ArcGIS v10.2 (ESRI, 2015) to obtain a census tract identifier. The dependent variable is preterm birth ($n = 12{,}157$), defined as infants born less than 37 weeks gestation. The individual-level characteristics of mothers and infants controlled for in statistical models are mother's race (black = 1, white = 0; other = 1, white = 0), age (continuous in years, centered on the mean 24 years), marital status (married = 1, not married = 0), and smoking history during pregnancy (yes = 1, no = 0). The generalized spatial dissimilarity index and concentrated poverty were calculated using race and poverty data from the American Community Survey (ACS) (2008–2012) (U.S. Census Bureau, 2014) at the census tract level.

33.2.1.1 Sample Size and Property of Exchangeability

The study utilized three-level multilevel models to conceptualize mothers and infants nested within neighborhoods, nested within urban areas. A general rule of thumb to estimate fixed parameters in two-level multilevel models is 30 neighborhoods with 30 individuals per neighborhood (Kreft, 2003), and to estimate cross-level interactions or variance and covariance components, 50 neighborhoods with 20 individuals per neighborhood (Hox, 1998). In this study, there were 16 urban areas with 1947 census tracts that met this criteria. Subramanian et al. (2003, p. 106) explain the property of exchangeability in multilevel models: "Neighborhoods are assumed to be randomly drawn from a common population. Residents of neighborhoods should therefore, be exchangeable with residents in all other neighborhoods to make inferences about the population. If residents in neighborhoods are dissimilar they should not be regarded as exchangeable and as such, should be treated as fixed effects" (as opposed to random effects that are allowed to vary across neighborhoods). Since this study assumes that Medicaid black mothers experience discriminatory resistance to social mobility, there is a low likelihood of exchangeability between black, white, and other mothers in these neighborhoods. Thus, neighborhood exposure for black mothers is modeled as fixed effects and random effects at the urban level only.

33.2.2 Methods

33.2.2.1 Calculation of Segregation Indices

The generalized spatial dissimilarity index $\breve{D}(m)$ was developed by Feitosa et al. (2007). It is a spatial version of the generalized dissimilarity index $D(m)$ created by Sakoda (1981) and informed by Wong (2002) and Reardon and O'Sullivan (2004). $\breve{D}(m)$ is a multigroup segregation index that estimates the global and local spatial dimensions of evenness and clustering. The index $\breve{D}(m)$ models "localities" as neighborhoods to account for the potential movement of residents across census tract boundaries. A kernel estimator is used to compute the local population intensity at each locality since the intensity of movement may vary by distance. At each census tract centroid j, a weighted average of the population groups is computed. The weights are given by a distance decay function and bandwidth of the kernel estimator, so that individuals of similar race are more likely to interact in closer tracts than tracts farther away. The local population intensity of a locality $j(\breve{L}_j)$ is expressed as

$$\breve{L}_j = \sum_{j=1}^{J} k(N_j), \tag{33.1}$$

where N_j is the total population in census tract j, J is the total number of census tracts in the study area, and k is the kernel estimator that estimates the influence of each census tract on locality j. The local population intensity of group m (black, white, or other) in locality $j(\breve{L}_{jm})$ is calculated by replacing the total population in tract $j(N_j)$ with the population of group m in tract $j(N_{jm})$ in Equation 33.1:

$$\breve{L}_{jm} = \sum_{j=1}^{J} k(N_{jm}). \tag{33.2}$$

Using the local population intensity value at each locality, the global generalized spatial dissimilarity index $\breve{D}(m)$ is calculated:

$$\breve{D}(m) = \sum_{j=1}^{J} \sum_{m=1}^{M} \frac{N_j}{2NI} \left|\breve{\tau}_{jm} - \tau_m\right|, \tag{33.3}$$

where

$$I = \sum_{m=1}^{M} (\tau_m)(1 - \tau_m) \tag{33.4}$$

and

$$\breve{\tau}_{jm} = \frac{\breve{L}_{jm}}{\breve{L}_j}. \tag{33.5}$$

In Equations 33.3 and 33.4, N is the total population in the study area, N_j is the total population in census tract j, τ_m is the proportion of group m in the study area $\breve{\tau}_{jm}$ is the local proportion of group m in locality j, J is the total number of census tracts in the study area, and M is the total number of population groups. In Equation 33.5, $\breve{L}_{jm}$ is the local population intensity of group m in locality j, and $\breve{L}_j$ is the local population intensity of locality j.

The local generalized spatial dissimilarity index $\breve{d}_j(m)$ is calculated by decomposing the global index:

$$\breve{d}_j(m) = \sum_{m=1}^{M} \frac{N_j}{2NI} \left|\breve{\tau}_{jm} - \tau_m\right|, \tag{33.6}$$

where the equation parameters are the same as in Equation 33.3. The local population intensities are estimated using a 0.25-mile bandwidth. Figure 33.1 shows the sensitivity of the local generalized spatial dissimilarity index to changes in the bandwidth parameter. At 0.25 miles, segregation is highest because of the spatial unevenness and clustering of similar racial groups in nearby neighborhoods. As distance and bandwidth size

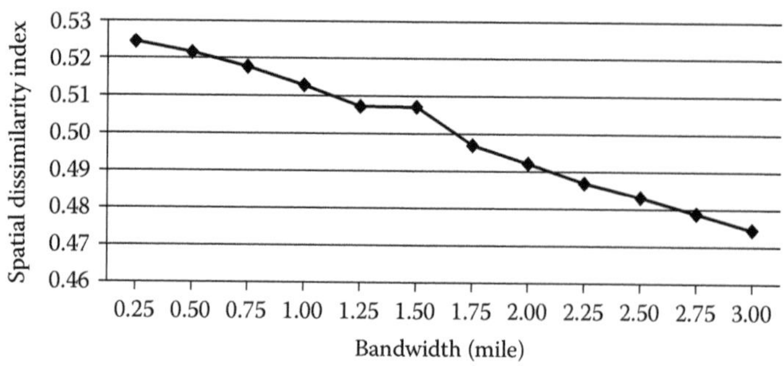

FIGURE 33.1
Global generalized spatial dissimilarity index by bandwidth in miles, lower Michigan, 2008–2011. Index calculated using white, black, and other racial groups in lower Michigan's 16 urban areas (American Community Survey 2008–2011). (From Feitosa, D.D., et al., *International Journal of Geographical Information Science* 21(3), 299–323, 2007.)

increase, the mixing of racial groups also increases, resulting in a decline in the segregation index. Selecting a bandwidth that conceptually define localities—herein referred to as neighborhoods—was important for this study. This segregation tool is freely available for use in ArcGIS (http://www.arcgis.com/) (ESRI, 2015).

The output from the model includes a local segregation index that ranges from 0 to 1, where 0 is the minimum degree of segregation and 1 the maximum degree. In addition, race-by-poverty measures are also calculated using a black–white local segregation index and the percentage of black and white poverty within neighborhoods. Race by poverty is not calculated for other racial groups because of their small numbers within neighborhoods. The segregation index is accompanied by a Z-score and p-value. High-segregated neighborhoods are defined as Z-score ≥ 1.65, which equals 0.57 on the index scale; medium-segregated neighborhoods are defined as Z-score ≥ 0.01 and <1.65, which equals 0.27 on the index scale; and racially mixed neighborhoods are defined as Z-score <0.01, which equals <0.27 on the index scale. If the proportion of black poverty is greater than the proportion of white poverty within high- or medium-segregated neighborhoods, then race by poverty is defined as high-black poverty = 1 and medium-black poverty = 1. If the proportion of white poverty is greater than the proportion of black poverty within high- or medium-segregated neighborhoods, then race by poverty is defined as high-white poverty = 1 or medium-white poverty = 1. All other neighborhoods are defined as racially mixed nonpoor = 0. These dichotomous measures of race by poverty are used to define exposure to neighborhood segregation and concentrated poverty in subsequent statistical analyses.

33.2.2.2 Descriptive Statistics

Rate ratios (RRs) and rate differences (RDs) in preterm birth and other characteristics of black, white, and other mothers are examined to assess the racial disparities in preterm birth by neighborhood types. The RR shows the strength of association between preterm birth and the neighborhood in which the mothers reside. The RD shows the excess preterm birth associated with neighborhood type exposure.

33.2.2.3 Multilevel Modeling

Three-level multilevel models are estimated in MLwiN v2.10 (Rashbash et al., 2012) using a logistic model with logit link for binomial response. The parameters are estimated using a first-order marginal quasi likelihood (MQL) iterative generalized least-squares (IGLS) procedure. The intercept and slope coefficients and accompanying standard errors are transformed into odds ratios (ORs) and 95% confidence intervals (CIs) for ease in interpretation.

First, a random intercept model is estimated to obtain the mean likelihood of preterm birth and the variability between urban and neighborhood levels (Model 1):

$$\begin{aligned}\text{logit } \pi_{ij} &= \beta_{0jk} + e_{ijk}\\ \beta_{0k} &= \beta_0 + v_{0k}\\ \beta_{0j} &= \beta_0 + u_{0j}\\ e_{ij} &\sim \mathrm{N}(0, \sigma_e^2)\\ v_{0j} &\sim \mathrm{N}(0, \sigma_{v0}^2)\\ u_{0j} &\sim \mathrm{N}(0, \sigma_{u0}^2),\end{aligned}$$

where β_{0k} and β_{0j} are the likelihood of preterm birth at the urban and neighborhood levels and e_{ij}, u_{0j}, and v_{0k} are the random effects or variation in preterm birth at level 1 maternal (e_{ij}), level 2 neighborhood (u_{0j}), and level 3 urban (v_{0j}). The random effects are assumed to follow a normal distribution with a mean of zero and variances $\sigma_e^2, \sigma_{u0}^2$, and σ_{v0}^2.

Second, a random intercept and slopes-as-outcomes model is estimated to obtain the likelihood that black mothers will have a preterm birth (Model 2) (equation not shown). Third, the likelihood of preterm birth for black mothers is estimated, controlling for the variation in proportion to black births across urban areas (Model 3):

$$\begin{aligned}\text{logit } \pi_{ij} &= \beta_{0jk} + \beta_1(\mathit{Black})_{ijk} + e_{ijk}\\ \beta_{0k} &= \beta_0 + v_{0k}\\ \beta_{0j} &= \beta_0 + u_{0j}\\ \beta_{1k} &= \beta_1 + v_{ik}.\end{aligned}$$

Fourth, a similar model is estimated, controlling for known maternal risk factors: age, married status, and smoking (Model 4) (equation not shown). Next, the effects of neighborhood types on the odds of preterm birth, controlling for age, married status, and smoking, are estimated (Model 5). Mother's black race is still allowed to vary across urban areas:

$$\begin{aligned}\pi_{ij} = \beta_{0j} &+ \beta_1(\mathit{Black})_{ij} + \beta_2(\mathit{MAge} - 24)_{ij} + \beta_3(\mathit{Married})_{ij} + \beta_4(\mathit{Smoke})_{ij}\\ &+ \beta_5(\mathit{HighBlackPov})_{ij} + \beta_6(\mathit{HighWhitePov})_{ij} + \beta_7(\mathit{MediumBlackPov})_{ij}\\ &+ \beta_8(\mathit{MediumWhitePov})_{ij}\\ \beta_{0k} &= \beta_0 + v_{0k}\\ \beta_{0j} &= \beta_0 + u_{0j}\\ \beta_{1k} &= \beta_1 + v_{ik}.\end{aligned}$$

Finally, in Model 6, interaction terms (high-black poverty*black and medium-black poverty*black) are added to the model to assess the impacts of maternal exposure to these neighborhoods on black preterm birth (equation not shown).

33.3 Results

The global generalized spatial dissimilarity index was 0.62, indicating high segregation in lower Michigan's 16 urban neighborhoods. Figure 33.2 is a map of the local generalized spatial dissimilarity index (Z-score range, −1.41 to 4.69) showing high segregation between

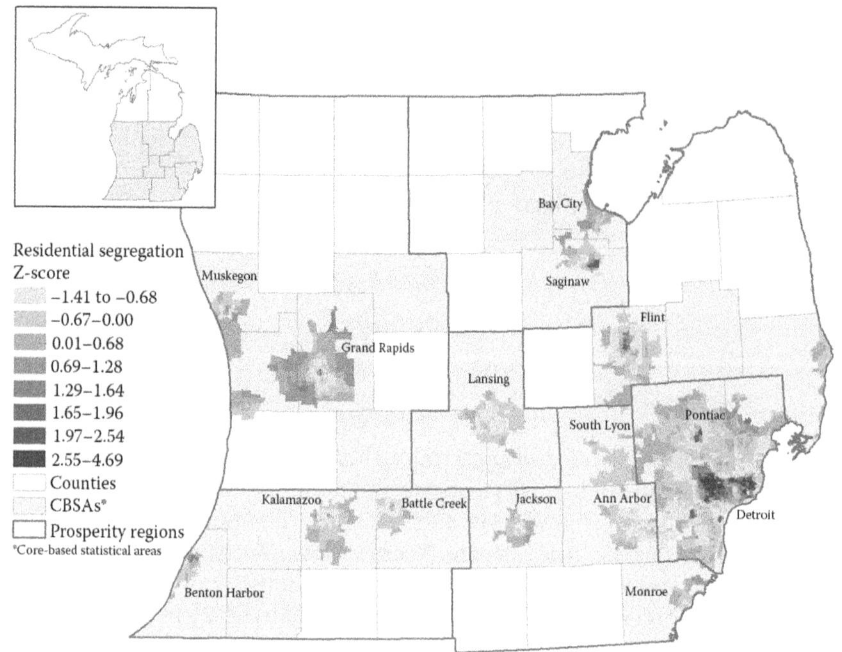

FIGURE 33.2
Local generalized spatial dissimilarity index: white, black, and other racial groups and reference map of urban areas in lower Michigan, 2008–2011. CBSA, core-based statistical area, U.S. Census Bureau, 2014.

blacks, whites, and others residing in neighborhoods in Flint, Grand Rapids, Pontiac, Saginaw, and Detroit and its northern and southeastern suburbs.

Figure 33.3 is a map of race by poverty with high-black poverty neighborhoods ($n = 160$, 8.2%) in the city centers of Benton Harbor, Flint, Grand Rapids, Muskegon, Pontiac, Saginaw, and Detroit and it southeastern suburbs. Medium-black poverty ($n = 197$, 10.1%) generally surrounded high-black poverty neighborhoods and was located in the cities of Ann Arbor and Kalamazoo and Detroit's northern and southeastern suburbs. High-white poverty neighborhoods ($n = 19$, <1.0%) were located in the city centers of Ann Arbor, Kalamazoo, Lansing and Detroit's southeastern suburbs. Medium-white poverty neighborhoods ($n = 218$, 11.2%) were in all urban areas except Benton Harbor and surrounded high-white poverty neighborhoods in Ann Arbor, Grand Rapids, Kalamazoo, and Lansing and otherwise located in suburbs or the outer periphery of urban areas. All other neighborhoods (1,353, 64.5%) were racially mixed nonpoor.

Figure 33.4 shows in more detail six high-black poverty neighborhoods and their locations in proximity to medium-black poverty neighborhoods in their city centers. These neighborhoods were generally enclosed to spaces between highways, railroad tracks, and natural barriers within which black residents were highly clustered.

Table 33.1 shows the characteristics of Medicaid mothers and infants by neighborhood type. A majority (52.0%) of Medicaid mothers lived in racially mixed nonpoor neighborhoods, followed by medium-white poverty (18.9%), high-black poverty (15.4%), medium-black poverty (8.5%), and high-white poverty (5.3%). In high-black poverty neighborhoods, the racial composition of Medicaid mothers was black (90.6%), white (7.7%), and other (1.8%). In high-white poverty neighborhoods, the racial composition of Medicaid mothers was white (75.8%), black (4.9%), and other (19.2%). These findings show

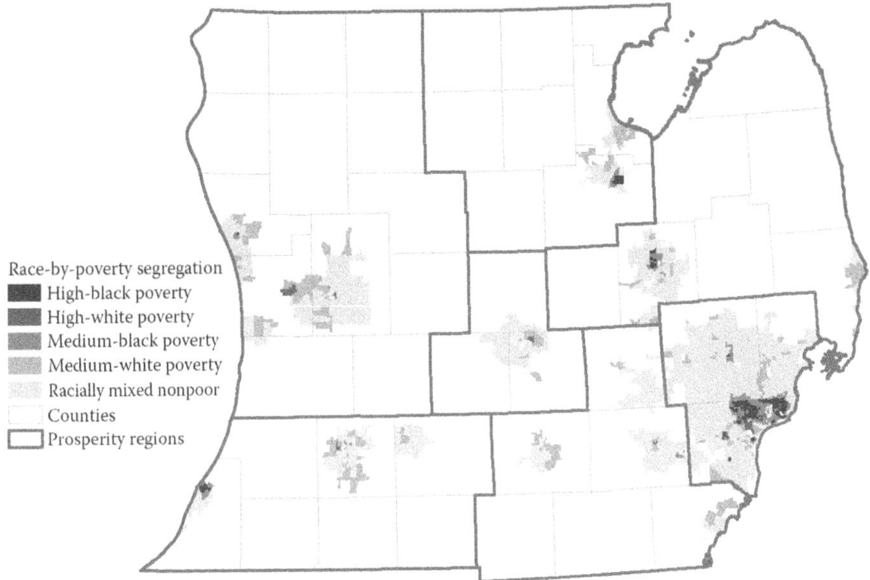

FIGURE 33.3
Local generalized spatial dissimilarity index: race-by-poverty neighborhood levels in urban areas, lower Michigan, 2008–2011. Index calculated using white, black, and other racial groups in lower Michigan's 16 urban areas (American Community Survey 2008–2011). The index was accompanied by a Z-score and p-value. High-segregated neighborhoods = Z-score ≥ 1.65 (~ 0.57 on index scale); medium-segregated neighborhoods = Z-score ≥ 0.01 and < 1.65 (~ 0.27 on index scale); and low-segregated neighborhoods = Z-score < 0.01 ($\sim< 0.27$ on index scale). If the proportion of black poverty was greater than the proportion of white poverty within high- or medium-segregated neighborhoods, then race by poverty was defined as high-black poverty or medium-black poverty. If the proportion of white poverty was greater than the proportion of black poverty within high- or medium-segregated neighborhoods, then race by poverty was defined as high-white poverty or medium-white poverty. All other neighborhoods were defined as racially mixed nonpoor.

that Medicaid black mothers are more highly clustered in high-black poverty neighborhoods than Medicaid white mothers in high-white poverty neighborhoods, where mothers are more racially mixed. In medium-black poverty neighborhoods, a majority of Medicaid mothers were black (73.9%), white (21.8%), and other (4.3%). In medium-white poverty neighborhoods, a majority of Medicaid mothers were white (78.6%), black (12.6%), and other (8.9%). These findings show that Medicaid white mothers are more likely to live in medium-black poverty neighborhoods than Medicaid black mothers in medium-white poverty neighborhoods.

The incidence of preterm birth was highest for Medicaid black mothers in all neighborhoods types, with the highest incidence in high-black poverty neighborhoods (13.5 per 100 live births). The rate ratio (RR) of preterm birth for Medicaid black mothers residing in racially mixed nonpoor versus high-black poverty neighborhoods was 0.8, and the rate difference (RD) was −2.1. The RR for Medicaid white mothers residing in racially mixed nonpoor versus high-white poverty neighborhoods was 1.2 (RD = 1.2). These findings show that Medicaid black mothers had a 20.0% reduction in preterm birth if they lived in racially mixed nonpoor compared to high-black poverty neighborhoods (2.1 excess preterm births per 100 live births). Why Medicaid white mothers had slightly more preterm births in

FIGURE 33.4
Local generalized spatial dissimilarity index: neighborhoods with high-black poverty, lower Michigan, 2008–2011. (a) Benton harbor, (b) Detroit, (c) Flint, (d) Grand Rapids, (e) Muskegon, and (f) Saginaw.

TABLE 33.1
Descriptive statistics of births by maternal race and race-by-poverty segregation for Medicaid mothers (N = 127,116) residing in neighborhoods in 16 urban areas, lower Michigan, 2008–2011

	High-black poverty		High-white poverty		Medium-black poverty		Medium-white poverty		Racially mixed nonpoor		Total	
Census tracts	(n = 160)		(n = 19)		(n = 197)		(n = 218)		(n = 1353)		(n = 1947)	
Characteristics	No.	(%)	No.	(%)	No.	(%)	No.	(%)	No.	(%)	No.	(%)
Black												
Preterm	2390	13.5	37	11.3	1015	12.7	342	11.3	1930	11.4	5714	12.4
Married	1551	8.8	65	19.8	856	10.7	413	13.7	2530	15.0	22,429	11.8
Smoke	3809	21.5	59	17.9	1740	21.8	704	23.3	3315	19.6	5415	21.0
Total	17,712	90.6	329	4.9	7970	73.9	3017	12.6	16,887	25.5	45,915	36.1
Age[a]	24.0	(5.8)	24.3	(5.4)	24.1	(5.7)	23.1	(5.3)	24.1	(5.6)	24.1	(5.7)
White												
Preterm	136	9.1	343	6.8	190	8.1	1504	8.0	3610	8.0	5783	8.0
Married	425	28.4	3716	73.2	776	33.1	7521	40.0	16,355	36.3	28,793	39.6
Smoke	553	37.0	542	10.7	814	34.7	5806	30.8	15,695	34.9	23,410	32.2
Total	1496	7.7	5077	75.8	2347	21.8	18,827	78.6	45,007	68.1	72,754	57.2
Age[a]	25.0	(5.7)	26.5	(5.9)	25.5	(5.6)	25.3	(5.6)	25.4	(5.4)	25.4	(5.5)
Other												
Preterm	25	7.2	83	6.4	38	8.2	187	8.8	327	7.7	660	7.8
Married	185	53.2	696	54.0	216	46.8	1175	55.4	2238	53.0	4510	53.4
Smoke	43	12.4	55	4.3	56	12.1	190	9.0	499	11.8	843	10.0
Total	348	1.8	1288	19.2	462	4.3	2123	8.9	4226	6.4	8447	6.7
Age[a]	26.4	(6.0)	27.0	(6.1)	26.8	(6.1)	26.9	(5.9)	27.0	(5.9)	26.9	(5.9)
Grand total	19,556	15.4	6694	5.3	10,779	8.5	23,967	18.9	66,120	52.0	127,116	100.0

[a]Mother's mean age (standard deviation).

racially mixed nonpoor than high-white poverty neighborhoods may be explained by the tertiary level of neonatal hospital care in Ann Arbor and Lansing that Medicaid mothers have access to. The RR of preterm birth for Medicaid black mothers residing in racially mixed nonpoor versus medium-black poverty neighborhoods was 0.9 (RD = −1.3). The RR for Medicaid white mothers residing in racially mixed nonpoor versus medium-white poverty neighborhoods was 1.0 (RD = −0.1). Thus, black mothers had a 10.0% reduction in preterm birth if they lived in racially mixed nonpoor versus medium-black poverty neighborhoods. There was virtually no difference for white mothers living in racially mixed nonpoor or medium-white poverty neighborhoods.

Within high-black poverty neighborhoods, the Medicaid black–white preterm birth rate ratio and rate difference were 1.5 and 4.4, respectively. Within high-white poverty neighborhoods, the Medicaid black–white preterm rate ratio and rate difference were 1.7 and 1.8, respectively. Thus, within high-black poverty neighborhoods, Medicaid black mothers had a 50.0% increase in preterm births (4.4 excess preterm babies per 100 live births). In high-white poverty neighborhoods, Medicaid black mothers had a 70.0% increase in preterm births compared to white mothers (1.8 excess preterm births per 100 live births). These findings demonstrate that within high-black poverty neighborhoods, Medicaid black mothers had a higher incidence of preterm births than Medicaid white mothers, which may be attributed, in part, to their limited social mobility.

The individual-level risk factors for preterm birth also varied by race and neighborhood type. The mean age of Medicaid black mothers across all neighborhood types was 24.1 years, compared to 25.4 years for Medicaid white and 26.9 years for other mothers. The youngest ages observed for Medicaid black, white, and other mothers were in high-black poverty neighborhoods. In high-black poverty neighborhoods, Medicaid black mothers were least likely to be married (8.8%) compared to Medicaid white (28.4%) and other (53.2%) mothers, although Medicaid white mothers had a higher percentage who smoked (37.0%) compared to Medicaid black (21.5%) and other (12.4%) mothers. Marriage rates were highest among Medicaid black (19.8%) and Medicaid white (73.2%) mothers in high-white poverty neighborhoods, where smoking rates were also lowest (black 17.9%, white 10.7%). In racially mixed nonpoor neighborhoods, approximately 15.0% of Medicaid black mothers were married, compared to 36.3% of Medicaid white and other (53.0%) mothers. In these same neighborhoods, smoking rates were highest among Medicaid white mothers (34.9%), compared to Medicaid black (19.6%) and other (6.4%) mothers.

Table 33.2 shows the results from the multilevel models. Model 1 shows that there was significant variation in preterm birth incidence across urban areas ($x^2 = 3.67$, 1 degree of freedom [d.f.]) and neighborhoods within urban areas ($x^2 = 58.19$ 1 d.f.). Model 2 shows that across the 16 urban areas, Medicaid black mothers were at increased odds (OR = 1.64, 95% CI 1.57–1.71) of preterm birth compared to Medicaid white mothers. Model 3 shows that the odds of preterm birth for Medicaid black mothers decreased (OR = 1.50, 95% CI 1.39–1.62) compared to Medicaid white mothers, controlling for the variation in proportion of Medicaid black births across the urban areas. Controlling for the variation in proportion of Medicaid black births also explained the variation in preterm birth across urban areas and neighborhoods within those urban areas. Model 4 shows the adjusted odds ratio (AOR) of preterm birth for Medicaid black mothers, controlling for individual-level risk factors, with smoking being the strongest risk factor (OR = 1.25, 95% CI 1.19–1.30) and marriage protective (OR = 0.83, 95% CI 0.79–0.88) of preterm birth. In Model 5, the neighborhood types were added to the model. The odds of preterm birth for Medicaid black mothers decreased further (AOR = 1.47, 95% CI 1.37–1.58), controlling for individual-level risk factors and neighborhood types. Medicaid mothers residing in high-black poverty versus racially mixed nonpoor neighborhoods had the highest odds of preterm birth (AOR = 1.16, 95% CI 1.08–1.23). There were significant differences between the probability of preterm birth

TABLE 33.2
Multilevel modeling of race-by-poverty segregation effects on preterm birth for Medicaid Mothers residing in urban neighborhoods, lower Michigan, 2008–2011

	β	Standard error (SE)	Odds ratio	95% Lower CI	Upper CI
Model 1					
Intercept	−2.28	0.03	0.10	0.10	0.11
Model 2[a]					
Intercept	−2.45	0.03	0.09	0.08	0.09
Black race	0.49	0.02	1.64	1.57	1.71
Other race	−0.01	0.04	0.99	0.91	1.08
Model 3[b]					
Intercept	−2.42	0.03	0.09	0.08	0.09
Black race	0.41	0.04	1.50	1.39	1.62
Other race	0.01	0.04	1.01	0.93	1.09
Model 4					
Intercept	−2.48	0.04	0.08	0.08	0.09
Black race	0.42	0.04	1.52	1.41	1.64
Other race	0.04	0.04	1.04	0.96	1.14
Age	0.02	0.01	1.02	1.00	1.04
Married	−0.18	0.03	0.83	0.79	0.88
Smoked	0.22	0.02	1.25	1.19	1.30
Model 5[c]					
Intercept	−2.49	0.04	0.08	0.08	0.09
Black race	0.39	0.04	1.47	1.37	1.58
High-black poverty	0.15	0.03	1.16	1.08	1.23
High-white poverty	−0.07	0.05	0.93	0.84	1.03
Medium-black poverty	0.06	0.04	1.07	0.99	1.15
Medium-white poverty	0.01	0.03	1.01	0.95	1.07
Model 6[c]					
Intercept	−2.48	0.04	0.08	0.08	0.09
High-black poverty* black race	0.09	0.04	1.10	1.02	1.18
Medium-black poverty* black race	0.01	0.04	1.01	0.93	1.10

	Test statistic[d]		
Variation	**Urban (*k*)**	**Neighborhood (*j*)**	**Black (*k*)**
Model 1	3.67*	58.193*	—
Model 2	3.90*	3.69*	—
Model 3	3.57	0.93	1.82
Model 4	3.49	0.76	1.64
Model 5	3.70	0.94	1.12
Model 6	3.68	0.87	1.24

[a]Black race estimated as fixed effects only.
[b]Black race estimated as fixed and random effects at the urban level.
[c]Models 5 and 6 controlling for black and other race, mother's age, married status, and smoked during pregnancy.
[d]Wald test statistic is approximately chi-square distributed on 1 d.f.; $^{*}p < 0.05$.
*Medicaid mothers of other racial groups vs. white not significant.

in high-black poverty and high-white poverty neighborhoods ($p = 0.000$), and high-black poverty and medium-black poverty neighborhoods ($p = 0.034$). There were not significant differences in the probability of preterm birth in high-white poverty and medium-white poverty ($p = 0.157$) or medium-black poverty and medium-white poverty ($p = 0.206$) neighborhoods. The findings demonstrate the importance of studying high-black poverty neighborhoods to better understand the high rates of preterm birth in those areas. Model 6 shows the effect of exposure to high-black poverty neighborhoods on preterm birth for Medicaid black mothers is 1.10 times that for Medicaid white mothers (AOR = 1.10, 95% CI 1.02–1.18). This interaction was significant. The effect of exposure to medium-black poverty neighborhoods on preterm birth for Medicaid black mothers was not significantly different than that of white mothers (AOR = 1.01, 95% CI 0.93, 1.10).

33.4 Discussion

In lower Michigan's 16 urban areas, two-thirds (62.3%) of Medicaid black mothers lived in segregated and poor neighborhoods. In contrast, two-thirds (60.2%) of white mothers and 50.0% of other mothers resided in racially mixed nonpoor neighborhoods. Medicaid black mothers were more highly clustered within high-black poverty and medium-black poverty neighborhoods than Medicaid white mothers in high-white poverty and medium-white poverty neighborhoods. This study showed that the incidence of preterm birth for Medicaid black mothers was more than twice as high as that for Medicaid white mothers, which was in part explained by the variation in Medicaid black births across the 16 urban areas. Individual-level risk factors for preterm birth were significant but did not substantially explain the Medicaid black–white disparity. The effect of exposure to high-black poverty neighborhoods on preterm birth was significantly higher for Medicaid black mothers than for Medicaid white mothers, but the black–white disparity was relatively small, suggesting that these neighborhoods are detrimental to both races—with a slight disadvantage for black mothers. This study assumed that Medicaid black mothers residing in high-black poverty neighborhoods would be susceptible to the untoward effects because of their real and perceived barriers to social mobility—a societal form of racism. Williams (1996) writes that racism can affect health at the individual and institutional levels. At the individual level, racism is a stressor that can lead to behavioral and physiological changes. At the institutional level, that is, neighborhood segregation, poverty is a form of racial discrimination that can result in the unequal distribution of employment, income, and resources and amenities necessary for a healthy pregnancy. Maternal exposures at both levels can operate independently and interactively. Exposure to high-black poverty neighborhoods may be viewed as a distal exposure and over the life course of a women can lead to proximal changes in behaviors and biological processes that ultimately contribute to preterm birth.

Two examples of behavioral and physiological pathways by which exposure to high-black poverty neighborhoods may contribute to the high rates of preterm birth among Medicaid black mothers include age at first sexual encounter and the risk of sexually transmitted diseases and discriminatory stressors and the risk of chronic and pregnancy-related hypertension. A recent study by Biello et al. (2013) found that black adolescents residing in highly segregated metropolitan areas—defined by the dimension of centralization—had a higher odds (OR = 1.3, 95% CI 1.11–1.53) of first sexual encounter than white adolescents also residing in these areas (mean age for black adolescents = 15.2 years and white adolescents = 16.2 years). Early sexual encounters increase the risk of sexually transmitted diseases and early pregnancy (Biello et al., 2013). There is evidence that sexually transmitted diseases, particularly syphilis and gonorrhea, increase the risk of preterm birth (Goldenberg et al., 2008). Other infectious diseases that may contribute to preterm birth include bacterial vaginosis and chorioamnionitis (inflammation of fetal membranes). In high-black poverty neighborhoods where adolescents are clustered, infectious diseases have the potential to rapidly spread while remaining relatively confined within the boundaries of segregation and limited social mobility. A few studies have estimated the indirect effects of chronic diseases, including hypertension and pregnancy-related hypertension, in the segregation and adverse birth outcome relationship (Grady and Ramírez, 2008). The potential pathways by which segregation may increase hypertension prevalence among pregnant mothers is through stress associated with racial discrimination and lack of high-quality preconceptual and prenatal health care (Greer et al., 2013; Jones, 2013; Kershaw et al., 2015). The racial composition of neighborhoods may potentially affect health care utilization through cultural preferences, social networks, higher rates of perceived discrimination, and low rates of trust in medical providers (Gaskin et al., 2012). Gaskin et al. (2012) also found that in black-segregated neighborhoods, there are fewer primary care providers, nurses, physical assistants, and midwives than in white neighborhoods (OR $= 0.52$, $p < 0.001$), controlling for individual and socioeconomic risk factors. Future research should continue to explore pathways by which highly segregated neighborhoods interact with behavioral and physiological changes and limited health care access that may contribute to preterm birth, as well as other diseases and conditions in maternal and child health.

The limitations of this study at the individual level are the lack of medical information on the mother's health, her prenatal health care, and the method of infant delivery; that is, births that are medically induced are considered a preterm birth in this study. Medicaid mothers often have high morbidity that may lead to higher rates of medically induced labor. There was also a lack of information on mother's education and occupation, which would influence her income and potential access to resources and amenities helpful to reduce the risk of preterm birth. At the neighborhood level, the errors in the American Community Survey estimates of race and poverty at the census tract level were not included in the calculation of the race-by-poverty neighborhood types. In addition, the local spatial segregation index modeled segregation as localities using a 0.25-mile buffer from the centroid of each census tract. There may still be small pockets of black-segregated and poor neighborhoods that were not captured at this geographic scale. Finally, Michigan has recently created prosperity regions across the state (Figures 33.1 and 33.2) from within which economic opportunities and health services are coordinated. The impacts of these institutional and large-scale activities on maternal health and preterm birth are not yet well understood. Research should consider addressing these limitations in future segregation and health studies.

33.5 Conclusions

Sociospatial epidemiological research focuses on understanding why certain populations experience poorer health outcomes or premature mortality than other populations within and across geographic areas. To reduce racial disparities in preterm birth, there is a need to address the high rate among black mothers. Racial residential segregation is a social structure that discriminately imposes limited social mobility upon individuals because of their black race. Not including segregation in studies of racial disparities in health will lead to an overestimation of individual-level risk factors (Acevedo-Garcia and Osypuk, 2008) that would otherwise, in part, be explained by the segregated environment in which blacks reside.

References

Acevedo-Garcia D, Osypuk TL. Invited Commentary: Residential Segregation and Health—The Complexity of Modeling Separate Social Contexts. *American Journal of Epidemiology* 2008;168(11):1255–58.

Bell JF, Zimmerman FJ, Mayer JD, Almgren GR, Huebner CE. Birth Outcomes among Urban African-American Women: A Multilevel Analysis of the Role of Racial Residential Segregation. *Social Science and Medicine* 2006;63(12):303–45.

Biello KB, Ickovics J, Niccolai L, Lin H, Kershaw T. Racial Differences in Age at First Sexual Intercourse: Residential Racial Segregation and the Black-White Disparity among U.S. Adolescents. *Public Health Reports* 2013;128:23–128.

Boone SD, Baumgartner KB, Baumgartner RN, Conner AE, Pinkston CM, Rai SN, Riley EC, et al. Associations between CYP19A1 Polymorphisms, Native American Ancestry, and Breast Cancer Risk and Mortality: The Breast Cancer Health Disparities Study. *Cancer Causes Control*, 2014;25(11):1461–71.

CDC (Centers for Disease Control and Prevention). CDC Healthy People 2020. Retrieved July 30, 2014 from http://wonder.cdc.gov/data2010/focraceg.htm.

CDC (Centers for Disease Control and Prevention). CDC WONDER Data Warehouse. Retrieved July 30, 2014 from http://wonder.cdc.gov/data2010/focraceg.htm.

Census Briefs. 2010. Overview of Race and Hispanic Origin: 2010. U.S. Census Bureau. Retrieved July 30, 2014, from http://www.census.gov/prod/cen2010/briefs/c2010br-02.pdf.

Chang VW. Racial Residential Segregation and Weight Status among US Adults. *Social Science and Medicine* 2006;63(5):1289–303.

Cooper RS, Kennelly JF, Durazo-Arvizu R. Relationship between Premature Mortality and Socioeconomic Factors in Black and White Populations of US Metropolitan Areas. *Public Health Reports* 2001;116(5):464–73.

Duncan OD, Duncan B. A Methodological Analysis of Segregation Indexes. *American Sociological Review* 1955;20:21–17.

Ellen IG. Is Segregation Bad for Your Health? The Case of Low Birthweight. *Brookings-Wharton Papers on Urban Affairs* 2000:203–29.

ESRI (Environmental Systems Research Institute). ArcGIS for Desktop. Retrieved July 15, 2015, from http://www.esri.com/softwre/arcgis-for-desktop.

Feitosa DD, Câmara, Monteiro MV, Koschitzki T, Silva MPS. Global and Local Spatial Indices of Urban Segregation. *International Journal of Geographical Information Science* 2007;21(3):299–323.

Fotheringham AS, Wong DWS. The Modifiable Areal Unit Problem in Multivariate Statistical Analysis. *Environment and Planning* 1991;23(7):1025–44.

Gaskin DJ, Dinwiddie GY, Chan KS, McCleary R. Residential Segregation and Disparities in Health Care Utilization. *Medical Care Research and Review* 2012;69(2)158–72.

Getis A, Ord JK. The Analysis of Spatial Association by Use of Distance Statistics. *Geographical Analysis* 1992;24(3):189–207.

Goldenberg RL, Culhane JF, Lams JD, Romero R. Preterm Birth 1 Epidemiology and Causes of Preterm Birth. *Lancet* 2008;371:75–84.

Grady SC. Racial Disparities in Low Birthweight and the Contribution of Residential Segregation: A Multilevel Analysis. *Social Science and Medicine* 2006;63(12):3013–29.

Grady SC, Darden JT. Spatial Methods to Study Local Racial Residential Segregation on Infant Health in Detroit, Michigan. *Annals of the Association of American Geographers* 2012;102(5):922–31.

Grady SC, Ramírez IJ. Mediating Medical Risk Factors in the Residential Segregation and Low Birthweight Relationship in New York City. *Health and Place* 2008;14(4):661–77.

Greer S, Kramer MR, Cook-Smith JN, Casper ML. Metropolitan Racial Residential Segregation and Cardiovascular Mortality: Exploring Pathways. *Journal of Urban Health: Bulletin of the New York Academy of Medicine* 2013;91(3):499–509.

Hox J. Multilevel Modeling: When and How. In Subramanian SV. Multilevel Methods for Public Health Research. In *Neighborhoods and Health*, Kawachi I, Berkman L (eds.), 65–111. New York: Oxford University Press, 2003.

Jones A. Segregation and Cardiovascular Illness: The Role of Individual and Metropolitan Socioeconomic Status. *Health and Place* 2013;22:56–67.

Kershaw KN, Osypuk TL, Phuong Do D, De Chavez PJ, Diez Roux AV. Neighborhood-Level Racial/Ethnic Residential Segregation and Incidence of Cardiovascular Disease. *Circulation* 2015;131:141–48.

Kreft IGG. Are Multilevel Techniques Necessary? An Overview, Including Simulation Studies. In Subramanian SV. Multilevel Methods for Public Health Research. In *Neighborhoods and Health*, Kawachi I, Berkman L (eds.), 65–111. New York: Oxford University Press, 2003

Kumar R, Tsai JH, Hong X, Gignous C, Pearson C, Ortiz K, Fu M, Pongracic JA, Burchard EG, Bauchner H, Wang X. African Ancestry, Early Life Exposure, and Respiratory Morbidity in Early Childhood. *Clinical and Experimental Allergy: Journal of the British Society for Allergy and Clinical Immunology* 2012;42(2):265–74.

LaVeist TA. Segregation, Poverty, and Empowerment: Health Consequences for African Americans. *Milbank Quarterly* 1993;71(1):41–64.

Massey DS, Denton NA. The Dimensions of Residential Segregation. *Social Forces* 1988;67(2):281–315.

Massey DS, Fischer MJ. How Segregation Concentrates Poverty. *Ethnic and Racial Studies* 2000; 23(4):670–91.

MDCH (Michigan Department of Community Health). Retrieved June 20, 2014, from http://www.mdch.gov.

Morrill RL. On the Measure of Spatial Segregation. *Geographic Research Forum* 1991; 11:25–36.

NCHS (National Center for Health Statistics). Retrieved July 30, 2014, from http://www.cdc.gov/nchs/.

Office of Management and Budget. Retrieved June 20, 2014, from http://www.whitehouse.gov/omb.

Polednak AP. Trends in US Black Infant Mortality, by Degree of Residential Segregation. *American Journal of Public Health* 1996;86(5):723–26.

Rashbash J, Steele F, Browne WJ, Goldstein H. *A User's Guide to MLwiN, v2.26*. Centre for Multilevel Modelling University of Bristol 2012.

Reardon SF, O'Sullivan D. Measures of Spatial Segregation. *Sociological Methodology* 2004;34(1):121–62.

Robert SA, Ruel E. Racial Segregation and Health Disparities between Black and White Older Adults. *Journal of Gerontology* 2006;61(4):S203–11.

Sakoda JM. A Generalized Index of Dissimilarity. *Demography* 1981;18(2):269–90.

Subramanian SV. Multilevel Methods for Public Health Research. In *Neighborhoods and Health*, Kawachi I, Berkman L (eds.), 65–111. New York: Oxford University Press, 2003.

U.S. Census Bureau American Community Survey 2008–2012. Retrieved July 15, 2014, from http://www.census.gov/acs/www/about_the_survey/american_community_survey.

U.S. Census Bureau, Core-Based Statistical Areas, Michigan 2014. Retrieved July 15, 2014 from https://www.census.gov/geo/maps-data/data/cbf/cbf_msa.html.

U.S. Census Bureau TIGER Products. Retrieved July 15, 2014, from http://www.census.gov/geo/maps-data/data/tiger.html.

Wilkes R, Iceland J. Hypersegregation in the Twenty-First Century. *Demography* 2004;41(1):23–36.

Williams DR. Race/Ethnicity and Socioeconomic Status: Measurement and Methodological Issues. *International Journal of Health Services* 1996;26(3):483–505.

Williams DR, Mohammed SA. Racism and Health II: A Needed Research Agenda for Effective Interventions. *American Behavioral Scientist* 2013;15(8):1200–26.

Wong DWS. Spatial Indices of Segregation. *Urban Studies* 1993;30:559–72.

Wong DWS. Modeling Local Segregation: A Spatial Interaction Approach. *Geographical and Environmental Modeling* 2002;6:81–97.

34

Sociospatial Epidemiology: Residential History Analysis

David C. Wheeler
Department of Biostatistics
Virginia Commonwealth University
Richmond, Virginia

Catherine A. Calder
Department of Statistics
The Ohio State University
Columbus, Ohio

CONTENTS

34.1 Introduction

In spatial analyses of chronic disease risk, the residential location at the time of diagnosis is often used as a surrogate for unknown environmental exposures, defined broadly to include household chemicals, pollutants, and radiation in environmental media, and lifestyle factors (see Chapters 2, 3, 6, 9, and 14). Residential location may be an appropriate proxy for relevant exposures in some situations; however, in other situations, it can be difficult to identify a true spatial signal in risk with only this surrogate for exposures. In particular, the combination of long disease latency and population mobility is a challenge for spatial epidemiological studies.

When a disease has a long latency, or lag time between exposure to an important risk factor and diagnosis of chronic disease, the relevance of the residential location at time of diagnosis may be minimal. For certain cancers, the latency period can be significant;

latency for cancers such as lung and bladder is thought to be between 20 and 30 years (Archer et al. 2004; Miyakawa et al. 2001), while mesothelioma is estimated to have a latency between 20 and 50 years (Lanphear and Buncher 1992). For these cancers, spatial epidemiologic studies need to consider the residential locations over study participants' lifetimes and allow for the possibility of different environmental exposures at each location.

Due to latency and population mobility, it is rational to expect that residential locations many years before diagnosis of disease are more relevant for risk than the location at time of diagnosis. Researchers in geography (Bentham 1988; Han et al. 2004; Sabel et al. 2009) and public health (Jacquez et al. 2005; Nuckols et al. 2011; Paulu et al. 2002; Vieira et al. 2005) have recognized the importance of population mobility when studying disease risk. Ignoring migration when studying health outcomes with long latencies can lead to exposure misclassification, biased risk estimates, and diminished study power (Tong 2000).

34.2 Population Mobility in the United States

Spatial analyses of epidemiologic studies conducted in the United States are challenged by the level of residential mobility in the population. It has been estimated that a person in the United States can expect to move 11.7 times in his or her lifetime based on the 2007 American Community Survey from the U.S. Census Bureau (U.S. Census Bureau 2014). According to census estimates, in 1996, the median duration of residence in the United States overall was 4.7 years (Schacter and Kuenzi 2002). Another complication is that the level of population mobility in the United States has changed over time. According to data from the 2013 Annual Social and Economic Supplement (ASEC) of the Current Population Survey (CPS) conducted by the U.S. Census Bureau, 11.7% of people aged 1 year or more living in the United States changed residences between 2012 and 2013 (Ihrke 2014). In contrast, the annual mover rate between 1998 and 1999 was 15.8%, and was 20.2% between 1985 and 1986 (U.S. Census Bureau 1987). The 5-year mover rate from 2005 to 2010 was 35.4%, but was 45.6% from 1970 to 1975 and 44.1% more recently, during 1990–1995 (Ihrke and Faber 2012).

In addition to different moving rates over time, there are clear differences in the rate of moving across different socioeconomic, demographic, and racial and ethnic groups in the United States. Not surprisingly, renters have much higher 5-year mover rates than property owners, for example, 65.6% versus 22.2%, respectively, in 2005–2010 (Ihrke and Faber 2012). Unemployed individuals were more mobile than employed individuals (47.7%–37.2%) during 2005–2010. Among those with a household income below the poverty line, the 5-year mover rate was 52.5%, which was much higher than the 5-year mover rate of 31.6% for those with a household income that was 150% or more of the poverty line. Among age groups, people in their mid-to-late twenties had the highest 5-year mover rate at 65.5% (Ihrke and Faber 2012). The two oldest age categories, 65–74 years and 75 years and over, had the lowest 5-year mover rates at 15.2% and 11.9%, respectively. Among racial and ethnic groups, Hispanics had the highest 5-year mover rate at 43.1%, followed by blacks with a rate of 42.9%. Whites had the lowest 5-year mover rate at 33.7%.

There have also been changes over time in the distance individuals have moved, as well as differences in distance moved by population groups. Of the 35.9 million moves between 2012 and 2013, 32.7% were between counties, while 64.4% were within a county. In 2010, 39% of 5-year moves were between different counties, while 61% of moves were in the same county (Ihrke and Faber 2012). In 2005, 48.4% of 5-year moves were between counties, and 51.6% were within county. In 2010, 13% of all 5-year moves were intercounty, but less than 50 miles, while 4.3% were 500 miles or greater. There is considerable variation in distance moved across education groups, with the percent of all 5-year moves in 2010 that are intercounty being 51.4% for those with a professional or graduate degree and only 29.3% for those not having a high school degree (Ihrke and Faber 2012). The implications of these characteristics of residential mobility in the United States are that using the residential location at time of diagnosis will likely be a poor surrogate for historic environmental exposures for many people enrolled in an epidemiologic study, and how poor of a surrogate it is depends on demographic and socioeconomic factors, as well as the timing of the study.

34.3 Residential Histories in a Case-Control Study

As an example of residential histories collected in an epidemiologic study, we describe the distributions of the residential locations collected in a case-control study of non-Hodgkin lymphoma (NHL). The National Cancer Institute (NCI) Surveillance, Epidemiology, and End Results (SEER) NHL study is a case-control study of 1321 cases aged 20–74 years that were diagnosed between July 1, 1998, and June 30, 2000, in four SEER cancer registries, including Detroit, Iowa, Seattle, and Los Angeles County. The study has been described previously (Wheeler et al. 2011). Briefly, population controls (1057) were selected from residents of the SEER areas using random digit dialing (<65 years of age) or Medicare eligibility files (65 and over) and were frequency matched to cases by age (within 5-year groups), sex, race, and SEER area. Among eligible subjects contacted for an interview, 76% of cases and 52% of controls participated in the study. The goal of the NCI-SEER NHL study was to investigate potential genetic and environmental risk factors for NHL.

Computer-assisted personal interviews were conducted during a visit to each subject's home to obtain lifetime residential and occupational histories, medical history, and information on demographics and risk factors. Written informed consent was obtained during the home visit, and human subject review boards approved the study at the NCI and at all participating institutions. Historic addresses were collected in a residential history section of an interviewer-administered questionnaire. Participants were mailed a residential calendar in advance of the interview and were requested to provide the complete address of every home in which they lived from birth to the current year, listing the years when they moved in and out of each address (De Roos et al. 2010). Interviewers reviewed the residential calendar with respondents and probed to obtain missing address information. Residential addresses were matched to geographic address databases to yield geographic coordinates.

Subjects reported residential addresses for the time period 1923–2001. There was substantial variation in the level of population mobility among subjects in the four study centers (Table 34.1). Detroit had the lowest mean level, with subjects enrolled there reporting 6.5 residential locations in the residential histories. Seattle had the highest population mobility, with a mean of 10.4 residential locations per subject. Over the 78 years of residential histories, this equates to an average of 7.5 years spent at each residential location. Interestingly, the most rural of the study centers, Iowa, had the second highest level of residential mobility, with a mean of 8.5 residential locations per subject. Los Angeles County had the second lowest level of population mobility, with a mean of 7.4, but the second highest variance in the number of residential locations per subject. There are also differences in the shapes of the distributions of the number of residential locations per subject across the study centers (Figure 34.1). For example, Los Angeles had relatively more subjects with less than 5 residential locations than in Detroit, but also had relatively more subjects with greater than 10 residential locations. The most common category across study centers was 5–10

TABLE 34.1
Mean, median, and variance of the number of residential locations reported for subjects enrolled in each of the four study centers in the NCI-SEER NHL study

	Number of residential locations		
Site	**Mean**	**Median**	**Variance**
Detroit	6.5	6	3.2
Los Angeles	7.4	7	4.2
Iowa	8.5	8	4.0
Seattle	10.4	10	5.1

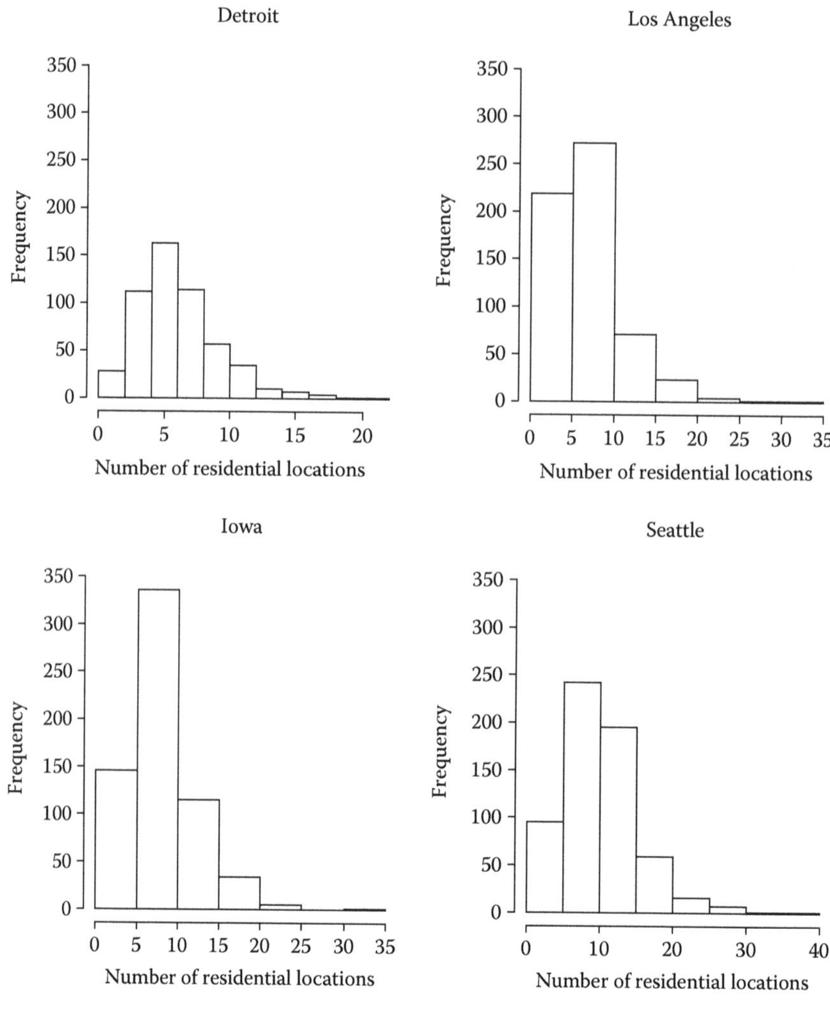

FIGURE 34.1
Distributions of the number of residential locations reported for subjects in each of the four study centers of the NCI-SEER NHL study.

residential locations per subject. This example of residential locations in the NCI-SEER NHL study demonstrates that subjects typically report many residential locations in their lifetime, and some subjects may report as many as 30 or more locations. Hence, one residential location at the time of study enrollment, which is the time of diagnosis for cases, is not likely to represent important environmental exposures over the life course.

34.4 Spatial Analysis of Disease Patterns with Residential Histories

In order to take into account long latencies and residential mobility, investigators increasingly are collecting and storing residential histories for subjects in case-control and cohort studies. Several approaches have been used in the spatial epidemiology literature to consider residential histories when analyzing disease patterns. Modeling the spatial variation in disease risk is often used in spatial epidemiology to describe a spatial pattern in risk and identify areas of statistically significant elevated risk (Diggle 2003; French and Wand 2004; Kelsall and Diggle 1998; Webster et al. 2006) (see Chapter 28). Using this approach, one may adjust for established risk factors or potential confounders that are collected in epidemiologic studies and then examine the unexplained risk for spatial patterns. When data on residential histories are available, one may use historic residential locations for individuals to study spatial risk over time. Residential histories can be informative for both the location and the timing of potential environmental exposures associated with residential locations, which is particularly useful when the mean latency for a disease with suspected environmental causes is unknown.

Several researchers have modeled spatial variation in cancer risk over time using a generalized additive model (GAM). Approaches differ in how residential histories are included in these models. One analysis approach has been to use all the residential locations collected for each subject in one model to estimate one risk surface (Vieira et al. 2005, 2009). This assumes that unmeasured environmental exposures at all residential locations up until diagnosis contribute equally to disease risk. Other approaches include using all the residential locations for each subject up until a time corresponding to a specific latency (e.g., 15 years) in one model (Vieira et al. 2005, 2009), or using all residential locations for a subject in a window of time (e.g., 11 years) beginning at a specific calendar year in a model (Vieira et al. 2008). In the latter case, overlapping time windows were used in a series of fitted models. These approaches consider only a subset of residential locations to be related to disease risk in one model. The above approaches ignore the dependence in the data by including multiple residential locations for each subject. An alternative approach would be to use a generalized additive mixed model (GAMM) to account for dependence in observations through an individual random effect. However, a practical concern with including multiple residential locations per subject is that a few highly mobile cases in a relatively small area could lead to the detection of a significant risk area that is based on only a few subjects.

Other approaches have considered only one residential location at a time for each subject in a GAM. Vieira et al. (2005, 2009) used only the residential location of longest duration in a GAM. Another proposed approach is to estimate a spatial risk surface for a single time lag using only the residential location at that time lag (Wheeler et al. 2011, 2012b). Analysis of deviance tests and Monte Carlo randomization have been used to identify the most significant time lag of exposure among those in a series of evaluated time lags. To better assess cumulative effects of unmeasured environmental exposures over space and time, other efforts have considered models that allowed for multiple residences per subject through spatial smoothing functions of residential locations at different times (Wheeler et al. 2012a). A limitation of these methods is that they do not use all the residential addresses collected in a residential history.

A different analysis method is that of Q-statistics (Jacquez et al. 2005), which consider residential histories of subjects in clustering analyses over time. Q-statistics are intended to indicate if a particular case is the center of clustering over time, if there is clustering of cases at a certain time, or if there is clustering of cases over the entire study period. However, use of Q-statistics requires an a priori knowledge of the k number of nearest-neighbor subjects to use in defining the relevant clustering size when searching over space, and this number is usually not known in advance of the analysis. In addition, it is not clear how relevant increasing spatially distant nearest neighbors are in the set of k nearest neighbors as time before study enrollment increases. Related work has applied the local spatial scan (Kulldorff 1997) in a spatial cluster analysis (see Chapters 2, 9, and 30) at particular times suggested by Q-statistics (Sloan et al. 2012).

34.5 Effect of Population Mobility on Modeling Spatial Risk: A Simulation Study

We conducted several simulation studies to evaluate the effect of population mobility on the ability of a generalized additive model to estimate the spatial variation in risk in a case-control study. We chose to use a GAM here because in a previous simulation study not considering residential histories (Young et al. 2010), a GAM with a spatial smoother was found to have greater power and sensitivity to detect areas of elevated risk than the popular local spatial scan statistic (Kulldorff 1997) (see Chapter 9). For each of our simulation studies, we generated many synthetic populations with realistic (based on national summaries) residential histories. Each individual in the synthetic population was assigned a case status based on a differential risk determined by whether the location of residence fell within a zone of elevated risk. For each simulated population, we then fit a series of GAMs to the case-control data associating individuals with their location of residence at different years (one fitted model per year). As we knew the year when residential location was most relevant to case status, we expected that the GAM corresponding to that year would best fit the data.

In our simulation studies, we considered a single study area with subjects that probabilistically changed residential locations within the study area over time. Note that the setting was a simplification of reality in that subjects could not leave the study area and new subjects could not enter the study area over time (i.e., the population was closed). The study region was taken to be the unit square, and $m = 1000$ subjects were assigned initial residential locations from a uniform distribution that covered the range of spatial coordinates in the study area. We defined a circle in the center of the study area with a radius of 0.25 to represent a zone of true elevated risk. The simulation studies differed according to the timing of the true elevated risk in the circle and the level of population mobility.

In the first simulation study, the zone of elevated risk existed 9 years prior to enrollment in the study. We assigned case status for individual i, Y_i, using a probability of 0.2 for those individuals who lived outside the elevated risk zone initially and a probability corresponding to an odds ratio (OR) of 3.5 for those individuals who lived in the elevated risk zone initially at time $t = -9$ years (9 years before study enrollment). An example of one realization from this scenario is shown in Figure 34.2, where cases are filled and controls are unfilled circles. We then allowed subjects to move randomly within the study area at a rate of 15% per year for 9 years until study enrollment (the annual mover rate in the United States was

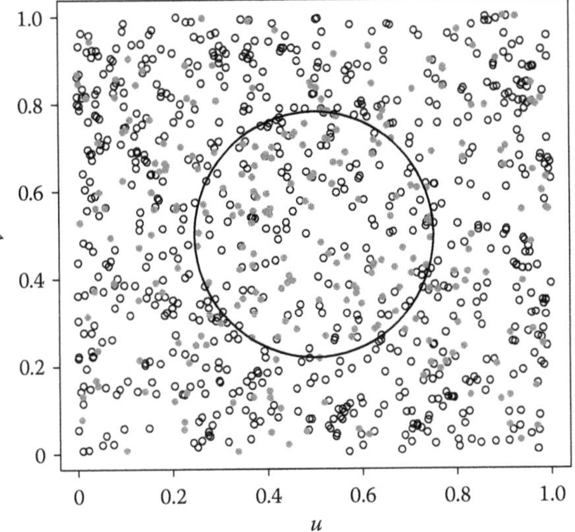

FIGURE 34.2
Case (filled circles) and control (unfilled circles) locations with a circular area of elevated risk 9 years before study enrollment for one data realization in the first simulation study.

approximately 15% in 2002). We denote the set of residential locations for individual i, $\{\mathbf{s}_{u} : t = 0, -1, \ldots, t_0\}$, where t_0 is the earliest year for which we consider the residential location relevant to disease status.

In the second and third simulation studies, we set case status for individuals using OR $= 1.0$ across the study area, including within the exposure zone, at time $t = 0$ (time of study enrollment). We then simulated historic annual population mobility over 9 years, with cases moving into the exposure zone at a rate of 5% and controls moving into the circle at a rate of 2% (study 2) or 1% (study 3) annually. Movement outside the circle within the study area was specified to be 10% annually. The fourth simulation study was based on simulation study 2, but cases and controls were moved outside the circle starting at time $t = -10$ at the rates of 5% and 2%, respectively. We also calculated the actual OR for the elevated risk circle each year.

The same modeling approach was used for all simulation studies. Defining the spatial risk function as $r(\mathbf{s})$ to be the probability that a person residing at location $\mathbf{s}$ will be a case, we express the general spatial log-odds function as $l(\mathbf{s}) = \log[r(\mathbf{s})/\{1 - r(\mathbf{s})\}]$. For each time $t = 0, \ldots, t_0$, we modeled the log-odds of disease through a generalized additive model as

$$\log[P(Y_i = 1)/P(Y_i = 0)] = \alpha + l_t(\mathbf{s}_{it}), \tag{34.1}$$

where α is an intercept and l_t is a locally weighted scatterplot smoother (LOESS) (Cleveland and Devlin 1988) over the subject locations centered at location $\mathbf{s}_{it}$ at a particular time t. We estimated the span parameter in LOESS for each GAM by minimizing the Akaike information criterion (AIC) (Akaike 1973).

To determine which temporal lag-specific GAM, denoted $M_0, \ldots, M_{t_0}$, best explained the spatial patterning of cases in each simulation study, we calculated the posterior probability that the model at year t was best supported by the data using

$$pr(M_t|Data) = \frac{\exp(-0.5 \times BIC_t)pr(M_t)}{\sum_i \exp(-0.5 \times BIC_i)pr(M_i)}, \tag{34.2}$$

where BIC_t is the Bayesian information criterion (Volinsky and Raftery 2000) for the model at time t, $pr(M_t)$ is the prior probability mass given the model, and the sum in the denominator is over all $i = 0, \ldots, t_0$. Previous work (Raftery 1995; Buckland et al. 1997) has shown that model weights based on the Bayesian information criterion are a computationally simple and effective frequentist approximation to the posterior probability that a given model is best, and the approach has been used previously to calculate model posterior probabilities for models with different exposure time lags (Schwartz et al. 2008). We assigned equal prior probability mass to each model so that no candidate model was favored in advance. The rationale for using this method of evaluation is that a study area containing a true area of elevated risk at a particular time, as reflected in the pattern of cases and controls, will have more spatial variation in risk, and therefore will be better fitted with a spatial model. The model fit should be better when the risk is more elevated in the area of increased disease incidence. In the simulation studies, we simulated 1000 realizations from the data-generating process and averaged the model posterior probabilities for each year over the realizations. In addition, we calculated the Monte Carlo error for the model posterior probability, where the Monte Carlo error was defined as the standard deviation of the model posterior probabilities divided by the square root of the number of simulations.

The results for simulation study 1 show that the model estimated based on the locations at $t = -9$ was best among the models with $t = 0, \ldots, -9$, with an average (across 1000 simulations) model posterior probability of 0.55 (Table 34.2). The model at time of study enrollment ($t = 0$) had a low average posterior probability of 0.02. In fact, after 2 years of population mobility, the average posterior probability was below 0.10, and it continued to trend to 0 with more sequences of residential moves. In simulation studies 2 and 3, the best model also was based on residential locations at $t = -9$, with average posterior probabilities of 0.43 and 0.73, respectively (Table 34.3). The actual OR associated with living in the exposure zone at this time was 1.98 in study 2 and 2.70 in study 3. The average posterior probabilities generally increased with the actual ORs for the area of historically elevated risk. The results from simulation study 4 were similar, where the largest mean model posterior probability was 0.35, corresponding to the model fitted based on the $t = -9$ residential locations, and was less than 0.01 at the time of study enrollment (Table 34.4). In all studies, the models at the time of study enrollment had very low posterior probability. The implication of these simulation studies is that using residential locations at the time

TABLE 34.2
Average model posterior probability and Monte Carlo error for generalized additive models fitted to 1000 simulated datasets in simulation study 1

Year	Model posterior probability	Monte Carlo error
0	0.023	0.002
−1	0.022	0.002
−2	0.023	0.002
−3	0.024	0.002
−4	0.030	0.002
−5	0.037	0.003
−6	0.052	0.003
−7	0.081	0.004
−8	0.154	0.056
−9	0.554	0.010

Note: The study has a true odds ratio of 3.5 in elevated risk area at time $t = -9$ and a 15% annual mover rate over a period of 9 years.

TABLE 34.3
Average model posterior probability and Monte Carlo error for generalized additive models fitted to 1000 simulated datasets in simulation studies 2 and 3

	Simulation study 2			Simulation study 3		
Year	**Model posterior probability**	**Monte Carlo error**	**Actual OR in circle**	**Model posterior probability**	**Monte Carlo error**	**Actual OR in circle**
0	0.040	0.014	1.01	0.001	0.001	1.01
−1	0.021	0.008	1.16	0.001	0.000	1.22
−2	0.018	0.005	1.29	0.000	0.000	1.41
−3	0.016	0.004	1.41	0.000	0.000	1.61
−4	0.022	0.004	1.52	0.001	0.000	1.80
−5	0.040	0.006	1.61	0.011	0.008	1.98
−6	0.069	0.011	1.71	0.013	0.003	2.16
−7	0.133	0.015	1.80	0.062	0.014	2.34
−8	0.208	0.017	1.89	0.186	0.021	2.52
−9	0.432	0.027	1.98	0.726	0.027	2.70

Note: The studies have a true odds ratio of 1.0 in the elevated risk area at time $t = 0$, a case annual mover rate of 5% into the circle, and a control mover rate of 2% (study 2) or 1% (study 3) into the circle. The actual OR associated with living in the exposure zone for each year is also listed.

TABLE 34.4
Average model posterior probability and Monte Carlo error for generalized additive models fitted to 1000 simulated datasets in simulation study 4

Year	Model posterior probability	Monte Carlo error	Actual OR in circle
0	0.006	0.002	1.03
−1	0.008	0.003	1.18
−2	0.011	0.003	1.31
−3	0.010	0.002	1.43
−4	0.019	0.006	1.53
−5	0.022	0.005	1.63
−6	0.034	0.005	1.73
−7	0.100	0.013	1.82
−8	0.206	0.019	1.91
−9	0.346	0.024	1.99
−10	0.101	0.012	1.86
−11	0.046	0.010	1.74
−12	0.026	0.006	1.63
−13	0.020	0.008	1.54
−14	0.001	0.003	1.46
−15	0.007	0.002	1.39
−16	0.010	0.003	1.31
−17	0.005	0.001	1.25
−18	0.006	0.002	1.19
−19	0.008	0.002	1.14

Note: The study has a true odds ratio of 1.0 in the elevated risk zone at time $t = 0$, a case annual mover rate of 5% into the zone and a control mover rate of 2% into the zone until time $t = -9$, and the same mover rates to outside the zone beginning at time $t = -10$. The actual OR associated with living inside the elevated risk zone for each year is also listed.

of study enrollment when there is historic population mobility is not likely to correctly find spatial variation in risk for diseases with long latencies and relevant environmental exposures, at least when compared to models based on the residential location most relevant for the elevated risk.

34.6 Bayesian Generalized Additive Model Analysis with Population Mobility: A Simulation and Example

In this example, we examine the effect of population mobility on the goodness of fit of a Bayesian generalized additive, or geoadditive model (Kammann and Wand 2003). French and Wand (2004) use a logistic GAM for modeling spatial variation in cancer risk while adjusting for covariates in a case-control study in Cape Code, Massachusetts (French and Wand 2004). Lawson (2013) specifies a Bayesian GAM for cancer count data in Ohio counties. Both use a spline to model the spatial risk surface while adjusting for study covariates. In our example, we use a Bayesian GAM with a spline to model spatial variation in disease risk for case-control data. A spline model is used as a practical alternative to the kriging-type model proposed by Diggle et al. (1998) to model spatial variation in risk for n binary data y_i (without controlling for study covariates), which requires inversion of an $n \times n$ matrix that is computationally infeasible for large datasets. The alternative we use is known as a reduced knot model or low-rank kriging, which has been used by Nychka et al. (1998) and French and Wand (2004).

The model is

$$\begin{aligned}\log[P(y_i = 1)/P(y_i = 0)] &= \beta_0 + \sum_{j=i}^{p} \beta_j x_{ij} + \sum_{k=1}^{K} \psi_k C\{\|\mathbf{s}_i - \boldsymbol{\kappa}_k\|\} \\ &= \mathbf{x}_i'\boldsymbol{\beta} + z_i'\boldsymbol{\psi}, \end{aligned} \tag{34.3}$$

where the px_j are covariates, and $\{\boldsymbol{\kappa}_1,\ldots,\boldsymbol{\kappa}_K\}$ are a set of K knot locations that are representative of the observed locations $\{\mathbf{s}_1,\ldots,\mathbf{s}_n\}$. We select knot locations using the efficient space-filling algorithm introduced by Johnson et al. (1990) that is implemented in the function cover.design() in the R package fields (Furrer et al. 2013). In the model, ψ_j is a Gaussian random effect, $z_i' = \{z_1,\ldots,z_K\}$, $\mathbf{z} = [C\{\|\mathbf{s}_i - \boldsymbol{\kappa}_k\|/\rho\}]_{1\le i\le n, 1\le k\le K}$, and $\|\mathbf{v}\| = \sqrt{\mathbf{v}^T\mathbf{v}}$. The covariance function is defined as

$$C\{d\} = (1+|d|)e^{-|d|}, \tag{34.4}$$

where ρ is the spatial range parameter and C is a member of the Matérn family of covariance functions (Stein 1999). Also, we define a square covariance matrix

$$\boldsymbol{\omega} = [C\{\|\boldsymbol{\kappa}_i - \boldsymbol{\kappa}_k\|/\rho\}]_{1\le i,k\le K}, \tag{34.5}$$

which is used in the joint random effect prior distribution

$$\psi \sim N(\mathbf{0}, \tau\boldsymbol{\omega}^{-1}). \tag{34.6}$$

To reduce computation burden, French and Wand (2004) select the range parameter in a low-rank kriging model to be the maximum interpoint distance,

$$\hat{\rho} = \max_{1\le i,j\le n} \|\mathbf{s}_i - \mathbf{s}_j\|. \tag{34.7}$$

Lawson (2013) follows the approach of French and Wand (2004) by fixing ρ to the maximum observed distance, but notes that it would be useful to estimate this parameter because it controls the degree of smoothing in the spatially structured component of the model. French and Wand also acknowledge that the smoothness of the estimated surface depends on the choice of ρ. Given the dependence of the results on the choice of ρ, we consider a spectrum of candidate values for ρ and select the best one according to the deviance information criterion (DIC) (Spiegelhalter et al. 2002).

The remaining prior distributions for the model parameters are as follows. The regression coefficient priors are vague normal: $\beta_0 \sim N(0, 0.0001), \beta_j \sim N(0, \tau_\beta)$, where $\tau_\beta \sim$ Gamma$(0.1, 0.0001)$ and $N(a, b)$ denotes the normal distribution with mean a and precision b. The prior on the precision τ for the random effects is specified using the standard deviation $\sigma = \sqrt{1/\tau}$, where $\sigma \sim$ Unif$(1, 10)$ and Unif(c, d) denotes the continuous uniform distribution on the interval (c, d). Similar priors were used for regression coefficients and the standard deviation for the random effects in a spatial spline model in Lawson (2013). The prior for the regression coefficient precision is similar to the one specified in Kelsall and Wakefield (2002).

To examine the effect of population mobility on the goodness of fit of the model, we used a simple simulation example. We generated data from a hypothetical case-control study with $n = 1000$ subjects living in a square study region with spatial coordinates $u \in (-1, 1)$ and $v \in (-1, 1)$. We considered four time periods. At time period 1, we randomly generated subject locations within the study region and assigned disease status. To create spatial variation in disease risk when assigning disease status, we defined a circular area of elevated risk (OR $= 3.5$) centered at $(0.5, 0.5)$ and a circular area of lowered risk (OR $= 0.5$) centered at $(-0.5, -0.5)$. The radius of each of these areas was set to 0.36 to cover 10% of the study area. We defined probability of disease to be 0.2 outside these areas. The simulated locations of cases (filled circles) and controls (unfilled circles) are shown in Figure 34.3, where there are more cases in the northeast and fewer in the southwest. To simulate population mobility, we randomly selected subjects to move randomly within the study region at a rate of 0.25 in time periods 2, 3, and 4. A move rate of 0.25 was selected as a realistic rate for multiyear

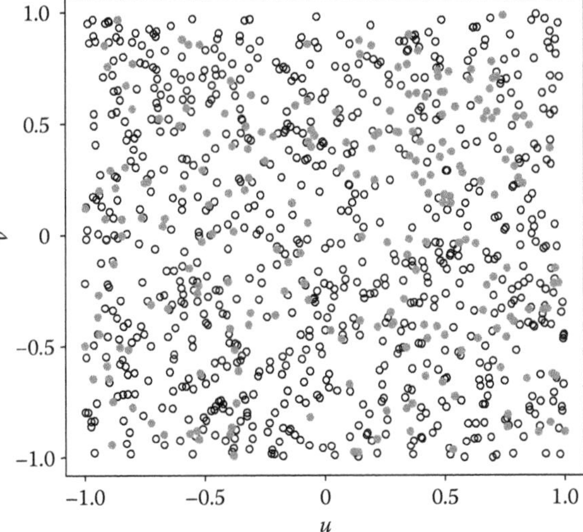

FIGURE 34.3
Case (filled circles) and control (unfilled circles) locations in time period 1 of simulation study.

periods in the United States. We suspected that this move rate would result in an effectively random pattern of cases and controls at time period 4. To represent covariates encountered in a study, we generated a continuous x_1 and a binary x_2. The covariates were by construction unrelated to the outcome.

We fitted Bayesian GAM models for the simulated data at each of the four time periods for candidate ρ values from 0.1 to the maximum distance of 2.76 by increments of 0.1. We used $K = 25$ knots in each model. We estimated the model parameters using Markov chain Monte Carlo (MCMC) implemented in the rjags package (Plummer 2013) in R. We used 1000 iterations for adaptation, 1000 iterations for burn-in, and 1000 iterations from three parallel chains with a thinning interval of 3 for sampling from the joint posterior distribution.

The DIC values for each of the time period models for each candidate ρ are shown in Figure 34.4. At the maximum distance (used as the choice of ρ by French and Wand [2004] and Lawson [2013]), the four model DIC values were nearly identical. The DIC values for the models for time periods 1, 2, 3, and 4 were 1064, 1065, 1066, and 1067, respectively, with the largest difference in model fit between time periods 1 and 4 with this spatial range value. The DIC curves for times 2–4 are similar and frequently overlap. In contrast, the DIC curve for time period 1 is separated from the others for almost all values of ρ, where the separation is greatest for small values of ρ. The lowest DIC (1041) is for the model for time period 1 with the minimum ρ of 0.1. This example demonstrates the importance of estimating ρ, particularly when the most important time for detecting spatial variation in risk is unknown.

We applied the above Bayesian GAM model to the Los Angeles County study site of the NCI-SEER NHL study. We fitted separate models using residential locations at the time of diagnosis and at times of interest of 5, 10, 15, and 20 years before diagnosis. In each model, we adjusted for subject age, gender, race, education level, and exposure to the insecticide chlordane through historic exposure to home termite treatment (Wheeler et al. 2011). For each of the five times, we evaluated a range of candidate ρ values in the models. Initially, we considered 20 equally spaced candidate ρ values in the range of 1% of the maximum

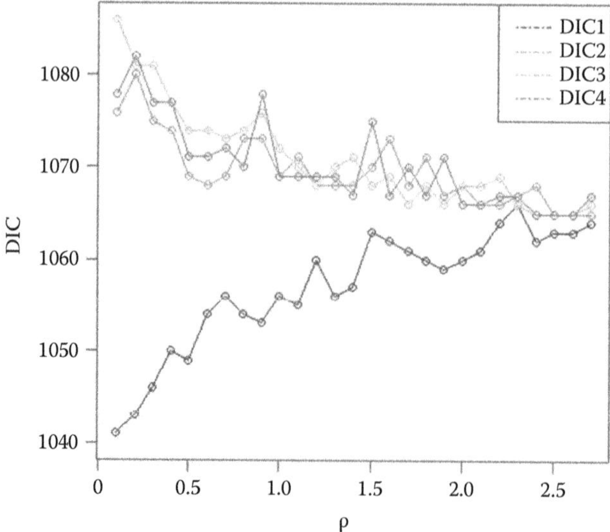

FIGURE 34.4
DIC as a function of ρ for each of the four time period models in the simulation study.

interpoint distance to the maximum distance of 108 (i.e., from 1.08 to 108 by increments of 5.6). Based on the output, we refined the candidate set to a range of 0.2–20 by increments of 1. To estimate parameters for each model using MCMC in rjags, we used 1000 iterations for adaptation, 1000 iterations for burn-in, and 1000 iterations from three parallel chains with a thinning interval of 3 for sampling from the joint posterior distribution.

With ρ set at the max distance, the model at the time of diagnosis had the highest DIC (501). The DIC values for the other times were all lower, with a lag of 10 years having the lowest DIC (489), and lags 20 (495), 15 (495), and 5 (492) years having similar DIC values. The DIC values for each value of ρ in the reduced candidate set in models for the five times are shown in Figure 34.5. For small values of ρ (i.e., 3–8), the model at a 20-year lag had a substantially lower DIC than the models for smaller time lags. The lowest DIC (469) for the 20-year lag model was at $\rho = 5.4$. Given the DIC results, we made predictions from the 20-year lag model at $\rho = 5.4$ over a 25×25 grid covering the study area and calculated the exceedance probabilities $q_j^c = \Pr(\theta_j > c)$ for the spatial odds ratio using the samples of ψ in the spatial component of the model in Equation 34.3. We defined the spatial odds at grid cell j as $\theta_j = \exp(z_j'\psi)$, used the mean spatial odds as the reference for the spatial odds ratios, and used $c = 1$ as the null risk value. We estimated the exceedance probabilities as $\hat{q}_j^c = \sum_{g=m+1}^{m+m_p} I(\theta_j^{(g)} > c)/G$, where $G = 1002$, and considered exceedance probability thresholds of $a = 0.90$ and 0.95 to indicate unusual risk areas, $\hat{q}_j^c > a$. The unusual risk areas at a time of 20 years before diagnosis using an exceedance probability threshold of 0.90 are identified as filled circles in the map in Figure 34.6, where case (square) and control (circle) locations have been jittered to mask the true locations. The area of unusual risk

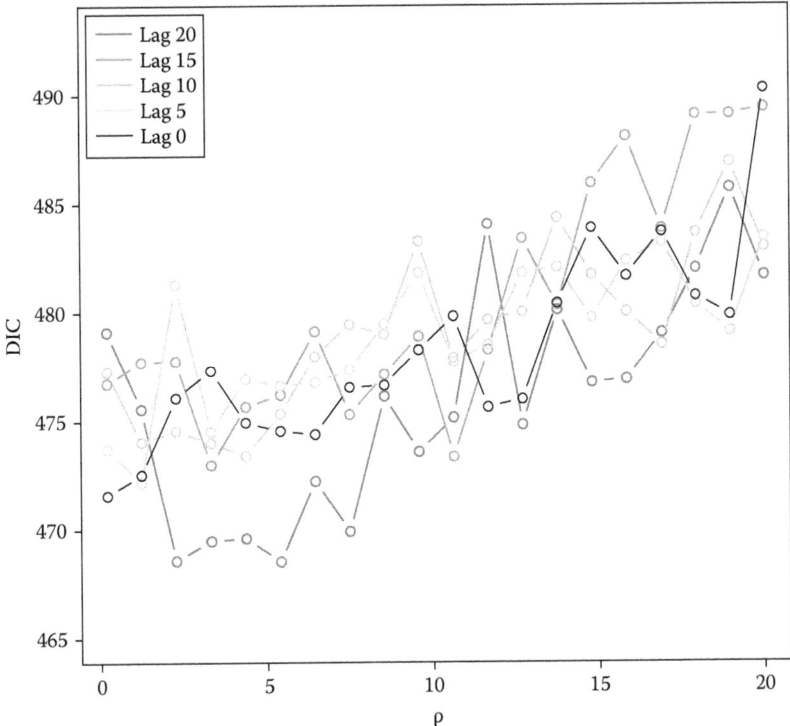

FIGURE 34.5
DIC as a function of ρ in the reduced candidate set in models for the five time periods considered in the NCI-SEER NHL Los Angeles study center.

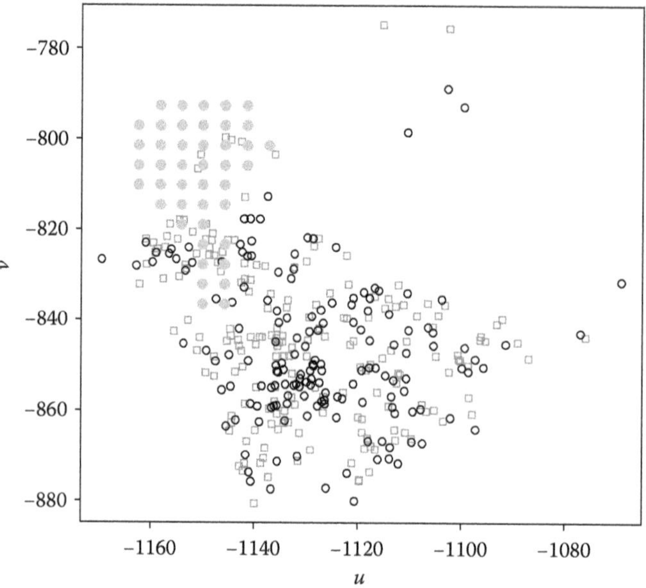

FIGURE 34.6
Jittered case (square) and control (unfilled circle) locations in Los Angeles County and unusual risk areas (filled circles) at a time of 20 years before diagnosis using an exceedance probability threshold of 0.90.

is in northwest Los Angeles County, which was identified previously as an area of elevated risk at a lag of 20 years using a different approach (Wheeler et al. 2011).

34.7 Spatial Scan Analysis with Residential Histories: An Example

As an example of using residential histories in a study of disease risk in a case-control study, we applied the local spatial scan (Kulldorff 1997) to the NCI-SEER NHL study using the SaTScan software (Kulldorff and Information Management Services 2009). We evaluated the local spatial scan statistic (see Chapter 9) for several time points of interest, including at time of diagnosis and 5, 10, 15, and 20 years before time of diagnosis. We conducted a purely spatial retrospective analysis, set the maximum spatial cluster size as 50% of the population at risk, used a Bernoulli probability model, and scanned for both circular and elliptical areas (Kulldorff et al. 2006) with high NHL rates in the four study areas of the NCI-SEER NHL study. For the Bernoulli model local spatial scan, the likelihood function for a search window of a particular size and location is

$$\left(\frac{c}{n}\right)^{c}\left(\frac{n-c}{n}\right)^{n-c}\left(\frac{C-c}{N-n}\right)^{C-c}\left(\frac{(N-n)-(C-c)}{N-n}\right)^{(N-n)-(C-c)}, \qquad (34.8)$$

where c is the observed number of cases within the search window, C is the total number of cases, n is the total number of cases and controls within the search window, and N is the total number of cases and controls in the study. This likelihood function is maximized over all window sizes and locations, and the window with the maximum likelihood is the most likely cluster. The distribution for the maximum likelihood ratio test statistic under the null hypothesis is generated using Monte Carlo randomization. The p-value for the most likely cluster is obtained by comparing the rank of the maximum likelihood for the observed data with the distribution of maximum likelihoods from the randomized datasets.

The p-values for the most likely cluster from each local spatial scan statistic are listed in Table 34.5. No statistically significant clusters were identified at the time of diagnosis (time lag = 0). There were significant clusters detected for Detroit and Los Angeles using time lags. For Detroit, a significant cluster was identified at lags of 15 and 20 years with a circular search window and at a lag of 15 years with an elliptical search window. The most likely cluster was marginally significant at a lag of 20 years with the elliptical window. In Los Angeles County, a statistically significant cluster was found at a time lag of 20 years with the elliptical window. The most likely cluster was marginally significant with the circular search window. These results are consistent with previous analyses, where Detroit and Los Angeles had the most significant unexplained spatial risk of NHL after adjusting for covariates in generalized additive models (Wheeler et al. 2011, 2012b). The most likely elliptical cluster at a lag of 20 years in Los Angeles is shown in Figure 34.7, and the most likely circular cluster at a lag of 20 years in Detroit is plotted in Figure 34.8. The residential locations shown in the maps have been jittered to mask the true locations. These clusters are located in previously identified areas of significantly elevated risk with the adjusted GAM approach (Wheeler et al. 2011). The cluster in Figure 34.7 also agrees with the unusual risk area identified in Figure 34.6 using the Bayesian GAM approach.

TABLE 34.5
Local spatial scan p-values for the most likely cluster from circular and elliptical search areas at five time points (time of diagnosis and lags of 5, 10, 15, and 20 years) in the four study centers of the NCI-NHL SEER study

Site	**Time lag**	**Circle p-value**	**Ellipse p-value**
Detroit	0	0.30	0.24
	5	0.25	0.40
	10	0.27	0.12
	15	0.02	0.04
	20	0.03	0.06
Los Angeles	0	0.79	0.76
	5	0.59	0.55
	10	0.26	0.30
	15	0.86	0.67
	20	0.09	0.05
Iowa	0	0.76	0.54
	5	0.48	0.74
	10	0.47	0.68
	15	0.86	0.50
	20	0.10	0.44
Seattle	0	0.67	0.47
	5	0.89	0.30
	10	0.54	0.47
	15	0.51	0.44
	20	0.75	0.54

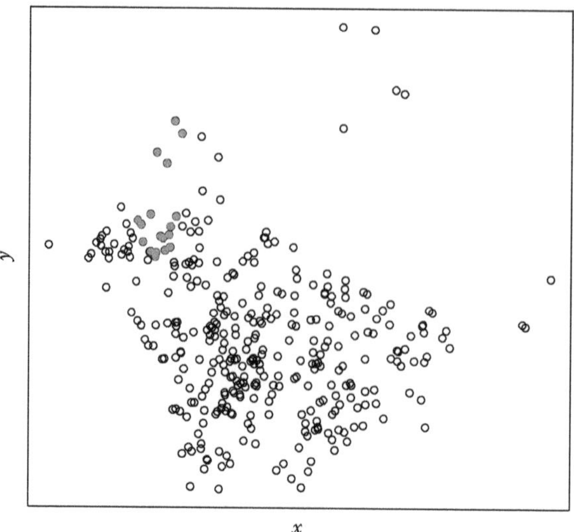

FIGURE 34.7
Subject locations and most likely elliptical cluster (filled circles) at a lag of 20 years in Los Angeles.

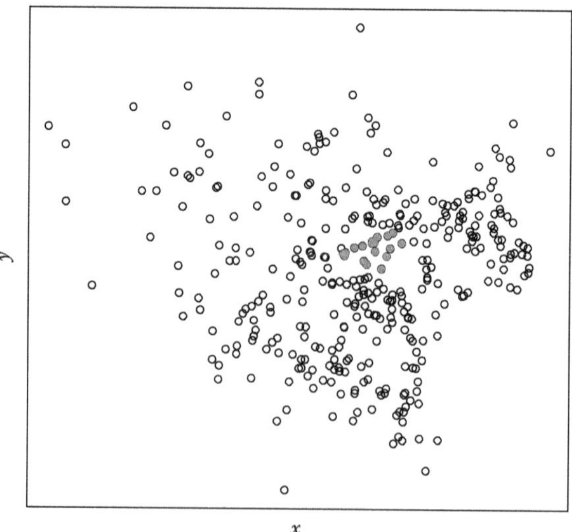

FIGURE 34.8
Subject locations and most likely elliptical cluster (filled circles) at a lag of 20 years in Detroit.

34.8 Future Directions: Cumulative Environmental Exposure Analysis

We have argued that there are issues with modeling spatial risk of a chronic disease with a long latency using the residential location at the time of diagnosis. Instead of using one location to model spatial risk to such diseases, one can use an approach to estimate life-course environmental exposures, which complements the increasingly popular vision in epidemiology of the "exposome" that seeks to characterize the totality of environmental exposures for disease risk (Rappaport and Smith 2010; Wild 2005). The rationale is that relevant cumulative environmental exposures could occur over multiple residential locations across time and models should attempt to encompass such life-course environmental exposures. In particular, there is a recent growth in interest in the public health literature on modeling early-life and life-course exposures (Murray et al. 2011; Richardson et al. 2013).

To illustrate modeling cumulative space–time risk, we consider the Los Angeles County study area of the NCI-SEER NHL study, which previously had the most significant unexplained spatial risk of NHL (Wheeler et al. 2011). The exposure of interest was the industrial solvent trichloroethylene (TCE), which has been found to be associated with NHL risk in occupational settings (Cocco et al. 2013; Mandel et al. 2006). We used the residential histories in the study to assign environmental TCE exposure using the closest facility in the Toxic Release Inventory (TRI), which is available from the Environmental Protection Agency (EPA). We assigned the total volume of TCE released each year from the closest facility to each of the residential locations for a period of 12 years prior to and including study enrollment.

We modeled the log-odds of NHL as

$$\log\left(\frac{P(Y_i=1)}{P(Y_i=0)}\right) = \mathbf{X}_i\beta + \sum_t \gamma(t)e_t(\mathbf{s}_{it}), \tag{34.9}$$

adjusting for the covariates gender and race in $\mathbf{X}$ and modeling the effect of historical exposure as a temporally varying weighted average. The sum in the second term on the right-hand side of (34.9) is over the historical period relevant to the study, and we specified the weighting function using a nonparametric cumulative exposure response curve,

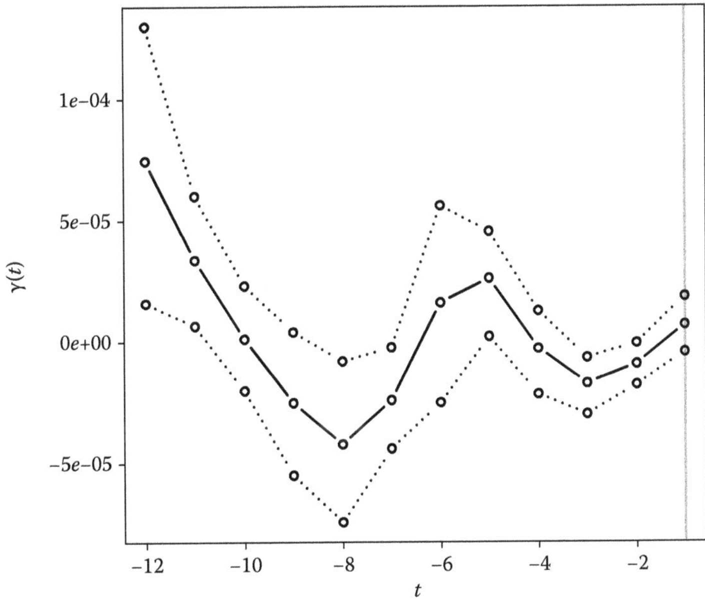

FIGURE 34.9
Estimated cumulative exposure response curve and 95% credible intervals for TCE and NHL in Los Angeles.

$$\gamma(t) = \sum_{j=1}^{J} b_j f_j(t), \quad \text{for } t = 0, \ldots, t_0, \tag{34.10}$$

to model the effect of TCE exposure e_t at time t. We used radial basis functions f_j with $J = 12$ knots and weights b_j. We estimated the model parameters using MCMC implemented in the rjags package (Plummer 2013) in R.

The estimated cumulative exposure response curve and 95% credible intervals show the most significant positive associations between TCE and NHL at years −12 and −11 (Figure 34.9). The effect of TCE is not significant at the time of diagnosis. TRI data are not available before 1987, so a limitation of this analysis is that it was not possible to estimate an effect for TCE exposure more than 12 years before study enrollment.

34.9 Discussion and Conclusions

In spatial analyses of disease risk, the residential location is typically used as a surrogate for unmeasured environmental exposures. However, residential mobility among subjects in epidemiologic studies makes the residential location at the time of study enrollment a poor surrogate for unknown environmental exposures for diseases with long latencies. Our simulation studies show that spatial models based on residential location at time of diagnosis are not likely to pick up true information about spatial patterns in disease when there is an area of true elevated risk in the past and residential mobility is at a level found in the recent history of the United States. Fortunately, investigators are increasingly collecting residential histories for subjects in cohort and case-control studies. Residential histories can be informative for both the location and the timing of potential environmental exposures associated with disease risk, and can also facilitate life-course analysis when linked to environmental exposure databases.

Acknowledgments

We would like to thank the following collaborators for providing access to the data from the National Cancer Institute Surveillance, Epidemiology, and End Results Interdisciplinary Case-Control Study of NHL: Mary Ward, Lindsay Morton, Patricia Hartge, Anneclaire De Roos, James Cerhan, Wendy Cozen, and Richard Severson. We also acknowledge Kabita Joshi for assistance with computer programming for the Bayesian GAM example, whose effort was supported by a grant from the American Cancer Society. David Wheeler's efforts were supported by an NCI grant (R03CA173823). Catherine Calder's efforts were supported in part by an NSF grant (DMS-1209161).

References

Akaike, Hirotogu. 1973. Information Theory and an Extension of the Maximum Likelihood Principle. In *International Symposium on Information Theory*, Budapest, pp. 267–281.

Archer, Victor E., Teresa Coons, Geno Saccomanno, and Dae-Yong Hong. 2004. Latency and the Lung Cancer Epidemic among United States Uranium Miners. *Health Physics* 87(5):480–489.

Bentham, Graham. 1988. Migration and Morbidity: Implications for Geographical Studies of Disease. *Social Science and Medicine* 26(1):49–54.

Buckland, Stephen T., Kenneth P. Burnham, and Nicole H. Augustin. 1997. Model Selection: An Integral Part of Inference. *Biometrics* 53:603–618.

Cleveland, William S., and Susan J. Devlin. 1988. Locally-Weighted Regression: An Approach to Regression Analysis by Local Fitting. *Journal of the American Statistical Association* 83(403):596–610.

Cocco, Pierluigi, Roel Vermeulen, Valeria Flore, Tinucia Nonne, Marcello Campagna, Mark P. Purdue, Aaron Blair, et al. 2013. Occupational Exposure to Trichloroethylene and Risk of Non-Hodgkin Lymphoma and Its Major Subtypes: A Pooled Interlymph Analysis. *Occupational and Environmental Medicine* 70(11):795–802.

De Roos, Anneclaire J., Scott Davis, Joanne S. Colt, Aaron Blair, Matthew Airola, Richard K. Severson, Wendy Cozen, James R. Cerhan, Patricia Hartge, John R. Nuckols, and Mary H. Ward. 2010. Residential Proximity to Industrial Facilities and Risk of Non-Hodgkin Lymphoma. *Environmental Research* 110:70–78.

Diggle, Peter J. *Statistical Analysis of Spatial Point Patterns.* 2nd ed. London: Edward Arnold, 2003.

Diggle, Peter J., J. A. Tawn, and R. A. Moyeed. 1998. Model-Based Geostatistics. *Applied Statistics* 47:299–326.

French, Jonathan L., and Matthew P. Wand. 2004. Generalized Additive Models for Cancer Mapping with Incomplete Covariates. *Biostatistics* 5(2):177–191.

Furrer, Reinhard, Douglas Nychka, and Stephen Sain. 2013. fields: Tools for Spatial Data. R package version 6.7.6. http://CRAN.R-project.org/package=fields.

Han, Daikwon, Peter A. Rogerson, Jing Nie, Matthew R. Bonner, John E. Vena, Dominica Vito, Paola Muti, Maurizio Trevisan, Stephen B. Edge, and Jo L. Freudenheim. 2004. Geographic Clustering of Residence in Early Life and Subsequent Risk of Breast Cancer (United States). *Cancer Causes and Control* 15:921–929.

Ihrke, David. 2014. Reason for Moving: 2012 to 2013. Current Population Reports, P20-574. Washington, DC: U.S. Census Bureau.

Ihrke, David K., and Carol S. Faber. 2012. Geographical Mobility: 2005 to 2010. Current Population Reports, P20-567. Washington, DC: U.S. Census Bureau.

Jacquez, Geoffrey M., Andy Kaufmann, Jaymie Meliker, Pierre Goovaerts, Gillian AvRuskin, and Jerome Nriagu. 2005. Global, Local and Focused Geographic Clustering for Case-Control Data with Residential Histories. *Environmental Health* 4(1):4.

Johnson, M. E., L. M. Moore, and D. Ylvisaker. 1990. Minimax and Maximin Distance Designs. *Journal of Statistical Planning and Inference* 26:131–148.

Kammann, E. E., and M. P. Wand. 2003. Geoadditive Models. *Applied Statistics* 52:1–18.

Kelsall, Julia E., and Peter J. Diggle. 1998. Spatial Variation in Risk of Disease: A Nonparametric Binary Regression Approach. *Applied Statistics* 47:559–573.

Kelsall, Julia E., and Jonathan Wakefield. 2002. Modelling Spatial Variation in Disease Risk: A Geostatistical Approach. *Journal of the American Statistical Association* 97:692–701.

Kulldorff, Martin. 1997. A Spatial Scan Statistic. *Communications in Statistics: Theory and Methods* 26:1487–1496.

Kulldorff, Martin, Lan Huang, Linda Pickle, and Luiz Duczmal. 2006. An Elliptic Spatial Scan Statistic. *Statistics in Medicine* 25:3929–3943.

Kulldorff, Martin, and Information Management Services, Inc. 2009. SaTScan v8.0: Software for the Spatial and Space-Time Scan Statistics. http://www.satscan.org/.

Lanphear, Bruce P., and C. Ralph Buncher. 1992. Latent Period for Malignant Mesothelioma of Occupational Origin. *Journal of Occupational Medicine* 34(7):718–721.

Lawson, Andrew B. *Bayesian Disease Mapping: Hierarchical Modeling in Spatial Epidemiology.* Boca Raton, FL: CRC Press, 2013.

Mandel, Jeffrey H., Michael A. Kelsh, Pamela J. Mink, Dominik D. Alexander, Renee M. Kalmes, Michal Weingart, Lisa Yost, and Michael Goodman. 2006. Occupational Trichloroethylene Exposure and Non-Hodgkin's Lymphoma: A Meta-Analysis and Review. *Occupational and Environmental Medicine* 63:597–607.

Miyakawa, Michiko, Masaaki Tachibana, Ayako Miyakawa, Katsumi Yoshida, Naoki Shimada, Masaru Murai, and Takefumi Kondo. 2001. Re-Evaluation of the Latent Period of Bladder Cancer in Dyestuff-Plant Workers in Japan. *International Journal of Urology* 8:423–430.

Murray, Emily T., Gita D. Mishra, Diana Kuh, Jack Guralnik, Stephanie Black, and Rebecca Hardy. 2011. Life Course Models of Socioeconomic Position and Cardiovascular Risk Factors: 1946 Birth Cohort. *Annals of Epidemiology* 21(8):589–597.

Nuckols, John R., Laura E. Beane Freeman, Jay H. Lubin, Matthew S. Airola, Dalsu Baris, Joseph D. Ayotte, Anne Taylor, et al. 2011. Estimating Water Supply Arsenic Levels in the New England Bladder Cancer Study. *Environmental Health Perspectives* 119(9): 1279–1285.

Nychka, Douglas, Perry Haaland, Michael O'Connell, and Stephen Ellner. FUNFITS, Data Analysis and Statistical Tools for Estimating Functions. In Douglas Nychka, Walter W. Piegorsch, and Lawrence H. Cox (eds.), *Case Studies in Environmental Statistics*, pp. 159–196, New York: Springer, 1998.

Paulu, Christopher, Ann Aschengrau, and David Ozonoff. 2002. Exploring Associations between Residential Location and Breast Cancer Incidence in a Case-Control Study. *Environmental Health Perspectives* 110(5):471–478.

Plummer, Martyn. 2013. rjags: Bayesian graphical models using MCMC. R package version 3-10. http://CRAN.R-project.org/package=rjags.

Raftery, Adrian E. 1995. Bayesian Model Selection in Social Research. *Sociological Methodology* 25:111–163.

Rappaport, Stephen M., and Martyn T. Smith. 2010. Environment and Disease Risks. *Science* 330:460–461.

Richardson, Douglas B., Nora D. Volkow, Mei-Po Kwan, Robert M. Kaplan, Michael F. Goodchild, and Robert T. Croyle. 2013. Spatial Turn in Health Research. *Science* 339:1390–1392.

Sabel, Clive E., Paul Boyle, Gillian Raab, Markku Löytönen, and Paula Maasilta. 2009. Modelling Individual Space-Time Exposure Opportunities: A Novel Approach to Unravelling the Genetic or Environmental Disease Causation Debate. *Spatial and Spatio-Temporal Epidemiology* 1(1):85–94.

Schacter, Jason P., and Jeffrey J. Kuenzi. 2002. Seasonality of Moves and the Duration and Tenure of Residence: 1996. Population Division Working Paper Series No. 69. Washington, DC: U.S. Census Bureau.

Schwartz, Joel, Brent Coull, Francine Laden, and Louis Ryan. 2008. The Effect of Dose and Timing of Dose on the Association between Airborne Particles and Survival. *Environmental Health Perspectives* 116(1):64–69.

Sloan, Chantel D., Geoffrey M. Jacquez, Carolyn M. Gallaghera, Mary H. Ward, Ole Raaschou-Nielsenf, Rikke Baastrup Nordsborgf, and Jaymie R. Meliker. 2012. Performance of Cancer Cluster Q-Statistics for Case-Control Residential Histories. *Spatial and Spatio-Temporal Epidemiology* 3(4):297–310.

Spiegelhalter, David J., Nicola G. Best, Bradley P. Carlin, and Angelika van der Linde. 2002. Bayesian Measures of Model Complexity and Fit. *Journal of the Royal Statistical Society: Series B* 64(4):583–639.

Stein, Michael L. *Statistical Interpolation of Spatial Data: Some Theory for Kriging.* New York: Springer, 1999.

Tong, Shilu. 2000. Migration Bias in Ecologic Studies. *European Journal of Epidemiology* 16:365–369.

U.S. Census Bureau. 1987. Current Population Reports, Series P-20, No. 420, Geographical Mobility: 1985. Washington, DC: U.S. Government Printing Office.

U.S. Census Bureau. 2014. Calculating Migration Expectancy Using ACS Data. Washington, DC: U.S. Government Printing Office. https://www.census.gov/hhes/migration/about/cal-mig-exp.html (accessed July 9, 2014).

Vieira, Verónica, Thomas Webster, Janice Weinberg, and Ann Aschengrau. 2008. Spatial-Temporal Analysis of Breast Cancer in Upper Cape Cod, Massachusetts. *International Journal of Health Geographics* 7:46.

Vieira, Verónica, Thomas Webster, Janice Weinberg, and Ann Aschengrau. 2009. Spatial Analysis of Bladder, Kidney, and Pancreatic Cancer on Upper Cape Cod: An Application of Generalized Additive Models to Case-Control Data. *Environmental Health* 8:3.

Vieira, Verónica, Thomas Webster, Janice Weinberg, Ann Aschengrau, and David Ozonoff. 2005. Spatial Analysis of Lung, Colorectal, and Breast Cancer on Cape Cod: An Application of Generalized Additive Models to Case-Control Data. *Environmental Health* 4:11.

Volinsky, Chris T., and Adrian E. Raftery. 2000. Bayesian Information Criterion for Censored Survival Models. *Biometrics* 56(1):256–262.

Webster, Thomas, Verónica Vieira, Janice Weinberg, and Ann Aschengrau. 2006. Method for Mapping Population-Based Case-Control Studies: An Application Using Generalized Additive Models. *International Journal of Health Geographics* 5:26.

Wheeler, David C., Anneclaire J. De Roos, James R. Cerhan, Lindsay M. Morton, Richard Severson, Wendy Cozen, and Mary H. Ward. 2011. Spatial-Temporal Cluster Analysis of Non-Hodgkin Lymphoma in the NCI-SEER NHL Study. *Environmental Health* 10:63.

Wheeler, David C., Lance A. Waller, Wendy Cozen, and Mary H. Ward. 2012a. Spatial-Temporal Analysis of Non-Hodgkin Lymphoma Risk Using Multiple Residential Locations. *Spatial and Spatio-Temporal Epidemiology* 3(2):163–171.

Wheeler, David C., Mary H. Ward, and Lance A. Waller. 2012b. Spatial-Temporal Analysis of Cancer Risk in Epidemiologic Studies with Residential Histories. *Annals of the Association of American Geographers* 102(5):1049–1057.

Wild, Christopher P. 2005. Complementing the Genome with an 'Exposome': The Outstanding Challenge of Environmental Exposure Measurement in Molecular Epidemiology. *Cancer Epidemiology, Biomarkers and Prevention* 14(8):1847–1850.

Young, Robin L., Janice Weinberg, Verónica Vieira, Al Ozonoff, Thomas F. Webster. 2010. A power comparison of generalized additive models and the spatial scan statistic in a case-control setting. *International Journal of Health Geographics* 9:37.

35

Spatiotemporal Modeling of Preterm Birth

Joshua L. Warren
Department of Biostatistics
Yale School of Public Health
Yale University
New Haven, Connecticut

Montserrat Fuentes
Department of Statistics
North Carolina State University
Raleigh, North Carolina

Amy H. Herring
UNC Gillings School of Global Public Health
The University of North Carolina at Chapel Hill
Chapel Hill, North Carolina

Peter H. Langlois
Birth Defects Epidemiology and Surveillance Branch
Texas Department of State Health Services
Austin, Texas

CONTENTS

35.1 Spatially Referenced Data

The analysis of spatiotemporal data has received increasing amounts of attention in recent years. This is due to the rising availability of spatially and temporally referenced data, mainly as a result of advancements in technology that allow researchers to more easily obtain location- and date-specific data records. A number of areas now regularly collect and analyze spatiotemporal data. Pollution monitors across a geographic domain collect information at fixed spatial locations during various times of the day. Public health studies often attempt to investigate the spread and severity of a disease outbreak using spatially referenced data. With census data, it is common to have access to information aggregated at the county or state level during discrete points in time. The ability to properly analyze spatiotemporal data is crucial in order to fully understand the data generating process and to ensure that the appropriate statistical inference is conducted. Failing to account for the spatiotemporal behavior of the process could lead to poor decisions regarding questions of interest in a particular study. This is especially important in the public health setting where researchers attempt to analyze spatiotemporally referenced data with the goal of improving the overall health of the population through informed lifestyle choices.

35.2 Health and Social Impact of Preterm Birth

Preterm birth (PTB), a delivery occurring before 37 completed weeks of gestation, is the leading cause of infant morbidity and mortality (Goldenberg et al., 2008). Over 35% of all infant deaths in 2009 were related to PTB complications, more than any other individual factor (Centers for Disease Control and Prevention, 2013). Survivors of a premature delivery often have a higher risk of neurological disabilities as they develop, including cerebral palsy. Additionally, PTB affects the developing lungs and intestines and can also lead to vision and hearing loss later in life. A general delay in social development, including learning, communication, self-care, and working with peers, is also common in developing premature children (March of Dimes, 2014c).

Along with the obvious impact on the health and development of the child, PTB also results in high medical costs due to lengthy hospital stays and continuing treatment. These costs are estimated to be $26.2 billion annually (Institute of Medicine, 2006). Currently in the United States, approximately 12% of all births result in premature delivery, approximately 500,000 each year (Centers for Disease Control and Prevention, 2013). The rates appear to be on the rise worldwide and range from 5% to 18% among countries where data have been collected. It is estimated that 1.1 million babies die each year of PTB complications, and that 75% of these deaths could have been prevented with relatively inexpensive treatment options (March of Dimes, 2014a). Known maternal risk factors include well-documented racial disparities, advanced age, smoking during the pregnancy, drinking high levels of alcohol during the pregnancy, illicit drug use during the pregnancy, giving birth to multiples, and having a previous PTB (Mayo Clinic, 2014). Though the prevalence is high, about half of the PTB cases are a result of spontaneous preterm labor that has unknown causes (March of Dimes, 2014b). This leads researchers to investigate the role of environmental exposures occurring during the pregnancy and their impact on PTB.

35.3 Standard Modeling Approaches for Preterm Birth Data

Statistical models have been developed to study the association between various exposures and PTB risk. The standard statistical models typically explore a single exposure during a defined period of the pregnancy or prepregnancy. Previously and currently investigated exposures include, but are not limited to, ambient air pollution (Bobak, 2000; Ritz et al., 2000), smoking (Kyrklund-Blomberg and Cnattingius, 1998; Shiono et al., 1986), alcohol and illicit drugs (Parazzini et al., 2003; Shiono et al., 1986, 1995), contaminated water and food (Ahmad et al., 2001; Savitz et al., 1995), and maternal occupational environment (Luke et al., 1995; Teitelman et al., 1990). We describe the basic framework for these standard PTB and exposure models in the following sections.

35.3.1 Preterm birth as a binary outcome

A majority of previous epidemiologic analyses of the association between PTB and a selected exposure have modeled the binary PTB outcome at the individual birth level (Ahmad et al., 2001; Bobak, 2000; Brauer et al., 2008; Gehring et al., 2011; Hansen et al., 2006; Huynh et al., 2006; Kyrklund-Blomberg and Cnattingius, 1998; Leem et al., 2006; Lin et al., 2001; Liu et al., 2003; Llop et al., 2010; Luke et al., 1995; Meis et al., 1998; Parazzini et al., 2003; Ponce et al., 2005; Ritz et al., 2000, 2007; Saurel-Cubizolles et al., 2004; Savitz et al., 1995; Shiono et al., 1986, 1995; Teitelman et al., 1990; Wilhelm and Ritz, 2003, 2005; Xu et al., 1995). This outcome for birth i is generally defined as

$$Y_i = \begin{cases} 1 & \text{if the completed weeks of gestation for birth } i \text{ at delivery is } {<}37; \\ 0 & \text{otherwise.} \end{cases}$$

Alternatively, the *very preterm* outcome is sometimes of interest due to the increased health problems associated with these births. In that case, 37 weeks is replaced by 32 weeks in the outcome definition.

The most common statistical method of analysis is a binary regression model that relates the probability of a birth occurring before 37 completed weeks of gestation with the exposure of interest, while possibly controlling for potential confounders and other covariates simultaneously. In general, these binary regression models are specified as

$$\begin{aligned} Y_i|p_i &\overset{\text{ind}}{\sim} \text{Bernoulli}\,(p_i) \\ g\,(p_i) &= \mathbf{x}_i^T \boldsymbol{\beta} + \mathrm{z}_i\theta, \end{aligned}$$

where Y_i is the previously defined binary preterm status of birth i, p_i is the probability that birth i results in a PTB, $g\,(.)$ is the selected link function relating the covariates and exposure to the individual probability, $\mathbf{x}_i$ is the vector of confounders or covariates specific to birth i, z_i is the main exposure of interest for birth i, $\boldsymbol{\beta}$ is the vector of regression parameters relating the covariates to the probability, and similarly, θ describes the association between the exposure and the probability of PTB.

Several link functions are available, with the two most common being the logit link, $g\,(p_i) = \ln\{p_i/\,(1-p_i)\}$, and the probit link, $g\,(p_i) = \Phi^{-1}\,(p_i)$, where $\ln\,(.)$ represents the natural logarithm function and $\Phi^{-1}\,(.)$ represents the inverse cumulative distribution function of the standard normal distribution. Logistic regression (logit link) is most popular in epidemiologic analyses due to the odds ratio interpretation of the regression parameters. Probit regression is most often used in Bayesian analyses because it results in semiconjugacy in the model, allowing for efficient sampling from the posterior distribution of the model parameters. Typically, both models result in similar statistical inference for the introduced parameters. See Chapter 7 in this volume for more information on Bayesian modeling and inference.

Models of this form most often represent the exposure as an average over a predefined period before or during the pregnancy for continuous exposures or as a categorical indicator of exposure thresholds for discrete variables. The purpose of these analyses is to conduct inference for θ and to determine if the examined exposure is associated with an increase or decrease in the probability of PTB for an individual.

35.3.2 Preterm birth counts

Another commonly implemented modeling option is a time series analysis of PTB counts. Time series models of the association between PTB counts and an exposure no longer describe probabilities on the individual level, but instead focus on the level of the geographic region in the study. A number of studies have used time series modeling to analyze the association between ambient air pollution exposure and PTB counts (Darrow et al., 2009; Jiang et al., 2007; Lee et al., 2008; Sagiv et al., 2005; Zhao et al., 2011). These models generally take the form originally introduced by Schwartz et al. (1996) such that

$$\begin{aligned} Y_{it}^*|\lambda_{it} &\overset{\text{ind}}{\sim} \text{Poisson}\left(\lambda_{it}\right) \\ \ln\left(\lambda_{it}\right) &= \mathbf{x}_{it}^T\boldsymbol{\beta} + \mathrm{z}_{it}\theta + O_{it}, \end{aligned}$$

where Y_{it}^* is the observed number of PTB cases in region i at time t, $\boldsymbol{\beta}$ and θ have been previously defined, $\mathbf{x}_{it}$ is a vector of covariates specific to region i at time t, z_{it} is the exposure specific to region i at time t, and O_{it} represents an offset term used to account for the number of individuals at risk in region i at time t. Often, smoothed functions of weather and time covariates are introduced instead of assuming the usual linear relationship that is displayed above. Extensions of the model have also been introduced to allow for region-specific random intercepts (Sagiv et al., 2005), while conditional autoregressive models can also be implemented to handle the spatial correlation existing between regions (Waller et al., 1997). The exposure of interest typically consists of lagged temporal averages preceding the modeled time point, while acute daily lagged exposure models have also been considered, specifically in the air pollution exposure setting.

35.4 Spatiotemporal Model for the Binary Preterm Birth Outcome

For certain time-varying exposures, assuming a constant association across the pregnancy period of interest and the geographic domain of the study may be inappropriate when modeling the binary PTB outcome. For example, averaging ambient air pollution exposure experienced throughout the pregnancy can potentially be misleading since certain periods of fetal development may be more important in terms of susceptibility to environmental insults. Similarly, assuming a constant association across all spatial locations of the study may be inadequate if exposure in certain areas is more or less associated with PTB. This is a particular concern with air pollution exposures, as the composition of particulate matter, for example, may vary substantially across different regions due to variation in sources. Spatial differences in the association of interest can indicate that important spatially referenced variables were omitted from the model, as well as suggest actual differences in the association that require further investigation. The standard epidemiologic models typically fail to allow for a fine temporal scale of exposure, and therefore are unable to identify susceptible windows of importance across the pregnancy where increased exposure more adversely affects PTB risk. These models also commonly assume a constant association across the geographic domain and analyze a single exposure. A model that has the ability to jointly handle multiple temporal windows and pollutants while also allowing the association to change spatially is needed to properly analyze the PTB outcome in some settings.

Warren et al. (2012b) introduced a model relating the binary PTB outcome to multiple exposures that allowed for ample spatiotemporal flexibility and the identification of weekly critical windows of importance across the pregnancy. The model is introduced in the Bayesian setting and uses the probit link function leading to semiconjugacy such that

$$\begin{aligned} Y_i|p_i &\overset{\text{ind}}{\sim} \text{Bernoulli}\left(p_i\right) \\ \Phi^{-1}\left(p_i\right) &= \mathbf{x}_i^T\boldsymbol{\beta} + \sum_{j=1}^{m}\sum_{k=1}^{\min\{d,ga_i\}} \mathrm{z}\left\{\boldsymbol{s}_i, t_i\left(k\right), j\right\}\theta\left(\boldsymbol{s}_i, j, k\right), \end{aligned} \quad (35.1)$$

where Y_i, p_i, $\Phi^{-1}(.)$, $\mathbf{x}_i$, and $\boldsymbol{\beta}$ have been previously described. Pregnancy period k is associated with a different calendar date for each woman due to differences in the date of conception. Therefore, we introduce $t_i(.)$, which maps an input pregnancy period to the appropriate calendar date of pregnancy for woman i and ensures that the correct exposures are selected. The exposure j at pregnancy period k corresponding to the calendar date of pregnancy for birth i at location $\boldsymbol{s}_i$ is represented by $\mathrm{z}\{\boldsymbol{s}_i, t_i(k), j\}$. The corresponding parameter that describes the association of interest is given by $\theta(\boldsymbol{s}_i, j, k)$. These spatially, temporally, and exposure-varying parameters describe the association between a single exposure at a specific location and pregnancy time period. This allows for the possibility that the association is changing among the different exposures, locations, and temporal windows. The general form of the model allows for multiple time-varying exposures (m) and flexibility when choosing the number of exposure periods of interest during the pregnancy (d). We prohibit exposures occurring after the pregnancy from affecting the probability of PTB through use of $\min\{d, ga_i\}$, where ga_i is the gestational age in the appropriate timescale for birth i (e.g., days, weeks, or months).

Through the introduction of the $\theta(\boldsymbol{s}_i, j, k)$ parameters, the model allows for the possibility of a changing association with each exposure across the pregnancy. This provides potential for the identification of temporal windows during the pregnancy where increased exposure more adversely affects the PTB outcome. The timescale of exposure (d) is completely dependent on the type and frequency of the exposure data. For example, air pollution data may be collected hourly in certain areas or can be aggregated to create monthly/trimester averages; either case is covered by the model.

A Gaussian process prior distribution is used that allows for the possibility that the parameters are more similar due to proximity in space, time, and more generally, between pollutants such that $\boldsymbol{\theta} \sim \text{MVN}\left(\mathbf{0}, \tau^2\Sigma_\theta\right)$, where $\boldsymbol{\theta}$ represents the entire vector of $\theta(\boldsymbol{s}_i, j, k)$ parameters and entries of Σ_θ are given as

$$\text{Corr}\left[\theta(\boldsymbol{s}, j, k), \theta(\boldsymbol{s}', j', k')\right] = \exp\left\{-\phi_1 \left\|\boldsymbol{s} - \boldsymbol{s}'\right\| - \phi_2|k - k'| - \phi_3 I(j \neq j')\right\}, \tag{35.2}$$

where $\text{Corr}(X, Y)$ represents the correlation between X and Y, $\phi_1 > 0$ controls the level of spatial smoothness between parameters, $\phi_2 > 0$ controls the level of temporal smoothness between parameters, and $\phi_3 > 0$ describes the correlation between parameters from different exposures. This relatively simple exponential structure allows for the possibility of separate degrees of shrinkage across parameters from different locations, time periods, and pollutants. While the number of introduced parameters is large and increasing as the number of locations, time periods, and pollutants increases, computationally, the covariance is tractable and is shown to have a Kronecker product form consisting of three smaller-dimension matrices (spatial, temporal, and pollutants).

We note that different prior distributions can be introduced for the $\theta(\boldsymbol{s}_i, j, k)$ parameters. The prior distribution specified by (35.2) assumes that the spatial process is isotropic, where the correlation between two locations depends only on the distance between the locations (constant mean previously specified). More generally, a stationary spatial process assumes a constant mean and that the correlation between two locations depends only on the separation vector between locations (distance and direction). In certain settings, these assumptions may be too restrictive or unjustified. In Warren et al. (2012a), a spatiotemporal kernel stick-breaking prior was introduced for these parameters, which allowed for a non-stationary spatiotemporal process to increase modeling flexibility. Similar modeling options may be appropriate when considering exposures that have a different temporal scale or spatial association with PTB. See Chapters 11 and 21 in this volume for more information on geostatistical approaches and Bayesian nonparametric modeling, respectively.

35.5 Case Study: Texas Air Pollution and Preterm Birth

In the presented case study, we analyze the binary PTB outcome obtained from birth records for all singleton, live births in Texas during 2002–2004. We investigate the association between PTB risk and exposure from multiple air pollutants, ozone (O_3), and particulate matter less than 2.5 μm ($PM_{2.5}$), experienced through the pregnancy. Texas is of interest spatially due to the differences in exposures seen across the counties. The state includes highly populated cities that produce relatively high levels of air pollution, as well as extremely rural areas with smaller overall levels. Along with differences in magnitude, the composition of $PM_{2.5}$ may also vary spatially due to the fact that $PM_{2.5}$ is a combination of individual pollutants that may also vary spatially. Two areas could have similar $PM_{2.5}$ levels while having different overall compositions. This spatial heterogeneity observed across counties requires a full spatial analysis in order to understand the association of interest.

The model is applied in two settings. First, we collapse the spatial component in (35.1) in order to investigate temporal changes in the PTB–exposure association across the pregnancy for a single county in Texas. Next, we apply the fully spatial version of the model to 13 counties in Texas, mapping each residence at delivery to its respective county centroid. The health dataset consists of full birth records for all singleton, live births in Texas from 2002 to 2004, geocoded to the residence at delivery. Available covariates include parental age group, race, education, and seasonality information. We consider six age groups for the mothers: 10–19, 20–24, 25–29, 30–34, 35–39, and 40+. We include two age groups for the fathers: <50 and 50+. For both mothers and fathers, we consider white (non-Hispanic), black (non-Hispanic), Hispanic, and other as the four race or ethnic groups in the analysis. To control for education level, the number of years of completed education is fit using a cubic B-spline with three degrees of freedom. To account for seasonality, we include the average temperature from the day of birth using a cubic B-spline with four degrees of freedom. The degrees of freedom are selected based on initial exploratory analyses and model fit comparisons.

Air quality system (AQS) monitoring data are available from the Environmental Protection Agency (EPA) and consist of a collection of ambient air pollution data from thousands of monitoring stations throughout the United States (http://www.epa.gov/airquality/airdata/ad_data_daily.html). The data are collected by the EPA, state, and local air pollution agencies and are used by the Office of Air Quality Planning and Standards and others in a number of air quality management functions (U.S. Environmental Protection Agency, 2013). The AQS data are spatially and temporally interpolated using kriging techniques in order to provide exposure predictions at the residences at delivery across the pregnancy. Maximum daily 8-hour average O_3 and daily average $PM_{2.5}$ weekly exposures from the first week of pregnancy through the entire pregnancy or week 36 are created, whichever week comes first.

Prior distributions for the remaining parameters are chosen to ensure that the observed data drive the inference rather than our prior beliefs. The correlation parameters that control the level of sharing across space, time, and pollutant (ϕ_1, ϕ_2, and ϕ_3, respectively) are assigned relatively uninformative uniform prior distributions such that $\phi_1 \sim \text{Uniform}(0.00001, 0.3)$, $\phi_2 \sim \text{Uniform}(0.00001, 5)$, and $\phi_3 \sim \text{Uniform}(0.01, 5)$. An inverse gamma prior distribution is used for the variance of the process such that $\tau^2 \sim \text{Inverse Gamma}(3, 1)$. Independent normal prior distributions centered at zero with large prior variances ($\sigma_\beta^2 = 10^9$) are selected for the regression parameters. In Warren et al. (2012b), a sensitivity analysis is carried out to examine a number of these choices.

The model is applied first in Harris County, Texas, with a focus on weekly $PM_{2.5}$ and O_3 ambient air pollution exposure during the pregnancy ($m = 2, d = 36$). As of July 1, 2009,

Harris County had the third largest county population in the United States with 4,070,989 estimated people (American Fact Finder). It includes the city of Houston, which provides a large amount of heterogeneity to our study population. Due to the smaller size of the geographic domain, we analyze the temporal associations across the pregnancy in terms of PTB and allow the spatial association to remain constant. Therefore, the general model in (35.1) is collapsed across space such that

$$\Phi^{-1}(p_i) = \mathbf{x}_i^T \boldsymbol{\beta} + \sum_{j=1}^{2} \sum_{k=1}^{\min\{36, ga_i\}} \mathrm{z}\{\boldsymbol{s}_i, t_i(k), j\} \theta(j,k),$$

where $\mathrm{z}\{\boldsymbol{s}_i, t_i(k), j\}$ represents the standardized average pollution exposure from pollutant j, at residence $\boldsymbol{s}_i$, on pregnancy week k corresponding to calendar week $t_i(k)$ for pregnancy i. The Gaussian process prior distribution in (35.2) is modified and assigned to the pollution parameters such that $\text{Cov}\{\theta(j,k), \theta(j',k')\} = \tau^2 \exp\{-\phi_2|k-k'| - \phi_3 I(j \neq j')\}$, where τ^2, ϕ_2, and ϕ_3 are previously defined in Section 35.4 and $\text{Cov}(X,Y)$ represents the covariance between X and Y. We refer to this as Model 1 hereafter. We consider the full spatiotemporal PTB model in the second data application.

For comparison purposes, we also fit two alternative models to the same data such that the following applies:

- *Model 1:* Temporal PTB probit regression model with Gaussian process prior distribution assigned to the pollution parameters
- *Model 2:* Standard epidemiologic PTB probit regression model jointly considering trimester 1 and 2 pollution exposure averages for $PM_{2.5}$ and O_3
- *Model 3:* Multiple probit regression model jointly considering all weekly $PM_{2.5}$ and O_3 exposure averages throughout the pregnancy

In Model 2, we consider the standard PTB or exposure model with first and second trimester averages from both pollutants included jointly, failing to account for the introduced multicollinearity. A more naive approach is applied in Model 3 by including all of the weekly exposures from both pollutants in a multiple probit regression model, also ignoring the present correlation between weekly and pollutant-specific parameters. This provides a crude way of carrying out the analysis, as we would expect the parameter estimates to be similar, with their respective posterior standard deviations possibly inflated due to the multicollinearity. We fit all three models using probit regression in the Bayesian setting for comparison purposes.

Table 35.1 displays the included covariate results from Model 1. Maternal covariates are more highly associated with PTB than paternal covariates. When compared with white (non-Hispanic) mothers, black (non-Hispanic) mothers have a higher probability of PTB in general. Mothers aged 35 and older are at a higher risk of having a PTB outcome than younger mothers. Males are more likely than females to be born prematurely. White (non-Hispanic) fathers are more likely to father preterm children than fathers who claim other as their race or ethnic group. Overall, these results agree with previously established PTB findings.

Figure 35.1a displays the ability of Model 1 to identify critical windows of importance across the pregnancy. The posterior median of each weekly pollution parameter is plotted against pregnancy week for each pollutant, along with the respective 95% credible intervals. The susceptible window for higher preterm probabilities covers the middle of the first trimester through the middle of the second trimester for $PM_{2.5}$. Week 14 has the largest estimated effect, with a posterior mean of 0.0327 (SD 0.007). This is interpreted as a one-unit increase in the standardized pollution exposure for week 14 leads to an increase in

TABLE 35.1
Included covariate results for Model 1

Covariate	Mean	SD	Percentiles 0.025	0.50	0.975
Intercept*	−1.515	0.279	−2.065	−1.513	−0.979
Maternal race					
Black vs. white*	0.149	0.067	0.018	0.149	0.280
Hispanic vs. white	−0.043	0.038	−0.117	−0.044	0.033
Other vs. white	0.086	0.070	−0.053	0.086	0.227
Paternal race					
Black vs. white	0.009	0.065	−0.118	0.010	0.135
Hispanic vs. white	−0.008	0.039	−0.084	−0.009	0.070
Other vs. white*	−0.190	0.073	−0.337	−0.189	−0.046
Maternal age group					
20–24 vs. 10–19*	−0.068	0.033	−0.132	−0.069	−0.003
25–29 vs. 10–19	−0.021	0.036	−0.091	−0.021	0.052
30–34 vs. 10–19	0.044	0.041	−0.036	0.045	0.126
35–39 vs. 10–19*	0.127	0.052	0.026	0.127	0.229
≥40 vs. 10–19*	0.308	0.090	0.128	0.309	0.478
Paternal age ≥50 vs. <50	−0.044	0.128	−0.302	−0.041	0.204
Maternal education					
Basis spline 1	−0.396	0.348	−1.053	−0.401	0.303
Basis spline 2	0.185	0.172	−0.140	0.180	0.529
Basis spline 3	0.022	0.204	−0.360	0.018	0.434
Paternal education					
Basis spline 1	0.160	0.307	−0.429	0.153	0.786
Basis spline 2	−0.003	0.150	−0.290	−0.005	0.296
Basis spline 3	−0.055	0.177	−0.389	−0.059	0.305
Female vs. male baby*	−0.069	0.021	−0.110	−0.069	−0.027

* Items have 95% credible intervals that do not include zero.

z-score of 0.0327 on average. An increase or decrease in z-score leads directly to an increase or decrease in the probability of PTB. Increased $PM_{2.5}$ exposure in weeks 4–22 increases the risk of PTB as each of the respective credible intervals fails to include zero. By combining the Markov Chain Monte Carlo (MCMC) output, we also estimate $P\left\{\boldsymbol{\theta}^{(1)}_{(window)} > \mathbf{0}|\mathbf{y}_{obs}\right\}$. This describes the relevant weekly effects jointly and allows inference for the critical windows to be carried out, as opposed to individual weekly analyses alone. This quantity is estimated to be 0.945 for the $PM_{2.5}$ results, with $\boldsymbol{\theta}^{(1)}_{(window)} = \{\theta(1,4), \ldots, \theta(1,22)\}^T$.

The O_3 results differ, with high exposure in the very early weeks of pregnancy appearing to increase the probability of PTB. Week 1 has the most drastic effect in terms of O_3 exposure, with a posterior mean of 0.0207 (SD 0.008). A similar interpretation exists for this effect. Increased O_3 exposure in weeks 1–5 increases the risk of PTB based on the associated credible intervals. For O_3, $P\left\{\boldsymbol{\theta}^{(2)}_{(window)} > \mathbf{0}|\mathbf{y}_{obs}\right\}$ is estimated to be 0.971, where $\boldsymbol{\theta}^{(2)}_{(window)} = \{\theta(2,1), \ldots, \theta(2,5)\}^T$, suggesting that the critical windows that are evident graphically are also impactful jointly.

The immediate benefits of Model 1 are seen when the results are directly compared to the results from the standard statistical methods. The graphical results from Models 1–3 are shown in Figure 35.1. While the plots from the standard methods do indicate some weeks or trimesters whose credible intervals do not include zero, a majority of the information is lost.

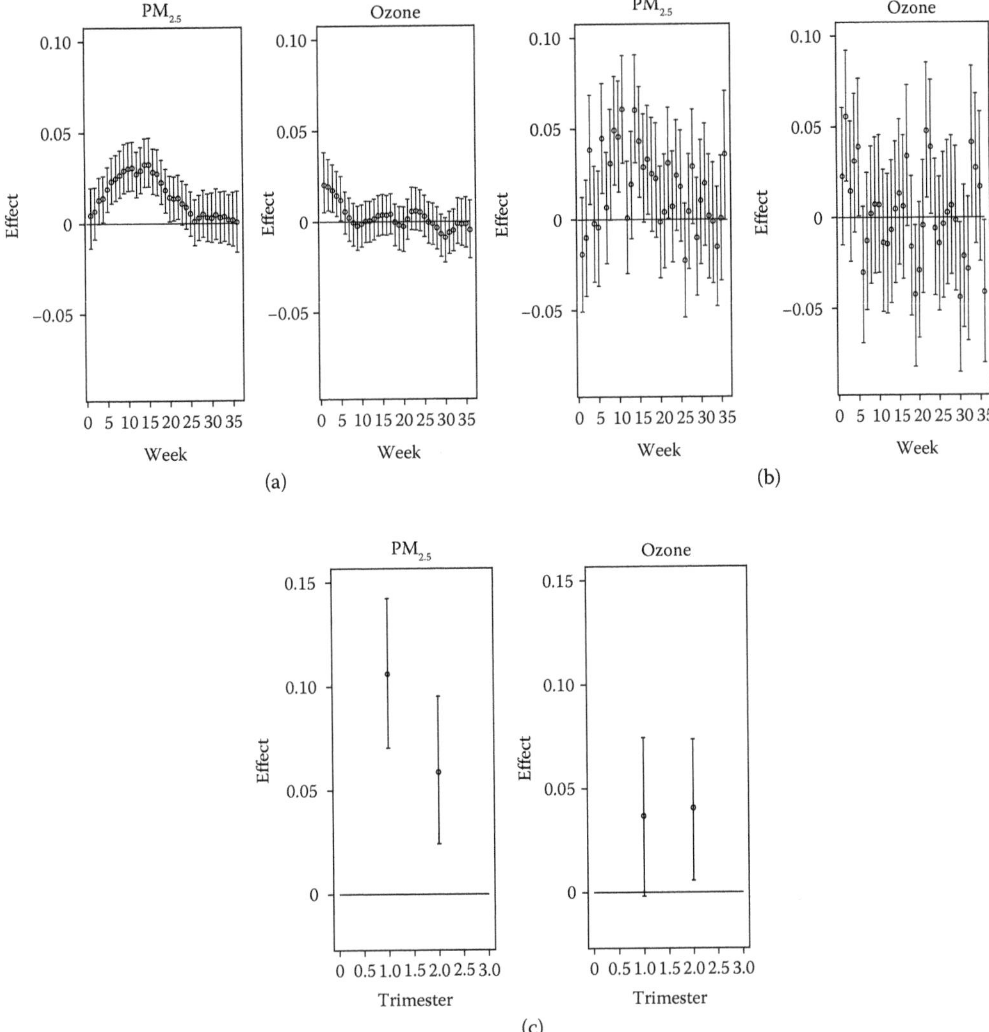

FIGURE 35.1
Susceptible windows of exposure results from Models 1–3. Posterior medians and 95% credible intervals are displayed. (a) Model 1: Temporal model. (b) Model 3: Multiple probit model. (c) Model 2: Trimester average model.

Model 3 is unable to uncover the true state of the windows of exposure. The uncertainty associated with the parameter estimates is much larger than that of Model 1, as the plots from both models are shown on the same scale. Model 2 does not include a fine-enough timescale to be as informative as Model 1.

The deviance information criterion (DIC) is used to carry out a more formal comparison of the models (Spiegelhalter et al., 2002). The DIC is based on the posterior distribution of the deviance statistic, $D(\boldsymbol{\gamma}) = -2\log f(\mathbf{y}|\boldsymbol{\gamma}) + 2\log h(\mathbf{y})$, where $f(\mathbf{y}|\boldsymbol{\gamma})$ represents the likelihood of the observed data given the $\boldsymbol{\gamma}$ vector of parameters and $h(\mathbf{y})$ is some standardizing function of the data alone. The posterior expectation of the deviance, $\overline{D}$, is used to describe the fit of the model to the data, while the effective number of parameters, p_D, is used to describe the complexity of the model. DIC is then defined as $DIC = \overline{D} + p_D$, with lower values of DIC representing a better model fit. The DIC clearly favors Model 1 (DIC 17246.3, p_D 42.4) when compared with Model 2 (DIC 17282.8, p_D 28.3) and Model 3 (DIC 17289.1, p_D 96.2).

The effective number of parameters gives insight into the ability of Model 1 to share information across pregnancy weeks. We include a total of 24 parameters in the $\boldsymbol{\beta}$ vector for the covariates of interest. For Model 2, we additionally include the 4 trimester parameters (2 trimester parameters for each pollutant), resulting in 28 total parameters. This is reflected in the effective number of parameters (p_D 28.3). In Model 3, we additionally include the 72 pollution parameters (36 weekly parameters for each pollutant) with exchangeable normal prior distributions, resulting in 96 total parameters. This is also reflected in the effective number of parameters (p_D 96.2). In Model 1, we additionally include the 72 pollution parameters with Gaussian process prior distribution, giving Model 1 the same number of parameters as Model 3. However, due to the ability of Model 1 to share information across pregnancy weeks, the effective number of parameters is reduced (p_D 42.4).

Next, we extend the spatial domain to include the 13 counties from Texas Department of State Health Services (TDSHS) Health Service Region 6, shown in Figure 35.2, and apply the spatiotemporal version of the model. We estimate the risk parameters for each county by introducing $\theta\{R(\boldsymbol{s}_i), j, k\}$, where $R(\boldsymbol{s}_i)$ represents the location of the county containing location $\boldsymbol{s}_i$. The specific location for a county is determined by the center of gravity of the births within the county. TDSHS Health Service Region 6 provides a good mix of urban and rural counties, which is useful for investigating possible changes in associations due to varying compositions of the pollutants across these counties. We work at the county level after examining within Harris County and finding that the association does not vary over the smaller spatial domain.

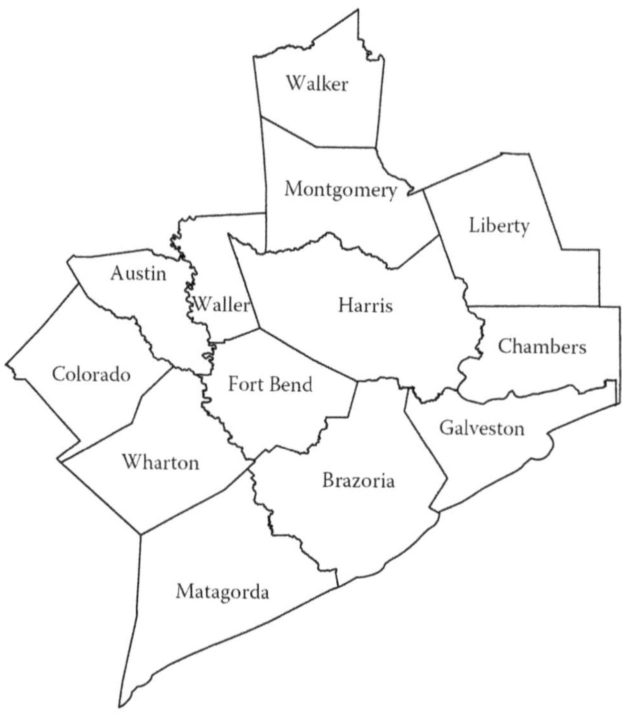

FIGURE 35.2
Texas Department of State Health Services (TDSHS) Health Service Region 6.

Two models are fit to the 13 county datasets in order to investigate the benefits of considering the spatial correlation between the risk parameters. Model 4 represents the spatiotemporal PTB probit regression model with Gaussian process prior distribution shown in (35.1) and (35.2). Model 5 ignores the spatial variability that exists between the risk effects by adapting (35.2) such that

$$\text{Cov}\left[\theta\left\{R\left(\boldsymbol{s}\right),j,k\right\},\theta\left\{R\left(\boldsymbol{s}'\right),j',k'\right\}\right] = \tau^2 \exp\left\{-\phi_2|k-k'| - \phi_3 I\left(j \neq j'\right)\right\}.$$

This structure assumes spatial independence of the risk parameters across the 13 counties.

Figure 35.3 displays the estimated posterior means and standard deviations of the average combined risk effects plotted across the 13 counties for the first trimester of pregnancy from Models 4 and 5. The first trimester average combined risk effect at location $R\left(\boldsymbol{s}\right)$ has the form $\overline{\theta}_1\left\{R\left(\boldsymbol{s}\right)\right\} = \frac{1}{13}\sum_{k=1}^{13}\left[\theta\left\{R\left(\boldsymbol{s}\right),1,k\right\}+\theta\left\{R\left(\boldsymbol{s}\right),2,k\right\}\right]$. This quantity gives some insight into the combined impact from both pollutants averaged over the first trimester on

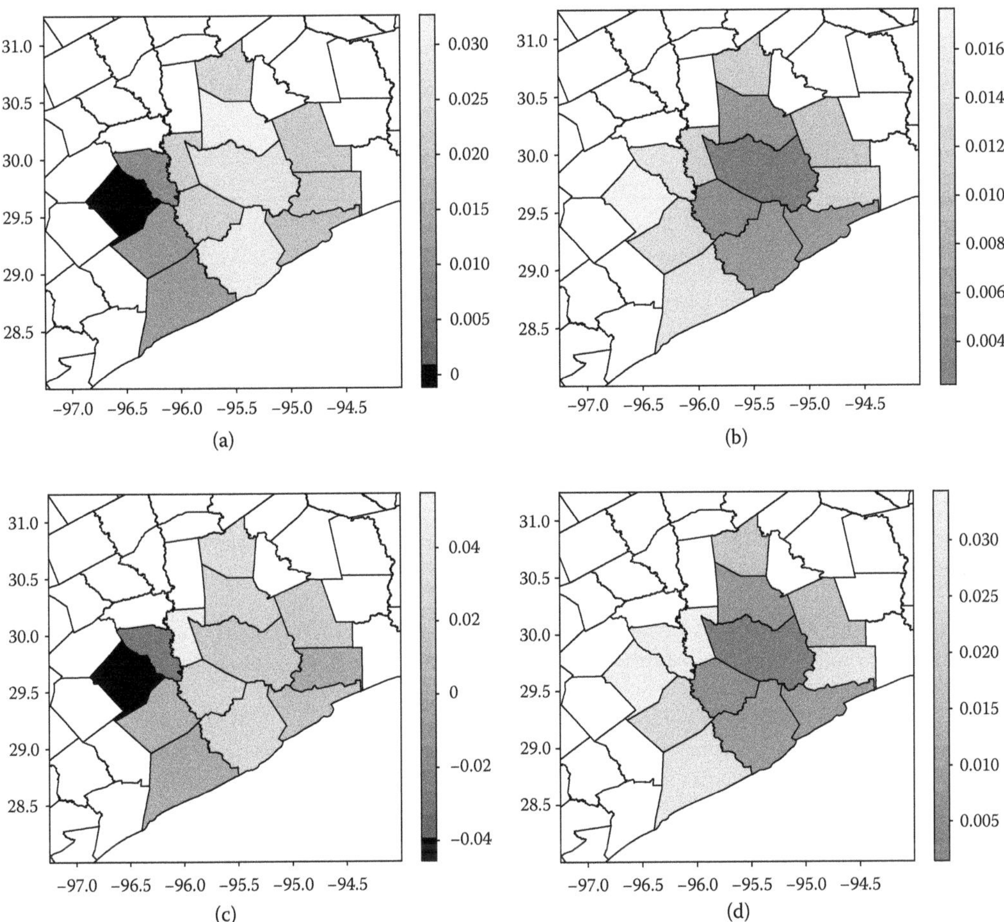

FIGURE 35.3
Posterior means and standard deviations of first trimester averages of the combined $PM_{2.5}$ and O_3 risk parameters by county for Model 4 (top row) and Model 5 (bottom row) model fits. (a) Posterior means: spatial (probit scale). (b) Posterior standard deviations: spatial. (c) Posterior means: nonspatial (probit scale). (d) Posterior standard deviations: nonspatial.

the probability of PTB at various spatial locations. The posterior standard deviation plots show the reduction in variation that occurs when we account for the spatial correlation between the pollution parameters. The posterior distributions of the risk parameters for Galveston, Liberty, and Walker Counties have 95% credible intervals that include zero for the Model 5 results, while being larger than zero once Model 4 is applied. Model 4 is able to identify these positive effects that are missed when space is ignored in Model 5. The DIC clearly favors Model 4 (DIC 23,857.4, p_D 61.4) over Model 5 (DIC 23,920.4, p_D 80.6).

Correctly characterizing the spatiotemporal dependence when modeling PTB data is important to fully understand the association between an exposure and PTB risk. The Texas air pollution application showed how the PTB risk can potentially vary across weeks of the pregnancy as well as geographically. Therefore, a statistical model for PTB data should allow for flexible spatiotemporal behavior between the risk parameters. The presented spatiotemporal model was shown to outperform the standard models and improve the understanding of the PTB–air pollution association. These application results further build the evidence supporting the link between air pollution and PTB while extending our knowledge regarding the specific periods during the pregnancy that have the greatest impact in terms of PTB risk. Applying similar spatiotemporal models to investigate alternative exposures could also prove beneficial for future PTB research efforts.

References

Ahmad, S. A., M. Sayed, S. Barua, M. H. Khan, M. Faruquee, A. Jalil, S. A. Hadi, and H. K. Talukder. (2001). Arsenic in drinking water and pregnancy outcomes. *Environmental Health Perspectives* 109(6), 629–631.

Bobak, M. (2000). Outdoor air pollution, low birth weight, and prematurity. *Environmental Health Perspectives* 108(2), 173–176.

Brauer, M., C. Lencar, L. Tamburic, M. Koehoorn, P. Demers, and C. Karr. (2008). A cohort study of traffic-related air pollution impacts on birth outcomes. *Environmental Health Perspectives* 116(5), 680–686.

Centers for Disease Control and Prevention. (2013). Preterm birth. http://www.cdc.gov/reproductivehealth/maternalinfanthealth/pretermbirth.htm (accessed June 1, 2014).

Darrow, L. A., M. Klein, W. D. Flanders, L. A. Waller, A. Correa, M. Marcus, J. A. Mulholland, A. G. Russell, and P. E. Tolbert. (2009). Ambient air pollution and preterm birth: a time-series analysis. *Epidemiology (Cambridge, Mass.)* 20(5), 689–698.

Gehring, U., A. H. Wijga, P. Fischer, J. C. de Jongste, M. Kerkhof, G. H. Koppelman, H. A. Smit, and B. Brunekreef. (2011). Traffic-related air pollution, preterm birth and term birth weight in the PIAMA birth cohort study. *Environmental Research* 111(1), 125–135.

Goldenberg, R. L., J. F. Culhane, J. D. Iams, and R. Romero. (2008). Epidemiology and causes of preterm birth. *Lancet* 371(9606), 75–84.

Hansen, C., A. Neller, G. Williams, and R. Simpson. (2006). Maternal exposure to low levels of ambient air pollution and preterm birth in Brisbane, Australia. *BJOG: An International Journal of Obstetrics and Gynaecology* 113(8), 935–941.

Huynh, M., T. J. Woodruff, J. D. Parker, and K. C. Schoendorf. (2006). Relationships between air pollution and preterm birth in California. *Paediatric and Perinatal Epidemiology* 20(6), 454–461.

Institute of Medicine. (2006). Preterm birth: causes, consequences, and prevention. https://www.iom.edu/Reports/2006/Preterm-Birth-Causes-Consequences-and-Prevention.aspx (accessed June 1, 2014).

Jiang, L., Y. Zhang, G. Song, G. Chen, B. Chen, N. Zhao, and H. Kan. (2007). A time series analysis of outdoor air pollution and preterm birth in Shanghai, China. *Biomedical and Environmental Sciences* 20(5), 426–431.

Kyrklund-Blomberg, N. B., and S. Cnattingius. (1998). Preterm birth and maternal smoking: risks related to gestational age and onset of delivery. *American Journal of Obstetrics and Gynecology* 179(4), 1051–1055.

Lee, S. J., S. Hajat, P. J. Steer, and V. Filippi. (2008). A time-series analysis of any short-term effects of meteorological and air pollution factors on preterm births in London, UK. *Environmental Research* 106(2), 185–194.

Leem, J.-H., B. M. Kaplan, Y. K. Shim, H. R. Pohl, C. A. Gotway, S. M. Bullard, J. F. Rogers, M. M. Smith, and C. A. Tylenda. (2006). Exposures to air pollutants during pregnancy and preterm delivery. *Environmental Health Perspectives* 114(6), 905–910.

Lin, M.-C., H.-F. Chiu, H.-S. Yu, S.-S. Tsai, B.-H. Cheng, T.-N. Wu, F.-C. Sung, and C.-Y. Yang. (2001). Increased risk of preterm delivery in areas with air pollution from a petroleum refinery plant in Taiwan. *Journal of Toxicology and Environmental Health Part A* 64(8), 637–644.

Liu, S., D. Krewski, Y. Shi, Y. Chen, and R. T. Burnett (2003). Association between gaseous ambient air pollutants and adverse pregnancy outcomes in Vancouver, Canada. *Environmental Health Perspectives* 111(14), 1773–1778.

Llop, S., F. Ballester, M. Estarlich, A. Esplugues, M. Rebagliato, and C. Iñiguez. (2010). Preterm birth and exposure to air pollutants during pregnancy. *Environmental Research* 110(8), 778–785.

Luke, B., N. Mamelle, L. Keth, F. Munoz, J. Minogue, E. Papiernik, and T. R. Johnson. (1995). The association between occupational factors and preterm birth: a United States nurses' study. *American Journal of Obstetrics and Gynecology* 173(3), 849–862.

March of Dimes. (2014a). Born too soon. http://www.marchofdimes.com/mission/global-preterm.aspx (accessed June 1, 2014).

March of Dimes. (2014b). Finding the causes of prematurity. http://www.marchofdimes.com/research/finding-the-causes-of-prematurity.aspx (accessed June 1, 2014).

March of Dimes. (2014c). Long-term health effects of premature birth. http://www.marchofdimes.com/baby/long-term-health-effects-of-premature-birth.aspx (accessed June 1, 2014).

Mayo Clinic. (2014). Premature birth: risk factors. http://www.mayoclinic.org/diseases-conditions/premature-birth/basics/risk-factors/con-20020050 (accessed June 1, 2014).

Meis, P. J., R. L. Goldenberg, B. M. Mercer, J. D. Iams, A. H. Moawad, M. Miodovnik, M. K. Menard, et al. (1998). The preterm prediction study: risk factors for indicated preterm births. *American Journal of Obstetrics and Gynecology* 178(3), 562–567.

Parazzini, F., L. Chatenoud, M. Surace, L. Tozzi, B. Salerio, G. Bettoni, and G. Benzi. (2003). Moderate alcohol drinking and risk of preterm birth. *European Journal of Clinical Nutrition* 57(10), 1345–1349.

Ponce, N. A., K. J. Hoggatt, M. Wilhelm, and B. Ritz. (2005). Preterm birth: the interaction of traffic-related air pollution with economic hardship in Los Angeles neighborhoods. *American Journal of Epidemiology* 162(2), 140–148.

Ritz, B., M. Wilhelm, K. J. Hoggatt, and J. K. C. Ghosh. (2007). Ambient air pollution and preterm birth in the environment and pregnancy outcomes study at the University of California, Los Angeles. *American Journal of Epidemiology* 166(9), 1045–1052.

Ritz, B., F. Yu, G. Chapa, and S. Fruin. (2000). Effect of air pollution on preterm birth among children born in Southern California between 1989 and 1993. *Epidemiology* 11(5), 502–511.

Sagiv, S. K., P. Mendola, D. Loomis, A. H. Herring, L. M. Neas, D. A. Savitz, and C. Poole. (2005). A time-series analysis of air pollution and preterm birth in Pennsylvania, 1997–2001. *Environmental Health Perspectives* 113(5), 602–606.

Saurel-Cubizolles, M. J., J. Zeitlin, N. Lelong, E. Papiernik, G. C. Di Renzo, and G. Bréart. (2004). Employment, working conditions, and preterm birth: results from the Europop case-control survey. *Journal of Epidemiology and Community Health* 58(5), 395–401.

Savitz, D. A., K. W. Andrews, and L. M. Pastore. (1995). Drinking water and pregnancy outcome in central North Carolina: source, amount, and trihalomethane levels. *Environmental Health Perspectives* 103(6), 592–596.

Schwartz, J., D. W. Dockery, and L. M. Neas. (1996). Is daily mortality associated specifically with fine particles? *Journal of the Air and Waste Management Association* 46(10), 927–939.

Shiono, P. H., M. A. Klebanoff, R. P. Nugent, M. F. Cotch, D. G. Wilkins, D. E. Rollins, J. C. Carey, and R. E. Behrman. (1995). The impact of cocaine and marijuana use on low birth weight and preterm birth: a multicenter study. *American Journal of Obstetrics and Gynecology* 172(1), 19–27.

Shiono, P. H., M. A. Klebanoff, and G. G. Rhoads. (1986). Smoking and drinking during pregnancy: their effects on preterm birth. *JAMA* 255(1), 82–84.

Spiegelhalter, D. J., N. G. Best, B. P. Carlin, and A. Van Der Linde. (2002). Bayesian measures of model complexity and fit. *Journal of the Royal Statistical Society: Series B (Statistical Methodology)* 64(4), 583–639.

Teitelman, A. M., L. S. Welch, K. G. Hellenbrand, and M. B. Bracke. (1990). Effect of maternal work activity on preterm birth and low birth weight. *American Journal of Epidemiology* 131(1), 104–113.

U.S. Environmental Protection Agency. (2013). Air data: basic information. http://www.epa.gov/airquality/airdata/ad_basic.html (accessed June 27, 2014).

Waller, L. A., B. P. Carlin, H. Xia, and A. E. Gelfand. (1997). Hierarchical spatio-temporal mapping of disease rates. *Journal of the American Statistical Association* 92(438), 607–617.

Warren, J., M. Fuentes, A. Herring, and P. Langlois. (2012a). Bayesian spatial–temporal model for cardiac congenital anomalies and ambient air pollution risk assessment. *Environmetrics* 23(8), 673–684.

Warren, J., M. Fuentes, A. Herring, and P. Langlois. (2012b). Spatial-temporal modeling of the association between air pollution exposure and preterm birth: identifying critical windows of exposure. *Biometrics* 68(4), 1157–1167.

Wilhelm, M., and B. Ritz. (2003). Residential proximity to traffic and adverse birth outcomes in Los Angeles County, California, 1994–1996. *Environmental Health Perspectives* 111(2), 207–216.

Wilhelm, M., and B. Ritz. (2005). Local variations in CO and particulate air pollution and adverse birth outcomes in Los Angeles County, California, USA. *Environmental Health Perspectives* 113(9), 1212–1221.

Xu, X., H. Ding, and X. Wang. (1995). Acute effects of total suspended particles and sulfur dioxides on preterm delivery: a community-based cohort study. *Archives of Environmental Health: An International Journal* 50(6), 407–415.

Zhao, Q., Z. Liang, S. Tao, J. Zhu, and Y. Du. (2011). Effects of air pollution on neonatal prematurity in Guangzhou of China: a time-series study. *Environmental Health* 10(2).

Index

B

C

D

E

H

I

M

N

S

Milton Keynes UK
Ingram Content Group UK Ltd.
UKHW051901071024
449327UK00025B/2047